Statistics for Business and Economics

Statistics for Business and Economics

Second Edition

David R. Anderson
University of Cincinnati

Dennis J. Sweeney
University of Cincinnati

Thomas A. Williams
Rochester Institute of Technology

West Publishing Company
St. Paul New York Los Angeles San Francisco

A workbook has been developed to assist you in mastering the concepts presented in the text. The workbook briefly summarizes the content of the chapter and reviews key terms and their definitions. All the formulas relevant to the chapter are shown in summary display form. You are shown step by step solutions to problems not shown in the text, then challenged to solve similar problems to reinforce your understanding. The workbook is available from your local bookstore under the title *Student Workbook to Accompany Statistics for Business and Economics,* Second Edition, prepared by Mohammad Ahmadi of the University of Tennessee at Chattanooga.

If you cannot locate it in the bookstore, ask your bookstore manager to order it for you.

Printed in the United States of America

Library of Congress Cataloging in Publication Data
Anderson, David Ray, 1941–
Statistics for business and economics.

Bibliography: p.
Includes index.
1. Commercial statistics. 2. Economics—Statistical methods. 3. Statistics. I. Sweeney, Dennis J. II. Williams, Thomas Arthur, 1944– . III. Title.
HF1017.A6 1984 519.5 83-21621
ISBN 0-314-77811-X
ISBN: International Edition 0-314-77822-5

To Lynn, Cherri, and Robbie

Preface

The purpose of this book is to provide students, primarily in the fields of business administration and economics, with a sound conceptual introduction to the field of statistics and its many applications. The text is applications oriented and has been written with the needs of the nonmathematician in mind.

The mathematical prerequisite is a course in college algebra. All of the text material end problems have been developed with this prerequisite in mind. The one exception is that a knowledge of differential calculus is necessary to understand the derivation of the normal equations for linear regression in the appendices to Chapters 14 and 15.

APPLICATIONS AND METHODOLOGY INTEGRATED

Applications of the statistical methodology are an integral part of the organization and presentation of material. Statistical techniques are introduced in conjunction with problem scenarios where the techniques are helpful. The discussion and development of each technique is thus centered around an application setting with the statistical results providing insights to decisions and solutions to problems.

Although the book is applications oriented, we have taken care to provide a sound methodological development. Throughout the text we have utilized notation that is generally accepted for the topic being covered. Thus, students will find that the text provides good preparation for the study of more advanced statistical material. A bibliography that should prove useful as a guide to further study has been included as an appendix.

CHANGES IN SECOND EDITION

The first edition of this text was entitled *Introduction to Statistics;* the title of this edition has been changed to *Statistics for Business and Economics.* We feel this title better reflects the orientation of the material and the types of applications presented. The current edition contains several new features.

Chapter-Ending Applications

To further emphasize the applications of statistics, cases supplied by practitioners from business and government have been added at the end of each chapter. Each application describes an actual organization and its current usage of the statistical methodology introduced in the chapter. We feel these applications help motivate the student to learn the material and provide an appreciation for some of the ways statistical methods are used in practice. The table that follows provides a guide to the companies and types of applications included.

An Overview of Chapter-Ending Applications

Chapter	*Chapter Title*	*Organization*	*Application*
1	Introduction	Kings Island Inc.	Consumer Profile Sample Survey
2	Descriptive Statistics I. Tabular and Graphical Approaches	The Colgate Palmolive Company	Quality Assurance for Heavy Duty Detergents
3	Descriptive Statistics II. Measures of Location and Dispersion	St. Luke's Hospital	Time Spent in Hospice Program
4	Introduction to Probability	Morton Thiokol Inc.	Evaluation of Customer Service Testing Program
5	Probability Distributions	Xerox Corporation	Performance Test of an On-Line Computerized Publication System
6	Special Probability Distributions	The Burroughs Corporation	Credit Cards for Banking
7	Sampling and Sampling Distributions	Mead Corporation	Estimating the Value of Mead Forest Ownership
8	Interval Estimation	Thriftway Inc.	Sampling for Estimation of LIFO Inventory
9	Hypothesis Testing	Harris Corporation	Testing for Defective Plating
10	Statistical Inference About Means and Proportions with Two Populations	Pennwalt Corporation	Evaluation of New Drugs
11	Inferences About Population Variances	U.S. General Accounting Office	Water Pollution Control
12	Tests of Goodness of Fit and Independence	United Way	Determining Community Perceptions of Charities
13	Experimental Design and Analysis of Variance	Burke Marketing Services Inc.	New Product Design
14	Simple Linear Regression and Correlation	Monsanto Company	Estimation of Effect of Feed Additive on Poultry Weight Gain
15	Multiple Regression	Champion International Corporation	Control of Pulp Bleaching Process
16	Time Series Analysis and Forecasting	The Cincinnati Gas & Electric Company	Forecasting Demand for Electricity
17	Indexes and Index Numbers	State of New York Department of Commerce	Index of Business Activity
18	Nonparametric Methods	West Shell Realtors	Comparison of Real Estate Prices across Neighborhoods
19	Decision Theory	Ohio Edison Company	Choice of Best Type of Particulate Control Equipment

Major Revisions

The two chapters on REGRESSION AND CORRELATION ANALYSIS have been substantially revised and now include discussions of residual analysis, the concept of multicollinearity, adding and deleting variables, and the general linear model. The role of computer packages in performing regression analysis is also highlighted.

The chapter on ANALYSIS OF VARIANCE has been expanded with an experimental design emphasis and now includes the completely randomized, randomized block, and factorial experimental designs.

The chapters on SAMPLING DISTRIBUTIONS AND HYPOTHESIS TESTING have been revised to provide better coverage of the central limit theorem and to describe the use of "p-values" in hypothesis testing.

The chapter on BASIC PROBABILITY has been revised and now includes additional examples and illustrations.

New Problems

Over 200 new problems have been added.

CHAPTER PEDAGOGY

What you will learn in this chapter is a note to students at the beginning of each chapter. On this page the concepts the student is expected to master are presented as a guide to the student's study of the material.

A summary of the **Key Formulas** has been provided at the end of each chapter. This along with the chapter summary and glossary is designed to aid the student in review and to provide a quick reference.

FLEXIBILITY

There is a reasonable amount of instructor flexibility possible in selecting material to satisfy specific course needs. As an illustration, a possible outline for a two-quarter sequence in introductory statistics is given below:

Possible Two-Quarter Course Outline

First Quarter	*Second Quarter*
Introduction (Chapter 1)	Hypothesis Testing (Chapter 9)
Descriptive Statistics (Chapters 2 and 3)	Two Population Cases (Chapter 10)
Introduction to Probability (Chapter 4)	Inferences About Population Variances (Chapter 11)
Probability Distributions (Chapter 5)	Tests of Goodness of Fit and Independence (Sections 12.1 and 12.2 of Chapter 12)
Special Probability Distributions (Chapter 6)	Analysis of Variance (Chapter 13)
Sampling and Sampling Distributions (Chapter 7)	Regression and Correlation (Chapters 14 and 15)
Interval Estimation (Chapter 8)	

Other possibilities exist for such a course, depending upon the time available and the background of the students. However, it is probably not possible to cover all the material in one semester or in two quarters unless some of the topics have been previously studied.

ANCILLARIES

Accompanying the text is a complete package of support materials. These include an

INSTRUCTOR'S MANUAL
STUDY GUIDE
TEST BANK
DEMONSTRATION PROBLEMS AND LECTURE NOTES

The instructor's manual, prepared by the authors, includes learning objectives and completely worked solutions to all the problems. The study guide, developed by Mohammad Ahmadi (University of Tennessee at Chattanooga), provides an additional source of problems and explanations for the students. The test bank, prepared by Edward Fagerlund (Wichita State University), provides a series of multiple choice questions and problems that will aid in the preparation of exams. Finally, the set of demonstration problems and lecture notes, prepared by the authors, can be used as a complete set of transparency masters and/or lecture notes. All of these demonstration problems are new illustrations of the material in the text.

We believe that the applications orientation of the text, combined with the package of support materials, provides an ideal basis for introducing students to statistics and statistical applications.

ACKNOWLEDGMENT

We owe a debt to many of our colleagues and friends for their helpful comments and suggestions in the development of this manuscript. Among these are:

Harry Benham
John Bryant
George Dery
Gopal Dorai
Edward Fagerlund
Nicholas Farnum
Ben Isselhardt
Jeffrey Jarrett
Thomas McCullough
Al Palachek
Ruby Ramirez
Tom Ryan
Willban Terpening
Hiroki Tsurumi
J. E. Willis
Donald Williams

Our associates from business and industry who supplied the applications made a major contribution. They are recognized individually by a credit line on the first page of each application.

We are also indebted to our editor, Mary C. Schiller, and others at West Publishing Company for their editorial counsel and support during the preparation of this text. Finally, we would like to express our appreciation to Phyllis Trosper and Janice Bruegge for their typing and secretarial support.

David R. Anderson
Dennis J. Sweeney
Thomas A. Williams

February 1984

Contents

Statistics for Business and Economics

1 Introduction

What you will learn in this chapter:

- there are two interpretations of the term statistics
- the difference between a population and a sample
- the advantages of sampling
- the process of statistical inference
- an overview of various business/economic applications of statistical analysis

Contents

The next time you read a newspaper, look for items such as the following:

46% of the people surveyed believe that the president is doing a good job in foreign affairs.
The average selling price of a new house is $74,500.
The unemployment rate is 8.9%.
New car sales are up 2.4% over last year.

The numerical facts—or data—in these news items (46%, $74,500, 8.9%, 2.4%) commonly are referred to as statistics. In everyday usage, then, the term *statistics* refers to numerical facts or data.

The field, or subject, of statistics involves much more than simply the calculation and presentation of numerical data. In a broad sense the subject of statistics involves the study of how numerical facts or data are collected, how they are analyzed, and how they are interpreted. A major reason for collecting, analyzing, and interpreting data is to provide managers with the information needed to make effective decisions. In this text we shall be oriented toward the use of statistics in a decision-making context.

1.1 THE POPULATION AND THE SAMPLE

Let us consider the case of a major political party that would like to estimate the percentage of voters currently favoring its presidential candidate. How could the party develop such an estimate?

There are approximately 100 million voters in the United States. For the above problem, then, this group of approximately 100 million voters is called the population. In general, we define the population as follows:

Definition of Population

A population is the collection of all items of interest in a particular study.

In theory, every voter could be contacted and asked if he or she preferred the party's candidate. Such a procedure would be called a census.

We can see that attempting to contact 100 million voters is impractical from both time and cost perspectives. Instead, let us suppose that the political party selects a subset of 1,500 voters believed to be representative of the 100 million voters. This subset of voters is referred to as a sample.

Definition of Sample

A sample is a portion of the population selected to represent the whole population.

Furthermore, suppose that of the 1,500 voters actually contacted, 600 favor the party's candidate. Expressing this result as a percentage, we find that (600/1,500) × 100%, or 40%, of the individuals in the sample favor the candidate. This percentage found for the sample could be used as an estimate of the percentage of all voters that favor the candidate.

Many situations have characteristics similar to those of the above illustration in that there exists a large group (individuals, voters, households, products, customers, etc.) about which information is being sought. Because of time, cost, or other considerations, however, data are collected only from a small portion of the group. As defined above, the larger group of items in a particular study is called the *population,* and the smaller group, the group actually contacted, is called the *sample.*

1.2 DATA SUMMARIZATION

In many statistical studies we are interested only in summarizing a set of data in order to present it in a more convenient or more easily interpreted form. For example, suppose that a sales manager has monthly sales figures for 200 of the firm's salespersons. In order to provide the manager with information about sales performance, the data could be summarized and presented in tabular fashion, as shown in Table 1.1.

TABLE 1.1 Summary of Monthly Sales for 200 Salespersons

Monthly Sales Volume (units sold)	*Number of Salespersons*
40 but less than 50	3
50 but less than 60	17
60 but less than 70	52
70 but less than 80	68
80 but less than 90	30
90 but less than 100	22
100 but less than 110	8
Total	200

From Table 1.1 we see that the most frequent sales volumes occur in the 70-but-less-than-80 interval. In addition, we see that $52 + 68 + 30 = 150$ salespersons sold 60 but less than 90 units and that only 8 of the 200 salespersons (4%) were able to sell 100 or more units. A number of additional observations can be made with the data summarized and presented in the tabular form of Table 1.1.

As another means of summarizing the data we could add the sales volumes for each salesperson and divide by 200 to compute an average sales volume. Suppose that doing so yields an average of 76 units per salesperson. While this average of 76 units provides a summarization of the data in a single numerical value, the tabular presentation in Table 1.1 provides more information about the variability in the data. Graphical approaches can also be used to summarize a data set. The study of methods for data summarization is referred to as *descriptive statistics.*

1.3 STATISTICAL INFERENCE AND PROBABILITY

Much of statistics is concerned with analyzing sample data in order to learn about characteristics of a population. In order to make our discussion more concrete, let us consider a situation in which a production manager has to decide whether or not to send a recently completed production run to the warehouse. Suppose that the run consists of 2,000 items. Suppose also that the manager follows a policy of sending the entire batch to the warehouse if no more than 3% of the items are defective; otherwise the batch is reworked. In this situation the population consists of the 2,000 items in the production run. The characteristic of interest is the percentage of items that are defective. Because of time and cost considerations the production manager has concluded that it is not practical to inspect every item in the population. Instead, the decision has been made to take a sample of 150 items. If more than 3% are defective in the sample, the batch will be reworked.

Let us suppose that the sample of 150 items is taken and that 3 defective items are found. An estimate of the percent defective in the population is $(3/150) \times 100\% = 2\%$. According to the production manager's policy, the entire batch would be accepted and shipped to the warehouse.

The process of estimating the percent defective in the population based on the percent defective in the sample is one example of the use of statistical inference in a decision-making context. Whenever we make a conclusion or inference about an entire population based on sample results we have to recognize the following: since the results are based on an analysis of only a small part of the population, they will not be exactly the same as if the entire population had been used. Hence it is desirable to provide some type of indication of how good the sample results are likely to be in terms of estimating the population characteristics. This is where probability plays an important role in statistical inference.

When statisticians make statements about the precision of their estimates, a measure of uncertainty is included. For instance, a statistician might state that the population percent defective is 2% with a possible error of $\pm.5\%$. Thus an interval estimate of 1.5% to 2.5% is provided. With the help of probability theory, the statistician can state how likely it is that the interval estimation procedure leads to an interval that contains the actual proportion defective in the population. By applying probability concepts to the analysis of data in a sample we will learn how to provide estimates of the characteristics of a population, including probabilistic statements about the quality or precision of the estimates.

1.4 SOME ILLUSTRATIVE APPLICATIONS

In this section we present briefly some applications which illustrate how statistics is used in business and economics.

Accounting Application

Suppose that a mail-order firm is interested in collecting information concerning recent ordering practices of its customers. In particular, the accounts receivable manager has asked for information about the dollar amounts billed on recent orders. By sampling the billing records for these orders, the accountant can provide the manager with estimates of the average amount billed per order, the percentage of orders less than $10.00, and so on.

Marketing Application

In an attempt to measure consumer acceptance of a new product, a firm has identified several test market areas throughout the country and has selected a sample of potential customers in each. The individuals in the sample are asked in interviews whether or not they prefer the new product to the product they currently are using. An analysis of the sample data will enable management to estimate the percentage of all potential customers who will prefer the new product.

Finance Application

A federal agency would like to identify the current practices and trends in home mortgages. To gather data pertinent to the problem, a statistician selected a sample of 100 savings and loan associations. The average initial mortgage issued by these 100 savings and loan associations was $60,000. Comparing this figure to similar figures for previous years, an inference could be made about the trend in home mortgages and the potential future demand for home mortgage loans.

Production Application

Many manufacturing and production applications of statistics involve quality control. Consider a manufacturing firm that produces several lines of soaps, shampoos, and toothpastes. Each product will have specifications for the filling weight of the container. Suppose that for a particular production line containers are to be filled with 12 ounces of product. While the automatic filling mechanism is very accurate, it can at times under- or overfill.

Periodically, a quality control inspector takes a sample of the containers from the production line. By weighing the sampled containers the inspector can estimate the filling characteristics of the production line. If under- or overfilling is suspected, the line can be shut down for repair; on the other hand, if the filling weights are meeting the product specifications the production process will be allowed to continue.

Personnel Application

In the area of wage and salary administration, consider the situation of a personnel manager who initiates a survey of starting salaries for various positions. By offering starting salaries that are within industry standards, the firm can remain competitive in recruiting new employees.

Numerous other applications are common, involving auditing, valuation of inventories, measurement of advertising effectiveness, purchasing behavior, manufacturing processes, stockholder opinions, and so on. It is the numerous uses and applications of statistics that make statistical procedures important and valuable in a decision-making context.

Applications of statistics such as those described above are an integral part of this text. These examples provide an overview of the breadth of statistical applications; however, without some understanding of the specific type of statistical technique that is used, it is impossible to discuss in any detail how the statistical methodology is actually applied. In this and subsequent chapters we have asked practitioners from the fields of business and economics to provide chapter-ending applications that serve to illustrate the material that has just been covered. We believe that these actual applications of statistics will provide you with a good appreciation of the importance of statistics in several different types of decision-making situations. Table 1.2 provides an overview of these chapter-ending applications.

TABLE 1.2 An Overview of Chapter-ending Applications

Chapter	*Chapter Title*	*Organization*	*Application*
1	Introduction	Kings Island Inc.	Consumer Profile Sample Survey
2	Descriptive Statistics I. Tabular and Graphical Approaches	The Colgate Palmolive Company	Quality Assurance for Heavy Duty Detergents
3	Descriptive Statistics II. Measures of Location and Dispersion	St. Luke's Hospital	Time Spent in Hospice Program
4	Introduction to Probability	Morton Thiokol Inc.	Evaluation of Customer Service Testing Program
5	Probability Distributions	Xerox Corporation	Performance Test of an On-Line Computerized Publication System
6	Special Probability Distributions	The Burroughs Corporation	Credit Cards for Banking
7	Sampling and Sampling Distributions	Mead Corporation	Estimating the Value of Mead Forest Ownership
8	Interval Estimation	Thriftway Inc.	Sampling for Estimation of LIFO Inventory
9	Hypothesis Testing	Harris Corporation	Testing for Defective Plating
10	Statistical Inference About Means and Proportions with Two Populations	Pennwalt Corporation	Evaluation of New Drugs
11	Inferences About Population Variances	U.S. General Accounting Office	Water Pollution Control
12	Tests of Goodness of Fit and Independence	United Way	Determining Community Perceptions of Charities
13	Experimental Design and Analysis of Variance	Burke Marketing Services Inc.	New Product Design
14	Simple Linear Regression and Correlation	Monsanto Company	Estimation of Effect of Feed Additive on Poultry Weight Gain
15	Multiple Regression	Champion International Corporation	Control of Pulp Bleaching Process
16	Time Series Analysis and Forecasting	The Cincinnati Gas & Electric Company	Forecasting Demand for Electricity
17	Indexes and Index Numbers	State of New York Department of Commerce	Index of Business Activity
18	Nonparametric Methods	West Shell Realtors	Comparison of Real Estate Prices across Neighborhoods
19	Decision Theory	Ohio Edison Company	Choice of Best Type of Particulate Control Equipment

Summary

Although the term *statistics* in everyday usage refers to numerical facts or data, the field of statistics is much broader. It involves the collection, analysis, and interpretation of data. Data summarization procedures have been developed to aid in the presentation and interpretation of data. The application of these procedures is usually called descriptive statistics. Chapters 2 and 3 are devoted to this topic.

Two key components of nearly all statistical studies are the population and the sample. Statistical inference involves the process of making inferences about the population and its characteristics based on sample results. Probability plays an important role in statistical inference by enabling the statistician to make statements concerning the precision of the results and the likelihood of error. Because of the numerous applications of statistics nearly every college student in business and economics is required to take a course in statistics. A brief introduction to a few of these applications was presented in Section 1.4.

Glossary

Population—The collection of all items (individuals, households, products, customers, etc.) of interest in a particular study.

Sample—A portion of the population selected to represent the whole population. Data collected from the sample provide the basis for making inferences about the characteristics of the population.

Supplementary Exercises

1. Give two examples of statistical inference. That is, find two examples of cases where sample results have been used to make inferences about a population and its characteristics.
2. What is descriptive statistics? Give two examples of its use.
3. A firm is interested in testing the advertising effectiveness of a new television commercial. As part of the test, the commercial is shown on a 6:30 P.M. local news program in Denver, Colorado. Two days later a market research firm conducts a telephone survey to obtain information on recall rates (percentage of viewers who recall seeing the commercial) and impressions of the commercial.
a. What is the population for this study?
b. What is the sample for this study?
c. Would you prefer a sample or a census? Explain.
4. The quality control department of a large manufacturing firm is responsible for maintaining product specifications for a variety of production line operations.
a. List some of the information that the quality control department might want in order to determine whether or not product specifications are being met.
b. Why would the firm be interested in sampling concepts from the area of statistics?
5. The Nielsen organization is well known for the studies it conducts on television viewing.
a. What is the Nielsen organization attempting to measure?
b. What is the population?
c. What would be a sample for such a study?
d. Why is a sample preferred to a census?
e. What kinds of decisions or actions are taken based on the Nielsen statistical studies?
6. A survey of starting salaries for 1983 college graduates with degrees in business administra-

tion was conducted in the summer of 1983. The survey reported an average annual starting salary of $18,000. This survey result was based on a nationwide sample of 400 college graduates who had accepted job offers during the spring of 1983.

a. What is the population for this study?

b. What would be a good estimate of the average starting salary for the population?

c. How much of a margin for error would you want to allow to be reasonably sure that your estimate in part b "plus or minus" the "margin for error" included the average starting salary for the population?

Application

Kings Island, Inc.*

Kings Island, Ohio

Kings Island Family Entertainment Center is a 1600 acre year-round recreational, sports, and shopping complex located in southwestern Ohio. The Kings Island theme park, one of America's top parks, provides rides, entertainment, and other attractions, drawing nearly 3 million people annually. Since its opening in 1972, over 30 million people have enjoyed the world of fun at Kings Island. In addition to the theme park, the Kings Island Entertainment Center includes the National Football Foundation's College Football Hall of Fame; The Jack Nicklaus Sports Center, site of both professional golf and professional tennis tournaments; The Resort Inn by Kings Island; the Kings Island Campground; and the Factory Outlet Mall. In short, Kings Island Entertainment Center is unmatched in the Midwest for entertainment, sports, shopping and relaxation.

The Taft Broadcasting Company owns and operates the Kings Island theme park. Taft entered the theme park industry in 1969 with the purchase of Coney Island, a century-old amusement park on the banks of the Ohio River. Coney was sold to Taft because of the threat of competition from the new theme parks and the annual flooding which took place at Coney Island. Three years later, Taft replaced the old Coney Island park with Kings Island.

The Kings Island theme park contains six theme areas. They are as follows:

International Street—A colorful European boulevard of shops and restaurants with five European buildings—Italian, French, Spanish, Swiss, and German—encircling the Royal Fountain. A 330-foot tower, a one-third size replica of the Eiffel Tower in Paris, is the Kings Island landmark located at the end of International Street.

Wild Animal Habitat—More than 350 wild animals from four continents roam freely on a 100 acre preserve. Air-conditioned monorail trains silently transport 2,000 guests per hour on the 2 mile journey.

Oktoberfest—The unforgettable sights, sounds, and pageantry of Germany's famed Oktoberfest tradition are celebrated all season long in this area of the park. The centerpiece of Oktoberfest is a new $2.5 million Festhaus featuring

*The authors are indebted to Bill Mefford, Manager of Marketing Communications, and Tom Russell, Market Representative in Charge of Planning and Analysis, for providing this application.

a hearty international menu and an original 30 minute musical production entitled World Cabaret.

Coney Island—A depiction of the famous amusement park that was located on the Ohio River for almost 100 years. It features the rides and attractions that made the 1920 amusement parks so popular. Roller coasters, Flying Carpets, Dodgems, the Scrambler, a double Ferris wheel, and numerous games and arcades are some of the attractions in this area of the park.

Rivertown—This scenic area depicts what life was like in the Ohio riverboat days, from thrilling action as seen from the Kings Island and Miami Valley Railroad to boat rides that shoot the rapids. The awesome Beast, however, is the main attraction. The biggest, "baddest" wooden roller coaster anywhere, it covers 7,400 feet of track at speeds of 65 mph; its two longest vertical drops are 135 and 141 feet.

Hanna-Barbera Land—This is a storybook kingdom that is brought to life for children and their families. The popular Enchanted Voyage boat ride and the Beastie roller coaster are in this area. In addition, an outdoor theatre with a stage show entitled "Yogi's Picnic" features Yogi Bear, Scooby Doo, Huckleberry Hound, and Ranger Smith.

Children enjoy a visit to the Yogi Bear Fountain in Kings Island Storybook Kingdom of Hannah Barbera Land

Kings Island is committed to offering the best possible quality family entertainment and recreational activities for its visitors. In the future the park will continue to expand, offering more new rides and attractions. Coming soon will be the first roller coaster of its kind in the United States, a standing loop roller coaster in which standing riders will coast over a 2,000-foot track with a 66-foot high vertical loop and a 540-degree loop over a lake. For visitors to the Midwest, Kings Island is a must-see attraction for the young and old alike.

KINGS ISLAND RESEARCH GROUP

Located on the Kings Island grounds but away from the crowds are the administrative offices of Kings Island. Here executives, managers, and staff plan the strategies and make the business decisions that continue to make Kings Island a successful business venture.

A particularly important part of the Kings Island organization is the research group, which operates with a staff of 11 individuals. This group is devoted to learning about the Kings Island consumer's behavior, attitudes, perceptions, and preferences and providing marketing inferences to Kings Island management. Such information guides operating policies and future plans for the park.

The major work done by the research group is statistical in nature. A population of park visitors can be easily identified; but the time, cost and inconvenience of using a census to collect data about the characteristics and preferences of this population is unacceptable. Yet to be effective and successful, Kings Island management must be knowledgeable about the needs and interests of its customers. Thus on a routine basis samples of park visitors are selected. The results from the samples provide the desired estimates and information about the population of park visitors.

The research group designs questionnaires, selects samples, conducts interviews, and analyzes data that provide the desired consumer attitudes and characteristics information. In total, the research group with its numerous special studies and samples will interview in the neighborhood of 30,000 visitors annually. One such sample, known as the consumer profile study, is described below.

THE CONSUMER PROFILE SAMPLE SURVEY

What types of individuals come to Kings Island, what their ages are, what their family size is, how far they drive, whether or not they have visited before, and so on comprise important information for Kings Island management. Such information helps determine advertising messages, identifies trends or changes in park attendees over time, and tells Kings Island how it is drawing from market areas such as Cincinnati, Dayton, Columbus, Lexington, Louisville, and Indianapolis as well as 12 other market areas throughout the Midwest.

Samples of park visitors are taken throughout the day. As the visitors enter the park, interviewers ask sampled individuals to answer a short questionnaire about themselves and why they came to the park today. Interest, cooperation, and participation by the park visitors is very good, and the number of visitor responses is high.

A variety of questions are asked in the 2 or 3 minute consumer profile interview. Examples of data collected include home zip code, distance traveled to the park, type of admission ticket (group sales, special promotion, regular ticket, season pass, etc.), group size, number of previous visits to park, respondent's age, and so on. The interviewer also notes the day/time of the interview as well as the sex of the respondent.

Methods from descriptive statistics are used to summarize the sample results for Kings Island managers and the marketing department. Each piece of statistical data collected is of interest to someone in the organization. In particular, the home zip code provides an indicator of how each market area is doing in terms of sending visitors to the park. In short, the consumer profile sample is the way Kings Island learns about its visitors. A wide variety of plans, strategies, and decisions are based on the critical information provided by the sample results.

OTHER SAMPLES

In addition to the consumer profile sample, the research group also takes numerous special purpose samples to determine visitor attitudes toward a variety of park features, including food service, rides, shows, games, and so on. These attitudinal samples are usually taken as park visitors are leaving at the end of the day. Statistical summaries of these samples guide the park's Operations Department in determining improvements and/or modifications that Kings Island management may want to consider.

To the visitor, Kings Island is a vast area of fun and excitement far removed from the busy world of business. However, behind the scenes an effective and efficient business organization operates Kings Island, much like its counterparts elsewhere in the business world. Consumer research through the use of samples and statistical methods plays a critical role in providing the data and information that keeps Kings Island one of the country's top theme parks and family entertainment centers.

2 Descriptive Statistics I. Tabular and Graphical Approaches

What you will learn in this chapter:

- how to construct and interpret tabular data summarization procedures such as
  - frequency distributions
  - relative frequency distributions
  - cumulative frequency distributions
- how to construct and interpret graphical data summarization procedures such as
  - histograms
  - frequency polygons
  - ogives

Contents

The purpose of this chapter is to introduce several tabular and graphical procedures that commonly are used to summarize data. Tabular and graphical presentations of data often can be found in annual reports, newspaper articles, research studies, and so on. All of us are exposed to these presentations. Hence it is important to understand how they are prepared and what they mean.

2.1 EXAMPLE: DATA ON STARTING SALARIES

Suppose that the placement office at your college sent a questionnaire to a sample of recent graduates requesting information on job offers and starting salaries. As the questionnaires were returned a secretary could record each student's major and monthly starting salary. Table 2.1 shows a set of data collected in this fashion for business school graduates. The cents have been dropped, and only the number of dollars earned is shown.

If you stopped by the placement office to ask for information about starting salaries, what would your reaction be if someone handed you the data in the form of Table 2.1? Most of us would wish that someone had summarized the data in a more convenient and readily interpretable form. For instance, we might be interested in knowing how many of the starting salaries were between $1400 and $1500, how many were over $1600, and so on. In the following sections we shall use the data in Table 2.1 to illustrate some of the procedures available for summarizing data and presenting it in a readily interpretable form.

2.2 FREQUENCY DISTRIBUTION

A *frequency distribution* is one of the most useful ways of summarizing a set of data. A frequency distribution for the monthly starting salary data of Table 2.1 is shown in Table 2.2. The data have been grouped into eight separate classes, with each of the starting salaries belonging to one and only one class.

Before discussing how to construct such a frequency distribution, let us see what observations about the salary data can be made from Table 2.2. We note the following:

1. The highest starting salaries are in the interval of at least $2,000 but less than $2,100, with 5 of 80 graduates earning salaries at this level.
2. The lowest starting salaries are in the interval of at least $1,300 but less than $1,400, with 8 of the 80 graduates earning salaries at this level.
3. The most frequently occurring starting salaries are in the interval of at least $1,600 but less than $1,700, with 19 of the 80 graduates earning salaries at this level.
4. Almost half of the graduates (36 of 80) have starting monthly salaries of at least $1,500 but less than $1,700.

Other relevant observations are possible. We see that the potential value of a frequency distribution is that it provides some insights about the data that cannot be obtained easily by viewing the data in the form shown in Table 2.1.

A frequency distribution can be used also to summarize the starting salary data for subdivisions of the original data set. For example, the frequency distribution for the

TABLE 2.1 Monthly Starting Salaries for Recent Business School Graduates

Graduate	*Major*	*Monthly Salary (dollars)*	*Graduate*	*Major*	*Monthly Salary (dollars)*
1	Finance	1,550	41	Finance	1,590
2	Management	1,310	42	Accounting	1,570
3	Management	1,575	43	Accounting	2,015
4	Marketing	1,675	44	Management	1,620
5	Accounting	1,585	45	Finance	1,860
6	Marketing	1,590	46	Marketing	1,625
7	Management	1,580	47	Management	2,000
8	Accounting	1,475	48	Marketing	1,850
9	Management	1,300	49	Finance	1,640
10	Finance	1,650	50	Marketing	1,900
11	Accounting	1,565	51	Accounting	1,450
12	Marketing	1,320	52	Accounting	1,815
13	Finance	1,750	53	Marketing	1,440
14	Accounting	1,725	54	Management	1,420
15	Marketing	1,650	55	Management	1,550
16	Management	1,740	56	Accounting	1,550
17	Accounting	1,650	57	Accounting	1,660
18	Accounting	1,875	58	Accounting	1,760
19	Marketing	1,620	59	Marketing	1,550
20	Finance	1,550	60	Management	1,650
21	Management	1,380	61	Management	1,775
22	Marketing	1,730	62	Finance	2,025
23	Management	1,640	63	Marketing	1,450
24	Accounting	2,000	64	Management	1,425
25	Marketing	1,400	65	Management	1,820
26	Management	1,325	66	Management	1,900
27	Accounting	1,900	67	Accounting	1,700
28	Marketing	1,600	68	Management	1,900
29	Accounting	1,600	69	Accounting	1,475
30	Accounting	1,555	70	Accounting	1,850
31	Marketing	1,700	71	Management	1,500
32	Marketing	1,380	72	Finance	1,620
33	Management	1,620	73	Management	1,600
34	Accounting	1,650	74	Finance	1,580
35	Accounting	2,000	75	Accounting	1,705
36	Finance	1,455	76	Management	1,780
37	Accounting	1,625	77	Management	1,400
38	Management	1,340	78	Accounting	1,550
39	Accounting	1,530	79	Accounting	1,390
40	Finance	1,410	80	Management	1,600

TABLE 2.2 Frequency Distribution of Monthly Salaries for Business School Graduates

Monthly Starting Salary (dollars)	*Number of Graduates*
1,300 but less than 1,400	8
1,400 but less than 1,500	11
1,500 but less than 1,600	17
1,600 but less than 1,700	19
1,700 but less than 1,800	10
1,800 but less than 1,900	6
1,900 but less than 2,000	4
2,000 but less than 2,100	5
Total	80

starting salaries of accounting majors is given in Table 2.3. If you are an accounting major, you probably are more interested in this frequency distribution than in the one presented in Table 2.2. Use Table 2.3 to make some observations about the starting salaries of the accounting majors who are represented in this set of data.

With our previous discussion as background we now present a definition of a frequency distribution:

Definition of Frequency Distribution

A frequency distribution is a tabular summary of a set of data showing the frequency (or number) of items in each of several nonoverlapping classes.

In constructing frequency distributions there are no "hard and fast" rules. The objective is to present the data in a tabular format so that it contains as much information as possible and does not mislead the reader. Important considerations are

TABLE 2.3 Frequency Distribution of Monthly Starting Salaries for Accounting Majors

Monthly Starting Salary (dollars)	*Number of Graduates*
1,300 but less than 1,400	1
1,400 but less than 1,500	3
1,500 but less than 1,600	7
1,600 but less than 1,700	5
1,700 but less than 1,800	4
1,800 but less than 1,900	3
1,900 but less than 2,000	1
2,000 but less than 2,100	3
Total	27

choosing the proper width for each class, determining the number of classes, and choosing class limits.

The choice of class width and number of classes cannot be made independently. Larger class widths mean fewer classes, and vice versa. Generally, larger data sets require more classes. Determining the exact number of classes and choosing a class width involves a tradeoff between too few and too many classes. For example, suppose that only two classes ("1,300 but less than 1,700" and "1,700 but less than 2,100") had been used for the frequency distribution of monthly salary offers (see Table 2.2). We would have obtained the frequency distribution given in Table 2.4. In Table 2.4 we

TABLE 2.4 **Two-Class Frequency Distribution of Monthly Starting Salaries for Business School Graduates**

Monthly Starting Salary (dollars)	*Number of Graduates*
1,300 but less than 1,700	55
1,700 but less than 2,100	25
Total	80

have 55 people in the "1,300 but less than 1,700" class and 25 in the "1,700 but less than 2,100" class. Such a frequency distribution provides too much grouping of the data at the expense of showing its dispersion. We obtain no information about how the 55 salaries are spread over the first interval and no information about how the other 25 salaries are spread over the other interval.

At the other extreme we could have a separate class for each offer received. This, of course, would not provide enough summarization of the data.

Two guidelines for selection of class width and number of classes are given below:

1. Use between 5 and 20 classes for the frequency distribution.
2. Make the classes of equal width.

Occasionally something unique in the data will cause us to deviate from these guidelines, but they are appropriate most of the time.

If the lowest and highest values in a data set are known, the following expression often is helpful in determining both the width of the class interval and the number of classes desired:

$$\text{Approximate Number of Classes} = \frac{\text{Highest Value} - \text{Lowest Value}}{\text{Width of Class}}. \tag{2.1}$$

For the starting salary data of Table 2.2, the lowest starting salary is $1,300 (graduate 9) and the highest starting salary is $2,025 (graduate 62). Using (2.1) with a trial class width of 100 shows that (2,025 − 1,300)/100 = 7.25. Rounding up, we find that eight classes would be required for the frequency distribution. Other class widths may be considered in (2.1); the decision on the class width and the number of classes is up to the user. We felt that for our data set the eight classes of width $100 offered a reasonable compromise between too many and too few classes.

In choosing class limits we seek to ensure that each item of data belongs to one and only one class. The frequency distributions shown in Tables 2.2 and 2.3 both have intervals defined as "$1,300 but less than $1,400," "$1,400 but less than $1,500," and so on, as opposed to $1,300–$1,400, $1,400–$1,500, and so on. The reason for this is that the latter form leaves some confusion as to which class a value of $1,400 belongs. The definition of class limits of "$1,300 but less than $1,400," "$1,400 but less than $1,500," and so on makes it clear that each data value belongs to one and only one class. For example, the $1,400 data value belongs in the "$1,400 but less than $1,500" class.

Class Midpoint

The midpoint of a class is the value that falls in the middle of the class interval. It is the average of the class limits. Thus for the class "$1,300 but less than $1,400" we obtain a midpoint of $(1,300 + 1,400)/2 = 1,350$. The other class midpoints for our frequency distribution of monthly salary data are $1,450, $1,550, and so on. In performing further computations, analysts usually treat the class midpoint as if it represented the average value for items in that class.

Determining the Frequencies

We have made decisions for the class width, the number of classes, and the class limits. The only task remaining is to refer to the data set and count the number of data values belonging to each class. This is accomplished easily by preparing a tally sheet like the one shown in Table 2.5. We consider the 80 data items in Table 2.1 one at a time. The tally sheet now can be used to record the class interval within which each data item falls.

TABLE 2.5 Tally Sheet for the Frequency Distribution of Monthly Starting Salaries for Business School Graduates

Monthly Starting Salary (dollars)	*Tally*	*Frequency*
1,300 but less than 1,400	𝍸 III	8
1,400 but less than 1,500	𝍸 𝍸 I	11
1,500 but less than 1,600	𝍸 𝍸 𝍸 II	17
1,600 but less than 1,700	𝍸 𝍸 𝍸 IIII	19
1,700 but less than 1,800	𝍸 𝍸	10
1,800 but less than 1,900	𝍸 I	6
1,900 but less than 2,000	IIII	4
2,000 but less than 2,100	𝍸	5
Total		80

The frequency distributions we have been discussing thus far have been based on numerical or quantitative data. A frequency distribution may also be based on qualitative data, however. For example, using the information in Table 2.1 we can develop a qualitative frequency distribution based upon the type of major. Summarizing these data leads to the frequency distribution shown in Table 2.6.

TABLE 2.6 Frequency Distribution of Type of Major for Business School Graduates

Major	*Frequency*
Accounting	27
Finance	12
Management	25
Marketing	16
Total	80

2.3 RELATIVE FREQUENCY DISTRIBUTION

Until now we have been expressing the frequency of each class in terms of the total number of data items in the data set that fall within that class. Sometimes we may find it of interest to consider also the relative frequency of items in each class. The ***relative frequency*** is simply the fraction or proportion of the total number of items belonging to the class. For a data set having a total of n observations, or items, the relative frequency of each class is given by

$$\text{Relative Frequency of a Class} = \frac{\text{Frequency of the Class}}{n}. \tag{2.2}$$

Using (2.2) we can compute a ***relative frequency distribution*** for the monthly starting salaries presented in Table 2.1. It is shown in Table 2.7. The relative frequency

TABLE 2.7 Relative Frequency Distribution of Monthly Starting Salaries for Business School Graduates

Monthly Starting Salary (dollars)	*Frequency*	*Relative Frequency*
1,300 but less than 1,400	8	8/80 = .1000
1,400 but less than 1,500	11	11/80 = .1375
1,500 but less than 1,600	17	17/80 = .2125
1,600 but less than 1,700	19	19/80 = .2375
1,700 but less than 1,800	10	10/80 = .1250
1,800 but less than 1,900	6	6/80 = .0750
1,900 but less than 2,000	4	4/80 = .0500
2,000 but less than 2,100	5	5/80 = .0625
Totals	80	1.0000

TABLE 2.8 Relative Frequency Distribution of Type of Major for Business School Graduates

Major	*Frequency*	*Relative Frequency*
Accounting	27	.3375
Finance	12	.1500
Management	25	.3125
Marketing	16	.2000
Totals	80	1.0000

distribution for type of major is shown in Table 2.8. Note that the relative frequency distributions enable us to make percentage statements about the data. For example, .0625, or 6.25%, of the graduates received starting salaries of at least $2,000 but less than $2,100, 45% of the graduates received starting salaries of at least $1,500 but less than $1,700, and accountants (33.75%) provided the largest percentage of graduates reporting starting salary information. Note that the sum of the relative frequencies for all classes is always 1.0.

2.4 CUMULATIVE FREQUENCY DISTRIBUTION

One of the variations of the basic frequency distribution, called the *cumulative frequency distribution,* is sometimes useful in summarizing a data set. The cumulative frequency distribution contains the same number of classes as the frequency distribution. However, for each class the cumulative frequency distribution shows the total number of data items with a value less than the upper limit. For example, the cumulative frequency distribution for our data on monthly starting salaries for business school graduates is given in Table 2.9.

TABLE 2.9 Cumulative Frequency Distribution for Monthly Starting Salaries of Business School Graduates

Monthly Starting Salary (dollars)	*Cumulative Frequency*
Less than 1,400	8
Less than 1,500	19
Less than 1,600	36
Less than 1,700	55
Less than 1,800	65
Less than 1,900	71
Less than 2,000	75
Less than 2,100	80

TABLE 2.10 Cumulative Relative Frequency Distribution of Monthly Starting Salaries of Business School Graduates

Monthly Starting Salary (dollars)	*Cumulative Relative Frequency*
Less than 1,400	.1000
Less than 1,500	.2375
Less than 1,600	.4500
Less than 1,700	.6875
Less than 1,800	.8125
Less than 1,900	.8875
Less than 2,000	.9375
Less than 2,100	1.0000

The classes for the cumulative frequency distribution are open ended in the sense that there are no lower class limits. If a frequency distribution is available, a cumulative frequency distribution can be computed quickly by just summing the class frequencies. The cumulative frequency distribution provides another way of viewing the data. For example, we quickly see from Table 2.9 that 55 graduates make less than $1,700 per month. Of these, 36 make less than $1,600, 19 make less than $1,500, and 8 make less than $1,400. This same information can be obtained from the frequency distribution, but not without some extra computations.

As a final point we note that a *cumulative relative frequency distribution* is one that shows the fraction of items with value less than the upper class limit. It can be developed from the relative frequency distribution in the same way that a cumulative frequency distribution is developed from the frequency distribution. It can be developed also by dividing the cumulative frequencies by the number of items in the data set. Table 2.10 contains the cumulative relative frequency distribution of monthly starting salaries for business school graduates.

EXERCISES

1. Using the data in Table 2.1, develop the frequency distribution of monthly starting salaries for the following:

a. Finance majors.
b. Marketing majors.
c. Management majors.

Using your frequency distributions, make some comments about the monthly starting salaries of the business school graduates in this study.

2. Using the data in Table 2.1 or the frequency distributions developed in Exercise 1, develop the relative frequency distribution for the following:

a. Accounting majors.
b. Finance majors.
c. Marketing majors.
d. Management majors.

3. Using the data in Table 2.1 or the frequency distributions developed in Exercise 1, develop the cumulative frequency distribution for the following:

a. Finance majors.
b. Marketing majors.

4. The data given below show the number of automobiles arriving at a toll booth during 20 intervals of 10 minutes duration each:

26	26	38	24
32	22	15	33
19	27	21	28
16	20	34	24
27	30	31	33

a. Develop a frequency distribution for the data.
b. Develop a relative frequency distribution for the data.
c. Develop a cumulative frequency distribution for the data.
d. Develop a cumulative relative frequency distribution for the data.

5. A doctor's office has studied the waiting times for patients who arrive at the office with a request for emergency service. The following data were collected over a 1 month period (the waiting times are in minutes):

2, 5, 10, 12, 4, 4, 5, 17, 11, 8, 9, 8, 12, 21, 6, 8, 7, 13, 18, 3.

Starting with 0 and using a class width of 5,

a. Show the frequency distribution.

b. Show the relative frequency distribution.

c. Show the cumulative frequency distribution.

d. Show the cumulative relative frequency distribution.

e. What proportion of patients needing emergency service have a waiting time of less than 10 minutes?

6. A survey of 250 company employees asks each employee to indicate the one-way mileage from his or her home to work. A partial relative frequency distribution is shown below:

Miles	*Relative Frequency*
0 but less than 5	.10
5 but less than 10	.22
10 but less than 15	
15 but less than 20	.18
20 but less than 25	.16

a. Complete the relative frequency distribution.

b. How many employees live 10 but less than 15 miles from work?

c. Show the frequency distribution for the data.

2.5 GRAPHICAL PRESENTATIONS OF FREQUENCY DISTRIBUTIONS

The concepts of frequency, relative frequency, cumulative frequency, and cumulative relative frequency distributions have been introduced as tabular summary procedures. They can be used to increase our understanding of the information contained in a data set. Graphical summaries often provide additional insight about the nature of a data set. The most common graphical representations of data sets are the histogram, frequency polygon, and ogive.

Histogram

The most common form of graphical presentation of a frequency distribution is the *histogram.* A histogram is constructed by placing the class intervals on the horizontal axis of a graph and the frequencies on the vertical axis. Each class is shown on the graph by drawing a rectangle whose base is the class interval and whose height is the corresponding frequency for the class.* The histogram for the frequency distribution of monthly starting salaries is shown in Figure 2.1.

The histogram provides a "picture" of how the data are distributed from low values to high. From Figure 2.1 we see quickly that starting salaries in the $1,500 to $1,700 range occur most frequently, with the frequencies decreasing for salaries below and above this level.

*When class intervals are required to be unequal because of some particular feature of the data set, the method of constructing a histogram should be modified. References providing a discussion of the necessary modifications are given in Appendix A.

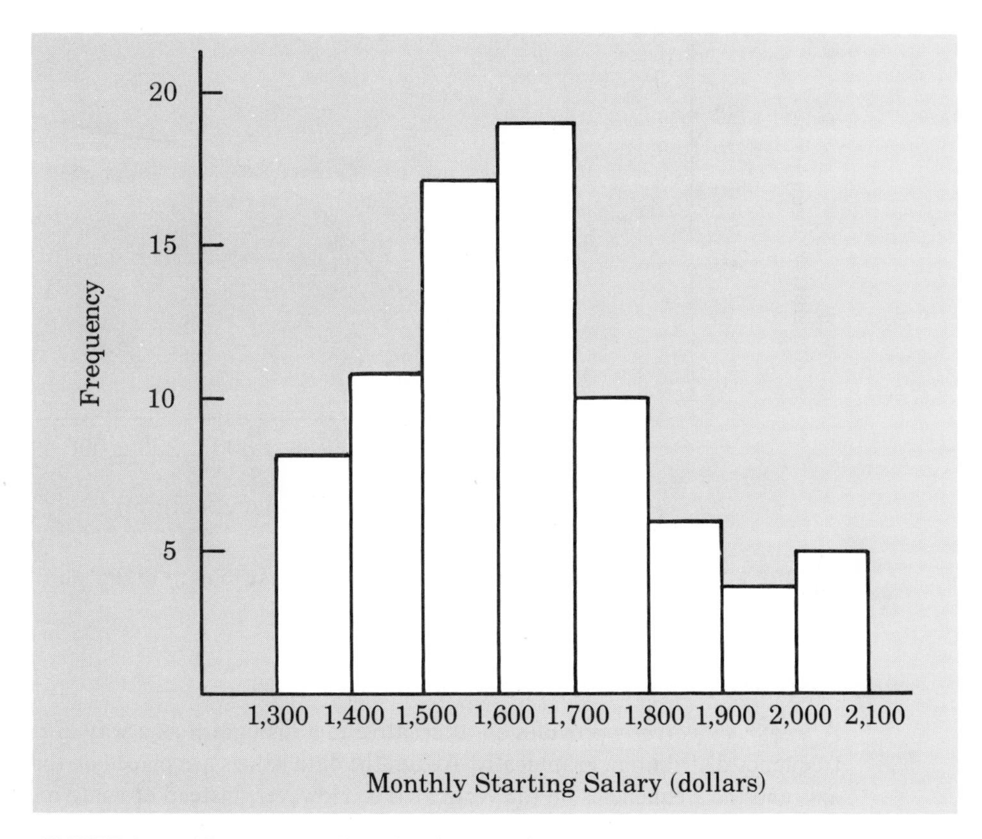

FIGURE 2.1 Histogram of Monthly Starting Salaries of Business School Graduates

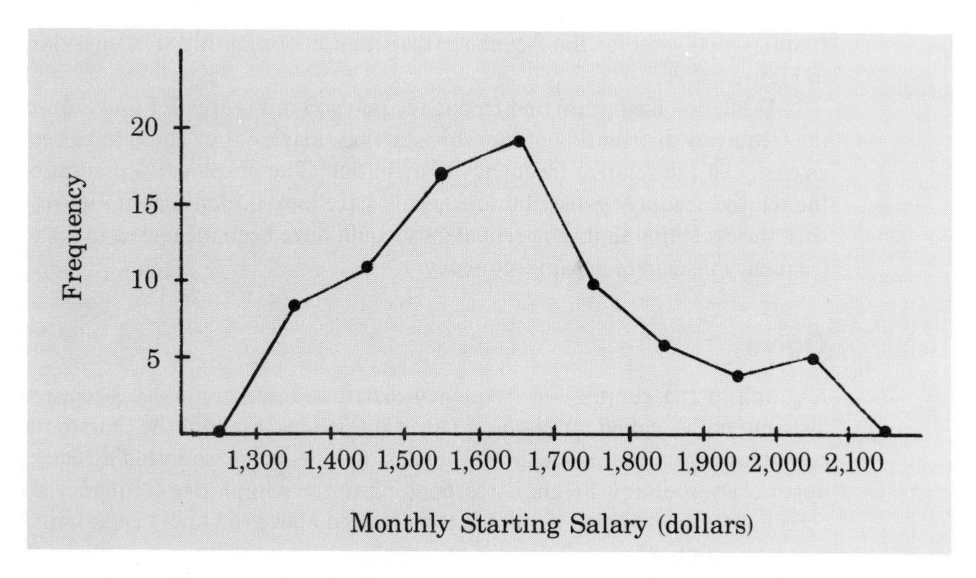

FIGURE 2.2 Frequency Polygon for the Monthly Starting Salaries of Business School Graduates

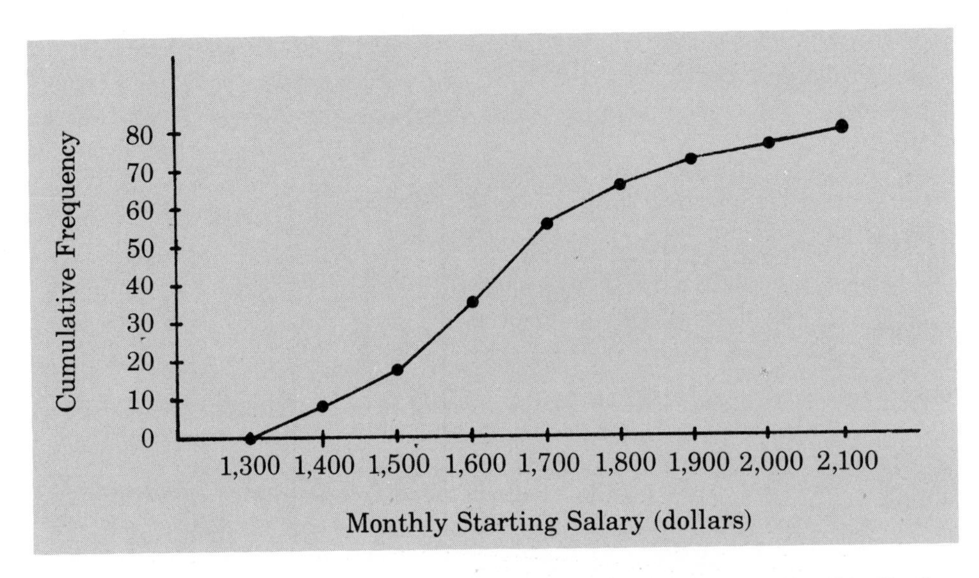

FIGURE 2.3 Ogive for the Cumulative Frequency Distribution of the Monthly Starting Salaries of Business School Graduates

Frequency Polygon

A *frequency polygon* provides an alternative to a histogram as a way of presenting a frequency distribution graphically. Again, the data values are placed on the horizontal axis and the frequencies on the vertical axis. However, instead of using rectangles, as with the histogram, we find the class midpoints on the horizontal axis and then plot points directly above the class midpoints at a height corresponding to the frequency of the class. Classes of zero frequency are added at each end of the frequency distribution so that the frequency polygon touches the horizontal axis at both ends of the graph. The frequency polygon is then formed by connecting the points with straight lines. The frequency polygon for the frequency distribution of monthly starting salaries is shown in Figure 2.2.

While the histogram and frequency polygon in Figures 2.1 and 2.2 were based on the frequency distribution of monthly starting salaries, they could have been based just as easily on the relative frequency distribution. The graphical presentations based on the relative frequency distributions would have looked identical to Figures 2.1 and 2.2 with the exception that the vertical axis would have been measured in terms of relative frequency instead of actual frequency.

Ogive

A graph of the cumulative frequency distribution or cumulative relative frequency distribution is called an *ogive*. The data values are on the horizontal axis and cumulative frequencies are on the vertical axis. A point is plotted directly above each upper class limit at a height corresponding to the cumulative frequency at that upper class limit. One additional point is then plotted above the lower class limit for the first class at a height of zero. These points are then connected by straight lines. The straight lines allow one to approximate the cumulative frequency between class limits by interpolating. For instance, the cumulative frequency at a monthly starting salary of $1,350 is approximately 4. The ogive for the distribution of monthly starting salaries is shown in Figure 2.3. The points for this ogive are based on the cumulative frequency distribution shown in Table 2.9.

EXERCISES

7. Using the information on monthly starting salaries for accounting majors given in Table 2.3, do the following:

a. Construct a histogram.
b. Construct a frequency polygon.
c. Construct an ogive.

8. Using the data in Table 2.1 or the frequency distribution developed in Exercise 1, do the following:

a. Construct a histogram for the monthly starting salaries of finance majors.
b. Construct a frequency polygon for the monthly starting salaries of finance majors.
c. Construct an ogive for the monthly starting salaries of finance majors.

9. Develop a histogram, frequency polygon, and ogive for the automobile arrival data in Exercise 4.

10. Show the following graphical presentations for the data on waiting times of emergency patients as given in Exercise 5:

a. A histogram of the frequency distribution.
b. A frequency polygon of the relative frequency distribution.
c. An ogive of the cumulative relative frequency distribution.
d. Use your answer to part c to show the proportion of patients needing emergency service who have a waiting time of less than 10 minutes.

11. RC Radio Corporation collects data on the monthly sales for its car telephone units. Over the past 12 months the sales have been as follows:

Month	*Sales*	*Month*	*Sales*
Jan	80	July	88
Feb	115	Aug	91
Mar	82	Sept	89
Apr	102	Oct	95
May	94	Nov	105
June	90	Dec	108

Beginning with a value of 80, develop a frequency distribution for these data using interval widths of 10. Also show the histogram of the relative frequency distribution.

12. National Airlines accepts flight reservations by phone. Shown below are the call durations (in minutes) for a sample of 20 phone reservations. Show histograms of the frequency and relative frequency distributions for the data.

2.1	4.8	5.5	10.4
3.3	3.5	4.8	5.8
5.3	5.5	2.8	3.6
5.9	6.6	7.8	10.5
7.5	6.0	4.5	4.8

Summary

A set of data, even if modest in size, is often difficult to interpret directly in the form in which it is gathered. Tabular and graphical statistical procedures provide ways of organizing and summarizing the data such that they are more easily used and interpreted. The concepts of frequency distribution, relative frequency distribution, cumulative frequency distribution, and cumulative relative frequency distribution were introduced as tabular methods of summarizing data. These distributions can also be presented graphically through the use of histograms, frequency polygons, and ogives. The purpose of the descriptive statistical procedures presented in this chapter is to provide tabular and graphical procedures which facilitate the interpretation of data.

Glossary

Frequency distribution—A tabular summary of a set of data showing the frequency or number of items in each of several nonoverlapping classes.

Class—A grouping of data items for purposes of constructing a frequency distribution.

Class width—The length of the interval forming a class.

Class midpoint—The middle value of a particular class. It is given by the average of the class limits.

Relative frequency distribution—A tabular summary of a set of data showing classes of the data and the fraction or proportion of the items belonging to each class.

Cumulative frequency distribution—A tabular summary of a set of data showing for each class the total number of data items with value less than the upper class limit.

Cumulative relative frequency distribution—A tabular summary of a set of data showing for each class the fraction of data items with value less than the upper class limit.

Histogram—A graphical presentation of a frequency or relative frequency distribution.

Frequency polygon—A graphical presentation of a frequency or relative frequency distribution.

Ogive—A graphical presentation of a cumulative frequency or cumulative relative frequency distribution.

Key Formulas

Approximate Number of Classes

$$\frac{\text{Highest Value} - \text{Lowest Value}}{\text{Width of Class}} \tag{2.1}$$

Relative Frequency of a Class

$$\frac{\text{Frequency of the Class}}{n} \tag{2.2}$$

Supplementary Exercises

13. Dinner check amounts for La Maison's French Restaurant are shown below:

42.65	36.12	52.90	44.26	52.00
34.10	39.86	29.40	48.75	82.00
38.40	44.50	79.80	74.45	71.81
46.62	56.12	63.00	63.06	59.42

a. Construct frequency and relative frequency distributions for the data.
b. Construct a cumulative relative frequency distribution for the data.

14. The following data represent quarterly sales volumes for 40 selected corporations:

17,864,000	15,065,000	42,200,000	13,523,000
49,747,000	20,510,000	5,520,000	7,985,000
3,624,000	11,556,000	1,855,000	9,023,000
3,804,000	5,933,000	23,900,000	6,145,000
9,232,000	2,979,000	1,059,000	42,789,000
5,143,000	33,380,000	20,779,000	6,145,000
2,141,000	17,768,000	18,017,000	42,800,000
5,090,000	41,626,000	12,003,000	6,840,000
3,669,000	37,738,000	40,765,000	21,946,000
13,614,000	39,914,000	7,846,000	25,837,000

a. Construct a frequency distribution to summarize these data. Use a class width of $5,000,000.
b. Develop a relative frequency distribution for the data.
c. Construct a cumulative frequency distribution for the data.
d. Construct a cumulative relative frequency distribution for the data.

15. Use the data in Exercise 14 for the following:
a. Construct a histogram as a graphical representation of the data.
b. Construct a frequency polygon for the data.
c. Construct an ogive for the data.

16. The data below show home mortgage loan amounts handled by a particular loan officer in a savings and loan company. Use a frequency distribution, relative frequency distribution, and histogram to help summarize these data:

20,000	38,500	33,000	27,500	34,000
12,500	25,999	43,200	37,500	36,200
25,200	30,900	23,800	28,400	13,000
31,000	33,500	25,400	33,500	29,200
39,000	38,100	30,500	45,500	30,500
52,000	40,500	51,600	42,500	44,800

17. Morrison Communications, Inc. periodically reviews sales personnel performance records. One member of the sales force has had the following weekly sales volume (units sold) over the past quarter, or 13 weeks. Use a relative frequency distribution, a cumulative frequency distribution, and a frequency polygon to summarize these sales data:

13, 19, 20, 17, 21, 27, 9, 15, 22, 18, 18, 23, 20.

18. Given below are the closing prices at week's end for 40 common stocks:

$7\frac{1}{2}$	$16\frac{1}{4}$	$19\frac{3}{4}$	$7\frac{3}{8}$	$10\frac{1}{8}$
$5\frac{3}{4}$	$7\frac{7}{8}$	$6\frac{7}{8}$	$24\frac{5}{8}$	$17\frac{1}{8}$
$12\frac{3}{4}$	11	5	$24\frac{3}{4}$	$34\frac{3}{4}$
$14\frac{3}{4}$	$10\frac{3}{8}$	$35\frac{1}{8}$	$20\frac{1}{4}$	$12\frac{5}{8}$
42	$10\frac{3}{8}$	57	$19\frac{3}{4}$	28
$63\frac{7}{8}$	$10\frac{3}{4}$	$7\frac{7}{8}$	$17\frac{5}{8}$	48
$17\frac{1}{4}$	$9\frac{3}{4}$	44	$11\frac{1}{8}$	20
$41\frac{3}{4}$	$16\frac{5}{8}$	$21\frac{1}{4}$	$8\frac{5}{8}$	$16\frac{3}{8}$

a. Construct frequency and relative frequency distributions for these data.
b. Construct cumulative frequency and cumulative relative frequency distributions for these data.

19. Use the data in Exercise 18 for the following:
a. Construct a histogram for the data. Plot relative frequency on the vertical axis.

b. Construct a frequency polygon for the data.
c. Construct an ogive for the data.

20. The grade point averages for 30 students majoring in economics are given below:

2.21	3.01	2.68	2.68	2.74
2.60	1.76	2.77	2.46	2.49
2.89	2.19	3.11	2.93	2.38
2.76	2.93	2.55	2.10	2.41
3.53	3.22	2.34	3.30	2.59
2.18	2.87	2.71	2.80	2.63

a. Construct a relative frequency distribution for the data.
b. Construct a cumulative relative frequency distribution for the data.

21. Use the data in Exercise 20 for the following:
a. Construct a histogram for the data. Plot relative frequency on the vertical axis.
b. Construct an ogive for the data.

22. Hospital records show the following number of days of hospitalization for 20 patients:

5	7	7	15
21	15	22	10
10	6	8	18
14	5	7	8
3	8	4	10

a. Construct frequency and relative frequency distributions for the data.
b. Construct a cumulative relative frequency distribution for the data.
c. Construct a histogram.

23. Points scored by the winning team in 25 A.C.C. (Atlantic Coast Conference) basketball games are shown below:

86	79	74	72	91
82	64	75	72	74
63	80	78	95	82
86	77	73	69	72
81	85	92	62	90

a. Construct frequency and relative frequency distributions for the data.
b. Construct a cumulative relative frequency distribution for the data.
c. Construct a histogram.

24. The duration of 20 long distance telephone calls is shown below (time recorded in minutes):

10.5	5.0	15.3	16.8	9.2
4.2	12.6	7.8	11.5	12.6
20.2	27.5	8.9	12.2	18.2
14.5	14.0	5.5	15.5	8.9

a. Construct a frequency distribution and a relative frequency distribution for the data.
b. Construct a histogram.

25. The numbers of television sets sold by Globe TV Sales are shown below. Data show weekly sales for a 15-week period:

13	10	6	9	10
11	5	8	3	9
10	7	8	10	6

a. Construct a frequency distribution and a relative frequency distribution for the data.
b. Construct a histogram.

The Colgate-Palmolive Company*

New York, New York

The Colgate-Palmolive Company dates back to 1806, or three years after the Louisiana Purchase. It started as a small shop in New York City that made and sold candles and soap. The business grew and prospered. Colgate and Company merged with the Palmolive Company in the late 1920s to form the present-day corporation, which now has annual sales in excess of $5 billion.

Colgate products can be found around the globe, with international operations in over 55 countries. While best known for its traditional product line of soaps, detergents, and toothpastes, subsidiary operations include the Kendall Company, Riviana Foods, Etonic, Bike Athletic Company, and others.

Colgate operates three large factories in the United States, with locations in Jersey City, New Jersey; Jeffersonville, Indiana; and Kansas City, Kansas. The Jersey City and Jeffersonville plants are area landmarks, with giant clocks that face the downtown metropolitan areas of nearby New York City and Louisville, Kentucky, respectively.

Most of Colgate's products are in the low-price, high-volume consumer market, which is known for its intense competition. Since the possibility of increasing a product's price is limited by the competition, Colgate's profitability is determined largely by the efficiency of its management techniques and production operations.

The use of statistics touches most areas of the business. Market research, forecasting, and quality control, among other areas, frequently use statistical procedures. In the following example we look at a quality assurance application at the point of manufacture.

QUALITY ASSURANCE FOR HEAVY DUTY DETERGENTS

The slogan on Colgate's crest reads "Quality Products since 1806," and every department is touched by the demands of this central business principle. The Quality Assurance and Improvement Department within Colgate devotes full time toward achieving this goal and includes a wide variety of special services. At each manufacturing plant, a group of chemists, inspectors, engineers, and managers are involved with product quality levels. The data that are collected must be communicated to others

*The authors are indebted to Mr. William R. Fowle, Manager of Quality Assurance, Colgate-Palmolive Company for providing this application.

throughout the organization. The format of the data summary is often vital to achieving the desired results.

A variety of statistical techniques are used. Relative frequency distributions and graphical techniques such as histograms are some of the most useful tools for communicating data and ideas. As an example of the use of these statistical techniques, consider the production of the familiar heavy duty detergent used for home laundries. Television advertisements claiming the relative merits of the competing brands are widely seen throughout the country.

The regular size carton of the detergent has a stated weight of 20 ounces. In the manufacturing process great pains are taken to ensure that the label specifications of 20 ounces of detergent per carton are maintained. However, meeting the 20-ounce-per-carton weight is not the only aspect of quality assurance addressed in the manufacture of this product. Of particular concern, from the point of view of quality, is the density of the detergent powder that is placed in the carton. Even with rigid quality control standards in the powder production process, at times the powder varies in its weight per unit volume. For example, if the weight of the powder is on the heavy side (a high specific gravity), it will not take as much powder to reach the 20-ounce-per-carton weight limit, and the company can be faced with the problem of filling cartons with 20 ounces but having the carton appear slightly underfilled when it is opened by the user.

To reduce this problem and maintain the quality standards, the powder is sampled periodically prior to being placed in the cartons. When the powder reaches an unacceptably high density or specific gravity, corrective action is taken to reduce the specific gravity of the powder before the filling operation is permitted to resume.

Repeated samples provide more and more data about the specific gravity of powder. At some point, various parties in the company are interested in knowing how the powder production process is doing in terms of meeting density guidelines. Tabular

Statistical summaries of product weights aid a Colgate-Palmolive management team during a meeting on quality assurance

TABLE 2A.1 Relative Frequency Distribution Showing the Specific Gravity of Heavy Duty Detergent (Based on 150 Sample Results)

Specific Gravity	*Relative Frequency*
.27 but less than .29	.02
.29 but less than .31	.18
.31 but less than .33	.50
.33 but less than .35	.21
.35 but less than .37	.06
.37 but less than .39	.02
.39 but less than .41	.01

and graphical summaries provide convenient ways to present the data to production, quality assurance and management personnel. Table 2A.1 shows a relative frequency distribution for the specific gravity of 150 samples taken over a 1 week period. Figure 2A.1 shows a histogram of these sample data. Note that the specific gravity of the powder varies, with a specific gravity of .32 occurring most frequently. The undesirably high specific gravity occurs around .40. Thus the summaries show that the operation is meeting its quality guidelines, with practically all the data showing values less than .40. Production management personnel would be pleased with the quality aspect of the powder product as indicated by these statistical summaries.

In cases where the relative frequency distribution and/or histogram summaries do not support the above conclusion, managers and quality assurance personnel begin to closely monitor the powder production process. Engineers may be consulted on ways of reducing the specific gravity to a more satisfactory level. After making any change in the process, data are collected and summarized in similar tabular and graphical

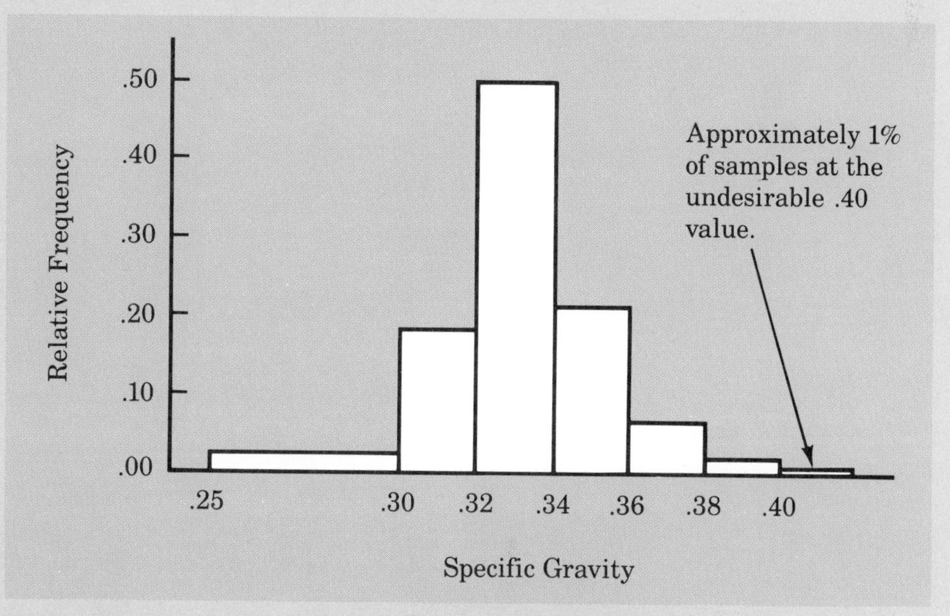

FIGURE 2A.1 Histogram Showing Relative Frequency of Specific Gravity of Heavy Duty Detergent

forms to determine how the modifications are affecting the quality of the product. The engineers' work continues until the statistical summaries show that the high quality level is once again obtained. The use of tabular and graphical methods of descriptive statistical summaries is essential in communicating data to the engineers, inspectors, and managers whose job it is to assure quality products at Colgate-Palmolive Company.

A SUMMARY NOTE

While there are many other examples of the use of statistics that could be shared, many of the ideas presented above would be the same. Statistical methods such as the descriptive tabular and graphical summaries presented above provide a basis for communicating data that help guide the decision-making process. It deserves mention here that people who are selected for promotion within our company are often good communicators, and those working with data are often good at presenting statistical results to management. It may seem simple when others do it effectively, but it is not always easy to pull together data and present them in a form that is understandable and easily grasped. This is a talent that improves with practice.

3 Descriptive Statistics II. Measures of Location and Dispersion

What you will learn in this chapter:

- how to compute and interpret the mean, median, and mode for a set of data
- what a percentile is and how it is computed
- how to compute and interpret measures of dispersion, such as the range, variance, standard deviation, and coefficient of variation
- how to compute descriptive statistics from grouped data
- Chebyshev's theorem and how to use it

Contents

In Chapter 2 we discussed tabular and graphical approaches to summarizing a set of data. These methods are very effective in written reports and as visual aids when presentations must be made to a group of people. Nonetheless, numerical values often are preferred for summarizing a data set. This is true particularly when the data set is a sample and is being used as a basis for making statistical inferences about a population.

Several numerical measures of location and dispersion for a data set will be introduced in this chapter. We will show how to compute these measures from the original ungrouped data set and then from the grouped data available in a frequency distribution. The most important numerical measures introduced are the mean, the variance, and the standard deviation. Numerical measures that are computed for a population are called *population parameters;* when they are computed for a sample, they are called *sample statistics.*

3.1 MEASURES OF LOCATION

In Chapter 2 we introduced an example that involved the monthly starting salary data for a sample of 80 recent business school graduates. For convenience, we have separated the data by majors; the result is shown in Table 3.1. To keep the numerical calculations at a minimum, in this chapter we will deal primarily with the sample of 12 finance majors and the sample of 25 management majors.

Our objective now is to use the data in Table 3.1 to develop numerical measures that describe the general nature of the entire data set as well as the general nature of the smaller data sets for each major. Since each of the data sets for the four majors is itself a sample, the numerical measures we compute for these groups will be called sample statistics. In this section we will focus on the use of the mean, median, and mode as measures of central location. We then will illustrate how percentiles can be used to describe the location of other values in the data set.

Mean

Perhaps the most important numerical measure is the *mean,* or simply "average" value, of the data. For example, a finance major would probably be interested in the mean starting salary for the sample of finance graduates. With knowledge of the mean he or she is better able to identify what constitutes a "good" or "poor" offer. The mean provides a good measure of central location for a data set.

Referring to the sample data in Table 3.1, let us compute the mean starting salary for the sample of 12 finance graduates. We let x_1 denote the monthly starting salary of the first finance graduate in the sample, x_2 denote the monthly starting salary of the second finance graduate, and so on. Thus $x_1 = 1{,}550$, $x_2 = 1{,}650, \ldots, x_{12} = 1{,}580$. (The symbol x_1 is pronounced "x sub 1," x_2 is pronounced "x sub 2," and so on.) The mean monthly starting salary for the sample of 12 finance graduates is computed as shown below, with the sample mean denoted by $\bar{x}$ (pronounced "x bar"):

$$\text{Mean Monthly Starting Salary} = \bar{x} = \frac{x_1 + x_2 + \cdots + x_{12}}{12}$$

$$= \frac{1{,}550 + 1{,}650 + \cdots + 1{,}580}{12}$$

$$= \frac{19{,}680}{12} = 1{,}640.$$

TABLE 3.1 **Monthly Starting Salaries for Recent Business School Graduates**

Finance Majors

1. 1,550	7. 1,590
2. 1,650	8. 1,860
3. 1,750	9. 1,640
4. 1,550	10. 2,025
5. 1,455	11. 1,620
6. 1,410	12. 1,580

Management Majors

1. 1,310	10. 1,340	18. 1,820
2. 1,575	11. 1,620	19. 1,900
3. 1,580	12. 2,000	20. 1,900
4. 1,300	13. 1,420	21. 1,500
5. 1,740	14. 1,550	22. 1,600
6. 1,380	15. 1,650	23. 1,780
7. 1,640	16. 1,775	24. 1,400
8. 1,325	17. 1,425	25. 1,600
9. 1,620		

Accounting Majors

1. 1,585	10. 1,555	19. 1,550
2. 1,475	11. 1,650	20. 1,660
3. 1,565	12. 2,000	21. 1,760
4. 1,725	13. 1,625	22. 1,700
5. 1,650	14. 1,530	23. 1,475
6. 1,875	15. 1,570	24. 1,850
7. 2,000	16. 2,015	25. 1,705
8. 1,900	17. 1,450	26. 1,550
9. 1,600	18. 1,815	27. 1,390

Marketing Majors

1. 1,675	9. 1,700
2. 1,590	10. 1,380
3. 1,320	11. 1,625
4. 1,650	12. 1,850
5. 1,620	13. 1,900
6. 1,730	14. 1,440
7. 1,400	15. 1,550
8. 1,600	16. 1,450

A shorthand way of writing the above formula for the sample mean is

$$\bar{x} = \sum_{i=1}^{12} x_i \bigg/ 12.$$

In this formula the numerator denotes the sum of the x_i values starting with $i = 1$ and ending with $i = 12$. That is,

$$\sum_{i=1}^{12} x_i = x_1 + x_2 + \cdot \cdot \cdot + x_{12}.$$

The symbol Σ is the summation sign.

The usual notation for the number of data items in a sample is n. The general formula for the sample mean is as follows:

Sample Mean

$$\bar{x} = \frac{\Sigma x_i}{n}, \tag{3.1}$$

where n = sample size.

Note that in the above formula we have not indicated the starting value for i below the summation sign, Σ, nor the ending value for i above Σ. This abbreviated notation means that one is to sum over all values that the subscript i can assume. A summary of the summation notation and operations used in this text is contained in Appendix C.

The value of summary statistics such as the sample mean can be realized when we start to make additional interpretations about a data set. For example, suppose that someone computed a sample mean of $1,590 for the 25 management majors. We see that the sample mean for management majors is lower than for the finance graduates. Note here that we do not want to say that the mean starting salary for the population of *all* finance majors is $50 per month greater than the mean starting salary for the population of *all* management majors. Rather this difference existed for the 12 finance majors and 25 management majors included in our sample. We are careful not to say *all* finance and management majors simply because we did not see all. We saw just a sample. In later chapters on statistical inference the focus will be on what conclusions we can draw about differences in population parameters—such as the population means—given observed differences in sample statistics—such as sample means.

Equation (3.1) shows how the mean is computed for a sample of n items. The formula for computing the mean of a population is similar, but we use different notation to indicate that we are dealing with the entire population. The number of items in the population is denoted by N, and the symbol for the population mean is the Greek letter μ (pronounced "mū"):

Population Mean

$$\mu = \frac{\Sigma x_i}{N}, \tag{3.2}$$

where N = population size.

Median

The *median* is another statistical measure of central location for a set of data. The median for a set of data is that value which falls in the middle when the data items are arranged in ascending order. If there is an odd number of data items, the median is the middle item. If the number of data items is even, some confusion exists as to what is the middle value. We follow the convention of defining the median to be the average of the middle two values in this case. A more formal definition is given below:

Definition of Median

If there is an odd number of items in the data set, the median is the value of the middle item when all items are arranged in ascending order.

If there is an even number of items in the data set, the median is the average value of the two middle items when all items are arranged in ascending order.

Let us apply the above definition to compute the median for the sample of monthly starting salaries for the 12 finance graduates. Arranging the 12 items in ascending order provides the following list:

$$1{,}410 \;\; 1{,}455 \;\; 1{,}550 \;\; 1{,}550 \;\; 1{,}580 \;\; \underbrace{1{,}590 \;\; 1{,}620}_{\text{Middle Two Values}} \;\; 1{,}640 \;\; 1{,}650 \;\; 1{,}750 \;\; 1{,}860 \;\; 2{,}025.$$

Since $n = 12$ is even, we have identified the middle two items. The median for the sample is the average of these two values:

$$\text{Median} = \frac{1{,}590 + 1{,}620}{2} = 1{,}605.$$

While the mean is the more commonly used measure of central location, there are a number of situations in which the median is a better measure. The mean is influenced by extreme values in a data set, but the median is not. For instance, suppose that one of the finance graduates had earned a starting salary of \$10,000 per month (maybe the individual's father owns the company). Changing the highest monthly starting salary from \$2,025 to \$10,000 and recomputing the mean, we obtain $\bar{x} = \$2{,}305$. The median, however, is unchanged, since \$1,590 and \$1,620 are still the middle two items. Clearly, with the extremely high starting salary included in the data set the median provides a better measure of central location than the mean. We can generalize to say that whenever there are extreme values in a data set the median is usually a better measure of central location than the mean.

We might be interested in comparing starting salaries for the different majors on the basis of the sample median. For the management majors there is an odd number of data items ($n = 25$). Thus the median is the middle, or 13th, item when the 25 items

TABLE 3.2 Calculation of Median for Sample of 25 Management Majors

1. 1,300	10. 1,550	18. 1,650
2. 1,310	11. 1,575	19. 1,740
3. 1,325	12. 1,580	20. 1,775
4. 1,340	*13. 1,600 Median*	21. 1,780
5. 1,380	14. 1,600	22. 1,820
6. 1,400	15. 1,620	23. 1,900
7. 1,420	16. 1,620	24. 1,900
8. 1,425	17. 1,640	25. 2,000
9. 1,500		

are arranged in order of their magnitude. In Table 3.2 the starting salaries for management majors are shown in ascending order. The median for this sample is $1,600. By comparing the median for management majors with that for finance majors we find a $5 difference in favor of the finance majors.

Although both the mean and median are used as measures of central location, more people are familiar with the mean and, for this reason, like to work with it. As pointed out above, however, when extreme values are present the median is a better measure of central location. Thus it is usually best to present both measures for the reader's benefit.

Mode

A third statistical measure, the *mode,* is sometimes used as a measure of central location. The mode is defined as follows:

Definition of Mode

The mode of a set of data is the value that occurs with greatest frequency.

Referring to the sample of finance graduates, we see that the only monthly starting salary value that occurs more than once is 1,550. Since this value, occurring with a frequency of 2, has the greatest frequency in the data set, it is the mode.

Checking the frequency of starting salary values for the management graduates shows that the greatest frequency, again 2, occurs at three different values: 1,600, 1,620, and 1,900. This means technically that the management data set has three different modes (i.e., it is multimodal). It is perhaps best, however, not to report the modes for this sample on the basis that the modes of 1,600, 1,620, and 1,900 would not do a very good job of describing the central location of the data. In fact, most practitioners would probably prefer to limit discussion to the mean and median for all of the starting salary data.

The type of data set for which the mode is considered a good measure of location is qualitative data. With our data on recent graduates, the student's major is considered qualitative data. There are 12 finance majors, 25 management majors, 27 accounting majors, and 16 marketing majors. Hence the mode, or most frequently occurring major, is accounting. For this type of data set it obviously makes no sense to speak of the mean or median type of major.

Let us look at another illustration of the use of the mode as a measure of location for qualitative data. Suppose that a manufacturer markets a product in three different package designs. Using the three package designs as the qualitative data items and measuring frequency of purchase, the modal package design would be the one most frequently purchased. Knowledge of the modal package design could be important managerial information. Other qualitative data sets where the mode is the preferred measure of central location are clothing styles, color preference, furniture styles, etc.

Percentiles

The statistical measure referred to as a *percentile* offers a means for identifying the location of values in the data set that are not necessarily central location values. This measure provides information regarding how the data items are spread over the interval from the lowest to the highest values. Hence percentiles can also be viewed as measures of dispersion or variability in the data set. In large data sets that do not have numerous repeat values, the *p*th percentile is a value that divides the data set into two parts. Approximately *p*% of the items take on values less than the *p*th percentile; approximately $(100 - p)$% of the items take on greater values.

We now present a formal definition of the *p*th percentile:

Definition of Percentile

The *p*th percentile of a data set is a value such that *at least p* percent of the items take on this value or less and *at least* $(100 - p)$ percent of the items take on this value or more.

Admission test scores for colleges and universities are frequently reported in terms of percentiles. For instance, suppose an applicant has a raw score of 54 on the verbal portion of an admissions test. It may not be readily apparent how this student performed relative to other students taking the same test. However, if the raw score of 54 corresponds to the 70th percentile, it is easily seen that approximately 70% of the students had a score less than this individual and approximately 30% scored better.

Again refer to the monthly starting salary data for finance graduates. Suppose that we are interested in the upper end of the salary scales. Specifically, assume that we are interested in the 90th percentile. This information will provide a good measure of top salaries, since approximately only 10% of the graduates received a starting salary at the 90th percentile value or above. The starting monthly salary data for the finance graduates are shown below in ascending order:

1,410, 1,455, 1,550, 1,550, 1,580, 1,590, 1,620, 1,640, 1,650, 1,750, 1,860, 2,025.

11 of 12 (91.7%)   2 of 12 (16.7%)

We see that the 11th item, a value of $1,860, satisfies the definition of the 90th percentile. At least 90% of the items are at this value or less (actually 91.7%), and at least 10% of the items are at this value or more (actually 16.7%). Thus the 90th percentile for the sample of finance majors is $1,860.

If two items in the data set satisfy the conditions required for the definition of the *p*th percentile, our convention will be to define the *p*th percentile as being halfway between the two values. For example, for the finance major data set, both the sixth and seventh data values, $1,590 and $1,620, satisfy the definition of the 50th percentile. Following our convention, we would designate $1,605 as the 50th percentile. Note that this is also the median for the finance major data set. Recall that in the discussion of the median we found the median divided the items ordered by magnitude into two equal parts. Thus in terms of percentiles the median is the 50th percentile. At times the 25th percentile and/or the 75th percentile may be of particular interest. These two percentiles are referred to as the first quartile and third quartile, respectively, in that they divide the data set into quarters.

In the above discussion we presented a conceptual framework for determining the *p*th percentile of a data set. Although you can use the definition to calculate the *p*th percentile, the following procedure is easier:

Calculating the pth Percentile

Step 1: Arrange the data values in ascending order.
Step 2: Compute an index i as follows:

$$i = \left(\frac{p}{100}\right) n,$$

where p is the percentile of interest and n is the number of data values.
Step 3: (a) If *i is not an integer,* the next integer value greater than i will denote the position of the pth percentile.
(b) If *i is an integer,* the pth percentile is the average of the data values in positions i and $i + 1$.

As an illustration of this procedure, let us determine the 90th percentile for the finance graduate data set.

Step 1: Arrange the 12 data values in ascending order:

1,410, 1,455, 1,550, 1,550, 1,580, 1,590, 1,620, 1,640, 1,650, 1,750, 1,860, 2,025.

Step 2:

$$i = \left(\frac{90}{100}\right) 12 = 10.8.$$

Step 3: Since i is not an integer, the position of the 90th percentile is the next integer value greater than 10.8, the 11th position.
Returning to the data presented in ascending order, we see that the 90th percentile corresponds to the 11th data value, or $1,860.

As another illustration of the above procedure, let us reconsider the calculation of the 50th percentile for the finance major data set. Applying step 2, we obtain

$$i = \left(\frac{50}{100}\right) 12 = 6.$$

Since i is an integer, step 3b states that the 50th percentile is the average of the 6th and 7th data values; thus the 50th percentile is ($1,590 + $1,620)/2 = $1,605.

EXERCISES

1. Use the data in Table 3.1 to compute the mean, median, and mode for accounting and marketing graduates. What observations or summary statements can you make after identifying these central location measures for the four different business administration majors? Be careful here to limit your statements to the sample data; we are not ready to draw inferences about all accounting graduates, all finance graduates, etc.
2. Compute the 90th percentile for the monthly starting salaries of management, accounting, and marketing graduates listed in Table 3.1. What observations can you make based on a comparison of the four 90th percentile values for the different majors?
3. The numbers of defective parts observed on 16 different days are shown below:

11, 14, 18, 14, 21, 17, 13, 21, 25, 19, 17, 13, 28, 13, 17, 18.

Compute the mean, median, mode, and 90th percentile for these data.

4. A bowler has the following scores for six games:

182, 168, 184, 190, 170, 174.

Using these data as a sample, compute the following descriptive statistics:

a. Mean.
b. Median.
c. Mode.
d. 75th percentile.

5. Exercise 11 in Chapter 2 described the RC Radio Corporation, with monthly sales data for car telephone units as shown below:

80, 115, 82, 102, 94, 90, 88, 91, 89, 95, 105, 108.

Compute the mean, median, and mode for monthly sales volumes.

6. A sample of 15 college seniors showed the following credit hours taken during the final term of the senior year:

15, 21, 18, 16, 18, 21, 19, 15, 14, 18, 17, 20, 18, 15, 16.

What are the mean, median, and mode for credit hours? Compute and interpret the 70th percentile for these data.

7. The data given below show the number of automobiles arriving at a toll booth during 20 intervals of 10 minutes duration. Compute the mean, median, mode, first quartile, and third quartile for the data.

26	26	38	24
32	22	15	33
19	27	21	28
16	20	34	24
27	30	31	33

8. In automobile mileage and gasoline consumption testing, 13 automobiles were road tested 300 miles in both city and country driving conditions. The following data were recorded for miles per gallon performance:

City: 16.2, 16.7, 15.9, 14.4, 13.2, 15.3, 16.8, 16.0, 16.1, 15.3, 15.2, 15.3, 16.2.
Country: 19.4, 20.6, 18.3, 18.6, 19.2, 17.4, 17.2, 18.6, 19.0, 21.1, 19.4, 18.5, 18.7.

Use the mean, median, and mode to make a statement about the difference in performance for city and country driving.

9. A survey of television viewing habits among college students provided the following data on viewing time in hours per week:

14, 9, 12, 4, 20, 26, 17, 15, 18, 15, 10, 6, 16, 15, 8, 5.

a. Compute the mean, median, and mode.
b. Compute the 10th and 80th percentiles.

3.2 MEASURES OF DISPERSION

Whenever data are collected, whether for a sample or a population, it is desirable to consider the variability or dispersion in the data values. For example, assume that you are a purchasing agent for a large manufacturing firm. You regularly place orders with two different suppliers. Both suppliers indicate that approximately 10 working days are required to fill your orders. After several months of operation you find that the mean number of days required to fill orders is indeed averaging around 10 days for both of the suppliers. The histograms summarizing the number of working days required to fill orders from each supplier are shown in Figure 3.1. Although the mean

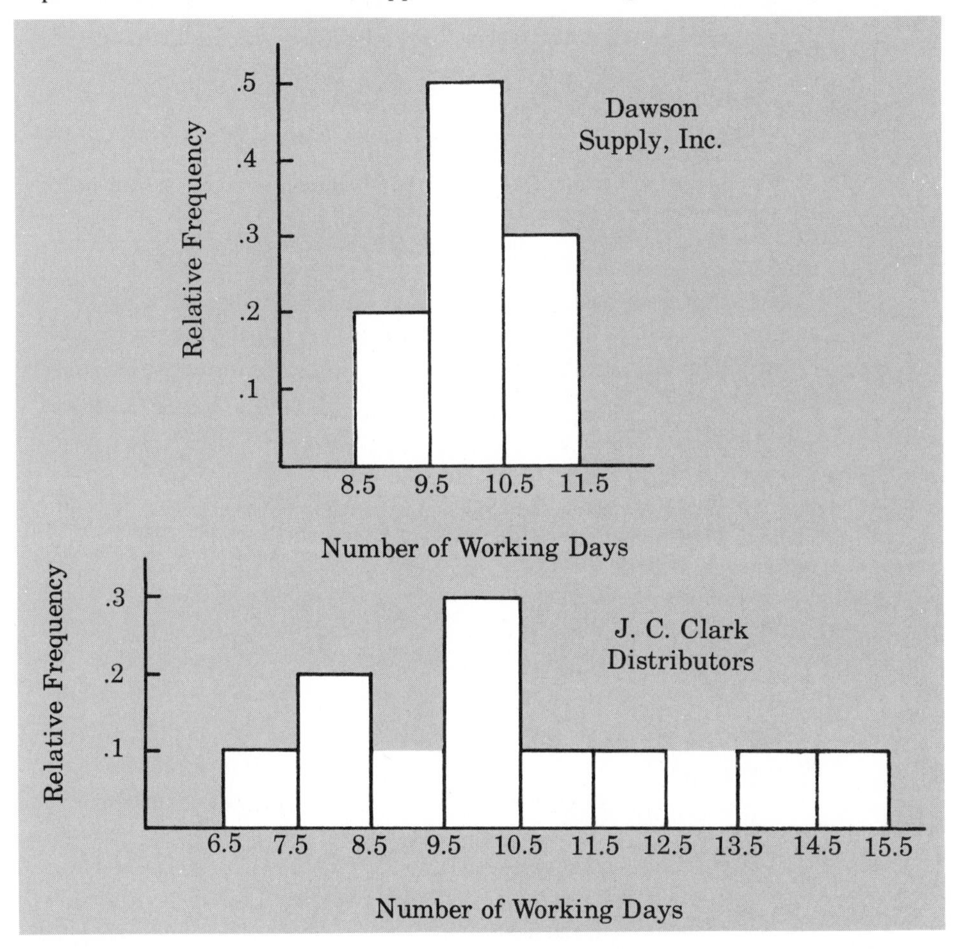

FIGURE 3.1 Historical Data Showing the Number of Days Required to Fill Orders

number of days required to fill orders is roughly 10 for both suppliers, do both suppliers possess the same degree of reliability in terms of making delivery on schedule? Note the variability or dispersion in the data. Which supplier would you prefer to order from?

For most firms, receiving materials and supplies on schedule is an important part of the purchasing agent's responsibility. At times the earlier than expected deliveries shown by the Clark firm might indicate that the purchasing agent was doing an excellent job. However, a few of the 14 or 15 day deliveries could be disastrous in terms of keeping a work force busy and production on schedule. The "average" or mean delivery time is a major consideration for a purchasing agent. Nonetheless, the above example illustrates a situation where the dispersion or variability in the delivery times is at least as important, perhaps even an overriding consideration.

We turn now to a discussion of some commonly used numerical measures of the variability or dispersion in a set of data.

Range

Perhaps the simplest measure of variability for a set of data is the *range:*

Definition of Range

The range for a set of data is the difference between the largest and smallest values in the set.

Let us refer to the data on monthly starting salaries for finance majors in Table 3.1. The largest starting salary is \$2,025, and the smallest is \$1,410. The range is then 2,025 − 1,410 = 615. For the management majors the range is 2,000 − 1,300 = 700.

While the range is the easiest of the statistical measures of dispersion to compute, it is not widely used. The reason is that the range is based on only two of the items in the data set and thus is influenced too much by extreme values. Suppose, as we did in the previous section, that one of the finance majors obtained a starting salary of \$10,000. The range would then be 10,000 − 1,410 = 8,590. This large value for the range would not be very descriptive of the variability in the data, since 11 of the 12 starting salaries are closely grouped between 1,410 and 1,860.

A measure sometimes used to overcome the difficulties of the range is the *interquartile range.* This measure is simply the difference between the third and first quartiles. For the data on finance major monthly starting salaries, the third quartile is (1,650 + 1,750)/2 = 1,700. The first quartile is (1,550 + 1,550)/2 = 1,550. Hence the interquartile range is 1,700 − 1,550 = 150 and would not change if the highest starting salary were \$10,000 instead of \$2,025. While this measure is not as sensitive to extreme values as the range, it also is based on only two data values.

Most statisticians prefer a measure of variability that involves all of the items in a data set. The measures discussed in the remainder of this section do involve all the data items. They are the most commonly used measures of variability.

Variance

A key step in developing a measure of variability that includes all the data items involves the computation of the differences between the data values and the mean for the data set. The difference between x_i and the mean ($\bar{x}$ *for a sample,* μ for a population) is called a *deviation about the mean.* Since we are seeking a descriptive statistical measure that summarizes the variability or dispersion in the entire data set, we want to consider the deviation of each data value about the mean. Thus for a sample size n and data values $x_1, x_2, x_3, \ldots, x_n$, we will need to compute the deviations $(x_1 - \bar{x}), (x_2 - \bar{x}), (x_3 - \bar{x}), \ldots, (x_n - \bar{x})$.

In the previous section we worked with the data on monthly starting salaries for business school graduates. Let us return to the sample of finance graduates and consider the deviations about the mean for this sample. Recall that the sample mean is $\bar{x} = 1{,}640$. A summary of the data, including the values of the deviations from the sample mean, is shown in Table 3.3.

We might think of summarizing the dispersion in a data set by computing the average deviation about the mean. However, a little reflection based upon a study of

TABLE 3.3 Deviation about the Mean of Finance Majors' Monthly Starting Salaries

Finance Graduate	*Monthly Salary* (x_i)	$\bar{x}$	*Deviation from Sample Mean* ($x_i - \bar{x}$)
1	1,550	1,640	−90
2	1,650	1,640	10
3	1,750	1,640	110
4	1,550	1,640	−90
5	1,455	1,640	−185
6	1,410	1,640	−230
7	1,590	1,640	−50
8	1,860	1,640	220
9	1,640	1,640	0
10	2,025	1,640	385
11	1,620	1,640	−20
12	1,580	1,640	−60
Totals	19,680	19,680	0

Table 3.3 would lead us to discard that idea—the sum of the deviations about the mean for the finance majors is equal to zero. This is true for any data set; that is,

$$\Sigma (x_i - \bar{x}) = 0.$$

The positive and negative deviations cancel each other out. Therefore the average deviation can provide no information about the variability in a data set. Hence if we are to use the deviations from the mean as a measure of dispersion, we must find another approach.

One possibility would be to eliminate the negative values by taking the absolute value of each deviation. We could then compute the average absolute deviation as a measure of variability. While this measure is sometimes used, the one most often used is based on squaring the deviations to eliminate the negative values. The squared

TABLE 3.4 Summary of the Calculation of the Sum of Squared Deviations for the Sample of Finance Graduate Monthly Starting Salaries

Finance Graduate	*Monthly Salary* (x_i)	*Deviation from Sample Mean* ($x_i - \bar{x}$)	*Squared Deviation* $(x_i - \bar{x})^2$
1	1,550	−90	8,100
2	1,650	10	100
3	1,750	110	12,100
4	1,550	−90	8,100
5	1,455	−185	34,225
6	1,410	−230	52,900
7	1,590	−50	2,500
8	1,860	220	48,400
9	1,640	0	0
10	2,025	385	148,225
11	1,620	−20	400
12	1,580	−60	3,600
Total			318,650

deviations and their sum for the sample of finance majors are shown in Table 3.4. The average of the squared deviations is

$$\text{Average Squared Deviation} = \frac{318{,}650}{12} = 26{,}554.17.$$

The average of the squared deviations for a data set representing a population is given a special name in statistics. It is called the *variance*. The population variance is denoted by the Greek symbol σ^2 (pronounced "sigma squared"). Given a population of N items and using μ to represent the population mean, we present a formula for the population variance:

Population Variance

$$\sigma^2 = \frac{\Sigma (x_i - \mu)^2}{N}. \tag{3.3}$$

In many statistical applications, the data set we are working with is a sample. When we compute a measure of variability for the sample, we often are interested in using the sample statistic obtained as an estimate of the population parameter, σ^2. At this point it might seem that the average of the squared deviations in the sample would provide a good estimate of the population variance. However, statisticians have found that the average squared deviation for the sample has the undesirable feature that it tends to underestimate the population variance σ^2. Because of this tendency toward underestimation we say it provides a biased estimate.

Fortunately, it can be shown that if the sum of the squared deviations in the sample is divided by $n - 1$, and not n, then the resulting sample statistic will provide an unbiased estimate of the population variance. For this reason the *sample variance* is not defined to be the average squared deviation in the sample. Rather, it is denoted by s^2 and is defined as follows:

Sample Variance

$$s^2 = \frac{\Sigma (x_i - \bar{x})^2}{n - 1}. \tag{3.4}$$

In later chapters we will frequently use the sample variance as an estimate of the population variance. When we do so we will use the estimate provided by (3.4).

Let us now compute the sample variance for the data on monthly starting salaries of finance majors. Table 3.4 gives the total squared deviations as 318,650. With $n - 1 = 11$ we obtain

$$s^2 = \frac{\Sigma (x_i - \bar{x})^2}{n - 1} = \frac{318{,}650}{11} = 28{,}968.18.$$

While admittedly it is difficult to obtain an intuitive feel for the meaning of the numerical value 28,968.18, we can note that larger variances could only be obtained from data sets with larger deviations about the mean and therefore more dispersion. Applying (3.4) to the sample of monthly starting salary values for the 25 management graduates provides a sample variance of $s^2 = 39{,}820.83$. Again, while this numerical value is not easy to interpret by itself, we can use it to compare the sample variances for the two groups of graduates. We restrict our comments specifically to the two sample data sets: we see that the monthly starting salary figures for management graduates show a larger degree of variability or dispersion than do the salary figures for the finance graduates.

Before concluding this subsection on the use of variance as a measure of dispersion, we provide shortcut formulas for computing the variance of a data set. The shortcut formulas are obtained after some algebraic manipulation of the numerator in (3.4). The derivation is shown in the chapter appendix.

Sample Variance (Shortcut Formula)

$$s^2 = \frac{\Sigma x_i^2 - n\bar{x}^2}{n-1}. \tag{3.5}$$

The advantage of working with (3.5) over (3.4) is that the step of calculating and squaring each deviation is not necessary. Formula (3.5) can be modified easily to compute a population variance. The appropriate formula is shown below:

Population Variance (Shortcut Formula)

$$\sigma^2 = \frac{\Sigma x_i^2 - N\mu^2}{N}. \tag{3.6}$$

Using the data on finance graduates, we show the calculations for the short-cut method of computing the sample variance in Table 3.5. Note that (3.4) and (3.5) provide the same results.

Standard Deviation

The *standard deviation* of a data set is defined to be the positive square root of the variance. Following the notation we adopted for a sample variance and a population variance, we will use s to denote a sample standard deviation and σ to denote a population standard deviation. The value of the standard deviation is derived from the value of the variance in the following manner:

Standard Deviation

$$\text{Sample Standard Deviation} = s = \sqrt{s^2}. \tag{3.7}$$

$$\text{Population Standard Deviation} = \sigma = \sqrt{\sigma^2}. \tag{3.8}$$

Obviously, the standard deviation is also a measure of dispersion, since the square root of a larger variance will provide a larger standard deviation. The sample standard deviations of the starting salaries of finance and management graduates are shown below:

Type of Graduate	*Sample Variance*	*Sample Standard Deviation*
Finance	$s^2 = 28{,}968.18$	$s = \sqrt{28{,}968.18}$
		$= 170.20$
Management	$s^2 = 39{,}820.83$	$s = \sqrt{39{,}820.83}$
		$= 199.55$

TABLE 3.5 Summary of the Calculation of the Sample Variance for the Sample of Finance Graduates' Monthly Starting Salaries (Shortcut Formula)

Finance Graduate	*Monthly Salary* (x_i)	x_i^2
1	1,550	2,402,500
2	1,650	2,722,500
3	1,750	3,062,500
4	1,550	2,402,500
5	1,455	2,117,025
6	1,410	1,988,100
7	1,590	2,528,100
8	1,860	3,459,600
9	1,640	2,689,600
10	2,025	4,100,625
11	1,620	2,624,400
12	1,580	2,496,400
Totals	19,680	32,593,850

$$\text{Sample Variance } s^2 = \frac{32{,}593{,}850 - 12(1{,}640)^2}{11}$$

$$= \frac{318{,}650}{11} = 28{,}968.18$$

A comparison of the sample standard deviations leads to the conclusion we previously obtained when comparing sample variances: that is, the data in the sample of management graduates show a greater degree of variability.

What is gained, then, by converting the variance to its corresponding standard deviation? Note that the units of the values being summed in the variance calculation are squared. For example, the sample variance for the monthly starting salaries of the finance graduates was $s^2 = 28{,}968.18$ (units of dollars squared). The fact that the units for variance are squared is the biggest reason why it is difficult to obtain an intuitive appreciation for the numerical value of the variance. Since the standard deviation is simply the square root of the variance, dollars squared in the variance is converted to dollars in the standard deviation. The standard deviation of finance majors' starting salaries is 170.20 dollars. In other words, the standard deviation is measured in the same units as the original data. For this reason the standard deviation often is more easily compared to the mean and other statistics which are measured in the units of the original data.

Coefficient of Variation

In some situations we may be more interested in a relative measure of the variability in a data set than the absolute measure provided by the standard deviation or variance. For example, a standard deviation of 1 inch would be considered very large for a batch of motor mount bolts used in automobiles. However, a standard deviation of 1 inch would be considered small for the length of telephone poles. The difficulty is that when the means for data sets differ greatly we do not get an accurate picture of the relative variability in the two data sets by comparing the standard deviations. A measure of variability that overcomes these difficulties is the *coefficient of variation*. The formula for computing the coefficient of variation is given by (3.9):

Coefficient of Variation

$$\frac{\text{Standard Deviation}}{\text{Mean}} \times 100\%. \tag{3.9}$$

For sample data the coefficient of variation is $(s/\bar{x}) \times 100\%$, and for a population it is $(\sigma/\mu) \times 100\%$. For the finance graduate sample the coefficient of variation is $(170.20/1{,}640) \times 100\% = 10.38\%$. In words, we could say that the standard deviation of the sample is 10.38% of the value of the sample mean.

The coefficient of variation may be helpful in comparing the relative variation in several data sets that have different means and different standard deviations. However, we caution that this measure should be used only with data sets involving all, or nearly all, positive values. We can see why by referring to (3.9). Values equal to or near zero for the mean could be obtained when both positive and negative values are present in the data set. A value near zero for the mean could cause the coefficient of variation to be very large even with a small standard deviation.

Chebyshev's Theorem

Oftentimes in statistical studies we are interested in specifying the percentage of items in a data set that lie within some specified interval when only the mean and standard deviation for the data set are known. The Russian mathematician Chebyshev provided a means for determining the percentage of data items within a specified distance of the mean. Chebyshev's theorem makes a statement about the fraction of items in a data set lying within a given number of standard deviations of the mean.

Chebyshev's Theorem

For any set of data and any value of k greater than or equal to 1, at least $1 - 1/k^2$ of the values in the data set must be within plus or minus k standard deviations of the mean.

In applying Chebyshev's theorem we treat every data set as if it were a population, and the formula for a population standard deviation is used.

To illustrate Chebyshev's theorem, let us consider the finance graduates. For this data set we have a mean of $1,640. We treat the data set itself as the total population of interest. We can refer to Table 3.4 to obtain the sum of squared deviations for use in computing the population standard deviation. It is given by

$$\sigma = \sqrt{\frac{318{,}650}{12}} = 162.95.$$

Applying Chebyshev's theorem with $k = 2$ standard deviations, the theorem says that at least $1 - 1/(2^2)$ of the values in the data set, or 75%, must be within plus or minus two standard deviations of the mean. Thus 75% of the values must be between $1{,}640 - 2(162.95) = \$1{,}314.10$ and $1{,}640 + 2(162.95) = \$1{,}965.90$. A check of the original data verifies Chebyshev's prediction, since 11 of 12, or 92%, of the values actually fall within the interval \$1,314.10 to \$1,965.90. Other values of k could be selected and different intervals computed.

In closing, we note that Chebyshev's theorem provides loose bounds on the percentage of data items in an interval. While Chebyshev's theorem guaranteed that at least 75% of the finance starting salaries were in the interval \$1,314.10 to \$1,965.90, we found that there were actually 92% in this interval. In later chapters we will see that when more knowledge about the data set is available (i.e., more than just knowing the mean and standard deviation), tighter bounds can often be computed.

EXERCISES

10. The Davis Manufacturing Company has just completed five weeks of operation using a new process that is supposed to increase productivity. The number of parts produced each week is shown below:

410, 420, 390, 400, 380.

Compute the sample variance and sample standard deviation using the definition of sample variance (3.4) as well as the shortcut formula (3.5).

11. Use the data in Table 3.1 to compute the variance, standard deviation, and range for the monthly starting salaries of accounting and marketing graduates. Compare your results to the measures of dispersion for the finance and management graduates. Comment on your findings.

12. Assume that the data used to construct the histograms of the number of days required to fill orders for Dawson Supply, Inc. and J. C. Clark Distributors (see Figure 3.1) are as follows:

Dawson Supply Days for Delivery: 10, 10, 9, 10, 11, 11, 9, 11, 10, 10.
Clark Distributors Days for Delivery: 8, 10, 14, 7, 10, 11, 10, 8, 15, 12.

Use the standard deviation and range to support our earlier observation that Dawson Supply provides the more consistent and reliable delivery schedules.

13. In Exercise 4, a bowler's scores for six games were as follows:

182, 168, 184, 190, 170, 174.

Using these data as a sample, compute the following descriptive statistics:

a. Range.
b. Variance.
c. Standard deviation.
d. Coefficient of variation.

14. Compute the range, variance, and standard deviation for the sample of credit hours taken by college seniors in their final term. The data are as follows:

15, 21, 18, 16, 18, 21, 19, 15, 14, 18, 17, 20, 18, 15, 16.

15. Given below are the yearly household incomes (in dollars) for ten families in Grimes, Iowa:

10,648	17,416
6,517	13,555
14,821	9,226
152,936	11,800
18,527	12,222

a. Compute the range as a measure of variability.
b. Compute the interquartile range as a measure of variability.
c. Compute the standard deviation as a measure of variability.
d. Which of the above measures do you feel is the best measure of variability in the data? Why?

16. A production department uses a sampling procedure to test the quality of newly produced items. The department employs the following decision rule at an inspection station: If a sample of 14 items has a variance of more than .01, the production line must be shut down for repairs. Suppose that the following data have just been collected:

3.43	3.45	3.43
3.48	3.52	3.50
3.39	3.48	3.41
3.38	3.49	3.45
3.51	3.50	

Should the production line be shut down? Why or why not?

17. The following times were recorded by the quarter-mile and mile runners of a university track team (times are in minutes):

Quarter-mile times: .92, .98, 1.04, .90, .99.
Mile times: 4.52, 4.35, 4.60, 4.70, 4.50.

After viewing this sample of running times, one of the coaches commented that the quarter-milers turned in the more consistent times. Use the standard deviation and the coefficient of variation to summarize the variability in the data sets. Does the use of the coefficient of variation measure indicate that the coach's statement should be qualified?

18. The monthly charges for credit card holders at Schip's Department Store have a population mean of $250 and a population standard deviation of $100. Use Chebyshev's theorem to answer the following questions:
a. What can be said about the percentage of the card holders who will have monthly charges between $100 and $400?
b. What can be said about the percentage of the card holders who will have monthly charges between 0 and $500?
c. Provide a range of credit card charges that will include at least 80% of all credit card customers.

19. During a recent football season, a major conference reported that the average attendance for its conference games was 45,000. The standard deviation in the attendance figure was σ = 4,000. Use Chebyshev's theorem for the following:
a. Develop an interval that contains the attendance figures for at least 75% of the games.
b. Develop an interval that contains the attendance figures for at least $8/9$ of the games.
c. The commissioner claims that at least 90% of the games had attendances of between 29,000 and 61,000. Is this statement warranted given the information we have?

3.3 MEASURES OF LOCATION AND DISPERSION FOR GROUPED DATA

The statistical measures we have presented for the central location and dispersion of data sets are computed using the individual data values. The computational procedures we have discussed in the last two sections provide the most common methods for computing measures of central location and dispersion. However, in some situations the data are available only in grouped or frequency distribution form. In these cases special procedures are used in order to obtain approximations to the common measures of central location and dispersion.

In cases where both the individual data values and a grouped form of the data are available, we recommend using the individual data values to compute the summary statistical measures. The reason for this is that we do not expect the calculations based on the ungrouped and grouped data to provide exactly the same numerical results. In fact, when data are grouped, some details about the data set are lost. Thus at best we can expect the summary measures from grouped data to be a good approximation to what would have been obtained if the entire data set of individual values had been used.

Let us consider the procedures available for computing measures of central location and dispersion for a data set presented in a frequency distribution format. To illustrate the calculations involved we will use the frequency distribution of starting salaries for the finance graduates. This distribution is presented in Table 3.6.

In Table 3.6 we have added an extra column to show the midpoint for each class.

TABLE 3.6 Frequency Distribution of Monthly Starting Salaries for Finance Graduates

Starting Monthly Salary (dollars)	*Frequency*	*Class Midpoint*
1,400 but less than 1,500	2	1,450
1,500 but less than 1,600	4	1,550
1,600 but less than 1,700	3	1,650
1,700 but less than 1,800	1	1,750
1,800 but less than 1,900	1	1,850
1,900 but less than 2,000	0	1,950
2,000 but less than 2,100	1	2,050
Total	12	

As you will recall from Chapter 2, the class midpoint (the average of the class limits) is used as an approximation of the mean of the items in that class. We note that the midpoint will be a good approximation if the data items are spread fairly evenly over the class interval. It will not be so good otherwise. These class midpoints play an important role in computing measures of location and dispersion for grouped data.

Mean

First, let us consider how to compute the mean monthly starting salary for the finance graduates. If the individual data values were available we would simply add them all up and divide by $n = 12$. Since they are not, we want to approximate the sum of the 12 starting salaries by using the information in Table 3.6. Let us begin by attempting to approximate the sum for each class.

In the first class the midpoint, 1,450, is assumed to represent the mean of the items in that class. Since there are two items in the class, an approximation of the sum of the items in that class is $2 \times 1{,}450 = 2{,}900$. Using M_1 as the notation for the midpoint of class 1 and f_1 as the notation for the frequency of class 1, we see that $f_1 M_1$

approximates the sum for class 1. In general we approximate the sum of data values in class i as follows:

$$\text{Approximate Sum of Data Values in Class } i = f_i M_i.$$

Thus the following calculations approximate the sums of data values in each of the classes:

$$\begin{aligned}
\text{Class 1:} \quad & f_1 M_1 = 2(1{,}450) = 2{,}900 \\
\text{Class 2:} \quad & f_2 M_2 = 4(1{,}550) = 6{,}200 \\
\text{Class 3:} \quad & f_3 M_3 = 3(1{,}650) = 4{,}950 \\
\text{Class 4:} \quad & f_4 M_4 = 1(1{,}750) = 1{,}750 \\
\text{Class 5:} \quad & f_5 M_5 = 1(1{,}850) = 1{,}850 \\
\text{Class 6:} \quad & f_6 M_6 = 0(1{,}950) = 0 \\
\text{Class 7:} \quad & f_7 M_7 = 1(2{,}050) = 2{,}050
\end{aligned}$$

Totaling the sums for all classes yields an approximation to the sum of all the data items.

Once an approximation of the sum of all the data items is obtained, the mean is computed by dividing this sum by the total number of data items. We show below the formula for computing the sample mean from grouped data:

Sample Mean for Grouped Data

$$\bar{x} = \frac{\Sigma f_i M_i}{n}, \tag{3.10}$$

where

f_i = frequency of class i,
M_i = midpoint of class i.

Applying (3.10) to the data for finance graduates, we find

$$\begin{aligned}
\bar{x} = \frac{\Sigma f_i M_i}{12} &= \frac{f_1 M_1 + f_2 M_2 + f_3 M_3 + f_4 M_4 + f_5 M_5 + f_6 M_6 + f_7 M_7}{12} \\
&= \frac{2{,}900 + 6{,}200 + 4{,}950 + 1{,}750 + 1{,}850 + 0 + 2{,}050}{12} \\
&= \frac{19{,}700}{12} \\
&= 1{,}641.67.
\end{aligned}$$

The sample mean for finance graduates was found earlier (working with the entire data set) to be 1,640. Thus we see that the sample mean computed from the grouped data is a good approximation.

Median

The method of approximating the median from grouped data can be understood best by referring to the histogram. In Figure 3.2 we show the histogram for starting salaries of finance majors. The median is the value that divides the area of the histogram into

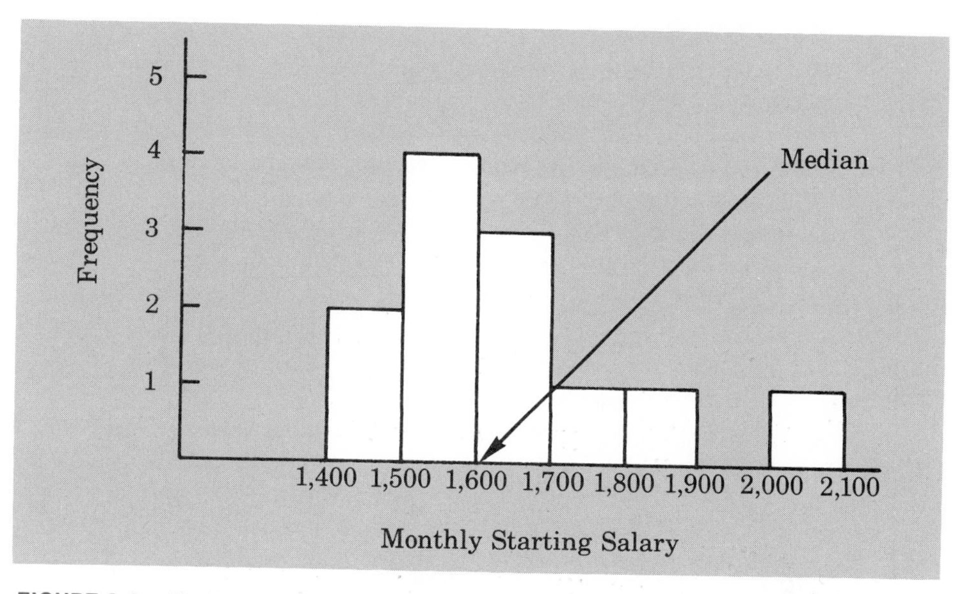

FIGURE 3.2 Median Divides the Area of the Histogram Into Two Equal Parts

two equal parts (50% in each half). The area of the first bar is 2(100) = 200, that of the second is 4(100) = 400, the third 3(100) = 300, the fourth 1(100) = 100, the fifth 1(100) = 100, the sixth 0(100) = 0, and the seventh 1(100) = 100. Thus, the total area of the histogram is 200 + 400 + 300 + 100 + 100 + 0 + 100 = 1,200. A starting salary of $1,600 is the value that divides the histogram into two parts each with an area of 600. Figure 3.2 shows 50% of the area to the left of the median and 50% of the area to the right. Earlier we found the median for the ungrouped data to be 1,605. Thus we have a good approximation.

To develop an algebraic formula for finding the median in a sample of n items, we choose as the median the approximate location of the item $n/2$. We first identify the *median class* as the class containing data item $n/2$. We then define $n_m = n/2$ − (number of items in classes *preceding* the median class). Using f_m to represent the frequency of the median class we find that n_m/f_m represents the fraction of the distance one must travel through the median class to reach the median. Letting L represent the lower class limit of the median class and W its width, we obtain an algebraic formula for computing the median from grouped data:

Median for Grouped Data

$$\text{Median} = \text{L} + \frac{n_m}{f_m} W. \tag{3.11}$$

Applying this formula to the data for finance graduates, we obtain

$$\text{Median} = 1{,}600 + (4/4)(100) = 1{,}600.$$

This is the same as the result we obtained working with the histogram.

Percentiles

Equation (3.11) can be modified slightly to find any percentile of interest. For example, the 90th percentile for the finance graduate data would be located at item (.90)12 = 10.8. Item 10.8 is interpreted as being in the "1,800 but less than 1,900" class, and this class, rather than the median class, becomes the focus of our analysis. With the notation of (3.11) now referring to the "1,800 but less than 1,900" class, we interpret the 90th percentile as being .8 units into this class. Thus the 90th percentile is computed from the grouped data as follows:

$$\text{90th Percentile} = 1{,}800 + \frac{.8}{1}(100) = 1{,}880.$$

This also compares favorably with the actual 90th percentile of 1,860 from the original ungrouped data set.

Mode

To find the mode in a grouped data set, we start by finding the modal, or most frequent, class. The mode of the data set is then approximated by the midpoint of this modal class. The frequency distribution of Table 3.6 shows the "1,500 but less than 1,600" class to be the modal class. Using the midpoint, we obtain a mode for the salary data of 1,550. Coincidentally, this turns out to be the same as the modal value found for the ungrouped data.

Variance

The easiest approach to computing the variance for a set of grouped data is to use a slightly altered form of the short-cut formula for the variance as provided in (3.5). Since we no longer have the individual data values, we treat the class midpoint as being representative of the data in the class. Then, just as we did with the sample mean calculation for grouped data, we weight each midpoint by the frequency of its corresponding class. Using this approach, (3.5) is modified as follows:

Sample Variance for Grouped Data

$$s^2 = \frac{\Sigma f_i M_i^2 - n\bar{x}^2}{n-1}. \qquad (3.12)$$

The calculation of the sample variance for the grouped data of Table 3.6 is summarized in Table 3.7.

The variance based on the formula for grouped data is slightly larger than the sample variance of 28,968.18 computed from the individual values. However, it is a good approximation.

TABLE 3.7 **Computation of the Variance of Monthly Starting Salaries of Finance Graduates Using Grouped Data**

Starting Monthly Salary (dollars)	*Frequency* (f_i)	*Midpoint* (M_i)	f_iM_i	M_i^2	$f_iM_i^2$
1,400 but less than 1,500	2	1,450	2,900	2,102,500	4,205,000
1,500 but less than 1,600	4	1,550	6,200	2,402,500	9,610,000
1,600 but less than 1,700	3	1,650	4,950	2,722,500	8,167,500
1,700 but less than 1,800	1	1,750	1,750	3,062,500	3,062,500
1,800 but less than 1,900	1	1,850	1,850	3,422,500	3,422,500
1,900 but less than 2,000	0	1,950	—	3,802,500	—
2,000 but less than 2,100	1	2,050	2,050	4,202,500	4,202,500
Totals	12		19,700		32,670,000

$$\bar{x} = \frac{19{,}700}{12} = 1{,}641.67 \qquad s^2 = \frac{32{,}670{,}000 - 12(1{,}641.67)^2}{11}$$

$$= \frac{329{,}035.33}{11}$$

$$= 29{,}912.30$$

Standard Deviation

The standard deviation computed from grouped data is simply the square root of the variance computed from grouped data. For the salary data on the sample of finance graduates, the sample standard deviation computed from grouped data is $s = \sqrt{29{,}912.30} = 172.95$. This differs only slightly from the value of 170.20 we obtained using the individual data items.

Before closing this section on computing measures of location and dispersion from grouped data, we note that the formulas in this section were presented only for data sets constituting a sample. Population summary measures are computed in a similar manner. We give the grouped data formulas for a population mean and variance below:

Population Mean for Grouped Data

$$\mu = \frac{\Sigma f_i M_i}{N}. \tag{3.13}$$

Population Variance for Grouped Data

$$\sigma^2 = \frac{\Sigma f_i M_i^2 - N\mu^2}{N}. \tag{3.14}$$

EXERCISES

20. The frequency distribution for the monthly starting salaries of all business graduates in the sample is shown as follows:

Monthly Starting Salaries (dollars)	*Frequency*
1,300 but less than 1,400	8
1,400 but less than 1,500	11
1,500 but less than 1,600	17
1,600 but less than 1,700	19
1,700 but less than 1,800	10
1,800 but less than 1,900	6
1,900 but less than 2,000	4
2,000 but less than 2,100	5
Total	80

Using the frequency distribution, compute the mean, median, mode, variance, and standard deviation for the sample of 80 graduates.

21. Waiting times for patients requesting emergency service at a doctor's office are summarized in the following frequency distribution:

Waiting Time (minutes)	*Frequency*
0 but less than 5	4
5 but less than 10	8
10 but less than 15	5
15 but less than 20	2
20 but less than 25	1
Total	20

Using the above grouped data, compute the following:

a. Mean.
b. Median.
c. Mode.
d. 75th percentile
e. Variance.
f. Standard deviation.

22. The following frequency distribution of grades for the first examination in Business Statistics was posted on the department bulletin board:

Examination Grade	*Frequency*
90 but less than 100	6
80 but less than 90	15
70 but less than 80	22
60 but less than 70	11
50 but less than 60	5
40 but less than 50	3
Total	62

Treating these data as a sample, identify the mean, median, mode, variance, and standard deviation.

23. A service station has recorded the following frequency distribution for the number of gallons of gasoline purchased by a sample of its customers.

Gasoline (gallons)	*Frequency*
0 but less than 5	74
5 but less than 10	192
10 but less than 15	280
15 but less than 20	105
20 but less than 25	23
25 but less than 30	6
Total	680

Compute the standard measures of central location and dispersion for these grouped data. If the service station expects to service about 120 cars on a given day, what is an estimate of the total number of gallons of gasoline that will be sold?

24. Consider the following frequency distribution for automobile repair costs for an insurance company's minor claims category:

Repair Cost (dollars)	*Frequency*
0 but less than 100	10
100 but less than 200	28
200 but less than 300	60
300 but less than 400	70
400 but less than 500	52
Total	220

Compute the mean, median, mode, variance, and standard deviation for these grouped data.

3.4 ROLE OF THE COMPUTER IN DESCRIPTIVE STATISTICS

We mentioned in Chapter 1 that the computer is playing an increasingly important role in statistical analysis. Statisticians, especially when working with large data sets, often rely on the computer to prepare data summaries, frequency distributions, etc. There are a number of computer packages available for use in data analysis. Each of these has somewhat different requirements for data input, and each presents the output in a slightly different format. Thus we shall not attempt to provide detailed instructions on how to use any particular package. Rather, we provide output from a standard package to illustrate what is available.

The data on monthly starting salaries for the 80 business school graduates were inputted to a standard data analysis program. Shown in Table 3.8 is a portion of the computer output. Note that the first column identifies the classes in the form 1,300–1,400, 1,400–1,500, and so on. It is understood that a salary of $1,400 goes in the second class, a salary of $1,500 goes in the third class, and so on. The second column, headed "code," is merely a numerical identification of the classes. The third column contains the class frequencies, and the fourth column contains the relative frequencies. These columns provide the frequency and relative frequency distributions, respectively. The fifth column provides a cumulative frequency distribution in percentage form.

TABLE 3.8 Computer Output for a Standard Data Analysis Computer Package

Category Label	*Code*	*Absolute Frequency*	*Relative Frequency (PCT)*	*Cumulative Frequency (PCT)*
1300–1400	1	8	10.0	10.0
1400–1500	2	11	13.8	23.8
1500–1600	3	17	21.3	45.0
1600–1700	4	19	23.8	68.8
1700–1800	5	10	12.5	81.3
1800–1900	6	6	7.5	88.8
1900–2000	7	4	5.0	93.8
2000–2100	8	5	6.3	100.0
Total		80	100.0	

Variable Salary

Mean = 1626.688 Std Dev = 181.2508
Variance = 32851.86
Range = 725.0000 Minimum = 1300.0000 Maximum = 2025.000
Sum = 130135

Valid Cases = 80

Beneath the tabular presentation are many of the numerical summary measures discussed in this chapter. The mean is given as a measure of central location, and many of the measures of dispersion we discussed are given. The number of "valid cases" is our sample size of 80.

This sample of the output available from a computer package provides an indication of how the computer can assist the analyst or decision maker. Manuals are available to assist anyone interested in using these packages, and most computer installations have some type of data analysis package available.

Summary

In this chapter we have introduced several statistical measures that are used to describe the location and dispersion of a data set. Unlike the tabular and graphical procedures for summarizing data, the measures introduced in this chapter summarize the data in terms of numerical values. When the numerical values obtained are for a sample, they are called sample statistics. When they are for a population, they are called population parameters.

As measures of central location, we defined the mean, median, and mode for both sample and population data sets. Then, the concept of a percentile was used to describe the location of other values in the data set. Next, we presented the range, interquartile range, variance, standard deviation, and coefficient of variation as statistical measures of variability or dispersion in a data set. Finally, we described how these summary measures could be computed for grouped as well as ungrouped data. However, we recommended using the measures based on the individual data values unless the grouped format was the only manner in which the data were available.

Glossary

Population parameter—A numerical value used as a summary measure for a population of data (e.g., the population mean, μ, and the population variance, σ^2).
Sample statistic—A numerical value used as a summary measure for a sample (e.g., the sample mean, $\bar{x}$, and the sample variance, s^2).
Mean—A measure of the central location of a data set. It is computed by summing all the values in the data set and dividing by the number of items.
Median—A measure of central location of a data set. It is the value which splits the data set into two equal groups—one with values greater than or equal to the median, and one with values less than or equal to the median.
Mode—A measure of central location of a data set, defined as the most frequently occurring data value.
Percentile—A value such that at least $p\%$ of the items in the data set are less than or equal to it and at least $(100 - p)\%$ of the items are greater than or equal to it. The median is the 50th percentile, the first quartile is the 25th percentile, and the third quartile is the 75th percentile.
Range—A measure of dispersion for a data set, defined to be the difference between the highest and lowest values.
Interquartile range—A measure of dispersion for a data set, defined to be the difference between the third and first quartiles.
Variance—A measure of dispersion for a data set, found by summing the squared deviations of the data values about the mean and then dividing the total by N if the data set is a population or by $n - 1$ if the data set is from a sample.
Standard deviation—A measure of dispersion for a data set, found by taking the square root of the variance.
Coefficient of variation—A measure of relative dispersion for a data set, found by dividing the standard deviation by the mean and multiplying by 100%.
Chebyshev's theorem—A theorem which allows the use of knowledge of the standard deviation and mean to draw conclusions about the fraction of data items within k standard deviations of the mean.
Grouped data—Data available in class intervals as summarized by a frequency distribution. Individual values of the original data are not recorded.
Median class—The class in a frequency distribution that contains the median.
Modal class—The class in a frequency distribution with the highest frequency.

Key Formulas

Sample Mean

$$\bar{x} = \frac{\Sigma x_i}{n} \qquad (3.1)$$

Population Mean

$$\mu = \frac{\Sigma x_i}{N} \qquad (3.2)$$

Population Variance

$$\sigma^2 = \frac{\Sigma (x_i - \mu)^2}{N} \tag{3.3}$$

Sample Variance

$$s^2 = \frac{\Sigma (x_i - \overline{x})^2}{n - 1} \tag{3.4}$$

Sample Variance (Shortcut Formula)

$$s^2 = \frac{\Sigma x_i^2 - n\overline{x}^2}{n - 1} \tag{3.5}$$

Population Variance (Shortcut Formula)

$$\sigma^2 = \frac{\Sigma x_i^2 - N\mu^2}{N} \tag{3.6}$$

Standard Deviation

$$\text{Sample Standard Deviation} = s = \sqrt{s^2} \tag{3.7}$$

$$\text{Population Standard Deviation} = \sigma = \sqrt{\sigma^2} \tag{3.8}$$

Coefficient of Variation

$$\left(\frac{\text{Standard Deviation}}{\text{Mean}}\right) \times 100\% \tag{3.9}$$

Sample Mean for Grouped Data

$$\overline{x} = \frac{\Sigma f_i M_i}{n} \tag{3.10}$$

Median for Grouped Data

$$\text{Median} = L + \frac{n_m}{f_m} W \tag{3.11}$$

Sample Variance for Grouped Data

$$s^2 = \frac{\Sigma f_i M_i^2 - n\overline{x}^2}{n - 1} \tag{3.12}$$

Population Mean for Grouped Data

$$\mu = \frac{\Sigma f_i M_i}{N} \tag{3.13}$$

Population Variance for Grouped Data

$$\sigma^2 = \frac{\Sigma f_i M_i^2 - N\mu^2}{N} \tag{3.14}$$

Supplementary Exercises

25. A sample of six recent home mortgage loans showed the following interest rates:

12.5, 13.2, 11.2, 13.0, 12.0, 12.5.

Compute the following descriptive statistics for the data set:

a. Mean.
b. Median.
c. Mode.
d. 25th percentile.
e. Range.
f. Interquartile range.
g. Variance.
h. Standard deviation.
i. Coefficient of variation.

26. The following data show home mortgage loan amounts handled by a particular loan officer in a savings and loan association:

20,000	38,500	33,000	27,500	34,000
12,500	25,900	43,200	37,500	36,200
25,200	30,900	23,800	28,400	13,000
31,000	33,500	25,400	33,500	20,200
39,000	38,100	30,500	45,500	30,500
52,000	40,500	51,600	42,500	44,800

Find the mean, median, and mode for these data.

27. Calculate the variance, standard deviation, and range for the mortgage amounts shown in Exercise 26. Conversion of the data to 1,000s (e.g., 38,500 being listed as 38.5) may ease the burden of having to work with large numerical values.

28. Morrison Communications, Inc. periodically reviews sales personnel performance records. One member of the sales force has had the following weekly sales volume (units sold) over the past quarter, or 13 weeks:

13, 19, 20, 17, 21, 27, 9, 15, 22, 18, 18, 23, 20.

a. Compute the mean, median, and mode.
b. Compute the first and third quartiles.
c. Compute the range and interquartile range.
d. Compute the standard deviation and coefficient of variation.

29. A sample of ten stocks on the New York Stock Exchange shows the following price/earnings ratios:

9, 4, 6, 7, 3, 11, 4, 6, 4, 7.

Using the above data, compute the mean, median, mode, range, variance, and standard deviation.

30. A sample of recent oil drilling locations shows oil found at the following depths (feet):

1,500 1,200 1,600 1,700 1,500 2,000.

Compute the mean, median, mode, range, variance, and standard deviation for the drilling depth data.

31. The number of patients treated at the Morton Hospital emergency room per day are shown below. Data are from a random sample of 12 days.

45, 50, 36, 59, 28, 42, 55, 67, 33, 35, 40, 50.

Compute the mean, median, mode, range, variance, and standard deviation for these data.

32. National Airlines accepts flight reservations by phone. Shown below are the durations (in minutes) for a sample of 20 phone reservations:

2.1	4.8	5.5	10.4
3.3	3.5	4.8	5.8
5.3	5.5	2.8	3.6
5.9	6.6	7.8	10.5
7.5	6.0	4.5	4.8

Using the following descriptive statistical methods, summarize these data:

a. Mean.
b. Median.
c. Mode.
d. Range.
e. Variance.
f. Standard deviation

33. Soft-drink purchases at the Wrigley Field concession stands show the following 1-day totals:

Drink	*Units Purchased*
Cola	4,553
Diet cola	2,125
Uncola	1,850
Orange soda	1,288
Root beer	1,572

What is the mode for the above sample data?

34. Light bulbs manufactured by a well known electrical equipment firm are known to have a mean life of 800 hours, with a standard deviation of 100 hours.

a. What percentage of the light bulbs will have a life of 600 to 1,000 hours?
b. What percentage of the light bulbs will have a life of 550 to 1,050 hours?
c. Provide an interval for light bulb life that will be true for at least 50% of the light bulbs.

35. Daily volume for the stock market over a 6 month period showed a mean of 40 million shares with a standard deviation of 7 million shares.

a. What percentage of the days during the 6 month period showed volumes between 30 and 50 million shares?
b. What percentage of the days during the 6 month period showed volumes between 20 and 60 million shares?
c. Provide an interval for daily volume that must include at least two-thirds of the days.

36. A frequency distribution for the duration of 20 long distance telephone calls is shown below:

Call Duration	*Frequency*
4 but less than 8	4
8 but less than 12	5
12 but less than 16	7
16 but less than 20	2
20 but less than 24	1
24 but less than 28	1
Total	20

Compute the mean, median, mode, variance, and standard deviation for the above data.

37. Dinner check amounts at the La Maison French Restaurant have the following frequency distribution:

Dinner Check (dollars)	*Frequency*
25 but less than 35	2
35 but less than 45	6
45 but less than 55	4
55 but less than 65	4
65 but less than 75	2
75 but less than 85	2
Total	20

Compute the mean, median, mode, variance, and standard deviation for the above data.

38. Automobiles traveling on the New York State Thruway are checked for speed by a state police radar system. A frequency distribution of speeds is shown below:

Speed (miles per hour)	*Frequency*
40 but less than 45	10
45 but less than 50	40
50 but less than 55	150
55 but less than 60	175
60 but less than 65	75
65 but less than 70	15
70 but less than 75	7
75 but less than 80	3
Total	475

a. What is the mean speed of the automobiles traveling on the New York State Thruway?
b. What are the median and modal speeds?
c. Compute the variance and the standard deviation.
d. What is the value of the 90th percentile? What is its meaning?

APPENDIX TO CHAPTER 3: DERIVATION OF SHORTCUT FORMULA FOR VARIANCE

In (3.1) we presented the formula for the sample mean:

$$\bar{x} = \frac{\Sigma x_i}{n}. \tag{3.1}$$

Multiplying both sides of (3.1) by n, we obtain (3.15):

$$n\bar{x} = \Sigma x_i. \tag{3.15}$$

Equation (3.15) is useful in developing a shortcut formula for the variance.

The formula for sample variance is

$$s^2 = \frac{\Sigma(x_i - \bar{x})^2}{n - 1}. \tag{3.4}$$

We will show how the shortcut formula for sample variance, (3.5), can be developed after several algebraic manipulations of (3.4):

$$\begin{aligned} s^2 &= \frac{\Sigma(x_i - \bar{x})^2}{n-1} \\ &= \frac{\Sigma(x_i^2 - 2x_i\bar{x} + \bar{x}^2)}{n-1} \\ &= \frac{\Sigma x_i^2 - \Sigma 2\bar{x}x_i + \Sigma\bar{x}^2}{n-1} \\ &= \frac{\Sigma x_i^2 - 2\bar{x}\Sigma x_i + n\bar{x}^2}{n-1} \qquad \text{by (3.15)} \\ &= \frac{\Sigma x_i^2 - 2\bar{x}(n\bar{x}) + n\bar{x}^2}{n-1} \\ &= \frac{\Sigma x_i^2 - 2n\bar{x}^2 + n\bar{x}^2}{n-1} \\ &= \frac{\Sigma x_i^2 - n\bar{x}^2}{n-1}. \end{aligned}$$

Therefore

$$s^2 = \frac{\Sigma x_i^2 - n\bar{x}^2}{n-1}. \tag{3.5}$$

St. Luke's Hospitals Hospice Care Program

St. Louis, Missouri*

The Hospice Program at St. Luke's Hospitals is based on the belief that people with advanced disease can live and die among family and friends in familiar surroundings—pain free and alert. We provide supportive care for the physical, emotional and spiritual needs of these people and their families.

Hospice offers an integrated set of services which care for people and their families when the doctor has said there is nothing more that can be done to cure their disease. The goal of Hospice is to enable dying persons to live the remainder of their lives at home among loved ones and familiar surroundings as free from pain and other symptoms of terminal illness as possible. Hospice helps these people to use their time and energy to deal with living and to retain maximum freedom of personal choice.

The word "hospice" or "hospes" dates back as early as the first century A.D., and was used to describe way-stations or shelters for travelers making long and difficult journeys. In the nineteenth century these way-stations became houses of sanctuary for the sick, wounded and dying. More recently the hospice movement can trace its beginnings to Sydeham, England, where in 1967, St. Christopher's Hospice admitted its first patient.

The first American hospice, The Connecticut Hospice, opened in 1974. By December 1982 it is estimated that over 1000 hospices will be in operation or in various stages of development, representing one hospice in every state and major metropolitan center.

St. Luke's Hospitals Hospice Care Program is proud to share in such a rich tradition of service to the terminally ill and their families. Since our opening in August of 1979, St. Luke's Hospice has served over 300 patients/families. In the first eight months, 20 patients/families were served; in the second eight months, 45 patients/families were served. We currently have 20 patients/families in the program on an ongoing basis.

The criteria for entering the program are:

1. The patient must have a limited life expectancy.
2. The patient must be able to participate in decision-making.
3. There must be a willing and able care-giving person (or persons) at home.
4. The patient/family unit must live within the general referral area of St. Luke's Hospitals.

*The authors are indebted to Ricki O'Meara, Hospice Coordinator, for St. Luke's Hospital for providing this application.

Often patients in St. Luke's Hospice program are able to live at home among loved ones and in familiar surroundings

Patients and families are referred to Hospice from many sources such as physicians, social workers, pastors, visiting nurses and friends. A patient or family member should discuss Hospice with the patient's physician who may indeed be the first person to suggest Hospice care. If they agree that Hospice is appropriate, the physician contacts either the Hospice Office or the Medical Director. If there is agreement between the physicians, the members of the family (and the patient if possible) are invited to the Hospice Office to talk with the Coordinator and the Assistant Coordinator. This is an opportunity to learn about the patient and family needs, to explain the Hospice concept, to see the Unit, meet the nurses, etc., and to begin the important two-way communication process that underlies the caring relationships which are the basis of Hospice. If the patient is unable to come to the hospital, the Hospice Coordinator will talk with the patient and family at home or in another hospital.

After the Hospice Medical Director and the interviewing team members have determined that the referred patient/family unit is appropriate for Hospice care, the patient is given a "consent to Hospice Care" form to sign when and if he/she and the family decide to enter the program. This form states (among other things), "I understand that in St. Luke's Hospice program I will not be given treatment directed at the cure of my disease, but rather I will be given symptomatic and supportive care directed at relieving my pain and suffering and keeping me comfortable and alert in my final days. . . . I hereby refuse all medical and surgical treatments of care, the sole purpose of which is to sustain my life process, such as cardiac resuscitation or the use of mechanical life support systems."

If severe pain or other distressing symptoms need to be controlled, the patient enters the Hospice Unit. A treatment plan is carefully devised by the whole team with the help of the patient and family. The patient's comfort is all-important. Whatever aids the comfort and meets the needs of the patient is employed, including the kinds and amount of food, time for eating, bathing, etc. Quietly and carefully the physical, psychological, spiritual and social needs of the ill person are ministered to, all directed toward pain-free, conscious comfort. Once this has been achieved, and the family members have become familiar with giving whatever care the patient will need at home, the Visiting Nurse who will give professional assistance in the home comes to meet the patient and family in the Hospice Unit and participates in the plan for home care. Volunteers are assigned to each patient/family unit to assist in whatever ways are needed. If it should become necessary the patient can be readmitted into the Hospice Unit. Patients are free to withdraw from the Hospice Program at any time should they wish to do so.

Finally, because of team work and open communication, the people in the Hospice create a community of genuine caring, not only for patients and family members, but for each other. It is in fact a community where love is acted out in service and support for all. Patients and family members give love as well as receive it in the mutuality of our common humanity.

STATISTICAL INFORMATION FOR HOSPICE

In the coordination and administration of the St. Luke's Hospice program, reports containing statistical summaries help the program's administrators better understand the ongoing operation of the program. Frequently, the statistical information provides a basis for operating policy decisions as well as a basis for planning for the future.

Examples of statistical summaries include the number of admissions to the Hospice program, the number of beds used by Hospice patients, the number of discharges, the percent occupancy, the ratio of patient deaths to total number of patients in the program, the number of readmissions, the total number of patient-days of care for the program and so on. Reports with these statistical summaries are prepared for the Hospice Coordinator. In addition, the reports are prepared for St. Luke's Vice President who uses the statistical summaries to better understand the Hospice program's contribution to the Hospital's total health care services. The Hospice report plus reports from other units throughout the Hospital help St. Luke's administrators to make decisions involving policy, planning and budget which best contribute to St. Luke's mission of offering quality medical services to its patients.

DESCRIPTIVE STATISTICS FOR LENGTH OF STAY IN HOSPICE

Patients admitted to the Hospice program are terminally ill. As such, the life expectancy and the corresponding length of stay in the Hospice program are expected to be relatively short. However, administrators who monitor the program operation and plan for the future in terms of number of beds, number of patients that can be admitted, and so on need periodic information showing the number of days the patients stay in the Hospice program.

As an illustration of the type of statistical information, Table A3.1 shows a frequency distribution providing the number of days in the Hospice program for a sample of 54 patients during 1982. While the frequency distribution is useful, the

Table A3.1 Length of Stay in Hospice for 1982 (Based on Sample of 54 Patients)

Length of Stay (Days)	*Frequency*
0 but less than 25	20
25 but less than 50	12
50 but less than 75	7
75 but less than 100	3
100 but less than 125	5
125 but less than 150	2
150 but less than 175	3
Over 175	2
	54

following descriptive statistics provided helpful summary information for the administrators.

Mean stay:	59.5 days
Median stay:	38 days
Mode stay:	4 days

Note that on average, the number of days a patient is in Hospice is about 2 months (59.5 days). However, the median shows half of the patients are in the program approximately a month (38 days) or less. The mode is consistent with the belief that the terminally ill Hospice patients have a relatively short stay in the program. The mode of 4 days is based on the fact that the most frequent length of stay of 4 days occurred for 7 of the 54 patients in the sample. The standard deviation of 75.7 days provides a measure of variability in the length of stay for the patients. Year-to-year comparisons of the standard deviations provide insight to changes in the variability of patient stays.

Without a doubt, Hospice provides a much-needed and valuable patient/family service for terminally ill patients. Descriptive statistics such as the length of stay statistics shown above provide program administrators with information essential for decision making and planning which will maximize the impact and contribution of the Hospice program.

4 Introduction to Probability

What you will learn in this chapter:

- an appreciation for the role probability plays in the decision-making process
- the three methods commonly used for assigning probabilities to outcomes
- how to use the basic laws of probability to compute the probability of an event
- how to use Bayes' theorem
- some important terms such as experiment, sample space, event, mutually exclusive events, independent events, and conditional probability

Contents

Throughout our lives we are faced with decision-making situations which involve an uncertain future. Perhaps you will be asked for an analysis of one of the following situations involving uncertainty:

1. What is the "chance" that sales will decrease if the price of the product is increased?
2. What is the "likelihood" that the new assembly method will increase productivity?
3. How "likely" is it that the project will be completed on time?
4. What are the "odds" that the new investment will be profitable?

The subject matter most useful in effectively dealing with such uncertainties is contained under the heading of probability. In everyday terminology, probability can be thought of as a numerical measure of the "chance" or "likelihood" that a particular event will occur. For example, if we consider the event "rain tomorrow," we understand that when the television weather report indicates "a near-zero probability of rain" there is almost no chance of rain. However, if a 90% probability of rain is reported, we know that it is very likely or almost certain that rain will occur. A 50% probability indicates that rain is just as likely to occur as not.

Probability values are always assigned on a scale from 0 to 1. A probability "near" 0 indicates that the event is very unlikely to occur; a probability "near" 1 indicates that the event is almost certain to occur. Other probabilities between 0 and 1 represent varying degrees of likelihood that the event will occur. Figure 4.1 depicts this view of probability.

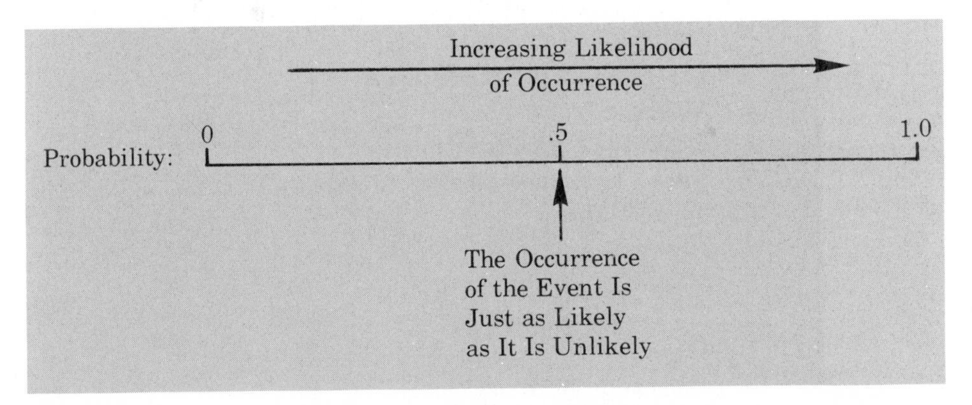

FIGURE 4.1 Probability as a Numerical Measure of the Likelihood of Occurrence

Probability is important in decision making because it provides a mechanism for measuring, expressing, and analyzing the uncertainties associated with future events. In this chapter we introduce the fundamental concepts of probability and begin to illustrate their use as decision-making tools. In subsequent chapters we will extend these basic notions of probability and demonstrate the important role that probability plays in statistical inference.

4.1 EXPERIMENTS AND SAMPLE SPACE

Using the terminology of probability, we define an *experiment* to be any process which generates well defined outcomes. By this we mean that on any single repetition of the experiment *one and only one* of the possible experimental outcomes will occur. Several examples of experiments and their associated outcomes are as follows:

Experiment	*Experimental Outcomes*
Toss a coin	Head, tail
Select a part for inspection	Defective, nondefective
Conduct a sales call	Purchase, no purchase
Roll a die	1, 2, 3, 4, 5, 6
Play a football game	Win, lose, tie

The first step in analyzing a particular experiment is to carefully define the experimental outcomes. When we have defined *all* possible experimental outcomes, we have identified the *sample space* for the experiment. That is, the sample space is defined as the set of all possible experimental outcomes. Any one particular experimental outcome is referred to as a *sample point* and is an element of the sample space.

Let us consider the experiment of tossing a coin, as mentioned above. The experimental outcomes are defined by the upward face of the coin—a head or a tail. If we let S denote the sample space, we can use the following notation to describe the sample space and sample points for the coin-tossing experiment:

$$S = \{\text{Head}, \text{Tail}\}.$$

Listing of the Sample Points

Using this notation, the experiment of selecting a part for inspection would have a sample space with sample points as follows:

$$S = \{\text{Defective}, \text{Nondefective}\}.$$

Finally, suppose that we consider the experiment of rolling a die, where the experimental outcomes are defined as the number of dots appearing on the upward face of the die. In this experiment, the numerical values 1, 2, 3, 4, 5, and 6 represent the possible experimental outcomes or sample points. Thus the sample space is denoted

$$S = \{1,2,3,4,5,6\}.$$

Let us extend our discussion of experiments, sample points, and sample spaces to a slightly more involved illustration. Consider an experiment of tossing two coins, with the experimental outcomes defined in terms of the pattern of heads and tails appearing on the upward faces of the two coins. How many experimental outcomes (sample points) are possible for this experiment?

Before attempting to answer this question, let us introduce a rule which often is helpful in determining the number of sample points for an experiment consisting of multiple steps:

A Counting Rule for Multiple-Step Experiments

If an experiment can be described as a sequence of k steps in which there are n_1 possible results on the first step, n_2 possible results on the second step, and so on, then the total number of experimental outcomes is given by $(n_1)(n_2) \cdots (n_k)$.

Looking at the experiment of tossing two coins as a sequence of first tossing one coin ($n_1 = 2$) and then tossing the other coin ($n_2 = 2$), we can see from the counting rule that there must be (2)(2) = 4 distinct sample points or experimental outcomes. These sample points are as follows:

Sample Point or Experimental Outcome	*First Coin*	*Second Coin*
1	Head	Head
2	Head	Tail
3	Tail	Head
4	Tail	Tail

If we use H to denote a head and T to denote a tail, (H,H) indicates the sample point with a head on the first coin and a head on the second coin. Similarly, (H,T) indicates the sample point with a head on the first coin and a tail on the second coin. Continuing this notation, we can describe the sample space S for the two-coin-tossing experiment as follows:

$$S = \{(H,H),(H,T),(T,H),(T,T)\}.$$

A graphical device that is helpful in visualizing an experiment and enumerating sample points is a *tree diagram.* Figure 4.2 provides the tree diagram for the two-coin-tossing experiment. Step 1 corresponds to tossing the first coin, step 2 corresponds to tossing the second coin, and each of the points on the right-hand end of the tree corresponds to a sample point or experimental outcome.

Let us now see how the concepts introduced thus far can be used in the analysis of the capacity expansion problem faced by the Kentucky Power and Light Company. We will begin by showing how the company's situation can be viewed as an experiment. Then we will attempt to define the appropriate sample points and the sample space for the experiment.

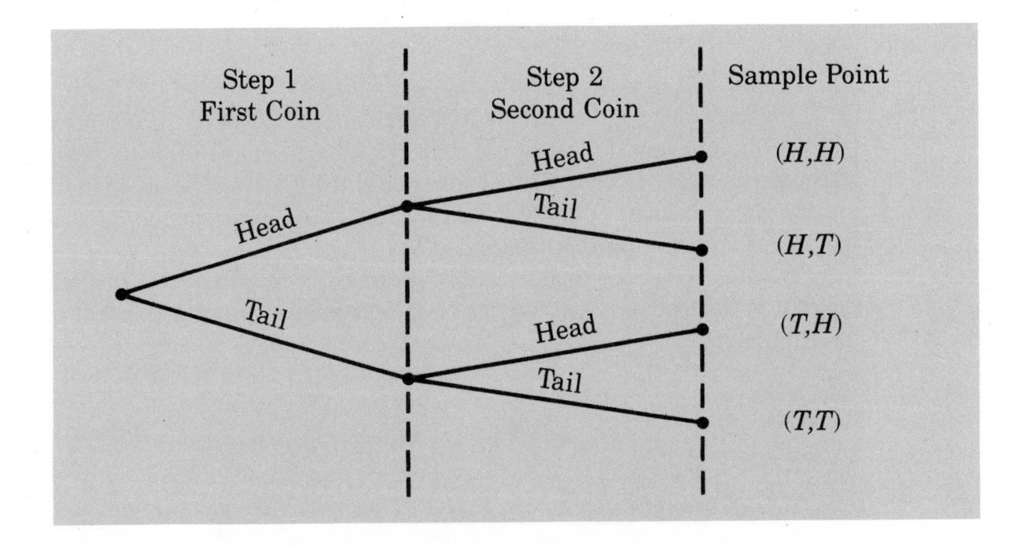

FIGURE 4.2 Tree Diagram for the Experiment of Tossing Two Coins

The Kentucky Power and Light Problem

The Kentucky Power and Light Company (KP&L) currently is starting work on a project designed to increase the generating capacity of one of its plants in Northern Kentucky. The project is divided into two sequential stages: stage 1 (design) and stage 2 (construction). While each stage will be scheduled and controlled as closely as possible, management cannot predict beforehand the exact elapsed time for each stage of the project. However, an analysis of similar construction projects over the past 3 years has shown completion times for the design stage of 2, 3, or 4 months and completion times for the construction stage of 6, 7, or 8 months. Thus management has decided to use these figures as the estimated completion times for the current project. In addition, because of the critical need for additional power, management has set a goal of 10 months for the total project completion time. Hence the entire project will be completed late if the total elapsed time to complete both stages exceeds 10 months.

Since there are three possible completion times for each stage of the project, and since the project involves a sequence of two stages, the counting rule for multiple-step experiments can be applied to determine that there is a total of (3)(3) = 9 experimental

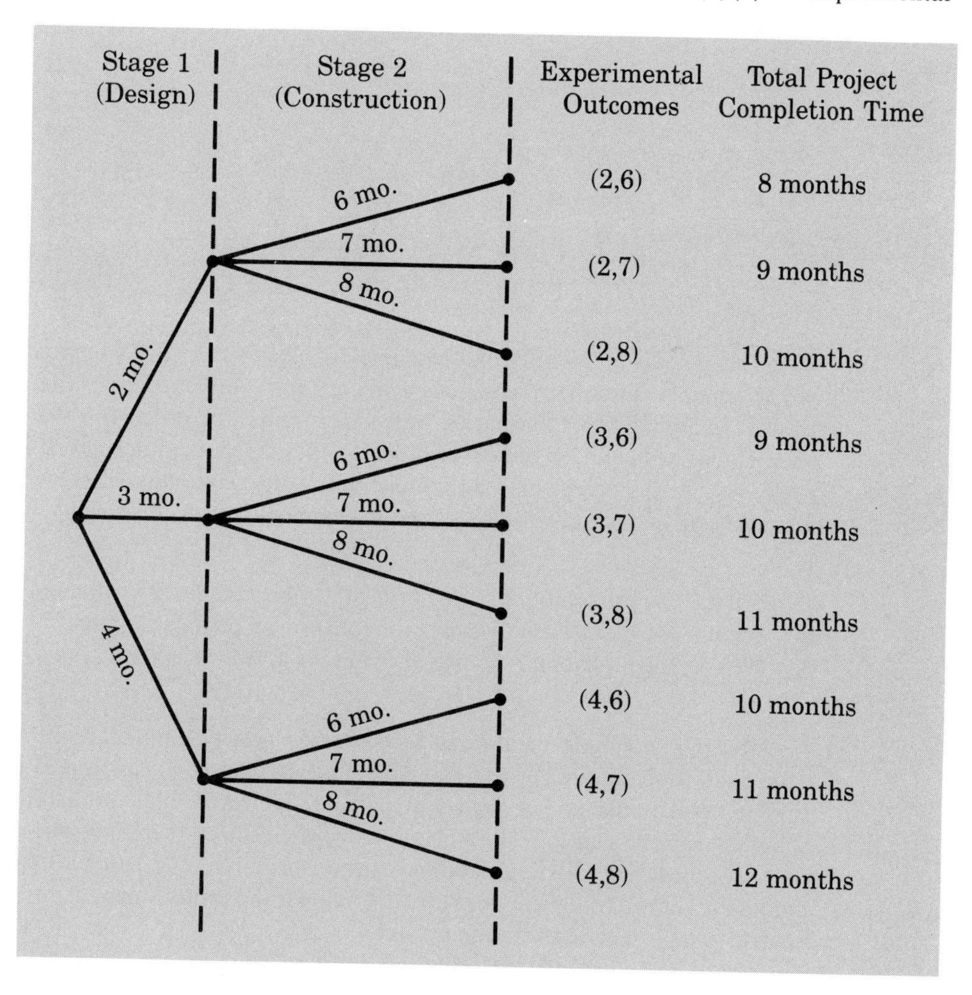

FIGURE 4.3 Tree Diagram for the KP&L Project

outcomes or sample points. To describe the experimental outcomes we will use a two-number notation. The first number will provide the completion time of stage 1 (design), and the second number will provide the completion time of stage 2 (construction). For example, the experimental outcome denoted by (2,6) would indicate the outcome with stage 1 being completed in *2 months* and stage 2 being completed in *6 months.* Table 4.1 uses this notation to summarize the nine experimental outcomes for the KP&L problem. Figure 4.3 shows a tree diagram indicating how these outcomes or sample points occur.

TABLE 4.1 Listing of Experimental Outcomes or Sample Points for the KP&L Problem

Completion Time (months)			
Stage 1 (Design)	*Stage 2 (Construction)*	*Notation for Experimental Outcome*	*Total Project Completion Time (months)*
2	6	(2,6)	8
2	7	(2,7)	9
2	8	(2,8)	10
3	6	(3,6)	9
3	7	(3,7)	10
3	8	(3,8)	11
4	6	(4,6)	10
4	7	(4,7)	11
4	8	(4,8)	12

The process of identifying the sample space for the KP&L project helps the project manager visualize the various outcomes and completion times that are possible. However, probability information will be essential in order for the manager to understand the likelihood of the various outcomes. From the information in Figure 4.3 we know that the project will be completed in 8 to 12 months. However, the project manager will undoubtedly be interested in questions such as the following:

1. What is the probability that the project can be completed in 10 months or less?
2. What is the most likely or most probable project completion time?
3. What is the probability the project will take a year (12 months) to complete?

Recall that management's goal is to have the project completed within 10 months. Thus high probabilities for early and/or on-time completions will give the manager confidence that his or her goal will be met. However, high probabilities for late completions may lead to revised planning for the project, including corrective actions such as scheduling overtime, adding to the workforce, and so on. In any case, the probability information will be critical in helping the project manager understand the uncertainties associated with the project.

In order to provide the desired probability information for the KP&L problem, we need to consider how probability values can be assigned to the various experimental outcomes. This is the topic of the next section.

EXERCISES

1. In a quality control process an inspector selects a completed part for inspection. The inspector then determines whether the part has a major defect, a minor defect, or no defect. Consider the selection and classification of the part as an experiment. List the sample points for the experiment.

2. Consider the experiment consisting of conducting three sales calls. On each of the calls there will be either a purchase or no purchase.

a. Construct a tree diagram for this experiment.

b. Identify each sample point and the sample space. How many sample points are there?

c. How many sample points would there be if the experiment consisted of four sales calls?

3. In the city of Milford applications for zoning changes go through a two-step process: a review by the planning commission and a final decision by the city council. At step 1 the planning commission will review the zoning change request and make a positive or negative recommendation concerning the change. At step 2 the city council will review the planning commission's recommendation and then vote to approve or to disapprove the zoning change. In some instances the city council vote has agreed with the planning commission's recommendation. However, in other instances the council vote has been the opposite of the planning commission's recommendation. Consider the application for a zoning change that has just been submitted by the developer of an apartment complex. Consider the application process as an experiment.

a. How many sample points are there for this experiment? List the sample points.

b. Construct a tree diagram for the experiment.

4. An investor has two stocks: stock A and stock B. Each stock may increase in value, decrease in value, or remain unchanged. Consider the experiment as the investment in the two stocks.

a. How many experimental outcomes are possible?

b. Show a tree diagram for the experiment.

c. How many of the experimental outcomes result in an increase in value for at least one of the two stocks?

d. How many of the experimental outcomes result in an increase in value for both of the stocks?

5. Consider the experiment of rolling a pair of dice. Each die has six possible results (the number of dots on its face).

a. How many sample points are possible for this experiment?

b. Show a tree diagram for the experiment.

c. How many experimental outcomes provide a sum of 7 for the dots on the dice?

6. Many states design their automobile license plates such that space is available for up to six letters or numbers.

a. If a state decides to use only numerical values for the license plates, how many different license plate numbers are possible? Assume that 000000 is an acceptable license plate number, although it will be used only for display purposes at the license bureau. (Hint: use the counting rule.)

b. If the state decides to use two letters followed by four numbers, how many different license plate numbers are possible? Assume that the letters I and O will not be used because of their similarity to numbers 1 and 0.

c. Would larger states, such as New York or California, tend to use more or fewer letters in license plates? Explain.

4.2 ASSIGNING PROBABILITIES TO EXPERIMENTAL OUTCOMES

We now have an understanding of the concept of an experiment and of the sample space as the set of all experimental outcomes. We are now ready to see how probabilities for the experimental outcomes (sample points) can be determined. Recall the discussion at the beginning of this chapter. The probability of an experimental outcome was said to be a numerical measure of the likelihood that the experimental outcome would occur. In the assigning of probabilities to the experimental outcomes there are various acceptable approaches; however, regardless of the approach taken the following two basic requirements must be satisfied:

1. The probability values assigned to each experimental outcome (sample point) must be between 0 and 1. That is, if we let E_i indicate the experimental outcome i and $P(E_i)$ indicate the probability of this experimental outcome, we must have

$$0 \leq P(E_i) \leq 1 \qquad \text{for all } i. \tag{4.1}$$

2. The sum of *all* of the experimental outcome probabilities must be 1. For example, if a sample space has k sample points (experimental outcomes), we must have

$$P(E_1) + P(E_2) + \cdots + P(E_k) = \sum P(E_i) = 1. \tag{4.2}$$

Any method of assigning probability values to the experimental outcomes which satisfies these two requirements and results in a reasonable numerical measure of the likelihood of the outcome is acceptable. In practice, one of the following three methods can be used:

1. Classical method.
2. Relative frequency method.
3. Subjective method.

Classical Method

To illustrate the classical method of assigning probabilities, let us again consider the experiment of flipping a coin. On any one flip, we will observe one of two experimental outcomes: "head" or "tail." It would seem reasonable to assume that the two possible outcomes are equally likely. Therefore since one of the two equally likely outcomes is "head," we logically should conclude that the probability of observing "head" is $\frac{1}{2}$, or .50. Similarly, the probability of observing "tail" is also .50. When the assumption of equally likely outcomes is used as a basis for assigning probabilities, the approach is referred to as the *classical method.* If an experiment has n possible outcomes, the classical approach would assign a probability of $1/n$ to each experimental outcome.

As another illustration of the classical method, consider again the experiment of rolling a die. In Section 4.1 we described the sample space and sample points for this experiment with the following notation:

$$S = \{1,2,3,4,5,6\}.$$

It would seem reasonable to conclude that the six possible outcomes are equally likely, and hence each outcome is assigned a probability of $\frac{1}{6}$. Thus, if $P(1)$ denotes the probability that one dot appears on the upward face of the die, then $P(1) = \frac{1}{6}$. Similarly, $P(2) = \frac{1}{6}$, $P(3) = \frac{1}{6}$, $P(4) = \frac{1}{6}$, $P(5) = \frac{1}{6}$, and $P(6) = \frac{1}{6}$. Note that this probability assignment satisfies the two basic requirements for assigning probabilities. In fact, requirements (4.1) and (4.2) will always be satisfied when the classical method is used, since each of the n sample points is assigned a probability of $1/n$.

The classical method was developed originally in the analysis of gambling problems, where the assumption of equally likely outcomes often is reasonable. In many business problems, however, this assumption is not valid. Hence alternative methods of assigning probabilities are required.

Relative Frequency Method

As an illustration of the relative frequency method, consider a firm that is preparing to market a new product. In order to estimate the probability that a customer will purchase the product, a test market evaluation has been set up wherein salespeople will call on potential customers. For each sales call conducted, there are two possible outcomes: the customer purchases the product or the customer does not purchase the product. Since there is no reason to assume that the two experimental outcomes are equally likely, the classical method of assigning probabilities would be inappropriate.

Suppose that in the test market evaluation of the product, 400 potential customers were contacted; 100 actually purchased the product, but 300 did not. In effect, then, we have repeated the experiment of contacting a customer 400 times and have found that the product was purchased 100 times. Thus we might decide to use the relative frequency of the number of customers that purchased the product as an estimate of the probability of a customer making a purchase. Hence we could assign a probability of $100/400 = .25$ to the experimental outcome of purchasing the product. Similarly, $300/400 = .75$ could be assigned to the experimental outcome of not purchasing the product. This approach to the assigning of probabilities is referred to as the *relative frequency method.*

Subjective Method

The classical and relative frequency methods cannot be applied to all situations where probability assessments are desired. For example, there are many situations where the experimental outcomes are not equally likely and where relative frequency data are unavailable. For example, consider the next football game that the Pittsburgh Steelers will play. What is the probability that the Steelers will win? The experimental outcomes of a win, a loss, or a tie are not necessarily equally likely. Also, since the teams involved have not played several times previously this year, there are no relative frequency data available that are relevant to this upcoming game. Thus if we want an estimate of the probability of the Steelers winning, we must use a subjective opinion of its value.

With the subjective method to assign probabilities to the experimental outcomes, we may use any data available, our experience, intuition, etc. However, after we consider all available information, a probability value that expresses the *degree of belief* that the experimental outcome will occur must be specified. This method of assigning probability is referred to as the *subjective method.* Since subjective probability expresses a person's "degree of belief," it is personal. Different people can be expected to assign different probabilities to the same event. Nonetheless, care must be taken when using the subjective method to ensure that requirements (4.1) and (4.2) are satisfied. That is, regardless of a person's "degree of belief," the probability value assigned to each experimental outcome must be between 0 and 1 and the sum of all the experimental outcome probabilities must be 1.

Even in situations where either the classical or relative frequency approach can be applied, management may want to provide subjective probability estimates. In such cases, the best probability estimates often are obtained by combining the estimates from the classical or relative frequency approaches with the subjective probability estimates.

Probabilities for the KP&L Problem

To perform further analysis for the KP&L problem, we must develop probabilities for each of the nine experimental outcomes listed in Table 4.1. Based on experience and judgment, management concluded that the experimental outcomes were not equally likely. Hence the classical approach to assigning probabilities could not be used. Management then decided to conduct a study of the completion times for similar projects undertaken by KP&L over the past three years. The results of a study of 40 similar projects are summarized in Table 4.2.

After reviewing the results of the study, management decided to employ the

TABLE 4.2 Completion Results for 40 KP&L Projects

Completion Time (Months)			*Number of Past Projects Having These Completion Times*
Stage 1	*Stage 2*	*Sample Point*	
2	6	(2,6)	6
2	7	(2,7)	6
2	8	(2,8)	2
3	6	(3,6)	4
3	7	(3,7)	8
3	8	(3,8)	2
4	6	(4,6)	2
4	7	(4,7)	4
4	8	(4,8)	6
		Total	40 projects

TABLE 4.3 Probability Assignments for the KP&L Problem Based on the Relative Frequency Method

Sample Point	*Project Completion Time*	*Probability of Sample Point*
(2,6)	8 months	$P(2,6) = 6/40 = .15$
(2,7)	9 months	$P(2,7) = 6/40 = .15$
(2,8)	10 months	$P(2,8) = 2/40 = .05$
(3,6)	9 months	$P(3,6) = 4/40 = .10$
(3,7)	10 months	$P(3,7) = 8/40 = .20$
(3,8)	11 months	$P(3,8) = 2/40 = .05$
(4,6)	10 months	$P(4,6) = 2/40 = .05$
(4,7)	11 months	$P(4,7) = 4/40 = .10$
(4,8)	12 months	$P(4,8) = 6/40 = .15$
		Total 1.00

relative frequency approach to assigning probabilities. Management could still have provided subjective probability estimates, but it was felt that the current project was quite similar to the 40 previous projects. Thus the relative frequency approach to assigning probabilities was judged best in this case.

In using the data in Table 4.2 to compute probabilities we note that outcome (2,6)—stage 1 completed in 2 months and stage 2 completed in 6 months—occurred 6 times in the 40 projects. Thus we can use the relative frequency approach to assign a probability of $6/40 = .15$ to this outcome. Similarly, outcome (2,7) also occurred in 6 of the 40 projects, providing a $6/40 = .15$ probability. Continuing in this manner, we obtain the probability assignments for the sample points of the KP&L project shown in Table 4.3. Note that $P(2,6)$ represents the probability of the sample point (2,6), $P(2,7)$ represents the probability of the sample point (2,7), and so on.

EXERCISES

7. Consider the experiment of selecting a card from a deck of 52 cards.
a. How many sample points are possible?
b. Which method (classical, relative frequency, or subjective) would you recommend for assigning probabilities to the sample points?
c. What are the probability assignments?
d. Show that your probability assignments satisfy the two basic requirements for assigning probabilities.

8. A small-appliance store in Madeira has collected data on refrigerator sales for the last 50 weeks. The data are presented below:

Number of Refrigerators Sold	*Number of Weeks*
0	6
1	12
2	15
3	10
4	5
5	2
	50

Suppose that we are interested in the experiment of observing the number of refrigerators sold in 1 week of store operations.
a. How many sample points (experimental outcomes) are there?
b. Which approach would you recommend for assigning probabilities to the sample points?
c. Assign probabilities to the sample points and verify that your assignments satisfy the two basic requirements.

9. Strom Construction has made a bid on two contracts. The owner has identified the possible outcomes and subjectively assigned probabilities as follows:

Experimental Outcome	*Obtain Contract 1*	*Obtain Contract 2*	*Probability*
1	Yes	Yes	.15
2	Yes	No	.15
3	No	Yes	.30
4	No	No	.25

a. Are these valid probability assignments? Why or why not?
b. What would have to be done to make the probability assignments valid?

4.3 EVENTS AND THEIR PROBABILITIES

Until now we have used the term "event" much as it would be used in everyday language. However, at this point we introduce the formal definition of an event as it relates to probability. This definition is as follows:

Definition of an Event
An *event* is a collection of sample points.

For an example, let us return to the KP&L problem and assume that the project manager is interested in the event that the entire project can be completed in 10 months or less. Referring to Table 4.3, we see that six sample points (2,6), (2,7), (2,8), (3,6), (3,7), and (4,6) provide a project completion time of 10 months or less. Let C denote the event that the project is completed in 10 months or less; we write

$$C = \{(2,6),(2,7),(2,8),(3,6),(3,7),(4,6)\}.$$

Event C is said to occur if *any one* of the six sample points shown above appears as the experimental outcome.

Other events that might be of interest to KP&L management include the following:

L = the event that the project is completed in *less* than 10 months;

M = the event that the project is completed in *more* than 10 months.

Using the information in Table 4.3 we see that these events consist of the following sample points:

$$L = \{(2,6),(2,7),(3,6)\},$$
$$M = \{(3,8),(4,7),(4,8)\}.$$

A variety of additional events can be defined for the KP&L problem, but in each case the event must be identified as a collection of sample points for the experiment.

Given the probabilities of the sample points (experimental outcomes) shown in Table 4.3, we can use the following definition to compute the probability of any event that KP&L management might want to consider:

Definition of the Probability of an Event
The probability of any event is equal to the sum of the probabilities of the sample points in the event.

Using this definition, we calculate the probability of a particular event by adding the probabilities of the experimental outcomes that make up the event. We can now compute the probability that the project will take 10 months or less to complete. Since this event is given by $C = \{(2,6),(2,7),(2,8),(3,6),(3,7),(4,6)\}$, the probability of event C

is shown below (note that P is used to denote the probability of the corresponding event or sample point):

$$P(C) = P(2,6) + P(2,7) + P(2,8) + P(3,6) + P(3,7) + P(4,6).$$

Refer to the sample point probabilities in Table 4.3; we have

$$P(C) = .15 + .15 + .05 + .10 + .20 + .05 = .70.$$

Similarly, since the event that the project is completed in less than 10 months is given by $L = \{(2,6),(2,7),(3,6)\}$, the probability of this event is given by

$$\begin{aligned} P(L) &= P(2,6) + P(2,7) + P(3,6) \\ &= .15 + 1.5 + .10 = .40. \end{aligned}$$

Finally, if the project is completed in more than 10 months, we have $M = \{(3,8),(4,7),(4,8)\}$ and thus

$$\begin{aligned} P(M) &= P(3,8) + P(4,7) + P(4,8) \\ &= .05 + .10 + .15 = .30. \end{aligned}$$

Using the above probability results, we can now tell KP&L management that there is a .70 probability that the project will be completed in 10 months or less, a .40 probability that the project will be completed in less than 10 months, and a .30 probability the project will be completed in more than 10 months. This procedure of computing event probabilities can be repeated for any event of interest to the KP&L management.

Any time that we can identify all the sample points of an experiment and assign the corresponding sample point probabilities, we can use the definition of this section to compute the probability of an event of interest to a decision maker. However, in many experiments the number of sample points is large and the identification of the sample points, as well as determining their associated probabilities, becomes extremely cumbersome if not impossible. In the remaining sections of this chapter we present some basic probability relationships that can be used to compute the probability of an event without requiring knowledge of sample point probabilities. These probability relationships require a knowledge of the probabilities for some events in the experiment. Probabilities of other events are then computed directly from the known probabilities using one or more of the probability relationships.

EXERCISES

10. Use the Kentucky Power and Light Company sample point and sample point probabilities in Table 4.3 to solve the following problems:

a. The design stage (stage 1) will run over budget if it takes 4 months to complete. List the sample points in the event the design stage is over budget.

b. What is the probability that the design stage is over budget?

c. The construction stage (stage 2) will run over budget if it takes 8 months to complete. List the sample points in the event the construction stage is over budget.

d. What is the probability that the construction stage is over budget?

e. What is the probability that both stages are over budget?

11. In Exercise 7 we considered the experiment of selecting a card from a deck of 52 cards. Each card corresponded to a sample point with a 1/52 probability.

a. List the sample points in the event an ace is selected.

b. List the sample points in the event a club is selected.

c. List the sample points in the event a face card (jack, queen, or king) is selected.
d. Find the probabilities associated with each of the events in parts a, b, and c.

12. Suppose that a manager of a large apartment complex provides the following subjective probability estimate about the number of vacancies that will exist next month:

Vacancies	*Probability*
0	.05
1	.15
2	.35
3	.25
4	.10
5	.10

List the sample points in each of the following events and provide the probability of the event:
a. No vacancies.
b. At least four vacancies.
c. Two or fewer vacancies.

13. Consider the experiment of rolling a pair of dice. Suppose that we are interested in the sum of the face values showing on the dice.
a. How many sample points are possible? (Hint: use the counting rule.)
b. List the sample points.
c. What is the probability of obtaining a value of 7?
d. What is the probability of obtaining a value of 9 or greater?
e. Since there are six possible even values (2, 4, 6, 8, 10, and 12) and only five possible odd values (3, 5, 7, 9, and 11), the dice should show even values more often than odd values. Do you agree with this statement? Explain.
f. What method did you use to assign the probabilities shown above?

14. A marketing manager is attempting to assign probability values to the possible profits and losses resulting from a new product. Relying on subjective probabilities, the manager's probability estimates are as follows:

$$P(\text{profit over } \$10{,}000) = .25,$$
$$P(\text{profit from } \$0 \text{ to } \$10{,}000) = .50,$$
$$P(\text{loss}) = .15.$$

What advice would you offer before the manager uses these estimates to perform further probability calculations?

15. The manager of a furniture store sells from zero to four china hutches each week. Based on past experience, the following probabilities are assigned to sales of zero, one, two, three, or four hutches:

$$P(0) = .08$$
$$P(1) = .18$$
$$P(2) = .32$$
$$P(3) = .30$$
$$P(4) = \underline{.12}$$
$$1.00$$

a. Are these valid probability assignments? Why or why not?
b. Let A be the event that two or fewer are sold in one week. Find $P(A)$.
c. Let B be the event that four or more are sold in one week. Find $P(B)$.

4.4 SOME BASIC RELATIONSHIPS OF PROBABILITY

Complement of an Event

Given an event A, the *complement* of A is defined to be the event consisting of all sample points that are *not* in A. The complement of A is denoted $\overline{A}$. Figure 4.4 provides

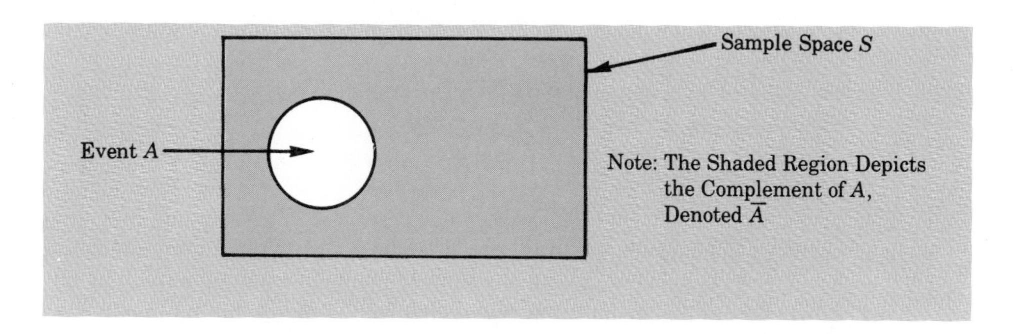

FIGURE 4.4 Complement of Event A

a diagram, known as a *Venn diagram,* which illustrates the concept of a complement. The rectangular area represents the sample space for the experiment and as such contains all possible sample points. The circle represents event A and contains only the sample points that belong to A. The shaded region of the diagram contains all sample points not in event A, which is by definition the complement of A.

In any probability application, event A and its complement $\overline{A}$ must satisfy

$$P(A) + P(\overline{A}) = 1. \tag{4.3}$$

Solving for $P(A)$, we obtain the following result:

Computing Probability Using the Complement

$$P(A) = 1 - P(\overline{A}). \tag{4.4}$$

Equation (4.4) shows that the probability of an event A can be easily computed if the probability of its complement, $P(\overline{A})$, is known.

As an example, consider the case of a sales manager who, after reviewing sales reports, states that 80% of new customer contacts result in no sale. By letting A denote the event of a sale and $\overline{A}$ denote the event of no sale, the manager is stating that $P(\overline{A}) = .80$. Using (4.4), we see that

$$P(A) = 1 - P(\overline{A}) = 1 - .80 = .20,$$

which shows that there is a .20 probability that a sale will be made on a new customer contact.

In another example, a purchasing agent states that there is a .90 probability that a supplier will send a shipment that is free of defective parts. Using the complement, we can conclude that there is a $1 - .90 = .10$ probability that the shipment will contain defective parts.

Addition Law

The addition law is a helpful probability relationship when we have two events and are interested in knowing the probability that at least one of the events occurs. That is, with events A and B we are interested in knowing the probability that event A or event B or both occur.

Before we present the addition law, we need to discuss two concepts concerning the combination of events: the *union* of events and the *intersection* of events.

Given two events A and B, the union of A and B is defined as follows:

Definition of the Union of Two Events

The *union* of A and B is the event containing *all* sample points belonging to *A or B or both*. The union is denoted $A \cup B$.

The Venn diagram shown in Figure 4.5 depicts the union of events A and B. Note that the shaded region contains all the sample points in event A as well as all the sample points in event B. The fact that the circles overlap indicates that there are some sample points contained in both A and B.

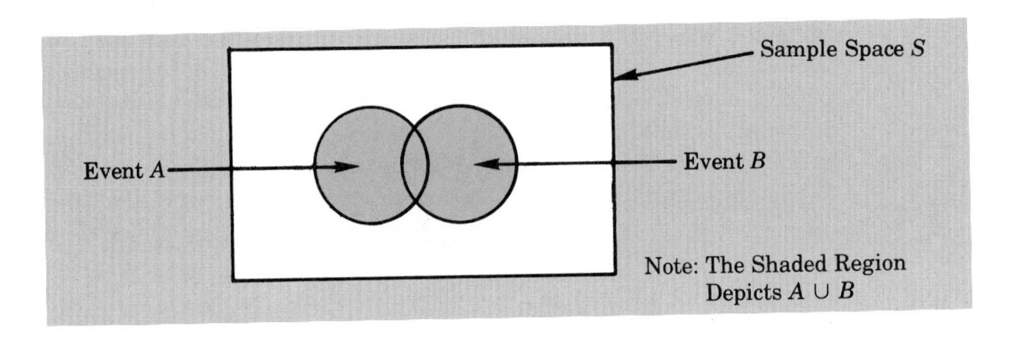

FIGURE 4.5 Union of Events A and B ($A \cup B$)

The definition of the intersection of two events A and B is shown below:

Definition of the Intersection of Two Events

Given two events A and B, the *intersection* of A and B is the event containing the sample points belonging to *both A and B*. The intersection is denoted $A \cap B$.

The Venn diagram depicting the intersection of the two events is shown in Figure 4.6. The area where the two circles overlap is the intersection; it contains the sample points that are in both A and B.

Let us now continue with a discussion of the addition law. The addition law provides a way to compute the probability of event A or B or both occurring. In other words, the addition law is used to compute the probability of the union of two events, $A \cup B$.

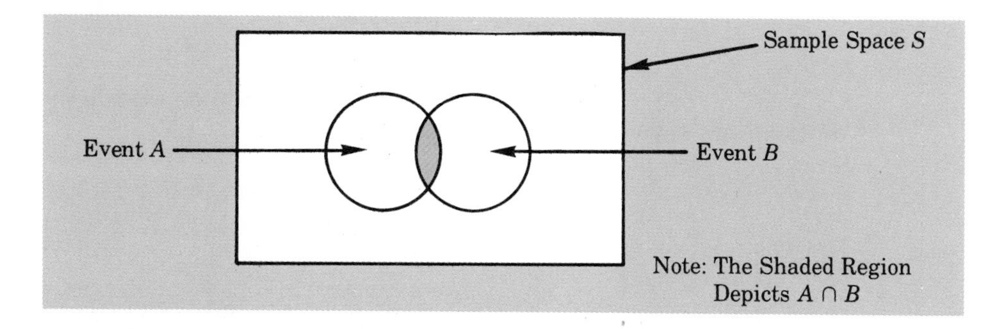

FIGURE 4.6 Intersection of Events A and B ($A \cap B$)

Addition Law

$$P(A \cup B) = P(A) + P(B) - P(A \cap B). \tag{4.5}$$

To obtain an intuitive understanding of the addition law, note that the first two terms in the addition law, $P(A) + P(B)$, account for all the sample points in $A \cup B$. However, since the sample points in the intersection $A \cap B$ are in both A and B, when we compute $P(A) + P(B)$ we are in effect counting each of the sample points in $A \cap B$ twice. We correct for this by subtracting $P(A \cap B)$.

In order to present an application of the addition law, let us consider the case of a small assembly plant with 50 employees. Each worker is expected to complete work assignments on time and in such a way that the assembled product will pass a final inspection. On occasion, some of the workers fail to meet the performance standards by completing work late and/or assembling defective products. At the end of a performance evaluation period, the production manager found that 5 of the 50 workers had completed work late, 6 of the 50 workers had assembled defective products, and 2 of the 50 workers had both completed work late *and* assembled defective products. Let

L = the event that the work is completed late,

D = the event that the assembled product is defective.

The above relative frequency information leads to the following probabilities:

$$P(L) = 5/50 = .10,$$
$$P(D) = 6/50 = .12,$$
$$P(L \cap D) = 2/50 = .04.$$

After reviewing the performance data, the production manager decided to assign a poor performance rating to any employee whose work was either late or defective; thus the event of interest is $L \cup D$. What is the probability that the production manager assigned an employee a poor performance rating?

Note that the probability question is about the union of two events. Specifically, we want to know $P(L \cup D)$. Here is where the addition law can be helpful. Using

(4.5), we have

$$P(L \cup D) = P(L) + P(D) - P(L \cap D).$$

Knowing values for the three probabilities on the right-hand side of the above expression, we can write

$$P(L \cup D) = .10 + .12 - .04 = .18$$

This tells us that there is a .18 probability that an employee will receive a poor performance rating.

As another example of the addition law, consider a recent study conducted by the personnel manager of a major computer software company. It was found that 30% of the employees that left the firm within 2 years did so primarily because they were dissatisfied with their salary, 20% left because they were dissatisfied with their work assignments, and 12% of the former employees said that *both* their salary and dissatisfaction with their work assignments were the primary reasons for leaving. What is the probability that an employee that leaves within 2 years does so because of being dissatisfied with salary, dissatisfied with the work assignment, or both?

Let

S = the event that the employee leaves due to salary,

W = the event that the employee leaves due to work assignment.

We have $P(S) = .30$, $P(W) = .20$, and $P(S \cap W) = .12$. Using (4.5), the addition law, we have

$$P(S \cup W) = P(S) + P(W) - P(S \cap W) = .30 + .20 - .12 = .38.$$

This shows that there is a .38 probability that an employee leaves because of salary or work assignment reasons.

Before we conclude our discussion of the addition law, let us consider a special case that arises for *mutually exclusive events:*

Definition of Mutually Exclusive Events

Two events are said to be *mutually exclusive* if the events have no sample points in common.

That is, events A and B are mutually exclusive if when one event occurs the other cannot occur. Thus a requirement for A and B to be mutually exclusive is that their intersection must contain no sample points. The Venn diagram depicting two mutually exclusive events A and B is shown in Figure 4.7. In this case $P(A \cap B) = 0$; hence the addition law can be written as follows:

Addition Law for Mutually Exclusive Events

$$P(A \cup B) = P(A) + P(B)$$

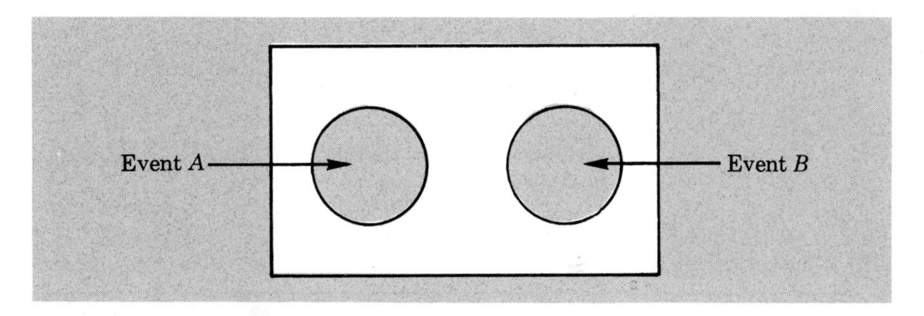

FIGURE 4.7 Mutually Exclusive Events

EXERCISES

16. Suppose that we have a sample space $S = \{E_1, E_2 E_3, E_4, E_5, E_6, E_7\}$, where $E_1, E_2, \ldots E_7$ denotes the sample points, and the following probability assignments for the sample points:

$$\begin{aligned} P(E_1) &= .05 \\ P(E_2) &= .20 \\ P(E_3) &= .20 \\ P(E_4) &= .25 \\ P(E_5) &= .15 \\ P(E_6) &= .10 \\ P(E_7) &= \underline{.05} \\ & \quad 1.00 \end{aligned}$$

Let

$$\begin{aligned} A &= \{E_1, E_4, E_6\}, \\ B &= \{E_2, E_4, E_7\}, \\ C &= \{E_2, E_3, E_5, E_7\}. \end{aligned}$$

a. Find $P(A)$, $P(B)$, and $P(C)$.
b. Find $(A \cup B)$ and $P(A \cup B)$.
c. Find $(A \cap B)$ and $P(A \cap B)$.
d. Are events A and C mutually exclusive?
e. Find $\bar{B}$ and $P(\bar{B})$.

17. An automotive store sells from zero to four car batteries each week. Based on past experience, the following probabilities are assigned to sales of zero, one, two, three, or four batteries.

$$\begin{aligned} P(0) &= .19 \\ P(1) &= .24 \\ P(2) &= .35 \\ P(3) &= .14 \\ P(4) &= \underline{.08} \\ & \quad 1.00 \end{aligned}$$

a. Are these valid probability assignments? Why or why not?
b. Let A be the event that two or fewer are sold in one week. Find $P(A)$.

c. Let B be the event that four or more are sold in one week. Find $P(B)$.
d. Are A and B mutually exclusive? Find $P(A \cap B)$ and $P(A \cup B)$.
18. Let

$$A = \text{the event that a person runs 5 miles or more per week,}$$
$$B = \text{the event that a person dies of heart disease,}$$
$$C = \text{the event that a person dies of cancer.}$$

Further, suppose that $P(A) = .01$, $P(B) = .25$, and $P(C) = .20$.
a. Are events A and B mutually exclusive? Can you find $P(A \cap B)$?
b. Are events B and C mutually exclusive? Find the probability that a person dies of heart disease or cancer.
c. Find the probability that a person dies from causes other than cancer.
19. During winter in Cincinnati, Mr. Krebs experiences difficulty in starting his two cars. The probability that the first car starts is .80, and the probability that the second car starts is .40. There is a probability of .30 that both cars start.
a. Define the events involved and use probability notation to show the probability information given above.
b. What is the probability that at least one car starts?
c. What is the probability that Mr. Krebs cannot start either of the two cars?
20. Consider an experiment where eight possible outcomes exist. We will denote the experimental outcomes as $E_1, E_2, \ldots, E_8$. Suppose the following events are defined:

$$A = (E_1, E_2, E_3, E_5),$$
$$B = (E_2, E_4, E_5, E_8).$$

Note that experimental outcomes E_6 and E_7 are in neither event A nor B.
Determine the sample points making up the following events:
a. $A \cup B$.
b. $A \cap B$.
c. $\bar{A}$.
d. Are A and B mutually exclusive events? Explain.
21. Let A be an event that a person's primary method of transportation to and from work is an automobile and B be an event that a person's primary method of transportation to and from work is a bus. Suppose that in a large city we find $P(A) = .45$ and $P(B) = .35$.
a. Are events A and B mutually exclusive? What is the probability that a person uses an automobile or a bus in going to and from work?
b. Find the probability that a person's primary method of transportation is something other than a bus.

Conditional Probability

In many probability situations it is important to be able to determine the probability of one event *given* that another event is known to have occurred. Suppose that we have an event A with probability $P(A)$. If we obtain new information or learn that another event, denoted B, has occurred, we will want to take advantage of this information in calculating the probability for event A.

This new probability of event A is written $P(A \mid B)$. The "|" is used to denote the fact that we are considering the probability of event A *given* the condition that event B has occurred. Thus the notation $P(A \mid B)$ is read "the probability of A given B."

As an illustration of the application of conditional probability, consider the situation of the promotional status of male and female officers of a major metropolitan police force in the eastern United States. The police force consists of 1200 officers, 960 men and 240 women. Over the past 2 years, 324 officers on the police force have been awarded promotions. The specific breakdown of promotions for male and female officers is shown in Table 4.4.

TABLE 4.4 Promotional Status of Police Officers Over the Past 2 Years

	Men	*Women*	*Total*
Promoted	288	36	324
Not Promoted	672	204	876
Total	960	240	1,200

After reviewing the promotional record, a committee of female officers raised a discrimination case on the basis that 288 male officers had received promotions but only 36 female officers had received promotions. The police administration has argued that the relatively low number of promotions for female officers is due not to discrimination but to the fact that there are relatively few female officers on the police force. Let us show how conditional probability could be used to analyze the discrimination charge.

Let

M = event an officer is a man,

W = event an officer is a woman,

A = event an officer is promoted.

Dividing the data values in Table 4.4 by the total of 1200 officers permits us to summarize the available information in the following probability values:

$P(M \cap A) = 288/1200 = .24$ = probability that an officer is a man *and* is promoted;

$P(M \cap \bar{A}) = 672/1200 = .56$ = probability that an officer is a man *and* is not promoted;

$P(W \cap A) = 36/1200 = .03$ = probability that an officer is a woman *and* is promoted;

$P(W \cap \bar{A}) = 204/1200 = .17$ = probability that an officer is a woman *and* is not promoted.

Since each of these values gives the probability of the intersection of two events, the probabilities are given the name of *joint probabilities*. Table 4.5, which provides a summary of the probability information for the police officer promotion situation, is referred to as a *joint probability table*.

The values in the margins of the joint probability table provide the probabilities of each event separately. That is, $P(M) = .80$, $P(W) = .20$, $P(A) = .27$, and $P(\bar{A}) = .73$. Thus we see that 80% of the force is male, 20% of the force is female, 27% of all officers received promotions, and 73% were not promoted. These probabilities are referred to as *marginal probabilities* because of their location in the margins of the joint probability table.

TABLE 4.5 Joint Probability Table for Promotions

Joint Probabilities Appear in the Body of the Table	*Men (M)*	*Women (W)*	*Total*
Promoted (A)	.24	.03	.27
Not Promoted ($\bar{A}$)	.56	.17	.73
Total	.80	.20	1.00

Marginal Probabilities Appear in the Margins of the Table

Let us begin the conditional probability calculations by computing the probability that an officer is promoted given that the officer is a man. In conditional probability notation we are attempting to determine $P(A \mid M)$. In order to calculate $P(A \mid M)$, we first realize that this notation simply means that we are considering the probability of the event A (promotion) given that the condition designated as event M (the officer is a man) is known to exist. Thus $P(A \mid M)$ tells us that we are now concerned only with the promotional status of the 960 male officers. Hence since 288 of the 960 male officers received promotions, the probability of being promoted given that the officer is a man is $288/960 = .30$. In other words, given that an officer is a man there has been a 30% chance of receiving a promotion over the past 2 years.

The above procedure was easy to apply in our illustration because the data values in Table 4.4 show the number of officers in each category. We now want to demonstrate how conditional probabilites such as $P(A \mid M)$ can be computed directly from probability information rather than the frequency data of Table 4.4.

We have shown that $P(A \mid M) = 288/960 = .30$. Let us now divide both the numerator and denominator of this fraction by 1,200, the total number of officers in the study. Thus,

$$P(A \mid M) = \frac{288}{960} = \frac{288/1200}{960/1200} = \frac{.24}{.80} = .30.$$

Hence we also see that the conditional probability $P(A \mid M)$ can be computed as $.24/.80$. Refer to the joint probability table shown in Table 4.5. Note in particular that .24 is the joint probability of A and M; that is, $P(A \cap M) = .24$. Also note that .80 is the marginal probability that an officer is a man; that is, $P(M) = .80$. Thus the conditional probability $P(A \mid M)$ can be computed as the ratio of the joint probability $P(A \cap M)$ to the marginal probability $P(M)$. That is,

$$P(A \mid M) = \frac{P(A \cap M)}{P(M)} = \frac{.24}{.80} = .30.$$

The fact that conditional probabilities can be computed as the ratio of a joint probability to a marginal probability provides the following general formula for conditional probability calculations for two events A and B.

Definition of Conditional Probability

$$P(A \mid B) = \frac{P(A \cap B)}{P(B)} \tag{4.6}$$

or

$$P(B \mid A) = \frac{P(A \cap B)}{P(A)}. \tag{4.7}$$

Let us return to the issue of discrimination against the female officers. The probabilities in Table 4.5 show that the probability of promotion of an officer is $P(A) = .27$ (regardless of whether that officer is male or female). However, the critical issue in the discrimination case involves the two conditional probabilities $P(A \mid M)$ and $P(A \mid W)$. That is, what is the probability of a promotion *given* that the officer is a man, and what is the probability of a promotion *given* that the officer is a woman? If these two probabilities are equal, there is no basis for a discrimination argument, since the chances of a promotion are the same for male and female officers. However, if the two conditional probabilities differ, there will be support for the position that male and female officers are treated differently when it comes to promotions.

We have already determined that $P(A \mid M) = .30$. Let us now use the probability values in Table 4.5 and the basic relationship of conditional probability (4.6) to compute the probability that an officer is promoted given that the officer is a woman; that is, $P(A \mid W)$. Using (4.6), we obtain

$$P(A \mid W) = \frac{P(A \cap W)}{P(W)} = \frac{.03}{.20} = .15.$$

What conclusions do you draw? The probability of a promotion given that the officer is a man is .30, twice the .15 probability of a promotion given that the officer is a woman. While the use of conditional probability does not in itself prove that discrimination exists in this case, the conditional probability values are support for the argument presented by the female officers.

Independent Events

In the above illustration we saw that $P(A) = .27$, $P(A \mid M) = .30$, and $P(A \mid W) = .15$. This shows that the probability of a promotion (event A) is affected or influenced by whether the officer is male or female. In particular, since $P(A \mid M) \neq P(A)$, we would say that events A and M are *dependent* events. That is, the probability of event A (promotion) is altered or affected by knowing whether or not M (the officer is a man) occurs. Similarly, with $P(A \mid W) \neq P(A)$, we would say that events A and W are *dependent* events. On the other hand, if the probability of event A were not changed by the existence of event M—that is, $P(A \mid M) = P(A)$—we would say that events A and M are *independent* events. This leads us to the following definition of the independence of two events:

Definition of Independent Events

Two events A and B are independent if

$$P(A \mid B) = P(A) \tag{4.8}$$

or

$$P(B \mid A) = P(B); \tag{4.9}$$

otherwise, the events are dependent.

Multiplication Law

While the addition law of probability is used to compute the probability of a union of two events, we can now show how the multiplication law can be used to find the probability of an intersection of two events. The multiplication law is based upon the definition of conditional probability. Using (4.6) and (4.7) and solving for $P(A \cap B)$, we obtain the *multiplication law:*

Multiplication Law

$$P(A \cap B) = P(B)P(A \mid B) \tag{4.10}$$

or

$$P(A \cap B) = P(A)P(B \mid A). \tag{4.11}$$

To illustrate the use of the multiplication law, consider a newspaper circulation department where it is known that 84% of the newspaper's customers subscribe to the daily edition of the paper. If we let D denote the event that a customer subscribes to the daily edition, $P(D) = .84$. In addition, it is known that the probability that a customer who already holds a daily subscription also subscribes to the Sunday edition (event S) is .75; that is, $P(S \mid D) = .75$. What is the probability that a customer subscribes to both the Sunday and daily editions of the newspaper? Using the multiplication law, we compute the desired $P(S \cap D)$ as follows:

$$P(S \cap D) = P(D)P(S \mid D) = .84\,(.75) = .63.$$

This tells us that 63% of the newspaper's customers take both the Sunday and daily editions.

Before concluding this section, let us consider the special case of the multiplication law when the events involved are independent. Recall that earlier in this section we defined independent events to exist whenever $P(A \mid B) = P(A)$ or $P(B \mid A) = P(B)$. Hence using (4.10) and (4.11) for the special case of independent events, the multiplication law becomes

Multiplication Law—Independent Events

$$P(A \cap B) = P(A)P(B) \tag{4.12}$$

Thus to compute the probability of the intersection of two independent events we simply multiply the corresponding probabilities. Note that the multiplication law for independent events provides another way to determine if A and B are independent. That is, if $P(A \cap B) = P(A)P(B)$, then A and B are independent; if $P(A \cap B) \neq P(A)P(B)$, then A and B are dependent.

As an application of the multiplication law for independent events, consider the situation of a service station manager who knows from past experience that 80 percent of the customers use a credit card when they purchase gasoline. What is the probability that the next two customers purchasing gasoline will each use a credit card? If we let

$$A = \text{the event that the first customer uses a credit card,}$$
$$B = \text{the event that the second customer uses a credit card,}$$

then the event of interest is $A \cap B$. Given no other information, it seems reasonable to assume that A and B are independent events. Thus

$$P(A \cap B) = P(A)P(B) = (.80)(.80) = .64.$$

EXERCISES

22. A Daytona Beach nightclub has the following data on the age and marital status of 140 customers:

		Marital Status	
		Single	*Married*
Age	Under 30	77	14
	30 or Over	28	21

a. Develop a joint probability table using the above data.
b. Use the marginal probabilities to comment on the age of customers attending the club.
c. Use the marginal probabilities to comment on the marital status of customers attending the club.
d. What is the probability of finding a customer who is single and under the age of 30?
e. If a customer is under 30, what is the probability that he or she is single?
f. Is marital status independent of age? Explain, using probabilities.

23. A survey of automobile ownership was conducted for 200 families in Houston. The results of the study showing ownership of automobiles of United States and foreign manufacture are summarized below:

		Do You Own a U.S. Car?		
		Yes	*No*	*Total*
Do you own a foreign car?	Yes	30	10	40
	No	150	10	160
Total		180	20	

a. Show the joint probability table for the above data.
b. Use the marginal probabilities to compare U.S. and foreign car ownership.
c. What is the probability that a family will own both a U.S. car and a foreign car?
d. What is the probability that a family owns a car, U.S. or foreign?
e. If a family owns a U.S. car, what is the probability that it also owns a foreign car?
f. If a family owns a foreign car, what is the probability that it also owns a U.S. car?
g. Are U.S. and foreign car ownership independent events? Explain.

24. The probability that Ms. Smith will get an offer on the first job she applies for is .5, and the probability that she will get an offer on the second job she applies for is .6. She thinks that the probability that she will get an offer on both jobs is .15.
a. Define the events involved, and use probability notation to show the probability information given above.
b. What is the probability that Ms. Smith gets an offer on the second job given that she receives an offer for the first job?
c. What is the probability that Ms. Smith gets an offer on at least one of the jobs she applies for?
d. What is the probability that Ms. Smith does not get an offer on either of the two jobs she applies for?
e. Are the job offers independent? Explain.

25. Shown below are data from a sample of 80 families in a midwestern city. The data shows the record of college attendance by fathers and their oldest sons.

		Son	
		Attended College	*Did Not Attend College*
Father	*Attended College*	18	7
	Did Not Attend College	22	33

a. Show the joint probability table.
b. Use the marginal probabilities to comment on the comparison between fathers and sons in terms of attending college.
c. What is the probability that a son attends college given that his father attended college?
d. What is the probability that a son attends college given that his father did not attend college?
e. Is attending college by the son independent of whether or not his father attended college? Explain, using probability values.

26. The Texas Oil Company provides a limited partnership arrangement whereby small investors can pool resources in order to invest in large scale oil exploration programs. In the exploratory drilling phase, locations for new wells are selected based on the geologic structure of the proposed drilling sites. Experience shows that there is a .40 probability of a type A structure present at the site given a productive well. It is also known that 50% of all wells are drilled in locations with type A structure. Finally, 30% of all wells drilled are productive.
a. What is the probability of a well being drilled in a type A structure *and* being productive?
b. If the drilling process begins in a location with a type A structure, what is the probability of having a productive well at the location?
c. Is finding a productive well independent of the type A geologic structure? Explain.

27. In a study involving a manufacturing process, 10% of all parts tested were defective, and 30% of all parts were produced on machine A. Given that a part was produced on machine A, there is a .15 probability that it is defective.
a. What is the probability that a part tested is both defective and produced by machine A?
b. If a part is found to be defective, what is the probability that it came from machine A?

c. Is finding a defective part independent of its being produced on machine A? Explain.
d. What is the probability of the part being either defective or produced by machine A?
e. Are the events "a defective part" and "produced by machine A" mutually exclusive events? Explain.

28. Assume that we have two events, A and B, which are mutually exclusive. Assume further that it is known that $P(A) = .30$ and $P(B) = .40$.
a. What is $P(A \cap B)$?
b. What is $P(A \mid B)$?
c. A student in statistics argues that the concepts of mutually exclusive events and independent events are really the same and that if events are mutually exclusive they must be independent. Do you agree with this statement? Use the probability information in this problem to justify your answer.
d. What general conclusion would you make about mutually exclusive and independent events given the results of this problem?

29. A purchasing agent has placed two rush orders for a particular raw material from two different suppliers A and B. If neither order arrives in 4 days the production process must be shut down until at least one of the orders arrives. The probability that supplier A can deliver the material in 4 days is .55. The probability that supplier B can deliver the material in 4 days is .35.
a. What is the probability that both suppliers deliver the material in 4 days? Since two separate suppliers are involved, we are willing to assume independence.
b. What is the probability that at least one supplier delivers the material in 4 days?
c. What is the probability the production process is shut down in 4 days because of a shortage in raw material (that is, both orders are late)?

4.5 BAYES' THEOREM

In the discussion of conditional probability we indicated that revising probabilities when new information is obtained is an important phase of probability analysis. Often, we begin our analysis with initial or *prior* probability estimates for specific events of interest. Then, from sources such as a sample, a special report, a product test, and so on we obtain some additional information about the events. Given this new information, we update the prior probability values by calculating revised probabilities, referred to as *posterior probabilities. Bayes' theorem* provides a means for making these probability calculations. The steps in this probability revision process are shown in Figure 4.8.

As an application of Bayes' theorem, consider a manufacturing firm that receives shipments of parts from two different suppliers. Let A_1 denote the event that a part is from supplier 1 and A_2 denote the event that a part is from supplier 2. Currently, 65% of the parts purchased by the company are from supplier 1, while the remaining 35% are from supplier 2. Thus if a part is selected at random, we would assign the prior probabilities $P(A_1) = .65$ and $P(A_2) = .35$.

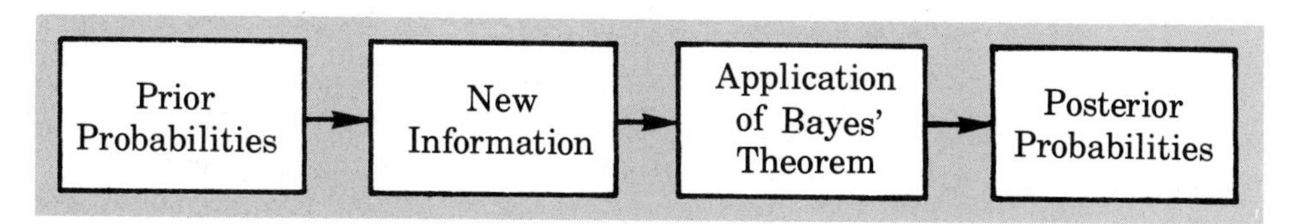

FIGURE 4.8 Probability Revision Using Bayes' Theorem

The quality of the purchased parts varies with the source of supply. Based upon historical data, the quality ratings of the two suppliers are as shown in Table 4.6. Thus if we let G denote the event that a part is good and B denote the event that a part is defective or bad, the information in Table 4.6 provides the following conditional probability values:

$$P(G \mid A_1) = .98, \qquad P(B \mid A_1) = .02,$$
$$P(G \mid A_2) = .95, \qquad P(B \mid A_2) = .05.$$

With the prior probabilities $P(A_1) = .65$ and $P(A_2) = .35$ and the above conditional probability information, we are ready to show how Bayes' theorem can be used to revise the prior probabilities in light of new information. For example, assume that the parts from the two suppliers are used in the firm's manufacturing process and that a machine breaks down because it attempts to process a defective or bad part. Given the information that the part causing the problem is bad, what is the probability that the part is from each of the suppliers? That is, letting B denote the bad parts, what are the posterior probabilities $P(A_1 \mid B)$ and $P(A_2 \mid B)$?

TABLE 4.6 Historical Quality Level of Two Parts Suppliers

	Percentage Good Parts	*Percentage Bad Parts*
Supplier 1	98%	2%
Supplier 2	95%	5%

In order to see how Bayes' theorem is applied and at the same time understand why the formula for Bayes' theorem works, let us look closely at the calculation of the probability $P(A_1 \mid B)$. $P(A_1 \mid B)$ is the probability the part came from supplier 1 given that the part is bad.

Since the probability that we are seeking, $P(A_1 \mid B)$, is a conditional probability, we can start with the definition of conditional probability expressed in (4.6):

$$P(A_1 \mid B) = \frac{P(A_1 \cap B)}{P(B)}.$$

Next, using the multiplication law of (4.11), we can substitute $P(A_1)P(B \mid A_1)$ for $P(A_1 \cap B)$. Thus we have

$$P(A_1 \mid B) = \frac{P(A_1)P(B \mid A_1)}{P(B)}. \qquad (4.13)$$

The above equation is one form of Bayes' theorem.

In our example, $P(A_1) = .65$ and $P(B \mid A_1) = .02$. However, $P(B)$, the probability of a bad part, must be determined before we can compute the desired probability $P(A_1 \mid B)$. Figure 4.9 should be helpful in seeing how to compute $P(B)$. In this figure we show that there are two outcomes that correspond to the part being bad: supplier 1 provides the part and it is bad $(A_1 \cap B)$, or supplier 2 provides the part and it is bad $(A_2 \cap B)$. Since these two events are mutually exclusive, the probability of event B is given by

$$P(B) = P(A_1 \cap B) + P(A_2 \cap B).$$

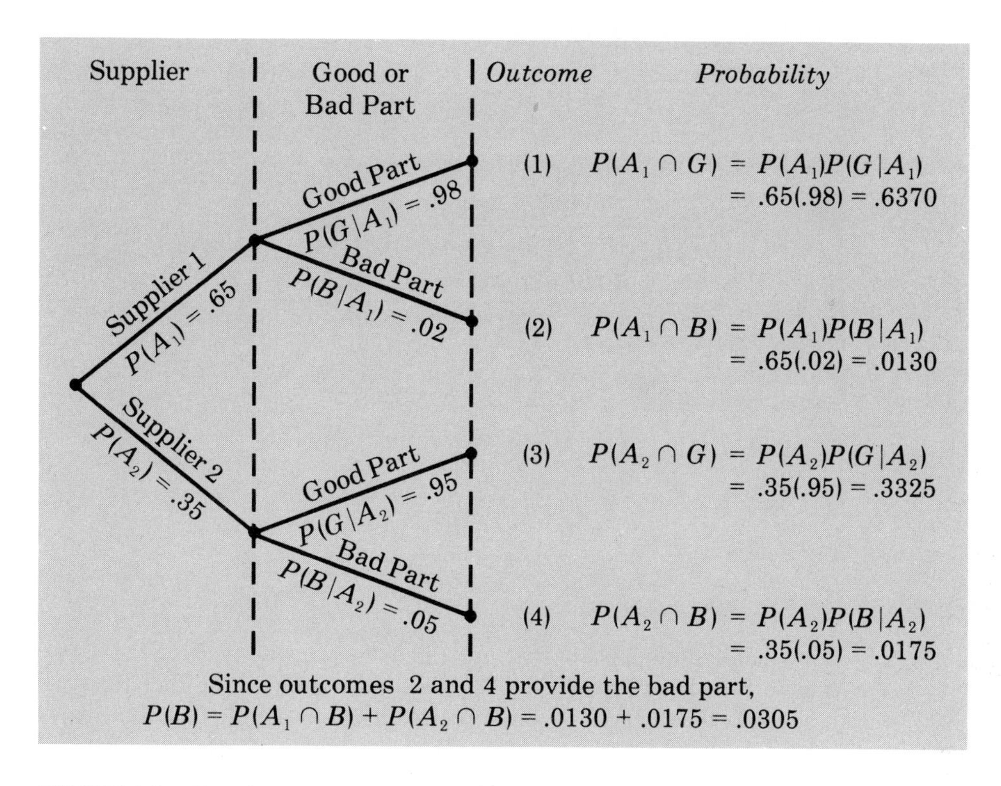

FIGURE 4.9 Tree Diagram of the Bayes' Theorem Computations for the Two-Supplier Problem

Returning to the multiplication law and referring to Figure 4.9, we see that

$$P(A_1 \cap B) = P(A_1)P(B \mid A_1)$$

and

$$P(A_2 \cap B) = P(A_2)P(B \mid A_2).$$

Therefore

$$P(B) = P(A_1)P(B \mid A_1) + P(A_2)P(B \mid A_2). \tag{4.14}$$

Using the probability values for our problem, we obtain

$$\begin{aligned} P(B) &= .65(.02) + .35(.05) \\ &= .0130 + .0175 = .0305. \end{aligned}$$

This probability calculation tells us that we should assign a .0305 probability to the event that a part is bad.

Substituting the expression for $P(B)$ as shown in (4.14) into (4.13) provides the following expression for Bayes' theorem with two events A_1 and A_2:

$$P(A_1 \mid B) = \frac{P(B \mid A_1)P(A_1)}{P(B \mid A_1)P(A_1) + P(B \mid A_2)P(A_2)}. \tag{4.15}$$

A similar argument can be used to show that $P(A_2 \mid B)$ can be computed using the following form of Bayes' theorem:

$$P(A_2 \mid B) = \frac{P(B \mid A_2)P(A_2)}{P(B \mid A_1)P(A_1) + P(B \mid A_2)P(A_2)}. \tag{4.16}$$

Using (4.15) and the probability values provided in our example, we have

$$\begin{aligned} P(A_1 \mid B) &= \frac{P(B \mid A_1)P(A_1)}{P(B \mid A_1)P(A_1) + P(B \mid A_2)P(A_2)} \\ &= \frac{(0.2)(.65)}{(.02)(.65) + (.05)(.35)} = \frac{.0130}{.0130 + .0175} \\ &= \frac{.0130}{.0305} = .426. \end{aligned}$$

In addition, using (4.16), we find $P(A_2 \mid B)$ as follows:

$$\begin{aligned} P(A_2 \mid B) &= \frac{(.05)(.35)}{(.02)(.65) + (.05)(.35)} \\ &= \frac{.0175}{.0130 + .0175} = \frac{.0175}{.0305} = .574. \end{aligned}$$

Note that in this application we initially had a probability of .65 that a part selected at random was from supplier 1. However, given information that the part is bad, the probability that the part is from supplier 1 drops to .426. In fact, if the part is bad, there is a better than 50–50 chance that the part came from supplier 2; that is, $P(A_2 \mid B) = .574$.

The Tabular Approach

A tabular approach helpful in conducting the Bayes' theorem calculations is shown in Table 4.7. The computations shown there are conducted as follows:

Step 1: Prepare the following three columns:

Column 1—The list of all mutually exclusive events that can occur in the problem.
Column 2—The prior probabilities for the events.
Column 3—The conditional probabilities of the new information *given* each event.

Step 2: In column 4 compute the joint probabilities for each event and the new information B. These joint probabilities are found by multiplying the values in column 2 by the corresponding values in column 3—that is, $P(A_i \cap B) = P(A_i)P(B \mid A_i)$.

Step 3: Sum the joint probability column to find the probability of the new information, $P(B)$. Thus we see that in the above example there is a .0130 probability of a bad part and supplier 1 and there is a .0175 probability of a bad part and supplier 2. Thus since these are the only two ways in which a bad part can be obtained, the sum .0130 + .0175 shows that there is an overall probability of .0305 of finding a bad part from the combined shipments of both suppliers.

Step 4: In column 5 compute the posterior probabilities using the basic relationship of conditional probability

$$P(A_i \mid B) = \frac{P(A_i \cap B)}{P(B)}.$$

Note that the joint probabilities $P(A_i \cap B)$ are found in column 4, while the probability $P(B)$ appears as the sum of column 4.

TABLE 4.7 Summary of Bayes' Theorem Calculations for the Two-Supplier Problem

(1) Events A_i	(2) Prior Probabilities $P(A_i)$	(3) Conditional Probabilities $P(B \mid A_i)$	(4) Joint Probabilities $P(A_i \mid B)$	(5) Posterior Probabilities $P(A_i \mid B)$
A_1	.65	.02	.0130	.0130/.0305 = .426
A_2	.35	.05	.0175	.0175/.0305 = .574
	1.00		$P(B)$ = .0305	1.000

As a final note, we can generalize Bayes' theorem to the case where there are n mutually exclusive events $A_1, A_2, \ldots, A_n$ and where one of the n events must occur when the experiment is conducted. In such a case Bayes' theorem for the computation of any posterior probability $P(A_i \mid B)$ appears as follows:

Bayes' Theorem

$$P(A_i \mid B) = \frac{P(B \mid A_i)P(A_i)}{P(B \mid A_1)P(A_1) + P(B \mid A_2)P(A_2) + \cdots + P(B \mid A_n)P(A_n)} \tag{4.17}$$

for $i = 1, 2, \ldots n$.

With prior probabilities $P(A_1), P(A_2), \ldots, P(A_n)$ and the appropriate conditional probabilities $P(B \mid A_1), P(B \mid A_2), \ldots, P(B \mid A_n)$, Eq. (4.17) can be used to compute the posterior probability of an event $A_1, A_2, \ldots, A_n$.

EXERCISES

30. The prior probabilities for events A_1, A_2, and A_3 are $P(A_1) = .20$, $P(A_2) = .50$, and $P(A_3) = .30$. The conditional probabilities of event B given A_1, A_2, and A_3 are $P(B \mid A_1) = .50$, $P(B \mid A_2) = .40$, and $P(B \mid A_3) = .30$.

a. Compute $P(B \cap A_1)$, $P(B \cap A_2)$, and $P(B \cap A_3)$.

b. Apply Bayes' theorem, Eq. (4.17), to compute the posterior probability $P(A_2 \mid B)$.

c. Use the tabular approach to applying Bayes' theorem to compute $P(A_1 \mid B)$, $P(A_2 \mid B)$, and $P(A_3 \mid B)$.

31. A consulting firm has submitted a bid for a large research project. The firm's management initially felt there was a 50–50 chance of getting the bid. However, the agency to which the bid was submitted has subsequently requested additional information on the bid. Past experience indicates that on 75% of the successful bids and 40% of the unsuccessful bids the agency requested additional information.

a. What is your prior probability the bid will be successful (i.e., prior to receiving the request for additional information)?
b. What is the conditional probability of a request for additional information given that the bid will ultimately be successful?
c. Compute a posterior probability that the bid will be successful given that a request for additional information has been received.

32. A local bank is reviewing its credit card policy with a view toward recalling some of its credit cards. In the past approximately 5% of cardholders have defaulted, and the bank has been unable to collect the outstanding balance. Thus management has established a prior probability of .05 that any particular cardholder will default. The bank has further found that the probability of missing one or more monthly payments for those customers who do not default is .20. Of course the probability of missing one or more payments for those who default is 1.
a. Given that a customer has missed a monthly payment, compute the posterior probability that the customer will default.
b. The bank would like to recall its card if the probability that a customer will default is greater than .20. Should the bank recall its card if the customer misses a monthly payment? Why or why not?

33. In a major eastern city, 60% of the automobile drivers are 30 years of age or older, and 40% of the drivers are under 30 years of age. Of all drivers 30 years of age or older, 4% will have a traffic violation in a 12 month period. Of all drivers under 30 years of age, 10% will have a traffic violation in a 12 month period. Assume that a driver has just been charged with a traffic violation; what is the probability that the driver is under 30 years of age?

34. A certain college football team plays 55% of their games at home and 45% of their games away. Given that the team has a home game, there is a .80 probability that it will win. Given that the team has an away game, there is a .65 probability that it will win. If the team wins on a particular Saturday, what is the probability that the game was played at home?

Summary

In this chapter we have introduced basic probability concepts and illustrated how probability analysis can be used to provide helpful decision-making information. We described how probability can be interpreted as a numerical measure of the likelihood that an event will occur. In addition, we saw that the probability of an event could be computed either by summing the probabilities of the experimental outcomes (sample points) comprising the event or by using the relationships established by laws of probability. For cases where additional information is available, we showed how Bayes' theorem could be used to obtain revised or posterior probabilities.

Glossary

Probability—A numerical measure of the likelihood that an event will occur.
Experiment—Any process which generates well defined outcomes.
Sample space—The set of all possible sample points (experimental outcomes).
Sample points—The individual outcomes of an experiment.
Tree diagram—A graphical device helpful in defining sample points of an experiment involving multiple steps.
Basic requirements of probability—Two principles or requirements which restrict the manner in

which probability assignments can be made:

a. For each experimental outcome E_i we must have $0 \leq P(E_i) \leq 1$.

b. If there are k experimental outcomes, then $\Sigma P(E_i) = 1$.

Classical method—A method of assigning probabilities which assumes that the experimental outcomes are equally likely.

Relative frequency method—A method of assigning probabilities based upon experimentation or historical data.

Subjective method—A method of assigning probabilities based upon judgment.

Event—A set consisting of a collection of sample points or experimental outcomes.

Union of events *A* and *B*—The event containing all sample points that are in A, in B, or in both.

Venn diagram—A graphical device for representing symbolically the sample space and operations involving events.

Intersection of *A* and *B*—The event containing all sample points that are in both A and B.

Mutually exclusive events—Events that have no sample points in common; that is, $A \cap B$ is empty and $P(A \cap B) = 0$.

Complement of event *A*—The event containing all sample points that are not in A.

Addition law—A probability law used to compute the probability of a union, $P(A \cup B)$. It is $P(A \cup B) = P(A) + P(B) - P(A \cap B)$. For mutually exclusive events, since $P(A \cap B) = 0$, it reduces to $P(A \cup B) = P(A) + P(B)$.

Conditional probability—The probability of an event given that another event has occurred. The conditional probability of A given B is $P(A \mid B) = P(A \cap B) \mid P(B)$.

Multiplication law—A probability law used to compute the probability of an intersection, $P(A \cap B)$. It is $P(A \cap B) = P(A)P(B \mid A)$ or $P(A \cap B) = P(B)P(A \mid B)$. For independent events it reduces to $P(A \cap B) = P(A)P(B)$.

Independent events—Two events A and B where $P(A \mid B) = P(A)$ or $P(B \mid A) = P(B)$; that is, the events have no influence on each other.

Prior probabilities—Initial estimates of the probabilities of events.

Posterior probabilities—Revised probabilities of events based on additional information.

Bayes' theorem—A method used to compute posterior probabilities.

Key Formulas

Computing Probability Using the Complement

$$P(A) = 1 - P(\bar{A}). \tag{4.4}$$

Addition Law

$$P(A \cup B) = P(A) + P(B) - P(A \cap B). \tag{4.5}$$

Definition of Conditional Probability

$$P(A \mid B) = \frac{P(A \cap B)}{P(B)} \tag{4.6}$$

or

$$P(B \mid A) = \frac{P(A \cap B)}{P(A)} \tag{4.7}$$

Multiplication Law

$$P(A \cap B) = P(B)P(A \mid B) \tag{4.10}$$

or

$$P(A \cap B) = P(A)P(B \mid A) \tag{4.11}$$

Multiplication Law for Independent Events

$$P(A \cap B) = P(A)P(B) \tag{4.12}$$

Bayes' Theorem

$$P(A_i \mid B) = \frac{P(B \mid A_i)P(A_i)}{P(B \mid A_1)P(A_1) + P(B \mid A_2)P(A_2) + \cdots + P(B \mid A_n)P(A_n)} \tag{4.17}$$

for $i = 1, 2, \ldots n$.

Supplementary Exercises

35. A financial manager has just made two new investments—one in the oil industry, and one in municipal bonds. After a one-year period, each of the investments will be classified as either successful or unsuccessful. Consider the making of the two investments as an experiment.

a. How many sample points exist for this experiment?
b. Show a tree diagram and list the sample points.
c. Let O = the event that the oil investment is successful and M = the event that the municipal bond investment is successful. List the sample points in O and in M.
d. List the sample points in the union of the events $(O \cup M)$.
e. List the sample points in the intersection of the events $(O \cap M)$.
f. Are events O and M mutually exclusive? Explain.

36. Consider an experiment where eight experimental outcomes exist. We will denote the experimental outcomes as $E_1, E_2, \ldots, E_8$. Suppose that the following events are identified:

$$\text{A} = \{E_1, E_2, E_3\},$$
$$\text{B} = \{E_2, E_4\}$$
$$\text{C} = \{E_1, E_7, E_8\}$$
$$\text{D} = \{E_5, E_6, E_7, E_8\}.$$

Determine the sample points making up the following events:

a. $A \cup B$.
b. $C \cup D$.
c. $A \cap B$.
d. $C \cap D$.
e. $B \cap C$.
f. $\overline{A}$.
g. $\overline{D}$.
h. $A \cup \overline{D}$.
i. $A \cap \overline{D}$.
j. Are A and B mutually exclusive?
k. Are B and C mutually exclusive?

37. Referring to Exercise 36 and assuming that the classical method is an appropriate way of

establishing probabilities, find the following probabilities:

a. $P(A)$, $P(B)$, $P(C)$, and $P(D)$.
b. $P(A \cap B)$.
c. $P(A \cup B)$.
d. $P(A \mid B)$.
e. $P(B \mid A)$.
f. $P(B \cap C)$.
g. $P(B \mid C)$.
h. Are B and C independent events?

38. In a particular resort area on the west coast of Florida the probability of the sun's shining on a given day is .80 and the probability of rain is .10. In addition, the probability that the resort experiences both sunshine and rain during the same day is .05. Assume that the experiment involves the weather possibilities during a 1 day period.

a. Are sun and rain mutually exclusive events? Explain.
b. Are sun and rain independent events?
c. We know that it rained on a given day. What is the probability that the resort also had sunshine during that day?

39. A salesperson makes contact with two customers. Each contact results in one of three outcomes: a sale, a request for a return call later, or no sale. Consider the contact of the two customers as an experiment.

a. How many sample points are possible?
b. Show a tree diagram and list the sample points.
c. Let A = the event that contact with customer 1 results in a sale and B = the event that contact with customer 2 results in a sale. List the sample points in event A, and then list the sample points in event B.
d. List the sample points in $A \cap B$.
e. List the sample points in $A \cup B$.

40. Suppose that $P(A) = .30$, $P(B) = .25$, and $P(A \cap B) = .20$.

a. Find $P(A \cup B)$, $P(A \mid B)$, and $P(B \mid A)$.
b. Are events A and B independent? Why or why not?

41. Suppose that $P(A) = .40$, $P(A \mid B) = .60$, and $P(B \mid A) = .30$.

a. Find $P(A \cap B)$ and $P(B)$.
b. Are events A and B independent? Why or why not?

42. Suppose that $P(A) = .60$, $P(B) = .30$, and events A and B are mutually exclusive.

a. Find $P(A \cup B)$ and $P(A \cap B)$.
b. Are events A and B independent?
c. Can you make a general statement about whether or not mutually exclusive events can be independent?

43. A market survey of 800 people found the following facts about the ability to recall a television commercial for a particular product and the actual purchase of the product:

	Could Recall Television Commercial	*Could Not Recall Television Commercial*	*Total*
Purchased the Product	160	80	240
Had Not Purchased the Product	240	320	560
Total	400	400	800

Let T be the event of the person recalling the television commercial and B the event of buying or purchasing the product.

a. Find $P(T)$, $P(B)$, and $P(T \cap B)$.

b. Are T and B mutually exclusive events? Use probability values to explain.
c. What is the probability that a person who could recall seeing the television commercial has actually purchased the product?
d. Are T and B independent events? Use probability values to explain.
e. Comment on the value of the commercial in terms of its relationship to purchasing the product.

44. A research study investigating the relationship between smoking and heart disease in a sample of 1000 men over 50 years of age provided the following data:

	Smoker	*Nonsmoker*	*Total*
Record of Heart Disease	100	80	180
No Record of Heart Disease	200	620	820
Total	300	700	1000

a. Show a joint probability that summarizes the results of this study.
b. What is the probability a man over 50 years of age is a smoker and has a record of heart disease?
c. Compute and interpret the marginal probabilities.
d. Given that a man over 50 years of age is a smoker, what is the probability that he has heart disease?
e. Given that a man over 50 years of age is a nonsmoker, what is the probability that he has heart disease?
f. Does the research show that heart disease and smoking are independent events? Use probability to justify your answer.
g. What conclusion would you draw about the relationship between smoking and heart disease?

45. A large consumer goods company has been running a television advertisement for one of its soap products. A survey was conducted. On the basis of this survey probabilities were assigned to the following events:

B = individual purchased the product,
S = individual recalls seeing the advertisement,
$B \cap S$ = individual purchased the product and recalls seeing the advertisement.

The probabilities assigned were $P(B) = .20$, $P(S) = .40$, and $P(B \cap S) = .12$. The following problems relate to this situation:
a. What is the probability of an individual's purchasing the product given that the individual saw the advertisement? Does seeing the advertisement increase the probability the individual will purchase the product? As a decision maker, would you recommend continuing the advertisement (assuming that the cost is reasonable)?
b. Assume that those individuals who do not purchase the company's soap product buy from its competitors. What would be your estimate of the company's market share? Would you expect that continuing the advertisement will increase the company's market share? Why or why not?
c. The company has also tested another advertisement and assigned it values of $P(S) = .30$ and $P(B \cap S) = .10$. What is $P(B \mid S)$ for this other advertisement? Which advertisement seems to have had the bigger effect on customer purchases?

46. A large company has done a careful analysis of a price promotion that it is currently testing. Some 20% of the people in a large sample of individuals in the test market both were aware of the promotion and made a purchase. It was further found that 80% were aware of the promotion and that prior to the promotion 25% of all people in the sample were purchasers of the product.

a. What is the probability that a person will make a purchase given that he or she is aware of the price promotion?

b. Are the events "made a purchase" and "aware of the price promotion" independent? Why or why not?

c. On the basis of these results, would you recommend that the company introduce this promotion on a national scale? Why or why not?

47. Cooper Realty is a small real estate company located in Albany, New York and specializing primarily in residential listings. They have recently become interested in the possibility of determining the likelihood of one of their listings being sold within a certain number of days. An analysis of company sales of 800 homes for the previous years produced the data shown below:

	Days Listed Until Sold			
Initial Asking Price	*Under 30*	*31–90*	*Over 90*	*Totals*
Under $25,000	50	40	10	100
$25,000–50,000	20	150	80	250
$50,000–75,000	20	280	100	400
Over $75,000	10	30	10	50
Totals	100	500	200	800 ↑ Total Homes Sold

a. If A is defined as the event that a home is listed for over 90 days before being sold, estimate the probability of A.

b. If B is defined as the event that the initial asking price is under $25,000, estimate the probability of B.

c. What is the probability of $A \cap B$?

d. Assuming that a contract has just been signed to list a home that has an initial asking price of less than $25,000, what is the probability the home will take Cooper Realty more than 90 days to sell?

e. Are events A and B independent?

48. In the evaluation of a sales training program, a firm found that of 50 salespersons making a bonus last year, 20 had attended a special sales training program. The firm has 200 salespersons. Let B = the event that a salesperson makes a bonus and S = the event a salesperson attends the sales training program.

a. Find $P(B)$, $P(S \mid B)$, and $P(S \cap B)$.

b. Assume that 40% of the salespersons have attended the training program. What is the probability that a salesperson makes a bonus given that the salesperson attended the sales training program, $P(B \mid S)$?

c. If the firm evaluates the training program in terms of the effect it has on the probability of a salesperson's making a bonus, what is your evaluation of the training program? Comment on whether B and S are dependent or independent events.

49. A company has studied the number of lost-time accidents occurring at its Brownsville, Texas plant. Historical records show that 6% of the employees had lost-time accidents last year. Management believes that a special safety program will reduce the accidents to 5% during the current year. In addition, it is estimated that 15% of those employees having had lost-time accidents last year will have a lost-time accident during the current year.

a. What percentage of the employees will have lost-time accidents in both years?

b. What percentage of the employees will have at least one lost-time accident over the 2 year period?

50. In a study of television viewing habits among married couples, a researcher found that for a popular Saturday night program 25% of the husbands viewed the program regularly and 30% of the wives viewed the program regularly. The study found that for couples where the husband watches the program regularly 80% of the wives also watch regularly.

a. What is the probability that both the husband and wife watch the program regularly?

b. What is the probability that at least one—husband or wife—watches the program regularly?

c. What percentage of married couples do not have at least one regular viewer of the program?

51. A statistics professor has noted from past experience that students who do the homework for the course have a .90 probability of passing the course. On the other hand, students who do not do the homework for the course have a .25 probability of passing the course. The professor estimates that 75% of the students in the course do the homework. Given a student who passes the course, what is the probability that she or he completed the homework?

52. A salesman for Business Communication Systems, Inc. sells automatic envelope-addressing equipment to medium and small size businesses. The probability of making a sale to a new customer is .10. During the initial contact with a customer, sometimes the salesperson will be asked to call back later. Of the 30 most recent sales, 12 were made to customers who initially told the salesperson to call back later. Of 100 customers who did not make a purchase, 17 had initially asked the salesperson to call back later. If a customer asks the salesperson to call back later, should the salesperson do so? What is the probability of making a sale to a customer who has asked the salesperson to call back later?

53. Migliori Industries, Inc. manufactures a gas-saving device for use on natural gas forced-air residential furnaces. The company is currently trying to determine the probability that sales of this product will exceed 25,000 units during next year's winter sales period. The company believes that sales of the product depend to a large extent on the winter conditions. Management's best estimate is that the probability that sales will exceed 25,000 units if the winter is severe is .8. This probability drops to .5 if the winter conditions are moderate. If the weather forcast is .7 for a severe winter and .3 for moderate conditions, what is Migliori's best estimate that sales will exceed 25,000 units?

54. The Dallas IRS auditing staff is concerned with identifying potential fraudulent tax returns. From past experience they believe that the probability of finding a fraudulent return given that the return contains deductions for contributions exceeding the IRS standard is 0.20. Given that the deductions for contributions do not exceed the IRS standard, the probability of a fraudulent return decreases to .02. If 8% of all returns exceed the IRS standard for deductions due to contributions, what is the best estimate of the percentage of fraudulent returns?

55. An oil company has purchased an option on land in Alaska. Preliminary geologic studies have assigned the following prior probabilities:

$$P(\text{high quality oil}) = .50,$$
$$P(\text{medium quality oil}) = .20,$$
$$P(\text{no oil}) = .30.$$

a. What is the probability of finding oil?

b. After 200 feet of drilling on the first well, a soil test is taken. The probabilities of finding this particular type of soil are as follows:

$$P(\text{soil} \mid \text{high quality oil}) = .20,$$
$$P(\text{soil} \mid \text{medium quality oil}) = .80,$$
$$P(\text{soil} \mid \text{no oil}) = .20.$$

How should the firm interpret the soil test? What are the revised probabilities, and what is the new probability of finding oil?

56. In the setup of a manufacturing process, a machine is either correctly or incorrectly adjusted. The probability of a correct adjustment is .90. When correctly adjusted, the machine operates with a 5% defective rate. However, if it is incorrectly adjusted, a 75% defective rate occurs.

a. After the machine starts a production run, what is the probability that a defect is observed when one part is tested?
b. Suppose that the one part selected by an inspector is found to be defective. What is the probability that the machine is incorrectly adjusted? What action would you recommend?
c. Before your recommendation in part b above was followed, a second part is tested and found to be good. Using your revised probabilities from part b as the most recent prior probabilities, compute the revised probability of an incorrect adjustment given that the second part is good. What action would you recommend now?

57. The Wayne Manufacturing Company purchases a certain part from three suppliers *A, B,* and *C*. Supplier *A* supplies 60% of the parts, *B* 30%, and *C* 10%. The quality of parts is known to vary among suppliers, with *A, B,* and *C* parts having .25%, 1%, and 2% defective rates, respectively. The parts are used in one of the company's major products.
a. What is the probability that the company's major product is assembled with a defective part?
b. Whan a defective part is found, which supplier is the likely source?

Application

MORTON THIOKOL

Morton Thiokol, Inc.*

Chicago, Illinois

Morton Norwich combined with Thiokol Corporation in 1982 to form the current company, Morton Thiokol, Inc. From a salt business first started in Chicago in 1848, Jay Morton named his firm the Morton Salt Company in 1910. In 1914 the Morton Girl and slogan "When it rains, it pours" established Morton Salt as a recognized name to the consumer. In the following years, Morton added specialty chemicals businesses and Texize, a manufacturer of consumer household products, to its rapidly growing salt business.

The slogan "When it rains, it pours," established Morton Salt as a recognized name to the consumer

Thiokol Corporation began in 1928 as a manufacturer of specialty polysulfide polymers that found widespread use as sealants. Other specialty chemical businesses that serve plastic, electronic and other industries were added later. As an outgrowth of the expanding chemical business, the company became involved in the manufacture of solid rocket propellants and today builds a variety of rocket motors including the boosters for the Space Shuttle and satellites used in the United States space programs.

The combination of Morton and Thiokol has resulted in a company with strong businesses in salt and household products, rocket motors, and specialty chemicals. In particular, the specialty chemicals group now consists of several decentralized manufacturing operations serving their particular market segment.

*The authors are indebted to Michael Haskell of Morton Thiokol's Carstab subsidiary for providing this application.

PROBABILITY ANALYSIS GUIDES MANAGEMENT DECISION MAKING

Managers at the various divisions throughout Morton Thiokol make decisions in light of uncertain outcomes. Often the managers' "feel" for the chances associated with the outcomes and/or subjective estimates of probabilities suggest a specific decision or course of action. Issues such as how much inventory to maintain, forecasts of demand,

The Morton Thiokol Corporation builds a variety of rocket motors including the boosters for the Space Shuttle

and estimates about order quantities from specific customers are instances where managers use probability considerations as part of the decision making process. In the following example we describe how probability considerations aided a customer service decision at Carstab Corporation, a subsidiary of Morton Thiokol.

Carstab Corporation provides a variety of specialty chemical products for its customers. Because of the diversity of customer applications, customers differ substantially in terms of the unique specifications they require for the product. One approach to servicing customers would be for Carstab to wait until a customer order is received and then make the product to the exact customer specifications. A disadvantage of this approach is that the customer's order would have to wait until Carstab could schedule production for the special order. In some instances this would require an undesirable waiting period for the customer. In addition, this approach also has the disadvantage of requiring Carstab to make relatively short production runs for specific customer orders thus losing the economic advantage of larger production runs. In order to avoid these disadvantages, Carstab makes relatively large production runs for many of its products and holds goods in inventory awaiting customers' orders. In doing this the company realizes that not all of its inventory will meet the unique specification requirements for all customers.

In one instance, a customer made small but repeated orders for an expensive catalyst product used in its chemical processing. Because of the nature of its operation, the customer placed unique specifications on the product. Some, but not all of the lots produced by Carstab would meet the customer's exact specifications.

The customer agreed to test each lot as it was received to determine whether or not the catalyst would perform the desired function. Carstab agreed to ship lots to the customer with the understanding the customer would perform the test and return the lots that did not pass the customer's specification test. The problem encountered was that only 60% of the lots sent to the customer would pass the customer's test. This meant that although the product was still good and usable to other customers, approximately 40% of the shipments sent to this customer were being returned.

Carstab explored the possibility of duplicating the customer's test and only shipping lots that passed the test. However, the test was unique to this one customer; and it was infeasible to purchase the expensive testing equipment needed to perform the customer's test.

Therefore, in order to improve the customer service Carstab chemists designed a new test, one that was believed to indicate whether or not the lot would eventually pass the customer's test. The question was would the Carstab test increase the probability that a lot shipped to the customer would pass the customer's test. The probability information sought was what is the probability that a lot will pass the customer's test given it has passed the company's test.

A sample of lots were tested under both the customer's procedure and the company's proposed procedure. Results were that 55% of the lots passed the company's test and 50% of the lots passed both the customer's and the company's test. In probability notation, we have

A = the event the lot passes the customer's test

B = the event the lot passes the company's test

where

$$P(B) = .55 \text{ and } P(A \cap B) = .50$$

The probability information sought was the conditional probability $P(A \mid B)$ which was given by

$$P(A \mid B) = \frac{P(A \cap B)}{P(B)} = \frac{.50}{.55} = .909$$

Prior to the company's test the probability a lot would pass the customer's test was .60. However, the new results showed that given a lot passed the company's test it had a .909 probability of passing the customer's test. This was good supporting evidence for the use of the test prior to shipment. Based on this probability analysis, the preshipment testing procedure was implemented at the company. Immediate results showed an improved level of customer service. A few lots were still being returned; however, the percentage was greatly reduced. The customer was more satisfied and return shipping costs were reduced.

As seen in this example, probability did not make the decision for the manager. However, some basic probability considerations provided important decision-making information and were a significant factor in the decision to implement the new testing procedure which resulted in improved service to the customer.

5 Probability Distributions

What you will learn in this chapter:

- what random variables are and how they are used
- what is meant by a probability distribution of a random variable
- how to distinguish between a discrete and a continuous random variable
- how to compute and interpret the expected value and variance of a random variable
- how probabilities are computed for a uniform probability distribution

Contents

In this chapter we continue the study of probability by introducing the concept of a random variable and its probability distribution. We first will study discrete random variables and see how to compute their expected values and variances. Then, we consider decisions based upon an analysis of two or more discrete random variables where the resulting probability distribution is referred to as a joint probability distribution. The chapter concludes with an introduction to continuous random variables and their probability distributions.

5.1 RANDOM VARIABLES

Recall that in Chapter 4 we defined an experiment as any process which generates well-defined outcomes. We now want to concentrate on the process of assigning *numerical values* to the experimental outcomes. This is where the notion of a random variable comes into play.

For any particular experiment a random variable can be defined such that each possible experimental outcome generates exactly one numerical value for the random variable. For example, if we consider the experiment of selling automobiles for one day at a particular dealership, we could elect to describe the experimental outcomes in terms of the *number* of cars sold. In this case, if x = number of cars sold, x is referred to as a random variable. The particular numerical value that the random variable takes on depends upon the outcome of the experiment.* That is, we will not know the specific value of the random variable until we have observed the experimental outcome. For example, if on a given day three cars were sold, the random variable would take on the value 3. If on another day (a repeat of the experiment) four cars were sold, the random variable would take on the value 4. We define a random variable as follows:

Random Variable

A random variable is a numerical description of the outcome of an experiment.

Some additional examples of experiments and associated random variables are given in Table 5.1. While many experiments such as those listed in Table 5.1 have experimental outcomes which lend themselves quite naturally to numerical values, others do not. Note, for example, the experiment of tossing a coin one time. The

TABLE 5.1 Examples of Random Variables

Experiment	*Random Variable* (x)	*Possible Values for the Random Variable*
Make 100 sales calls	Total number of sales	0, 1, 2, . . . , 100
Inspect a shipment of 70 radios	Number of defective radios	0, 1, 2, . . . , 70
Work 1 year on a project to build a new library	Percentage of project completed after 6 months	$0 \leq x \leq 100$

*More advanced texts on probability make a distinction between the random variable, denoted X, and the values it can take on, denoted x. In this text we will use x for both purposes.

experimental outcome will be either heads or tails, neither of which has a natural numerical value. However, we still may want to express the outcomes in terms of a random variable. Thus we need a rule that can be used to assign a numerical value to each of the experimental outcomes. One possibility is to let the random variable $x = 1$ if the experimental outcome is heads and $x = 0$ if the experimental outcome is tails. While the numerical values for the random variable x are arbitrary (we could have used 3 and 4), they are acceptable in terms of the definition of a random variable. Namely, x is a random variable because it describes the experimental outcomes numerically.

A random variable may be classified as either discrete or continuous depending upon the specific numerical values it can have. A random variable that may only take on a finite or countable number of different values is referred to as a *discrete random variable.* The number of units sold, the number of defects observed, the number of customers that enter a bank during one day of operation, and so on, are examples of discrete random variables. On the other hand, random variables such as weight, time, temperature, and so forth, which may take on any value in a certain interval or collection of intervals, are referred to as *continuous random variables.* For instance, the third random variable in Table 5.1 (percentage of project completed after 6 months) is a continuous random variable because it may take on any value in the interval from 0 to 100 (for example, 56.33, 64.22, etc.).

EXERCISES

1. Consider the experiment of tossing a coin twice.
a. List the experimental outcomes.
b. Define a random variable that represents the number of heads occurring on the two tosses.
c. Show what value the random variable would assume for each of the experimental outcomes.

2. Listed is a series of experiments and associated random variables. In each case identify the values that the random variable can take on and state whether the random variable is discrete or continuous.

	Experiment	*Random Variable* (x)
a.	Take a 20-question examination	Number of questions answered correctly
b.	Observe cars arriving at a tollbooth for 1 hour	Number of cars arriving at tollbooth
c.	Audit 50 tax returns	Number of returns containing errors
d.	Observe an employee's work	Number of nonproductive hours in an 8-hour work day
e.	Weigh a shipment of goods	Number of pounds

5.2 DISCRETE PROBABILITY DISTRIBUTIONS

In order to demonstrate the use of a discrete random variable, let us consider the sales of automobiles at DiCarlo Motors, Inc. in Saratoga, New York. The owner of DiCarlo Motors is interested in the daily sales volume for automobiles. Suppose we let x be a random variable denoting the number of cars sold on a given day. Sales records show that five is the maximum number of cars that DiCarlo has ever sold during one day. Since the owner believes that the previous history of sales adequately represents what will occur in the future, we would expect the random variable x to take on one of the numerical values 0, 1, 2, 3, 4, or 5. The possible values of the random variable are finite; thus we would classify x as a discrete random variable.

Once we have defined an appropriate discrete random variable for a particular situation, we can turn our attention to determining the probability associated with each possible value of the random variable. In the DiCarlo Motors problem we are interested in determining the probabilities of x being 0, 1, 2, 3, 4, or 5. In other words, we would like to know the probabilities associated with each possible daily sales volume for automobiles.

Suppose that in checking DiCarlo's sales records we find that over the past year the firm has been open for business on exactly 300 days. The sales volumes generated and the frequency of their occurrence are summarized in Table 5.2. With these

TABLE 5.2 Cars Sold per Day at DiCarlo Motors

Sales Volume	*Number of Days*
No sales	54
Exactly one car	117
Exactly two cars	72
Exactly three cars	42
Exactly four cars	12
Exactly five cars	3
Total	300

historical data available, the owner of DiCarlo Motors feels the relative frequency method will provide a reasonable means of assessing the probabilities for the random variable x. First, however, we need to define the notation used to represent these probabilities. The *probability function,* denoted $f(x)$, provides the probability that the random variable x takes on some specific value. For example, since on 54 of the 300 days of historical data DiCarlo Motors did not sell any cars and since no sales corresponds to $x = 0$, we assign to $f(0)$ the value $54/300 = .18$. Similarly, since $f(1)$ denotes the probability that x takes on the value 1, we assign to $f(1)$ the value $117/300 = .39$. After computing the relative frequencies for the other possible values of x, we can develop a table of x and $f(x)$ values, as shown in Table 5.3. This table is one way of representing the *probability distribution* of the random variable x.

TABLE 5.3 Probability Distribution for the Number of Cars Sold per Day

x	$f(x)$
0	.18
1	.39
2	.24
3	.14
4	.04
5	.01
Total	1.00

We can also represent the probability distribution of x graphically. In Figure 5.1 the values of the random variable x are shown on the horizontal axis. The probability that x takes on these values is shown on the vertical axis. For many discrete random variables the probability distribution can also be represented as a formula that provides $f(x)$ for every possible value of x.

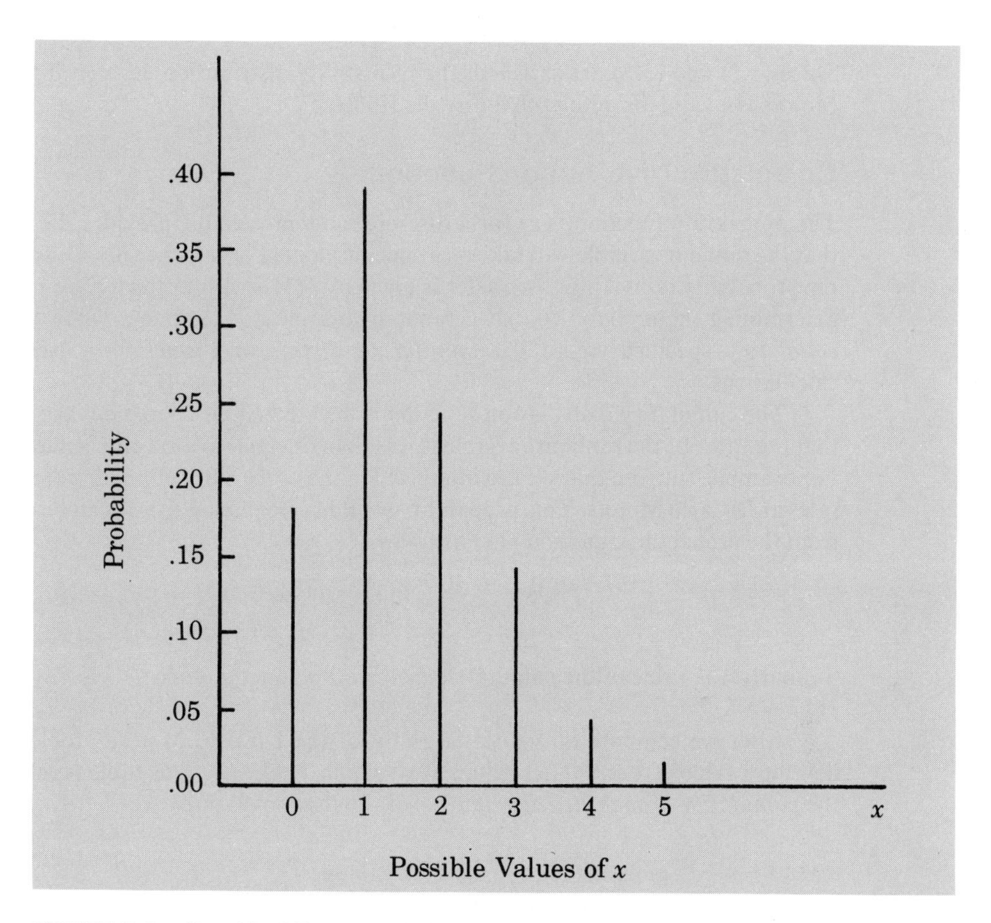

FIGURE 5.1 Graphical Representation of the Probability Distribution for Number of Cars Sold Per Day

In the development of the probability distribution for a discrete random variable, the following two conditions must always be satisfied:

Required Conditions for a Discrete Probability Distribution

$$f(x) \geq 0, \tag{5.1}$$

$$\Sigma f(x) = 1. \tag{5.2}$$

Condition (5.1) is the requirement that the probabilities associated with each value of

x must be greater than or equal to zero, while (5.2) indicates that the sum of the probabilities for all values of the random variable x must be equal to one. Referring to Table 5.3 for the DiCarlo Motors probability distribution, we see that the probabilities are all greater than or equal to zero. In addition, we note that

$$\begin{aligned}\Sigma f(x) &= f(0) + f(1) + f(2) + f(3) + f(4) + f(5)\\ &= .18 + .39 + .24 + .14 + .04 + .01 = 1.00.\end{aligned}$$

Since (5.1) and (5.2) are satisfied, the probability distribution developed for DiCarlo Motors is a valid discrete probability distribution.

Cumulative Distribution Function

The probability function $f(x)$ for a discrete random variable provides the probability that the random variable will take on a specific value. For instance, for DiCarlo Motors the probability of making three sales is given by $f(3) = .14$. Often we are interested in determining the probability that a random variable will assume a value *less than or equal to* a specified value. The *cumulative distribution function* is used for these calculations.

The cumulative distribution function is denoted $F(x)$. It provides the probability that the value of the random variable is less than or equal to a specific value of interest. For example, suppose that we are interested in the probability of making three or fewer sales at DiCarlo Motors. This probability would be denoted $F(3)$. It can be determined from the probability function $f(x)$ as follows:

$$\begin{aligned}F(3) &= f(0) + f(1) + f(2) + f(3)\\ &= .18 + .39 + .24 + .14 = .95.\end{aligned}$$

Similarly, the value of the cumulative distribution function for $x = 1$ is $F(1) = f(0) + f(1) = .57$.

After we compute all values of $F(x)$ for the DiCarlo Motors problem, we can develop a table of x and $F(x)$ values as shown in Table 5.4. This table is referred to as the *cumulative probability distribution* of the random variable x.

TABLE 5.4 Cumulative Probability Distribution

x	$F(x)$
0	.18
1	.57
2	.81
3	.95
4	.99
5	1.00

EXERCISES

3. A stock broker has given the following probability estimates for the price of Mills Corporation stock at the end of next week: $P(\$22) = .10$, $P(\$23) = .40$, $P(\$24) = .30$, $P(\$25) = .20$.

a. Identify an appropriate probability function and specify a probability distribution for the price of the stock at the end of next week.

b. Draw a graph of the probability distribution.
c. Show that the probability distribution satisfies (5.1) and (5.2).

4. QA Properties is considering making an offer to purchase an apartment building. Management has subjectively assessed a probability distribution for x, the purchase price:

x	$f(x)$
\$148,000	.20
\$150,000	.40
\$152,000	.40

a. Determine if this is a proper probability distribution [check (5.1) and (5.2)].
b. Find the cumulative distribution function $F(x)$. What is $F(152{,}000)$?
c. What is the probability that the apartment house can be purchased for \$150,000 or less?

5. The cleaning and changeover operation for a production system requires from 1 to 4 hours, depending upon the specific product that will begin production. Let x be a random variable indicating the time in hours required to make the changeover. The following probability function can be used to compute the probability associated with any changeover time x:

$$f(x) = \frac{x}{10} \qquad \text{for } x = 1, 2, 3, \text{ or } 4.$$

a. Show that the probability function meets the required conditions of (5.1) and (5.2).
b. What is the probability that the changeover will take 2 hours?
c. What is the probability that the changeover will take more than 2 hours?
d. Graph the probability distribution for the changeover times.

6. MRA is a new company that began operation in the spring of 1983. Shown below is the firm's profit probability distribution for the first 6 months of operation (the negative value shows a loss):

Profit in 1,000's x	$f(x)$
−100	.10
0	.20
50	.30
100	.25
150	.10
200	

a. What is the value of $f(200)$? What is your interpretation of this value?
b. What is the probability that MRA will be profitable?
c. What is the probability that MRA will make at least \$100,000?
d. Show the cumulative distribution function, $F(x)$.

5.3 EXPECTED VALUE AND VARIANCE

Expected Value

Once we have constructed the probability distribution for a random variable, we often want to compute the mean or expected value of the random variable. The *expected value* of a discrete random variable is a weighted average of all possible values of the random variable, where the weights are the probabilities associated with the values. The mathematical expression for computing the expected value of a discrete random variable x is shown below:

Expected Value of a Discrete Random Variable

$$E(x) = \mu = \Sigma x f(x). \tag{5.3}$$

As can be seen from the above expression, both the notations $E(x)$ and μ are used to refer to the expected value of a random variable.

Equation (5.3) tells us that in order to compute the expected value of a discrete random variable we must multiply each value of the random variable by the corresponding value of its probability function. We then add the resulting terms. The calculation of the expected value of the random variable (number of daily sales) for DiCarlo Motors is shown in Table 5.5. We see that 1.50 is the expected value of the number of cars sold per day.

TABLE 5.5 Expected Value of Random Variable

x	$f(x)$	$x f(x)$
0	.18	$0(.18) = .00$
1	.39	$1(.39) = .39$
2	.24	$2(.24) = .48$
3	.14	$3(.14) = .42$
4	.04	$4(.04) = .16$
5	.01	$5(.01) = .05$
		$E(x) = \mu = \Sigma x f(x) = 1.50$

The expected value of a random variable can be thought of as a mean or "average" value. That is, for experiments that can be repeated numerous times, the expected value can be interpreted as the "long run" average value for the random variable. However, the expected value is not necessarily the number that we think the random variable will take on the next time the experiment is conducted. In fact, it is impossible for DiCarlo to sell exactly 1.5 cars on any given day. But if we envision selling cars at DiCarlo Motors for many days into the future, the expected value of 1.5 cars provides a good estimate of the mean or average daily sales volume.

It is important to note that often probability problems involve nonrepeatable experiments. That is, unlike the DiCarlo Motors daily sales problem, many experiments are "one time only" occurrences. For example, consider the situation of a firm considering the development of a major shopping center where the random variable of interest is the return on investment. In this type of experiment we cannot interpret the expected value of the random variable as the "long run" average, since the experiment could not be repeated a large number of times. Nonetheless, if the decision maker is willing to think in terms of the experiment being "hypothetically" repeated a large number of times, the "long run average" interpretation can be made.

The expected value can be important to the manager from both planning and decision-making points of view. For example, suppose that DiCarlo Motors will be open 60 days during the next 3 months. How many cars can the owner expect to sell during this time? While we cannot specify the exact sales for any given day, the expected value of 1.50 cars provides an expected sales estimate of 60(1.50) = 90 cars for the next 3 month period. In terms of setting sales quotas and/or planning orders, the expected value may provide helpful decision-making information.

Variance

While the expected value gives us an idea of the average or central value for the random variable, often we would also like a measure of the dispersion or variability of the possible values of the random variable. For example, if the values of the random variable range from quite large to quite small, we would want a "large" value for our measure of variability. On the other hand, if the values of the random variable show only modest variation, we would want a relatively "small" value. Just as we used variance in Chapter 3 to summarize the dispersion in a data set, we now want to use the variance measure to summarize the variability in the values of a random variable. The mathematical expression for the variance of a discrete random variable is shown below:

Variance of a Discrete Random Variable

$$\text{Var}(x) = \sigma^2 = \Sigma(x - \mu)^2 f(x) \tag{5.4}$$

As (5.4) shows, an essential part of the variance formula is a *deviation,* $x - \mu$, which measures how far a particular value of the random variable is from the expected value or mean, μ. In computing the variance of a random variable, the deviations are squared and then weighted by the corresponding value of the probability function. The sum of these weighted squared deviations for all values of the random variable is referred to as the variance. The calculation of the variance for the number of daily sales in the DiCarlo Motors problem is summarized in Table 5.6. We see that the variance for the

TABLE 5.6 Calculation of Variance

x	$x - \mu$	$(x - \mu)^2$	$f(x)$	$(x - \mu)^2 f(x)$
0	0 − 1.50 = −1.50	2.25	.18	2.25(.18) = .4050
1	1 − 1.50 = − .50	.25	.39	.25(.39) = .0975
2	2 − 1.50 = .50	.25	.24	.25(.24) = .0600
3	3 − 1.50 = 1.50	2.25	.14	2.25(.14) = .3150
4	4 − 1.50 = 2.50	6.25	.04	6.25(.04) = .2500
5	5 − 1.50 = 3.50	12.25	.01	12.25(.01) = .1225
				$\sigma^2 = \Sigma(x - \mu)^2 f(x) = 1.2500$

number of cars sold per day is 1.25. As we saw in Chapter 3, a related measure of variability is the *standard deviation,* σ, which is defined simply as the positive square root of the variance. For DiCarlo Motors the standard deviation of the number of cars sold per day is

$$\sigma = \sqrt{1.25} = 1.118.$$

Recall from Chapter 3 that for the purpose of easier managerial interpretation the standard deviation may be preferred over the variance because it is measured in the same units as the random variable ($\sigma = 1.118$ cars sold per day). The variance (σ^2) is measured in squared units and is thus more difficult for a manager to interpret.

EXERCISES

7. The McCormick Hardware Store places an order for riding lawn mowers each February. The following probability distribution for the demand is assumed:

Demand	*Probability*
0	0.10
1	0.15
2	0.30
3	0.20
4	0.15
5	0.10

a. If the store orders three riding lawn mowers, what is the probability of selling all three?
b. What is the expected demand for the lawn mowers?
c. What is the variance in the demand for the lawn mowers? What is the standard deviation?
8. Glazer's Winton Woods apartment complex has 80 two-bedroom apartments. The number of apartment air-conditioner units that must be replaced during the summer season has the probability distribution shown below:

Air Conditioners Replaced	*Probability*
0	0.30
1	0.35
2	0.20
3	0.10
4	0.05

a. What is the expected number of air-conditioner units that will be replaced during a summer season?
b. What is the variance in the number of air-conditioner replacements?
c. What is the standard deviation?
9. A roulette wheel at a Las Vegas casino has 18 red numbers, 18 black numbers, and 2 green numbers. Assume that a $5 bet is placed on the black numbers. If a black number comes up, the player wins $5; otherwise the player loses $5.
a. Let x be a random variable indicating the player's *net* winnings on one bet. Show the probability distribution for x.
b. What is the expected winnings? What is your interpretation of this value?
c. What is the variance in the winnings? What is the standard deviation?
d. If a player places 100 bets of $5 each, what is the expected winnings? Comment on why casinos like a high volume of betting.
10. The probability distribution for collision insurance claims paid by the Newton Automobile Insurance Company is as follows:

Claim	*Probability*
$ 0	0.90
$ 200	0.04
$ 500	0.03
$1,000	0.01
$2,000	0.01
$3,000	0.01

a. Use the expected collision claim amount to determine the collision insurance premium that would allow the company to break even on the collision portion of the policy.
b. The insurance company charges an annual rate of $130 for the collision coverage. What is the expected value of the collision policy for the policyholder? Why does the policyholder purchase a collision policy with this expected value?
11. The number of dots on the face of a die has the following probability function:

$$f(x) = 1/6 \qquad \text{for } x = 1, 2, 3, 4, 5, \text{ or } 6.$$

a. Show that this probability function possesses the properties of all probability distributions.
b. Draw a graph of the probability distribution.
c. What is the expected value? What is the interpretation of this value?
d. What are the variance and the standard deviation for the number of dots on the face of a die?
12. The demand for a product of Carolina Industries varies greatly from month to month. Based on the past 2 years of data, the following probability distribution shows the company's monthly demand:

Unit Demand:	***300***	***400***	***500***	***600***
Probability:	.20	.30	.35	.15

a. If the company places monthly orders based on the expected value of the monthly demand, what should Carolina's monthly order quantity be for this product?
b. Assume that each unit demanded generates $70 in revenue and that each unit ordered costs $50. How much will the company gain or lose in a month if it places an order based on your answer to part a and where the actual demand for the item is 300 units?
13. What are the variance and the standard deviation for the number of units demanded in Exercise 12?
14. The J. R. Ryland Computer Company is considering a plant expansion that will enable the company to begin production of a new computer product. The company's president must determine whether to make the expansion a medium-scale or large-scale project. An uncertainty involves the demand for the new product, which for planning purposes may be low demand, medium demand, or high demand. The probability estimates for the demands are .20, .50 and .30, respectively. Letting x indicate the annual profit in 1,000's, the firm's planners have developed profit forecasts for the medium-scale and large-scale expansion projects:

	Medium-Scale Expansion Profits		*Large-Scale Expansion Profits*	
Demand	x	$f(x)$	x	$f(x)$
Low	50	.20	0	.20
Medium	150	.50	100	.50
High	200	.30	300	.30

a. Compute the expected value for the profit associated with the two expansion alternatives. Which decision is preferred for the objective of maximizing the expected profit?
b. Compute the variance for the profit associated with the two expansion alternatives. Which decision is preferred for the objective of minimizing the risk or uncertainty?

5.4 LINEAR FUNCTIONS AND SUMS OF RANDOM VARIABLES

Expected Value of a Linear Function of a Random Variable

Sometimes a decision maker is more interested in some function of a random variable than in the variable itself. For instance, DiCarlo might be more interested in his expected daily profit than in expected daily sales. Suppose DiCarlo makes a $300 profit from each car sold and, in addition, makes a fixed daily profit of $200 from a long-term leasing contract. Thus the daily profit for the combined sales and leasing operation is

$$\text{Daily profit} = 300x + 200.$$

This equation is an example of a linear function of the random variable x. In general, a linear function of a random variable x can be represented as $ax + b$, with a and b constants. In such cases, the following equation can be used to compute the expected value of the linear function:

Expected Value of a Linear Function of a Random Variable

$$E(ax + b) = aE(x) + b, \tag{5.5}$$

where a and b are constants.

We can apply (5.5) with $a = 300$, $b = 200$, and $E(x) = 1.5$ to compute the expected daily profit for the combined sales and leasing operations at DiCarlo Motors:

$$\begin{aligned} E(\text{daily profit}) &= E(300x + 200) \\ &= 300E(x) + 200 = 300(1.5) + 200 = 450 + 200 = 650. \end{aligned}$$

Variance of a Linear Function of a Random Variable

As with the expected value, we are sometimes interested in computing the variance of a linear function of a random variable. Equation (5.6) can be used to find the variance of a linear function of the random variable x:

Variance of a Linear Function of a Random Variable

$$\text{Var}(ax + b) = a^2\,\text{Var}(x), \tag{5.6}$$

where a and b are constants.

As an application of this formula, let us compute the variance for daily profit for the combined sales and leasing operations at DiCarlo Motors. We can apply (5.6) with $a = 300$, $b = 200$, and Var $(x) = 1.25$:

$$\begin{aligned}\text{Var(daily profit)} &= \text{Var}(300x + 200) \\ &= (300)^2\,\text{Var}(x) = 90{,}000(1.25) = 112{,}500.\end{aligned}$$

The standard deviation of daily profit is $\sqrt{112{,}500} = 335.41$. We can now advise DiCarlo that the expected daily profit is \$650, with a standard deviation of \$335.41.

Expected Value of The Sum of Random Variables

Another expected value calculation commonly occurring is that of computing the expected value of a sum of random variables. For instance, suppose DiCarlo has a second dealership in Albany. Given in Table 5.7 is the probability distribution for the random variable y, the daily sales at the Albany dealership, as well as the computation

TABLE 5.7 Probability Distribution for Daily Sales

y	$f(y)$	$yf(y)$
0	.05	.00
1	.10	.10
2	.30	.60
3	.25	.75
4	.15	.60
5	.10	.50
6	.05	.30
	1.00	$E(y) = 2.85$

of the expected value of y. Suppose that DiCarlo is interested in the expected daily sales for both dealerships. In this case we would want to calculate the expected value of the sum of the random variables x and y [that is, $E(x + y)$]. Equation (5.7) shows that the expected value of the sum of two random variables is given by the sum of expected values:

Expected Value of the Sum of Two Random Variables

$$E(x + y) = E(x) + E(y). \qquad (5.7)$$

For DiCarlo Motors the expected total daily sales from both dealerships is

$$\begin{aligned}E(x + y) &= E(x) + E(y) \\ &= 1.50 + 2.85 = 4.35.\end{aligned}$$

Equation (5.7) can be extended to provide a method for computing the expected value of the sum of any number of random variables. The result is that the expected value of the sum of any number of random variables is equal to the sum of their individual expected values. For example, if DiCarlo Motors had a third dealership and if we let z represent the daily sales at the third dealership, then the expected value of the total daily sales for all three dealerships would be $E(x + y + z) = E(x) + E(y) + E(z)$.

Variance of the Sum of Independent Random Variables

In order to compute the variance of total sales at two of DiCarlo's dealerships, we need an expression for computing the variance of a sum of two random variables. Equation (5.8) provides the variance for the sum of two independent random variables:

Variance of the Sum of Two Independent Random Variables

$$\text{Var}(x + y) = \text{Var}(x) + \text{Var}(y). \tag{5.8}$$

In the previous chapter we said that two events were independent if the occurrence of one of the events did not affect the probability of the other event occurring. Similarly, two random variables are said to be independent if the value that one takes on does not affect the probabilities associated with the value that the other can take on. A more detailed discussion of independent random variables will be presented in Section 5.5. Note that (5.8) applies only when the random variables are independent.

As an illustration of (5.8) let us compute the variance of DiCarlo's total sales at the Saratoga and Albany dealerships. The variance of daily sales at Saratoga is 1.25 and at Albany is 2.13. If DiCarlo feels that the sales at the dealerships are independent, the variance of total sales is given by

$$\begin{aligned}\text{Var}(\text{total sales}) &= \text{Var}(\text{Saratoga sales}) + \text{Var}(\text{Albany sales}) \\ &= 1.25 + 2.13 = 3.38.\end{aligned}$$

Note that the standard deviation of total sales is given by $\sigma = \sqrt{3.38} = 1.84$. Thus we see that the standard deviation of total sales is not equal to the sum of the individual standard deviations.

Equation (5.8) can be generalized to provide a method for computing the variance of the sum of any number of independent random variables. The result is that the variance of the sum is equal to the sum of the variances. For example, if DiCarlo Motors had a third dealership and we let z be the daily sales at the third dealership, then the variance of total sales is $\text{Var}(\text{total sales}) = \text{Var}(x) + \text{Var}(y) + \text{Var}(z)$. In order to use this approach, however, sales at all three dealerships must be independent.

EXERCISES

15. Consider the following probability distribution for the random variable x:

x	$f(x)$
10	.20
20	.40
30	.25
40	.15
	1.00

a. Find the expected value of x.
b. Find the variance and standard deviation.
c. If $y = 3x + 5$, find the expected value, variance, and standard deviation for y.

16. The probability distribution for DiCarlo's Albany dealership is repeated below:

y	$f(y)$
0	.05
1	.10
2	.30
3	.25
4	.15
5	.10
6	.05
	1.00

a. If DiCarlo makes $400 per sale at this dealership and has no other income, compute the expected daily profit.
b. Compute the variance and standard deviation of daily sales at the Albany dealership.

17. The demand for a product is defined by the random variable x and its probability distribution as shown below:

Demand x	$f(x)$
100	.50
200	.30
300	.20

a. What is the expected demand for this product?
b. What is the variance in the demand for this product?
c. Production costs include a $100 fixed setup cost and a $5 variable cost for each unit produced. Let y be a random variable indicating total cost, where $y = 100 + 5x$; what is the expected value and variance of y?

18. A brokerage firm has offices in Houston and Dallas. The daily commissions at the Houston office show $E(x) = \$2,000$ and a standard deviation of $300, while daily commissions at Dallas show $E(y) = \$1,500$ and a standard deviation of $400.
a. What are the total expected daily commissions for the two offices?
b. Assuming that daily commissions at the two offices are independent, determine the standard deviation of total sales for the two offices.

5.5 JOINT PROBABILITY DISTRIBUTIONS

Although the probability distributions studied so far have involved only one random variable, many decisions are based upon an analysis of two or more random variables. In problem situations that involve two or more random variables, the resulting probability distribution is referred to as a *joint probability distribution.* In addition, if the problem involves only two random variables, the resulting joint probability distribution is said to be a *bivariate distribution.* If it involves two or more random variables, it is called a *multivariate distribution.* The purpose of this section is to provide an introduction to joint probability distributions.

TABLE 5.8 Sales Data for 300 Days at DiCarlo Motors

Number of Salespersons Working	*Number of Cars Sold* 1	2	3	*Totals*
0	15	34	5	54
1	33	78	6	117
2	23	45	4	72
3	17	14	11	42
4	2	7	3	12
5	0	2	1	3
Totals	90	180	30	300

To illustrate the concept of a joint probability distribution, consider the data shown in Table 5.8 for the 300 days of sales at DiCarlo Motors. Note that these data show not only the number of cars sold each day but also the number of salespersons working on the day of the sale. For example, on 15 of the 300 days there were no cars sold and one salesperson was working, on 33 days one car was sold and one salesperson was working, and so on. Recall that we previously defined the random variable x as the number of cars sold on a given day. Now, let us define the random variable y to be the number of salespersons working on a given day. Using these two random variables, we define $f(x,y)$ to be the joint probability function of x and y. For example, $f(0,1)$ denotes the probability that $x = 0$ (no cars sold) and $y = 1$ (one salesperson working). Using the sales data shown in Table 5.8 we could use the relative frequency method to assign probabilities. Thus we would assign $f(0,1)$ the probability of $15/300 = .050$, since on 15 of the 300 days we observed no sales and one salesperson working. Similarly, $f(1,1) = 33/300 = .11$, and so on. The joint probabilities for the remaining values of x and y are shown in Table 5.9, which provides the joint probability distribution of the random variables x and y.

TABLE 5.9 Joint Probability Distribution of x and y

Number of Salespersons Working on the Day of Sale (y)	*Number of Cars Sold (x)* 1	2	3	*Totals*
0	.050	.113	.017	.18
1	.110	.260	.020	.39
2	.077	.150	.013	.24
3	.056	.047	.037	.14
4	.007	.023	.010	.04
5	.000	.007	.003	.01
Totals	.300	.600	.100	1.00

The individual probability distributions for the random variables x and y are given in the margins of Table 5.9. For example, we see that the row totals provide the probability distribution of x, the number of cars sold. That is, the probability of zero cars sold is .18, the probability of one car sold is .39, and so on. In a similar fashion, the column totals provide the probability distribution of y, the number of salespersons working on the day of the sale. The probability of 1 salesperson working is .30, the probability of 2 salespersons working is .60, and the probability of 3 salespersons working is .10. As we saw in Chapter 4, the separate probability distributions are found in the margins of the table and are referred to as the *marginal probability distributions.*

Note that the marginal probability distribution for x is identical to the probability distribution for x that we presented before bringing y into the analysis (see Table 5.3). We will use the notation $f_x(x)$ and $f_y(y)$ for the marginal probability functions of x and y, respectively. Thus $f_x(1)$ will denote the probability $x = 1$ and $f_y(1)$ will denote the probability $y = 1$. The marginal probability distributions for x and y are repeated in Table 5.10.

TABLE 5.10 Marginal Probability Distributions for x and y

x	$f_x(x)$	y	$f_y(y)$
0	.18	1	.30
1	.39	2	.60
2	.24	3	.10
3	.14		1.00
4	.04		
5	.01		
	1.00		

When we deal with two or more random variables we shall also follow the practice of placing a subscript on μ and σ to identify the random variable we are referring to. For example μ_x and σ_x refer to the mean and standard deviation of x, while μ_y and σ_y denote the mean and standard deviation for the random variable y.

Independence of Two Discrete Random Variables

In Section 5.4 we introduced the concept of two independent random variables. A formal definition of independence for two discrete random variables is shown below:

Definition of Independence for Two Discrete Random Variables

Two discrete random variables x and y are said to be independent if $f(x,y) = f_x(x) f_y(y)$ for all possible values of x and y. Otherwise the two random variables are said to be dependent.

For the DiCarlo Motors problem we see from Table 5.9 that $f(0,1) = .050$. According to the definition of independence, if x and y are independent, then $f(0,1) = f_x(0) f_y(1)$. From Table 5.10 we see that $f_x(0) = .18$ and $f_y(1) = 0.30$. Thus $f_x(0) f_y(1) = (.18)(.30) = .054$. Hence since $f(0,1) = 0.050$ does not equal $f_x(0) f_y(1)$, we must conclude that x and y are dependent. In order to conclude that the two random variables are dependent all we need do is to find one example where $f(x,y)$ is not equal to $f_x(x) f_y(y)$. However, in order for independence to be proved this relationship must hold for all possible values of x and y.

EXERCISES

19. Shown below is the joint probability distribution for two random variables x and y:

x	y = 5	y = 10	Totals
10	.12	.08	.20
20	.30	.20	.50
30	.18	.12	.30
Totals	.60	.40	1.00

a. Find $f(10,10)$, $f(30,5)$, and $f(20,5)$.
b. Specify the marginal probability distributions for x and y.
c. Compute the mean and variance for x and y.
d. Are x and y independent random variables? Justify your answer.

20. In order to determine whether or not the decision to purchase a new product is dependent upon an individual seeing a commercial for the new product, an experiment was conducted using the random variables x and y as defined below:

$$x = \begin{cases} 0 & \text{if the individual does not purchase} \\ 1 & \text{if the individual does purchase} \end{cases}$$

$$y = \begin{cases} 0 & \text{if the individual did not see the commercial} \\ 1 & \text{if the individual did see the commercial} \end{cases}$$

The joint probability distribution for x and y is shown below:

x	y = 0	y = 1	Totals
0	.52	.28	.80
1	.08	.12	.20
Totals	.60	.40	1.00

a. Are x and y independent random variables?
b. Compute the mean and standard deviation of x and y.
c. Should the company consider continuing the commercial? Justify your answers.

5.6 CONTINUOUS PROBABILITY DISTRIBUTIONS

Until now we have been concerned with probability distributions for discrete random variables, i.e., random variables that can take on a finite or countable number of values. In this section we will introduce probability distributions for continuous random variables. Recall from Section 5.1 that random variables which can take on any value in a certain interval or collection of intervals are said to be continuous. Some examples of continuous random variables are as follows:

1. The *number of ounces* of soup placed in a can labeled "8 ounces."
2. The *flight time* of an airplane traveling from Chicago to New York.

3. The *lifetime* of the picture tube in a new color television set.
4. The *drilling depth* required to reach oil in an offshore drilling operation.

In order to understand the nature of continuous random variables more fully, let us consider the first example given above. Suppose that in checking the filling weights for the cans of soup we find that one can has 8.2 ounces and another 8.3 ounces. Other cans could weigh 8.25 ounces, 8.225 ounces, etc. In fact, the actual weight can be any numerical value from 0 ounces for an empty can to, say, 10.00 ounces for a can filled to capacity. Since there are an infinite number of values in this interval, we can no longer list each value of the random variable and then identify its associated probability. In fact, for continuous random variables we will need to introduce a new method for computing the probabilities associated with the values of the random variable.

Let us consider the random variable x which represents the total flight time of an airplane traveling from Chicago to New York. Assume that the minimum time is 2 hours and that the maximum time is 2 hours and 20 minutes. Thus in terms of minutes the flight time can be any value in the interval from 120 minutes to 140 minutes (e.g., 124 minutes, 125.48 minutes, etc.). Since the random variable x can take on any value from 120 to 140 minutes, x is a continuous rather than a discrete random variable. Let us assume in addition that sufficient actual flight data are available to conclude that the probability of a flight time between 120 and 121 minutes is the same as the probability of a flight time within any other 1 minute interval up to and including 140 minutes. With every 1 minute interval being equally likely, the random variable x is said to have a *uniform probability distribution.* The function, referred to as the *probability density function,* which describes the uniform probability distribution for the flight time random variable is

$$f(x) = \begin{cases} 1/20 & \text{for } 120 \le x \le 140, \\ 0 & \text{elsewhere.} \end{cases}$$

A graph of this probability density function is shown in Figure 5.2. In general, the uniform probability density function for a random variable x is

$$f(x) = \begin{cases} \dfrac{1}{b-a} & \text{for } a \le x \le b, \\ 0 & \text{elsewhere.} \end{cases} \tag{5.9}$$

In the total flight time example, $a = 120$ and $b = 140$.

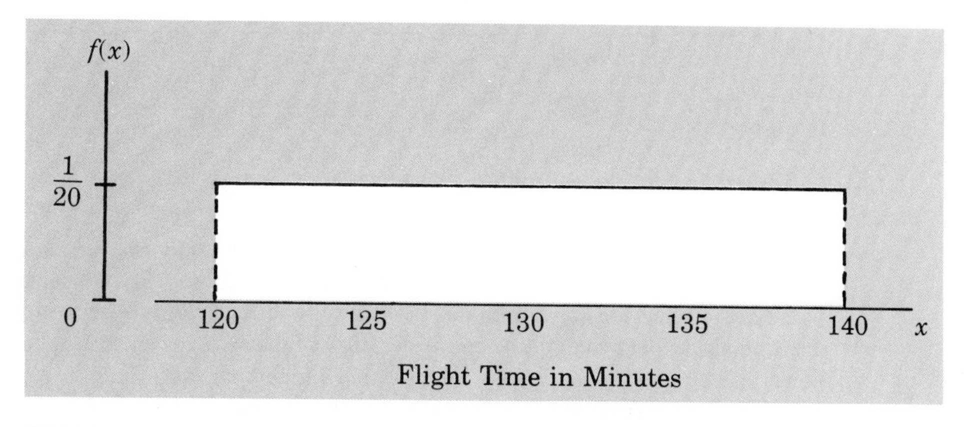

FIGURE 5.2 Uniform Probability Density Function for Flight Time

In the graph of the probability density function, $f(x)$ shows the height or value of the function at any particular value of x. Because we have a *uniform* probability density function the height or value of the function is the same for each value of x between 120 and 140. That is, $f(x) = 1/20$ for all values of x between 120 and 140. The probability density function $f(x)$, unlike the probability function for a discrete random variable, represents the height of the function at any particular value of x and *not* probability. Recall that for each value of a discrete random variable (say, $x = 2$), the probability function yielded the probability of x having *exactly* that value [for example, $f(2)$]. However, since a continuous random variable has an infinite number of possible values, we can no longer attempt to identify the probability for each specific value of x. Rather we must consider probability only in terms of the likelihood that a random variable has a value within a *specified interval.* For example, in our flight time problem an acceptable probability question is, "What is the probability that the flight time is between 120 and 130 minutes?" That is, what is $P(120 \leq x \leq 130)$? Since the flight time must be between 120 and 140 minutes and since the probability was described as being uniform over this interval, we feel comfortable saying $P(120 \leq x \leq 130) = .50$. Indeed, as we shall see, this is correct.

Area as a Measure of Probability

Let us make an observation about the graph shown in Figure 5.3. Specifically, note the *area under the graph of* $f(x)$ in the interval from 120 to 130. The region is rectangular in shape and the area of a rectangle is simply the width times the height. With the width of the interval equal to $130 - 120 = 10$ and the height of the graph $f(x) = 1/20$, we have area = width × height = $10(1/20) = 10/20 = .50$.

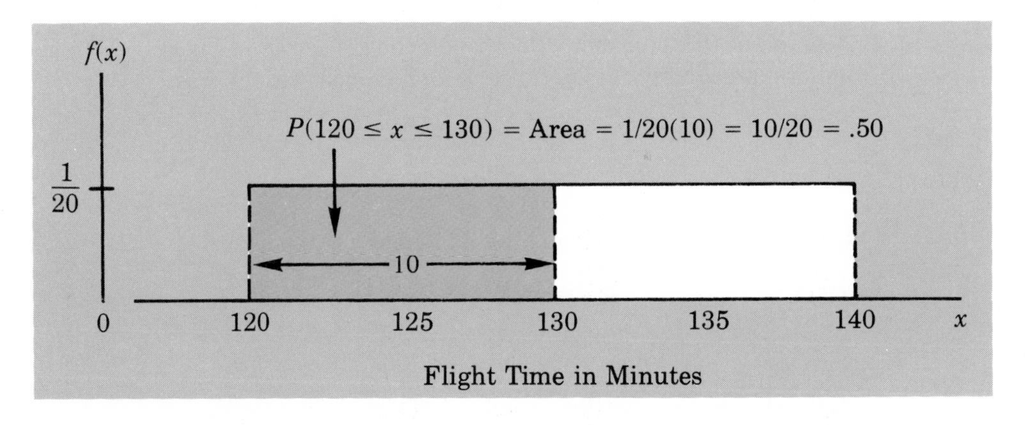

FIGURE 5.3 Area Provides Probability of Flight Time

What observation can you make about the area under the graph of $f(x)$ and probability? They are identical! Indeed, this is true for all continuous random variables. Namely, once a probability density function $f(x)$ has been identified for a continuous random variable, then the probability that x takes on a value between some lower value x_1 and some higher value x_2 can be found by computing the *area* under the graph of $f(x)$ over the interval x_1 to x_2.

Once we have the appropriate probability distribution and accept the interpretation of area as probability, we can answer any number of probability questions. For example, what is the probability of a flight time between 128 and 136 minutes? The width of the interval is $136 - 128 = 8$. With the uniform height of $1/20$, we see $P(128 \leq x \leq 136) = 8/20 = .40$.

Note that $P(120 \leq x \leq 140) = 20(1/20) = 1$. That is, the total area under the $f(x)$ curve is equal to one. This property holds for all continuous probability distributions and is the analog of the condition that the sum of the probabilities has to equal one for a discrete probability distribution. For a continuous probability distribution we must also require that $f(x) \geq 0$ for all values of x. This is the analog of the requirement that $f(x) \geq 0$ for discrete probability distributions.

When we deal with continuous random variables and probability distributions, two major differences stand out as compared to the treatment of their discrete counterparts:

1. We no longer talk about the probability of the random variable taking on a particular value. Instead we talk about the probability of the random variable taking on a value within some given interval.
2. The probability of the random variable taking on a value within some given interval from x_1 to x_2 is defined to be the area under the graph of the probability density function between x_1 and x_2. This implies that the probability that a continuous random variable takes on any particular value exactly is zero, since the area under the graph of $f(x)$ at a single point is zero.

The calculation of the expected value $E(x)$ and variance $\text{Var}(x)$ for a continuous random variable is analogous to that for a discrete random variable. However, since the computational procedure involves integral calculus, we leave the derivation of the appropriate formulas to more advanced texts.

For the uniform continuous probability distribution introduced in this section the formulas for the expected value and variance are

$$E(x) = \frac{a + b}{2},$$

$$\text{Var}(x) = \frac{(b - a)^2}{12}.$$

In the above formulas a is the smallest value and b is the largest value that the random variable may take on.

Applying these formulas to the uniform probability distribution for flight times from Chicago to New York, we obtain

$$E(x) = \frac{(120 + 140)}{2} = 130,$$

$$\text{Var}(x) = \frac{(140 - 120)^2}{12} = 33.33.$$

The standard deviation of flight times can be found by taking the square root of the variance. Thus $\sigma = 5.77$ minutes.

EXERCISES

21. The travel time for a truck traveling from Davenport to Iowa City is uniformly distributed between 70 and 90 minutes.

a. Give a mathematical expression for the probability density function.
b. Compute the probability that the truck will make the trip in 75 minutes or less.
c. What is the probability that the trip will take longer than 82 minutes?
d. Find the expected travel time and its standard deviation.
e. What is the probability that the trip will take exactly 80 minutes?

22. The total time to process a loan application is uniformly distributed between 3 and 7 days.

a. Give a mathematical expression for the probability density function.
b. What is the probability that the loan application will be processed in fewer than 3 days?
c. Compute the probability that a loan application will be processed in 5 days or less.
d. Find the expected processing time and its standard deviation.

23. Bus arrival times at a particular location are uniformly distributed between 2:10 and 2:25 P.M., with 2:10 listed as the scheduled arrival time.

a. Show a graph of the probability density function for the bus arrival times.
b. What is the expected value of the arrival time?
c. The bus is considered delayed when it arrives more than 5 minutes after the scheduled arrival time. What is the probability that the bus will be considered delayed?
d. If a person arrives at the bus stop at 2:17, what is the probability that the person will still catch the bus?

Summary

In this chapter the concept of a random variable and its probability distribution was introduced. We saw that random variables are used to provide numerical descriptions of experimental outcomes. Such a numerical description is necessary if computations such as computing expected values and variances are to be performed. We have seen that probability distributions for discrete random variables can be represented by tables or graphs that show the values of the random variable together with the associated probabilities. In such cases the probability function $f(x)$ provides probabilities for each value of the random variable. Joint probability distributions were introduced to provide a numerical description of experimental outcomes involving two or more random variables. The chapter concluded with an introduction to continuous random variables. Specifically, the uniform probability distribution was introduced as an example of a continuous probability distribution.

Glossary

Random variable—A numerical description of the outcome of an experiment.

Discrete random variable—A random variable that can take on only a finite or countable number of values.

Continuous random variable—A random variable that may take on any value in an interval or collection of intervals.

Discrete probability distribution—A table, graph, or equation describing the values of the random variable and the associated probabilities.

Cumulative distribution function—A function $F(x)$ which gives the probability that a random variable can take on a certain value or less.

Expected value—A weighted average of the values of the random variable, where the probability function provides the weights. If an experiment can be repeated a large number of times, the expected value can be interpreted as the "long run average."

Variance—A measure of the dispersion or variability in the random variable. It is a weighted average of the squared deviations from the mean μ.

Standard deviation—The positive square root of the variance.

Independent random variables—Two random variables are independent if the value that one takes on does not affect the probabilities for the values that the other random variable can take on.

Joint probability distribution—A probability distribution for two or more random variables. For two discrete random variables the probability function is denoted $f(x,y)$ and gives the probability that x and y take on particular values simultaneously.

Bivariate distribution—A joint probability distribution involving two random variables.

Multivariate distribution—A joint probability distribution that involves two or more variables.

Marginal probability distribution—A term used when dealing with two or more random variables. It refers to the separate probability distribution for each of the individual random variables.

Uniform probability distribution—A continuous probability distribution where the probability that the random variable will assume a value in any interval of equal length is the same for each interval.

Probability density function—The function that describes the probability distribution of a continuous random variable.

Key Formulas

Expected Value of a Discrete Random Variable

$$E(x) = \mu = \Sigma x f(x) \tag{5.3}$$

Variance of a Discrete Random Variable

$$\text{Var}(x) = \sigma^2 = \Sigma (x - \mu)^2 f(x) \tag{5.4}$$

Expected Value of a Linear Function of a Random Variable

$$E(ax + b) = aE(x) + b \tag{5.5}$$

Variance of a Linear Function of a Random Variable

$$\text{Var}(ax + b) = a^2 \text{Var}(x) \tag{5.6}$$

Expected Value of the Sum of Two Random Variables

$$E(x + y) = E(x) + E(y) \tag{5.7}$$

Variance of the Sum of Two Independent Random Variables

$$\text{Var}(x + y) = \text{Var}(x) + \text{Var}(y) \tag{5.8}$$

Supplementary Exercises

24. A chain of rental stores would like to determine the number of exercise bicycles to stock at each of its stores. Assume that the probability distribution of daily demand at a particular store is as follows:

Daily Demand x	$f(x)$
0	.15
1	.30
2	.40
3	.10
4	.05

a. Compute the expected value of daily demand.
b. If the daily rental cost for a bicycle is $20 per day, what is the expected value of daily bicycle revenue?

25. The number of defective units returned to a manufacturer varies from week to week. Assume that the number of defective units returned (x) has the following probability distribution:

x	$f(x)$
0	.10
1	.15
2	.30
3	.25
4	.10
5	.10

a. What are the mean and the variance of the number of units returned?
b. If the cost to replace a defective unit is $125, what is the expected weekly cost to replace defective units?

26. Which of the following are and which are not probability distributions? Explain.

x	$f(x)$
0	.20
1	.30
2	.25
3	.35

y	$f(y)$
0	.25
2	.05
4	.10
6	.60

z	$f(z)$
−1	.20
0	.50
1	−.10
2	.40

27. The number of weekly lost-time injuries at a particular plant has the following probability distribution:

Number of Injuries x	$f(x)$
0	.05
1	.20
2	.40
3	.20
4	.15

a. Compute the expected value.
b. Compute the variance.

28. Assume that the plant in Exercise 27 initiated a safety training program and that the number of lost-time injuries during the 20 weeks following the training program was as follows:

Number of Injuries	*Number of Weeks*
0	2
1	8
2	6
3	3
4	1
	20

a. Develop a probability distribution for weekly lost-time injuries based on these data.
b. Compute the expected value and the variance and use both to evaluate the effectiveness of the safety training program.

29. The Hub Real Estate Investment stock is currently selling for $16 per share. An investor plans to buy shares and hold the stock for 1 year. Let x be the random variable indicating the price of the stock after 1 year. The probability distribution for x is shown below:

Price of Stock (x)	$f(x)$
16	.35
17	.25
18	.25
19	.10
20	.05

a. Show that the above probability distribution possesses the properties of all probability distributions.
b. What is the expected price of the stock after 1 year?
c. What is the expected gain per share of the stock over the 1 year period? What percent return on the investment is reflected by this expected value?
d. What is the variance in the price of the stock over the 1 year period?
e. Another stock with a similar expected return has a variance of 3. Which stock appears to be the better investment in terms of minimizing risk or uncertainty associated with the investment? Explain.

30. The budgeting process for a midwestern college resulted in expense forecasts for the coming year (in 1,000,000's) of $9, $10, $11, $12, and $13. Since the actual expenses were unknown, the following respective probabilities were assigned: .3, .2, .25, .05, and .2.
a. Show the probability distribution for the expense forecast.
b. What is the expected value of the expenses for the coming year?
c. What is the variance in the expenses for the coming year?
d. If income projections for the year are estimated at $12 million, comment on the financial position of the college.

31. Exercise 5 provided a probability function for x, the hours required to change over a production system, as follows:

$$f(x) = \frac{x}{10} \quad \text{for } x = 1, 2, 3, \text{ or } 4.$$

a. What is the expected value of the changeover time?
b. What is the variance of the changeover time?

32. A retailer has shelf space for two units of a highly perishable item which must be disposed of at the end of the day if it is not sold. Each unit costs \$2.50 and sells for \$5.00. Demand probabilities are as follows: $P(\text{demand} = 0) = .40$, $P(\text{demand} = 1) = .20$, and $P(\text{demand} = 2) = .40$. Let y be a random variable indicating daily profit if a retailer stocks two units each day. Let x be a random variable indicating daily profit if a retailer stocks one unit each day.
a. Show the probability distributions for y and x.
b. Using the expected values of y and x, determine whether the retailer would be better off stocking one or two units per day.
33. Consider the following joint probability distribution of x = time of arrival and y = traffic conditions, where x and y take on values as follows:

Arrival Time	*Value of x*	*Traffic Conditions*	*Value of y*
Early	1	Light	1
On-time	2	Moderate	2
Late	3	Heavy	3

The joint probability distribution of x and y is given below.

		y	
x	*1*	*2*	*3*
1	.12	.06	.02
2	.25	.15	.10
3	.04	.06	.20

a. Is this joint probability distribution a valid probability distribution? Explain.
b Are the random variables x and y independent? Explain.
34. In an office building the waiting time for an elevator is found to be uniformly distributed between 0 minutes and 5 minutes.
a. What is the probability density function $f(x)$ for this uniform distribution?
b. What is the probability of waiting longer than 3.5 minutes?
c. What is the probability that the elevator arrives in the first 45 seconds?
d. What is the probability of a waiting time between 1 and 3 minutes?
e. What is the expected waiting time?
35. The time required to complete a particular assembly operation is uniformly distributed between 30 and 40 minutes.
a. What is the mathematical expression for the probability density function?
b. Compute the probability that the assembly operation will require more than 38 minutes to complete.
c. If management wants to set a time standard for this operation, what time should be selected such that 70% of the time the operation will be completed within the time specified?
d. Find the expected value and standard deviation for the assembly time.
36. A particular make of automobile is listed as weighing 4,000 pounds. Because of weight differences due to the options ordered with the car, the actual weight varies uniformly between 3,900 and 4,100 pounds.
a. What is the mathematical expression for the probability density function?
b. What is the probability that the car will weigh less than 3,950 pounds?
37. The number of ounces of soup placed in a can labeled as containing "8 ounces" is actually uniformly distributed between 7.98 and 8.02 ounces. What is the probability that more than 8.01 ounces of soup are in a can?

XEROX

Xerox Corporation*

Stamford, Connecticut

Xerox Corporation is in the information products and systems business worldwide. As a major part of this business, it develops, makes and markets xerographic copiers and duplicators; facsimile transceivers; electrostatic printers; processor memory disks, drives and high-speed terminals; electronic typewriters; information-processing products, office information systems; electronic printing systems; automatic labeling, binding and mailing machines, and xeroradiographic devices.

MULTINATIONAL DOCUMENTATION & TRAINING SERVICES

Multinational Documentation & Training Services (MD&TS) provides customers with timely, cost effective, and high quality communication services. In this regard, MD&TS provides four basic services:

1. *Documentation.* Quality documentation assures that the intent of service or product design is upheld, marketing and service goals are maintained, and that the entire organization is striving toward the same end result. MD&TS provides a variety of documentation services such as: Operation Manuals; Installation Instructions; Preventive Maintenance Procedures; Testing and Troubleshooting Procedures.
2. *Training.* MD&TS has the experience and the facilities to provide training programs that use a variety of proven instructional methods. These programs can be implemented at any of our 28 modern training facilities, the customer location, or through correspondence.
3. *Translation.* Documentation and training requirements that are multinational must also be multilingual. MD&TS can provide translation services for technical English materials into a number of different foreign languages. Our highly skilled linguistic specialists ensure grammatical and connotative correctness, as well as technical accuracy.
4. *Publishing.* MD&TS offers complete publishing capabilities with over 100 communication and media specialists. The quality of the publishing service provided is assured by following a policy whereby every document is scrutinized and validated for technical accuracy, consistency, grammar, and print quality.

*The authors are indebted to Soterios M. Flouris, Manager, Systems Development and Maintenance, Webster, New York, for providing this application.

Our professionals have many years of experience in the design, development and implementation of multinational documentation and training materials. We believe this experience enables us to provide the most flexible, efficient, and cost-effective documentation and training services available today.

PERFORMANCE TEST SIMULATION

The professional writers and translators working for MD&TS use an on-line computerized publication system. Management of MD&TS was interested in determining the effect of different system configurations (for example, type of computer, maximum number of on-line users, and so forth) on system performance. Specifically, for a given system configuration, management was interested in determining the following:

1. The probability of a user being refused access by the system because of an excess number of users.
2. The probability of any specific number of users being on the system simultaneously.

To determine the above probabilities, a computer simulation model was developed. The purpose of the computer simulation model was to provide a representation of different system configurations. Through a series of computer runs the behavior of the simulation model—and hence each system configuration—was studied. The operating characteristics of the simulation models were then used to make inferences about the operating characteristics of different system configurations.

In order to build the simulation model it was necessary to determine a probability distribution for the following two random variables:

1. The length of time a user is on the system, referred to as the on time per session.
2. The length of time between one user session and the next user session, referred to as the idle time per session.

Based upon a survey of users, the probability distribution of on time per session was approximated* as shown in Table A5.1 and Figure A5.1, where x is the random

TABLE A5.1 Probability Distribution of On Time per Session

x	$f(x)$
10	.05
20	.06
30	.08
40	.20
50	.25
60	.20
70	.08
80	.06
90	.02

*The actual distribution used in the simulation study has been modified to protect proprietary information and to simplify the discussion.

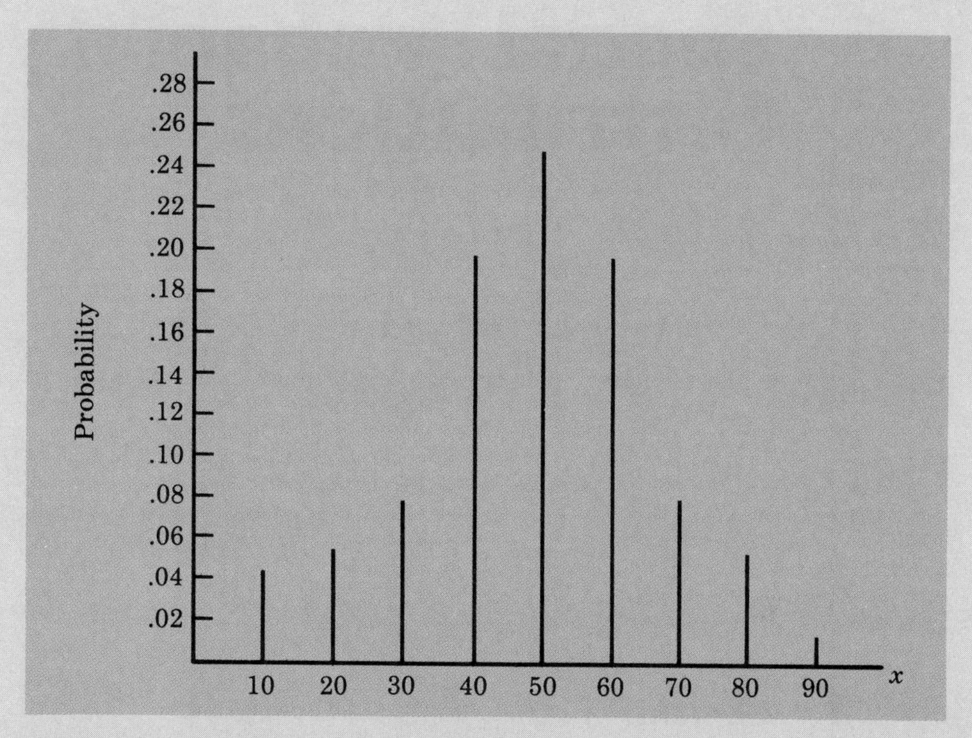

FIGURE A5.1 Graphical Representation of the Probability Distribution of On Time Per Session

variable indicating on time in minutes. Using the data in Table 1, the expected value and variance of on time per session were calculated with $E(x) = 48.8$ and $Var(x) = 336.2$. The expected value shows that the mean or "average" length of time a user spends on the system is 48.8 minutes. Another probability distribution was developed for the random variable indicating idle time per session.

In the computer simulation model, whenever the simulation indicated that a user was to begin operation on the system, it was necessary to generate a value of the random variable from the on time per session distribution. Similarly, whenever the simulation indicated that a user had completed a session, it was necessary to generate a value of the random variable from the idle time distribution. Although the details of the simulation model are somewhat complex and beyond the scope of this application, the concept of a random variable and its probability distribution was an important part of this simulation study.

RESULTS

The probability distributions for on time per session and idle time per session were key components in the simulation model developed to investigate the effect of different system configurations. The results from the simulation study helped MD&TS to determine a system configuration that would ensure a near zero probability that a user would be refused access to the system.

6 Special Probability Distributions

What you will learn in this chapter:

- the advantages of using special probability distributions to compute probabilities
- when it is appropriate to use special probability distributions, such as the binomial, Poisson, and normal probability distributions
- how to use probability functions and/or tables to compute probabilities
- when it is advantageous to use the Poisson and normal probability distributions to approximate the binomial probability distribution

Contents

In this chapter we will introduce three probability distributions that have been successfully applied in a wide variety of decision-making situations. Two of the distributions, the binomial and the Poisson, apply to discrete random variables. The third, the normal distribution, applies to continuous random variables. The purpose of this chapter is to show the types of situations in which these distributions can be applied.

6.1 THE BINOMIAL DISTRIBUTION

In this section we consider a class of experiments possessing the following characteristics:

1. The overall experiment can be described in terms of a sequence of n identical experiments, which are called *trials*.
2. Two outcomes are possible on each trial. We refer to one outcome as a *success* and the other as a *failure*.
3. The probabilities of the two outcomes do not change from one trial to the next.
4. The trials are independent.

For example, consider the situation of an insurance salesperson who contacts ten different families. The outcome associated with visiting each family can be referred to as a success if the family purchases an insurance policy and a failure if not. If the probability of selling a policy is assumed to be the same for each family, and if the decision to purchase or not purchase insurance by one family is not affected by the decision of any other family, then we have an example that exhibits the above four characteristics.

Experiments which satisfy conditions 2, 3, and 4 are said to be generated by a *Bernoulli process*. In addition, if condition 1 is satisfied (there are n trials), we say we have a *binomial experiment*. An important discrete random variable associated with the binomial experiment is the number of successful outcomes in the n trials. If we let x denote the value of this random variable, then x can have a value of $0, 1, 2, 3, \ldots, n$, depending on the number of successes observed in the n trials. The probability distribution associated with this random variable is called the *binomial probability distribution*.

In cases where the binomial distribution is applicable, a mathematical formula can be used to compute the probability associated with any possible value of the random variable. We now show how such a formula can be developed in the context of an illustrative problem.

The Nastke Clothing Store Problem

As an illustration of the binomial probability distribution, let us consider the event that a customer who enters the Nastke Clothing Store makes a purchase. To keep the problem relatively small, let us restrict our attention to the next three customers who enter the store. If, based on past experience, the store manager estimates the probability that any one customer will make a purchase to be .30, what is the probability that exactly two of the next three customers make a purchase? What is the probability that none of the three make a purchase?

We first want to demonstrate that three customers entering the clothing store and electing whether or not to make a purchase can be viewed as a binomial experiment. Checking the four requirements for a binomial experiment, we see the following:

1. The experiment can be described as a sequence of three identical trials, one trial for each of the three customers that will enter the store.
2. Two outcomes—the customer makes a purchase (success) or the customer does not make a purchase (failure)—are possible for each trial or customer.
3. The probabilities of the purchase (.30) and no purchase (.70) outcomes are assumed to be the same for all customers.
4. The purchase decision of each customer is independent of the decisions of the other customers.

Thus if we define the random variable x as the number of customers making a purchase (i.e., the number of successes in the three trials), we see that the requirements of the binomial probability distribution have been satisfied.

In Figure 6.1 we show a tree diagram of the Nastke problem with three customers entering the store. A success (S) indicates the customer made a purchase, while a failure (F) indicates no purchase. Let us now attempt to determine the probability that exactly two of the three customers will make a purchase. Looking at Figure 6.1, we see that there are only three possible outcomes in which exactly two successes occur; (S,S,F), (S,F,S), and (F,S,S). Since these three outcomes represent three mutually exclusive events, we can compute the probability of two successes by adding the

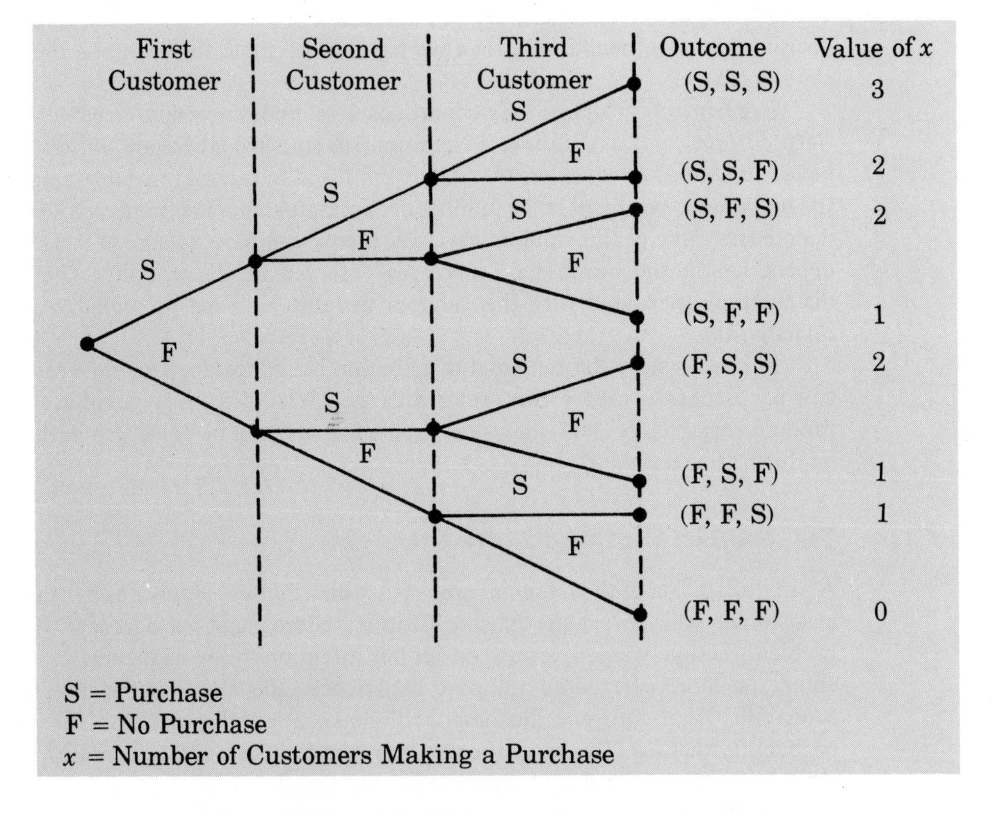

FIGURE 6.1 Tree Diagram for the Nastke Clothing Store Problem: S = Purchase, F = No Purchase, x = Number of Customers Making a Purchase

probabilities of the three experimental outcomes. Let us start with the outcome (S,S,F). To observe this outcome the first customer must make a purchase, the second customer must make a purchase, and the third customer must leave without making a purchase. Since these events are independent, the probability of their joint occurrence must equal the product of their respective probabilities. Hence the probability of observing this outcome must equal $P(\text{SSF}) = P(\text{S})P(\text{S})P(\text{F}) = (.30)(.30)(.70) = (.30)^2(.70) = .063$.

Similar computations for the other two outcomes yielding exactly two successes provide the following results:

Outcome	*Probability*
(S,F,S)	$P(\text{S})P(\text{F})P(\text{S}) = (.30)(.70)(.30) = (.30)^2(.70) = .063$
(F,S,S)	$P(\text{F})P(\text{S})P(\text{S}) = (.70)(.30)(.30) = (.30)^2(.70) = .063$

Again, since the three experimental outcomes corresponding to two successes are mutually exclusive, we can apply the addition law (Chapter 4) to obtain

$$\text{Probability of two successes} = .063 + .063 + .063 = .189.$$

We can now generalize the above development to the case of a binomial experiment involving n trials, with p representing the probability of success on any one trial and $(1 - p)$ the probability of failure on any one trial. Given a binomial experiment involving n trials, the probability of obtaining any one sequence of outcomes resulting in exactly x successes is given by

$$p^x(1 - p)^{n-x}. \tag{6.1}$$

The exponent on p represents the number of successes, and the exponent on $1 - p$ represents the number of failures in the sequence of n trials. For the Nastke problem with $n = 3$ and $p = .30$, the probability of one outcome with exactly $x = 2$ successes is given by (6.1) as

$$(.30)^2(.70)^1 = .063.$$

Since there may be more than one experimental outcome corresponding to x successes, in order to compute the probability of x success in n trials we must multiply (6.1) by the number of experimental outcomes providing exactly x successes in n trials. This number can be computed from the following formula.*

Number of Experimental Outcomes Providing Exactly x Successes in n Trials

$$\frac{n!}{x!(n - x)!} \tag{6.2}$$

where

$$n! = n(n - 1)(n - 2) \cdots (2)(1) \tag{6.3}$$

with 0! defined to be 1.

*This is the formula commonly used to determine the number of combinations of n objects selected x at a time. For the binomial experiment, this combinatorial formula provides the number of experimental outcomes having x successes in n trials.

For the Nastke problem with $n = 3$ and $x = 2$, (6.2) would have told us there were

$$\frac{3!}{2!1!} = \frac{(3)(2)(1)}{(2)(1)(1)} = \frac{6}{2} = 3$$

experimental outcomes providing exactly two successes. Multiplying (6.1) by (6.2) would have told us that the probability of exactly two purchases is 3(.063) = .189.

Combining (6.1) and (6.2) into one equation provides the mathematical formula for the binomial probability function:

Binomial Probability Function

$$f(x) = \frac{n!}{x!(n-x)!} p^x (1-p)^{n-x}, \qquad x = 0, 1, \ldots, n. \qquad (6.4)$$

Equation (6.4) shows that if an experiment possesses the binomial properties, by specifying the number of trials (n) and the probability of success on each trial (p) we can compute the probability for any particular number of successes (x). Table 6.1 depicts in tabular form the probability distribution of x for the Nastke Clothing Store

TABLE 6.1 Probability Distribution for the Number of Customers Making a Purchase

x	$f(x)$
0	$\frac{3!}{0!3!}(.30)^0(.70)^3 = .343$
1	$\frac{3!}{1!2!}(.30)^1(.70)^2 = .441$
2	$\frac{3!}{2!1!}(.30)^2(.70)^1 = .189$
3	$\frac{3!}{3!0!}(.30)^3(.70)^0 = .027$
	1.000

(Note: 0! = 1)

problem. Note that for all four values of the random variable (0,1,2, and 3), (6.4) was used to compute the corresponding probability. A graph of the probability distribution is shown in Figure 6.2.

If we consider any variation of the Nastke experiment, such as ten customers rather than three entering the store, the binomial probability function given by (6.4) is still applicable. For example, the probability of making exactly four sales to ten potential customers entering the store is

$$f(4) = \frac{10!}{4!6!}(.30)^4(.70)^6 = .2001.$$

This is a binomial experiment with $n = 10$, $x = 4$, and $p = .30$.

With use of (6.4) tables have been developed which provide the probability of x successes in n trials for a binomial experiment. These tables are generally easier and

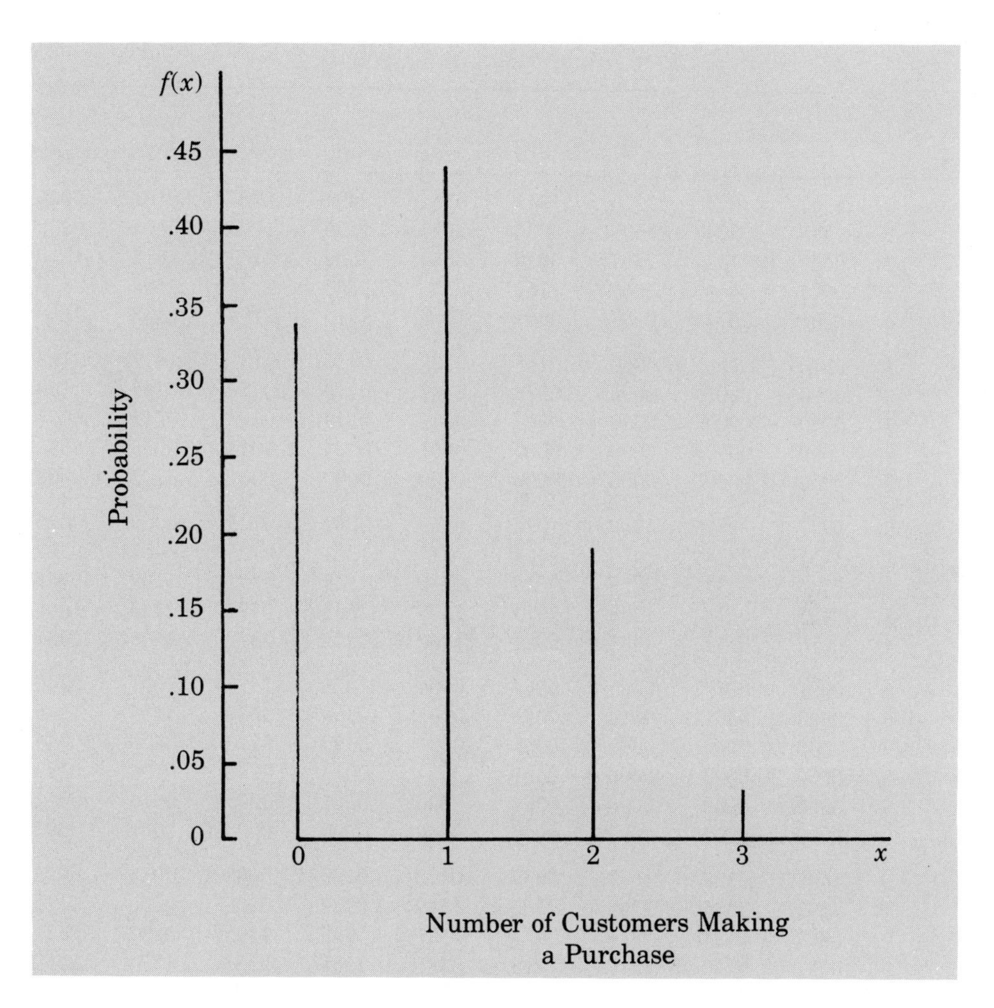

FIGURE 6.2 Graphical Representation of the Probability Distribution of x for the Nastke Clothing Store Problem

quicker to use than equation (6.4), especially when the number of trials involved is large. Such a table of binomial probability values is provided as Table 5 of Appendix B. We have included a portion of this table in Table 6.2. In order to use this table it is necessary to specify the values of n, p, and x for the binomial experiment of interest. Check the use of this table by using it to verify the probability of four successes in ten trials for the Nastke Clothing Store problem. Note that the value of $f(4) = .2001$ can be read directly from the table of binomial probabilities, making it unnecessary to perform the calculations required by (6.4).

The Expected Value and Variance for the Binomial Probability Distribution

Given the data in Table 6.1, we can use (5.3) to compute the expected value or expected number of customers making a purchase:

$$\mu = \Sigma x f(x) = 0(.343) + 1(.441) + 2(.189) + 3(.027) = .9.$$

TABLE 6.2 Selected Values of the Binomial Probability Tables
Example: $n = 10$, $x = 4$, $p = 0.30$; $f(4) = .2001$.

		(p)									
n	x	.05	.10	.15	.20	.25	.30	.35	.40	.45	.50
9	0	.6302	.3874	.2316	.1342	.0751	.0404	.0207	.0101	.0046	.0020
	1	.2985	.3874	.3679	.3020	.2253	.1556	.1004	.0605	.0339	.0176
	2	.0629	.1722	.2597	.3020	.3003	.2668	.2162	.1612	.1110	.0703
	3	.0077	.0446	.1069	.1762	.2336	.2668	.2716	.2508	.2119	.1641
	4	.0006	.0074	.0283	.0661	.1168	.1715	.2194	.2508	.2600	.2461
	5	.0000	.0008	.0050	.0165	.0389	.0735	.1181	.1672	.2128	.2461
	6	.0000	.0001	.0006	.0028	.0087	.0210	.0424	.0743	.1160	.1641
	7	.0000	.0000	.0000	.0003	.0012	.0039	.0098	.0212	.0407	.0703
	8	.0000	.0000	.0000	.0000	.0001	.0004	.0013	.0035	.0083	.0176
	9	.0000	.0000	.0000	.0000	.0000	.0000	.0001	.0003	.0008	.0020
10	0	.5987	.3487	.1969	.1074	.0563	.0282	.0135	.0060	.0025	.0010
	1	.3151	.3874	.3474	.2684	.1877	.1211	.0725	.0403	.0207	.0098
	2	.0746	.1937	.2759	.3020	.2816	.2335	.1757	.1209	.0763	.0439
	3	.0105	.0574	.1298	.2013	.2503	.2668	.2522	.2150	.1665	.1172
	4	.0010	.0112	.0401	.0881	.1460	**.2001**	.2377	.2508	.2384	.2051
	5	.0001	.0015	.0085	.0264	.0584	.1029	.1536	.2007	.2340	.2461
	6	.0000	.0001	.0012	.0055	.0162	.0368	.0689	.1115	.1596	.2051
	7	.0000	.0000	.0001	.0008	.0031	.0090	.0212	.0425	.0746	.1172
	8	.0000	.0000	.0000	.0001	.0004	.0014	.0043	.0106	.0229	.0439
	9	.0000	.0000	.0000	.0000	.0000	.0001	.0005	.0016	.0042	.0098
	10	.0000	.0000	.0000	.0000	.0000	.0000	.0000	.0001	.0003	.0010
11	0	.5688	.3138	.1673	.0859	.0422	.0198	.0088	.0036	.0014	.0005
	1	.3293	.3835	.3248	.2362	.1549	.0932	.0518	.0266	.0125	.0054
	2	.0867	.2131	.2866	.2953	.2581	.1998	.1395	.0887	.0531	.0269
	3	.0137	.0710	.1517	.2215	.2581	.2568	.2254	.1774	.1259	.0806
	4	.0014	.0158	.0536	.1107	.1721	.2201	.2428	.2365	.2060	.1611
	5	.0001	.0025	.0132	.0388	.0803	.1321	.1830	.2207	.2360	.2256
	6	.0000	.0003	.0023	.0097	.0268	.0566	.0985	.1471	.1931	.2256
	7	.0000	.0000	.0003	.0017	.0064	.0173	.0379	.0701	.1128	.1611
	8	.0000	.0000	.0000	.0002	.0011	.0037	.0102	.0234	.0462	.0806
	9	.0000	.0000	.0000	.0000	.0001	.0005	.0018	.0052	.0126	.0269
	10	.0000	.0000	.0000	.0000	.0000	.0000	.0002	.0007	.0021	.0054
	11	.0000	.0000	.0000	.0000	.0000	.0000	.0000	.0000	.0002	.0005

Note that we could have obtained this same expected value simply by multiplying n by p: $np = 3(.30) = .9$. For the special case of a binomial probability distribution, the expected value of the random variable is given by

Expected Value for the Binomial

$$\mu = np \tag{6.5}$$

Thus it is not necessary to carry out the detailed calculations required by (5.3) in order to compute the expected value.

Suppose that during the next month Nastke's Clothing Store expects 1,000 customers to enter the store. What is the expected number of customers who will make a purchase? The answer is $\mu = np = (1000)(.3) = 300$. Thus in order to increase the expected number of sales Nastke's must induce more customers to enter the store and/or somehow increase the probability that any individual customer will make a purchase after entering.

For the special case of a binomial distribution the variance of the random variable can be computed as follows:

Variance for the Binomial

$$\sigma^2 = np(1 - p). \tag{6.6}$$

For the Nastke Clothing Store problem with three customers, we see that the variance and standard deviation for the number of customers making a purchase are

$$\sigma^2 = np(1 - p) = 3(.3)(.7) = .63.$$

$$\sigma = \sqrt{.63} = .79.$$

Exercise 3 below asks you to use the probability distribution of Table 6.1 and the formula for calculating the variance of a random variable to check this result.

EXERCISES

1. When a particular machine is functioning properly only 1% of the items produced are defective. Assume that the machine is functioning properly in answering the following questions:

a. If two items are examined, what is the probability that one is defective?
b. If five items are examined, what is the probability that none are defective?
c. What is the expected number of defective items in a sample of 200?
d. What is the standard deviation of the number of defective items in a sample of 200?

2. At a particular university it has been found that 20% of the students withdraw without completing the introductory statistics course. Assume that 20 students have registered for the course this quarter.

a. What is the probability that two or fewer will withdraw?
b. What is the probability that exactly four will withdraw?
c. What is the probability that more than three will withdraw?
d. What is the expected number of withdrawals?

3. For the special case of a binomial random variable we stated that the variance measure could be computed from the formula $\sigma^2 = np(1 - p)$. For the Nastke Clothing Store problem data in Table 6.1 we found $\sigma^2 = np(1 - p) = .63$. Use the general definition of variance for a discrete random variable, Equation (5.4), and the data in Table 6.1 to verify that the variance is in fact .63.

4. National Oil Company conducts exploratory oil drilling operations in the southwestern United States. In order to fund the operation, investors form partnerships which provide the financial support necessary to drill a fixed number of oil wells. Each well drilled is classified as a "producer" well or a "dry" well. Past experience shows that this type of exploratory operation provides producer wells for 15% of all wells drilled. A newly formed partnership has provided the financial support for drilling at 12 exploratory locations.

a. What is the probability that all 12 wells will be producer wells?

b. What is the probability that all 12 wells will be dry wells?

c. What is the probability that exactly one well will be a producer well?

d. In order to make the partnership venture profitable, at least three of the exploratory wells must be producer wells. What is the probability that the venture will be profitable?

5. Military radar and missile detection systems are designed to warn a country against enemy attacks. A probability question deals with the ability of the detection system to identify the attack and perform the warning correctly. Assume that a particular detection system has a .90 probability of detecting a missile attack. Answer the following questions using the binomial probability distribution.

a. What is the probability that a single detection system will detect an attack?

b. If two detection systems are installed in the same area and operate independently, what is the probability that at least one of the systems will detect the attack?

c. If three systems are installed, what is the probability that at least one of the systems will detect the attack?

d. Would you recommend that multiple detection systems be operated simultaneously? Explain.

6. Assume that the binomial distribution applies for the case of a college basketball player shooting free throws. Late in a basketball game, a team will sometimes foul intentionally in the hope that the player shooting the free throws will miss and the team committing the foul will get the ball. Assume that the best player on the opposing team has a .82 probability of making a free throw and that the worst player has a .56 probability of making a free throw.

a. What are the probabilities that the best player makes 0, 1, and 2 points if fouled and given two free throws?

b. What are the probabilities that the worst player makes 0, 1, and 2 points if fouled and given two free throws?

c. Does it make sense for a coach to have a preset plan as to which player to intentionally foul late in a basketball game? Explain.

6.2 THE POISSON DISTRIBUTION

In this section we will consider a discrete random variable which is often useful when dealing with the number of occurrences of an event over a specified interval of time or space. For example, the random variable of interest might be the number of arrivals at a carwash in 1 hour, the number of repairs needed in 10 miles of highway, or the number of leaks in 100 miles of pipeline. If the two assumptions shown below are satisfied, it can be shown that the resulting probability function of the random variable of interest is given by (6.7), referred to as the *Poisson probability function:*

1. The probability of an occurrence of the event is the same for any two intervals of equal length.
2. The occurrence or nonoccurrence of the event in any interval is independent of the occurrence or nonoccurrence in any other interval.

Poisson Probability Function

$$f(x) = \frac{\mu^x e^{-\mu}}{x!} \qquad \text{for } x = 0, 1, 2, \ldots, \tag{6.7}$$

where

μ = expected value or average number of occurrences in an interval,

e = 2.71828.

Before we consider a specific example to see how the Poisson distribution can be applied, note that (6.7) shows that there is no limit to the number of possible values that a Poisson random variable can take on. That is, although x is still a discrete random variable with $x = 0, 1, 2, \ldots,$ the Poisson random variable has no specific upper limit.

A Poisson Example Involving Time Intervals

Suppose that we are interested in the number of arrivals at the drive-in teller window of a bank during a 15-minute period on weekday mornings. If we can assume that the probability of a car arriving is the same for any two time periods of equal length and that the arrival or nonarrival of a car in any time period is independent of the arrival or nonarrival in any other time period, the Poisson probability function is applicable. Assuming that these assumptions are satisfied and that an analysis of historical data shows that the average number of cars arriving in a 15 minute period of time is 10, the following probability function applies:

$$f(x) = \frac{10^x e^{-10}}{x!} \qquad \text{for } x = 0, 1, 2, \ldots.$$

If management wanted to know the probability of exactly five arrivals in 15 minutes, we would set $x = 5$ and thus obtain*

$$\begin{matrix}\text{Probability of exactly} \\ \text{five arrivals in 15 minutes}\end{matrix} = f(5) = \frac{10^5 e^{-10}}{5!} = .0378.$$

Although the above probability was determined by evaluating the probability function with $\mu = 10$ and $x = 5$, it is often easier to refer to tables for the Poisson probability distribution. These tables provide probabilities for specific values of x and μ. We have included such a table as Table 7 of Appendix B. For convenience we have reproduced a portion of this table as Table 6.3. Note that in order to use the table of Poisson probabilities we need know only the values of x and μ. Thus from Table 6.3 we see that the probability of five arrivals in a 15-minute period is found by locating the value in the row of the table corresponding to $x = 5$ and the column of the table corresponding to $\mu = 10$. Hence we obtain $f(5) = .0378$.

*Values of $e^{-\mu}$ can be found in Table 6 of Appendix B.

TABLE 6.3 Selected Values of the Poisson Probability Tables
Example: $\mu = 10, x = 5; f(5) = .0378$

	μ									
x	9.1	9.2	9.3	9.4	9.5	9.6	9.7	9.8	9.9	10
0	.0001	.0001	.0001	.0001	.0001	.0001	.0001	.0001	.0001	.0000
1	.0010	.0009	.0009	.0008	.0007	.0007	.0006	.0005	.0005	.0005
2	.0046	.0043	.0040	.0037	.0034	.0031	.0029	.0027	.0025	.0023
3	.0140	.0131	.0123	.0115	.0107	.0100	.0093	.0087	.0081	.0076
4	.0319	.0302	.0285	.0269	.0254	.0240	.0226	.0213	.0201	.0189
5	.0581	.0555	.0530	.0506	.0483	.0460	.0439	.0418	.0398	.0378
6	.0881	.0851	.0822	.0793	.0764	.0736	.0709	.0682	.0656	.0631
7	.1145	.1118	.1091	.1064	.1037	.1010	.0982	.0955	.0928	.0901
8	.1302	.1286	.1269	.1251	.1232	.1212	.1191	.1170	.1148	.1126
9	.1317	.1315	.1311	.1306	.1300	.1293	.1284	.1274	.1263	.1251
10	.1198	.1210	.1219	.1228	.1235	.1241	.1245	.1249	.1250	.1251
11	.0991	.1012	.1031	.1049	.1067	.1083	.1098	.1112	.1125	.1137
12	.0752	.0776	.0799	.0822	.0844	.0866	.0888	.0908	.0928	.0948
13	.0526	.0549	.0572	.0594	.0617	.0640	.0662	.0685	.0707	.0729
14	.0342	.0361	.0380	.0399	.0419	.0439	.0459	.0479	.0500	.0521
15	.0208	.0221	.0235	.0250	.0265	.0281	.0297	.0313	.0330	.0347
16	.0118	.0127	.0137	.0147	.0157	.0168	.0180	.0192	.0204	.0217
17	.0063	.0069	.0075	.0081	.0088	.0095	.0103	.0111	.0119	.0128
18	.0032	.0035	.0039	.0042	.0046	.0051	.0055	.0060	.0065	.0071
19	.0015	.0017	.0019	.0021	.0023	.0026	.0028	.0031	.0034	.0037
20	.0007	.0008	.0009	.0010	.0011	.0012	.0014	.0015	.0017	.0019
21	.0003	.0003	.0004	.0004	.0005	.0006	.0006	.0007	.0008	.0009
22	.0001	.0001	.0002	.0002	.0002	.0002	.0003	.0003	.0004	.0004
23	.0000	.0001	.0001	.0001	.0001	.0001	.0001	.0001	.0002	.0002
24	.0000	.0000	.0000	.0000	.0000	.0000	.0000	.0001	.0001	.0001

A Poisson Example Involving Length or Distance Intervals

Let us illustrate the variety of applications where the Poisson probability distribution is useful. Suppose that we are concerned with the occurrence of major defects in a section of highway 1 month after resurfacing. We will assume that the probability of a defect is the same for any two intervals of equal length. We assume also that the occurrence or nonoccurrence of a defect in any one interval is independent of the occurrence or nonoccurrence in any other interval. Thus the Poisson probability distribution can be applied to this situation.

Suppose we learn that major defects 1 month after resurfacing occur at the average rate of two per mile. Let us find the probability that there will be no major defects in a particular 3 mile section of the highway. Since we are interested in an interval with a length of 3 miles, μ = (2 defects/mile)(3 miles) = 6 represents the expected number of major defects over the 3 mile section of highway. Thus by using (6.7) or Table 7 in Appendix B we see that the probability of no major defects is

0.0025. Thus it is very unlikely that there will be no major defects in the 3 mile section. In fact, there is a $1 - .0025 = .9975$ probability of at least one major defect in the highway section.

Poisson Approximation of Binomial Probabilities

Binomial probability tables are usually not available for large values of n, the number of trials. For example, note that Table 5 of Appendix B goes up only to $n = 20$. Thus in cases where we are interested in computing binomial probabilities for large values of n, some other means must be used. It can be shown that if n becomes larger and larger and p becomes smaller and smaller in such a way that np remains a constant, then the Poisson distribution with $\mu = np$ provides an approximation to the binomial distribution. A common rule of thumb is that the Poisson approximation will be good as long as $p \leq .05$ and $n \geq 20$. Of course the approximation can still be used when the rule of thumb is not satisfied, but the approximation will not be as good.

Let us illustrate the use of the Poisson distribution to approximate the binomial with a case where $n = 20$ and $p = .05$. A certain grocery store manager reports that 5% of the customers make purchases in excess of $100. The store manager would like to know the probability that no customers in a sample of 20 will make purchases in excess of $100. Using the binomial probability tables or (6.4) with $n = 20$ and $p = .05$, we find the probability of no successes to be .3585. In order to use the Poisson approximation we set $\mu = np = (20)(.05) = 1$ and $x = 0$. The Poisson probability tables (Table 7 of Appendix B) give a probability of no successes of 0.3679. In Table 6.4 we show the correct binomial probabilities and compare them with the ones obtained from the Poisson approximation for the other values of x.

TABLE 6.4 Comparison of Binomial Probabilities with Poisson Approximation for $n = 20$ and $p = .05$

x	*Binomial Probability*	*Poisson Approximation*	*Difference*
0	.3585	.3679	−.0094
1	.3774	.3679	.0095
2	.1887	.1839	.0048
3	.0596	.0613	−.0017
4	.0133	.0153	−.0020
5	.0022	.0031	−.0009
6	.0003	.0005	−.0002

In comparing the probabilities given in Table 6.4 we see that the Poisson approximation is quite good. In the worst case the probabilities differ by slightly less than .01. For larger values of n and smaller values of p the Poisson approximation would be even more accurate.

As one final example of the use of the Poisson approximation to the binomial, let us reconsider the illustration above involving the grocery store manager. This time, however, assume that $n = 100$. Note that since the binomial tables go only to $n = 20$, we need to use the Poisson approximation to determine the probability that no customers in a sample of 100 will make purchases in excess of $100. Referring to Table 7 of Appendix B with $\mu = np = (100)(.05) = 5$ and $x = 0$, we obtain a probability of .0067.

EXERCISES

7. A certain restaurant has a reputation for good food. Restaurant management boasts that on a Saturday night groups of customers arrive at the rate of 15 groups every half hour.
a. What is the probability that 5 minutes will pass with no customers arriving?
b. What is the probability that eight groups of customers will arrive in 10 minutes?
c. What is the probability that more than five groups will arrive in a 10 minute period of time?
8. During rush hours accidents occur in a particular metropolitan area at the rate of two per hour. The morning rush period lasts for 1 hour and 30 minutes and the evening rush period lasts for 2 hours.
a. On a particular day what is the probability that there will be no accidents during the morning rush period?
b. What is the probability of two accidents during the evening rush period?
c. What is the probability of four or more accidents during the morning rush period?
d. On a particular day what is the probability there will be no accidents during both the morning and evening rush periods?
9. An accounting firm has come to expect 1% of a company's accounts receivable balances to be in error. A sample of 150 accounts has been selected for audit.
a. What is the probability that none of the accounts selected will contain errors? (Hint: use the Poisson approximation.)
b. What is the probability that four or more of the accounts will contain errors?
c. What is the probability that exactly two accounts will contain errors?
10. Airline passengers arrive randomly and independently at the passenger screening facility at a major international airport. The mean arrival rate is 10 passengers per minute.
a. What is the probability of no arrivals in a 1 minute period?
b. What is the probability three or fewer passengers arrive in a 1 minute period?
c. What is the probability of no arrivals in a 15 second period?
d. What is the probability of at least one arrival in a 15 second period?
11. For a given model of hand calculators, 3% of the calculators will fail within the first 30 days of operation and be returned to the manufacturer for repair. Assume that there is a batch of 120 calculators:
a. What is the expected number of calculators that will fail in the first 30 days of operation?
b. What is the probability that at least two will fail?
c. What is the probability that exactly three will fail?
12. The Oakland Bell Telephone Company subcontracts for delivery of its new telephone books. The telephone book delivery service has a record of delivering books to 97% of the names in the telephone book. The company decides to check this record by randomly calling numbers from the telephone book to see if the new telephone books have been delivered. Answer the following questions assuming the 97% delivery rate is correct:
a. Out of 100 calls, what is the probability that exactly three will not have received the new telephone books?
b. Out of 50 calls, what is the probability that at least two will not have received the new telephone books?

6.3 THE NORMAL DISTRIBUTION

In the previous chapter we introduced the notion of a continuous random variable and its probability distribution by considering the uniform probability distribution.

Perhaps the most important probability distribution used to describe a continuous random variable is the *normal probability distribution.* The normal probability distribution is applicable in a great many practical situations and is an integral part of many of the statistical procedures discussed later in this text. Its probability density function has the form of the "bell-shaped" curve shown in Figure 6.3.

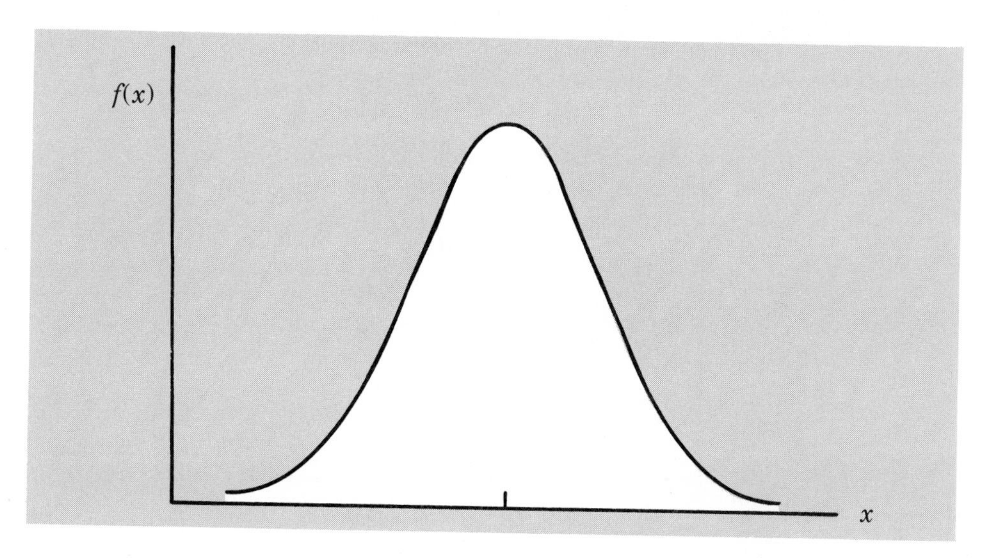

FIGURE 6.3 Bell-Shaped Curve of the Normal Probability Distribution

The mathematical form of the probability density function for the uniform distribution was fairly simple: $f(x) = 1/(b - a)$ for $a \leq x \leq b$. This resulted in a horizontal line for its graph. The mathematical function which provides the bell-shaped curve of the normal probability density function is more complex. The mathematical formula is given by (6.8):

Normal Probability Density Function

$$f(x) = \frac{1}{\sigma\sqrt{2\pi}} e^{-(x-\mu)^2/2\sigma^2} \quad \text{for } -\infty < x < \infty, \tag{6.8}$$

where

μ = mean or expected value of the random variable x,
σ^2 = variance of the random variable x,
σ = standard deviation of the random variable x,
π = 3.14159,
e = 2.71828.

Recall that for a continuous random variable $f(x)$ is the height of the curve at a particular value of x. Thus once the mean (μ) and either the standard deviation (σ) or variance (σ^2) are specified, (6.8) can be used to draw the graph or curve for the corresponding normal distribution. For example, let us consider a normal distribution with a mean of 50 and a standard deviation of 10. Substituting $\mu = 50$ and $\sigma = 10$ into (6.8) provides the following probability density function:

$$f(x) = \frac{1}{(10)\sqrt{2\pi}} e^{-(x-50)^2/2(100)}. \tag{6.9}$$

Figure 6.4 shows the graph of this particular normal probability distribution.

Figure 6.5 shows two other normal distributions, one with $\mu = 50$ and $\sigma = 15$ and

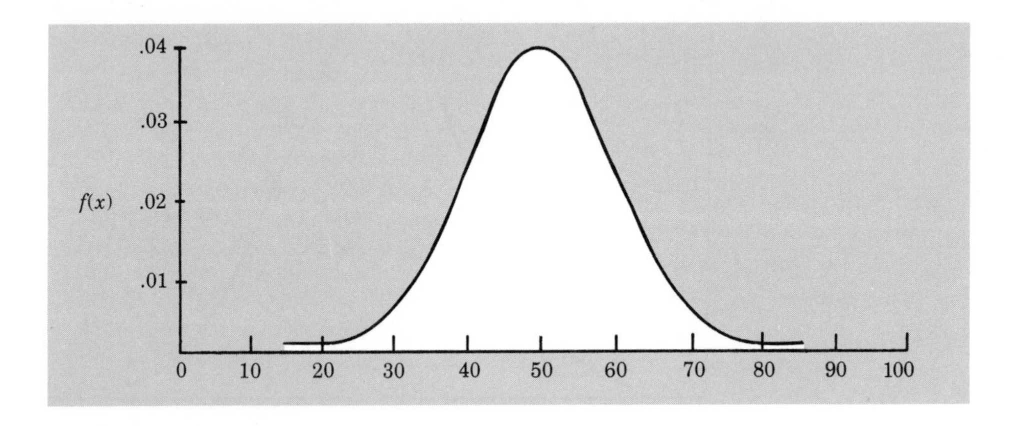

FIGURE 6.4 Normal Distribution with $\mu = 50$ and $\sigma = 10$

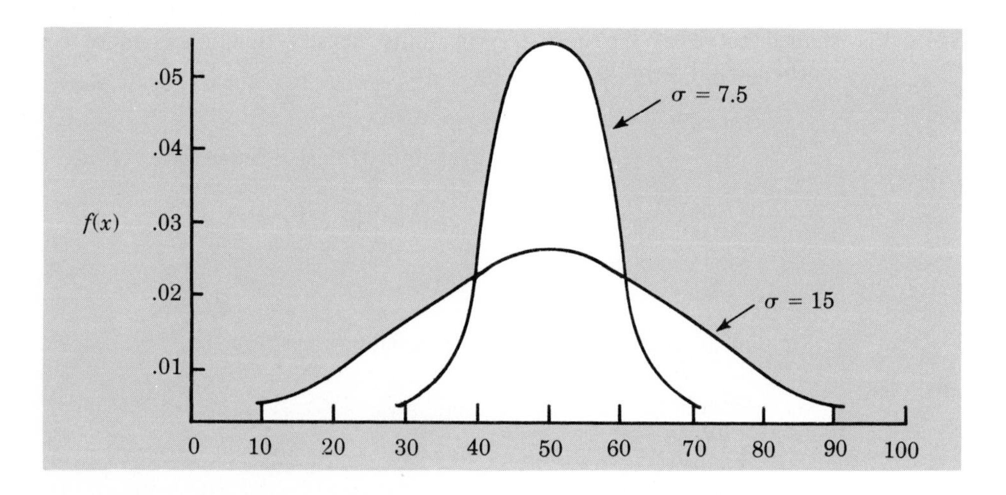

FIGURE 6.5 Other Normal Distributions with $\mu = 50$

another with $\mu = 50$ and $\sigma = 7.5$. Note in particular the effect that the standard deviation σ has on the general shape of the normal curve. A larger standard deviation tends to flatten and broaden the curve. This, of course, is what we should expect, since larger values of σ indicate a larger variability in the values of the random variable.

In order to present the procedure used to compute probabilities [areas under the $f(x)$ curve] associated with a random variable having a normal distribution, we must first introduce the standard normal distribution.

The Standard Normal Distribution

A random variable which has a normal distribution with a mean of 0 and a standard deviation of 1 is said to have a *standard normal distribution.* We use the letter z to designate this particular normal random variable. Thus for a standard normal random variable z we let $\mu = 0$ and $\sigma = 1$ in (6.8) in order to determine its probability density function. For the standard normal distribution we have

$$f(z) = \frac{1}{\sqrt{2\pi}} e^{-z^2/2}. \tag{6.10}$$

The graph of the standard normal distribution is shown in Figure 6.6. Note that it has the same general appearance as other normal distributions, but with the special properties of $\mu = 0$ and $\sigma = 1$.

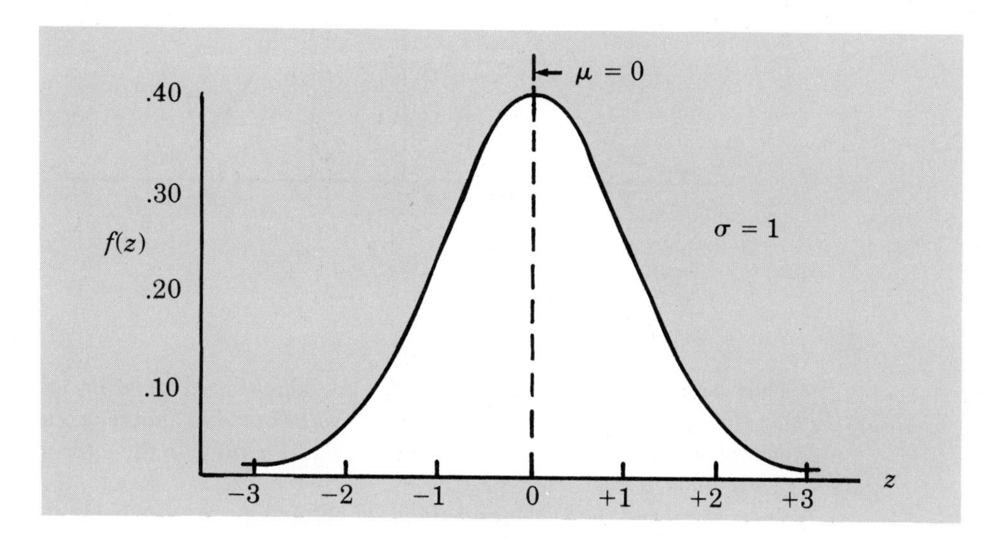

FIGURE 6.6 Standard Normal Distribution

Now let us recall the procedure for finding probabilities associated with a continuous random variable. We wish to determine the probability of the random variable having a value in a specified interval from a to b. Thus we have to find the area under the curve [given by the probability density function $f(x)$] in the interval from a to b. In Chapter 5 we saw that finding probabilities, or "areas under the curve," for a uniform distribution was relatively easy. All we had to do was multiply the width of the interval by the height of the curve. However, finding areas under the normal distribution curve appears at first glance to be much more difficult, since the height of the curve varies. The mathematical technique for obtaining these areas is beyond the scope of the text, but fortunately tables are available which provide the areas or probability values for the standard normal distribution. Table 6.5 is such a table of areas. This table is also available as Table 1 of Appendix B.

Let us see how Table 6.5 is used to find areas or probabilities. First note that values of z appear in the left-hand column, with the second decimal value of z appearing in the top row. For example, for a z value of 1.00 we find the 1.0 in the left-hand column and .00 in the top row. Then by looking in the body of the table we find a value of .3413 corresponding to the 1.00 value for z. The value .3413 is the area under the curve between the mean ($z = .00$) and $z = 1.00$. This is shown graphically in Figure 6.7.

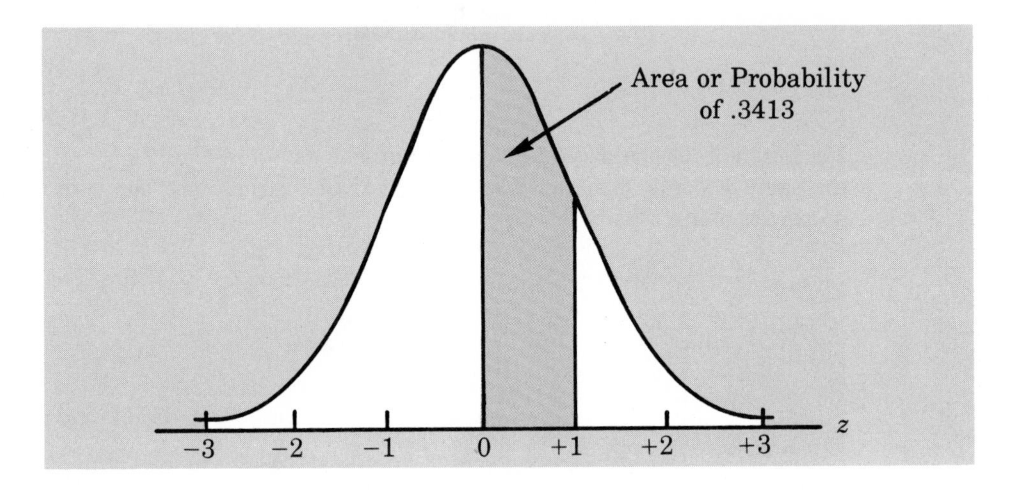

FIGURE 6.7 Probability of z Between 0.00 and + 1.00

Thus we see that the values in Table 6.5 provide *the area under the curve between the mean* ($z = 0.00$) *and any specified positive value of* z. For another example, we can use the table to find that the area or probability of a z value in the interval $z = .00$ to $z = 1.25$ is .3944.

Suppose that we want the probability of obtaining a z value between $z = -1.00$ and $z = 1.00$. We already have used Table 6.5 to find that the probability of a z value between $z = .00$ and $z = 1.00$ is 0.3413. Note now that the normal distribution is symmetric. That is, the shape of the curve to the left of the mean is the mirror image of the shape of the curve to the right of the mean. Thus the probability of a z value between $z = .00$ and $z = -1.00$ is the same as that between $z = .00$ and $z = 1.00$, that is, .3413. Hence the probability of a z value between $z = -1.00$ and $z = 1.00$ must be $.3413 + .3413 = .6826$. This is shown graphically in Figure 6.8.

Similarly, we can find that the probability of a z value between -2.00 and $+2.00$ is $.4772 + .4772 = .9544$, while the probability of a z value between -3.00 and $+3.00$ is $.4986 + .4986 = .9972$. Since we know that the total probability or total area under

TABLE 6.5 Areas or Probabilities for the Standard Normal Distribution

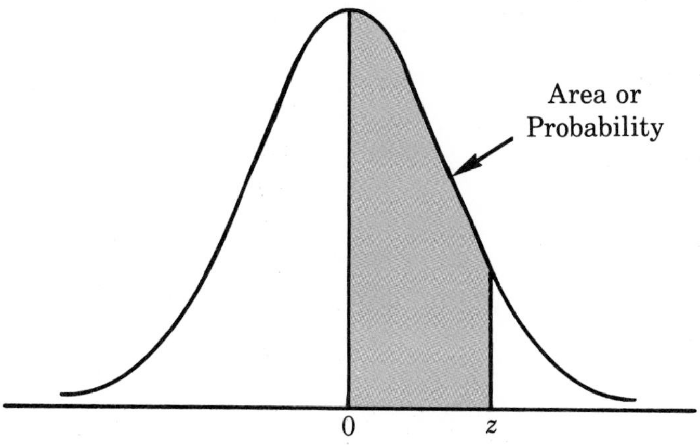

z	.00	.01	.02	.03	.04	.05	.06	.07	.08	.09
.0	.0000	.0040	.0080	.0120	.0160	.0199	.0239	.0279	.0319	.0359
.1	.0398	.0438	.0478	.0517	.0557	.0596	.0636	.0675	.0714	.0753
.2	.0793	.0832	.0871	.0910	.0948	.0987	.1026	.1064	.1103	.1141
.3	.1179	.1217	.1255	.1293	.1331	.1368	.1406	.1443	.1480	.1517
.4	.1554	.1591	.1628	.1664	.1700	.1736	.1772	.1808	.1844	.1879
.5	.1915	.1950	.1985	.2019	.2054	.2088	.2123	.2157	.2190	.2224
.6	.2257	.2291	.2324	.2357	.2389	.2422	.2454	.2486	.2518	.2549
.7	.2580	.2612	.2642	.2673	.2704	.2734	.2764	.2794	.2823	.2852
.8	.2881	.2910	.2939	.2967	.2995	.3023	.3051	.3078	.3106	.3133
.9	.3159	.3186	.3212	.3238	.3264	.3289	.3315	.3340	.3365	.3389
1.0	.3413	.3438	.3461	.3485	.3508	.3531	.3554	.3577	.3599	.3621
1.1	.3643	.3665	.3686	.3708	.3729	.3749	.3770	.3790	.3810	.3830
1.2	.3849	.3869	.3888	.3907	.3925	.3944	.3962	.3980	.3997	.4015
1.3	.4032	.4049	.4066	.4082	.4099	.4115	.4131	.4147	.4162	.4177
1.4	.4192	.4207	.4222	.4236	.4251	.4265	.4279	.4292	.4306	.4319
1.5	.4332	.4345	.4357	.4370	.4382	.4394	.4406	.4418	.4429	.4441
1.6	.4452	.4463	.4474	.4484	.4495	.4505	.4515	.4525	.4535	.4545
1.7	.4554	.4564	.4573	.4582	.4591	.4599	.4608	.4616	.4625	.4633
1.8	.4641	.4649	.4656	.4664	.4671	.4678	.4686	.4693	.4699	.4706
1.9	.4713	.4719	.4726	.4732	.4738	.4744	.4750	.4756	.4761	.4767
2.0	.4772	.4778	.4783	.4788	.4793	.4798	.4803	.4808	.4812	.4817
2.1	.4821	.4826	.4830	.4834	.4838	.4842	.4846	.4850	.4854	.4857
2.2	.4861	.4864	.4868	.4871	.4875	.4878	.4881	.4884	.4887	.4890
2.3	.4893	.4896	.4898	.4901	.4904	.4906	.4909	.4911	.4913	.4916
2.4	.4918	.4920	.4922	.4925	.4927	.4929	.4931	.4932	.4934	.4936
2.5	.4938	.4940	.4941	.4943	.4945	.4946	.4948	.4949	.4951	.4952
2.6	.4953	.4955	.4956	.4957	.4959	.4960	.4961	.4962	.4963	.4964
2.7	.4965	.4966	.4967	.4968	.4969	.4970	.4971	.4972	.4973	.4974
2.8	.4974	.4975	.4976	.4977	.4977	.4978	.4979	.4979	.4980	.4981
2.9	.4981	.4982	.4982	.4983	.4984	.4984	.4985	.4985	.4986	.4986
3.0	.4986	.4987	.4987	.4988	.4988	.4989	.4989	.4989	.4990	.4990

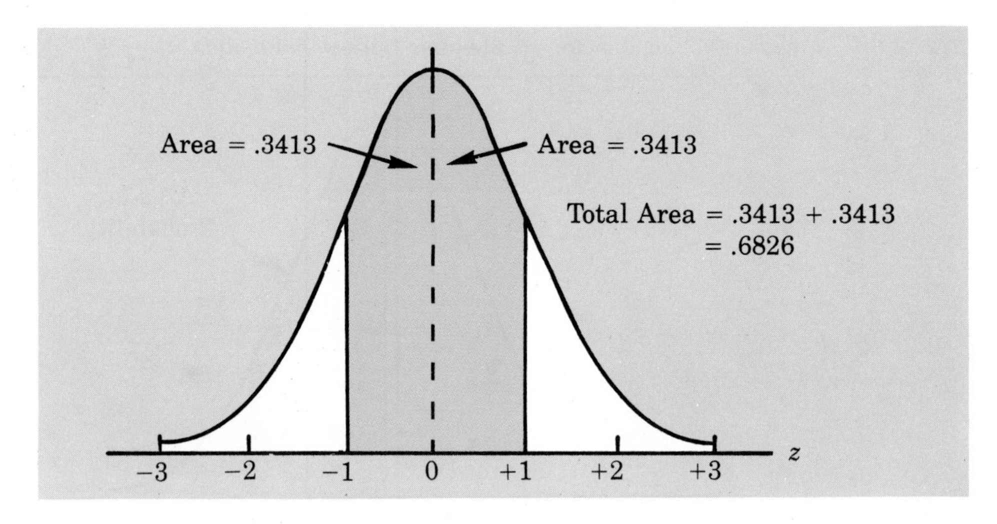

FIGURE 6.8 Probability of z Between −1.00 and 1.00

the curve for any continuous random variable must be 1.0000, the probability of .9972 tells us the value of z will almost always fall between −3.00 and +3.00. Note that the figures depicting the standard normal distribution show this graphically.

Look now at two final examples of computing areas for the standard normal distribution. Let us find the probability that z is greater than 2.00 and the probability that z is between 1.00 and 2.00. In the first case, we see from Table 6.5 that the area between $z = .00$ and $z = 2.00$ is .4772. Since .5000 is the total area above the mean, the area above $z = 2.00$ must be $.5000 - .4772 = .0228$. This is shown graphically in Figure 6.9. Again, because of the fact that the normal distribution is symmetric, the probability of obtaining a value of z less than $z = -2.00$ must also be .0228.

To find the probability that z is between 1.00 and 2.00, we first note that the area between the mean $z = 0$ and $z = 2.00$ is .4772. The area between the mean $z = 0$ and

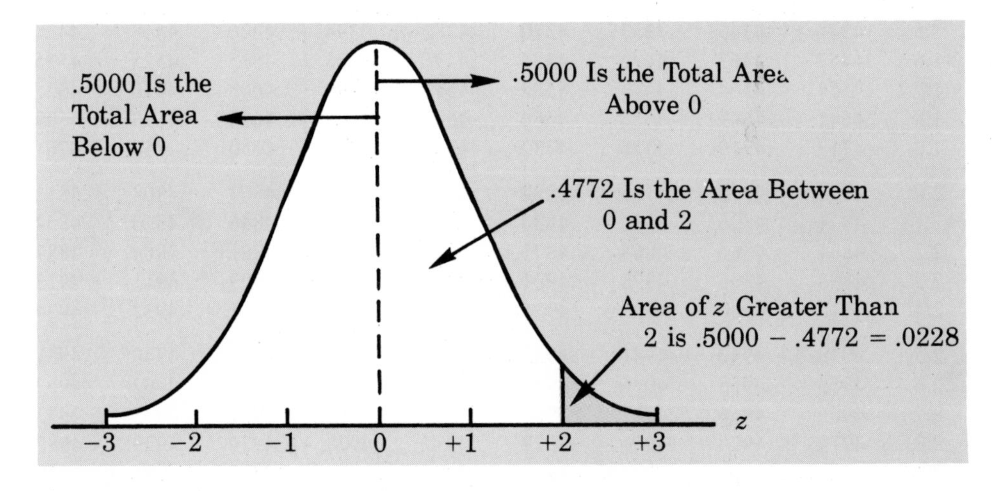

FIGURE 6.9 Probability of z Greater Than 2.00

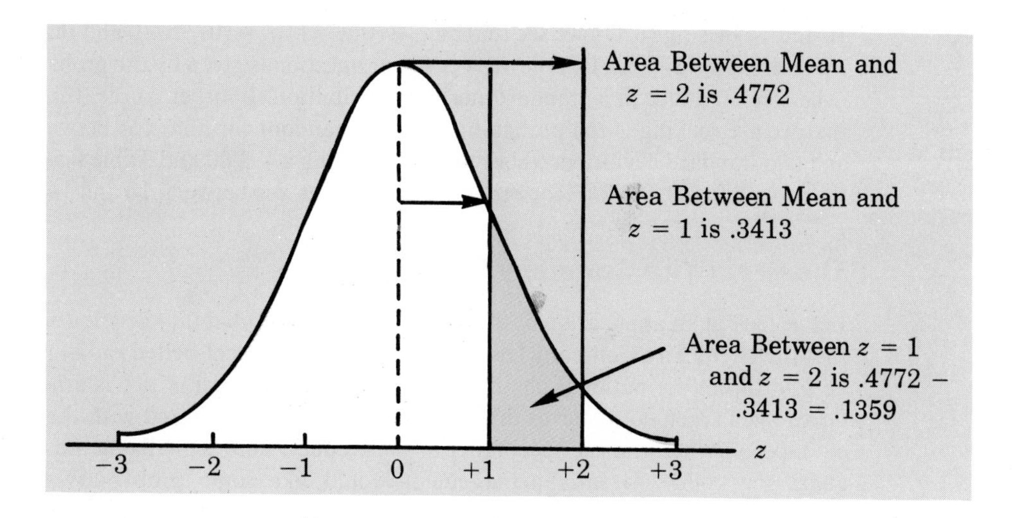

FIGURE 6.10 Probability of z Between 1.00 and 2.00

$z = 1.00$ is .3413. Thus the area between $z = 1.00$ and $z = 2.00$ must be $.4772 - .3413 = .1359$. This result is shown graphically in Figure 6.10. Note again that because of the symmetry of the normal distribution .1359 is also the probability of obtaining a value of z between $z = -1.00$ and $z = -2.00$.

Computing Probabilities for Any Normal Distribution by Converting to the Standard Normal Distribution

The reason that we have been discussing the standard normal distribution so extensively is that probabilities for any normal distribution can be computed by first converting to the standard normal distribution. Thus when we have a normal distribution with any mean μ and any standard deviation σ, we can answer probability questions about this distribution by converting to the standard normal distribution. We then use Table 6.5 and the appropriate z values to find the area or probability values. The formula used to convert any normal random variable x with mean μ and standard deviation σ to the standard normal distribution is

$$z = \frac{x - \mu}{\sigma}. \tag{6.11}$$

Note that a value of x equal to its mean μ results in $z = (\mu - \mu)/\sigma = 0$. Thus we see that x equal to its mean μ corresponds to z at its mean, 0. Now suppose that x is one standard deviation above its mean; that is, $x = \mu + \sigma$. We apply (6.11) and see that the corresponding z value is $z = (\mu + \sigma - \mu)/\sigma = \sigma/\sigma = +1$. Thus we see that a value of x that is one standard deviation above the mean is equivalent to $z = 1$. In other words, we can interpret the z value as *the number of standard deviations that an x value is from its mean μ.*

To see how this conversion enables us to compute probabilities for any normal distribution, let us consider an example. Suppose that we have a normal distribution with $\mu = 10$ and $\sigma = 2$. What is the probability that the random variable x is between

10 and 14? Using (6.11) we see that at $x = 10$, $z = (10 - 10)/2 = 0$ and that at $x = 14$, $z = (14 - 10)/2 = 2$. Thus the answer to our question is given by the probability that z is between 0 and 2 in a standard normal distribution. In other words, the probability that we are seeking is the probability that the random variable x is between its mean and two standard deviations above the mean. Using $z = 2.00$ and Table 6.5, we see that the probability is .4772. Hence the probability that x is between 10 and 14 is .4772.

The Grear Tire Company Problem

Let us look at an application of the use of the normal probability distribution. Suppose that the Grear Tire Company has just developed a new steel-belted radial tire that will be sold through a national chain of discount stores. Since the tire is a new product, Grear's management believes that the mileage guarantee offered with the tire will be an important factor in the acceptance of the product. Before finalizing the tire mileage guarantee policy, Grear's management would like some probability information concerning the number of miles the tires will last.

From actual road tests with the tires Grear's engineering group has estimated the mean tire mileage at $\mu = 36{,}500$ miles and the standard deviation at $\sigma = 5{,}000$. In addition, the data collected indicate that a normal distribution is a reasonable assumption.

Using the normal distribution, what percentage of the tires can be expected to last more than 40,000 miles? In other words, what is the probability that the tire mileage will exceed 40,000? This question can be interpreted as trying to find the area of the shaded region in Figure 6.11.

At $x = 40{,}000$ we have

$$z = \frac{x - \mu}{\sigma} = \frac{40{,}000 - 36{,}500}{5{,}000} = \frac{3{,}500}{5{,}000} = .70.$$

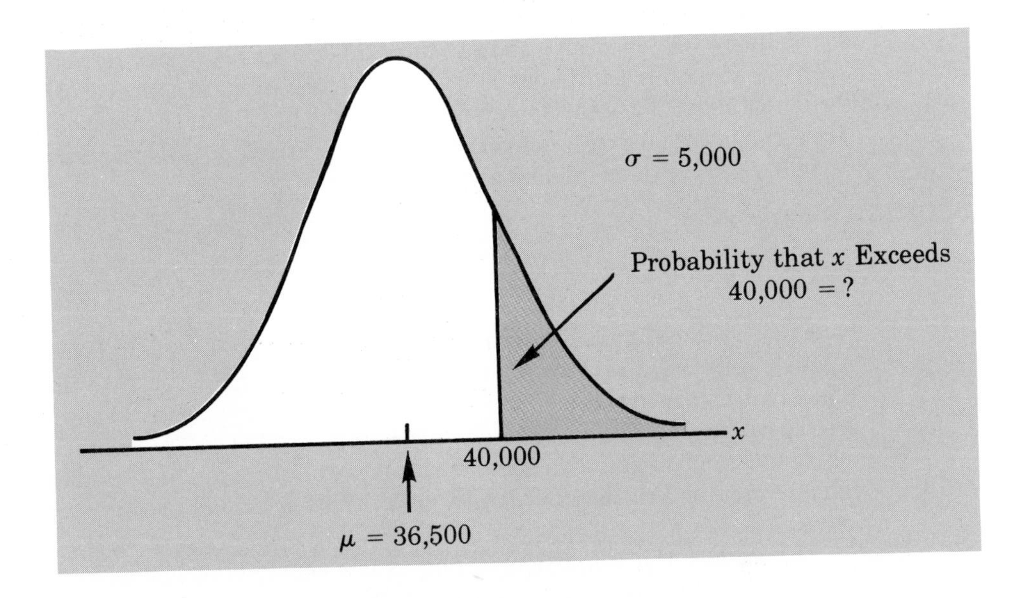

FIGURE 6.11 **Grear Tire Company Tire Mileage**

Using Table 6.5 we see that the area between the mean and $z = .70$ is .2580. Thus $.5000 - .2580 = .2420$ is the probability that x will exceed 40,000. We can conclude that about 24.2% of the tires will exceed 40,000 in mileage.

Let us now assume that Grear is considering a guarantee that will provide a discount on a new set of tires if the original tires do not exceed the mileage stated in the guarantee. What should the guarantee mileage be if Grear would like no more than 10% of the tires to be eligible for the discount guarantee? This question is interpreted graphically in Figure 6.12. According to Figure 6.12, 40% of the area must be between

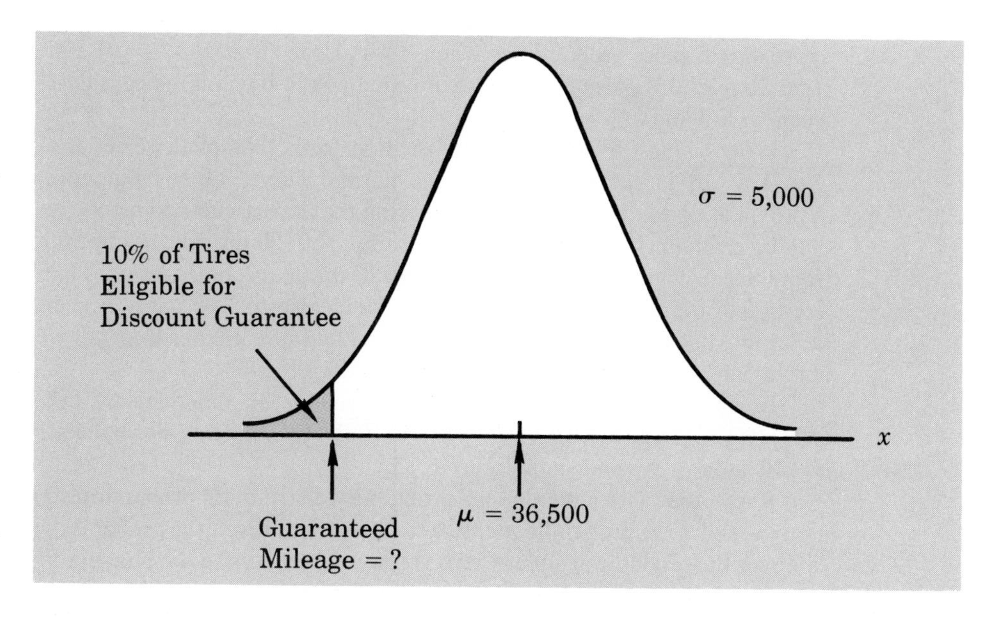

FIGURE 6.12 Grear's Discount Guarantee

the mean and the unknown guarantee mileage. We look up .4000 in the body of Table 6.5 and see that this area occurs at approximately 1.28 standard deviations *below the mean.* That is, $z = -1.28$. To find the mileage (x) corresponding to $z = -1.28$ we have

$$z = \frac{x - \mu}{\sigma} = -1.28,$$

$$x - \mu = -1.28\sigma,$$

$$x = \mu - 1.28\sigma.$$

or

$$x = 36{,}500 - 1.28(5{,}000) = 30{,}100.$$

We see that a guarantee of 30,100 miles will meet the requirement that approximately 10% of the tires will be eligible for the guarantee. Perhaps with this information the firm will set its tire mileage guarantee policy at 30,000 miles.

Again we see the important role that probability distributions play in providing decision-making information. Namely, once a probability distribution is established for a particular problem situation, it can be used to rather quickly and easily provide probability data about the problem. While the data do not make a decision recommendation directly, they do provide information that helps the decision maker better understand the problem. Ultimately this information may assist the decision maker in reaching a good decision.

Normal Approximation of Binomial Probabilities

As discussed in the previous section, binomial probability tables for large values of n usually are not available. We saw that the Poisson distribution could be used to approximate those probabilities when n was large (that is, $n \geq 20$) and p was small (that is, $p \leq .05$). A normal approximation to the binomial is considered acceptable when $np \geq 5$ and $n(1 - p) \geq 5$.

When using the normal approximation to the binomial we set $\mu = np$ and $\sigma = \sqrt{np(1 - p)}$ in the definition of the normal curve. Let us illustrate the normal approximation to the binomial by supposing that a particular company has a history of making errors in 10% of its invoices. A sample of 100 invoices has been taken, and we would like to compute the probability that 12 invoices contain errors. That is, we would like to find the binomial probability of 12 successes in 100 trials. Since the binomial tables in Appendix B are not tabulated for values of n greater than 20, we will use the normal approximation to compute the desired probability.

In applying the normal approximation to the binomial we set $\mu = np = (100)(.1) = 10$ and $\sigma = \sqrt{np(1 - p)} = \sqrt{(100)(.1)(.9)} = 3$. A normal distribution with $\mu = 10$ and $\sigma = 3$ is shown in Figure 6.13.

Recall that with a continuous probability distribution probabilities are computed as areas under the probability density function. As a result the probability of any single value for the random variable is zero. Thus to approximate the binomial probability of

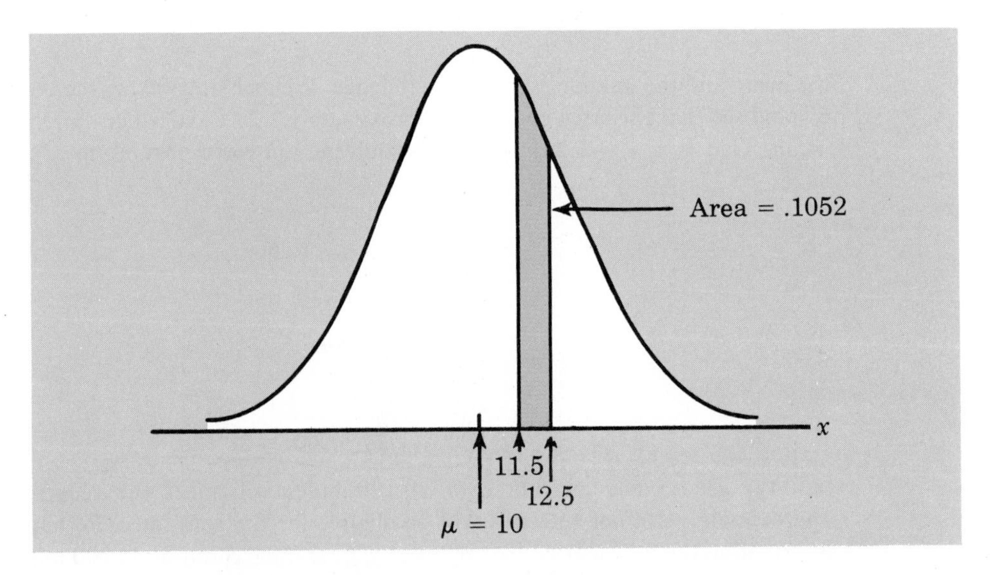

FIGURE 6.13 **Normal Approximation to a Binomial Probability Distribution with $n = 100$ and $p = .10$. Probability of 12 Errors in 100 Trials Is Approximately .1052**

12 successes we must compute the area under the corresponding normal curve between 11.5 and 12.5. The 0.5 that we added and subtracted from 12 is called a *continuity correction factor*. It is introduced because a continuous distribution is being used to approximate a discrete distribution. Thus $P(12)$ for the *discrete* binomial distribution is approximated by $P(11.5 \leq x \leq 12.5)$ for the *continuous* normal distribution.

Converting to the standard normal distribution in order to compute $P(11.5 \leq x \leq 12.5)$, we have

$$z = \frac{x - \mu}{\sigma} = \frac{12.5 - 10.0}{3} = .83 \qquad \text{at } x = 12.5,$$

$$z = \frac{x - \mu}{\sigma} = \frac{11.5 - 10.0}{3} = .50 \qquad \text{at } x = 11.5.$$

From Table 6.5 we find the area under the curve (in Figure 6.13) between 10 and 12.5 is .2967. Similarly, the area under the curve between 10 and 11.5 is .1915. Therefore the area between 11.5 and 12.5 is .2967 − .1915 = .1052. The normal approximation to the probability of 12 successes in 100 trials thus is .1052.

For another illustration, suppose that we want to compute the probability of 13 or fewer errors in our sample of 100 invoices. Figure 6.14 shows the area under the normal curve which approximates this probability. Note that the use of the continuity correction factor results in the value of 13.5 being used to compute the desired probability. The z value corresponding to $x = 13.5$ is

$$z = \frac{13.5 - 10.0}{3.0} = 1.17.$$

Table 6.5 shows that the area under the curve between 10 and 13.5 is .3790. Hence the shaded portion of the graph in Figure 6.14 represents an area of .3790 + .5000 = .8790.

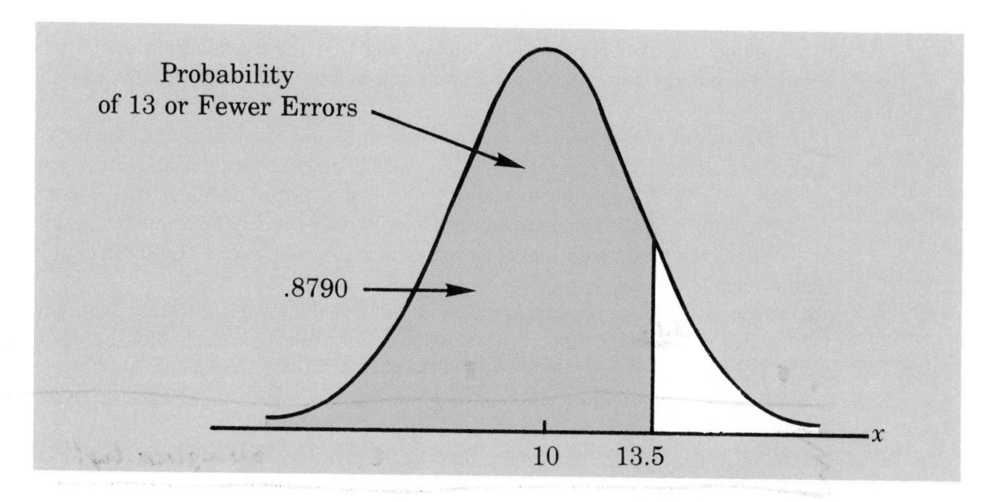

FIGURE 6.14 **Normal Approximation to a Binomial Probability Distribution with n = 100 and p = .10. Probability of 13 or Fewer Errors in 100 Trials Is Approximately .8790**

EXERCISES

13. The demand for a new product is assumed to be normally distributed with $\mu = 200$ and $\sigma = 40$. Letting x be the number of units demanded, find the following:

a. $P(180 \leq x \leq 220)$.
b. $P(x \geq 250)$.
c. $P(x \leq 100)$.
d. $P(225 \leq x \leq 250)$.

14. The Webster National Bank is reviewing its service charge and interest-paying policies on checking accounts. The bank has found that the average daily balance on personal checking accounts is \$550.00, with a standard deviation of \$150.00. In addition, the average daily balances have been found to be normally distributed.

a. What percentage of personal checking account customers carry average daily balances in excess of \$800.00?
b. What percentage of the bank's customers carry average daily balances below \$200.00?
c. What percentage of the bank's customers carry average daily balances between \$300.00 and \$700.00?
d. The bank is considering paying interest to customers carrying average daily balances in excess of a certain amount. If the bank does not want to pay interest to more than 5% of its customers, what is the minimum average daily balance it should be willing to pay interest on?

15. General Hospital's patient account division has compiled data on the age of accounts receivables. The data collected indicate that the age of the accounts follows a normal distribution, with $\mu = 28$ days and $\sigma = 8$ days.

a. What portion of the accounts are between 20 and 40 days old? $[P(20 \leq x \leq 40)]$?
b. The hospital administrator is interested in sending reminder letters to the oldest 15% of accounts. How many days old should an account be before a reminder letter is sent?
c. The hospital administrator would like to give a 5% discount to those accounts which pay their balance by the 21st day. What percentage of the accounts will receive the discount?

16. The time required to complete a final examination in a particular college course is normally distributed, with a mean of 80 minutes and a standard deviation of 10 minutes. Answer the following questions:

a. What is the probability of completing the exam in 1 hour or less?
b. What is the probability a student will complete the exam in more than 60 minutes but less than 75 minutes?
c. Assume that the class has 60 students and that the examination period is 90 minutes in length. How many students do you expect will be unable to complete the exam in the allotted time?

17. The useful life of a computer terminal at a university computer center is known to be normally distributed, with a mean of 3.25 years and a standard deviation of .5 years.

a. Historically 22% of the terminals have had a useful life less than the manufacturer's advertised life. What is the manufacturer's advertised life for the computer terminals?
b. What is the probability that a computer terminal will have a useful life of at least 3 but less than 4 years?

18. From past experience, the management of a well known fast food restaurant estimates that the number of weekly customers at a particular location is normally distributed, with a mean of 5,000 and a standard deviation of 800 customers.

a. What is the probability that on a given week the number of customers will be 4,760 to 5,800?
b. What is the probability of more than 6,500 customers?
c. For 90% of the weeks the number of customers should exceed what amount?

19. In order to obtain cost savings, a company is considering offering an early retirement incentive for its older management personnel. The consulting firm that designed the early retirement program has found that approximately 22% of the employees qualifying for the program will select early retirement during the first year of eligibility. Assume that the company offers the early retirement program to 50 of its management personnel.

a. What is the expected number of employees who will elect early retirement in the first year?

b. What is the probability at least 8 but not more than 12 employees will elect early retirement in the first year?

c. What is the probabiilty that 15 or more employees will select the early retirement option in the first year?

d. For the program to be judged successful, the company believes that it should entice at least 10 management employees to elect early retirement in the first year. What is the probability that the program is successful?

20. Suppose that 54% of a large population of registered voters favor the Democratic candidate for state senator. A public opinion poll uses randomly selected samples of voters and asks each person in the sample his or her preference: the Democratic candidate or the Republican candidate. The weekly poll is based on the response of 100 voters.

a. What is the expected number of voters who will favor the Democratic candidate?

b. What is the variance in the number of voters who will favor the Democratic candidate?

c. What is the probability that the poll will show that *less than* 50% of the voters favor the Democratic candidate when in fact 54% of the population of registered voters favor the candidate? That is, what is the probability that 49 or fewer individuals in the sample express support for the Democratic candidate?

21. Thirty percent of the students at a particular university attended Catholic high schools. A random sample of 50 of this university's students has been taken. Use the normal approximation to the binomial probability distribution to answer the following questions:

a. What is the probability that exactly 10 of the students selected attended Catholic high schools?

b. What is the probability that 20 or more of the students attended Catholic high schools?

c. What is the probability that the number of students from Catholic high schools is between 10 and 20 inclusively?

Summary

This chapter was devoted to a discussion of the binomial, Poisson, and normal probability distributions.

The binomial probability distribution is used for problems where it is necessary to determine the probability of obtaining x successes in an experiment consisting of n trials. The following assumptions must be satisfied for the binomial distribution to be applicable:

1. The experiment can be described in terms of a sequence of n identical trials.
2. Two outcomes are possible on each trial, one called success, the other failure.
3. The probabilities of success and failure do not change from trial to trial.
4. The trials are independent.

If assumptions 3 and 4 are nearly (but not exactly) satisfied, probabilities can still be calculated using the binomial probability distribution. However, they should then be considered only approximations to the true probabilities.

The Poisson probability distribution is used when it is desired to determine the probability of obtaining x occurrences over an interval of time or space. The following assumptions were required for the Poisson distribution to be applicable:

1. The probability of an occurrence of the event is the same for any two intervals of equal length.

2. The occurrence or nonoccurrence of the event in any interval is independent of the occurrence or nonoccurrence in any other interval.

Both the binomial and Poisson distributions were discrete probability distributions.

The final section of the chapter was devoted to the normal probability distribution. It is a continuous probability distribution that is regarded by many individuals as the most important probability distribution in statistical analysis. We saw how the standard normal distribution could be used to compute probabilities for any particular normally distributed random variable.

Glossary

Binomial probability distribution—The probability distribution for a discrete random variable. It is used to compute the probability of x successes in n trials.

Poisson probability distribution—The probability distribution for a discrete random variable. It is used to compute the probability of x occurrences of an event over a specified interval.

Normal probability distribution—A continuous probability distribution. Its probability density function is bell shaped and determined by the mean μ and standard deviation σ.

Standard normal distribution—A normal distribution with a mean of 0 and a standard deviation of 1.

Continuity correction factor—A value of .5 that is added and/or subtracted from a value of x when a continuous probability distribution (e.g., the normal) is used to approximate a discrete probability distribution (e.g., the binomial).

Key Formulas

Number of Experimental Outcomes Providing Exactly x Successes in n Trials

$$\frac{n!}{x!(n-x)!} \tag{6.2}$$

Binomial Probability Function

$$f(x = \frac{n!}{x!(n-x)!} p^x(1-p)^{n-x}, \qquad x = 0, 1, \ldots, n \tag{6.4}$$

Expected Value for the Binomial

$$\mu = np \tag{6.5}$$

Variance for the Binomial

$$\sigma^2 = np(1-p) \tag{6.6}$$

Poisson Probability Function

$$f(x) = \frac{\mu^x e^{-\mu}}{x!} \text{ for } x = 0, 1, 2, \ldots \tag{6.7}$$

Converting to the Standard Normal Distribution

$$z = \frac{x - \mu}{\sigma} \tag{6.11}$$

Supplementary Exercises

22. A firm estimates the probability of employee disciplinary problems on a particular day to be .10.

a. What is the probability that the company experiences 5 days without a disciplinary problem?

b. What is the probability of exactly 2 days with disciplinary problems in a 10 day period?

c. What is the probability of at least 2 days with disciplinary problems in a 20 day period?

23. Suppose that a salesperson makes a sale on 20% of the calls.

a. If the salesperson makes three calls per day, what is the probability of more than three sales in a 5 day week?

b. If the salesperson works 50 weeks per year and makes a commission of $100 per sale, how many sales can be expected annually? What is the salesperson's expected annual income?

24. The salesperson in Exercise 23 is being asked by the sales manager to make one extra call per day. If the salesperson increases the calls from three to four per day, what is the probability of making more than three sales per week? How much of an increase can be expected in annual income?

25. In an audit of a company's billings, an auditor randomly selects five bills. If 3% of all bills contain an error, what is the probability that the auditor will find the following:

a. Exactly one bill in error?

b. At least one bill in error?

26. A salesperson contacts eight potential customers per day. From past experience we know that the probability of a potential customer making a purchase is .10.

a. What is the probability the salesperson makes *exactly* two sales in a day?

b. What is the probability the salesperson makes *at least* two sales in a day?

c. What percentage of the days will the salesperson not make a sale?

d. What is the expected number of sales per day? Over a 5 day week, how many sales are expected?

27. A manufacturing process produces parts that are classified as either defective or acceptable. If the probability that the process produces a defective part is .10, how many defective parts would you expect to find in a lot of 500 parts? What is the variance of the number of defective parts in the lot?

28. Cars arrive at a carwash at the average rate of 15 cars per hour. If the number of arrivals per hour follows a Poisson distribution, what is the probability of 20 or more arrivals during any given hour of operation?

29. A new automated production process has been experiencing an average of 1.5 breakdowns per day. Because of the cost associated with a breakdown, management is concerned about the possibility of having three or more breakdowns during a given day. Assume that the number of breakdowns per day follows a Poisson distribution. What is the probability of observing three or more breakdowns?

30. The regional director of the Small Business Administration (SBA) in Pennsylvania is concerned about the number of SBA-sponsored businesses that end as failures. If the average number of failures per month is ten, what is the probability that exactly four SBA-sponsored businesses will fail during a given month? Assume that the number of SBA-sponsored businesses failing per month follows a Poisson distribution.

31. The arrivals of customers at a bank follow the Poisson distribution. Answer the following questions assuming a mean arrival rate of three per minute.
a. What is the probability of exactly three arrivals in a 1 minute period?
b. What is the probability of at least three arrivals in a 1 minute period?
32. During the registration period at a local university, students consult advisors with questions about course selection. A particular advisor noted that during the registration period an average of eight students per hour ask questions, although the exact arrival times of the students were random in nature. Use the Poisson distribution to answer the following questions:
a. What is the probability that exactly eight students come in for consultation during a particular 1 hour period?
b. What is the probability that three students come in for consultation during a particular ½ hour period?
33. A soup company markets eight varieties of homemade soup throughout the Eastern states. The standard-size soup can holds a maximum of 11 ounces, while the label on each can advertises contents of 10¾ ounces. The extra ¼ ounce is to allow for the possibility of the automatic filling machine placing more soup than the company actually wants in a can. Past experience shows that the number of ounces placed in a can is approximately normally distributed, with a mean of 10¾ and a standard deviation of 0.1 ounces. What is the probability that the machine will attempt to place more than 11 ounces in a can, causing an overflow to occur?
34. The sales of High-Brite Toothpaste are believed to be approximately normally distributed, with a mean of 10,000 tubes per week and a standard deviation of 1,500 tubes per week.
a. What is the probability that more than 12,000 tubes will be sold in any given week?
b. In order to have a .95 probability that the company will have sufficient stock to cover the weekly demand, how many tubes should be produced?
35. Points scored by the winning team in NCAA college football games are approximately normally distributed, with a mean of 24 and a standard deviation of 6.
a. What is the probability that a winning team in a football game scores between 20 and 30 points [that is, $P(20 \leq x \leq 30)$]?
b. How many points does a winning team have to score to be in the highest 20% of scores for college football games?
36. Ward Doering Auto Sales is considering offering a special service contract that will cover the total cost of any service work required on leased vehicles. From past experience the company manager estimates that yearly service costs are approximately normally distributed, with a mean of $150 and a standard deviation of $25.
a. If the company offers the service contract to customers for a yearly charge of $200, what is the probability that any one customer's service costs will exceed the contract price of $200?
b. What is Ward's expected profit per service contract?
37. The attendance at football games at a certain stadium is normally distributed, with a mean of 45,000 and a standard deviation of 3,000.
a. What percentage of the time should attendance be between 44,000 and 48,000?
b. What is the probability of exceeding 50,000?
c. Eighty percent of the time the attendance should be at least how many?
38. Assume that the test scores from a college admissions test are normally distributed, with a mean of 450 and a standard deviation of 100.
a. What percentage of the people taking the test score between 400 and 500?
b. Suppose that someone receives a score of 630. What percentage of the people taking the test score better? What percentage score worse?
c. If a particular university will not admit anyone scoring below 480, what percentage of the persons taking the test would be acceptable to the university?
39. The lifetime of a color television picture tube is normally distributed, with a mean of 7.8 years and a standard deviation of 2 years.
a. What is the probability that a picture tube will last more than 10 years?

b. If the firm guarantees the picture tube for 2 years, what percentage of the television sets sold will have to be replaced because of picture tube failure?

c. If the firm is willing to replace the picture tubes in a maximum of 1% of the television sets sold, what guarantee period can be offered for the television picture tubes?

40. A machine fills containers with a particular product. The standard deviation of filling weights is known from past data to be .6 ounces. If only 2% of the containers hold less than 18 ounces, what is the mean filling weight for the machine? That is, what must μ equal? Assume the filling weights have a normal distribution.

41. Forty-five percent of the residents in a township who are of voting age are not registered to vote.

a. On a street with ten people of voting age, what is the probability that five are not registered to vote?

b. On a street with ten people of voting age, what is the probability that two or less are not registered to vote?

c. In a neighborhood having 84 people of voting age, what is the probability that between 30 and 40 people are not registered to vote? That is, find $P(30 \leq x \leq 40)$.

42. A Myrtle Beach Resort Hotel has 120 rooms. In the spring months, hotel room occupancy is approximately 75%. Use the normal approximation to the binomial distribution to answer the following questions:

a. What is the probability that at least half the rooms are occupied on a given day?

b. What is the probability that 100 or more rooms are occupied on a given day?

c. What is the probability that 70 or fewer rooms are occupied on a given day?

43. It is known that 30% of all customers of a major national charge card pay their bills in full before any interest charges are incurred. Answer the following questions for a group of 150 credit card holders:

a. What is the probability that between 40 and 60 customers pay their account charges before any interest charges are incurred? That is, find $P(40 \leq x \leq 60)$.

b. What is the probability that 30 or fewer customers pay their account charges before any interest charges are incurred?

Application

Burroughs

The Burroughs Corporation*

Rochester, New York

The business of the Burroughs Corporation is information management. The company was founded on the invention of one of the first information processing devices—the Burroughs adding machine—and has grown to serve all the major aspects associated with the recording, computation, editing, processing and communication of information. Established as the American Arithmometer Company in St. Louis, Missouri in 1886, the Company moved its operations to Detroit, Michigan in 1904, and changed its name to Burroughs one year later. Through decades of marketing growth and technological developments, Burroughs has evolved as a multinational corporation with a full range of products and a reputation for innovation and reliability in information management. Burroughs employs nearly 67,000 people worldwide.

The Office Products Group (OPG) is one of the major operating groups of the corporation and is headquartered in Rochester, New York. The Office Products Group designs, engineers, manufactures, and markets a range of business forms, checks, office supplies, and document encoding, signing, and protective equipment, as well as accounting systems and credit cards.

The Quality Assurance Department, located in the Office Products Group, makes extensive use of the concepts involving probability and probability distributions in order to establish tolerances for new manufacturing processes and to determine probabilities that existing manufacturing processes exceed known tolerances. This type of analysis frequently involves the application of the normal probability distribution. One such application is described below.

CREDIT CARDS FOR BANKING

One of the products manufactured by OPG is a plastic credit card that is used in Burroughs automatic bank teller machines. Some of the banks that were using these machines were having problems with the credit cards being rejected in the equipment. An analysis of the rejected credit cards showed that these cards did not meet the product specifications for length and height as shown in Table 6A.1. A process study was started in an attempt to determine the cause of the credit card rejection problem.

*The authors are indebted to Frank C. Garcia of Burroughs Corporation, Rochester, New York, for providing this application.

TABLE 6A.1 Credit Card Specifications

	Minimum Dimension	*Maximum Dimension*
Length	3.365 in	3.375 in
Height	2.123 in	2.127 in

RESULTS OF PROCESS STUDY

In the manufacturing process, the plastic cards are cut to their final dimensions from larger plastic sheets using a die cutting machine. Four different dies can be used, but all dies are designed to produce cards that meet the same product specifications. Approximately 250 cards from each of the four dies were sampled during typical manufacturing runs. The height and length of each card were accurately measured. The results of this study are presented in Table 6A.2.

TABLE 6A.2 Height and Length Characteristics of Cards Produced by Four Dies

		Mean	*Standard Deviation*
Die 1	Length	3.367 in	.0010 in
	Height	2.123 in	.0018 in
Die 2	Length	3.368 in	.0014 in
	Height	2.125 in	.0021 in
Die 3	Length	3.367 in	.0010 in
	Height	2.128 in	.0032 in
Die 4	Length	3.366 in	.0007 in
	Height	2.124 in	.0015 in

The height and length of the cards produced by each die were assumed to follow a normal distribution. The probabilities that the height and length of the cards produced by these dies would not meet product specifications were calculated. For example, the probability of die 1 producing a card with a length less than the minimum acceptable length of 3.365 inches was calculated as follows:

1. Assume a normal distribution for length of cards produced by die 1 with a mean of 3.367 inches and a standard deviation of .0010 inches (see Table 6A.2).
2. The minimum acceptable length is 3.365 inches; the z value corresponding to 3.365 is

$$z = \frac{x - \mu}{\sigma} = \frac{3.365 - 3.367}{.001} = -2.$$

3. Using the table for the standard normal distribution the corresponding probability of a length as small as 3.365 inches or smaller is .5000 − .4772 = .0228, or approximately .023.

A view of OPG Headquarters located in Rochester, New York

Thus approximately 2.3% of the cards manufactured using die 1 will have a length less than the minimum length specification. The probability values for the four dies are summarized in Table 6A.3.

The probabilities that the cards would not meet the length specifications were relatively low for all four dies. However, probabilities that the cards would not meet the height specifications were unacceptably large for all four dies. For example, die 3 showed a .622 probability of exceeding the maximum height specification. Inspection of die 3 revealed that the cutting edges were dull; thus this die was removed from production for sharpening. Although the inspection of the other three dies indicated that they were in good physical condition, they still could not meet the required height specifications. Further investigation revealed that the entire die cutting operation had to be upgraded to meet the height dimension specifications. As a result, new process equipment was installed. Tests and probability calculations showed that the new equipment would be able to produce cards that met all dimensional specifications.

TABLE 6A.3 Probabilities of Not Meeting Product Specifications for the Four Dies

	Minimum Length	*Maximum Length*	*Minimum Height*	*Maximum Height*
Die 1	.023	*	.500	.013
Die 2	.016	*	.171	.171
Die 3	.023	*	.059	.622
Die 4	.076	*	.251	.023

*Denotes a probability value less than .01.

7 Sampling and Sampling Distributions

What you will learn in this chapter:

- the importance of sampling
- how to select a simple random sample
- how results from samples can be used to provide estimates of population parameters
- what a sampling distribution is
- what the central limit theorem is and the important role it plays in statistics
- other sampling techniques, such as stratified sampling, cluster sampling, systematic sampling, convenience sampling, and judgment sampling

Contents

As stated in Chapter 1, the purpose of statistical inference is to provide information about a *population* based upon a portion of the population called a *sample*. Let us begin our discussion of statistical inference by citing typical situations where sampling is conducted in order to provide a manager or decision maker with information about a population.

1. A tire manufacturer is considering producing a new tire which is designed to provide an increase in mileage over the firm's current line of tires. In order to determine whether or not to market the new tire, management needs an estimate of the mean or expected number of miles provided by the new tires. Since testing all new tires is impossible, a sample of 120 new tires is selected and tested. The mean tire life for the sample of tires is 36,500 miles. Thus 36,500 miles is used as an estimate of the mean tire life for the population of new tires.
2. Members of a political party are considering supporting a particular candidate for election to the United States Senate. In order to decide whether or not to enter the candidate in the upcoming primary election, party leaders need an estimate of the proportion of registered voters that favor the candidate. The time and cost associated with contacting every individual in the population of registered voters are prohibitive. Thus a sample of 400 registered voters is selected. If 160 of the 400 voters indicate a preference for the candidate, a reasonable estimate of the proportion of the population of voters that favor the candidate is $160/400 = .40$.
3. A manufacturing firm must decide which of two proposed assembly methods should be used for a particular new product. Management wants to know if there is a difference between the average or mean number of parts that can be produced per day using the two methods. During a testing period, sample production runs of 20 days are used to collect data on the daily output of each of the two assembly methods. The sample results show an average production of 550 units per day for method A and an average production of 500 units per day for method B. Thus an estimate of the difference in the mean daily outputs for the two methods is $550 - 500 = 50$ units per day, with method A providing the higher output.

From the preceding examples we can see how sampling and the subsequent sample results can be used to develop estimates of population characteristics. Note that in the tire mileage example, collecting the data on tire life requires the wearout or destructive testing of the tires. Clearly it is not feasible to test every tire in the population; a sample is the only realistic way to obtain the desired tire mileage information. In the example involving the primary election, it is theoretically possible to contact every registered voter in the population; however, the time and cost involved in doing so would be prohibitive. Thus a sample of registered voters is preferred.

The above examples point out some of the reasons for using samples. However, it is important to realize that sample results only provide *estimates* of the values of the population characteristics. That is, in the previous examples we do not expect the sample mean of 36,500 miles to be the *exact* mean mileage for all tires in the population; neither do we expect *exactly* 40% of the voter population to favor the candidate, or exactly 50 units per day to be the difference in the two assembly methods. We cannot expect this simply because the sample only contains a portion of the population. However, we are inclined to believe that with proper sampling methods the sample results will provide "good" estimates of the population characteristics. But how "good" can we expect the sample results to be? Fortunately, statistical procedures are available for answering this question.

In this chapter we introduce the simple random sampling method of selecting samples. We also will show how data obtained from a simple random sample can be used to compute estimates of population characteristics such as means, proportions, variances, etc. In addition, we will introduce the important concept of a sampling distribution. As we shall show, knowledge of the appropriate sampling distribution is what enables us to make statements about the "goodness" of the sample results.

7.1 THE ELECTRONICS ASSOCIATES SAMPLING PROBLEM

Electronics Associates, Inc. (EAI) is an international company that manufactures a diverse line of products in plants located throughout the United States, Canada, and Europe. The firm's director of personnel has been assigned the task of developing a group profile for the company's 2,500 managers. This group includes department heads, plant superintendents, and division managers. The group characteristics that are to be identified include the mean annual salary and the proportion of managers that have completed the company's management training program.

With the 2,500 managers considered the population for this study, we can find the annual salary and the training program status for each element or individual in the population by referring to the firm's personnel records. Let us assume that this has been done and that we have obtained a list of all 2,500 managers along with the corresponding annual salary and management training program participation information.

Using the formulas for a population mean and a population standard deviation that were presented in Chapter 3, we can compute the mean and standard deviation of annual salary for the population of 2,500 EAI managers. Assume that these calculations have been performed with the following results:

$$\text{Population mean } \mu = \frac{\Sigma x_i}{2{,}500} = \$31{,}800,$$

$$\text{Population standard deviation } \sigma = \sqrt{\frac{\Sigma (x_i - \mu)^2}{2{,}500}} = \$4{,}000.$$

Furthermore, assume that a review of the 2,500 records shows that 1,500 managers have completed the training program. Letting p denote the proportion of the population that has completed the training program, we see that $p = 1500/2500 = .60$.

In common statistical terminology, the characteristics of the population such as μ, σ, and p are referred to as *parameters* of the population. The question we would like to consider is how the firm's director of personnel could have obtained values or estimates of these population parameters by using a sample of managers rather than all 2,500 in the population.

Clearly the time and the cost required to develop a profile for 30 managers would be substantially less than those for the entire population. If the personnel director could be assured that a sample of 30 managers would provide adequate information about the population of 2,500 managers, working with a sample would be preferred to working with the entire population. Let us explore the possibility of using a sample for the EAI study by first considering how we could identify a sample of 30 managers.

7.2 SIMPLE RANDOM SAMPLING

The prime objective of sampling is to select a sample that is *representative* of the population. There are several types of samples that may be selected from a population; one of the most common of these is a *simple random sample*. The definition of a simple random sample and the process of selecting such a sample depends upon whether the population involved is *finite* or *infinite* in size. Since the EAI sampling problem introduced in the previous section involves a finite population of 2,500 managers, we consider simple random samples from finite populations first.

Sampling From Finite Populations

A simple random sample of size n from a finite population of size N is defined as follows:

Definition of a Simple Random Sample (Finite Population)

A simple random sample of size n from a finite population of size N is a sample selected such that each possible sample of size n has the same probability of being selected.

Let us demonstrate the concept of a simple random sample by considering a small-scale problem where we wish to select a simple random sample of two items from a population of five items. The example involves a regional sales manager who has a sales force of five people selling mobile telephone units to both private and commercial customers. Assume that the sales manager decides to use the number of units sold by two members of the sales force as an indicator of how the five member sales force is progressing toward meeting its quarterly sales quota. The population of five salespersons consists of individuals named Adams, Baker, Collins, Davis, and Edwards. Let us use the first letter of each salesperson's name as an abbreviation, with A for Adams, B for Baker, and so on. Doing so provides the population of size $N = 5$ denoted by the letters A, B, C, D, and E.

With this population of five we note that there are *several* different ways of selecting a sample consisting of two members of the sales force. For example, we might select a sample consisting of A (Adams) and E (Edwards); however, B (Baker) and E (Edwards) also could form the sample, C (Collins) and D (Davis) could form the sample, and so on. Actually there are a total of ten different possible samples of size $n = 2$ that could be selected: AB, AC, AD, AE, BC, BD, BE, CD, CE, and DE. The sample that we ultimately select must be *one* of these ten possible samples.

Now recall the definition of a simple random sample from a finite population: A simple random sample is a sample selected so that each possible sample has the same probability of being selected. In our illustration a simple random sample of size 2 can be selected by making sure each of the ten different samples has an *equal probability* of being selected. Thus to select the sample, we could write each of the ten possible samples on a standard sized piece of paper. Mixing the ten pieces of paper thoroughly and then selecting one piece of paper would provide the sample of size 2. Since each possible sample had the same $1/10$ probability of being selected, the specific sample selected would be referred to as a *simple random sample*.

While listing all possible samples of size n, writing each possible sample on a piece of paper, and then selecting one piece of paper does indeed provide a method for identifying a simple random sample, this process becomes cumbersome and impractical as the population size increases. For the EAI problem, where a sample of size 30 is to be drawn from a population of 2,500 EAI managers, this process would be so time-consuming that we would find it easier to take a complete census rather than to try to identify all possible samples of 30 managers.* Thus we need a better way to identify a simple random sample from a finite population.

Fortunately there is a relatively easy and straightforward procedure for identifying a simple random sample from a finite population without listing all possible samples. This practical way of selecting a simple random sample enables us to choose, or select, the items for the sample *one at a time.* At each selection we make sure that each of the items remaining in the population has the *same probability* of being selected for the sample. Sampling n items in this fashion will satisfy the definition of a simple random sample from a finite population.

Let us demonstrate this method by referring to the EAI sampling problem. First, we will assume that the 2,500 EAI managers have been numbered sequentially (i.e., 1, 2, 3, ..., 2,499, 2,500) in the order that they appear in the EAI personnel file. We could then write each number from 1 to 2,500 on equal sized pieces of paper. The 2,500 pieces of paper could then be placed in a hat and mixed thoroughly. We would begin the process of identifying managers for the sample by reaching into the hat and selecting one piece of paper *randomly.* The number on the chosen piece of paper would correspond to one of the numbered managers in the file of 2,500 managers, who is thus selected for the sample. The remaining 2,499 pieces of paper are thoroughly mixed again, after which another piece of paper is selected. This second number corresponds to another EAI manager to be included in the sample. The process continues until 30 managers have been selected from the population. The 30 managers identified in this manner form a simple random sample from the population.

In this procedure note that we took care not to place a selected (sampled) piece of paper back into the hat after it was drawn. Thus we are selecting a simple random sample *without replacement.* Certainly we could have followed the sampling procedure of *replacing* each sampled item before selecting subsequent items. This form of sampling, referred to as sampling *with replacement,* would have made it possible for some items to appear in the sample more than once. While sampling with replacement is a valid way of identifying a simple random sample, sampling *without replacement* is the sampling procedure currently used most often. Whenever we refer to simple random sampling from a finite population, we will make the assumption that the sampling is done without replacement.

This procedure for selecting a simple random sample of 30 EAI managers requires the labeling of 2,500 pieces of paper. In practice, tables of random numbers can be used to provide the same results much more easily. Tables of random numbers are available from a variety of handbooks† that contain page after page of random numbers. We have included one such page of random numbers in Table 8 of Appendix B. A portion of this page of random numbers is also shown in Table 7.1. The first line of

*According to the area of mathematics referred to as combinatorial analysis, the number of samples of size n that can be selected from a population of size N when sampling without replacement is given by the number of *combinations* of N items taken n at a time. For the EAI problem with $N = 2{,}500$ and $n = 30$, combinatorial analysis tells us there are in the neighborhood of 2.75×10^{69} possible samples!

†For example, The Rand Corporation, *A Million Random Digits with 100,000 Normal Deviates.* New York: The Free Press, 1955.

TABLE 7.1 Random Numbers

63271	59986	71744	51102	15141	80714	58683	93108	13554	79945
88547	09896	95436	79115	08303	01041	20030	63754	08459	28364
55957	57243	83865	09911	19761	66535	40102	26646	60147	15702
46276	87453	44790	67122	45573	84358	21625	16999	13385	22782
55363	07449	34835	15290	76616	67191	12777	21861	68689	03263
69393	92785	49902	58447	42048	30378	87618	26933	40640	16281
13186	29431	88190	04588	38733	81290	89541	70290	40113	08243
17726	28652	56836	78351	47327	18518	92222	55201	27340	10493
36520	64465	05550	30157	82242	29520	69753	72602	23756	54935
81628	36100	39254	56835	37636	02421	98063	89641	64953	99337
84649	48968	75215	75498	49539	74240	03466	49292	36401	45525
63291	11618	12613	75055	43915	26488	41116	64531	56827	30825
70502	53225	03655	05915	37140	57051	48393	91322	25653	06543
06426	24771	59935	49801	11082	66762	94477	02494	88215	27191
20711	55609	29430	70165	45406	78484	31639	52009	18873	96927
41990	70538	77191	25860	55204	73417	83920	69468	74972	38712
72452	36618	76298	26678	89334	33938	95567	29380	75906	91807
37042	40318	57099	10528	09925	89773	41335	96244	29002	46453
53766	52875	15987	46962	67342	77592	57651	95508	80033	69828
90585	58955	53122	16025	84299	53310	67380	84249	25348	04332
32001	96293	37203	64516	51530	37069	40261	61374	05815	06714
62606	64324	46354	72157	67248	20135	49804	09226	64419	29457
10078	28073	85389	50324	14500	15562	64165	06125	71353	77669
91561	46145	24177	15294	10061	98124	75732	00815	83452	97355
13091	98112	53959	79607	52244	63303	10413	63839	74762	50289

this table of random numbers begins as follows:

63271 59986 71744 51102 15141 80714

The digit appearing in any one of the above positions is a random selection of the digits 0, 1 . . . ., 9 with each digit having an equal chance of occurring. The grouping of the numbers into sets of five is simply for the convenience of making the table easy to read.

Let us see how the numbers in this random number table can be used to select a simple random sample of 30 EAI managers. As we did using the pieces of paper, we want to select numbers from 1 to 2,500 such that every number has an equal chance of being selected. Since the largest number in our population, 2,500, has four digits, we will select random numbers from the table in sets or groups of four digits. While we could select four-digit numbers from any portion of the random number table, suppose we start by using the first row of random numbers appearing in Table 7.1. The four-digit grouping of the random numbers provides

6327 1599 8671 7445 1102 1514 1807

Since the numbers in the table are random, the above four-digit numbers are all equally probable or equally likely.

We can now use the equally likely random numbers to give each element in the population an equal chance of being included in the sample. The first number, 6,327, is greater than 2,500. It does not correspond to an element in the population, and thus it is discarded. The second number, 1,599, is between 1 and 2,500. Thus the first individual selected for the sample is number 1,599 on the list of EAI managers. Continuing the process, we ignore 8,671 and 7,445 before identifying individuals 1,102, 1,514, and 1,807 as the next managers to be included in the sample. This process of selecting managers continues until the desired simple random sample of size 30 has been obtained. We note that with this random number procedure for simple random sampling, a random number previously used to identify an item for the sample may reappear in the random number table. Since we want to select the simple random sample *without replacement,* previously used random numbers are ignored because the corresponding element is already included in the sample.

As a final comment, note that random numbers can be selected from anywhere in the random number table. We chose to use the first row of the table in the above example. However, we could have started at any other point in the table and continued in any direction. Once the arbitrary starting point is selected, it is recommended that a predetermined systematic procedure, such as reading across rows or down columns, be used to determine the subsequent random numbers.

Sampling from Infinite Populations

To this point we have restricted our attention to selecting a simple random sample from a finite population. Most sampling situations in business and economics involve finite populations, but there are situations in which the population is either infinite or so large that for practical purposes it must be treated as infinite. In sampling from an infinite population we must give a new definition for a simple random sample. Since the items cannot be numbered, we must use a different process for selecting items for the sample.

Let us consider a situation which can be viewed as requiring a simple random sample from an infinite population. Suppose we want to estimate the average time between placing an order and receiving food for customers arriving at a fast food restaurant during the 11:30 A.M. to 1:30 P.M. lunch period. If we consider the population as being all possible customer visits, we see that it would be next to impossible to specify a finite limit on the number of possible visits. In fact, if we view the population as being all customer visits that could *conceivably* occur during the lunch period, we can consider the population as being infinite. Our task is now to select a simple random sample of n customers from this population. With this situation in mind we now state the definition of a simple random sample from an infinite population:

Definition of a Simple Random Sample (Infinite Population)

A simple random sample from an infinite population is a sample selected such that the following conditions are satisfied:

1. Each item selected comes from the same population.
2. Each item is selected independently.

For our problem of selecting a simple random sample of customer visits at a fast food restaurant, we find that the first condition defined above is satisfied by any customer visit occurring during the 11:30 A.M. to 1:30 P.M. lunch period while the restaurant is operating with its regular staff under "normal" operating conditions. The second condition is satisfied by ensuring that the selection of a particular customer does not influence the selection of any other customer.

A well known fast food restaurant has implemented a simple random sampling procedure for just such a situation. The sampling procedure is based on the fact that some customers will present discount coupons which provide special prices on sandwiches, drinks, french fries, and so on. Whenever a customer presents a discount coupon, the *next* person is selected for the sample. Since the customers present discount coupons in a random and independent fashion, the firm is satisfied that the sampling plan satisfies the two conditions for a simple random sample from an infinite population.

As a final note, let us reconsider the finite population case when simple random sampling with replacement is conducted. With this sampling procedure each item is returned to the population after being sampled; thus the same item may be selected several times for the sample. Samples selected in this manner have the following two properties:

1. Each item is selected from the same population.
2. At each trial each item has an equal probability of being selected, so that the items selected are independent.

Since these are the conditions for a simple random sample from an infinite population, we see that simple random sampling *with replacement from a finite population* is equivalent to simple random sampling from an *infinite population.*

EXERCISES

1. Assume that the simple random sample for the EAI study had been based on the seventh column of the five-digit random numbers shown in Table 7.1. Ignoring the first digit in the column and moving down the column, identify the first five EAI manager numbers that will be selected for the simple random sample. Note that this procedure begins with the four-digit random number 8683.

2. A student government organization is interested in estimating the proportion of students who favor a mandatory "pass-fail" grading policy for elective courses. A list of names and addresses of the 645 students during the current quarter is available from the registrar's office. Using row 10 of Table 7.1 and moving across the row from left to right, identify the first ten students who would be selected by simple random sampling. When every digit in row 10 is used the three-digit random numbers begin with 816, 283, and 610.

3. In this section we described a population of five salespersons selling mobile telephone units to private and commercial customers. The individuals in the population were identified by the letters A, B, C, D, and E. The ten possible samples of size two that could be selected were AB, AC, AD, AE, BC, BD, BE, CD, CE, and DE. Using the 15th row of random numbers in Table 7.1, use the random number procedure to select a simple random sample of size 2 from the population of size 5.

a. What sample is selected?

b. What sample would have been selected if the random numbers in row 20 had been used?

4. Schuster's Interior Design, Inc. specializes in a variety of home decorating services for its clients. During the previous year the firm provided major decorating consultation for 875 homes. Schuster's management was interested in obtaining information about customer satisfaction 6 to 12 months after the project was complete. To obtain this information, the firm decided to sample

30 of the 875 clients and interview the group to learn about client satisfaction and ways that Schuster might improve its service. Begin in column 10 of Table 7.1 with the three-digit random number 945. Moving down the column, identify the first ten clients that would be included in the sample. Assume that the 875 clients are numbered sequentially in the order in which the decorating projects were conducted.

5. Haskell Public Opinion Poll, Inc. conducts telephone surveys concerning a variety of political and general public interest issues. The households included in the survey are identified by taking a simple random sample from telephone directories in selected metropolitan areas. The telephone directory for a major Midwest area contains 853 pages with 400 lines per page.

a. Describe a two-stage random selection procedure that could be used to identify a simple random sample of 200 households. The selection process should involve first selecting a page at random and then selecting a line on the sampled page. Use the random numbers in Table 7.1 to illustrate this process. Select your own arbitrary starting point in the table.

b. What would you do if the line selected in part a was clearly inappropriate for the study (that is, the line provided the phone number of a business, restaurant, etc.)?

6. Indicate whether the populations listed below should be considered finite or infinite:

a. All the registered voters in the state of California.

b. All the television sets that could be produced by the Allentown, Pennsylvania plant of the TV-M Company.

c. All orders that could be processed by a mail-order firm.

d. All emergency telephone calls that could come into a local police station.

e. All of the items that were produced on the second shift on May 17th.

7. Read the Kings Island consumer profile sample survey application at the end of Chapter 1.

a. Assume that the Kings Island research group treats the population of consumer visits as an infinite population. Is this acceptable? Explain.

b. Assume that immediately after completing an interview with a consumer, the interviewer returns to the entrance gate and begins counting individuals as they enter the park. The 25th individual counted is selected as the next person to be sampled for the survey. After completing this interview, the interviewer returns to the entrance and again selects the 25th individual entering the park. Does this sampling process appear to provide a simple random sample? Explain.

7.3 POINT ESTIMATION

Now that we know how to select a simple random sample, let us proceed with the Electronics Associates problem. We will assume that a simple random sample of 30 managers has been selected and that the corresponding data on annual salary and management training program participation are as shown in Table 7.2. The notation x_1, x_2, etc. is used to denote the annual salary of the first manager in the sample, the second manager in the sample, and so on. Participation in the management training program is indicated by a "yes" in the management training program column.

In order to estimate the population mean and the population standard deviation for the annual salary of EAI managers, we simply calculate the corresponding sample statistics: the sample mean and sample standard deviation. Using the formulas for a sample mean and a sample standard deviation as presented in Chapter 3 and the data in Table 7.2, we have for the sample mean

$$\bar{x} = \frac{\Sigma x_i}{n} = \frac{954{,}420}{30} = \$31{,}814.00$$

TABLE 7.2 Annual Salary and Training Program Status for a Simple Random Sample of 30 Managers

Annual Salary (dollars)	*Management Training Program?*
x_1 = 29,094.30	Yes
x_2 = 33,263.90	Yes
x_3 = 29,643.50	Yes
x_4 = 29,894.90	Yes
x_5 = 27,621.60	No
x_6 = 35,924.00	Yes
x_7 = 29,092.30	Yes
x_8 = 31,404.40	Yes
x_9 = 30,957.70	Yes
x_{10} = 35,109.70	Yes
x_{11} = 25,922.60	Yes
x_{12} = 37,268.40	No
x_{13} = 35,688.80	Yes
x_{14} = 31,564.70	No
x_{15} = 36,188.20	No
x_{16} = 31,766.00	Yes
x_{17} = 32,541.30	No
x_{18} = 24,980.00	Yes
x_{19} = 31,932.60	Yes
x_{20} = 32,973.00	Yes
x_{21} = 25,120.90	Yes
x_{22} = 31,753.00	Yes
x_{23} = 34,391.80	No
x_{24} = 30,164.20	No
x_{25} = 32,973.60	No
x_{26} = 30,241.30	No
x_{27} = 32,793.90	No
x_{28} = 30,979.40	Yes
x_{29} = 35,860.90	Yes
x_{30} = 37,309.10	No

and for the sample standard deviation

$$s = \sqrt{\frac{\Sigma(x_i - \bar{x})^2}{n - 1}} = \sqrt{\frac{325{,}009{,}260}{29}} = \$3{,}347.72.$$

In addition, by computing the proportion of managers in the sample who have responded "yes" we can estimate the proportion of managers in the population who have completed the management training program. Table 7.2 shows that 19 of the 30 managers in the sample have completed the training program. Thus the sample proportion, denoted $\bar{p}$, is given by

$$\bar{p} = \frac{19}{30} = .63.$$

This value is used as the estimate of the population proportion.

We have just completed the statistical procedure called *point estimation.* In point estimation we use the data from the sample to compute a value of a sample statistic that serves as an estimate of a population parameter. Using the terminology of point estimation, we would refer to $\bar{x}$ as the *point estimator* of the population mean μ, s as the *point estimator* of the population standard deviation σ, and $\bar{p}$ as the *point estimator* of the population proportion p. The actual numerical value obtained for $\bar{x}$, s, or $\bar{p}$ in a particular sample is called the *point estimate* of the parameter. Thus according to the EAI sample of 30 managers $31,814.00 is the point estimate of μ, $3,347.72 is the point estimate of σ, and .63 is the point estimate of p. Table 7.3 provides a summary of the sample results and compares the point estimates to the actual values of the population parameters.

TABLE 7.3 Summary of Point Estimates Obtained from a Simple Random Sample of 30 EAI Managers

Population Parameter	*Parameter Value*	*Point Estimator*	*Point Estimate*
μ = Population mean annual salary	$31,800.00	$\bar{x}$ = Sample mean annual salary	$31,814.00
σ = Population standard deviation for the annual salary	$4,000.00	s = Sample standard deviation for the annual salary	$3,347.72
p = Population proportion having completed the management training program	.60	$\bar{p}$ = Sample proportion having completed the management training program	.63

EXERCISES

8. A simple random sample of 5 months of sales data provides the following:

Month	1	2	3	4	5
Units Sold	94	100	85	94	92

a. What is a point estimate of the mean number of units sold per month?
b. What is a point estimate of the standard deviation for the population?

9. The California Highway Patrol maintains records showing the time between an accident report being received and an officer arriving at the accident scene. A simple random sample of ten records shows the following times in minutes:

12.6, 3.4, 4.8, 5.0, 6.8, 2.3, 3.6, 8.1, 2.5, 10.3.

a. What is a point estimate of the mean time between accident report and officer arrival?
b. What is a point estimate of the standard deviation of time between accident report and officer arrival?

10. Nine nurses selected at random from a large hospital were asked if they believed that the hospital nursing department was understaffed. The responses are shown below:

Nurse	1	2	3	4	5	6	7	8	9
Response	No	Yes	Yes	No	No	Yes	No	No	No

Develop a point estimate of the proportion of all hospital nurses that believe that an understaffing of nurses exists.

11. Develop a point estimate of the proportion of the pages in this text that contain a "figure" or a "table." That is, if a randomly selected page has at least one figure or at least one table, the response is yes; otherwise the response is no. Use the random numbers in Table 7.1 to select a simple random sample of 20 pages. Compare your point estimate with those of others in the class.

7.4 INTRODUCTION TO SAMPLING DISTRIBUTIONS

In the previous section we saw how a simple random sample of 30 managers could serve as the basis for developing point estimates of the mean and standard deviation of annual salary of all EAI managers as well as the proportion of the managers that have completed the company's management training program. In this section we consider the point estimates that would be observed if additional simple random samples, each of size 30, were selected. The resulting analysis will introduce the important concept of a sampling distribution.

Suppose we were to select another simple random sample of 30 EAI managers. Let us assume that this has been done and that an analysis of the sample data provides the following:

$$\text{Sample Mean } \bar{x} = \$32{,}669.70,$$
$$\text{Sample Standard Deviation } s = \$4{,}239.07,$$
$$\text{Sample Proportion } \bar{p} = .70.$$

These results show that different values of $\bar{x}$, s, and $\bar{p}$ have been obtained with the second sample. In general, this is to be expected because most likely this second simple random sample will not contain the exact same 30 managers that were in the first sample. Let us imagine carrying out the same process of selecting a new simple random sample of 30 managers over and over again, each time computing values of $\bar{x}$, s, and $\bar{p}$. We could begin thus to identify the variety of values that these point estimators can take on. To illustrate this, we repeated the simple random sampling process for the EAI managers until we obtained 500 samples of 30 managers each and their corresponding $\bar{x}$, s, and $\bar{p}$ values. A portion of the results is shown in Table 7.4. Table 7.5 shows the frequency distribution for the 500 $\bar{x}$ values. Figure 7.1 shows the relative frequency histogram for the $\bar{x}$ results.

TABLE 7.4 Values of $\bar{x}$, s, and $\bar{p}$ from 500 Simple Random Samples of 30 Managers Each

Sample Number	*Sample Mean* $\bar{x}$	*Sample Standard Deviation* s	*Sample Proportion* $\bar{p}$
1	\$31,814.00	\$3,347.72	.63
2	\$32,669.70	\$4,239.07	.70
3	\$31,780.30	\$4,433.43	.67
4	\$31,587.90	\$3,985.32	.53
.	.	.	.
.	.	.	.
.	.	.	.
500	\$31,752.00	\$3,857.82	.50

TABLE 7.5 Frequency Distribution of $\bar{x}$ From 500 Simple Random Samples of 30 EAI Managers Each

Annual Salary	*Frequency*	*Relative Frequency*
$29,500 but less than $30,000	2	.004
$30,000 but less than $30,500	16	.032
$30,500 but less than $31,000	52	.104
$31,000 but less than $31,500	101	.202
$31,500 but less than $32,000	133	.266
$32,000 but less than $32,500	110	.220
$32,500 but less than $33,000	54	.108
$33,000 but less than $33,500	26	.052
$33,500 but less than $34,000	6	.012
Total	500	1.000

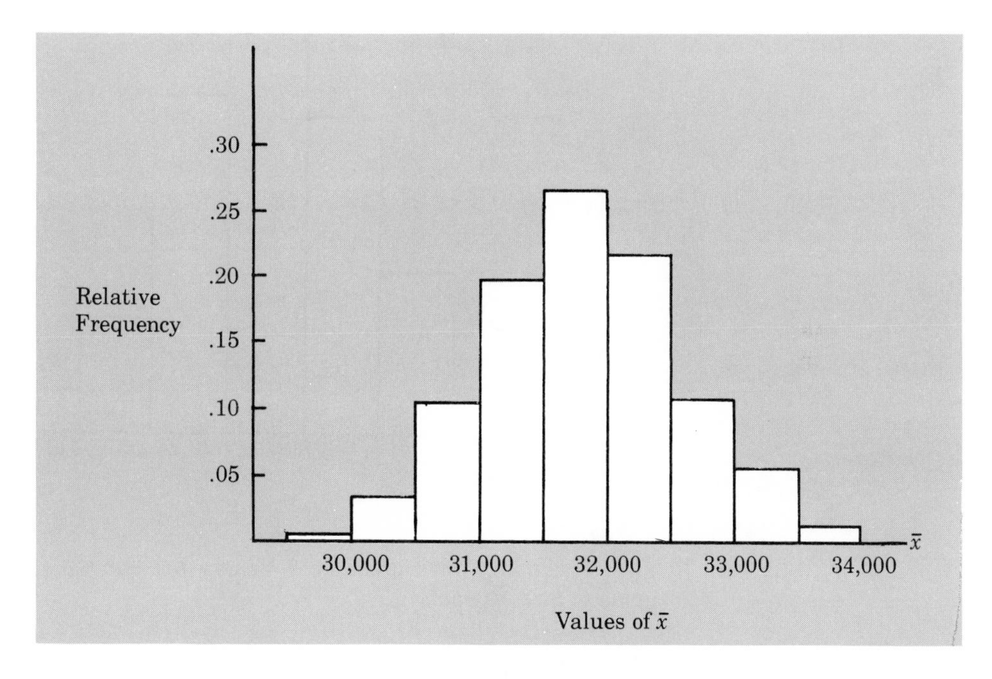

FIGURE 7.1 Relative Frequency Histogram of $\bar{x}$ Values for 500 Simple Random Samples of Size 30 Each

Recall that in Chapter 4 we defined a random variable as a numerical description of the outcome of an experiment. If we consider the simple random sampling process as an experiment, the sample mean $\bar{x}$ is the numerical description of the experimental outcome. Thus the sample mean $\bar{x}$ is a random variable. As a result $\bar{x}$, just like other random variables, has a mean or expected value, a variance, and a probability distribution. Since the various possible values of $\bar{x}$ are the result of different simple random *samples*, the probability distribution of $\bar{x}$ is called the *sampling distribution of*

$\bar{x}$. Knowledge of this sampling distribution and its properties will enable us to make probability statements about how close the sample mean $\bar{x}$ is to the population mean μ.

Let us return to Figure 7.1. We would need to enumerate every possible sample of 30 managers and compute each sample mean in order to determine completely the sampling distribution of $\bar{x}$. However, the histogram of 500 $\bar{x}$ values gives an approximation of this sampling distribution. From this approximation we observe the bell-shaped appearance of the distribution. We also note the fact that the mean of the 500 $\bar{x}$ values is near the population mean $\mu = \$31,800$. We will describe the properties of the sampling distribution of $\bar{x}$ more fully in the next section.

The 500 values of the sample standard deviation s and the 500 values of the sample proportion $\bar{p}$ are summarized by the relative frequency histograms in Figures 7.2 and 7.3. As in the case of $\bar{x}$, both s and $\bar{p}$ are random variables that provide

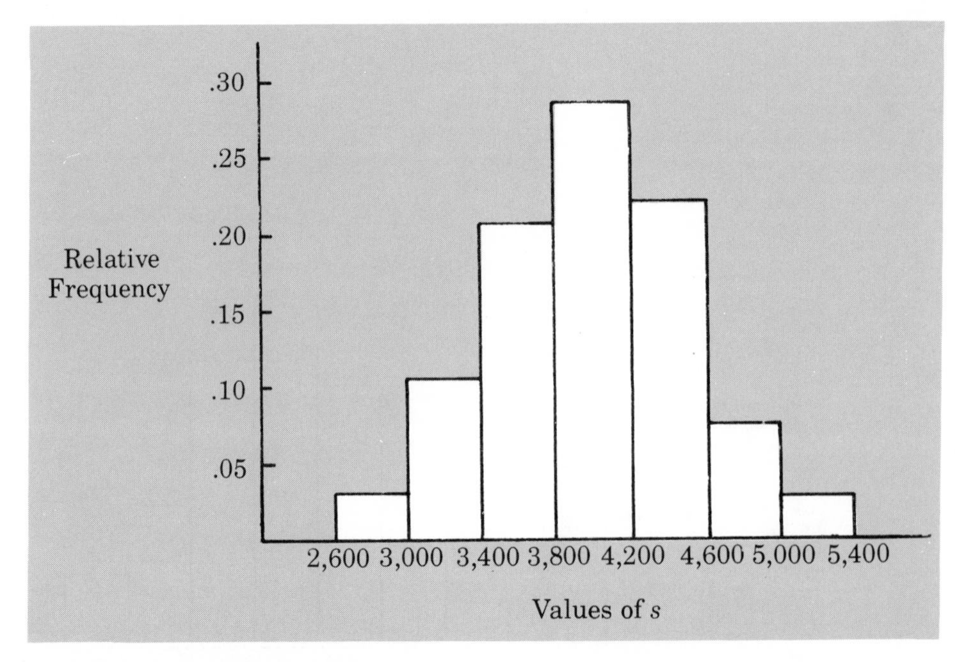

FIGURE 7.2 Relative Frequency Histogram of s Values for 500 Simple Random Samples of Size 30 Each

numerical descriptions of the outcome of a simple random sampling process. If every possible sample of size 30 were selected from the population and if a value of s and a value of $\bar{p}$ were computed for each sample, the resulting probability distributions would be called the sampling distribution of s and the sampling distribution of $\bar{p}$, respectively. The histograms of the 500 sample values shown in Figures 7.2 and 7.3 provide a general idea of the appearance of these two sampling distributions.

In closing this section let us note that in practice we will select only one simple random sample from the population. We would never actually carry out 500 repeated samples. The repeated sampling we did in this section was to illustrate that many different samples are possible and that the different samples generate a variety of values for the estimates of the population parameters. The probability distribution of

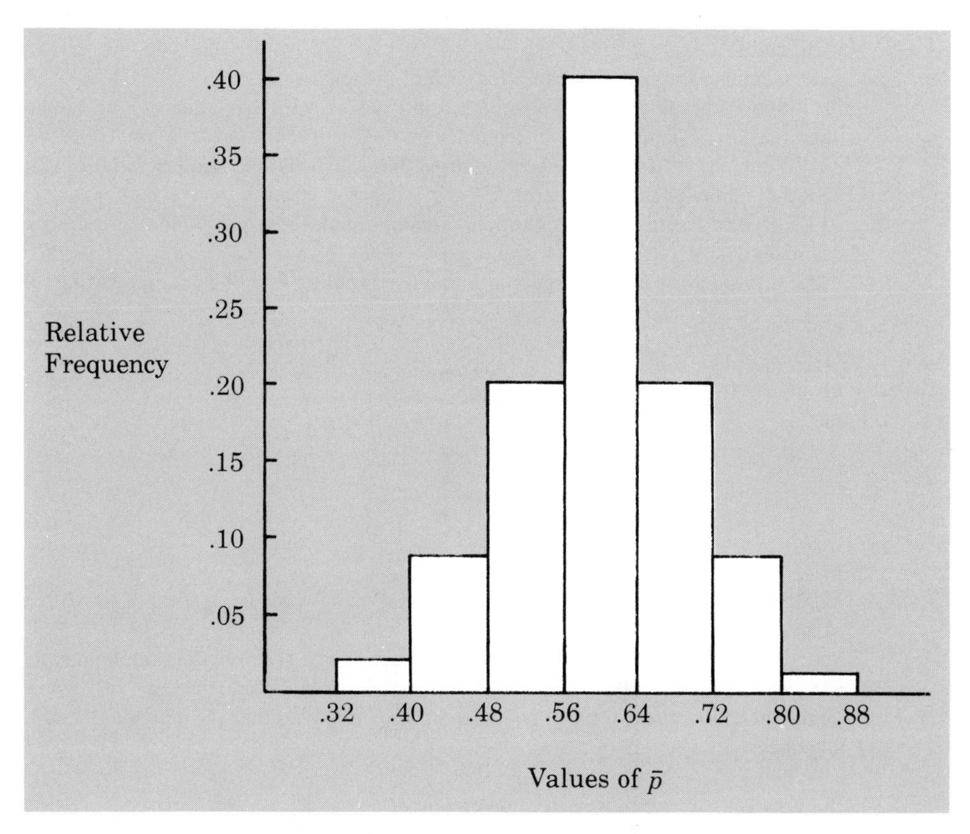

FIGURE 7.3 **Relative Frequency Histogram of $\bar{p}$ Values for 500 Simple Random Samples of Size 30 Each**

any particular sample statistic, such as $\bar{x}$, s, or $\bar{p}$, is called the sampling distribution of the statistic. In the next section we show that the characteristics of the sampling distribution of $\bar{x}$ are known even before we sample from the population. In Section 7.6 we will show the characteristics of the sampling distribution of $\bar{p}$. We will defer further discussion of the sample standard deviation s until we consider sampling distributions pertaining to sample variances, covered in Chapter 11.

EXERCISES

12. Return to the population of five salespersons selling mobile telephone units to private and commercial customers (Section 7.2). Assume that the number of units sold for each person is as follows:

Salesperson	*Units Sold*
Adams (A)	14
Baker (B)	20
Collins (C)	12
Davis (D)	8
Edwards (E)	16

There were a total of ten possible samples of size 2 given by AB, AC, AD, AE, BC, BD, BE, CD, CE, and DE.

a. Compute the sample mean $\bar{x}$ for *each* of the ten samples.

b. Show a graph of the ten possible sample means, with the values of $\bar{x}$ on the horizontal axis and the corresponding relative frequency on the vertical axis.

c. Your answer to part b shows a probability distribution of all possible $\bar{x}$ values. What is another name for this distribution?

13. Repeat Exercise 12 using simple random samples of size 3. Note that there are ten possible samples of size 3–ABC, ABD, and so on.

14. Consider the following population of five families, with the data indicating family size:

Family	*Family Size*
1	2
2	4
3	3
4	5
5	3

a. Simple random sampling with a sample size of 3 provides a total of ten different samples. List the ten possible samples.

b. Compute the mean, $\bar{x}$, for each sample and show a graph of the sampling distribution of $\bar{x}$.

15. Consider the following population of 25-year-old males, where a "yes" indicates that the individual has a life insurance policy and a "no" indicates that the individual does not have such a policy:

Individual	*Response*
1	Yes
2	No
3	No
4	Yes
5	No
6	Yes

a. Selecting simple random samples of size 4 provides a total of 15 possible samples. List the 15 samples.

b. Compute the proportion of "yes" responses for each sample, and show a histogram of the sampling distribution of $\bar{p}$.

7.5 SAMPLING DISTRIBUTION OF $\bar{x}$

In the previous section we saw that the sample mean, $\bar{x}$, is a random variable. The probability distribution of this random variable is referred to as the sampling distribution of $\bar{x}$. The purpose of this section is to describe the properties of the sampling distribution of $\bar{x}$, including the expected value or mean of $\bar{x}$, the standard deviation of $\bar{x}$, and the shape or form of the sampling distribution itself. As we shall see in Chapter 8, knowledge of the sampling distribution of $\bar{x}$ will enable us to make probability statements about the error involved when a sample mean $\bar{x}$ is used as a point estimator of a population mean μ. Let us begin by considering the mean of all possible $\bar{x}$ values, or, simply, the expected value of $\bar{x}$.

Expected Value of $\bar{x}$

As we saw in the EAI sampling problem, different simple random samples result in a variety of values for the sample mean (for example, \$31,814.00, \$32,669.70, \$31,780.30, \$31,587.90, and so on). When we realize that many different values of the random variable $\bar{x}$ are possible, we are often interested in the mean of all possible values of $\bar{x}$ that can be generated by the various simple random samples. This value can be provided by realizing that the mean of the $\bar{x}$ random variable is simply the expected value of $\bar{x}$. As a result, we are able to use concepts from sampling theory to calculate the expected value of $\bar{x}$; we need not actually go through the process of computing all possible values of $\bar{x}$ and then computing their mean. Let $E(\bar{x})$ represent the expected value of $\bar{x}$, or simply the mean of all possible $\bar{x}$ values and μ equal the population mean. It can be shown that when using simple random sampling these two values are the same. That is,

$$E(\bar{x}) = \mu. \tag{7.1}$$

This result, which is derived in the Appendix to this chapter, shows that with simple random sampling the expected value or mean for $\bar{x}$ is equal to the mean of the population. Refer to the EAI study and recall that in Section 7.1 we saw that the mean annual salary for the population of EAI managers was $\mu = \$31,800$. Thus according to (7.1) the mean of all possible sample means for the EAI study also is \$31,800.

In cases where the expected value of the estimator is equal to the population parameter, the estimator is said to be an *unbiased* estimator of the parameter. Using this terminology, (7.1) shows that for simple random sampling the sample mean $\bar{x}$ is an unbiased estimator of the population mean, μ.

Standard Deviation of $\bar{x}$

As we have stated, various simple random samples can be expected to generate a variety of $\bar{x}$ values. Let us now explore what sampling theory tells us about the standard deviation of all possible $\bar{x}$ values. We will use the following notation:

$$\begin{aligned} \sigma_{\bar{x}} &= \text{the standard deviation of all possible } \bar{x} \text{ values,} \\ \sigma &= \text{the population standard deviation,} \\ n &= \text{the sample size,} \\ N &= \text{the population size.} \end{aligned}$$

It can be shown that with simple random sampling the standard deviation of $\bar{x}$ depends upon whether the population is finite or infinite. The two expressions for the standard deviation of $\bar{x}$ are as follows:*

*The expression for $\sigma_{\bar{x}}$ in the finite population case is based on sampling without replacement. Although sampling with replacement is rarely conducted, we stated in Section 7.2 that sampling with replacement from a finite population possesses the same properties as sampling from an infinite population. Thus if a population is finite and if sampling is done with replacement, the expression of $\sigma_{\bar{x}}$ for the infinite population case would be appropriate.

Standard Deviation of $\bar{x}$

Finite Population	*Infinite Population*	
$\sigma_{\bar{x}} = \sqrt{\frac{N-n}{N-1}}\frac{\sigma}{\sqrt{n}}$	$\sigma_{\bar{x}} = \frac{\sigma}{\sqrt{n}}$	(7.2)

A derivation of the formulas for $\sigma_{\bar{x}}$ is discussed in the Appendix to this chapter. In comparing the two expressions in (7.2) we see that the factor $\sqrt{(N-n)/(N-1)}$ is required for the finite population but not for the infinite population case. This factor is commonly referred to as the *finite population correction factor*. In many practical sampling situations we find that the population involved, although finite, is "large" while the sample size is relatively "small." In such cases the finite population correction factor $\sqrt{(N-n)/(N-1)}$ is close to 1. As a result the difference between the values of the standard deviation of $\bar{x}$ for the finite and infinite population cases becomes negligible. When this occurs, $\sigma_{\bar{x}} = \sigma/\sqrt{n}$ becomes a very good approximation to the standard deviation of $\bar{x}$ even though the population is finite. As a general guideline or rule of thumb for computing the standard deviation of $\bar{x}$, we state the following:

Use the following expression to calculate the standard deviation of $\bar{x}$

$$\sigma_{\bar{x}} = \frac{\sigma}{\sqrt{n}} \qquad (7.3)$$

whenever (1) the population is infinite
or (2) the population is finite *and* the sample size is less than or equal to 5% of the population size. That is, $n/N \leq .05$.

In cases where $n/N > .05$, the finite population version of (7.2) should be used in the computation of $\sigma_{\bar{x}}$.

Now let us return to the EAI study and determine the standard deviation of all possible sample means that can be generated with samples of 30 EAI managers. Recall that in Section 7.1 we identified the population standard deviation for the annual salary data to be $\sigma = 4{,}000$. In this case the population is finite, with $N = 2{,}500$. However, with a sample size of 30, we have $n/N = 30/2{,}500 = .012$. Following the rule of thumb given above, we can ignore the finite population correction factor and use $\sigma_{\bar{x}} = \sigma/\sqrt{n}$ to compute the standard deviation of $\bar{x}$:

$$\sigma_{\bar{x}} = \frac{\sigma}{\sqrt{n}} = \frac{4{,}000}{\sqrt{30}} = 730.30.$$

In this example the value of the finite population correction factor is $\sqrt{(N-n)/(N-1)} = \sqrt{(2{,}500-30)/(2{,}500-1)} = .994$. Since this is very close to 1, it has a negligible effect on the value of the $\sigma_{\bar{x}}$. Thus we follow the common practice of ignoring the finite population correction factor, and we conclude that $\sigma_{\bar{x}} = 730.30$. Unless specifically noted, throughout the text we will be assuming that the population size is "large" and that the finite population correction factor is unnecessary.

Later we will see that the value of $\sigma_{\bar{x}}$ is helpful in determining how far the sample mean may be from the population mean. Because of the role that $\sigma_{\bar{x}}$ plays in computing possible estimation errors, $\sigma_{\bar{x}}$ is referred to as the *standard error of the mean.*

Central Limit Theorem

The final step in identifying the characteristics of the sampling distribution of $\bar{x}$ is to determine the form of the probability distribution of $\bar{x}$. We will consider two cases: one where the population distribution is unknown, and one where the population distribution is known to be normal.

For the situation where the population distribution is unknown, we rely on one of the most important theorems in statistics—the *central limit theorem.* A statement of the central limit theorem as it applies to the sampling distribution of $\bar{x}$ is as follows:*

Central Limit Theorem

In selecting simple random samples of size n from a population with mean μ and standard deviation σ, the probability distribution of the sample mean $\bar{x}$ approaches a normal distribution with mean μ and standard deviation $\sigma/\sqrt{n}$ as the sample size becomes large.

Figure 7.4 shows how the central limit theorem works for three different populations; in each case the population clearly is not normal. However, note what begins to happen to the sampling distribution of $\bar{x}$ as the sample size is increased. When the samples are of size 2 we see that the sampling distribution of $\bar{x}$ begins to take on an appearance different than the population distribution. For samples of size 5 we see all three sampling distributions beginning to take on a bell shaped appearance. Finally, the samples of size 30 show all three sampling distributions to be approximately normal. General statistical practice is to assume that regardless of the population distribution, the sampling distribution of $\bar{x}$ can be approximated by a normal probability distribution whenever the sample size is 30 or more. In effect the sample size of 30 is the rule of thumb that allows us to assume that the large sample conditions of the central limit theorem have been satisfied. This observation about the sampling distribution of $\bar{x}$ is so important that we restate it:

The sampling distribution of $\bar{x}$ can be approximated by a normal probability distribution whenever the sample size is large. The large sample size condition can be assumed for simple random samples of size 30 or more.

The central limit theorem is the key to identifying the form of the sampling distribution of $\bar{x}$ whenever the population distribution is unknown. However, we may encounter some sampling situations where the population is assumed or believed to have a normal distribution. When this condition occurs, it is not necessary to use the

*The theoretical proof of the central limit theorem requires independent observations or items in the sample. This condition exists for infinite populations or for finite populations where sampling is done with replacement. Although the central limit theorem does not directly address sampling without replacement from finite populations, general statistical practice has been to apply the findings of the central limit theorem in this situation provided that the population size is large.

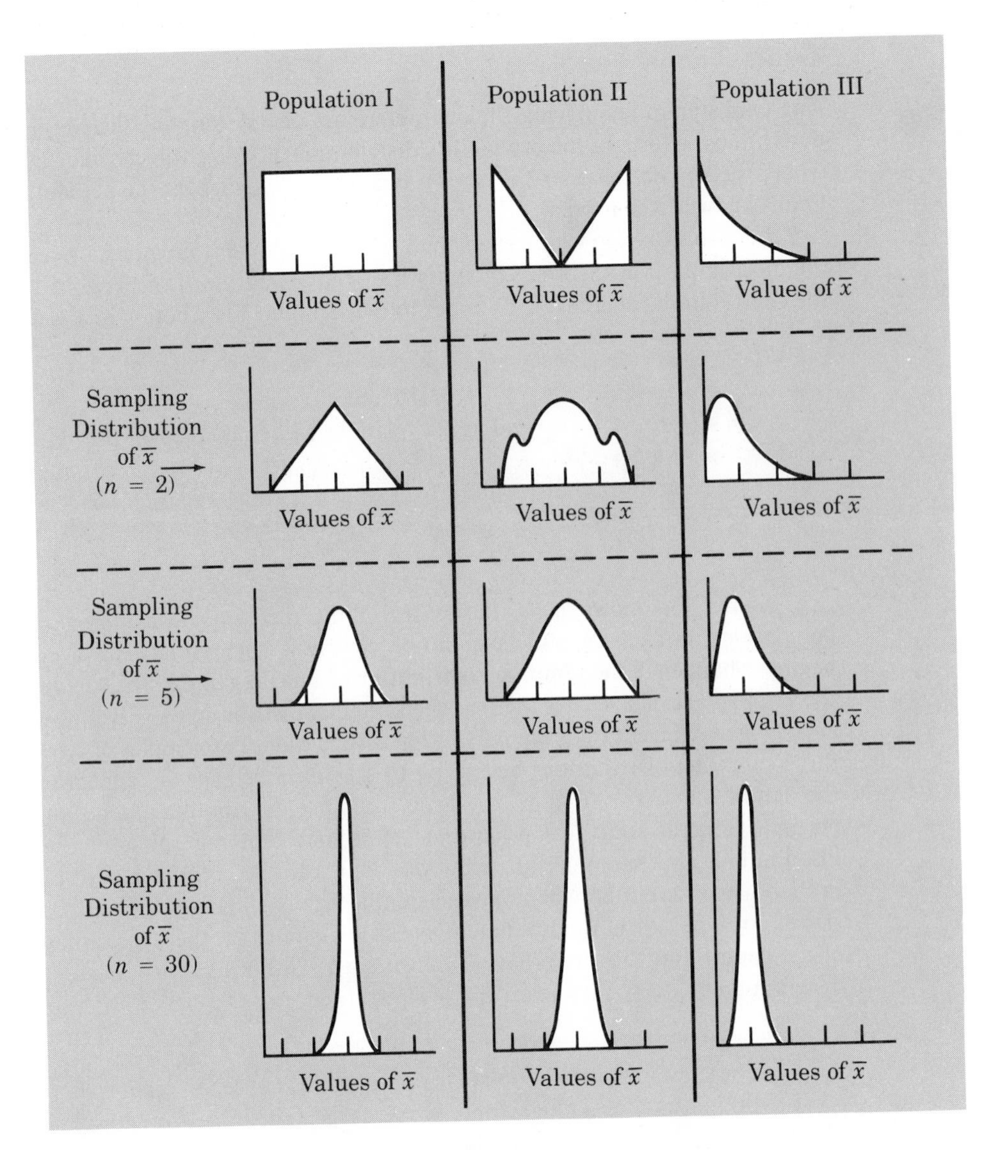

FIGURE 7.4 Illustration of the Central Limit Theorem for Three Populations

central limit theorem because the following result identifies the form of the sampling distribution of $\bar{x}$:

> Whenever the population has a normal probability distribution, the sampling distribution of $\bar{x}$ is normal for any sample size.

In summary, whenever we are using a large simple random sample (rule of thumb: $n \geq 30$), the central limit theorem enables us to conclude that the sampling distribution

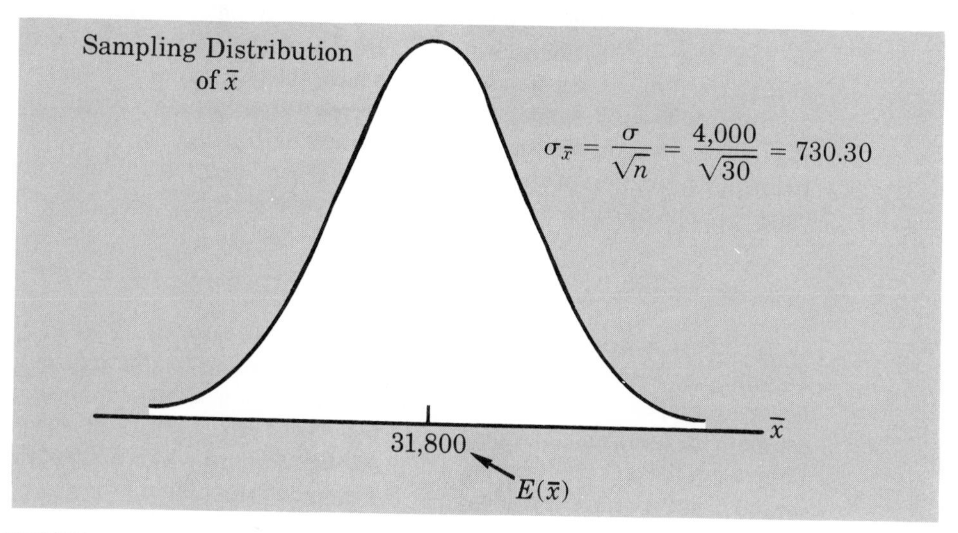

FIGURE 7.5 Sampling Distribution of $\bar{x}$ for the Mean Annual Salary of a Simple Random Sample of 30 Managers

of $\bar{x}$ can be approximated by a normal distribution. In cases where the simple random sample is small ($n < 30$), the sampling distribution of $\bar{x}$ can be considered normal only if we believe that the assumption of a normal distribution for the population is appropriate.

Sampling Distribution of $\bar{x}$ for the EAI Problem

Let us draw upon our knowledge of the sampling distribution of $\bar{x}$ to determine the properties of the sampling distribution of $\bar{x}$ for the EAI study. Recall from Section 7.1 that the population mean annual salary is $\mu = 31,800.00$ and that the population standard deviation is $\sigma = 4,000$. From the central limit theorem we know that with

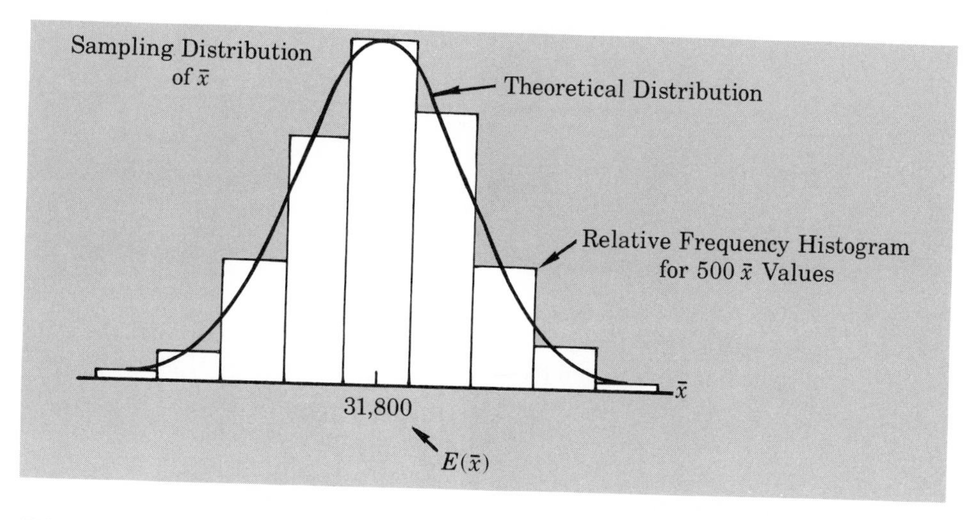

FIGURE 7.6 Comparison of the Theoretical Sampling Distribution of $\bar{x}$ and the Relative Frequency Histogram of $\bar{x}$ from 500 Simple Random Samples

simple random samples of 30 managers the distribution of all possible $\bar{x}$ values follows the sampling distribution shown in Figure 7.5. Although we do not have actual data available for all possible $\bar{x}$ values, we do have the 500 values of $\bar{x}$ that were obtained from the 500 simple random samples referred to in Section 7.4 (see Figure 7.1). In Figure 7.6 we compare the theoretical sampling distribution of $\bar{x}$ with the relative frequency histogram of the 500 $\bar{x}$ values that we have actually observed. Note how close the approximation is.

Practical Value of the Sampling Distribution of $\bar{x}$

Let us begin to demonstrate some of the practical reasons why we are interested in knowing about the sampling distribution of $\bar{x}$. Assume for the moment that we are going to sample 30 managers for the EAI study. When we use the sample mean to estimate the mean annual salary for the population of 2,500 EAI managers, what is the probability that the point estimate that we obtain will be within \$500 of the population mean? Since we have just identified the sampling distribution for $\bar{x}$, we will use this distribution to answer this question. Refer to the sampling distribution shown in Figure 7.7. We are asking about the probability that the sample mean is between \$31,300 and \$32,300. If the value of the sample mean $\bar{x}$ is in this interval, the point estimate will be within \$500 of the population mean. The probability is given by the shaded area of the sampling distribution shown in Figure 7.7. Since the sampling distribution is normal,

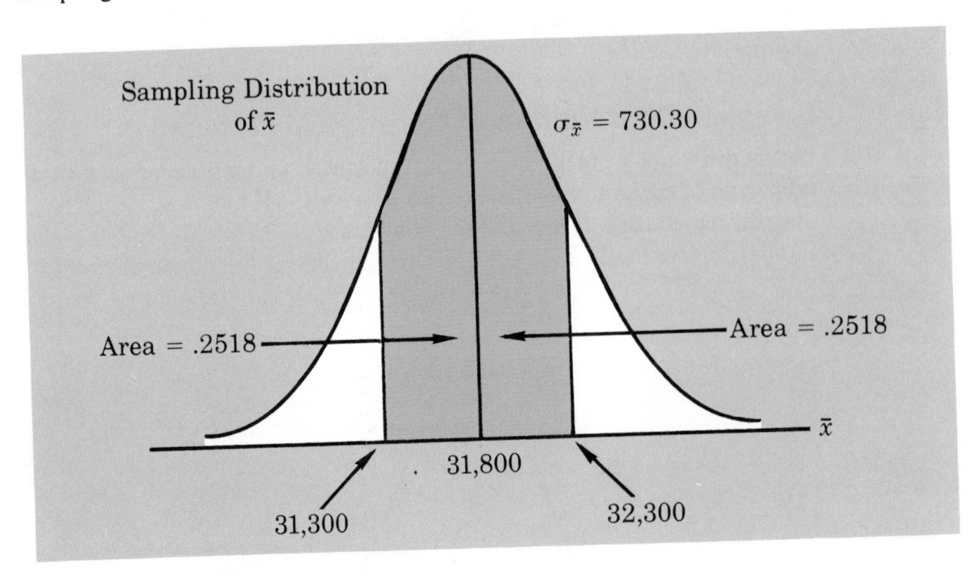

FIGURE 7.7 Sampling Distribution of $\bar{x}$ for the EAI Sampling Problem

with mean 31,800 and standard deviation 730.30, we can use the standard normal distribution table to find the area or probability. At $\bar{x} = 31{,}300$, we have

$$z = \frac{31{,}300 - 31{,}800}{730.30} = -.68.$$

Referring to Table 1 of Appendix B, we find an area between $z = 0$ and $z = -.68$ of .2518. Similar calculations for $\bar{x} = 32{,}300$ show an area between $z = 0$ and $z = +.68$ of

.2518. Thus the probability of the value of the sample mean being between 31,300 and 32,300 is .2518 + .2518 = .5036.

EXERCISES

16. The following data show the number of automobiles owned in a population of five families.

Family	*Number of Automobiles*
1	2
2	1
3	0
4	2
5	3

a. Compute the mean and standard deviation for the population using formulas for a population mean and population standard deviation as given in Chapter 3. See Eqs. (3.2) and (3.8).

b. List the ten possible samples of size 3 for this population.

c. Compute the sample mean for *each* sample and show a graph of the sampling distribution of $\bar{x}$.

d. Using the ten values of $\bar{x}$ found in part c, compute the mean and standard deviation of $\bar{x}$. Use Eqs. (3.2) and (3.8) with the ten values of $\bar{x}$ as the data.

e. Use the population mean and population standard deviation found in part a in Eqs. (7.1) and (7.2) in order to compute the mean and standard deviation of $\bar{x}$. Note that with $n/N = 3/5 = .60$, you must use the finite population version of (7.2).

f. Compare the results in part d with the results in part e. Which approach—listing and using all possible $\bar{x}$ values (part d) or the use of the equations for $E(\bar{x})$ and $\sigma_{\bar{x}}$ (part e)—provided the easier way to determine the mean and standard deviation of the $\bar{x}$ values?

17. Refer to the EAI sampling problem. Suppose that the simple random sample had contained 60 managers.

a. Sketch the sampling distribution of $\bar{x}$ when simple random samples of size 60 are used.

b. What happens to the sampling distribution of $\bar{x}$ if simple random samples of size 100 are used?

c. What general statement can you make about what happens to the sampling distribution of $\bar{x}$ as the sample size is increased? Does this seem logical? Explain.

18. In the EAI sampling problem, we showed that with a sample size of $n = 30$ there was .5036 probability of selecting a simple random sample having a sample mean within plus or minus $500 of the population mean.

a. What is the probability of an $\bar{x}$ value being within $500 of the population mean if a sample of size 60 is used?

b. Answer part a for a sample of size 100.

19. What important role does the central limit theorem serve whenever a sample mean $\bar{x}$ is used as a point estimator of a population mean μ?

20. During a complete review of 2 months of billings an accountant found the following values for the mean and standard deviation of the dollar amounts per billing: $\mu = \$22.00$ and $\sigma = \$7.00$. The company's controller believes that the accountant could have obtained very good estimates of the mean billing amount by taking a simple random sample of 50 billings. Assume that a simple random sampling procedure was conducted.

a. Explain how the sample mean billing $\bar{x}$ would have a sampling distribution.

b. Show the sampling distribution of $\bar{x}$.

c. What is the standard error of the mean?

d. What would happen to the sampling distribution of $\bar{x}$ if a sample size of 100 was considered?

21. An automatic machine used to fill cans of soup has the following characteristics: $\mu = 15.9$ ounces, $\sigma = .5$ ounces.

a. Show the sampling distribution of $\bar{x}$, where $\bar{x}$ is the sample mean for 40 cans selected randomly by a quality control inspector.

b. What is the probability of finding a sample of 40 cans with a mean $\bar{x}$ greater than 16 ounces?

22. In a population of 4,000 employees, a simple random sample of 40 employees is selected in order to estimate the mean age for the population.

a. Would you use the finite population correction factor in calculating the standard error of the mean? Explain.

b. If the population standard deviation is $\sigma = 8.2$ years, compute the standard error both with and without using the finite population correction factor. What is the rationale behind ignoring the finite population correction factor whenever $n/N \leq .05$?

23. What is the probability that the sample mean age of the employees in Exercise 22 will be within (plus or minus) 2 years of the population mean age? Use the population standard deviation of $\sigma = 8.2$ years and the sample size of 40.

24. A library checks out an average of 320 books per day, with a standard deviation of 75 books. Consider a sample of 30 days of operation, with $\bar{x}$ being the sample mean number of books checked out per day.

a. Show the sampling distribution of $\bar{x}$.

b. What is the standard error?

c. What is the probability that the sample mean for the 30 days will be between 300 and 340 books?

d. What is the probability the sample will show more than 325 books checked out per day?

7.6 SAMPLING DISTRIBUTION OF $\bar{p}$

As we observed in Section 7.4, the sample proportion $\bar{p}$ is a random variable that provides a numerical description of the outcome of a simple random sample. In order to determine how close the point estimate based on $\bar{p}$ is to the population parameter p, we need to understand the properties of the sampling distribution of $\bar{p}$: the expected value of $\bar{p}$, the standard deviation of $\bar{p}$, and the shape of the sampling distribution of $\bar{p}$.

Expected Value of $\bar{p}$

It can be shown by relying on the definition of expected value that the expected value of $\bar{p}$—that is, the mean of all possible values of $\bar{p}$—is as follows:

$$E(\bar{p}) = p, \tag{7.4}$$

where

$E(\bar{p})$ = the expected value of the random variable $\bar{p}$,

p = the population proportion.

This result shows that the mean of all possible $\bar{p}$ values is equal to the population proportion p. Since $E(\bar{p}) = p$, the sample proportion is an unbiased estimator of the population proportion.

Recall that in Section 7.2 we showed that $p = .60$ for the EAI population, where p was the proportion of the population of managers who had participated in the company's management training program. Thus the expected value of the random variable $\bar{p}$ is .60.

Standard Deviation of $\bar{p}$

The different random samples generate a variety of values for $\bar{p}$. We now are interested in determining the standard deviation of $\bar{p}$, which is referred to as the *standard error of the proportion.* Just as we found for the sample mean $\bar{x}$, the standard deviation of $\bar{p}$ depends upon whether the population is finite or infinite. The two expressions for the standard deviation of $\bar{p}$ are as follows:

Standard Deviation of $\bar{p}$

Finite Population | *Infinite Population*

$$\sigma_{\bar{p}} = \sqrt{\frac{N-n}{N-1}}\sqrt{\frac{p(1-p)}{n}} \qquad \sigma_{\bar{p}} = \sqrt{\frac{p(1-p)}{n}} \tag{7.5}$$

Comparing the two expressions in (7.5), we see that the finite population correction factor $\sqrt{(N-n)/(N-1)}$ is again required for the finite population case.

As was the case with the sample mean $\bar{x}$, we find that the difference between the above expressions for the finite population and the infinite population becomes negligible if a finite population is large compared to the sample size. We follow the same rule of thumb that we recommended for the sample mean. That is, if the population is infinite or if the population is finite with $n/N \leq .05$, we will simply use $\sigma_{\bar{p}} = \sqrt{p(1-p)/n}$. However, if the population is finite and if $n/N > .05$, the finite population correction factor must be used, as shown in (7.5). Again, unless specifically noted, throughout the text we will be assuming that the population size is "large" relative to the sample size and that the finite population correction factor is unnecessary.

For the EAI study the population proportion of managers that have participated in the management training program was $p = .60$. With $n/N = 30/2{,}500 = .012$, we ignore the finite population correction factor and compute the standard deviation of the $\bar{p}$ values as follows:

$$\sigma_{\bar{p}} = \sqrt{\frac{p(1-p)}{n}} = \sqrt{\frac{.60(1-.60)}{30}} = \sqrt{.008} = .089.$$

As we shall see in Chapter 8, knowledge of the value of $\sigma_{\bar{p}}$ will be helpful in determining the possible sampling error whenever the sample proportion $\bar{p}$ is used to estimate the population proportion p.

Form of the Sampling Distribution of $\bar{p}$

Now that we know the mean and standard deviation of $\bar{p}$, we want to consider the form of the sampling distribution of $\bar{p}$. Applying the central limit theorem as it relates to the

$\bar{p}$ random variable, we have the following:

> The sampling distribution of $\bar{p}$ can be approximated by a normal probability distribution whenever the sample size is large.

With $\bar{p}$, the sample size can be considered large whenever the following two conditions are satisfied:

$$np \geq 5,$$
$$n(1 - p) \geq 5.$$

To understand the rationale behind the above rule of thumb for a "large" sample size, first note that the population proportion p is equivalent to the probability of success associated with the binomial probability distribution. In fact, the exact sampling distribution of $\bar{p}$ is the binomial probability distribution. However, as we saw in Chapter 6, whenever n is large it is computationally convenient to use the normal distribution to approximate the binomial distribution. The above rule of thumb indicates when the normal approximation of the binomial distribution is appropriate and thus also when the normal approximation is appropriate for the sampling distribution of $\bar{p}$.

Recall that for the EAI sampling problem we know that the population proportion of managers having participated in the training program is $p = .60$. Let us use this value to determine the characteristics of the sampling distribution of $\bar{p}$ for the EAI study. With a simple random sample of size 30, we have $np = 30(.60) = 18$ and $n(1 - p) = 30(.40) = 12$. Thus according to the rule of thumb we conclude that the sampling distribution of $\bar{p}$ can be approximated by a normal distribution. Figure 7.8 shows this sampling distribution for the EAI study. Figure 7.9 shows the close

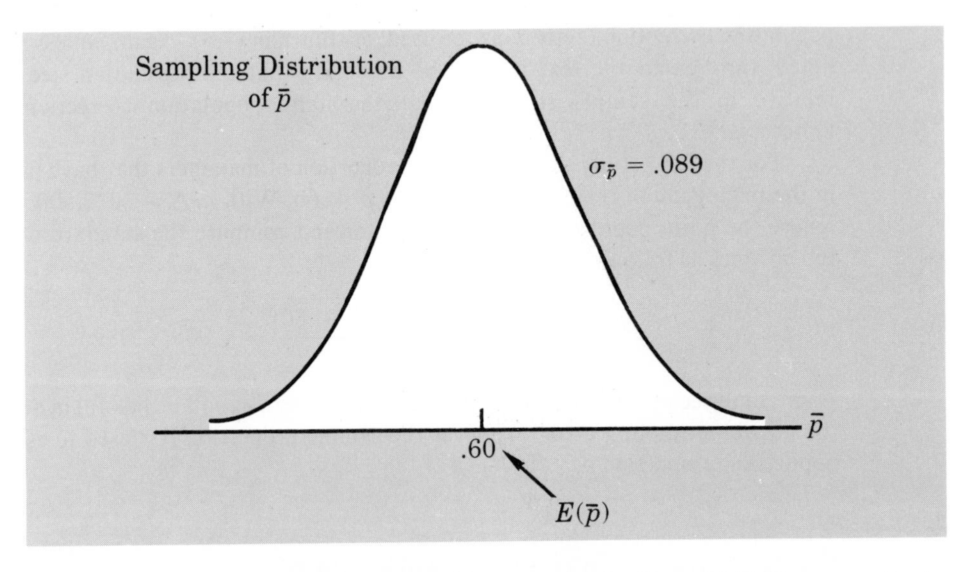

FIGURE 7.8 **Sampling Distribution of $\bar{p}$ for the Proportion of EAI Managers Having Participated in the Management Training Program**

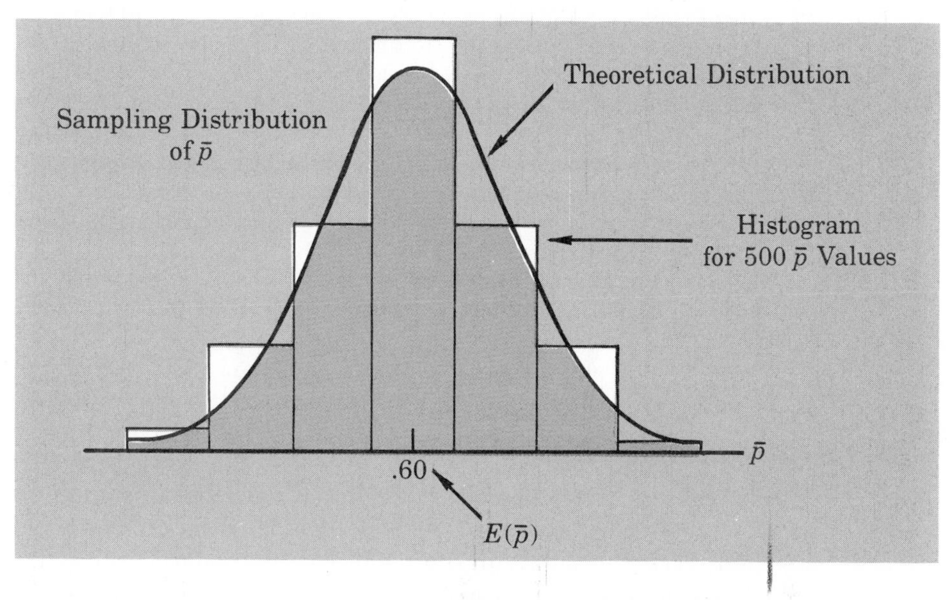

FIGURE 7.9 Comparison of the Theoretical Sampling Distribution of $\bar{p}$ and the Relative Frequency Histogram of $\bar{p}$ for 500 Simple Random Samples

agreement between this theoretical sampling distribution of $\bar{p}$ and the histogram of the 500 $\bar{p}$ values we obtained from the 500 repeated samples of size 30 (see Figure 7.3).

EXERCISES

25. The president of Doerman Distributors, Inc. believes that 30% of the firm's orders come from new or first-time customers. A simple random sample of 100 orders will be used to estimate the proportion of new or first-time customers. The results of the sample will be used to verify the president's claim of $p = .30$.

a. Assume that the president is correct with $p = .30$. What is the sampling distribution of $\bar{p}$ for this study?

b. What is the probability that the sample proportion $\bar{p}$ will be between .20 and .40?

c. What is the probability that the sample proportion will be within plus or minus .05 of the population proportion $p = .30$?

26. A particular county in West Virginia has a 9% unemployment rate. A monthly survey of 800 individuals is conducted by a state agency. This study provides the basis for monitoring the unemployment rate of the county.

a. Assume that $p = .09$. What is the sampling distribution of $\bar{p}$ when a sample of size 800 is used?

b. What is the probability that a sample proportion $\bar{p}$ of at least .08 will be observed?

27. What is the probability that the EAI estimate of the proportion of managers having completed the firm's training program, $\bar{p}$, is within $\pm.05$ of the population proportion $p = .60$? Use samples of size 30, size 60, and size 100.

28. Assume that 15% of the items produced in an assembly line operation are defective but that the firm's production manager is not aware of this situation. Assume further that 50 parts are tested by the quality assurance department in order to determine the quality of the assembly operation. Let $\bar{p}$ be the sample proportion defective found by the quality assurance test.

a. Show the sampling distribution for $\bar{p}$.

b. What is the probability that the sample proportion will be within $\pm.03$ of the population proportion defective?

c. If the test shows $\bar{p} = .10$ or more, the assembly line operation will be shut down to check for the cause of the defects. What is the probability that the sample of 50 parts will lead to the conclusion that the assembly line should be shut down?

29. A doctor believes that 80% of all patients having a particular disease will be fully recovered within 3 days after receiving a new drug.

a. A simple random sample of 20 medical records will be used to develop an estimate of the proportion of patients who were fully recovered within 3 days after receiving the drug. If a data analyst suggests using a normal probability distribution approximation for the sampling distribution of $\bar{p}$, what would you say? Explain.

b. If the sample of patient records is increased to 60, what is the probability that the sample proportion will be within $\pm.10$ of the population proportion? (Assume that the population proportion is .80.)

7.7 PROPERTIES OF POINT ESTIMATORS

In our discussions of the sampling distribution of $\bar{x}$ and the sampling distribution of $\bar{p}$, we stated that $E(\bar{x}) = \mu$ and $E(\bar{p}) = p$. Thus the point estimators $\bar{x}$ and $\bar{p}$ were described as being unbiased estimators of their corresponding population parameters μ and p. We delay a more complete discussion of the sampling distribution of the sample standard deviation s and sample variance s^2 until Chapter 11. However, it can be shown that $E(s^2) = \sigma^2$. Thus we would conclude similarly that the sample variance s^2 is an unbiased estimator of the population variance σ^2. In fact, when we first presented the formula for the sample variance in Chapter 3, we stated that s^2 was computed with $n - 1$ rather than n in the denominator. The reason for using $n - 1$ rather than n in the denominator was to obtain an unbiased estimator of the population variance. If we had used n in the denominator for the sample variance formula, the sample variance would have been a biased estimator. In this case, the sample variance would tend to underestimate the population variance.

Unbiasedness is a property that is desirable for point estimators. However, there are other properties of estimators that statisticians may consider when attempting to determine appropriate point estimators of population parameters. One such property is *consistency*. Loosely speaking, an estimator is said to be a consistent estimator if values of the point estimator tend to lie closer to the population parameter as the sample size becomes larger. In other words, larger sample sizes tend to provide better point estimates. Note here that for the point estimator $\bar{x}$ we showed that $\sigma_{\bar{x}} = \sigma/\sqrt{n}$. Since this expression shows that larger sample sizes result in a smaller standard deviation or standard error $\sigma_{\bar{x}}$, we would conclude that larger sample sizes tend to provide point estimates of $\bar{x}$ closer to the population mean. In this sense, we would say that $\bar{x}$ is a consistent estimator of μ.

Finally, assume that two unbiased point estimators can be used to estimate a population parameter. We would select the estimator with the smaller standard deviation or variance on the basis that it would tend to provide estimates closer to the population parameter. The estimator with the smaller standard deviation or variance would be considered the more efficient estimator. The ratio of the variances of the two estimators would provide their *relative efficiency*.

7.8 OTHER SAMPLING METHODS

We have described the simple random sampling procedure and developed the appropriate sampling distributions for $\bar{x}$ and $\bar{p}$ based on simple random sampling. It is important to realize that simple random sampling is not the only sampling method available, however. Sampling methods such as stratified random sampling, cluster sampling, and systematic sampling offer alternatives that in some situations have advantages over simple random sampling. Let us describe briefly some of the alternative sampling methods.

Stratified Random Sampling

Under stratified random sampling, the population is first divided into groups of elements called strata such that each item in the population belongs to one and only one stratum. The basis for forming the various strata, such as department, location, age, industry type, and so on, is up to the discretion of the designer of the sample. Best results, however, are obtained whenever the elements within each stratum are as much alike as possible. Figure 7.10 shows a diagram with the population divided into H strata.

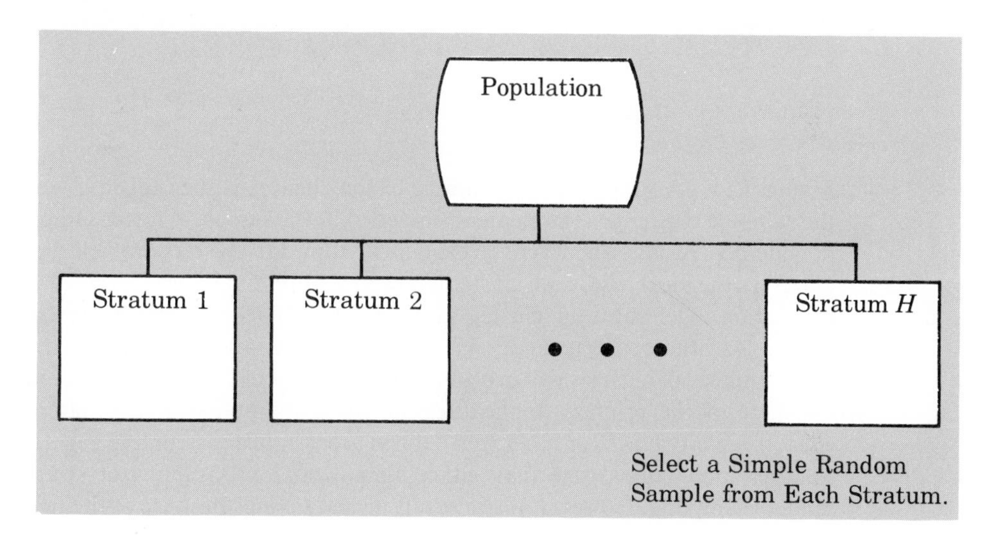

FIGURE 7.10 Diagram for Stratified Simple Random Sampling

After the strata are formed, a simple random sample is taken from every stratum. Formulas are available for combining the results for the individual samples into one estimate of the population parameter of interest. The formulas differ from those for simple random sampling. However, procedures are available for determining the sampling distribution of the estimator whenever a stratified random sampling procedure is used.

The value of stratified random sampling depends upon how homogeneous the elements are within the strata. If units within strata are *alike* (homogeneity), the strata will have low variances. Thus relatively small sample sizes can be used to obtain good estimates of the strata characteristics. If homogeneous strata exist, the stratified random sampling procedure will provide results similar to simple random sampling but will do so with a smaller total sample size.

Cluster Sampling

In cluster sampling the population is first divided into separate groups of elements called clusters. Each element of the population belongs to one and only one cluster (see

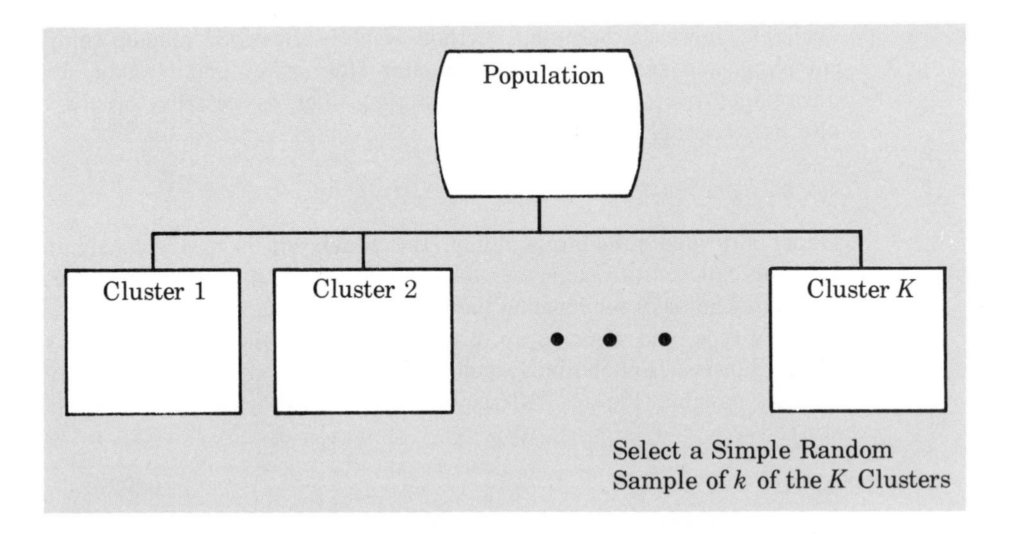

FIGURE 7.11 Diagram for Cluster Sampling

Figure 7.11). A simple random sample of the clusters is then taken. Elements within the sampled cluster are the elements included in the sample. Cluster sampling tends to provide best results whenever the elements within the clusters are heterogeneous (not alike). In the ideal case, each cluster is a representative small-scale version of the entire population. The value of cluster sampling depends upon how representative each cluster is of the entire population. If each cluster is alike in this regard, sampling a small number of clusters will provide good estimates of the population parameters.

One of the primary applications of cluster sampling is area sampling, where clusters are city blocks or other well defined areas. Cluster sampling generally requires a larger total sample size than either simple random sampling or stratified simple random sampling. However, it can result in cost savings because of the fact that when an interviewer is sent to a sampled cluster (or city-block location) many interviews or sample observations can be obtained in a relatively short time. As a result, a larger sample size may be obtainable with a significantly lower cost per element sampled and thus possibly a lower total cost.

Systematic Sampling

In some sampling situations, especially those with large populations, it is often time-consuming to select a simple random sample by first finding a random number and then counting or searching through the list of population items until the corresponding element is found. An alternative to a simple random sampling is a systematic sampling procedure. For example, if a sample size of 50 is desired from a population containing 5,000 elements, we might sample one element for every $5{,}000/50 = 100$ elements in the population. A systematic sample for this case would involve selecting randomly 1 of the first 100 elements from the population list. Other sample elements are identified by starting with the first sampled element and then selecting every 100th element that follows in the population list. In effect the sample of 50 is identified by moving systematically through the population and identifying every

100th element after the first randomly selected element. The sample of 50 usually will be easier to identify in this manner than would be the case if simple random sampling were used. Since the first element selected is a random choice, the assumption is usually made that a systematic sample has the properties of a simple random sample. This assumption is especially applicable when the list of the population elements is believed to be a random ordering of the elements.

Convenience Sampling

All of the sampling methods discussed thus far fall under the heading of *probability sampling* techniques. By this we mean that elements selected from the population have a known probability of being included in the sample. The advantage of probability sampling is that the sampling distribution of the appropriate sample statistic generally can be identified. Formulas such as the ones for simple random sampling presented earlier in this chapter can be used to determine the properties of the sampling distribution. Then the sampling distribution can be used to make probability statements about possible sampling errors associated with the sample results.

Convenience sampling falls under the heading of *nonprobability sampling* techniques. As the name implies, the sample is identified primarily by convenience. Items are included in the sample without prespecified or known probabilities of being selected. For example, a professor conducting research at a university may ask student volunteers to constitute a sample for the study simply because they are readily available and will often participate as subjects for little or no cost. In another example, a shipment of oranges may be sampled by an inspector who selects oranges haphazardly from among several crates. Labeling each orange and using a probability method of sampling would be impractical. Samples such as wildlife captures and volunteer panels for consumer research would also have to be considered convenience samples.

Convenience samples have the advantage of relatively easy sample selection and data collection; however, it is impossible to evaluate the "goodness" of the sample in terms of its ability to estimate characteristics of the population. A convenience sample may provide good results, or it may not. However, there is no statistically justified procedure that will allow a probability analysis and inference about the quality of the sample results. Nevertheless, at times you will see statistical methods designed for probability samples applied to a convenience sample. The researcher argues that the convenience sample may well provide a sample which may be treated as if it were a random sample. However, this argument cannot be supported, and we should be very cautious in interpreting convenience samples that are used to make inferences about populations.

Judgment Sampling

One additional nonprobability sampling technique is judgment sampling. In this situation the person most knowledgeable on the subject of the study selects individuals or other elements of the population that he or she feels are most representative of the population. Often this is a relatively easy way of selecting a sample. However, the quality of the sample results are dependent on the judgment of the person selecting the sample. Again great caution must be used in drawing conclusions based on judgment samples used to make inferences about populations.

Summary

This chapter has presented the important concepts of simple random sampling and sampling distributions. We demonstrated how a simple random sample could be selected and how the data collected for the sample could serve as the basis for developing point estimates of population parameters. Since different simple random samples provided a variety of different values for the point estimators, we saw that the point estimators such as $\bar{x}$ and $\bar{p}$ were random variables. The probability distribution of the random variable is called its sampling distribution. In particular, we described the sampling distributions of the sample mean, $\bar{x}$, and the sample proportion, $\bar{p}$.

In considering the characteristics of the sampling distributions, we stated that since $E(\bar{x}) = \mu$ and $E(\bar{p}) = p$, both $\bar{x}$ and $\bar{p}$ are unbiased estimators of their corresponding population parameters. After developing the standard error formulas for these estimators, we showed how the central limit theorem provided the basis for using a normal distribution to approximate the sampling distributions. Rules of thumb were provided for determining when "large" sample size conditions were satisfied and the normal distribution approximation was appropriate.

We then discussed some properties of point estimators, including unbiasedness, consistency, and relative efficiency. We concluded the chapter by describing sampling methods other than simple random sampling, including stratified, cluster, systematic, convenience, and judgment sampling.

Glossary

Parameter—A population characteristic, such as a population mean μ, a population standard deviation σ, a population proportion p, and so on.

Sample statistic—A sample characteristic, such as a sample mean $\bar{x}$, a sample standard deviation s, a sample proportion $\bar{p}$, and so on. The value of the sample statistic is used to estimate the value of the population parameter.

Sampling distribution—A probability distribution for all possible values of a sample statistic.

Simple random sampling—Finite population: a sample selected such that each possible sample of size n has the same probability of being selected. Infinite population: a sample selected such that each item comes from the same population and the successive items are selected independently.

Sampling without replacement—Once an item from the population has been included in the sample it is removed from further consideration and cannot be selected for the sample a second time.

Sampling with replacement—As each item is selected for the sample, it is returned to the population. It is possible that a previously selected item may be selected again and therefore appear in the sample more than once.

Point estimate—A single numerical value used as an estimate of a population parameter.

Point estimator—The sample statistic, such as $\bar{x}$, s, $\bar{p}$, etc., that provides the point estimate of the population parameter.

Finite population correction factor—The multiplier term $\sqrt{(N - n)/(N - 1)}$ that is used in the formulas for $\sigma_{\bar{x}}$ and $\sigma_{\bar{p}}$ whenever a finite population rather than an infinite population is being sampled. The generally accepted rule of thumb is to ignore the finite population correction factor whenever $n/N \leq .05$.

Standard error—The standard deviation of a point estimator.

Central limit theorem—A theorem that allows us to use the normal probability distribution to approximate the sampling distribution of $\bar{x}$ and $\bar{p}$ whenever the sample size is large.

Unbiased—A property of a point estimator that occurs whenever the expected value of the point estimator is equal to the population parameter it estimates.

Consistency—A property of a point estimator that occurs whenever larger sample sizes tend to provide point estimates closer to the population parameter.

Relative efficiency—The ratio of the variances of two point estimators of a given parameter. The estimator with the smaller variance is the more efficient.

Probability sample—A sample selected such that each element in the population has a known probability of being included in the sample. Simple random sampling, stratified random sampling, cluster sampling, and systematic sampling can be classified as methods leading to probability samples.

Nonprobability sample—A sample selected such that the probability of each element being included in the sample is unknown. Convenience and judgment samples are nonprobability samples.

Key Formulas

Standard Deviation of $\bar{x}$

Finite Population — *Infinite Population*

$$\sigma_{\bar{x}} = \sqrt{\frac{N-n}{N-1}}\frac{\sigma}{\sqrt{n}} \qquad \sigma_{\bar{x}} = \frac{\sigma}{\sqrt{n}} \tag{7.2}$$

Standard Deviation of $\bar{p}$

Finite Population — *Infinite Population*

$$\sigma_{\bar{p}} = \sqrt{\frac{N-n}{N-1}}\sqrt{\frac{p(1-p)}{n}} \qquad \sigma_{\bar{p}} = \sqrt{\frac{p(1-p)}{n}} \tag{7.5}$$

Supplementary Exercises

30. Nationwide Supermarkets has 2400 retail stores located in 32 states. At the end of each year a sample of 35 stores is selected for physical inventories. Results from the inventory samples are used in annual tax reports. Assume that the retail stores are listed sequentially on a computer printout. Begin at the bottom of the third column of the random numbers shown in Table 8 of Appendix B. Using four-digit random numbers beginning with 7762, read *up* the column to identify the first five stores to be included in the simple random sample.

31. Four college students are taking the following number of credit hours during the current term:

Student	*Number of Credit Hours*
Albert	15
Becky	17
Cindy	19
Dan	17

Treat the students as a population of size 4 and answer the following questions:

a. Compute the mean and standard deviation for the population.

b. If an estimate of the mean number of credit hours taken by these students is to be based on a sample of two students, how many samples are possible? List them.

c. Show a graph of the sampling distribution of $\bar{x}$.
d. Use (7.1) and (7.2) to compute the expected value and standard deviation of $\bar{x}$.

32. The sizes of the ten offices on the 12th floor of the new Crosley Tower Bank Building are as follows:

Office	*Size (sq. ft.)*	*Office*	*Size (sq. ft.)*
1	150	6	300
2	175	7	140
3	180	8	150
4	180	9	150
5	225	10	200

In a space allocation study, simple random samples of four offices will be selected in order to estimate the mean office size on each floor of the building.
a. Compute the mean and standard deviation for the population of ten offices.
b. Explain how the sampling distribution of $\bar{x}$ could be developed. Select two or three simple random samples to demonstrate that various samples will provide a variety of values for $\bar{x}$.
c. Using the mean and standard deviation of the population, determine the mean and standard deviation of the $\bar{x}$ values.

33. A study of the time from computer program submission until program return (i.e., "turnaround" time) was conducted at a university computer center. Assume that under standard operating conditions the population mean is 120 minutes, with a population standard deviation of 40 minutes.
a. Future studies of turnaround time are to be based on simple random samples of 30 programs. Show the sampling distribution of the sample mean turnaround time.
b. What is the probability that the sample mean for 30 programs will be less than 100 minutes? Over 125 minutes?

34. An electrical component is designed to provide a mean service life of 3,000 hours, with a standard deviation 800 hours. A customer purchases a batch of 50 components, which can be considered a simple random sample of the population of components. What is the probability that the mean life for the group of 50 components will be at least 2,750 hours? At least 3,200 hours?

35. In the EAI study the population of managers had annual salaries with $\mu = \$31,800$ and $\sigma = \$4,000$. Samples of size 30 provided a .5036 probability of selecting a sample with $\bar{x}$ within $\pm\$500$ of the population mean. How large a sample should be selected if the personnel director wishes the probability of a sample mean $\bar{x}$ being within $\pm\$500$ of μ to be .95?

36. Three firms have inventories that vary in size. Firm A has a population of 2,000 items, firm B has a population of 5,000 items, and firm C has a population of 10,000 items. The population standard deviation for the cost of the items is $\sigma = 144$. A statistical consultant recommends that each firm take a sample of 50 items from their respective populations in order to provide statistically valid estimates of the average cost per item. Management of the small firm states that since it has the smallest population, it should be able to obtain the data from a much smaller sample size than required by the larger firms. However, the consultant states that in order to obtain the same standard error and thus the same precision in the sample results, all firms should take the same sample size regardless of population size.
a. Using the finite population correction factor, compute the standard error for each of the three firms given a sample of size 50.
b. For each firm what is the probability that the sample mean $\bar{x}$ will be within ± 25 of the population mean, μ?

37. A survey reports its results by stating that the standard error of the mean was 20. The population standard deviation was 500.
a. How large was the sample used in this survey?
b. What is the probability that the estimate would be within ± 25 of the population mean?

38. A production process is checked periodically by a quality control inspector. The inspector selects simple random samples of 30 finished products and computes the sample mean product weights, $\bar{x}$. If test results over a long period of time show that 5% of the $\bar{x}$ values are over 2.1 pounds and 5% are under 1.9 pounds, what are the mean and the standard deviation for the population produced with this process?

39. The grade point average for all juniors at Strausser College has a standard deviation of .50.

a. A random sample of 20 students is to be used to estimate the population mean grade point average. What assumption is necessary in order to compute the probability of obtaining a sample mean within plus or minus .2 of the population mean?

b. Provided that this assumption can be made, what is the probability of $\bar{x}$ being within plus or minus .2 of the population mean?

c. If this assumption cannot be made, what would you recommend doing?

40. For a population where $p = .35$ is the proportion of persons having a college degree:

a. Explain how the sampling distribution of $\bar{p}$ results from random samples of size 80 being used to estimate the proportion of individuals having a college degree.

b. Show the sampling distribution for $\bar{p}$ in this case.

c. If the sample size is increased to 200, what happens to the sampling distribution of $\bar{p}$? Compare the standard error for the $n = 80$ and $n = 200$ alternatives.

41. A market research firm conducts telephone surveys with a 40% historical response rate. What is the probability that in a new sample of 400 telephone numbers at least 150 individuals will cooperate and respond to the questions? In other words, what is the probability of a sample proportion $\bar{p} \geq 150/400 = .375$?

42. A production run is not acceptable for shipment to customers if a sample of 100 items contains at least 5% defective. If a production run has a population proportion defective of $p = .10$, what is the probability that $\bar{p}$ will be at least .05?

43. The proportion of individuals insured by the All-Driver Automobile Insurance Company that have received at least one traffic ticket during a 5-year period is .15.

a. Show the sampling distribution of $\bar{p}$ if a random sample of 150 insured individuals is used to estimate the proportion having received at least one ticket.

b. What is the probability that the sample proportion will be within plus or minus .03 of the population proportion?

44. Historical records show that .50 of all orders placed at Big Burger Fast Food restaurants include a soft drink. With a simple random sample of 40 orders, what is the probability that between .45 and .55 of the sampled orders will include a soft drink?

45. Lori Jeffrey is a successful sales representative for a major publisher of college textbooks. Historically, Lori obtains a book adoption on 25% of her sales calls. Viewing her sales calls for 1 month as a sample of all possible sales calls, a statistical analysis of the data yields a standard error of the proportion of .0625.

a. How large was the sample used in this analysis? That is, how many sales calls did Lori make during the month?

b. Let $\bar{p}$ indicate the sample proportion of book adoptions obtained during the month. Show the sampling distribution $\bar{p}$.

c. Use the sampling distribution of $\bar{p}$. What is the probability that Lori will obtain book adoptions on 30% or more of her sales calls during the 1 month period?

APPENDIX TO CHAPTER 7: THE EXPECTED VALUE AND STANDARD DEVIATION OF $\bar{x}$

In this appendix we present the mathematical basis for the expressions for the expected value of $\bar{x}$, $E(\bar{x})$, as given by (7.1) and the standard deviation of $\bar{x}$, $\sigma_{\bar{x}}$, as given by (7.2).

Expected Value of $\bar{x}$

Assume a population with mean μ and variance σ^2. A simple random sample of size n is selected with individual observations denoted $x_1, x_2, \ldots, x_n$. A sample mean $\bar{x}$ is computed as follows:

$$\bar{x} = \frac{\Sigma x_i}{n}.$$

With repeated simple random samples of size n, $\bar{x}$ is a random variable that takes on different numerical values depending upon the specific n items selected. The expected value of the random variable $\bar{x}$, or the mean of all possible $\bar{x}$ values, is as follows:

$$\begin{aligned} \text{Mean of } \bar{x} = E(\bar{x}) &= E\left(\frac{\Sigma x_i}{n}\right) \\ &= \frac{1}{n}[E(x_1 + x_2 + \cdots + x_n)] \\ &= \frac{1}{n}[E(x_1) + E(x_2) + \cdots + E(x_n)]. \end{aligned}$$

Since for any x_i we have $E(x_i) = \mu$, we can write

$$\begin{aligned} E(\bar{x}) &= \frac{1}{n}(\mu + \mu + \cdots + \mu) \\ &= \frac{1}{n}(n\mu) = \mu. \end{aligned}$$

The above expression shows that the mean of all possible $\bar{x}$ values is the same as the population mean μ.

Standard Deviation of $\bar{x}$

Again assume a population with mean μ, variance σ^2, and a sample mean $\bar{x}$ given by

$$\bar{x} = \frac{\Sigma x_i}{n}.$$

With repeated simple random samples of size n, we know that $\bar{x}$ is a random variable that takes on different numerical values depending upon the specific n items selected. Shown below is the derivation of the expression for the standard deviation of the $\bar{x}$ values, $\sigma_{\bar{x}}$, for the case where the population is infinite. Since sampling with replacement from a finite population is equivalent to the infinite population case, the expression holds for this situation as well. The derivation of the expression for $\sigma_{\bar{x}}$ for a finite population when sampling is done without replacement is more difficult and is beyond the scope of this text.

Returning to the infinite population case, recall that a simple random sample from an infinite population consists of observations $x_1, x_2, \ldots, x_n$ that are independent. In Section 5.4 we provided the following formulas concerning the variance of random variables:

$$\text{Var}(ax) = a^2 \,\text{Var}(x), \tag{5.6}$$

where a is a constant and x is a random variable, and

$$\text{Var}(x + y) = \text{Var}(x) + \text{Var}(y), \tag{5.7}$$

where x and y are *independent* random variables. Using (5.6) and (5.7), we can develop the expression for the variance of the random variable $\overline{x}$ as follows:

$$\text{Var}(\overline{x}) = \text{Var}\left(\frac{\Sigma x_i}{n}\right) = \text{Var}\left(\frac{1}{n}\Sigma x_i\right).$$

Using (5.6) with $1/n$ viewed as the constant, we have

$$\begin{aligned}\text{Var}(\overline{x}) &= \left(\frac{1}{n}\right)^2 \text{Var}(\Sigma x_i) \\ &= \left(\frac{1}{n}\right)^2 \text{Var}(x_1 + x_2 + \cdots + x_n).\end{aligned}$$

With the infinite population case, the random variables $x_1, x_2, \ldots, x_n$ are independent. Thus (5.7) enables us to write

$$\text{Var}(\overline{x}) = \left(\frac{1}{n}\right)^2 \left[\text{Var}(x_1) + \text{Var}(x_2) + \cdots + \text{Var}(x_n)\right].$$

Since for any x_i we have $\text{Var}(x_i) = \sigma^2$, we have

$$\text{Var}(\overline{x}) = \left(\frac{1}{n}\right)^2 \underbrace{\left(\sigma^2 + \sigma^2 + \cdots + \sigma^2\right)}_{n \text{ items}}.$$

With n values of σ^2 in this expression, we have

$$\text{Var}(\overline{x}) = \left(\frac{1}{n}\right)^2 (n\sigma^2) = \frac{\sigma^2}{n}.$$

Taking the square root provides the formula for the standard deviation of $\overline{x}$ for the infinite population case:

$$\sigma_{\overline{x}} = \sqrt{\text{Var}(\overline{x})} = \frac{\sigma}{\sqrt{n}}.$$

This expression, which appeared as (7.2), shows that as the sample size n is increased the standard deviation of the various sample means $\overline{x}$ will decrease. In effect, the larger samples tend to provide better estimates of the population mean, such that as n increases there is less variation in the values of the different sample means.

Application

Mead Corporation*

Mead

Dayton, Ohio

Mead is a diversified paper and forest products company which manufactures paper, pulp, and lumber and converts paperboard into shipping containers and beverage carriers. The company's strong distribution capability is used to market many of its own products, including paper, school supplies, and stationery. Mead's Advanced Systems group develops business applications for the future. These involve storing, retrieving, and reproducing data through the innovative application of digital technology.

DECISION ANALYSIS AT MEAD CORPORATION

Decision Analysis is an internal consulting group located in a larger department known as Decision Support Applications (DSA). The principal thrust of the DSA department is to increase the productivity of Mead's human and computer resources by providing products and training which will enable Mead employees worldwide to use the computer to do their jobs more efficiently and/or effectively. DSA focuses on providing timely and relevant responses that will satisfy staff/line business needs in the following areas:

Word processing/office automation
Operations research/statistical analysis
Financial/ planning/modeling
Hotline assistance and training in the use of user-friendly computer products
Identification of end-user software
Consulting on efficient use of end-user-developed applications

To accomplish these goals, DSA is divided into four departments: Office Systems, Decision Analysis, Information Center, and Financial Modeling Coordination.

The major role of the Decision Analysis department is to provide quantitative support to decision makers throughout the corporation, including both line and staff management in the functional areas of operations, finance, accounting, and human resources. Consulting projects include not only analysis for one-time decision situa-

*The authors are indebted to Dr. Edward P. Winkofsky, Mead Corporation, Dayton, Ohio for providing this application.

tions, but also the development of support tools for periodic and recurring decision problems. The activities of the department can be divided into six major categories: financial analysis, data and statistical analysis, resource allocation modeling, simulation modeling, forecasting, and user-friendly planning models. Decision Analysis is also responsible for monitoring and disseminating to appropriate Mead management new developments in operations research and decision support methodologies. The emphasis in this effort is to identify those developments which would result in significant productivity benefits to the organization.

AN APPLICATION USING STRATIFIED SAMPLING

Mead maintains a continuous forest inventory (CFI) system which provides information concerning the stocking of a large portion of its timberland holdings. The CFI system consists of permanent plots of 1/5 or 1/7 acre which are systematically located throughout the Mead forest. On a periodic basis the trees on certain CFI plots are measured and general site information is gathered on each plot. These data are used to estimate the present volume and past growth of the forest and to project the future growth of the forest.

To identify the plots for measurement, a stratified sampling technique was used. First, the forest was divided into three strata. Then, based upon previous measurements, estimates of the variance of the volume of each stratum were developed. Using the estimated variance and the accuracy specified for each stratum, the sample size needed in each case was determined. For strategic reasons, the required accuracy was not identical across the strata.

Once the sample size was determined for each stratum, a random sample of plots within each stratum was selected. In order to determine whether CFI plots had been treated differently from the surrounding forest, for each CFI plot selected, two temporary plots—each located in close proximity to the permanent CFI plot—were located at randomly selected points. Decision Analysis provided tables of computer generated pseudorandom numbers to be used in the selection process. Specifically, these numbers were used to identify which permanent plots would be included in the sample and to define the location of the associated temporary plots.

Foresters throughout the organization participated in the field measurements. They gathered information on each tree of every selected plot, as well as general plot information. These data were entered in the field on computer generated plot reports. These reports provided space to enter the new data and showed the data from the previous measurements for the permanent plots. The data were collected by several two person teams, and an additional two persons acted as an audit team. As the plot information was collected, it was entered into a data base by CRT operators.

Decision Analysis generated a number of frequency distributions from the data base. These reports were used in part to edit the data base but also to estimate the proportions of the forest in each of several species groups. The proportions for each stratum were first determined and then weighted to provide an estimate for the overall forest. Since certain species groups are more valuable than others, these estimates provided an initial indication of the value of Mead's land holdings. A more accurate measure was then determined through volume estimates.

The volume of each tree was determined using several species-specific formulas common to forestry. Next, the volume estimates were compiled for each stratum and then weighted to provide the forest estimates. These reports provided management

A paper machine known as the "Spirit of Escanaba" is one of the largest and most modern paper machines in the world

with the necessary information to evaluate the Mead timberlands. Growth estimates are now in the process of being developed from these data. These estimates will be used in a large scale linear program which will develop long term harvest schedules.

ADVANTAGES TO MEAD

With advanced sampling techniques, a small number of measurements were used to estimate the value of the total Mead forest ownership. Stratified sampling allowed the sample to provide greater accuracy for a more critical area of the forest without increasing the sample size over the less critical areas. Decision Analysis assisted in the sample selection and played a major role in the development of the programs necessary to compute the required statistical and summary information. In addition, the group has developed a user-friendly front end to the data to allow Corporate and the divisional woodlands to make their own report requests. This system allows the user to select from over ten prespecified reports and plots to be included in the estimates, thus providing a facility to develop estimates for any subset of the total ownership. The information provided by these reports is important input for decisions involving the management of Mead's woodland assets.

8 Interval Estimation

What you will learn in this chapter:

- how to construct and interpret an interval estimate of a population mean and a population proportion
- what is meant by confidence level
- the concept of a sampling error
- how to use the *t* distribution to construct interval estimates
- how to determine the sample size for estimating a population mean or a population proportion

Contents

In Chapter 7 we saw how a simple random sample could be used to develop a point estimate of a population parameter. Unfortunately, a point estimate by itself does not provide information about how close the estimate is to the population parameter. As we saw in Chapter 7 we cannot expect the point estimate to provide the exact value of the population parameter. Thus we often desire a statistical statement about the quality or precision of the estimate. Interval estimation has been developed to provide this kind of information. Let us introduce the interval estimation procedure by considering a sampling study conducted by the Statewide Insurance Company.

8.1 INTERVAL ESTIMATION OF A POPULATION MEAN—σ KNOWN

The Statewide Insurance Company provides a variety of life, health, disability, and business insurance policies for customers located throughout the United States. As part of an annual review of life insurance policies, a simple random sample of 36 Statewide policyholders is selected. The corresponding 36 life insurance policies are reviewed in terms of the amount of the coverage, the cash value of the policy, the disability options, and so on. For the current policy review study the project manager has requested information on the ages of the life insurance policyholders. Table 8.1 shows the age data collected from the simple random sample of 36 policyholders. Let us use these data to develop an interval estimate of the mean age of the population of life insurance policyholders covered by Statewide.

TABLE 8.1 Ages of Life Insurance Policyholders from a Simple Random Sample of 36 Statewide Policyholders

Policyholder	*Age*	*Policyholder*	*Age*	*Policyholder*	*Age*
1	32	13	39	25	23
2	50	14	46	26	36
3	40	15	45	27	42
4	24	16	39	28	34
5	33	17	38	29	39
6	44	18	45	30	34
7	45	19	27	31	35
8	48	20	43	32	42
9	44	21	54	33	53
10	47	22	36	34	28
11	31	23	34	35	49
12	36	24	48	36	39

The estimation procedure that we develop in this section is based on the assumption that the value of the population standard deviation is *known.* For the Statewide study, previous studies on policyholder ages permit us to use a known population standard deviation of $\sigma = 7.2$ years. In Section 8.2 we will show the interval estimation procedure for a population mean when the value of the population standard deviation is unknown.

Let x_1 indicate the age of the first policyholder in the sample, x_2 the age of the second policyholder, and so on. The sample mean of the data in Table 8.1 provides a point estimate of the population mean μ. The calculation of the sample mean is

$$\bar{x} = \frac{\Sigma x_i}{n} = \frac{1{,}422}{36} = 39.5.$$

Thus a point estimate of the mean age of the population of Statewide life insurance policyholders is 39.5 years.

As we stated in Chapter 7, we cannot expect the value of a sample mean $\bar{x}$ to *exactly* equal the value of the population mean μ. Thus anytime a sample mean is used to provide a point estimate of a population mean, someone may ask, "How good is the estimate?" The "how good" question is a way of asking about the error involved when the value of $\bar{x}$ is used as a point estimate of the population mean μ. In general, we shall refer to the magnitude of the difference between an unbiased point estimator and the population parameter as the *sampling error*. For the case of a sample mean estimating a population mean, the sampling error would be expressed as follows:

$$\text{Sampling error} = |\bar{x} - \mu|. \tag{8.1}$$

Note, however, that even after a sample is selected and the sample mean is computed ($\bar{x} = 39.5$ in the Statewide example) we will not be able to use (8.1) to find the value of the sampling error. This is true because the population mean μ is still unknown. This inability to determine the value of the sampling error appears to be a serious limitation. However, we shall see that the sampling distribution of $\bar{x}$ developed in Chapter 7 can be used to make probability statements about the sampling error.

From the central limit theorem we know that whenever the sample size is large ($n \geq 30$) the sampling distribution of $\bar{x}$ can be approximated by a normal probability distribution with a mean μ and a standard deviation* $\sigma_{\bar{x}} = \sigma/\sqrt{n}$. For the Statewide Insurance study, with $\sigma = 7.2$ years and $n = 36$, this theorem enables us to conclude that the sampling distribution of $\bar{x}$ is approximately normal with a mean μ and a

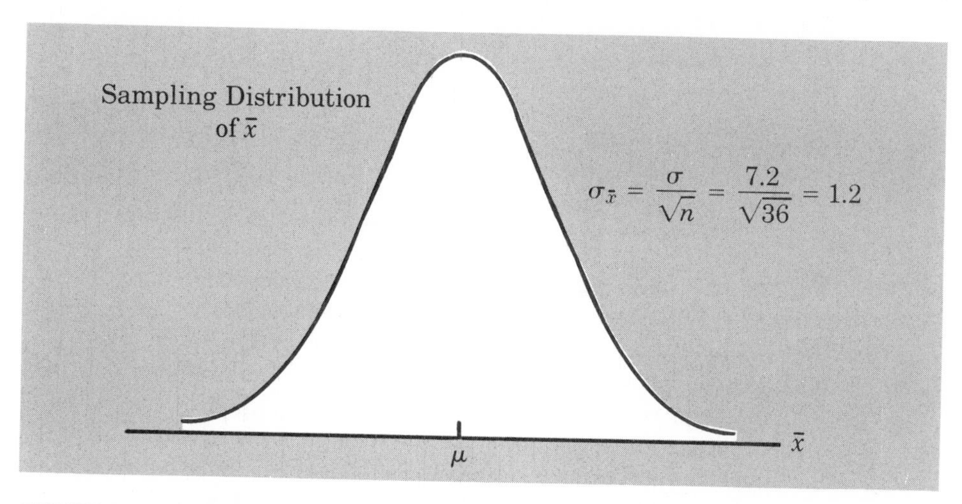

FIGURE 8.1 Sampling Distribution of the Sample Mean Age ($\bar{x}$) from Simple Random Samples of 36 Statewide Insurance Company Policyholders

*In this chapter we will be assuming that $n/N \leq .05$. Thus the finite population correction factor is not needed in the computation of $\sigma_{\bar{x}}$.

standard deviation $\sigma_{\bar{x}} = 7.2/\sqrt{36} = 1.2$ years. This sampling distribution is shown in Figure 8.1.

Although the population mean μ is unknown, the sampling distribution in Figure 8.1 shows how the $\bar{x}$ values are distributed around μ. In effect, this distribution is telling us about the possible differences between $\bar{x}$ and μ and as a result about the possible sampling error.

Probability Statements About the Sampling Error

From our study in Chapter 6 of the normal probability distribution we know that 95% of the values of a normally distributed random variable lie within plus or minus 1.96 standard deviations of the mean. Hence for the sampling distribution of $\bar{x}$ shown in Figure 8.1, 95% of all sample means lie within plus or minus $1.96\sigma_{\bar{x}}$ of the mean μ. Since $1.96\sigma_{\bar{x}} = 1.96(1.2) = 2.35$, we can state that 95% of all sample means lie within plus or minus 2.35 years of the population mean μ. The location of all the sample means that provide a sampling error of 2.35 years or less is shown in Figure 8.2. It is possible

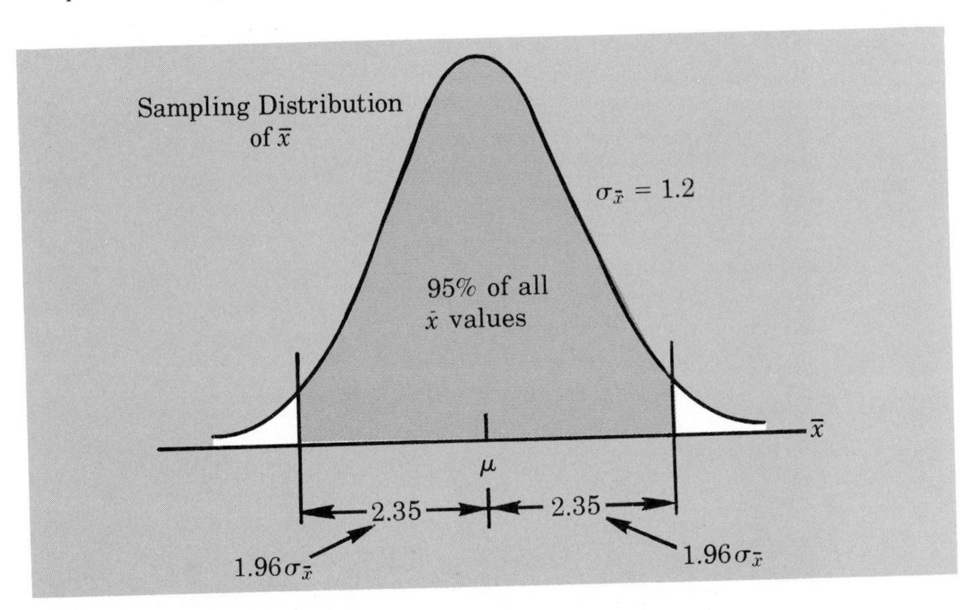

FIGURE 8.2 **Sampling Distribution of $\bar{x}$ Showing the Location of Sample Means That Cause a Sampling Error of 2.35 Years or Less**

for a sample mean to fall in one of the two tails of the sampling distribution, which would result in a sampling error greater than 2.35 years. However, we see from Figure 8.2 that the probability of this occurring is only $1 - .95 = .05$. Thus our knowledge of the sampling distribution of $\bar{x}$ enables us to make the following probability statement about the sampling error whenever a simple random sample of 36 Statewide policyholders is used to provide a point estimate of the mean age of the population:

> There is a .95 probability that the sample mean will provide a sampling error of 2.35 years or less.

The above probability statement about the sampling error is a statement of the *precision* of the estimate. If the project manager is not satisfied with this degree of precision, a larger sample size will be necessary. We shall discuss a procedure for determining the sample size necessary to obtain a desired precision in Section 8.3.

Note that in the above analysis the .95 probability used in the statement about the sampling error was arbitrary. Although a .95 probability is frequently used in making such statements, other probability values can be selected. Probabilities of .90 and .99 are popular alternatives. Let us consider what would have happened to the precision statement if a probability of .99 had been selected. Figure 8.3 shows the location of 99% of the sample means for the Statewide Insurance sampling problem. From the standard normal probability table (Table 1 of Appendix B) we find that 99% of the $\bar{x}$ values lie within plus or minus 2.575 standard deviations of the mean μ. Since $2.575\sigma_{\bar{x}} = 2.575(1.2) = 3.09$, we can make the following statement about the sampling

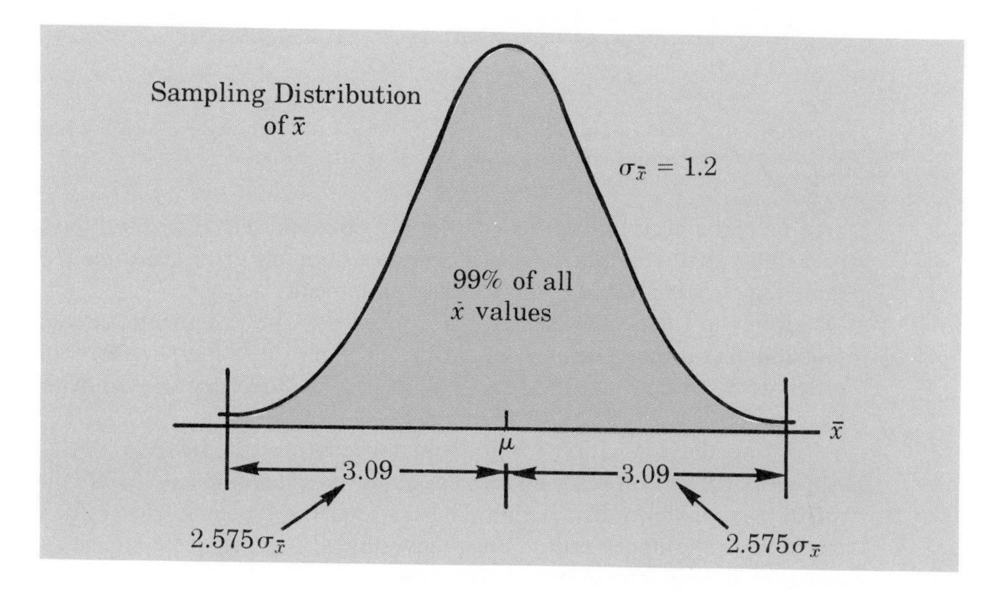

FIGURE 8.3 Sampling Distribution of $\bar{x}$ Showing the Location of 99% of the $\bar{x}$ Values

error whenever a simple random sample of 36 Statewide policyholders is used to provide a point estimate of the mean age of the population:

> There is a .99 probability that the sample mean will provide a sampling error of 3.09 years or less.

A similar calculation with a .90 probability shows that there is a .90 probability that the sample mean will provide a sampling error of $1.645\sigma_{\bar{x}} = 1.645(1.2) = 1.97$ years or less.

The above results show that there are various probability statements that can be made about the sampling error. They also show that there is a tradeoff between the probability specified and the stated limit on the sampling error. In particular, note that the higher probability statements possess larger values for the sampling error.

Let us now generalize the procedure we are using to make probability statements about the sampling error whenever a mean of a simple random sample is used to provide a point estimate of a population mean. We will use the Greek letter α ("alpha") to indicate the probability that a sampling error is *larger* than the sampling error mentioned in the precision statement. Refer to Figure 8.4. We see that $\alpha/2$ will be the

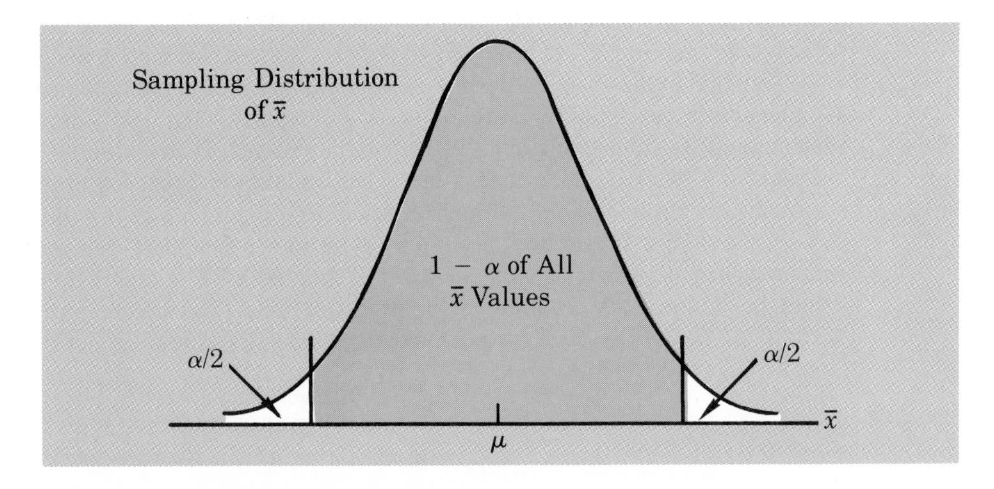

FIGURE 8.4 Areas of a Sampling Distribution of $\bar{x}$ Used to Make Probability Statements About the Sampling Error

area or probability in each tail of the distribution, and $1 - \alpha$ will be the area or probability that a sample mean will provide a sampling error *less than or equal to* the sampling error contained in the precision statement.

Refer to the Statewide Insurance example. The statement that there is a .95 probability that the value of a sample mean will provide a sampling error of 2.35 years or less is based on $\alpha = .05$ and $1 - \alpha = .95$. The area in each tail of the sampling distribution is $\alpha/2 = .025$ (see Figure 8.2).

Using the z notation for the standard normal random variable (Table 1 of Appendix B), we will place a subscript on the z value to denote the area in the *upper tail* of the probability distribution. Thus $z_{.025}$ will correspond to the z value with .025 of the area in the upper tail of the probability distribution. As can be found in the standard normal probability distribution table, $z_{.025} = 1.96$. If we desired a .99 probability statement, $\alpha = .01$ and we would be interested in an area of $\alpha/2 = .005$ in the upper tail of the distribution. Here, $z_{.005} = 2.575$.

Let $z_{\alpha/2}$ = the value for the standard normal random variable corresponding to an area of $\alpha/2$ in the upper tail of the distribution. Also let $\sigma_{\bar{x}}$ = the standard deviation of the sampling distribution of $\bar{x}$ (also called the standard error of the mean). We now have the following general procedure for making a probability statement about the sampling error whenever $\bar{x}$ is used to provide a point estimate of μ:

Probability Statement About the Sampling Error

There is a $1 - \alpha$ probability that the value of a sample mean will provide a sampling error of $z_{\alpha/2}\sigma_{\bar{x}}$ or less.

Calculating an Interval Estimate

We have the ability to make probability statements about the sampling error. We now can combine the point estimate with the probability information about the sampling error to obtain an *interval estimate* of the population mean. The rationale for the interval estimation procedure is as follows: We have already stated that there is a $1 - \alpha$

probability that the value of a sample mean will provide a sampling error of $z_{\alpha/2}\sigma_{\bar{x}}$ or less. This means that there is a $1 - \alpha$ probability that the sample mean or point estimate *will not miss* the population mean *by more than* $z_{\alpha/2}\sigma_{\bar{x}}$. Thus if we form an interval by subtracting $z_{\alpha/2}\sigma_{\bar{x}}$ from the sample mean $\bar{x}$ and then adding $z_{\alpha/2}\sigma_{\bar{x}}$ to the sample mean $\bar{x}$, we would have a $1 - \alpha$ probability of obtaining an interval that *includes* the population mean μ.

Let us return to the Statewide Insurance study. We previously stated that there is a .95 probability that the value of a sample mean provides a sampling error of 2.35 years or less. Look at the sampling distribution of $\bar{x}$ as shown in Figure 8.5 Let us consider some possible values of the sample mean $\bar{x}$ that could be obtained from different simple random samples of 36 policyholders. Remember, in each case we will form an interval estimate of the population mean by subtracting 2.35 from $\bar{x}$ and also by adding 2.35 to $\bar{x}$.

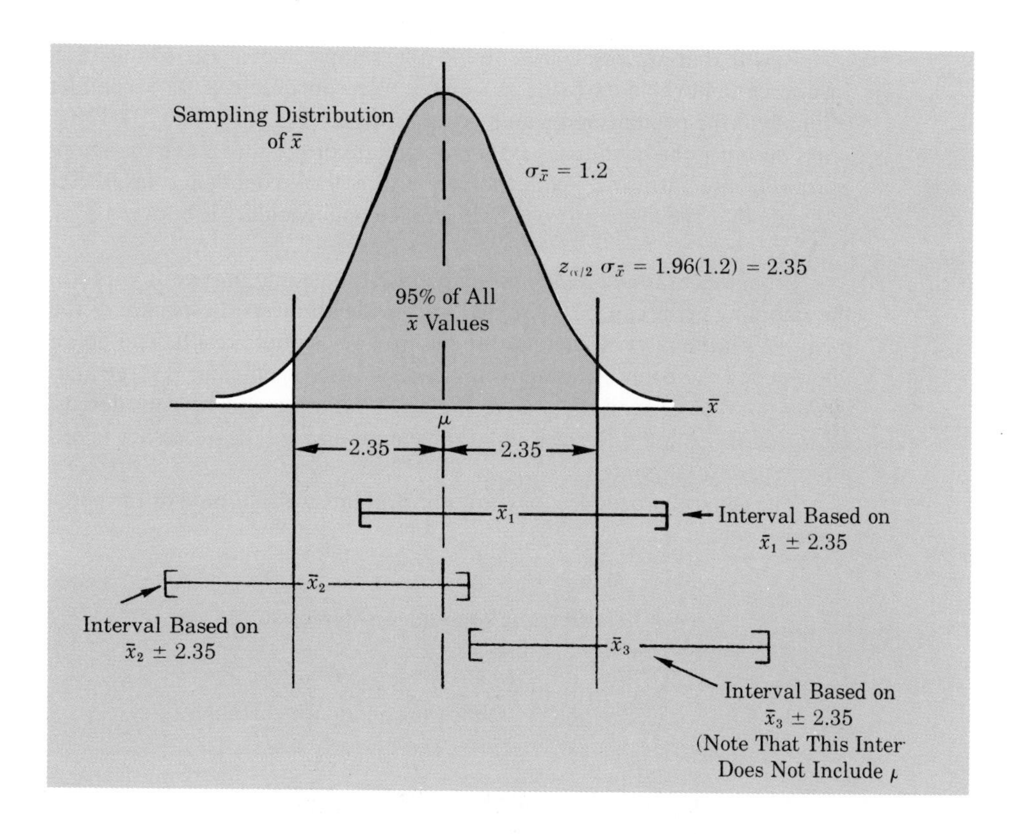

FIGURE 8.5 Intervals Formed from Selected Sample Means at Locations $\bar{x}_1$, $\bar{x}_2$, and $\bar{x}_3$

First, consider what happens if the sample mean turns out to have the value shown in Figure 8.5 as $\bar{x}_1$. Note that in this case the interval formed by subtracting 2.35 from $\bar{x}$ and adding 2.35 to $\bar{x}$ includes the population mean μ. Now consider what happens if the sample mean turns out to have the value shown in Figure 8.5 as $\bar{x}_2$. While this sample mean is different from the first sample mean, we see that the interval based on $\bar{x}_2$ also includes the population mean μ. However, the interval based on the sample mean denoted $\bar{x}_3$ does not include the population mean. The reason for this is that the sample mean $\bar{x}_3$ lies in a tail of the probability distribution at a distance further than 2.35 from μ. Thus subtracting and adding 2.35 to $\bar{x}_3$ forms an interval that does not include μ.

Now think of repeating the sampling process many times, each time computing the value of the sample mean and then forming an interval $\bar{x} - 2.35$ to $\bar{x} + 2.35$. Any sample mean ($\bar{x}$) that falls in the shaded region in Figure 8.5 will provide an interval that includes the population mean μ. Since 95% of the sample means are in the shaded region, 95% of all intervals that could be formed will include μ. As a result, we say that we are 95% confident that an interval constructed from $\bar{x} - 2.35$ to $\bar{x} + 2.35$ will include the population mean. In common statistical terminology the interval is referred to as a *confidence interval.* With 95% of the sample means leading to a confidence interval including μ, we say the interval is established at the 95% *confidence level.* The value .95 is referred to as the *confidence coefficient* for the interval estimate.

Recall that we previously found the sample mean age for 36 Statewide Life Insurance policyholders to be $\bar{x} = 39.5$. We can obtain a 95% confidence interval estimate of the population mean by computing 39.5 ± 2.35. Thus 37.15 years to 41.85 years becomes the confidence interval estimate of the mean age for the population of Statewide life insurance policyholders. At a 95% confidence level, Statewide can conclude that the mean age of life insurance policyholders is between 37.15 years and 41.85 years.

From this example we see how a point estimate and probability information about the sampling error can be combined to provide an interval estimate of the population mean. With the interval estimate the user of the sample results can obtain an idea of "how good" the point estimate is. If the confidence coefficient is large and the interval is rather small or modest in size, the point estimate can be considered a good one. However, if this is not the case, a larger sample size will be necessary in order to obtain a more precise estimate.

The general procedure for computing an interval estimate of the population mean is shown below:

Interval Estimate of a Population Mean

$$\bar{x} \pm z_{\alpha/2}\sigma_{\bar{x}}, \tag{8.2}$$

where $1 - \alpha$ is the confidence coefficient.

The central limit theorem played an important role in the development of the above interval estimation procedure. Specifically, with a large sample size ($n \geq 30$) the central limit theorem allowed us to conclude that the sampling distribution of $\bar{x}$ could be approximated by a normal probability distribution. This nomal distribution approximation is what makes the use of $z_{\alpha/2}$ appropriate in (8.2). However, what happens to the interval estimation procedure when the sample size is small ($n < 30$) and we cannot use the central limit theorem to justify the use of a normal sampling distribution? Recall that in Chapter 7 we pointed out that if the population has a normal distribution, the sampling distribution of $\bar{x}$ is normal regardless of the sample size. Thus if we are faced with a small sample size situation, we consider the possibility of the population having a normal distribution. If this appears to be a reasonable assumption, the sampling distribution of $\bar{x}$ will be normal regardless of the sample size. Thus with σ known, (8.2) can be used to compute an interval estimate of a population mean for this small sample size case. If we are unwilling to make the assumption of a normal population, the alternative is to increase the sample size to $n \geq 30$ and rely on the central limit theorem as the basis for the interval estimation given by (8.2).

Finally, note that whether the sample size is large or small, the interval estimation procedure we have been using in this section is based on the assumption that the population standard deviation σ is known. The procedure for computing an interval estimate of a population mean when σ is unknown is the topic of the next section.

EXERCISES

1. In an effort to estimate the mean amount spent per customer for dinner meals at a major Atlanta restaurant, data were collected for a sample of 49 customers over a 3-week period.

a. Assume a population standard deviation of $2.50. What is the standard error of the mean?

b. With a .95 probability, what statement can be made about the sampling error?

c. If the sample mean is $12.60, what is the 95% confidence interval estimate of the population mean?

2. On a final examination for a statistics course at a large university, 30 randomly selected papers were graded shortly after the exam was over. From the previous year's experience the standard deviation of test scores for the population of students was assumed known, with σ = 13.2.

a. With a .90 probability, what statement can be made about the sampling error when a sample of 30 exam scores is used to estimate the mean exam score for the population?

b. If the sample mean for the 30 papers is 72, what is the 90% confidence interval estimate of the mean exam score for the population?

c. Repeat parts a and b with a .95 probability and a 95% confidence interval.

3. A production filling operation has a historical standard deviation of 5.5 ounces. A quality control inspector periodically selects 36 containers at random and uses the sample mean filling weight to estimate the mean filling weight for the production process.

a. What is the standard error of the mean?

b. With .75, .90, and .99 probabilities, what statements can be made about the sampling error? What happens to the statement about the sampling error when the probability is increased? Why does this happen?

c. What is the 99% confidence interval estimate for the mean filling weight for the process if a sample mean is 48.6 ounces?

4. A sample of 64 customers at Ron and Ted's Service Station shows a mean number of gallons of gasoline purchased per customer of 13.6 gallons. If the population standard deviation is 3.0 gallons, what is the 95% confidence interval estimate of the mean number of gallons purchased per customer?

5. E. Lynn and Associates is an energy research firm that provides estimates of monthly heating costs for new homes based on style of house, square footage, insulation, and so on. The firm's service is used by both builders and potential buyers of new homes who wish advance information on heating costs. For winter months the variance in the home heating bills for residential homes in a certain area is believed to be 10,000. Assume that a sample of 36 homes in a particular subdivision will be used to estimate the mean monthly heating bills for all homes in this type of subdivision.

a. What is the standard error of the mean?

b. Show the sampling distribution for the sample mean heating bill.

c. At an 80% probability, what can be said about the sampling error? Show this probability on a graph of the sampling distribution in part b.
d. What is the 98% confidence interval for the mean monthly heating bill if the sample mean is $196.50?

8.2 INTERVAL ESTIMATION OF A POPULATION MEAN—σ UNKNOWN

In the previous section we showed that an interval estimate of a population mean μ is given by

$$\bar{x} \pm z_{\alpha/2}\sigma_{\bar{x}} \tag{8.2}$$

or with $\sigma_{\bar{x}} = \sigma/\sqrt{n}$, we have

$$\bar{x} \pm z_{\alpha/2}\frac{\sigma}{\sqrt{n}}. \tag{8.3}$$

A difficulty in using (8.3) is that in many sampling situations the value of the population standard deviation σ is *unknown.* In these instances we simply use the value of the sample standard deviation s as the point estimate of the population standard deviation σ. The estimator of the standard deviation of $\bar{x}$ can then be computed as follows:

$$\text{Estimator of } \sigma_{\bar{x}} = s_{\bar{x}} = \frac{s}{\sqrt{n}}. \tag{8.4}$$

The $s_{\bar{x}}$ notation is used for the estimator of $\sigma_{\bar{x}}$ just as s is used as the estimator of σ. With the above estimator of $\sigma_{\bar{x}}$, the interval estimation procedure depends upon whether the sample size is large or small. Let us first consider the interval estimation of a population mean when the sample size is large.

Large-Sample Case

If the sample size is large, the estimated standard error of the mean, $s_{\bar{x}}$, can be accepted as a good estimator of $\sigma_{\bar{x}}$. As a result, the confidence interval is given by the following expression:

Interval Estimate of a Population Mean (Large-Sample Case with σ Unknown)

$$\bar{x} \pm z_{\alpha/2}s_{\bar{x}}, \tag{8.5}$$

where $1 - \alpha$ is the confidence coefficient.

Thus with a large sample size the procedure for developing an interval estimate of a population mean with σ unknown follows the interval estimation procedure of Section 8.1. The only differences are that the sample standard deviation s is used to estimate the population standard deviation σ and that the estimated standard error of the mean, $s_{\bar{x}}$, is used in the interval estimation computation, as shown in (8.5). As a working rule, we shall conclude that the large sample size case is appropriate whenever the size of the random sample is 30 or more.

Small-Sample Case

If the sample size is small, it no longer is safe to simply use the estimated standard error of the mean, $s_{\bar{x}}$, in the interval estimation procedure as shown above. In fact, the ability to construct a confidence interval estimate now depends upon the distribution of the population.

Consider a sampling situation where the population has a normal distribution with an unknown σ. An interval estimate of the population mean can be based on a probability distribution known as the *t distribution.* The t distribution is actually a family of similar probability distributions, with a specific t distribution depending upon a parameter known as the *degrees of freedom.* That is, there is a unique t distribution with 1 degree of freedom, with 2 degrees of freedom, with 3 degrees of freedom, and so on. As the number of degrees of freedom increases, the difference between the t distribution and the standard normal distribution becomes smaller and smaller. Figure 8.6 shows t distributions with 10 and 20 degrees of freedom and their relationship to the standard normal probability distribution.

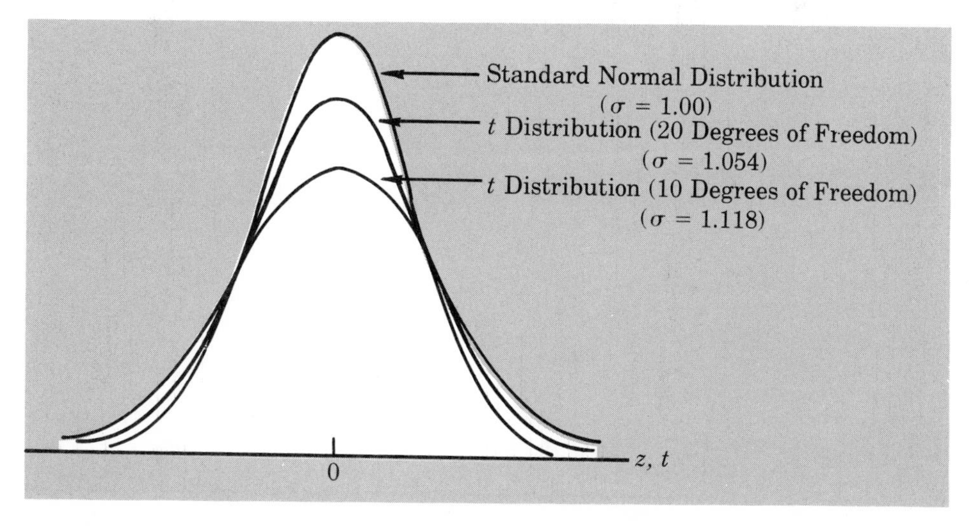

FIGURE 8.6 Comparison of the Standard Normal Distribution with *t* Distributions Having 10 and 20 Degrees of Freedom

We will use a subscript for t to indicate the area in the upper tail of the t distribution. For example, just as we used $z_{.025}$ to indicate the z value providing a .025 area in the upper tail of a standard normal probability distribution, we will use $t_{.025}$ to indicate a .025 area in the upper tail of the t distribution. At times we will refer to the upper tail t value as $t_{\alpha/2}$ with the $\alpha/2$ area in the tail of the t distribution as shown in Figure 8.7.

A table for the t distribution is provided in Table 2 of Appendix B. This table is also shown in Table 8.2. Note, for example, that for a t distribution with 10 degrees of freedom, $t_{.025} = 2.228$. Similarly, for a t distribution with 20 degrees of freedom, $t_{.025} = 2.086$.

Now that we have an idea of what the t distribution is, let us show how it is used to

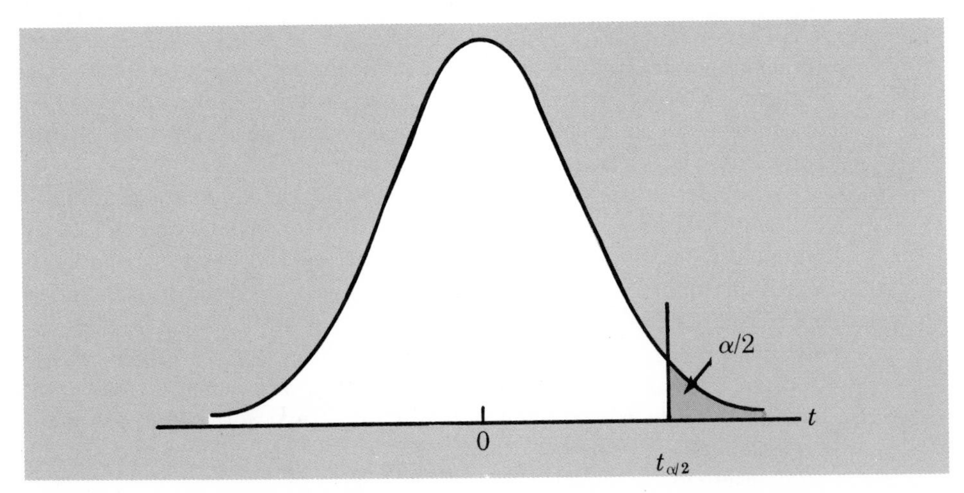

FIGURE 8.7 *t* Distribution with $\alpha/2$ Area or Probability in the Upper Tail

TABLE 8.2 *t* Distribution Tables for Areas in the Upper Tail. Example: With 10 Degrees of Freedom $t_{.025} = 2.228$

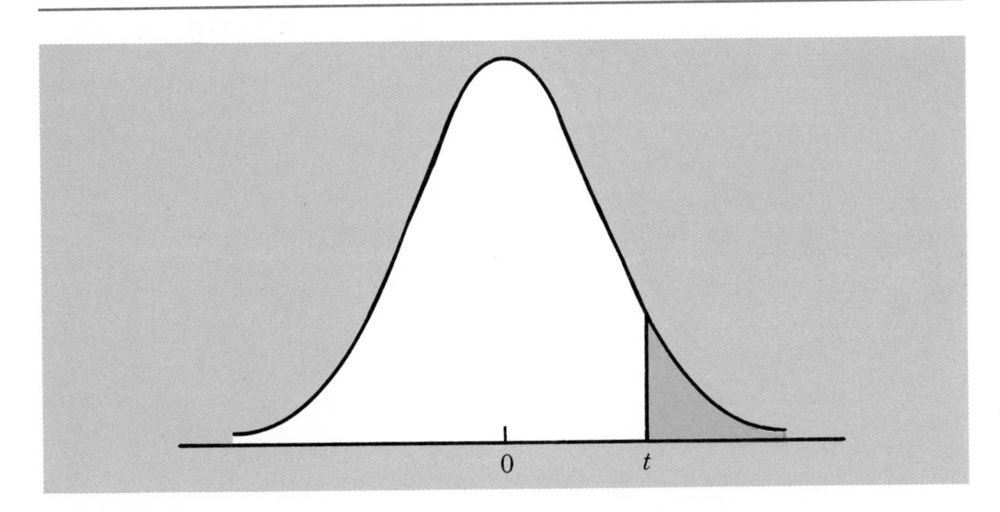

Degrees of Freedom	*Upper-Tail Areas (Shaded)* .10	.05	.025	.01	.005
1	3.078	6.314	12.706	31.821	63.657
2	1.886	2.920	4.303	6.965	9.925
3	1.638	2.353	3.182	4.541	5.841
4	1.533	2.132	2.776	3.747	4.604
5	1.476	2.015	2.571	3.365	4.032
6	1.440	1.943	2.447	3.143	3.707
7	1.415	1.895	2.365	2.998	3.499
8	1.397	1.860	2.306	2.896	3.355
9	1.383	1.833	2.262	2.821	3.250

TABLE 8.2 Continued

10	1.372	1.812	2.228	2.764	3.169
11	1.363	1.796	2.201	2.718	3.106
12	1.356	1.782	2.179	2.681	3.055
13	1.350	1.771	2.160	2.650	3.012
14	1.345	1.761	2.145	2.624	2.977
15	1.341	1.753	2.131	2.602	2.947
16	1.337	1.746	2.120	2.583	2.921
17	1.333	1.740	2.110	2.567	2.898
18	1.330	1.734	2.101	2.552	2.878
19	1.328	1.729	2.093	2.539	2.861
20	1.325	1.725	2.086	2.528	2.845
21	1.323	1.721	2.080	2.518	2.831
22	1.321	1.717	2.074	2.508	2.819
23	1.319	1.714	2.069	2.500	2.807
24	1.318	1.711	2.064	2.492	2.797
25	1.316	1.708	2.060	2.485	2.787
26	1.315	1.706	2.056	2.479	2.779
27	1.314	1.703	2.052	2.473	2.771
28	1.313	1.701	2.048	2.467	2.763
29	1.311	1.699	2.045	2.462	2.756
30	1.310	1.697	2.042	2.457	2.750
40	1.303	1.684	2.021	2.423	2.704
60	1.296	1.671	2.000	2.390	2.660
120	1.289	1.658	1.980	2.358	2.617
∞	1.282	1.645	1.960	2.326	2.576

develop an interval estimate of a population mean. Assume that the population has a normal probability distribution and that the sample standard deviation s is used as a point estimate of the population standard deviation σ. The following interval estimate now is applicable:

Interval Estimate of a Population Mean (Small-Sample Case with σ Unknown)

$$\bar{x} \pm t_{\alpha/2} s_{\bar{x}}, \tag{8.6}$$

where the t value is based on a t distribution with $n - 1$ degrees of freedom and where $1 - \alpha$ is the confidence coefficient.

The reason the number of degrees of freedom associated with the t value in (8.6) is $n - 1$ has to do with the use of s^2 as an estimate of the population variance σ^2. First recall that

$$s^2 = \frac{\Sigma(x_i - \bar{x})^2}{n - 1}.$$

Degrees of freedom here refers to the number of independent pieces of information that go into the computation of $\Sigma(x_i - \bar{x})^2$. The pieces of information involved in computing $\Sigma(x_i - \bar{x})^2$ are $x_1 - \bar{x}, x_2 - \bar{x}, \ldots, x_n - \bar{x}$. In Section 3.3 we indicated that $\Sigma(x_i - \bar{x}) = 0$ for any data set. Thus only $n - 1$ of the $x_i - \bar{x}$ values are independent; if we know $n - 1$ of the values, the remaining value can be determined exactly using the condition that all of the $x_i - \bar{x}$ values must sum to 0.

Let us demonstrate the use of the above interval estimate by considering the training program evaluation conducted by Scheer Industries. Scheer's director of manufacturing is interested in a computer assisted training program that can be used to train the firm's maintenance employees for machine repair operations. It is anticipated that the computer assisted training will reduce training time and training costs for Scheer employees. In order to evaluate the training method, the director of manufacturing has requested an estimate of the mean training time required using the computer assisted training technique.

Suppose that management has agreed to train 15 employees with the new approach. The data on training days required for each employee in the sample are shown in Table 8.3. The sample mean and sample standard deviation for these data are as follows:

$$\bar{x} = \frac{\Sigma x_i}{n} = \frac{808}{15} = 53.87 \text{ days},$$

$$s = \sqrt{\frac{\Sigma(x_i - \bar{x})^2}{n - 1}} = \sqrt{\frac{651.73}{14}} = 6.82 \text{ days}.$$

The point estimate of the mean training time for the population of employees is 53.87 days. We can obtain information about the precision of this estimate by developing an interval estimate of the population mean. Since the population standard deviation is unknown, we will use the sample standard deviation $s = 6.82$ days as the point estimate of σ. The corresponding estimate of the standard error of the mean is given by

$$s_{\bar{x}} = \frac{s}{\sqrt{n}} = \frac{6.82}{\sqrt{15}} = 1.76.$$

With the small sample size $n = 15$, we will look to (8.6) to make the interval estimate calculations. Assume that the population of training times is normal. The t distribution with $n - 1 = 14$ degrees of freedom is the appropriate probability distribution for the estimation procedure. Selecting a .95 confidence coefficient, we see from Table 8.2

TABLE 8.3 Machine Repair Training Time in Days for the Computer Assisted Instruction Program

Employee	*Time*	*Employee*	*Time*	*Employee*	*Time*
1	52	6	59	11	54
2	44	7	50	12	58
3	55	8	54	13	60
4	44	9	62	14	62
5	45	10	46	15	63

(Table 2 of Appendix B) that $t_{\alpha/2} = t_{.025} = 2.145$. Using (8.6), we have

$$\bar{x} \pm t_{.025} s_{\bar{x}},$$
$$53.87 \pm (2.145)(1.76),$$
$$53.87 \pm 3.78.$$

Thus the 95% confidence interval estimate of the population mean is 50.09 days to 57.65 days.

Strictly speaking, the above approach to interval estimation based on the use of the t distribution is applicable whenever the population being sampled has a normal probability distribution. However, statistical research has shown that (8.6) is applicable even if the population being sampled is not quite normal. That is, confidence intervals based on the t distribution provide good results provided that the population distribution does not differ extensively from a normal probability distribution. The fact that the t distribution can be used to provide satisfactory results for many possible population distributions is referred to as the *robustness* property of the t distribution.

As a final comment, we point out that the t distribution is not restricted to the small-sample situation. Actually the t distribution is applicable whenever the population is normal or near normal and whenever the sample standard deviation is used to estimate the population standard deviation. However, (8.5) shows that with a large sample the interval estimation procedure can be conducted without using the t distribution. Thus we generally do not consider the t distribution until we encounter a small-sample-size case.

EXERCISES

6. During a water shortage a water company randomly sampled residential water meters in order to monitor daily water consumption. On one particular day, a sample of 50 meters showed a sample mean of $\bar{x} = 240$ gallons and a sample standard deviation of $s = 45$ gallons. Provide a 90 percent confidence interval estimate of the mean water consumption for the population.

7. A simple random sample of 35 Metro buses shows a sample mean of 225 passengers carried per day per bus. The sample standard deviation is computed to be 60 passengers. Provide a 98% confidence interval estimate of the mean number of passengers carried per bus during a 1 day period.

8. In the testing of a new production method, 18 employees were randomly selected and asked to try the new method. The sample mean production rate for the 18 employees was 80 parts per hour. The sample standard deviation was 10 parts per hour. Provide 90% and 95% confidence interval estimates for the mean production rate for the new method.

9. The following data are family sizes from a simple random sample of households in a new test market area:

Household	*Family Size*	*Household*	*Family Size*
1	4	7	3
2	3	8	2
3	2	9	3
4	2	10	6
5	4	11	3
6	5	12	2

Provide a 95% confidence interval estimate for the mean family size for the population.

10. Sales personnel for Skillings Distributors are required to submit weekly reports listing the customer contacts made during the week. A sample of 61 weekly contact reports showed a mean of 22.4 customer contacts per week for the sales personnel. The sample standard deviation was 5 contacts.

a. Develop a 95% confidence interval estimate for the mean number of weekly customer contacts for the population of sales personnel.

b. Assume that the population of weekly contact data has a normal distribution. The t distribution can also be used to develop an interval estimate of the population mean. Use the t distribution to develop a 95% confidence interval for the mean number of weekly customer contacts.

c. Compare your answers for parts a and b. Comment on why in the large-sample case it is permissible to base interval estimates on the procedure used in part a even if the t distribution also is applicable.

11. A simple random sample of five people provided the following data on ages: 21, 25, 20, 18, and 21. Develop a 95% confidence interval for the mean age of the population being sampled.

12. Shown below are the duration (in minutes) for a sample of 20 telephone flight reservations:

2.1	4.8	5.5	10.4
3.3	3.5	4.8	5.8
5.3	5.5	2.8	3.6
5.9	6.6	7.8	10.5
7.5	6.0	4.5	4.8

a. What is the point estimate of the population mean time for flight reservation phone calls?

b. Develop a 95% confidence interval estimate of the population mean time.

8.3 DETERMINING THE SIZE OF THE SAMPLE

Recall that in Section 8.1 we were able to make the following probability statement about the sampling error whenever a sample mean was used to provide a point estimate of a population mean:

> There is a $1 - \alpha$ probability that the value of the sample mean will provide a sampling error of $z_{\alpha/2}\sigma_{\bar{x}}$ or less.

Note that $\sigma_{\bar{x}} = \sigma/\sqrt{n}$. We now can rewrite this statement as follows:

> There is a $1 - \alpha$ probability that the value of the sample mean will provide a sampling error of $z_{\alpha/2}(\sigma/\sqrt{n})$ or less.

From the above statement we see that the values of $z_{\alpha/2}$, σ, and the sample size n combine to determine the sampling error mentioned in the precision statement. Once we select a confidence coefficient or probability $1 - \alpha$, $z_{\alpha/2}$ can be determined. Since the population standard deviation σ is fixed and does not depend upon the sample result, we see that the probability statement about the sampling error is dependent solely upon the size of the sample, n. A larger sample size will provide a smaller sampling error and thus more precision. In fact, given values for $z_{\alpha/2}$ and σ, we can adjust the sample size n to provide any precision desired. The formula used to compute the required sample size is developed as follows.

Let E = the sampling error mentioned in the statement about the desired precision. We now have

$$z_{\alpha/2}\frac{\sigma}{\sqrt{n}} = E. \tag{8.7}$$

Using (8.7) to solve for $\sqrt{n}$, we have

$$\sqrt{n} = \frac{z_{\alpha/2}\sigma}{E}.$$

Squaring both sides of this expression, we obtain the following expression for the sample size

Sample Size for Interval Estimate of a Population Mean

$$n = \frac{(z_{\alpha/2})^2\sigma^2}{E^2}. \tag{8.8}$$

This sample size n will provide a precision statement with a $1 - \alpha$ probability that the sampling error will be *E or less.*

To see how (8.8) can now be used to determine a recommended sample size, let us return to the Scheer Industries problem. We showed in Section 8.2 that for a sample of 15 employees the 95% confidence interval was 53.87 days ± 3.78 days. Assume that after considering the interval estimate of 50.09 days to 57.65 days, Scheer's director of manufacturing is not satisfied with this degree of precision. Furthermore, assume that the director makes the following statement about the desired precision: "I would like a .95 probability that the value of the sample mean will provide a sampling error of 2 days or less." From the above statement we have $E = 2$ days. In addition, with $1 - \alpha$ specified at .95 we have $\alpha = .05$ and $z_{\alpha/2} = z_{.025} = 1.96$. Note that we are using the z value in our calculations of the sample size even though the original computations for the Scheer problem employed the t distribution. The reason for this is that since the sample size is yet to be determined, we are anticipating that it will be larger than 30. Thus (8.8) is based on a z value rather than a t value.

Refer to (8.8). We see that with $E = 2$ and $z_{.025} = 1.96$ we need a value for σ in order to compute the sample size that will provide the desired precision. Do we have a value of σ that we could use for the Scheer problem? Although σ is unknown, let us take advantage of the sample results reported in Section 8.2, where we found a sample standard deviation $s = 6.82$ days. With this value as a planning value for the population standard deviation σ, (8.8) shows that the following sample size should provide the director's desired precision:

$$n = \frac{(z_{\alpha/2})^2\sigma^2}{E^2} = \frac{(1.96)^2(6.82)^2}{2^2} = 44.67.$$

In cases where the computed n is a fraction, we round up to the next integer value, thus making the recommended sample size for the Scheer problem 45 employees. Since Scheer already has test data for 15 employees, an additional $45 - 15 = 30$ employees should be tested if the director wishes to obtain the desired precision of ±2 days at a 95% confidence level.

Note that in (8.8) the values of $z_{\alpha/2}$ and E follow directly from the statement about the desired precision. However, the value of the population standard deviation σ may or may not be known. In cases where σ is known we have no problem in determining the sample size. However, in cases where σ is unknown we see that we must at least have a preliminary or *planning value* for σ in order to compute the sample size. In the Scheer Industries example, we were fortunate to have a sample of 15 employees which provided a point estimate for σ at 6.82 days. In instances where this initial or preliminary sample is unavailable, we may be able to obtain a good approximation of σ from past data on "similar" studies. Without such past data we may have to use a judgment or "best-guess" value for σ. In any case, regardless of the source, we see from (8.8) that we must have a planning value for σ in order to determine a recommended sample size.

EXERCISES

13. What sample size would have been recommended for the Scheer Industries problem if the director of manufacturing had specified a .95 probability for a sampling error of 1.5 days or less? How large a sample would have been necessary if the precision statement had specified a .90 probability for a sampling error of 2 days or less?

14. In Section 8.1 the Statewide Insurance Company used a simple random sample of 36 policyholders to estimate the mean age of the population of policyholders. The resulting precision statement was reported to have a .95 probability that the value of the sample mean provided a sampling error of 2.35 years or less. This statement was based on a known population standard deviation of 7.2 years.

a. How large a simple random sample would have been necessary to reduce the sampling error statement to 2 years or less? To 1.5 years or less? To 1 year or less?

b. Would you recommend that Statewide attempt to estimate the mean age of the policyholders with $E = 1$ year? Explain.

15. Starting annual salaries for college graduates with business administration degrees are believed to have a standard deviation of approximately $2,000.00. Assume that a 95% confidence interval estimate of the mean annual starting salary is desired. How large a sample size should be taken if the size of the sampling error in the precision statement is to be

a. $500.00?

b. $200.00?

c. $100.00?

16. A national survey research firm has past data that indicate that the interview time for a consumer opinion study has a standard deviation of 6 minutes.

a. How large a sample should be taken if the firm desires a .98 probability of estimating the mean interview time to within 2 minutes or less?

b. Assume that the simple random sample you recommended in part a is taken and that the mean interview time for the sample is 32 minutes. What is the 98% confidence interval estimate for the mean interview time for the population of interviews?

17. A gasoline service station shows a standard deviation of $6.25 for the charges made by the credit card customers. Assume that the station's management would like an estimate of the mean gasoline bill for its credit card customers to within ±$1.00 of the actual population mean. For a 95% confidence level, how large a sample would be necessary?

8.4 INTERVAL ESTIMATION OF A POPULATION PROPORTION

In Section 8.2 we presented the Scheer Industries problem, which involved estimating the mean employee training time for a new machine-repair training program. In order to evaluate the program from a different perspective, management has requested that some measure of program quality be developed. In the past the degree of success of the training program has been measured by the score the employee obtains on a standard examination given at the end of the training program. From past experience the company has found that an individual scoring 75 or better on the examination has excellent chances of high performance on the job. After some discussion management has agreed to evaluate the program quality for the new training method based on the proportion of the employees that score 75 or better on the examination. Let us assume that Scheer implemented the sample size recommendation of the preceding section. Thus we now have a sample of 45 employees which can be used to develop an interval estimate for the proportion of the population that score 75 or better on the examination.

In Chapter 7 we learned that a sample proportion $\bar{p}$ was an unbiased estimator of a population proportion p and that the large sample approximation of the sampling distribution of $\bar{p}$ is normal as shown in Figure 8.8. Recall that the use of the normal

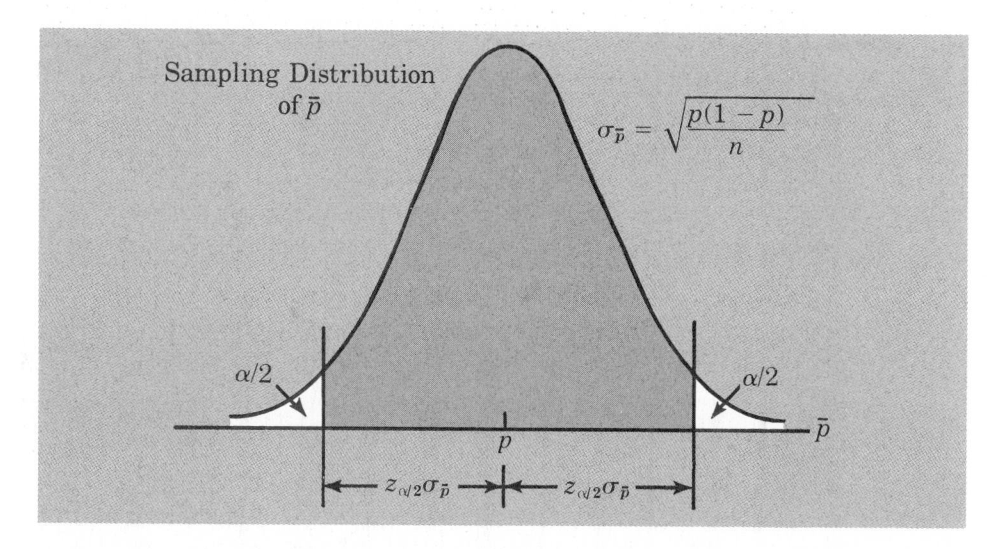

FIGURE 8.8 Normal Approximation of the Sampling Distribution of $\bar{p}$ When Both $np \geq 5$ and $n(1 - p) \geq 5$

distribution as an approximation of the sampling distribution of $\bar{p}$ was based on the assumption that both np and $n(1 - p)$ are 5 or more. We will be using our knowledge of the sampling distribution of $\bar{p}$ to make probability statements about the sampling error whenever a sample proportion $\bar{p}$ is used to estimate a population proportion p. In this case the sampling error is defined as the magnitude of the difference between $\bar{p}$ and p, written $|\bar{p} - p|$.

The probability statements that we can make about the sampling error for the proportion take the following form:

> There is a $1 - \alpha$ probability that the value of the sample proportion will provide a sampling error of $z_{\alpha/2}\sigma_{\bar{p}}$ or less.

The rationale for the above statement is the same as we used when the value of a sample mean was used as an estimate of a population mean. Namely, since we know that the sampling distribution of $\bar{p}$ can be approximated by a normal probability distribution, we can use the value of $z_{\alpha/2}$ and the value of the standard error of the proportion, $\sigma_{\bar{p}}$, to make the probability statement about the sampling error.

Once we see that the probability statement concerning the sampling error is based on $z_{\alpha/2}\sigma_{\bar{p}}$, we can subtract this value from $\bar{p}$ and add it to $\bar{p}$ in order to obtain an interval estimate of the population proportion. Such an interval estimate is given by

$$\bar{p} \pm z_{\alpha/2}\sigma_{\bar{p}}, \tag{8.9}$$

where $1 - \alpha$ is the confidence coefficient (applicable provided that $np \geq 5$ and $n(1 - p) \geq 5$).

One difficulty remains. In order to use (8.9) we need to compute $\sigma_{\bar{p}} = \sqrt{p(1 - p)/n}$. While the value of the sample size n is known, the value of the population proportion p is what we are trying to estimate. Thus its value is unknown. As a result we estimate $\sigma_{\bar{p}}$ by using the sample proportion $\bar{p}$. Such an estimator of $\sigma_{\bar{p}}$, denoted by $s_{\bar{p}}$, is given by

$$s_{\bar{p}} = \sqrt{\frac{\bar{p}(1 - \bar{p})}{n - 1}}. \tag{8.10}$$

The $n - 1$ term in the denominator is necessary to provide an unbiased estimate. The theoretical basis for this is similar to the rationale for using $n - 1$ in the denominator of s^2 in order to make the sample variance an unbiased estimator of the population variance.

With $s_{\bar{p}}$ as the estimator of $\sigma_{\bar{p}}$, the general expression for an interval estimate of a population proportion in the large sample case is as follows:

Interval Estimate of a Population Proportion

$$\bar{p} \pm z_{\alpha/2}s_{\bar{p}}, \tag{8.11}$$

where $1 - \alpha$ is the confidence coefficient.

Let us return to the Scheer Industries problem. Assume that in the sample of 45 employees who completed the new training program, 36 scored 75 or above on the examination. Thus the point estimate of the proportion in the population that score 75 or above on the examination is $\bar{p} = 36/45 = .80$. Using (8.11) and a .95 confidence coefficient, the interval estimate for the population proportion is given by

$$\bar{p} \pm z_{.025}s_{\bar{p}},$$

$$.80 \pm 1.96\sqrt{\frac{.80(1 - .80)}{44}},$$

$$.80 \pm .12.$$

Thus we see that at the 95% confidence level the interval estimate of the population proportion is .68 to .92.

Determining the Size of the Sample

Let us consider the question of how large the sample size should be in order to obtain a point estimate of a population proportion at a specified level of precision. The rationale for the sample size determination in developing interval estimates of p is very similar to the rationale used in Section 8.3 to determine the sample size for estimating a population mean.

Earlier in this section we provided the following probability statement about the sampling error:

> There is a $1 - \alpha$ probability that the value of the sample proportion will provide a sampling error of $z_{\alpha/2}\sigma_{\bar{p}}$ or less.

With $\sigma_{\bar{p}} = \sqrt{p(1 - p)/n}$, the sampling error in the above statement is based on the values of $z_{\alpha/2}$, the population proportion p, and the sample size n. For a given confidence coefficient $1 - \alpha$, $z_{\alpha/2}$ can be determined. Then, since the value of the population proportion is fixed, the sampling error mentioned in the precision statement is determined by the sample size n. Larger sample sizes again provide better precision.

Let E = the sampling error mentioned in the statement about the desired precision. We then have

$$z_{\alpha/2}\sqrt{\frac{p(1 - p)}{n}} = E.$$

Solving the above equation for n provides the following formula for sample size determination when estimating a population proportion:

Sample Size for Estimating a Population Proportion

$$n = \frac{(z_{\alpha/2})^2 p(1 - p)}{E^2}. \qquad (8.12)$$

Assume that the manager in the Scheer study has requested a sampling error of .10 or less with a .95 probability. This degree of precision specifies $z_{.025} = 1.96$ and $E = .10$.

In order to use (8.12) to find the necessary sample size, we need a planning value for the population proportion p. Obviously, p will never be known exactly, since it is the population parameter we are trying to estimate. Thus we will use past data, a preliminary sample, or judgment to determine a planning value for p. For the Scheer Industries example, we can use the sample proportion of .80 from the sample of 45 employees as the planning value for the population proportion p. Substituting this value for p into (8.12) provides the following sample size:

$$n = \frac{(1.96)^2 .80(1 - .80)}{(.10)^2} = 61.47, \qquad \text{or } 62.$$

Thus for our calculations a sample size of 62, or 17 more than the current sample of 45, would be necessary to meet the precision requirement of $\pm.10$ at a 95% confidence level.

In the above example we were fortunate to have the sample of 45 and the associated planning value of .80 available for p. In other cases it may be more difficult to determine an appropriate planning value for p. However, note that the numerator of (8.12) shows that the sample size is proportional to the quantity $p(1 - p)$. In Table 8.4 are shown some possible values for this quantity. To be on the safe or conservative side by always providing a sample size that will meet the precision requirement, we have to consider the largest possible value for $p(1 - p)$. Thus whenever there is trouble determining an appropriate planning value for p we suggest using $p = .50$. Checking (8.12) with other planning values for p will show that $p = .50$ provides the largest sample size recommendation. If the proportion is different than the .50 planning value,

Table 8.4 Possible Values for $p(1 - p)$

p	$p(1 - p)$
.10	(.10)(.90) = .09
.30	(.30)(.70) = .21
.40	(.40)(.60) = .24
.50	(.50)(.50) = .25 ← Largest Value for $p(1 - p)$
.60	(.60)(.40) = .24
.70	(.70)(.30) = .21
.90	(.90)(.10) = .09

the precision statement will be better than anticipated. However, in using this procedure at least we have guaranteed that the required level of precision will be obtained.

In the Scheer Industries sample size determination, a planning value of $p = .50$ would have provided the following sample size recommendation:

$$n = \frac{(1.96)^2 .50(1 - .50)}{(.10)^2} \approx 96.$$

The larger recommended sample size reflects the caution we took in using the conservative planning value for the population proportion.

EXERCISES

18. A simple random sample of 100 residents of Watkins Glen, New York resulted in 65 individuals stating that they would support a newly proposed water treatment facility. Develop a 95% confidence interval estimate of the proportion of all Watkins Glen residents that would support the new water treatment facility.

19. In a telephone followup survey of a new advertising campaign, 45 of 150 individuals contacted could recall the new advertising slogan associated with the product. Develop a 90% confidence interval estimate of the proportion in the population that will recall the advertising slogan.

20. A sample of 90 students at a particular college showed 27 students favoring pass–fail grades for elective courses.

a. What is the point estimate of the proportion of all students who would favor pass–fail grades for elective courses?

b. Provide a 90% confidence interval estimate of the proportion of the population of students who would favor pass–fail grades for elective courses.

21. In an election campaign, a campaign manager requests that a sample of voters be polled to determine the support for the candidate. From a sample of 120 voters, 64 express plans to support the candidate.

a. What is the point estimate of the proportion of the voters in the population who will support the candidate?

b. Develop and interpret the 95% confidence interval for the proportion of voters in the population who will support the candidate.

c. From the result from part b, is the campaign manager justified in feeling confident that the candidate has support of at least 50% of the voters? Explain.

d. How many voters should be sampled if we desired to estimate the population proportion with a sampling error of 5% or less? Continue to use the 95% confidence level.

22. A new cheese product is to be test marketed by giving a free sample to randomly selected customers and asking them to state whether or not they like the product. With a 98% confidence level and a target sampling error of .05 or less, what sample size would you recommend

a. If preliminary estimates are that approximately 35% of the individuals in the population will like the product?

b. If no information is available about the proportion in the population that will like the product?

23. A firm provides national survey and interview services designed to estimate the proportion of the population that have certain beliefs or preferences. Typical questions seek to find the proportion favoring gun control, the proportion favoring abortion, the proportion favoring a particular political candidate, and so on. Assume that all interval estimates of population proportions are conducted at the 95% confidence level. How large a sample size would you recommend if the firm desired the sampling error to be

a. 3% or less?

b. 2% or less?

c. 1% or less?

Summary

In this chapter we presented confidence interval estimation procedures for a population mean μ and a population proportion p. The purpose of developing an interval estimate of a population parameter is to provide the user of the sample results with a better understanding of the sampling error that may exist for a point estimate. If the width of an interval estimate is considered to be too large, the sample size can be increased in order to improve the precision of the estimate.

In cases where the population standard deviation is unknown, we showed how the sample standard deviation could be used in the development of the confidence interval for a population mean. If the sample size is large, $\bar{x} \pm z_{\alpha/2}s_{\bar{x}}$ can be used to compute the confidence interval. However, when the sample size is small, the t distribution is used to compute a confidence interval provided that the population is normal or at least near normal. For a t distribution with $n - 1$ degrees of freedom, the confidence interval estimate is given by $\bar{x} \pm t_{\alpha/2}s_{\bar{x}}$.

In addition, we showed how to determine the sample size so that the interval estimates of μ and p would possess a specified or desired level of precision.

Glossary

Interval estimate—An estimate of a population paramcter that provides an interval of values believed to contain the value of the parameter.

Sampling error—The magnitude of the difference between the point estimate, such as the sample mean $\bar{x}$, and the value of the population parameter it estimates, such as the population mean μ. In this case the sampling error is $|\bar{x} - \mu|$. In the case of the population proportion, the sampling error is given by $|\bar{p} - p|$.

Precision—A probability statement about the sampling error.

Confidence level—The confidence associated with the ability of an interval estimate to contain the value of the parameter of interest. For example, if an interval estimation procedure provides intervals such that 95% will include the population parameter, an interval estimate is said to be constructed at the 95% confidence level, and .95 is referred to as the *confidence coefficient*.

***t* distribution**—A family of probability distributions which can be used to develop interval estimates of a population mean whenever the population standard deviation is unknown and the population has a normal or near-normal distribution.

Degrees of freedom—A parameter of the t distribution that specifies the t distribution of interest. When the t distribution is used in the computation of an interval estimate of a population mean, the appropriate t distribution has $n - 1$ degrees of freedom, where n is the size of the simple random sample.

Key Formulas

Sampling Error

$$|\bar{x} - \mu| \tag{8.1}$$

Interval Estimate of a Population Mean

$$\bar{x} \pm z_{\alpha/2}\sigma_{\bar{x}} \tag{8.2}$$

Estimator of $\sigma_{\bar{x}}$

$$s_{\bar{x}} = \frac{s}{\sqrt{n}} \tag{8.4}$$

Interval Estimate of a Population Mean (Large-Sample Case with σ Unknown)

$$\bar{x} \pm z_{\alpha/2}s_{\bar{x}} \tag{8.5}$$

Interval Estimate of a Population Mean (Small-Sample Case with σ Unknown)

$$\bar{x} \pm t_{\alpha/2}s_{\bar{x}} \tag{8.6}$$

Sample Size for Interval Estimate of a Population Mean

$$n = \frac{(z_{\alpha/2})^2\sigma^2}{E^2} \tag{8.8}$$

Estimator of $\sigma_{\bar{p}}$

$$s_{\bar{p}} = \sqrt{\frac{\bar{p}(1 - \bar{p})}{n - 1}} \tag{8.10}$$

Interval Estimate of a Population Proportion

$$\bar{p} \pm z_{\alpha/2}s_{\bar{p}} \tag{8.11}$$

Sample Size for Interval Estimate of a Population Proportion

$$n = \frac{(z_{\alpha/2})^2 p(1 - p)}{E^2} \tag{8.12}$$

Supplementary Exercises

24. The Benson Property Management firm located in St. Louis would like to estimate the mean cost of repairing damages in apartments that are vacated by tenants. A sample of 36 vacated apartments resulted in a sample mean repair cost of $86.00, with a sample standard deviation of $12.25. Develop a 95% confidence interval to estimate the mean repair cost for the population of apartments.

25. The North Carolina Savings and Loan Association would like to develop an estimate of the mean size of home improvement loans granted by its member institutions. A sample of 100 loans granted by member institutions resulted in a sample mean of $3,400 and a sample standard deviation of $650. With these data develop a 98% confidence interval for the mean dollar amount of home improvement loans.

26. In a test of phone utilization, a firm recorded the length of time for phone calls handled by its main switchboard. A sample of 50 phone calls provided a sample mean of 8.9 minutes and a sample standard deviation of 5 minutes.

a. What is a 90% confidence interval estimate for the mean phone call duration?

b. What is a 99% confidence interval estimate?

27. Dailey Paints, Inc. implemented a long-term painting test study designed to check the wear resistance of its major brand of paint. The test consisted of painting eight houses in various parts of the United States and observing the number of months until signs of peeling were observed. The following data were obtained:

House	1	2	3	4	5	6	7	8
Time Until Signs of Peeling (months)	60	51	64	45	48	62	54	56

a. What is a point estimate of the mean number of months until signs of peeling are observed?

b. Develop a 95% confidence interval to estimate the population mean number of months until signs of peeling are observed.

c. Develop a 99% confidence interval for the population mean.

28. The directors of a university computer center have been studying the usage of the center's 30 computer terminals on Friday evenings. A sample of 5 weeks resulted in the following data:

Week	1	2	3	4	5
Number of Terminals in Use at 9:00 P.M.	12	18	21	15	9

Treat the data as being from a simple random sample and develop a 95% confidence interval estimate for the mean number of terminals in use on Friday evenings at 9:00 P.M.

29. The Atlantic Fishing and Tackle Company has developed a new synthetic fishing line. In order to estimate the breaking strength of this line, testers subjected six lengths of line to breakage testing. The following data were obtained:

Line	1	2	3	4	5	6
Breaking Strength (pounds)	18	24	19	21	20	18

Develop a 95% confidence interval estimate of the mean breaking strength of the new line.

30. Sample assembly times for a particular manufactured part were 8, 10, 10, 12, 15, and 17 minutes. If the mean of the sample is to be used to estimate the mean of the population of assembly times, provide a point estimate and a 90% confidence interval estimate of the population mean.

31. A utility company finds that a sample of 100 delinquent accounts yields an average amount owed of \$131.44, with a sample standard deviation of \$16.19. Develop a 90% confidence interval for the population mean amount owed.

32. In Exercise 31 a utility company sampled 100 delinquent accounts in order to estimate the mean amount owed by these accounts. The sample standard deviation was \$131.44. How large a sample should be taken if the company wants to be 90% confident that the estimate of the population mean will have a sampling error of \$10 or less?

33. Consider the Atlantic Fishing and Tackle Company problem presented in Exercise 29. How large a sample would be necessary in order to estimate the mean breaking strength of the new line with a .99 probability of a sampling error of 1 pound or less?

34. Mileage tests are conducted for a particular model of automobile. If the desired precision is stated such that there is to be .98 probability of a sampling error of 1 mile per gallon or less, how many automobiles should be used in the test? Assume that preliminary mileage tests indicate a standard deviation in miles per gallon for the automobiles to be 2.6 miles per gallon.

35. In developing patient appointment schedules, a medical center desires to estimate the mean time that a staff member spends with each patient. How large a sample should be taken if the precision of the estimate is to be ±2 minutes at a 95% level of confidence? How large a sample for a 99% level of confidence? Use a planning value for the population standard deviation of 8 minutes.

36. The New Orleans Beverage Company has been experiencing problems with the automatic machine that places labels on bottles. The company desires an estimate of the percentage of bottles that have improperly applied labels. A simple random sample of 400 bottles resulted in 18 bottles with improperly applied labels. Using these data, develop a 90% confidence interval estimate of the population proportion of bottles with improperly applied labels.

37. H. G. Forester and Company is a distributor of lumber supplies throughout the Southwest United States. Management of H. G. Forester would like to check a shipment of over 1 million pine boards in order to determine if excessive warpage exists for the boards. A sample of 50 boards resulted in the identification of 7 boards with excessive warpage. With these data develop a 95% confidence interval estimate of the proportion of boards defective in the whole shipment.

38. Consider the H. G. Forester and Company problem represented in Exercise 37. How large a sample would be required to estimate the proportion of boards with warpage to within ±.01 at a 95% confidence level?

39. A well known bank credit card firm is interested in estimating the proportion of credit card holders that carry a nonzero balance at the end of the month and incur an interest charge. Assume that the desired precision for the proportion estimate is ±3% at a 98% confidence level.

a. How large a sample should be recommended if it is anticipated that roughly 70% of the firm's cardholders carry a nonzero balance at the end of the month?

b. How large a sample would be recommended if no planning value for the population proportion could be specified?

40. A sample of 200 people were asked to identify their major source of news information; 110 stated that their major source was television news coverage.

a. Construct a 95% confidence interval for the proportion of the people in the population that consider television news their major source.

b. How large a sample would be necessary to estimate the population proportion with a sampling error of .05 or less at a 95% confidence level?

Application

Thriftway

Thriftway, Inc.*

Cincinnati, Ohio

Thriftway, Inc. was formed in 1959 to operate two small supermarkets. Today it is one of the larger local supermarket chains in the country and is among the top 50 nonpublic supermarket chains nationwide. The company retails groceries, meats, fresh produce, drugs, and general merchandise through 22 locations throughout Cincinnati and neighboring communities in southwestern Ohio and northern Kentucky.

Being in an inventory-intense business (the company carries in excess of 25,000 supermarket and nonfood items), Thriftway, Inc. made the decision to adopt the LIFO (last in–first out) method of inventory valuation. Under this accounting practice, the inventory on hand at the close of an accounting period is valued at the first price paid regardless of fluctuation affecting the actual cost of the total inventory. The reasons for adopting this method of valuation are many, the significant ones are as follows:

1. To better match current costs against current revenues, thereby minimizing the effect of radical price changes and their influence on profit or loss results.
2. To reduce income and thereby income taxes during periods of inflation. This in turn brings disposable cash generated from operations more in line with current income and allows for replacement of inventory at current costs.

*The authors are indebted to Mr. Kenneth R. Sayers, Controller, Thriftway, Inc. for providing this application.

ESTABLISHING AN ANNUAL LIFO INDEX

While we do not want to go into the complete details of the LIFO method of inventory valuation, realize that the LIFO computations require the company to establish a LIFO index for its inventory over the given year of interest. The desired LIFO index is based on two components shown in Figure 8A.1. Component 1 is referred to as the *base cost* for the December 31 inventory; component 2 is referred to as the *current cost* for the same inventory, For example, a base cost for a given mix of inventory items might

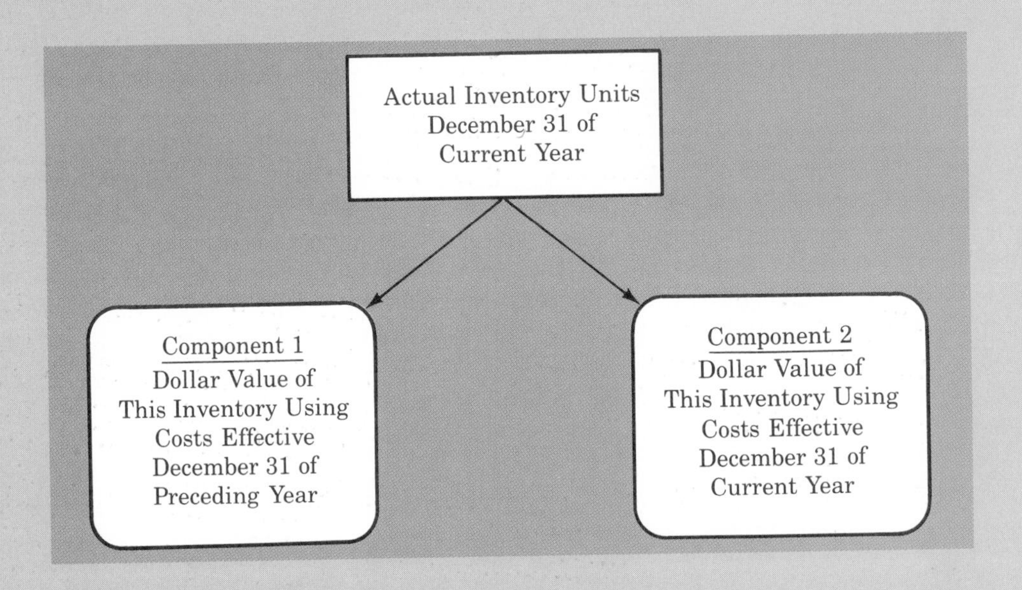

FIGURE 8A.1 Components of LIFO Index

be $1.4 million. However, using current costs, which are higher because they reflect the year's inflationary effect, might show this same mix of inventory items to cost $1.5 million. The LIFO index for the inventory is then computed by the ratio of the current cost to the base cost. Using these data, the LIFO index would be $1.5 million/$1.4 million = 1.071. Interpretation of this value is that the company's inventory at current costs contains a 7.1% increase in value due to the inflation occurring during the 1 year period.

In electing the LIFO method of inventory valuation, the company must have a way to establish its annual LIFO index. The straightforward way is to determine the actual number of units in inventory for every item (a census) carried at Thriftway's 22 locations. Then, using the current cost per unit as well as the preceding year's base cost per unit, the two components of the index can be determined. However, with over 25,000 items carried in inventory, the time and cost of conducting this census would be prohibitive. This is where sampling and the statistical estimation procedures you have been studying play an essential role at Thriftway.

SAMPLING AND INTERVAL ESTIMATION OF THE LIFO INDEX

Rather than taking the census of all 25,000 items in inventory, a random sample of 500 items is selected. A stratification procedure is used to ensure that the 500 items have proportional representation from the company's four inventory pools: basic grocery, alcoholic beverages, tobacco products, and health and beauty aids. The physical inventory counts for the sample items are taken during the last week of December. A clerical employee identifies the current cost per unit and base cost per unit for each item in the sample. Component 1 of the LIFO index is the dollar value of the *sample* inventory using the base costs, while component 2 is the dollar value of the *sample* inventory using current costs. The ratio of the sample current cost to the sample base cost provides the point estimate for the company's LIFO index.

For example, for a given year this sample index was 1.045. However, we realize that this index is only an estimate of the population's LIFO index. A statement about the sampling error and the associated interval estimate are essential in determining the goodness of the sample index. Using the sample results, the standard error was computed to be .01. With a 95% confidence level ($z = 1.96$), the maximum sampling error was approximately .02. Thus the interval of 1.025 to 1.065 provided the 95% confidence interval estimate for the actual LIFO index. The precision of the sample was judged acceptable, and the point estimate of the index, 1.045, was used in the LIFO computations.

The sample of 500 items (2% of all items) provided the time and cost savings that made the election of the LIFO inventory policy acceptable for Thriftway. Without sampling and the interval estimate showing the goodness of the sample index, Thriftway would have been unable to obtain the advantages of the LIFO policy.

9 Hypothesis Testing

What you will learn in this chapter:

- what a hypothesis is and how it is formulated
- how to use sample results to test hypotheses about a population mean or a population proportion
- how to determine the probability of making Type I and Type II errors
- how to use standardized test statistics and *p*-values to make hypothesis tests
- how to determine the sample size for hypothesis tests

Contents

Statistical inference is the process of drawing conclusions about population characteristics based on information contained in a sample. Clearly the point and interval estimation procedures introduced in Chapters 7 and 8 are forms of statistical inference. Another type of statistical inference is hypothesis testing. In hypothesis testing we begin by stating a hypothesis about a population characteristic. This hypothesis, called the *null hypothesis,* is assumed to be true unless sufficient evidence can be found in a sample to reject it. The situation is quite similar to that in a criminal trial. The defendant is assumed to be innocent; if sufficient evidence to the contrary is presented, however, the jury will reject this hypothesis and conclude that the defendant is guilty.

In statistical hypothesis testing, often the null hypothesis is an assumption about the value of a population parameter. A sample is selected from the population, and a point estimate is computed. By comparing the value of the point estimate to the hypothesized value of the parameter we draw a conclusion with respect to whether or not there is sufficient evidence to reject the null hypothesis. A decision is made and often a specific action is taken depending upon whether or not the null hypothesis about the population parameter is accepted or rejected.

The purpose of this chapter is to introduce hypothesis testing and to show how the testing procedure is accomplished. We begin by considering the product weight tests conducted for Hilltop Coffee, Inc.

9.1 DEVELOPING HYPOTHESES

The Federal Trade Commission (FTC) periodically tests the advertising claims of manufacturers in order to verify the manufacturer's statements about its products. A case currently under investigation concerns the container weights specified on the labels of coffee products manufactured by Hilltop Coffee, Inc. In order to demonstrate the hypothesis testing procedure, let us show how a test on label accuracy could be made for Hilltop's 3 pound can of coffee.

With no evidence to the contrary, we tentatively assume that the labels are correct. A hypothesis test is then designed to provide evidence regarding the truth of this claim. The mean filling weight for the population of 3 pound coffee cans could be greater than 3 pounds, exactly equal to 3 pounds, or less than 3 pounds. Let μ denote the population mean container weight. The three possibilities for μ, the conclusion, and the followup action for each are as follows:

Possible Value	*Conclusion and Action*
$\mu > 3$	Hilltop is exceeding label specification; *take no action*
$\mu = 3$	Hilltop is meeting label specification; *take no action*
$\mu < 3$	Hilltop is not meeting label specification; *file complaint and take action* to ensure that the firm complies with label specification

While there are three possibilities for the mean filling weight μ, we see that two of the alternatives lead to the same decision. No action will be taken against the firm if the mean filling weight is greater than or equal to 3 (that is, $\mu > 3$ or $\mu = 3$). In both of these cases the consumer is being treated fairly. However, if the mean filling weight is less than the label's claim of 3 pounds, a complaint will be filed with the company and corrective action will be taken to ensure that the company complies with the product label.

Given this situation, we formally state the hypotheses about the mean filling weight of the 3 pound can of coffee as follows:

Hypothesis	*Conclusion and Action*
$H_0: \mu \geq 3$	Hilltop okay; no action necessary
$H_1: \mu < 3$	Hilltop violating label specification; take appropriate followup action

For the Hilltop situation it is natural to assume that the company is innocent of mislabeling until sufficient evidence to the contrary is presented. H_0 is the notation we shall use for the *null hypothesis.* Thus for the Hilltop example the null hypothesis is that the company is meeting its product label specifications. With the null hypothesis assumed true, we select a sample from the population. If the sample results do not differ significantly from the assumed null hypothesis, we "accept" H_0 as being true. However, if the sample results do differ significantly from the null hypothesis, we "reject" H_0 and conclude that the *alternative hypothesis,* H_1, is true.

The hypothesis test that we have set up for Hilltop Coffee concerns the value of the population mean. Before we show how to conduct this particular hypothesis test, let us state in general the three forms for hypothesis tests concerning the value of a population mean. In the statements below μ_0 denotes a specific value for the population mean:

Form 1	*Form 2*	*Form 3*
$H_0: \mu \geq \mu_0$	$H_0: \mu \leq \mu_0$	$H_0: \mu = \mu_0$
$H_1: \mu < \mu_0$	$H_1: \mu > \mu_0$	$H_1: \mu \neq \mu_0$

In the Hilltop Coffee situation we are interested in conducting a hypothesis test of Form 1 with $\mu \geq 3$. Other problems require the hypotheses to be stated in Form 2 or Form 3. The following exercises are designed to provide practice in choosing the proper form for a hypothesis test.

EXERCISES

1. The manager of the Danvers-Hinton Resort Hotel believes that the mean guest bill is at least \$250. A sample of billing statement will be used to test the manager's claim. Which of the forms of the hypothesis (below) should be used for testing whether or not the manager's claim is correct? Explain.

a	*b*	*c*
$H_0: \mu = 250$	$H_0: \mu \leq 250$	$H_0: \mu \geq 250$
$H_1: \mu \neq 250$	$H_1: \mu > 250$	$H_1: \mu < 250$

2. A quality control inspector at Morgan Manufacturing Co. tests part dimensions for a machining operation. Specifications require the mean part diameter to be 2 inches. If a sample leads the quality control inspector to believe that part diameters are too large or too small, the machine will be shut down and readjusted. State the hypothesis test that could be used to determine whether or not the machine should be shut down.

3. The manager of an automobile dealership is considering a new bonus plan that may increase sales volume. Currently, the mean sales rate for sales personnel is four automobiles per month. A sample of sales personnel will be allowed to sell under the new bonus plan for 1 month. What hypothesis test is appropriate for testing the ability of the new bonus plan to increase the sales volume?

4. Because of production changeover time and costs a director of manufacturing must convince management that a proposed manufacturing method reduces costs before the new method can be implemented. The current production line operates with a mean cost of $220 per hour. A new production line has been proposed and a sample production period specified. What hypothesis test should be formulated in order to test whether or not the company should convert to the new production line?

9.2 ERRORS INVOLVED IN HYPOTHESIS TESTING

In any hypothesis-testing procedure we begin by tentatively assuming that the null hypothesis, H_0, is true. The ultimate aim or goal of the hypothesis-testing procedure is to reach a decision to either accept H_0 (confirming our tentative assumption) or reject H_0. A sample is taken, and this decision is made after a comparison of the sample results with the results that one would expect if the null hypothesis were true. Ideally we would like the hypothesis-testing procedure always to lead us to accept H_0 when H_0 is true and reject H_0 when H_0 is false. However, this will not always be the case.

For hypothesis tests concerning a population mean, the null hypothesis will be accepted or rejected based on whether or not the sample mean differs significantly from what we would expect if the null hypothesis were true. From our study of sampling distributions, we realize that the value of the sample mean cannot be expected to equal the population mean exactly. Indeed, it may differ substantially from the population mean. Hence there is a possibility that the hypothesis-testing procedure could lead to a wrong decision. A conclusion to reject H_0 when it is true or a conclusion to accept H_0 when it is false are errors that might be made. Let us consider these errors more closely and see how they are dealt with in the hypothesis-testing procedure.

First, consider the case where the null hypothesis H_0 reflects the true situation in the population. In this case, we hope that the hypothesis-testing procedure will lead to the conclusion to accept H_0. However, if, instead, the procedure leads us to reject H_0, we will be making what is called a *Type I error*. The Type I error is formally defined as follows:

Definition of Type I Error

The Type I error is the error of rejecting H_0 when it is true.

In the Hilltop coffee study, consider the case where the company is actually meeting or exceeding its label weight specifications on the average. That is, the null hypothesis H_0: $\mu \geq 3$ is true. We hope that the hypothesis-testing procedure will lead us to accept H_0. However, if it leads to the conclusion of rejecting H_0, a Type I error will be made. A complaint will be filed and action taken against Hilltop when in fact the company is meeting its label specifications and is not guilty of any wrongdoing.

Consideration of the case where a null hypothesis is actually false points to another possible hypothesis-testing error. If in fact H_0 is false and H_1 is true, we want the hypothesis-testing procedure to lead to rejecting H_0. However, if, instead, we make the error of accepting H_0, we are making what is called a *Type II error.* The Type II error is formally defined as follows:

Definition of Type II Error

The Type II error is the error of accepting H_0 when it is false.

Let us return to the Hilltop coffee example. Suppose that H_1 is true and $\mu < 3$. We would be making a Type II error if we made the decision to accept H_0. In this case the Type II error would lead to no action being taken to stop Hiltop's underfilling violations when in fact the company is not meeting its label specifications.

At first we might be a little disappointed to learn that a statistical hypothesis-testing procedure could lead to errors. However, we may be comforted to learn that the probabilities of making the two types of errors can be determined and controlled. With common statistical notation we denote the probabilities of the hypothesis-testing errors as follows:

$$\alpha \text{ ("alpha")} = \text{the probability of making a Type I error,}$$
$$\beta \text{ ("beta")} = \text{the probability of making a Type II error.}$$

Because of this notation the Type I error is sometimes referred to as the *α error* and the Type II error as the *β error.*

Figure 9.1 summarizes the conditions under which Type I and Type II errors are made. It also shows the situations where the hypothesis-testing procedure leads to correct decisions. The specific hypotheses shown in Figure 9.1 are from the Hilltop Coffee hypothesis-testing example.

Hypotheses:

H_0: $\mu \geq 3$

H_1: $\mu < 3$

	Situation in the Population	
	H_0 True ($\mu \geq 3$)	H_0 False ($\mu < 3$)
Accept H_0 (Conclude $\mu \geq 3$)	Correct Decision	Type II Error (β)
Reject H_0 (Conclude $\mu < 3$)	Type I Error (α)	Correct Decision

FIGURE 9.1 Type I and Type II Errors in Hypothesis Testing

The square in the upper left-hand corner of Figure 9.1 corresponds to the case when H_0 is true and the hypothesis test leads to the decision to accept H_0. In this case a correct decision is made. The lower left-hand corner corresponds to the case when the null hypothesis is actually true but the hypothesis testing procedure leads us to reject it. In this case a Type I error is made. Similarly, the square in the upper right-hand corner corresponds to making a Type II error, and the square in the lower right-hand corner corresponds to a correct decision.

9.3 ONE-TAILED HYPOTHESIS TESTS ABOUT A POPULATION MEAN

We now are familiar with how hypothesis tests about a population mean are formulated, and we have an understanding of the types of errors that can be committed in hypothesis testing. The decision rules used in conducting a hypothesis test are designed to control the probabilities of error.

Developing Decision Rules

Let us return to the Hilltop Coffee study. Assume that a sample of 36 coffee containers has been taken for the purpose of testing the manufacturer's claim that the cans contain an average of at least 3 pounds of coffee. From past studies on coffee can filling weights it is known that the population standard deviation is equal to .18 pounds (that is, $\sigma = .18$). Recall that the hypotheses were stated as follows:

$$H_0: \mu \geq 3,$$
$$H_1: \mu < 3.$$

Now suppose that the sample of 36 coffee cans provides a sample mean of $\bar{x} = 3.12$ pounds. Should we accept or reject H_0? Most of us would probably be inclined to accept H_0. On the other hand, suppose that a sample mean of $\bar{x} = 2.60$ pounds is observed; what should we do now? In this case, most of us would probably reject H_0 on the grounds that the sample mean is substantially less than 3 pounds. Generally, high values for the sample mean lead to acceptance of H_0, while low values for the sample mean lead to rejection of H_0. But we need to know how much lower than 3 the value of $\bar{x}$ must be before it is safe to reject H_0.

A job for the statistical analyst is developing a decision rule based on the value of the sample mean that specifies under what conditions H_0 will be accepted and under what conditions it will be rejected. The general form of the decision rule for the Hilltop coffee example is as follows:

Decision Rule

Accept H_0 if $\bar{x} \geq c$,

Reject H_0 if $\bar{x} < c$,

where c is called the *critical value* for the test.

Once the critical value c has been chosen, we have determined how much smaller than 3 the value of $\bar{x}$ must be before H_0 should be rejected. The choice of a value for c is made with the objective in mind of keeping the probability of a Type I error under control. To see how this is done, let us consider the following possible decision rule for

Hilltop Coffee:

$$\text{Accept } H_0 \text{ if } \bar{x} \geq 3,$$
$$\text{Reject } H_0 \text{ if } \bar{x} < 3.$$

This might appear to be a sensible decision rule on the grounds that 3 pounds is the label weight specification. If a sample results in a sample mean of 3 pounds or more, the rule leads to accepting H_0 and concluding that Hilltop is meeting the label specifications. However, a sample mean under 3 pounds provides a rejection of H_0 and the conclusion that Hilltop is underfilling its coffee cans.

Before we adopt the critical value of 3 and the above decision rule, let us determine the probability of making a Type I error under this decision rule. Recall that the Type I error occurs whenever H_0 is actually true and the sample results lead us to make the error of rejecting H_0. Thus to determine the probability of making a Type I error we have to consider the situation where H_0 is true. H_0 is true whenever $\mu \geq 3$, but for the moment assume that $\mu = 3$ and that Hilltop is meeting its label specifications exactly.

We have seen in the past two chapters that simple random samples taken from a population generate a sampling distribution for the sample mean $\bar{x}$. We further know that the mean of the sampling distribution is equal to the population mean μ, and its standard deviation is given by $\sigma_{\bar{x}} = \sigma / \sqrt{n}$. With our sample size of 36, the central limit theorem allows us to assume that the sampling distribution is normal. With $\sigma = .18$ and $n = 36$ we have $\sigma_{\bar{x}} = .18/\sqrt{36} = .03$. With our current assumption of $\mu = 3$, the sampling distribution of $\bar{x}$ for all possible samples of size 36 is shown in Figure 9.2. Thus with $\mu = 3$, so that H_0 is true, Figure 9.2 shows the distribution of all possible sample means that could be observed using simple random samples of 36 Hilltop coffee cans.

Let us now use the decision rule with a critical value of 3 to compute the probability of making a Type I error. This error probability is shown graphically in Figure 9.3. Note that 50% of the sample means that can be observed are actually below

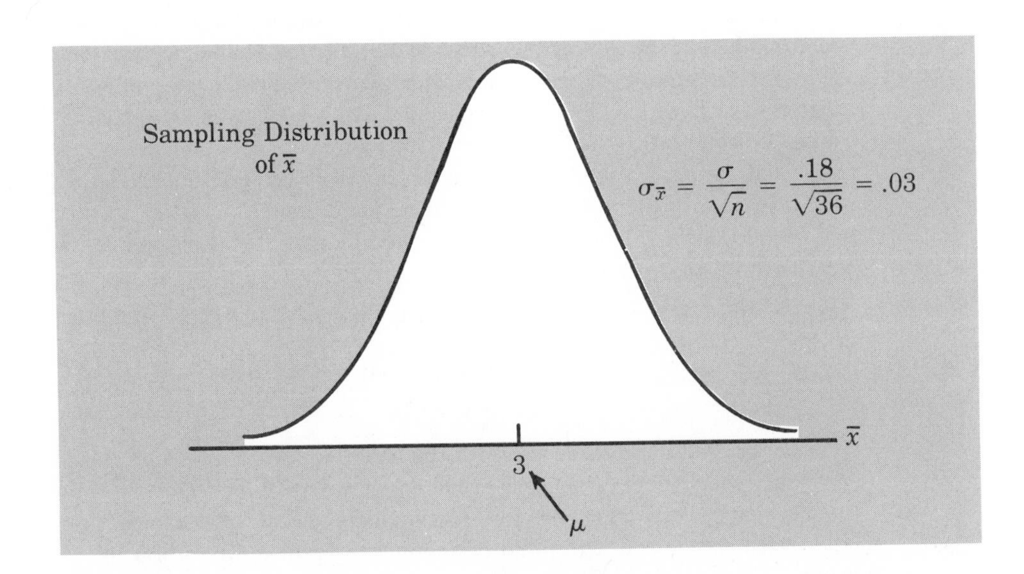

FIGURE 9.2 Sampling Distribution of $\bar{x}$ for the Hilltop Coffee Study When $\mu = 3$

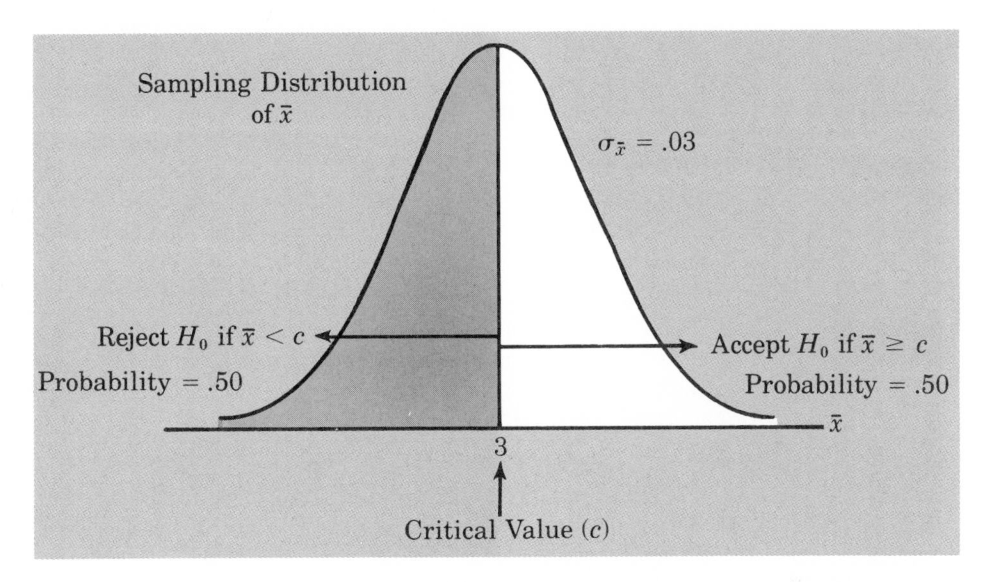

FIGURE 9.3 Sampling Distribution of $\bar{x}$ for the Hilltop Coffee Study Showing Probabilities of Accepting H_0 and Rejecting H_0 When $\mu = 3$ and Critical Value $c = 3$

3 pounds. Thus there is a .50 probability that we will reject H_0 and claim that Hilltop is not meeting its label specifications when in fact Hilltop is meeting the label specifications with an average filling weight of $\mu = 3$. Committing a Type I error could be a rather serious and embarrassing situation. We would be accusing Hilltop of violations and taking corrective action when Hilltop is not doing anything wrong. Hence a .50 probability for a Type I error is too high. To reduce this probability we will need to adjust the critical value of the decision rule.

In the above discussion, we first selected a critical value for our decision rule and then found the probability of making a Type I error. While we could try other critical values and then compute the probability of making the Type I error, in practice we find it much easier to first determine the maximum allowable probability of making a Type I error and then compute the corresponding critical value for the decision rule. Specifying the maximum allowable probability for the Type I error is up to the manager or director of the weight-testing program. However, because of the seriousness of making the error of falsely accusing a company that is meeting the product weight specification, the program director might make the following statement: "If the company is meeting its weight specifications exactly and $\mu = 3$, I would like a 99% chance of concluding that the company is meeting the specifications. While I do not want to wrongly accuse the company of underfilling its product, I am willing to live with a 1% chance of making this error."

In effect, the manager is saying that if $\mu = 3$ he or she wants a .99 probability of accepting H_0; thus the probability of rejecting H_0 and making a Type I error would be .01. Figure 9.4 shows the sampling distribution of $\bar{x}$ for the Hilltop Coffee study, with $\alpha = .01$ shown in the lower tail of the distribution. The corresponding critical value is computed as follows:

$$c = 3 - z_{.01}\sigma_{\bar{x}}.$$

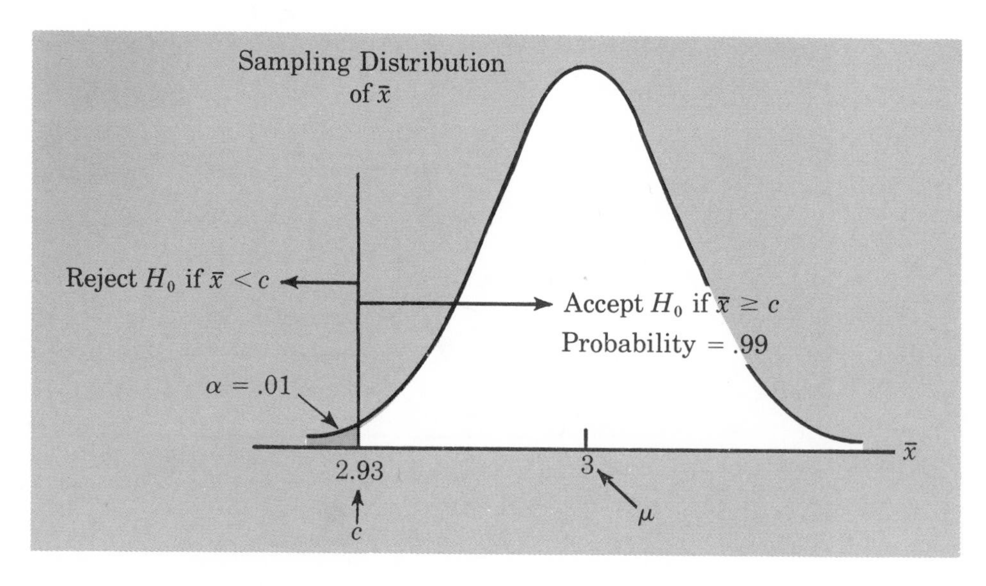

FIGURE 9.4 Sampling Distribution of $\bar{x}$ for the Hilltop Coffee Study Showing Critical Value Based on a .01 Probability of Type I Error

Here $z_{.01}$ refers to the standard normal probability distribution value with .01 area in the upper tail. We find from Table 1 of Appendix B that $z_{.01} = 2.33$. In addition, we found previously that $\sigma_{\bar{x}} = \sigma/\sqrt{n} = .03$ and thus

$$c = 3 - 2.33(.03) = 2.93.$$

With this critical value, the decision rule for the Hilltop Coffee study that provides a .01 probability of making a Type I error when $\mu = 3$ is as follows:

Hilltop Coffee Study Decision Rule

$$\text{Accept } H_0 \text{ if } \bar{x} \geq 2.93, \tag{9.1}$$

$$\text{Reject } H_0 \text{ if } \bar{x} < 2.93. \tag{9.2}$$

In establishing the critical value for the decision rule, we have been assuming that $\mu = 3$ so that H_0 was satisfied in the equality case. The reason for this is that the probability of committing a Type I error is at a maximum when equality holds. Thus when we choose a critical value to control the Type I error in the equality case we are controlling the maximum Type I error. To see this, suppose that the actual filling weights for the Hilltop Coffee cans were $\mu = 3.02$ pounds. The probability of making a Type I error with a critical value of 2.93 in this case is only .0013. This situation is shown in Figure 9.5. Other values of $\mu > 3$ will all make the probability of making a Type I error less than .01. Thus in establishing the critical value for a particular hypothesis-testing situation we always assume that the null hypothesis holds as an equality. This allows us to control the maximum probability of a Type I error.

The maximum probability of a Type I error is referred to as the *level of significance* for the hypothesis test. As in the Hilltop case, the committing of a Type I error often is serious. Thus it is common to choose small values such as $\alpha = .01$, $\alpha = .05$, or $\alpha = .10$ as levels of significance. In choosing the decision rule given by (9.1) and

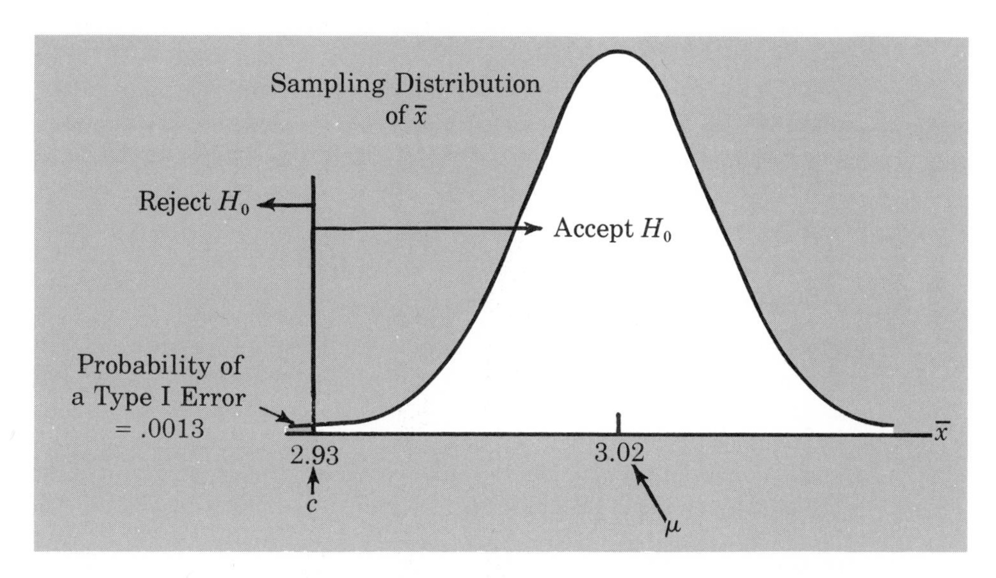

FIGURE 9.5 Probability of Making a Type I Error in the Hilltop Coffee Study When $\mu = 3.02$ and $c = 2.93$

(9.2) we have a level of significance of .01 for the Hilltop Coffee hypothesis test. Thus it is very unlikely that we will falsely accuse Hilltop of labeling violations.

We note that the above decision rule has been established without regard to the probability of making a Type II error. While we do not want to completely ignore the probability of making this error, in many cases (e.g., the Hilltop Coffee study) the cost of making a Type I error is much higher than that of making a Type II error. In these cases it is common practice to consider only the level of significance (i.e., the probability of a Type I error) in establishing the decision rule and its critical value. In Section 9.6 we will discuss the probability of making a Type II error. There we will see what adjustments we might recommend for the hypothesis testing procedure when the cost of making this error is high and the probability of this error is excessive.

Making the Decision

We are now ready to see how the decision is made after the sample results have been observed. The decision rule that we have established for the Hilltop Coffee study requires that we reject H_0 if $\bar{x} < 2.93$ and accept H_0 otherwise. Suppose that a sample of 36 containers yields a sample mean of $\bar{x} = 2.92$ pounds. Following the decision rule, we would reject H_0 and conclude that Hilltop is underfilling its product. The sampling distribution of $\bar{x}$ shown in Figure 9.4 shows that a sample mean of $\bar{x} = 2.92$ is a very unlikely occurrence if the population mean is $\mu = 3$. In fact, this value of the sample mean is so unlikely that we are inclined to believe that it did not come from this sampling distribution. Thus the null hypothesis is rejected, and the conclusion that it must have come from a sampling distribution with $\mu < 3$ is reached. Followup action to ensure that Hilltop complies with its label specifications would be undertaken.

Let us now consider another possible sample result, a sample mean of $\bar{x} = 2.97$ pounds. Again, following the decision rule, this time we would accept H_0, even with a sample mean below 3. While to some this may appear surprising, note that in Figure 9.4 the value $\bar{x} = 2.97$ pounds is greater than the critical value of 2.93 pounds. In fact, $\bar{x} = 2.97$ is a reasonable value for the sample mean when the population mean is actually 3, as assumed. Our conclusion is that we have insufficient evidence to claim that $\mu \geq 3$ is not true. Thus we make the decision to accept H_0. In this case no followup action would be taken toward Hilltop.

Summary of Statistical Decision Rules for One-Tailed Tests About a Population Mean

Since the null hypothesis for the Hilltop Coffee study was rejected only when the sample result $\bar{x}$ was in the lower tail of the sampling distribution, the hypothesis test can be classified as a *one-tailed test*. We now wish to generalize the hypothesis-testing procedure for one-tailed tests about a population mean. Let μ_0 indicate the value of the population mean appearing in the null hypothesis. The general form of the one-tailed test as encountered in the Hilltop Coffee study is as follows (see Figure 9.6):

Hypothesis Test About a Population Mean for a One-Tailed Test of the Form

$$H_0: \mu \geq \mu_0$$
$$H_1: \mu < \mu_0$$

Decision rule:

$$\begin{aligned} &\text{Accept } H_0 \text{ if } \bar{x} \geq c, \\ &\text{Reject } H_0 \text{ if } \bar{x} < c, \\ &\text{where } c = \mu_0 - z_\alpha \sigma_{\bar{x}}. \end{aligned} \qquad (9.3)$$

As we saw in Section 9.1, other forms of the null and alternative hypotheses are possible. A second form of the one-tailed test rejects the null hypothesis only when the sample result is in the upper tail of the sampling distribution. This one-tailed test and decision rule are as follows (see Figure 9.6):

Hypothesis Test About a Population Mean for a One-Tailed Test of the Form

$$H_0: \mu \leq \mu_0$$
$$H_1: \mu > \mu_0$$

Decision rule:

$$\begin{aligned} &\text{Accept } H_0 \text{ if } \bar{x} \leq c, \\ &\text{Reject } H_0 \text{ if } \bar{x} > c, \\ &\text{where } c = \mu_0 + z_\alpha \sigma_{\bar{x}}. \end{aligned} \qquad (9.4)$$

These hypothesis-testing procedures and critical values, as given by (9.3) and (9.4), are based upon the assumption that the population standard deviation σ is known. With σ known, the standard error of the mean, $\sigma_{\bar{x}} = \sigma/\sqrt{n}$, can be computed

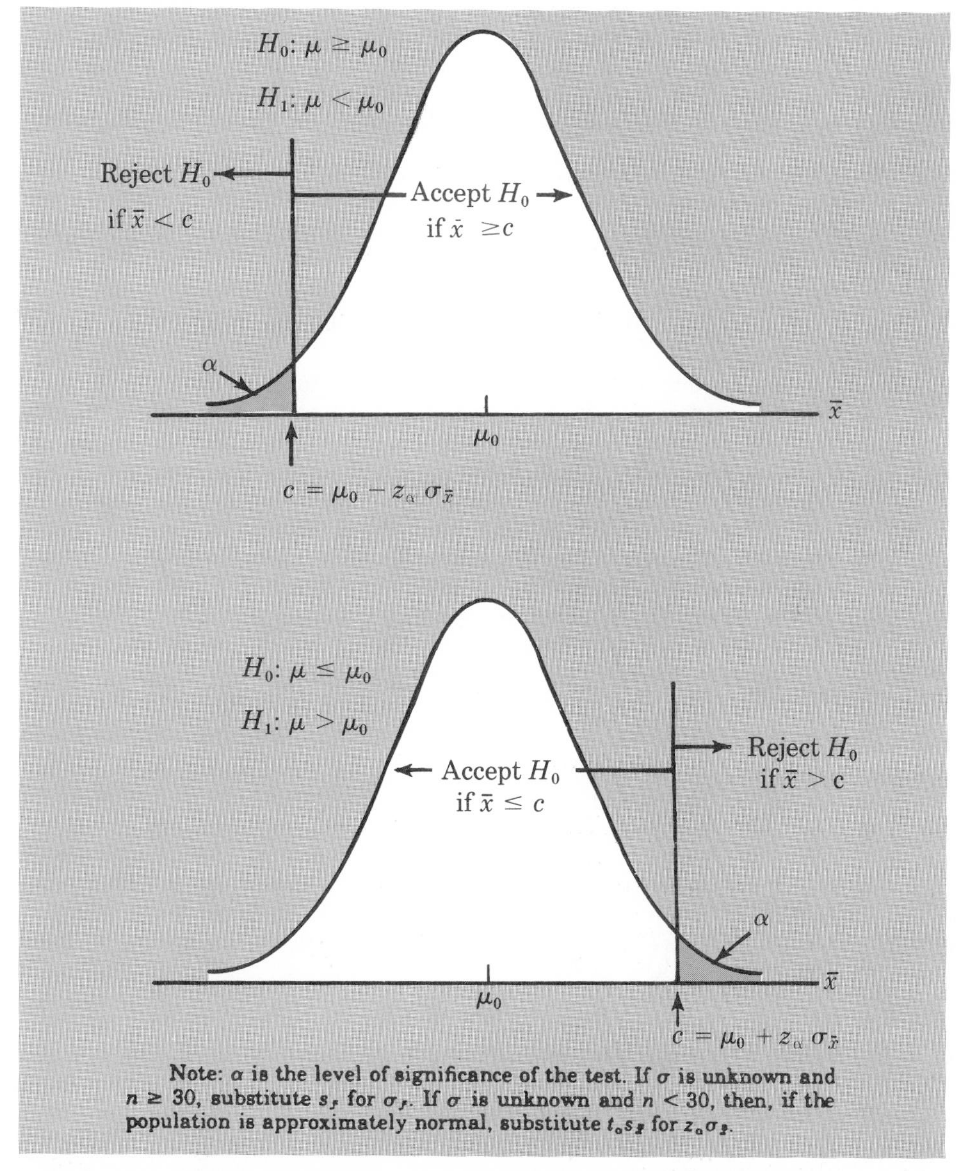

FIGURE 9.6 Summary of Statistical Decision Rules for One-Tailed Tests About a Population Mean

easily. With a sample of size $n \geq 30$, we are justified (the central limit theorem applies) in using the normal probability distribution to determine the appropriate z_α value. If the sample size $n < 30$, these procedures are appropriate if we are willing to assume that the population has a normal distribution.

But what happens to these hypothesis-testing procedures when the population standard deviation is unknown? First of all, we simply use the sample standard deviation s as an estimator of the population standard deviation σ. The standard error of the mean, $\sigma_{\bar{x}}$, is then estimated by $s_{\bar{x}} = s/\sqrt{n}$. With a sample size of $n \geq 30$, we substitute $s_{\bar{x}}$ for $\sigma_{\bar{x}}$ in (9.3) and (9.4) in order to determine the critical values for the test. For a sample size $n < 30$, we must make the assumption that the population has a normal distribution. In this instance the t distribution with $n - 1$ degrees of freedom applies, with $t_{\alpha}s_{\bar{x}}$ being substituted for $z_{\alpha}\sigma_{\bar{x}}$ in (9.3) and (9.4) to provide the appropriate critical values.

EXERCISES

5. The manager of the Danvers-Hinton Resort Hotel believes that the mean guest bill is at least $250.

a. With $\sigma = 50$, what is the decision rule for accepting or rejecting the manager's claim if a sample of 60 bills is to be the basis for making the decision? Use a .05 level of significance for the test.

b. If the sample mean is $\bar{x} = \$235$, what decision would you make?

6. The president of Fightmaster and Associates Real Estate, Inc. claims that the mean selling time of a residential home is 40 days or less after it is listed with the company. A sample of 50 recently sold residential homes shows a sample mean selling time of 45 days and a sample standard deviation of 20 days. Use a .02 level of significance and test the president's claim.

7. A long distance trucking firm believes that its mean weekly loss due to damaged shipments is $2,000 or less. A sample of 15 weeks of operations shows a sample mean weekly loss of $2,200, with a sample standard deviation of $500. Use a .05 level of significance and test the trucking firm's claim that the mean weekly loss is $2,000 or less.

8. Fowle Marketing Research, Inc. bases charges to a client on the assumption that telephone surveys can be completed with a mean time of 15 minutes or less. If a greater mean survey time is required, a premium rate is charged the client. Does a sample of 35 surveys that shows a sample mean of 17 minutes and a sample standard deviation of 4 minutes justify the premium rate? Test at a .01 level of significance.

9. New tires manufactured by a company in Findlay, Ohio are designed to provide a mean of at least 28,000 miles. Tests with 20 tires show a sample mean of 26,500 miles with a sample standard deviation of 1,000 miles. Use a .01 level of significance and test for whether or not there is sufficient evidence to reject the claim of a mean of at least 28,000 miles.

10. It is estimated that on average a housewife with a husband and two children works 55 hours or less per week on household related activities. Shown below are the actual hours worked by a sample of eight housewives:

58, 52, 64, 63, 59, 62, 62, 55.

Using a .05 level of significance, test the claim about the housewife work week.

11. Joan's Nursery specializes in custom designed landscaping for residential areas. The labor cost associated with a particular landscaping proposal is estimated based on the number of plantings of trees, shrubs, and so on associated with the project. For labor cost estimating purposes, management allows a maximum of 2 hours of labor time for the planting of a medium-size tree. Actual times from a sample of ten plantings during the past month are as follows (times in hours):

1.9, 2.1, 2.8, 3.0, 2.6, 2.5, 2.8, 3.2, 2.2, 2.5.

Using a .05 level of significance, test the claim that the mean tree planting time is 2 hours or less. What is your conclusion, and what recommendations would you consider making to management?

9.4 TWO-TAILED TESTS ABOUT A POPULATION MEAN

KMGM is a new rock radio station in Seattle, Washington. Advertising spots ("commercials") are sold on the basis of KMGM's target listening audience having an average age of 21 years. A particular concern of KMGM's station manager is whether or not the station is reaching this target listening audience. To answer the question KMGM has hired an independent radio and television survey firm that specializes in identifying listening and viewing audience characteristics. If the audience survey shows that KMGM is not reaching the desired audience, programming changes will be implemented.

A statistical hypothesis-testing procedure can be used to determine whether or not the station has a listening audience with an average age of 21 years. The hypotheses and the corresponding conclusions and actions are shown below:

Hypothesis	*Conclusion and Action*
H_0: $\mu = 21$	KMGM is reaching the target audience; no action necessary
H_1: $\mu \neq 21$	KMGM is not reaching the target audience; consider programming modifications or other corrective action

Let us assume that the audience characteristics are identified for a sample of 100 KMGM listeners. Also, suppose that from similar audience profile studies the population standard deviation is known to be $\sigma = 5$ years. Following the hypothesis-testing procedure developed in the previous section, we would first determine a maximum allowable probability of making a Type I error for the test. Suppose that we choose $\alpha = .05$ as the level of significance. We are saying we are willing to live with a .05 probability of rejecting H_0 if it is true. In other words, we are willing to tolerate a .05 probability that we will conclude that KMGM is not reaching its target audience when in fact it really is.

Next we make the usual (tentative) assumption that the null hypothesis is true and consider the corresponding sampling distribution of $\bar{x}$. This distribution is shown in Figure 9.7. Note that the standard error of the mean is given by $\sigma_{\bar{x}} = \sigma/\sqrt{n} = 5/\sqrt{100} = .5$. We now need to develop the decision rule that can be used to determine whether to accept or reject H_0. Since KMGM should reject the claim $\mu = 21$ when the sample

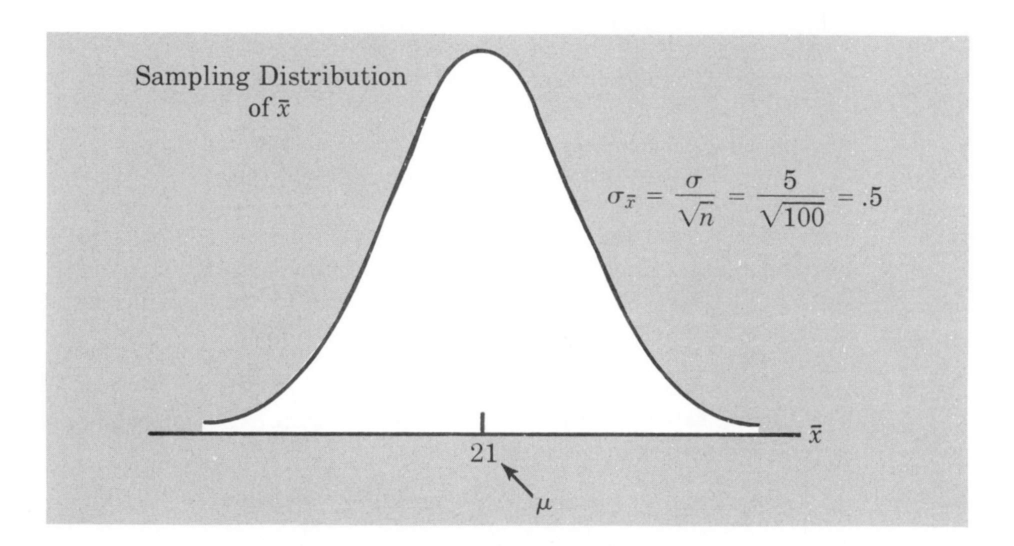

FIGURE 9-7 Sampling Distribution of $\bar{x}$ for the KMGM Study When $\mu = 21$

mean $\bar{x}$ is significantly larger than 21 (i.e., an older audience than desired) *or* when $\bar{x}$ is significantly smaller than 21 (i.e., a younger audience than desired), the decision rule for the test should have the following general form:

Decision Rule

Accept H_0 if $c_1 \leq \bar{x} \leq c_2$,
Reject H_0 if $\bar{x} < c_1$ or if $\bar{x} > c_2$,

where c_1 and c_2 are the critical values for the test.

The KMGM study is an example of a two-tailed hypothesis test. That is, H_0 will be rejected for values of the sample mean in either the lower tail or the upper tail of the sampling distribution. Both tails of the sampling distribution of $\bar{x}$ are involved in the test, and we have previously specified $\alpha = .05$; therefore we shall place an area of .025 in each tail of the sampling distribution in order to determine the c_1 and c_2 critical values. This situation is shown graphically in Figure 9.8. We shall adopt the convention of placing an area of $\alpha/2$ in each tail for all two-tailed hypothesis-testing situations. Thus the probability of falsely rejecting H_0 will be the same for each tail.

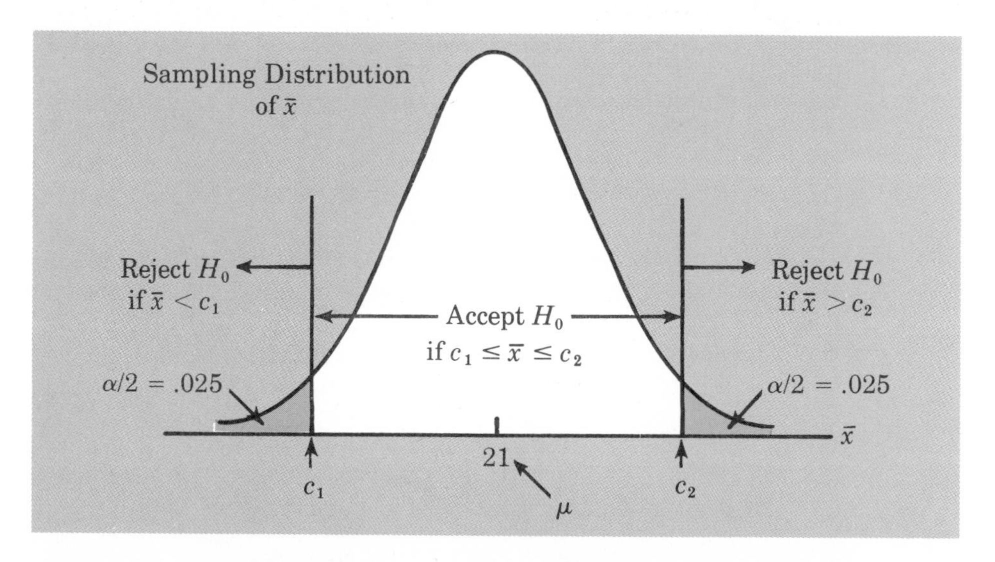

FIGURE 9.8 Sampling Distribution of $\bar{x}$ for the KMGM Study Showing Critical Values Based on a .05 Probability of a Type I Error

With $n = 100$, the sampling distribution of $\bar{x}$ can be approximated by a normal probability distribution. Therefore in order to cut off an area of .025 in the upper tail we use $z_{.025} = 1.96$. With $\sigma_{\bar{x}} = .5$ the critical values for the two-tailed test become

$$c_1 = 21 - (1.96)(.5) = 20.02,$$
$$c_2 = 21 + (1.96)(.5) = 21.98.$$

The decision rule for the KMGM study that provides a .05 probability of making a Type I error if H_0 is true is as follows:

KMGM Study Decision Rule

Accept H_0 if $20.02 \leq \bar{x} \leq 21.98$,
Reject H_0 if $\bar{x} < 20.02$ or if $\bar{x} > 21.98$.

Once the decision rule has been established for the test, the step of comparing the observed value of the sample to the critical values and reaching a conclusion becomes a rather simple task. For example, let us assume that the listening audience sample provided a sample mean age of 20.3 years. As the decision rule indicates, H_0 is accepted. In other words, we can conclude that the survey shows no results that should alarm KMGM about not meeting its target audience; no programming changes are needed. Actually, the hypothesis-testing procedure has not proven that $\mu = 21$, but there seems to be little evidence to the contrary.

Summary of Statistical Decision Rules for Two-Tailed Tests About a Population Mean

Let μ_0 indicate the value of the population mean appearing in the null hypothesis of a two-tailed test. The general form of the decision rule for this test is as follows (see Figure 9.9):

Hypothesis Test About a Population Mean for a Two-Tailed Test of the Form

$$H_0: \mu = \mu_0$$
$$H_1: \mu \neq \mu_0$$

Decision rule:

Accept H_0 if $c_1 \le \bar{x} \le c_2$,
Reject H_0 if $\bar{x} < c_1$ or if $\bar{x} > c_2$,

where
$$c_1 = \mu_0 - z_{\alpha/2}\sigma_{\bar{x}} \quad (9.5)$$
$$c_2 = \mu_0 + z_{\alpha/2}\sigma_{\bar{x}} \quad (9.6)$$

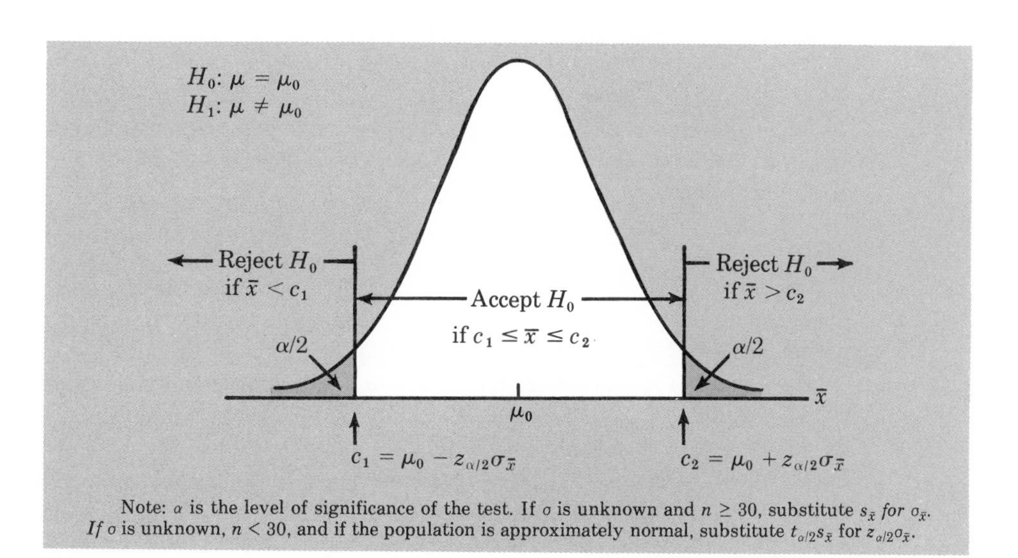

FIGURE 9.9 **Summary of Statistical Decision Rules for Two-Tailed Tests About a Population Mean**

If the population standard deviation is unknown, $s_{\bar{x}} = s/\sqrt{n}$ can be used as an estimator of $\sigma_{\bar{x}}$. There are two cases to consider: If the sample size $n \geq 30$, (9.5) and (9.6) can be used to compute the critical values, with $s_{\bar{x}}$ replacing $\sigma_{\bar{x}}$. If the sample size $n < 30$, the population must be assumed to have a normal distribution if we wish to conduct this test. In this case the t distribution with $n - 1$ degrees of freedom applies, with $t_{\alpha/2}s_{\bar{x}}$ replacing $z_{\alpha/2}\sigma_{\bar{x}}$ in (9.5) and (9.6) for determination of the appropriate critical values.

The Relationship Between Interval Estimation and Hypothesis Testing

In Chapters 8 and 9 we have discussed statistical procedures which can be used to make inferences about population parameters. In Chapter 8 we discussed interval estimation, while in Chapter 9 we have focused on hypothesis testing. Since both procedures require the use of a population parameter, a sample statistic, and the sampling distribution of the sample statistic, it may not be too surprising to learn that estimation and hypothesis testing are closely related statistical procedures.

In the case of inferences about a population mean, the interval estimation procedure in Chapter 8 was based on the assumption that the value of the parameter μ was unknown. Once the sample was selected and the sample mean $\bar{x}$ computed, we developed an interval around the value of $\bar{x}$ that had a good chance of including the parameter μ. The interval computed was referred to as a confidence interval with $1 - \alpha$ defined as the confidence coefficient for the estimate. The formula for the confidence interval given by (8.2) is repeated below:

$$\bar{x} \pm z_{\alpha/2}\,\sigma_{\bar{x}}. \tag{9.7}$$

The hypothesis-testing approach to statistical inference requires us to make a hypothesis and corresponding assumption about the value of the population parameter. In the case of the population mean, the two-tailed hypothesis test has the form

$$H_0: \mu = \mu_0,$$

$$H_1: \mu \neq \mu_0,$$

where μ_0 is the stated or hypothesized value for the population mean. Expressions (9.5) and (9.6) can be combined to show the acceptance region for the hypothesis with level of significance α:

$$\mu_0 \pm z_{\alpha/2}\,\sigma_{\bar{x}}. \tag{9.8}$$

A close look at (9.7) and (9.8) will provide insight into the relationship between the estimation and hypothesis-testing approaches to statistical inference. Note in particular that both procedures require the computation of the value $z_{\alpha/2}\,\sigma_{\bar{x}}$. Focusing on α, we see that a confidence level of $1 - \alpha$ for interval estimation corresponds to a level of significance of α in hypothesis testing. For example, a 95% confidence interval for estimation corresponds to a .05 level of significance for hypothesis testing. Furthermore, (9.7) and (9.8) show that since $z_{\alpha/2}\,\sigma_{\bar{x}}$ is the plus or minus value for both expressions, if $\bar{x}$ falls in the acceptance region defined by (9.8), the hypothesized value μ_0 will be in the confidence interval defined by (9.7). Conversely, if the hypothesized value μ_0 falls in the confidence interval defined by (9.7), the sample mean $\bar{x}$ will be in the acceptance region of the hypothesis H_0: $\mu = \mu_0$. These observations lead to the following procedure for using confidence interval results to draw hypothesis-testing conclusions:

A Confidence Interval Approach to Testing a Hypothesis of the Form

$$H_0: \mu = \mu_0$$
$$H_1: \mu \neq \mu_0$$

1. Select a simple random sample from the population and use the value of the sample mean $\bar{x}$ to develop the confidence interval

$$\bar{x} \pm z_{\alpha/2}\,\sigma_{\bar{x}}.$$

2. If the confidence interval contains the hypothesized value μ_0, accept H_0. If not, reject H_0.

Let us return to the KMGM study discussed earlier in this section to demonstrate the use of confidence intervals in hypothesis testing. The KMGM study investigating the mean age of 21 years for its listening audience resulted in the following two-tailed test:

$$H_0: \mu = 21,$$
$$H_1: \mu \neq 21.$$

In order to test this hypothesis with a level of significance of $\alpha = .05$, we would need to sample of 100 listeners provided a sample mean age of $\bar{x} = 20.3$. The value of the standard error of the mean was $\sigma_{\bar{x}} = .5$. Using these results along with $z_{.025} = 1.96$, the confidence interval becomes

$$\bar{x} \pm z_{\alpha/2}\,\sigma_{\bar{x}}$$
$$20.3 \pm (1.96)\,(.5)$$
$$20.3 \pm .98$$

or

$$19.32 \text{ to } 21.28.$$

This finding enables the radio station to conclude with 95% confidence that the mean age for the population of listeners is between 19.32 years and 21.28 years. Since the hypothesized value for the population mean $\mu_0 = 21$ is in the interval, the hypothesis-testing conclusion is that the null hypothesis H_0: $\mu = 21$ is accepted.

Note that this discussion and example have been devoted to two-tailed hypothesis tests about a population mean. However, the same confidence interval and hypothesis-testing relationship exists for other population parameters as well. In addition, the relationship can be expanded to make one-tailed tests about population parameters. However, this requires the development of one-sided confidence intervals. Exercise 18 will ask you to use such a one-sided confidence interval to make a conclusion about a one-tailed hypothesis test.

EXERCISES

12. A production line operates with a filling weight standard of 16 ounces per container. Overfilling or underfilling is a serious problem, and the production line should be shut down if either occurs. From past data σ is known to be .8 ounces. A quality control inspector samples 30 items every 2 hours and at that time makes the decision of whether or not to shut the line down for adjustment purposes.

a. With a .05 level of significance, what is the decision rule for the hypothesis testing procedure?

b. If a sample mean of $\bar{x} = 16.32$ ounces occurs, what action would you recommend?

c. If $\bar{x} = 15.82$ ounces, what action would you recommend?

13. An automobile assembly line operation has a scheduled mean completion time of 2.2 minutes. Because of the effect of completion time on both earlier and later assembly operations it is important to maintain the 2.2 minute standard. A random sample of 45 times shows a sample mean completion time of 2.39 minutes, with a sample standard deviation of .20 minutes. Use a .02 level of significance and test whether or not the operation is meeting its 2.2 minute standard.

14. Historically, evening long distance phone calls from a particular city have averaged 15.20 minutes per call. In a random sample of 25 calls, the sample mean time was 14.30 minutes per call, with a sample standard deviation of 5 minutes. Use this sample information to test whether or not there has been a change in the average duration of long distance phone calls. Use a .05 level of significance.

15. Refer to the production line example in Exercise 12.

a. Develop a 95% confidence interval for the population mean based on the sample mean of $\bar{x} = 16.32$.

b. What is your verbal interpretation of your answer to part a? Use this result to test the hypothesis H_0: $\mu = 16$.

c. Repeat parts a and b for the sample result $\bar{x} = 15.82$.

16. Refer to the long distance phone call application in Exercise 14.

a. Develop a 95% confidence interval for the population mean time per call based on the sample results.

b. Use your results from part a to test the hypothesis H_0: $\mu = 15.20$. What is your conclusion?

17. It is hypothesized that the average driving speed on the interstate highway system is 55 miles per hour. A sample of 72 automobiles shows the average driving speed of 59.5, with a sample standard deviation of 12 miles per hour.

a. What are the null and alternative hypotheses for this situation? Use a two-tailed test.

b. Develop a 95% percent confidence interval for the mean driving speed for the population of automobiles on the interstate highway system.

c. Use your confidence interval developed in part b to test the hypothesis in part a. What is your conclusion?

18. A company currently pays an average wage of \$9.00 per hour for its employees. The company is planning to build a new factory, and several locations are being considered. The availability of labor at a rate less than \$9.00 per hour is a major factor in the location decision. For one location, a sample of 36 workers showed a mean wage of $\bar{x} = \$8.50$ and a sample standard deviation of $s = \$.60$.

a. A one-sided confidence interval uses the sample results to establish either an upper limit or a lower limit for the value of the population parameter. For this exercise establish an upper 95% confidence limit for the hourly wage rate. The form of this one-sided confidence interval requires that we be 95% confident that the population mean is this value or less. What is the 95% confidence statement for this one-sided confidence interval?

b. Use the one-sided confidence interval result to test the hypothesis H_0: $\mu \geq 9$. What is your conclusion? Explain.

9.5 OTHER METHODS OF HYPOTHESIS TESTING

The decision rules we have presented for the one- and two-tailed hypothesis tests in Sections 9.3 and 9.4 require us to compare the observed sample mean $\bar{x}$ with a critical value(s) in order to reach the accept-or-reject conclusion. When the decision rules are stated in this form, $\bar{x}$ is called the *test statistic*. Sometimes the decision rules are stated in a different but equivalent form yet provide the same conclusions. This section describes two other forms that the decision rules can have. These forms involve the use of standardized test statistics and p-values.

The Use of Standardized Test Statistics

In Section 9.3 we showed that the critical value for a one-tailed test of H_0: $\mu \geq \mu_0$ was $c = \mu_0 - z_{\alpha/2}\sigma_{\bar{x}}$. The decision rule for this test can be written

$$\text{Accept } H_0 \text{ if } \bar{x} \geq \mu_0 - z_\alpha \sigma_{\bar{x}},$$
$$\text{Reject } H_0 \text{ if } \bar{x} < \mu_0 - z_\alpha \sigma_{\bar{x}}.$$

Working algebraically with the rule for accepting H_0, we have

$$\bar{x} - \mu_0 \geq -z_\alpha \sigma_{\bar{x}},$$
$$\frac{\bar{x} - \mu_0}{\sigma_{\bar{x}}} \geq -z_\alpha. \tag{9.9}$$

Similarly, the rule for rejecting H_0 can be written

$$\frac{\bar{x} - \mu_0}{\sigma_{\bar{x}}} < -z_\alpha. \tag{9.10}$$

Here $-z_\alpha$ is the critical value that cuts off an area α in the lower tail of the standard normal probability distribution. Since $z = (\bar{x} - \mu_0)/\sigma_{\bar{x}}$ is a standard normal random variable,* we say that (9.9) and (9.10) state the decision rule in standardized form.

Let us return to the Hilltop Coffee example discussed in Section 9.3 to show how the standardized form of the decision rules can be used to test the hypotheses

$$H_0: \mu \geq 3,$$
$$H_1: \mu < 3.$$

Previously we developed the decision rules for a .01 level of significance as follows:

$$\text{Accept } H_0 \text{ if } \bar{x} \geq 2.93,$$
$$\text{Reject } H_0 \text{ if } \bar{x} < 2.93.$$

The sampling distribution for $\bar{x}$ in the Hilltop study is shown in the upper part of Figure 9.10. The critical value of 2.93 and a sample mean value of $\bar{x} = 2.92$ are shown. From the decision rule we see that $\bar{x} = 2.92$ leads to rejection of the null hypothesis.

Now let us see how the test works in standardized form. With $\sigma_{\bar{x}} = .03$ and $-z_{.01} = -2.33$, (9.9) and (9.10) provide the decision rule:

$$\text{Accept } H_0 \text{ if } \frac{\bar{x} - 3}{.03} \geq -2.33,$$

$$\text{Reject } H_0 \text{ if } \frac{\bar{x} - 3}{.03} < -2.33.$$

*Chapter 6 contains a discussion of the standard normal random variable.

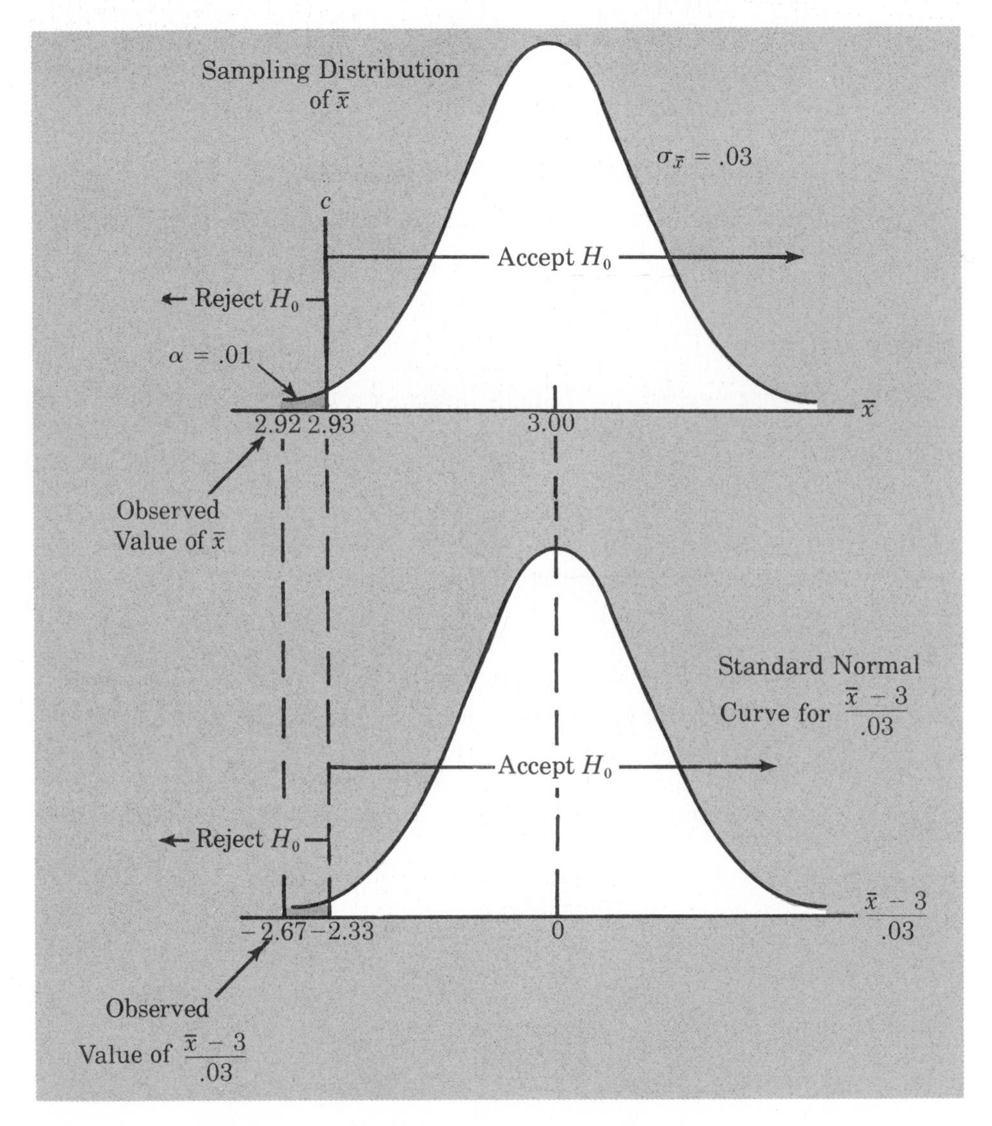

FIGURE 9.10 Sampling Distribution of $\bar{x}$ and of $(\bar{x} - 3)/.03$ for the Hilltop Coffee Study

With $\bar{x} = 2.92$, we have

$$\frac{\bar{x} - 3}{.03} = \frac{2.92 - 3.00}{.03} = \frac{-.08}{.03} = -2.67.$$

Since $-2.67 < -2.33$, we would reject using the standardized form of the decision rule, just as we did before.

Note that in the bottom part of Figure 9.10 the values of $(\bar{x} - 3)/.03$ corresponding to the values of $\bar{x}$ are shown. We see that -2.67 is in the same part of the rejection region for the standard normal curve that 2.92 is for the sampling distribution of $\bar{x}$.

Similarly, note the correspondence between the critical value of $c = 2.93$ and $-z_{.01} = -2.33$.

Whether we use $\bar{x}$ or $(\bar{x} - \mu_0)/\sigma_{\bar{x}}$ as the test statistic, the conclusion will be the same. The choice of decision rules is up to the user. One advantage of using the standardized form of the decision rule is that the critical values for the test can be read directly from the table of the standard normal probability distribution. Also, it is often the case when computers are used for data analysis that the value of $(\bar{x} - \mu)/\sigma_{\bar{x}}$ is printed as part of the output. Hence a hypothesis test can be conducted by simply comparing this value with the appropriate critical value from the normal tables.

The standardized decision rule for a two-tailed test requires us to use critical values of $-z_{\alpha/2}$ and $+z_{\alpha/2}$, since the two-tailed test has a rejection region of $\alpha/2$ in each tail of the distribution. The decision rule for the two-tailed test using the standardized test statistic becomes

$$\text{Accept } H_0 \text{ if } -z_{\alpha/2} \leq \frac{\bar{x} - \mu_0}{\sigma_{\bar{x}}} \leq z_{\alpha/2},$$

$$\text{Reject } H_0 \text{ if } \frac{\bar{x} - \mu_0}{\sigma_{\bar{x}}} < -z_{\alpha/2} \text{ or if } \frac{\bar{x} - \mu_0}{\sigma_{\bar{x}}} > z_{\alpha/2}.$$

In cases where the sample size is small, σ unknown, and the population normal, the t distribution with $n - 1$ degrees of freedom must be used for the standardized test statistic decision rule: t_α provides the critical value for a one-tailed test, and $t_{\alpha/2}$ provides the critical values for a two-tailed test.

The Use of *p*-Values

The hypothesis-testing procedures we have discussed thus far have utilized decision rules providing a known probability of making a Type I error. This probability, α, is referred to as the level of significance for the test.

Given a hypothesis and a particular sample result, a number called a *p-value* can be calculated. The p-value is the smallest value of α for which the given sample outcome would lead to accepting H_0. Thus if the p-value $< \alpha$ we would reject H_0. For example, if a sample result provides a p-value of .12, we know that there is a .12 probability of making a Type I error if we reject H_0. Now suppose that we desire a level of significance of $\alpha = .05$ for the test. What conclusion should we draw in terms of accepting or rejecting the null hypothesis H_0? The p-value tells us that a decision to reject would have a .12 probability of a Type I error. Since this is larger than the desired $\alpha = .05$, we would not want to reject H_0. Thus in this case the p-value leads us to accept H_0. On the other hand, suppose a particular sample result provides a p-value of .02. Since this p-value is less than $\alpha = .05$, we find that rejecting H_0 based on the sample evidence possesses a lower probability of a Type I error than the specified .05. Thus in this case the p-value leads us to reject H_0.

From the above discussion, we find that higher p-values indicate higher probabilities of making a Type I error. On the other hand, lower p-values indicate lower probabilities of making a Type I error. The following general relationship between the p-value and the level of significance α shows how the p-value can be used to make hypothesis-testing conclusions:

$$\text{Accept } H_0 \text{ if } p\text{-value} \geq \alpha,$$

$$\text{Reject } H_0 \text{ if } p\text{-value} < \alpha.$$

Let us return to the Hilltop coffee hypothesis test to show how p-values are computed for one-tailed tests. Specifically, we will compute the p-value associated with the sample result $\bar{x} = 2.92$. Figure 9.11 shows the sampling distribution of $\bar{x}$ and the location of the $\bar{x} = 2.92$ sample result under the assumption that the null hypothesis is true at $\mu = 3$. The p-value is given by the area in the tail of the sampling distribution below the $\bar{x} = 2.92$ value. The calculation of the p-value is as follows:

At $\bar{x} = 2.92$, we have

$$z = \frac{\bar{x} - \mu_0}{\sigma_{\bar{x}}} = \frac{2.92 - 3.00}{.03} = -2.67.$$

The table for the standard normal distribution shows the area between the mean and $z = -2.67$ to be .4962. Thus the area in tail of the sampling distribution must be $.5000 - .4962 = .0038$. Thus we have a p-value of .0038 associated with the $\bar{x} = 2.92$ sample result. Since the level of significance is $\alpha = .01$, we have a p-value less than α, leading to the rejection of the null hypothesis.

Figure 9.11 shows that a critical value of 2.93 corresponds to $\alpha = .01$. Since $\bar{x} = 2.92$ is less than 2.93, the p-value will be less than .01. Thus we see that the p-value approach will provide the same hypothesis-testing conclusion as the previous methods. Note here that use of the p-value is an alternative approach to reaching a hypothesis-testing conclusion. In addition, it has the advantage of providing probability information that indicates how close the accept or reject decision was. For example, if the level of significance for a particular test is .05, a p-value of .00001 would show the sample result to be an extreme value in the tail of the sampling distribution. Such a p-value would be a clear signal to reject the null hypothesis H_0. However, a p-value of .0495, being very close to $\alpha = .05$, shows that the hypothesis was just barely rejected. In fact, a p-value of .0495 shows that a level of significance of $\alpha = .01$ or .02 would have lead to the opposite conclusion: acceptance of the null hypothesis H_0. Finally, we see that a

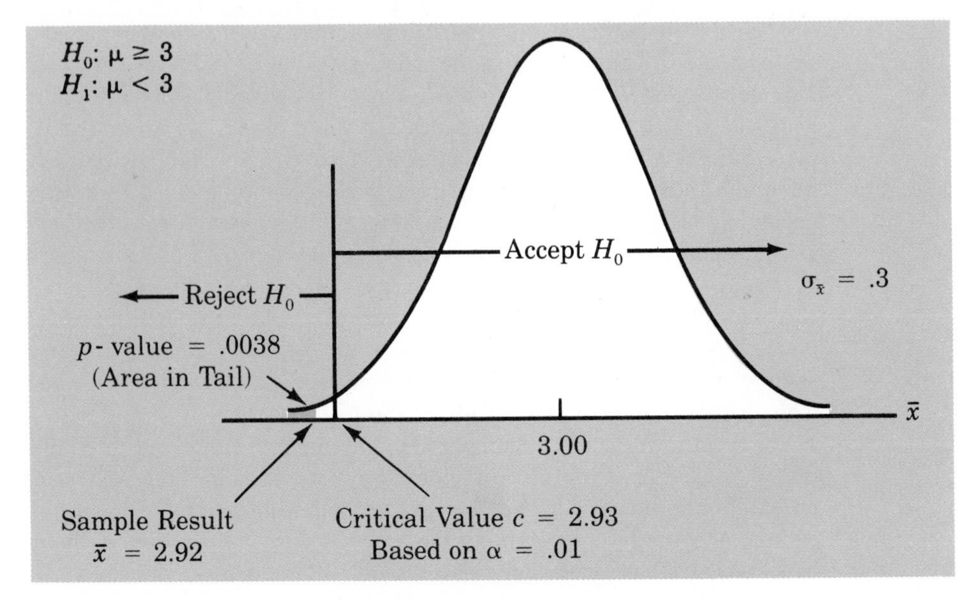

FIGURE 9.11 ***p*-Value for the Hilltop Coffee Hypothesis Based on Sample Result $\bar{x} = 2.92$**

p-value much larger than α, such as a p-value $= .3825$, indicates that the sample result is clearly in the accept region. Thus we see that comparing the p-value with the level of significance α enables us to make the accept/reject decision for the hypothesis and at the same time obtain an idea of how close the decision really was.

Current computer software for statistical analysis often provides p-values associated with the observed sample statistics. By comparing a p-value to the selected level of significance for the test, we can determine whether or not the hypothesis is accepted or rejected. Use of both the sample statistic and the standardized test statistic requires additional computation and/or looking up critical values in tables before the hypothesis-testing conclusion can be reached. In this regard, knowing how to use and interpret p-values can lead to an easy method for making hypothesis-testing conclusions.

Another advantage of using a p-value is that a researcher can simply report the p-value associated with the sample result for a particular test. The user of the research results can make his or her own choice of the preferred level of significance. By comparing the selected α to the reported p-value, the user of the results can easily reach the appropriate accept-or-reject decision. Thus one user might choose to reject and another to accept a certain hypothesis given the same sample results.

Before leaving this discussion of the p-value, consider the computation of p-values for a two-tailed hypothesis test. Let us return to the KMGM study where the two-tailed hypothesis testing situation was as follows:

$$H_0: \mu = 21,$$
$$H_1: \mu \neq 21.$$

The sampling distribution under the assumption that H_0 is true and the location of the observed sample result $\bar{x} = 20.3$ are shown in Figure 9.12. Let us proceed as before and

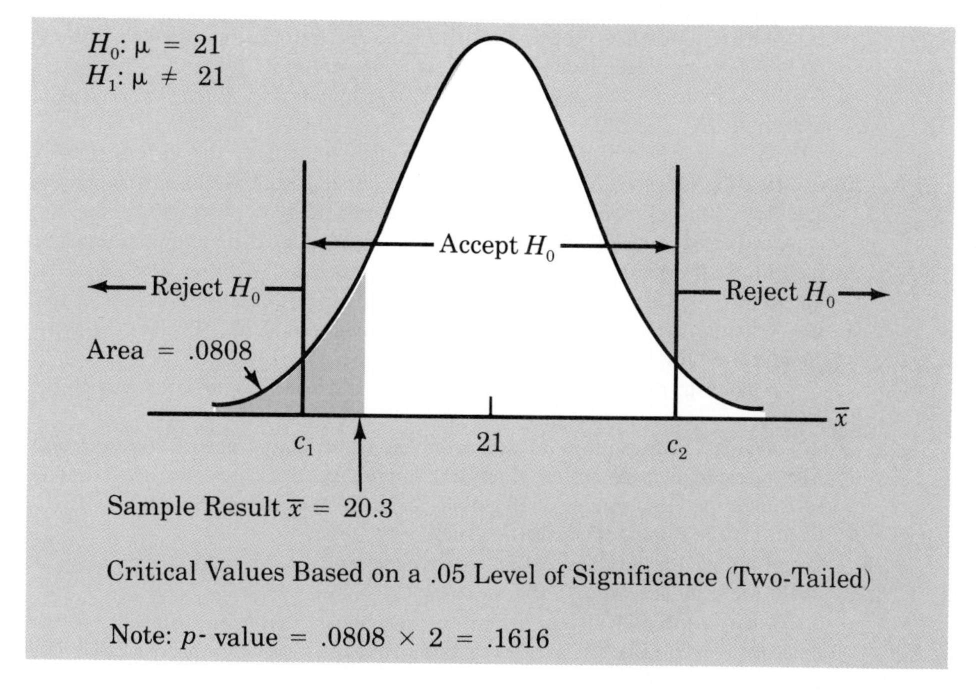

FIGURE 9.12 p-Value for the KMGM Hypothesis Based on a Sample Result of $\bar{x} = 20.3$

compute the p-value based on the sample result $\bar{x} = 20.3$. Referring to the sampling distribution in Figure 9.12, at $\bar{x} = 20.3$ we have

$$z = \frac{\bar{x} - \mu_0}{\sigma_{\bar{x}}} = \frac{20.3 - 21}{.5} = -1.40.$$

The table for the standard normal distribution shows the area between the mean and $z = -1.40$ to be .4192. Thus the area in the tail of the sampling distribution must be $.5000 - .4192 = .0808$. However, for the two-tailed hypothesis test this value is *not* the p-value for the test. Recall that the two-tailed test develops a two-tailed rejection region based on an area or probability of $\alpha/2$ in each tail. Because of this, the p-value for a two-tailed test is found by *doubling* the area or probability found in the above computations. That is, the p-value associated with the $\bar{x} = 20.3$ sample result is $.0808 \times 2 = .1616$. Since .1616 is greater than the level of significance $\alpha = .05$, the p-value calls for the acceptance of the hypothesis $\mu = 21$. Thus in computing the p-value for a two-tailed hypothesis test we must double the probability or area of one-tailed result in order to find the appropriate value.

EXERCISES

19. Develop standardized decision rules for an upper one-tailed hypothesis test when the population is normal, σ is unknown, and the sample size is small ($n < 30$).

20. Develop standardized decision rules for the two-tailed hypothesis test when σ is known and the sample size is large.

21. In Exercise 5 the manager of the Danvers-Hinton Resort Hotel believed that the mean guest bill was at least $250.

a. With a .05 level of significance, what is the standardized decision rule for this test?

b. With $\sigma = 50$, what is the test statistic value and the resulting conclusion based on a sample of 60 bills showing a sample mean amount of $235 per bill?

c. What is the p-value associated with this sample result? What is your conclusion about the hypothesis H_0: $\mu \geq 250$?

22. Stout Electric Company operates a fleet of trucks for its electrical service to the construction industry. Monthly mean maintenance costs have been $75 per truck. A random sample of 40 trucks shows a sample mean maintenance cost of $82.50 per month, with a sample standard deviation of $30. Management would like a test to determine whether or not the mean monthly maintenance cost has increased.

a. With a .05 level of significance, what is the standardized decision rule for this test?

b. What is your conclusion based on the sample mean of $82.50? Test the hypothesis with the standardized test statistic.

c. What is the p-value associated with this sample result? What is your conclusion based on the p-value?

23. Spread Easy paint is labeled as having a mean coverage of 400 square feet per gallon. Mean coverage more or less than 400 square feet indicates that the paint is not covering satisfactorily. For a sample of 20 gallons and a .05 level of significance

a. State the standardized decision rule for the test.

b. If sample results show $\bar{x} = 380$ square feet and $s = 25$ square feet, what is your conclusion about the paint coverage?

c. What is the difficulty in using the t-distribution table as it appears in the Appendix to determine the p-value? What can be said about the p-value in this exercise? What hypothesis-testing conclusion would be reached?

24. In Exercise 6 the president of Fightmaster and Associates Real Estate, Inc. claimed that the mean selling time of a residential home is 40 days or less. A sample of 50 homes showed a sample mean of 45 days and a sample standard deviation of 20 days; $\alpha = .02$ was used to make the test.

a. Test this hypothesis using a standardized test statistic. What is your conclusion?
b. What is the p-value associated with the sample result? What is your conclusion?

25. In exercise 17 it was hypothesized that the average driving speed on the interstate highway system was 55 miles per hour. The test was conducted as a two-tailed test. A sample of 72 automobiles showed an average driving speed of 59.5 miles per hour, with a sample standard deviation of 12 miles per hour.

a. Use a standardized test statistic to test this hypothesis. Use $\alpha = .05$.
b. What is the p-value associated with this sample result? What is your conclusion?

26. At Western University the historical mean scholarship examination score of entering students has been 900, with a standard deviation of 180. Each year a sample of applications is taken to see if the examination scores are at the same level as in previous years. The null hypothesis tested is H_0: $\mu = 900$. A sample of 200 students in this year's class shows a sample mean score of 935. Use a .05 level of significance.

a. Use a confidence interval estimation procedure to test this hypothesis.
b. Use a standardized test statistic to test this hypothesis.
c. What is the p-value for this test? What is your conclusion?

9.6 CALCULATING THE PROBABILITY OF TYPE II ERRORS

In the hypothesis-testing applications we have been discussing, the maximum allowable probability of making a Type I error has been specified. The critical value(s) for the decision rule were then set to control the Type I error when $\mu = \mu_0$. In effect, the entire hypothesis-testing procedure was developed without consideration for β, the probability of the Type II error. The approach we have followed is common and is justified on the grounds that the Type I error is often the more serious of the two errors. However, in many situations it also is important to control the probability of a Type II error. In this section we show how to compute the probabilities associated with Type II errors. In the next section we show how proper sample size selection can be used to control the probability of both Type I and Type II errors.

For purposes of illustrating the procedure for computing the probabilities associated with making a Type II error, let us return to the Hilltop Coffee study. The hypotheses for this application and the corresponding actions are restated below:

Hypothesis	*Conclusion and Action*
H_0: $\mu \geq 3$	Hilltop okay; no action necessary
H_1: $\mu < 3$	Hilltop violating label specification; take appropriate followup action

For a .01 level of significance the following decision rule was established:

$$\text{Accept } H_0 \text{ if } \bar{x} \geq 2.93,$$
$$\text{Reject } H_0 \text{ if } \bar{x} < 2.93.$$

The level of significance for the test specifies the probability of a Type I error at a maximum value of $\alpha = .01$. That is, if μ is any value greater than or equal to 3, then the probability of a Type I error does not exceed .01 (at $\mu = 3$ we have $\alpha = .01$). In Section

9.2 we defined the Type II error as the error of accepting H_0 when it is false. For the Hilltop Coffee problem this means claiming that Hilltop is meeting or exceeding its label specification (that is, concluding that $\mu \geq 3$) when in fact Hilltop is not meeting label specifications (that is, $\mu < 3$). Thus a Type II error would occur if Hilltop were underfilling the cans of coffee on the average but the sample mean $\bar{x}$ for the 36 cans turned out to be 2.93 pounds or more. The decision rule would lead to acceptance of H_0 when it was false.

In order to compute the probability of a Type II error we have to consider possible values of μ less than 3 pounds. For example, suppose that Hilltop is actually underfilling the 3 pound coffee cans by an average of 2 ounces. Since 2 ounces is equivalent to $2/16 = .125$ pounds, this degree of underfilling corresponds to a mean filling weight of $3 - .125 = 2.875$ pounds. If $\mu = 2.875$, we want to know the probability that a sample mean $\bar{x}$ will be greater than or equal to the critical value of 2.93, leading us to accept H_0. This probability is the probability of making a Type II error when the actual population mean is $\mu = 2.875$. The shaded area of the sampling distribution of $\bar{x}$ shown in Figure 9.13 indicates the probability of making a Type II error when $\mu = 2.875$.

Refer to Figure 9.13. What is the probability of obtaining a value of $\bar{x}$ greater than or equal to 2.93 and hence accepting H_0 when in fact H_1 is true with $\mu = 2.875$? Since the sampling distribution of $\bar{x}$ is approximately normal, we can use the standard normal probability distribution to answer this question. Computing the corresponding z value at $c = 2.93$, we have

$$z = \frac{2.930 - 2.875}{.03} = 1.83.$$

That is, the critical value is 1.83 standard deviations above the mean $\mu = 2.875$. The table of areas for the standard normal probability distribution shows the area in the tail above $z = 1.83$ to be $.5000 - .4664 = .0336$. Thus the probability of making a Type II error if $\mu = 2.875$ is $\beta = .0336$. Hence the probability is $1 - .0336 = .9664$ that Hilltop will be accused of violating label weight specifications if they are underfilling with $\mu = 2.875$.

Let us now consider the situation where Hiltop is underfilling its 3 pound coffee cans by an average of only 1 ounce ($1/16 = .0625$ pounds). This degree of underfilling shows a mean filling weight for the cans to be $3 - .0625 = 2.9375$. The sampling distribution of $\bar{x}$ when $\mu = 2.9375$ can be used to compute the probability of obtaining a value of $\bar{x}$ greater than 2.93 pounds and thus the probability of making a Type II error when $\mu = 2.9375$. Refer to Figure 9.14. We find the z value corresponding to the critical value of $c = 2.93$ to be

$$z = \frac{2.93 - 2.9375}{.03} = \frac{-.0075}{.03} = -.25.$$

The table of areas for the standard normal probability distribution shows the area above $z = -.25$ to be $.5000 + .0987 = .5987$. Thus the probability of a Type II error if Hilltop is actually underfilling with $\mu = 2.9375$ pounds is $\beta = .5987$. With this probability of making a Type II error we see that if Hilltop is underfilling the coffee cans at an average weight of 2.9875 pounds there is a sizable probability (.5987) that the hypothesis test will lead us to accept H_0 and conclude that Hilltop is meeting the label specifications when it is not.

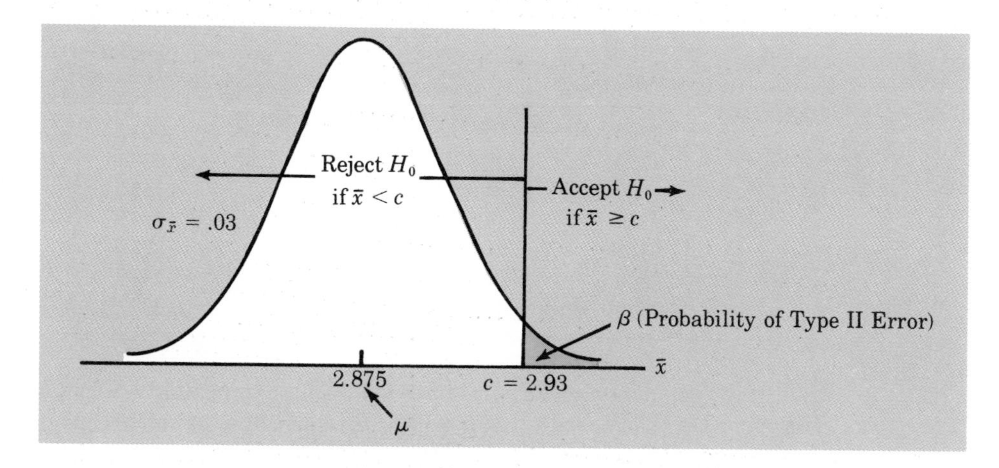

FIGURE 9.13 **Sampling Distribution of $\bar{x}$ and the Probability of Making a Type II Error When Actual $\mu = 2.875$ for the Hilltop Study**

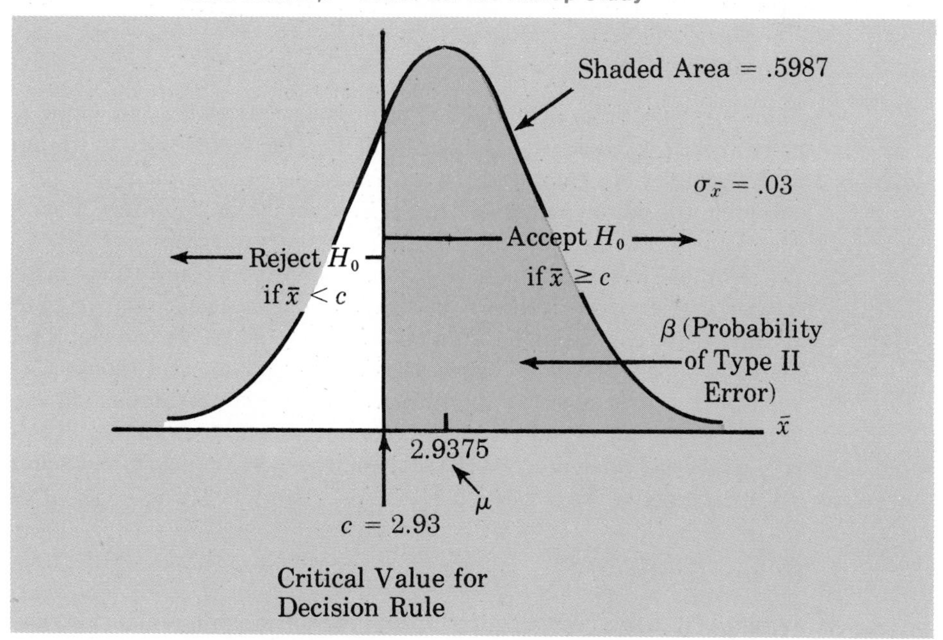

FIGURE 9.14 **Sampling Distribution of $\bar{x}$ and the Probability of Making a Type II Error When Actual $\mu = 2.9375$ for the Hilltop Study**

Let us pause for a moment and reflect on what we have learned about the probability of making a Type II error. First of all, we know that a Type II error can occur only when the null hypothesis is false. In the Hilltop case this corresponds to values of μ less than 3 pounds. Moreover, we learned that the probability of making a Type II error is not one unique value; it depends upon the actual value of μ. Thus by following the procedure we used to compute the probability of a Type II error when $\mu = 2.875$ and $\mu = 2.9375$ we can compute the probability of making a Type II error for other possible values of μ. The results are shown in Table 9.1. Note that as the actual

TABLE 9.1 **Type II Error Probabilities for the Hilltop Coffee Example**

If Actual μ *Is*	$z = \dfrac{2.93 - \mu}{.03}$	*Probability of Making a Type II Error* (β)
2.8500	2.67	.0038
2.8750	1.83	.0336
2.9000	1.00	.1587
2.9250	.17	.4325
2.9375	−.25	.5987
2.9500	−.67	.7486
2.9750	−1.50	.9332

value of the population mean gets closer to $\mu = 3$, the probability of making a Type II error increases. This tells us that if Hilltop is underfilling its coffee cans but is coming very close to meeting the label specification of 3 pounds, there is a high probability that we will not detect the label violation. Minor violations are more difficult to detect with our hypothesis test than major ones.

Operating Characteristic Curve

Table 9.1 shows the probabilities of making a Type II error by accepting H_0 when H_0 is actually false. Continuation of the calculations for values of $\mu \geq 3$ will provide a table of probabilities of accepting H_0 when H_0 is true. This is the probability of correctly accepting the null hypothesis. Table 9.2 shows the probabilities of accepting H_0 for values of $\mu \geq 3$. Note that for $\mu = 3$ the probability of accepting H_0 is $1 - \alpha$.

Plotting the probability of accepting H_0 for the possible values of the population mean μ provides what is known as the *operating characteristic curve* (OC curve) for the hypothesis test. Using the data in Tables 9.1 and 9.2 we can plot the OC curve for the Hilltop Coffee study, as shown in Figure 9.15. Note that for cases where the null hypothesis is true ($\mu \geq 3$) the OC curve provides the probabilities of correctly accepting the null hypothesis. However, for cases where the null hypothesis is false ($\mu < 3$) the OC curve provides the probabilities of accepting H_0 and hence committing a Type II error.

Power Curve

With the OC curve providing the probabilities of accepting H_0 for various values of the population mean μ, we might ask also about the probabilities of rejecting H_0 for various values of μ. Since we must either accept or reject H_0, the probability of accepting H_0

TABLE 9.2 **Probabilities of Accepting H_0 for the Hilltop Coffee Example When $\mu \geq 3$**

If Actual μ *Is*	$z = \dfrac{2.93 - \mu}{.03}$	*Probability of Accepting* H_0
3.00	−2.33	.9900
3.01	−2.67	.9962
3.02	−3.00	.9986
3.05	−4.00	1.0000

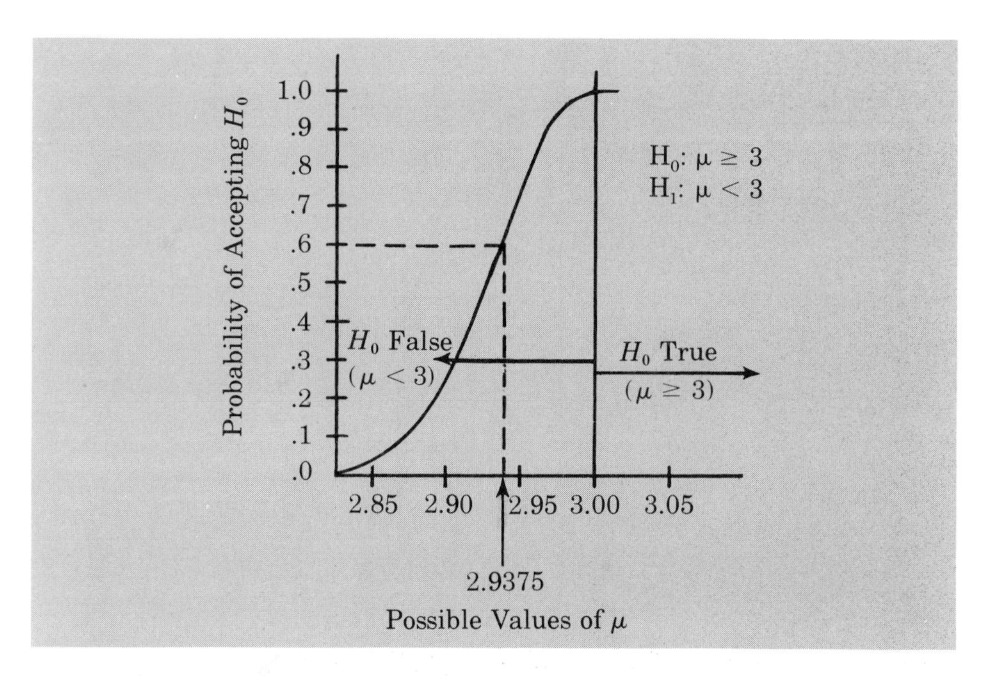

FIGURE 9.15 Operating Characteristic Curve for the Hilltop Study

and the probability of rejecting H_0 must sum to 1. Therefore the probability of rejecting H_0 for each possible value of μ is simply 1 minus the probability given by the OC curve. The probability of rejecting H_0 at a particular value of μ is called the *power* of the test, and the plot of these probabilities for various values of μ is called the *power curve.* The power curve for the Hilltop study is shown in Figure 9.16.

Note in using the power curve that for cases where the null hypothesis is true ($\mu \geq 3$) the power curve provides the probabilities of rejecting H_0 or, in effect, the probabilities associated with making a Type I error. However, for cases where the null hypothesis is false ($\mu < 3$) the power curve provides the probabilities of correctly rejecting H_0 when it should be rejected. Thus the higher the curve is in this region, the more power the hypothesis test has for correctly rejecting the null hypothesis. This is why it is called a power curve. In the Hilltop example these probabilities are the probabilities of detecting that Hilltop is violating its label specification when it is actually guilty of doing so.

One contribution of the hypothesis-testing procedure is that it provides the user with probability information about the ability to make correct as well as incorrect decisions. A complete summarization of these types of probabilities is contained in the OC and power curves. If the user finds these probabilities of being correct or in error satisfactory, the decision rule used to reach the hypothesis-testing conclusion should be a good one.

EXERCISES

27. In Exercise 8 Fowle Marketing Research, Inc. based charges to a client on the assumption that telephone surveys can be completed with a mean time of 15 minutes or less. If a greater mean survey time is required, a premium rate is charged to the client. Use a sample of 35

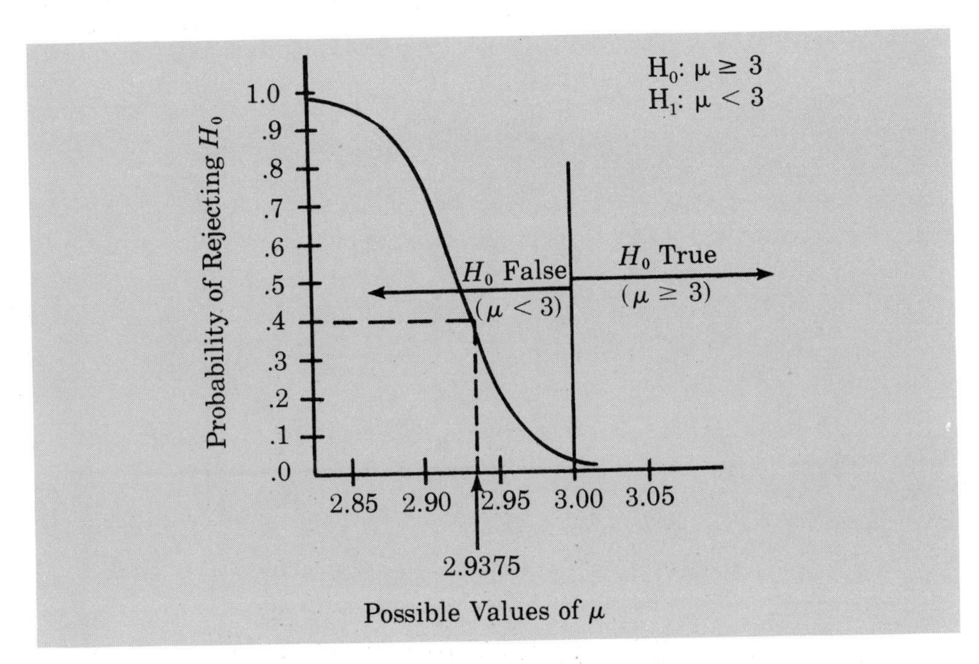

FIGURE 9.16 Power Curve for the Hilltop Study

surveys, a standard deviation of 4 minutes, and a level of significance of .01; the sample mean is used to test the hypothesis H_0: $\mu \le 15$.

a. What is your interpretation of the Type II error for this problem? What is its impact on the firm?

b. What is the probability of making a Type II error when the actual mean time is $\mu = 17$ minutes?

c. What is the probability of making a Type II error when the actual mean time is $\mu = 18$ minutes?

d. Sketch the general shape of the operating characteristic curve for this test. How does it differ from the operating characteristic curve for the Hilltop Coffee example? Why does this difference exist?

e. Sketch the general shape of the power curve for this test.

28. From the results of mileage testing, an automobile manufacturer claims that a new economy model will show at least 25 miles per gallon of gasoline.

a. Using a .02 level of significance and a sample test of 30 cars, what is the decision rule for the test to determine if the manufacturer's claim is supported? Assume that σ is known at 3 miles per gallon.

b. What is the probability of committing a Type II error if the actual mileage is 23 miles per gallon?

c. What is the probability of committing a Type II error if the actual mileage is 24 miles per gallon?

d. What is the probability of committing a Type II error if the actual mileage is 25.5 miles per gallon?

29. In the KMGM study in Section 9.4, the following hypotheses about the mean age of the station's listeners were tested:

$$H_0: \mu = 21,$$
$$H_1: \mu \ne 21.$$

a. What would it mean to KMGM to make a Type II error in this situation?
b. The population standard deviation was considered known at $\sigma = 5$, and the sample size was set at 100 listeners. What is the probability of accepting H_0 for each of the following possible values of μ:

$$23, 22.5, 22, 21.5, 21, 20.5, 20, 19.5, 19.$$

c. Show the operating characteristic curve for this problem. Why does it have a shape different from the operating characteristic curve in Figure 9.15?
d. Show the power curve for this hypothesis test.
e. What is the power at $\mu = 23$? What does this tell KMGM's management?

30. The production line operation referred to in Exercise 12 was tested for filling weight accuracy with the following hypotheses:

Hypothesis	*Conclusion and Action*
H_0: $\mu = 16$	Filling okay; keep running
H_1: $\mu \neq 16$	Filling off standard; stop and adjust machine

The sample size was 30 and the population standard deviation was considered known at $\sigma = .8$ $(\alpha = .05)$.
a. What would a Type II error mean in this situation?
b. What is the probability of making a Type II error when the machine is overfilling by .5 ounces?
c. What is the power of the statistical test when the machine is overfilling by .5 ounces?
d. Show the power curve for this hypothesis test. What information does it contain for the production manager?

31. Repeat parts b and c (Exercise 27) for the Fowle Marketing Research example if the firm selects a sample of 50 surveys. What observation can you make about how increasing the sample size affects the probability of making a Type II error?

32. Sparr Investments, Inc. specializes in tax deferred investment opportunities for its clients. Recently Sparr has offered a payroll deduction investment program for the employees of a particular company. Sparr estimates that the employees are currently averaging $100 or less per month in tax deferred investments. A sample of 40 employees will be used to test Sparr's hypothesis about the current level of investment activity among the population of employees. Assume that the monthly employee tax deferred investment amounts have a standard deviation of $75 and that a .05 level of significance will be used in the hypothesis test.
a. What is the Type II error in this situation?
b. What is the probability of the Type II error if the actual mean employee monthly investment is $120?
c. What is the probability of the Type II error if the actual mean employee monthly investment is $130?
d. Repeat parts b and c if a sample of size 80 employees is used.
e. Show the operating characteristic curves for both samples (sizes 40 and 80). What happens to this curve as the sample size is increased? Discuss.

9.7 DETERMINING SAMPLE SIZES FOR HYPOTHESIS TESTING

In Chapter 8 we discussed how sample size decisions could be made for interval-estimation problems once the user had specified the desired precision. Sample size decisions can be made for hypothesis-testing problems once the user of the results has specified allowable probabilities for making Type I and Type II errors. Let us show

how we could establish a sample size for the Hilltop Coffee hypothesis-testing problem if the director of the project had made the following statements about the error probabilities associated with the test:

Type I error statements: "If the company is filling coffee cans exactly to the weight specification $\mu = 3$ pounds, I am willing to risk a .01 probability of claiming that the firm is in violation when it is not."
Type II error statements: "If the company is underfilling coffee cans at an average of 1 ounce below the label weight (i.e., filling at $\mu = 2.9375$ pounds), I am willing to risk a .20 probability of not claiming that the firm is in violation when it really is."

Our problem now is to determine the sample size that should be recommended in order to construct a test which will satisfy the above statements about the allowable error probabilities for the test.

Before we get into the details of the sample size computation, let us make some observations about the error probability statements. First, note that the statement about the probability of a Type I error is simply a restatement of the level of significance or α value for the test. As we saw in the previous sections, we need to know α in order to establish the decision rule. Thus we will always have a specified probability of a Type I error. The statement about the probability of a Type II error is a new statement which must be provided in order to determine the appropriate sample size. Since we saw that this probability, β, depends upon the particular value of μ, we see that the β value specified must be accompanied by a definite statement about the value of μ.

The value of μ and the associated β used in the statement about a Type II error are completely up to the user of the test. The probabilities chosen for the two types of errors usually reflect the costs of the respective errors. If the cost of a Type I error is high, a small value for α should be chosen. If the cost of a Type II error is high, a small value for β should be chosen. In the Hilltop example, the Type II error condition is stated with reference to $\mu = 2.9375$. The director feels that he or she can risk a probability of error of $\beta = .20$ for this amount of violation. In Figure 9.14 we showed that the probability of a Type II error at $\mu = 2.9375$ was $\beta = .5987$. Since this β is based on a sample size of $n = 36$, we will need a substantial sample size increase in order to reduce the error probability to $\beta = .20$.

What sample size would you recommend for the Hilltop study in order to limit error probabilities to $\alpha = .01$ when the null hypothesis is true at $\mu = 3$ and to $\beta = .20$ when Hilltop is underfilling at $\mu = 2.9375$? In order to answer this question we consider two sampling distributions of $\bar{x}$, one with $\mu = 3$ and one with $\mu = 2.9375$. These two sampling distributions are shown together in Figure 9.17.

Recall that the general form of the decision rule for the Hilltop hypothesis test is

$$\text{Accept } H_0 \text{ if } \bar{x} \geq c,$$
$$\text{Reject } H_0 \text{ if } \bar{x} < c.$$

The locations of the sample mean values that lead to the acceptance and rejection of H_0 are indicated in Figure 9.17. From Figure 9.17 we see that what we need to do is choose a sample size such that when the critical value is computed the following statements are true: the rejection region for the upper sampling distribution will have an area of .01 or less in the lower tail, and the rejection region for the lower sampling distribution will have an area of .20 or less in the upper tail.

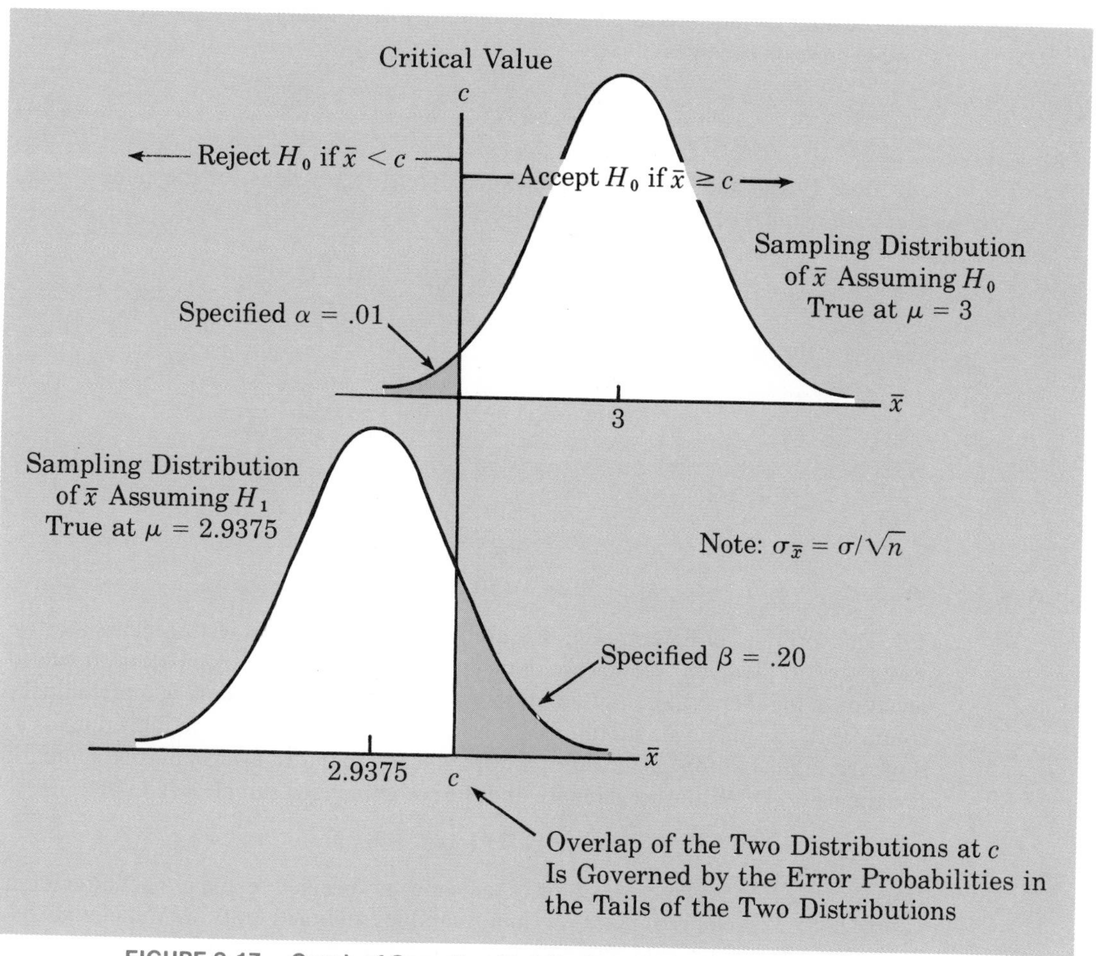

FIGURE 9.17 **Graph of Sampling Distributions Used to Determine the Sample Size for the Hilltop Coffee Example**

Recall that the population standard deviation for Hilltop coffee was assumed known at $\sigma = .18$ pounds. However, the standard error of the mean, $\sigma_{\bar{x}} = \sigma/\sqrt{n}$, cannot be determined, since we have not yet agreed on the sample size n. Thus leaving $\sigma/\sqrt{n}$ as the notation for the standard error of the mean we can use the sampling distribution with $\mu = 3$ to show that the critical value for the test must be as follows:

$$c = 3 - z_{.01}\frac{\sigma}{\sqrt{n}}.$$

With $z_{.01}$ given in the normal probability tables as $z_{.01} = 2.33$, we have

$$c = 3 - 2.33\frac{\sigma}{\sqrt{n}}. \qquad (9.13)$$

Next, using the sampling distribution corresponding to $\mu = 2.9375$, we can develop a second expression for the critical value of the test as follows:

$$c = 2.9375 + z_{.20}\frac{\sigma}{\sqrt{n}}.$$

With the table of the standard normal probability distribution showing $z_{.20} = .84$, the above expression becomes

$$c = 2.9375 + .84 \frac{\sigma}{\sqrt{n}}. \tag{9.14}$$

Since (9.13) and (9.14) provide two different expressions for the same critical value c, they must be equal. Thus we have

$$3 - 2.33 \frac{\sigma}{\sqrt{n}} = 2.9375 + .84 \frac{\sigma}{\sqrt{n}}.$$

Solving for the sample size n, we have

$$3 - 2.9375 = 2.33 \frac{\sigma}{\sqrt{n}} + .84 \frac{\sigma}{\sqrt{n}} = 3.17 \frac{\sigma}{\sqrt{n}}$$

$$.0625 \sqrt{n} = 3.17\sigma$$

$$\sqrt{n} = \frac{3.17}{.0625} \sigma = 50.72\sigma$$

$$n = (50.72)^2 \sigma^2.$$

This formula for determining the sample size in hypothesis testing shows that we must have a planning value (preliminary estimate) for the population standard deviation σ. In other situations we might have to rely on historical data or a preliminary sample with its sample standard deviation s in order to obtain a planning value for σ. However, in the Hilltop Coffee Study we are able to use the known population standard deviation $\sigma = .18$. With this value we find the recommended sample size to be

$$n = (50.72)^2(.18)^2 = 83.3.$$

Thus in order for the error probabilities to be met as specified by the project director, a sample of 83.3 is required. Since we cannot sample fractional units, we round up to 84. Rounding down would not quite permit us to meet the error probability requirements, whereas rounding up will permit us to make the error probabilities slightly smaller.

With $n = 84$, we can use (9.13) to compute the critical value c as follows:

$$c = 3 - 2.33 \frac{.18}{\sqrt{84}} = 2.954.$$

Thus for a sample mean from a sample of 84 coffee cans the following decision rule can be used:

Accept H_0 if $\bar{x} \geq 2.954$,

Reject H_0 if $\bar{x} < 2.954$.

General Outline for Sample Size Computation

We now have seen the detailed calculations necessary to determine the sample size for the Hilltop Coffee study. Let us outline the general steps that can be used to determine sample sizes:

1. Obtain statements about the allowable α value when the null hypothesis is assumed true and the allowable β value for a stated value of μ satisfying the alternate hypothesis.

2. Sketch a diagram showing the two sampling distributions of $\bar{x}$, one where the null hypothesis is assumed true and one where the μ from the alternate hypothesis is assumed true (see Figure 9.17).
3. In the above diagram show the critical value c for the test. Note that the two sampling distributions will "overlap" with the tail probabilities specified by the α and β values (again see Figure 9.17).*
4. Use $\sigma_{\bar{x}} = \sigma/\sqrt{n}$ and develop two equations for the critical value c, one based on each sampling distribution. These equations should have a form similar to (9.13) and (9.14).
5. Set the two equations in step 4 equal to each other and solve for n in terms of the population standard deviation σ.
6. Using a known or planning value for σ, solve for the desired sample size n.

EXERCISES

33. Suppose that the project director for the Hilltop Coffee study had asked for a .10 probability of not claiming that the firm was in violation when it really was underfilling by 1 ounce ($\mu = 2.9375$ pounds). What sample size would have been recommended?

34. A special industrial battery is manufactured with an estimated average use life of at least 400 hours. A hypothesis test is to be conducted with a .02 level of significance. If the batteries from a particular production run have an actual mean use life of 385 hours, the production manager would like a sampling procedure that erroneously concludes the batch is acceptable only 10% of the time. What sample size is recommended for the battery test? Use 30 hours as a planning value for the population standard deviation.

35. In the KMGM study, H_0: $\mu = 21$ was tested at a .05 level of significance. If KMGM's station manager would permit only a .15 probability of erroneously claiming that KMGM was reaching a listening audience with an average age of 21 when the true mean age was really 22, what sample size should have been selected? Assume $\sigma = 5$.

36. The automobile mileage study in Exercise 28 tested the following hypothesis:

Hypothesis	*Conclusion*
H_0: $\mu \geq 25$	Manufacturer's claim supported
H_1: $\mu < 25$	Manufacturer's claim rejected; average mileage per gallon less than stated

For $\sigma = 3$ and a .02 level of significance, what sample size would be recommended if the tester desired an 80% chance of detecting that μ was less than 25 miles per gallon when it was actually $\mu = 24$?

9.8 HYPOTHESIS TESTS ABOUT A POPULATION PROPORTION

In regard to testing hypotheses regarding a population mean, there are three forms that we can consider when testing hypotheses regarding the value of a population proportion. Using the symbol p to denote the true population proportion and p_0 to

*For upper-tail hypothesis testing situations the overlapping will occur between the upper tail of the sampling distribution when H_0 is assumed true and the lower tail of the other sampling distribution. In addition, if the hypothesis test is a two-tailed test, $\alpha/2$ rather than α will be in the tail of the sampling distribution when H_0 is assumed true.

denote a particular value for the proportion, we can write the three forms as follows:

Form 1	*Form 2*	*Form 3*
$H_0: p \geq p_0$ $H_1: p < p_0$	$H_0: p \leq p_0$ $H_1: p > p_0$	$H_0: p = p_0$ $H_1: p \neq p_0$

The first two forms require a one-tailed test about a population proportion, while the third form requires a two-tailed test. The specific form used depends upon the application of interest and the decision that must be made.

Let us illustrate the hypothesis-testing procedure for a population proportion by considering the problem faced by Byroni's Italian Restaurant. Although Byroni's currently operates as an eat-in-only restaurant, management is considering opening a carry-out operation. The carry-out service could be a profitable venture, but management is concerned about the potential for profits lost because not all individuals placing orders by phone actually pick up their orders. After considering the loss associated with carry-out orders, management believes that if at least 90% of the phone orders are picked up, the carry-out service will be profitable. However, if less than 90% of the orders are picked up, the carry-out service will show a loss. Management has decided to go ahead with the carry-out operation unless they can be convinced that fewer than 90% of the orders will be picked up. Let p indicate the proportion of carry-out orders that will be picked up. The following hypotheses could be formulated to help management determine whether or not to implement the carry-out service:

Hypothesis	*Conclusion and Action*
$H_0: p \geq .90$	Carry-out pickup satisfactory; go with carry-out service
$H_1: p < .90$	Carry-out pickup unsatisfactory; do not offer carry-out service

In order to determine the feasibility of the operation, management has decided to run a 6-week test of the carry-out service. At the end of the test period a decision will be made as to whether or not the carry-out service should be continued. During this period, a sample of 250 phone-in orders will be used to estimate the proportion of carry-out orders that are picked up. The results of the sample will be used to test the above hypotheses and help management arrive at a decision on the carry-out service.

As usual, we will begin the hypothesis-testing procedure by assuming that $p = .90$ and hence that H_0 is true. Using the sample proportion $\bar{p}$ to estimate p, we next consider the sampling distribution of $\bar{p}$. Since $\bar{p}$ is an unbiased estimator of p, we know that the mean of the sampling distribution of $\bar{p}$ is .90 if $p = .90$. In addition, we know from Chapter 7 that the standard error of the proportion is given by

$$\sigma_{\bar{p}} = \sqrt{\frac{p(1-p)}{n}}.$$

With the assumed value of $p = .90$ and a sample size of $n = 250$, the standard error of the proportion for the Byroni problem is

$$\sigma_{\bar{p}} = \sqrt{\frac{.90(1-.90)}{250}} = .019.$$

In Chapter 7 we saw that the sampling distribution of $\bar{p}$ can be considered normal if both np and $n(1 - p)$ are greater than or equal to 5. In the Byroni case $np = 250(.90) = 225$ and $n(1 - p) = 250(.10) = 25$; thus the normal distribution approximation is appropriate. The sampling distribution of $\bar{p}$ for the Byroni problem is shown in Figure 9.18. Let us assume that $\alpha = .05$ has been adopted as the level of significance for the test. With $z_{.05} = 1.645$, the critical value for the test is shown in Figure 9.19. The decision rule becomes

$$\text{Accept } H_0 \text{ if } \bar{p} \geq .869,$$
$$\text{Reject } H_0 \text{ if } \bar{p} < .869.$$

Once the decision rule has been determined, it is a relatively simple task to collect the data, compute the value of the point estimate $\bar{p}$, and compare that value to the critical value of the decision rule in order to arrive at a decision. Let us assume that of the 250 phone orders sampled, 220 were actually picked up. In this case $\bar{p} = 220/250 = .88$. Comparing this value of $\bar{p}$ to the critical value we accept H_0. Byroni's would therefore continue with the carry-out service.

We see that hypothesis tests can be conducted on a population proportion as well as a population mean. The procedure used for both situations is similar, with the difference being that the sampling distribution of $\bar{x}$ is used when the hypothesis involves a population mean and the sampling distribution of $\bar{p}$ is used when the hypothesis involves a population proportion. The assumption that the null hypothesis is true, the use of the level of significance to establish the critical value, and the comparison of the point estimate to the critical value are identical for both testing procedures.

Although we shall not discuss other aspects of hypothesis testing about a population proportion, realize that the concepts involved with the probability of Type II errors, the operating characteristics curve, the power curve, and the determination of sample size hold for the proportion case just as they did when the hypothesis testing focused on the population mean. A summary of the decision rules for hypothesis tests

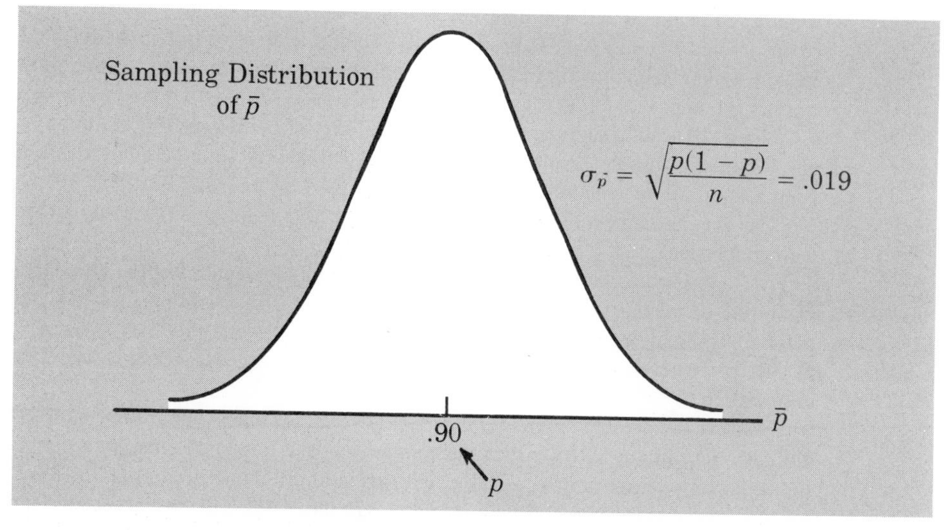

FIGURE 9.18 Sampling Distribution of $\bar{p}$ for the Byroni Carry-Out Decision with $p = .90$ and $n = 250$

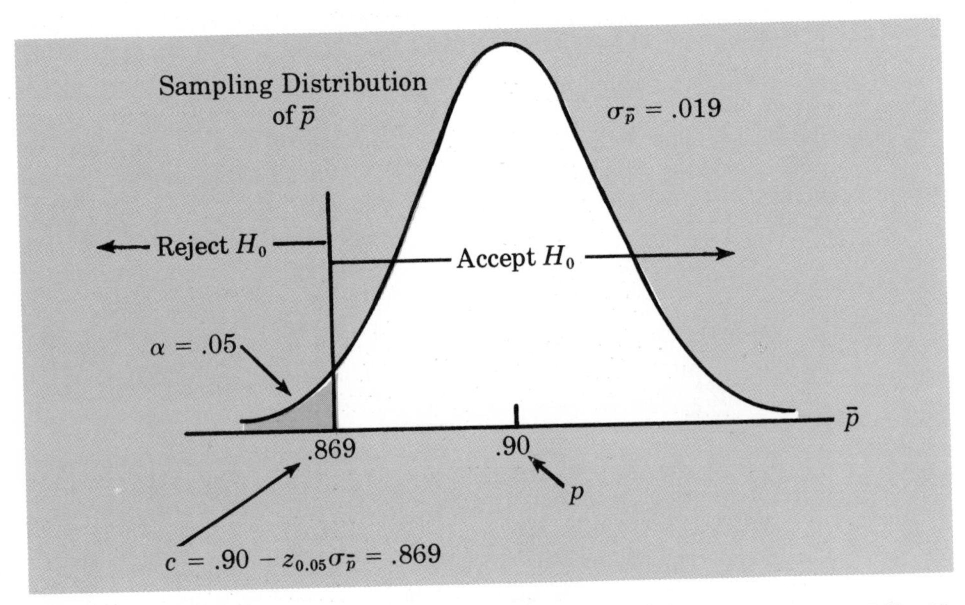

FIGURE 9.19 **Critical Value for the Byroni Hypothesis Test at a .05 Level of Significance**

about a population proportion is presented in Figure 9.20. We assume the large-sample case ($np \geq 5$ and $n(1 - p) \geq 5$) where the normal distribution is a good approximation to the sampling distribution of $\bar{p}$.

EXERCISES

37. The director of a college placement office claims that at least 80% of graduating seniors have made employment commitments 1 month prior to graduation. At a .05 level of significance, what is your conclusion if a sample of 100 seniors shows that 75 have actually made employment commitments 1 month prior to graduation? Should the director's claim be rejected?

38. A magazine claims that 25% of its readers are college students. A random sample of 200 readers is taken. It is found that 42 of these readers are college students.

a. Use a .10 level of significance to test the validity of the magazine's claim.

b. Using the sample results, develop a 90% confidence interval for the proportion of the population that are college students.

c. Use the interval estimate of part b to test the magazine's claim about its readers. What is your conclusion?

39. A new television series must prove that it has more than 25% of the viewing audience after its initial 13 week run in order to be judged successful. Assume that in a sample of 400 households, 112 were watching the series.

a. At a .10 level of significance, can the series be judged successful based on the sample information?

b. What is the p-value for the sample results? What is your hypothesis-testing conclusion?

40. An accountant believes that the company's cash flow problems are a direct result of the slow collection of accounts receivable. The accountant claims that at least 70% of the current accounts receivable are over 2 months old. A sample of 120 accounts receivable shows 78 over 2 months old. Test the accountant's claim at the .05 level of significance using a standardized test statistic.

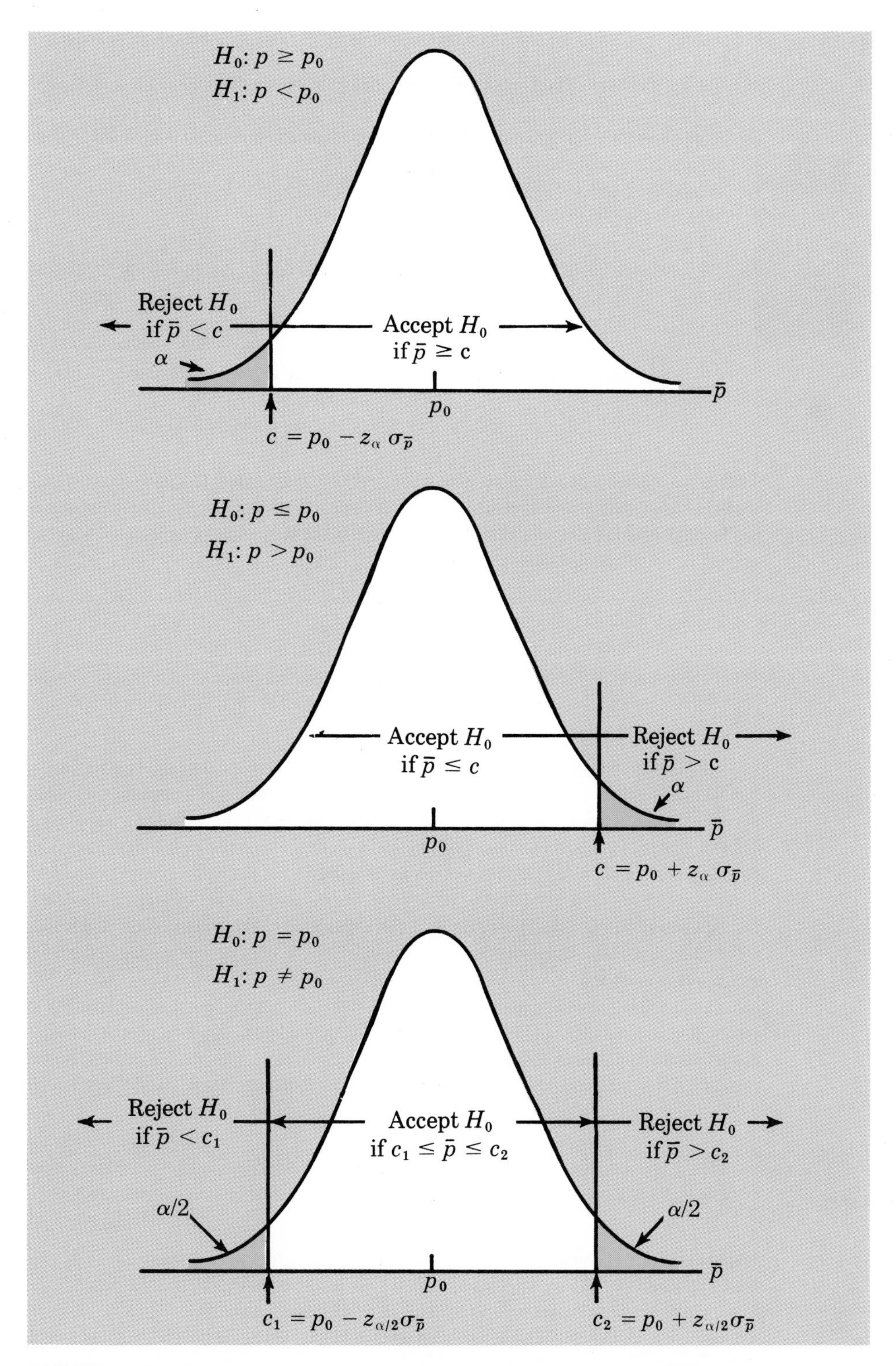

FIGURE 9.20 Summary of Statistical Decision Rules for Tests About a Population Proportion—Large-Sample Case [$np \geq 5$ and $n(1 - p) \geq 5$]

41. A supplier claims that at least 96% of the parts it supplies meet the product specifications. In a sample of 500 parts received over the past 6 months, 36 were defective. Test the supplier's claim at a .05 level of significance.

42. In testing the effectiveness of a new drug, a major drug firm administers the drug to a sample of 50. The hypothesis being tested is that the drug will prove effective for at least 80% of the people using the drug. In conducting the test, the hypothesis is supported if at least 36 of the 50 subjects experience positive test results.

a. What is the level of significance for the test?
b. What is the probability of the Type II error if $p = .75$?
c. What is the probability of the Type II error if $p = .70$?

43. A television station predicts election winners based on the following hypothesis test:

Hypothesis	*Conclusion*
H_0: $p \leq .50$	Candidate cannot be claimed a winner
H_1: $p > .50$	Candidate claimed a winner

The above hypotheses are tested with a .01 level of significance to guard against falsely claiming that a winner exists. On the other hand, if the true proportion for the candidate is .54, a .05 probability of not detecting that the candidate is the winner is permissible. What sample size is needed for this hypothesis-testing situation?

Summary

In this chapter we have introduced the statistical process of hypothesis testing by discussing hypothesis testing situations involving population means and population proportions. The hypothesis-testing procedure begins with a statement about the value of a population parameter. This statement is called the null hypothesis. Assuming that the null hypothesis is satisfied as an equality, we can then develop the sampling distribution of the appropriate point estimator. Using a maximum allowable probability of a Type I error as the level of significance, a critical value can be computed for a decision rule that can be used to make the decision to accept or reject H_0. Depending upon the acceptance or rejection of H_0, a conclusion is reached and a decision or action recommended.

Additional hypothesis-testing concepts, such as computing the probability of a Type II error, the use of the operating characteristic curve, and the use of the power curve, were discussed. In hypothesis-testing problems about a population mean we showed how a sample size could be determined to control the probabilities for both the Type I and Type II errors.

Glossary

Hypothesis—A statement about the value of a population parameter.
Null hypothesis—The hypothesis assumed true in the hypothesis-testing procedure. The sample results lead to the acceptance or rejection of this hypothesis.
Alternative hypothesis—The hypothesis assumed true if the null hypothesis is rejected.
Type I error—The error of rejecting H_0 when it is true.
Type II error—The error of accepting H_0 when it is false.

Critical value—A value which is compared with the test statistic to determine whether H_0 is to be accepted or H_0 is to be rejected.

Level of significance—The maximum probability of a Type I error that the user will tolerate in the hypothesis-testing procedure. The level of significance needs to be known in order to determine the critical value.

One-tailed test—A hypothesis test in which rejection of the null hypothesis occurs for values of the point estimator in one tail of the sampling distribution.

Two-tailed test—A hypothesis test in which rejection of the null hypothesis occurs for values of the point estimator in either tail of the sampling distribution.

***p*-value**—Given a sample result, the p-value is the smallest value of α for which H_0 would be rejected. The null hypothesis should be rejected only if the p-value is less than the level of significance for the test.

Operating characteristic curve—A graph of the probability of accepting H_0 for all possible values of the population parameter. In instances where the value of the population parameter is different than specified in the null hypothesis, the operating characteristic curve provides the probability of making a Type II error.

Power—The probability of rejecting H_0.

Power curve—A graph of the probability of rejecting H_0 for all possible values of the population parameter. In instances where the value of the population parameter is different than specified in the null hypothesis, the power curve provides the probability of correctly rejecting the null hypothesis.

Key Formulas

Critical Value for H_0: $\mu \geq \mu_0$, H_1: $\mu < \mu_0$

$$c = \mu_0 - z_\alpha \sigma_{\bar{x}} \tag{9.3}$$

Critical Value for H_0: $\mu \leq \mu_0$, H_1: $\mu > \mu_0$

$$c = \mu_0 + z_\alpha \sigma_{\bar{x}} \tag{9.4}$$

Note: In (9.3) and (9.4), if σ is unknown and $n \geq 30$, substitute $s_{\bar{x}}$ for $\sigma_{\bar{x}}$. If σ is unknown, $n < 30$, and the population is approximately normal, substitute $t_\alpha s_{\bar{x}}$ for $z_\alpha \sigma_{\bar{x}}$

Critical Value for H_0: $\mu = \mu_0$, H_1: $\mu \neq \mu_0$

$$c_1 = \mu_0 - z_{\alpha/2} \sigma_{\bar{x}} \tag{9.5}$$

$$c_2 = \mu_0 + z_{\alpha/2} \sigma_{\bar{x}} \tag{9.6}$$

Note: In (9.5) and (9.6), if σ is unknown and $n \geq 30$, substitute $s_{\bar{x}}$ for $\sigma_{\bar{x}}$. If σ is unknown, $n < 30$, and the population is approximately normal, substitute $t_{\alpha/2} s_{\bar{x}}$ for $z_{\alpha/2} \sigma_{\bar{x}}$.

Hypothesis Testing Using p-Values

Accept H_0 if p-value $\geq \alpha$

Reject H_0 if p-value $< \alpha$

Supplementary Exercises

44. The manager of the Keeton Department Store believes that the mean annual income of the store's credit card customers is at least $18,000 per year. A sample of 58 credit card customers shows a sample mean of $17,200 and a sample standard deviation of $3,000. At the .05 level of significance can the manager's claim be rejected?

45. The chamber of commerce of a Florida Gulf Coast community advertises area commercial property available at a mean cost of $25,000 or less per acre. Use a .05 level of significance and test this claim for a sample of 32 properties providing a sample mean of $26,000 per acre and a sample standard deviation of $2,500.

46. A bath soap manufacturing process is designed with the expectation that each batch prepared in the mixing department will produce a mean of 120 bars of soap per batch. Quantities over or under this standard are undesirable. A sample of ten batches shows the following numbers of bars of soap:

108, 118, 120, 122, 119, 113, 124, 122, 120, 123.

Use a .05 level of significance and test to see if the sample result supports the conclusion that the mean number of bars of soap per batch is 120.

47. The monthly rent for a two bedroom apartment in a particular city is reported to average $350. Assume we would like a test of the hypotheses $H_0 = 350$ versus $H_1{:}\mu \neq 350$. A sample of 36 two bedroom apartments is selected. The sample mean turns out to be $\bar{x} = \$338$, with a sample standard deviation of $s = \$40$.

a. Conduct this hypothesis test with a .05 level of significance.

b. Use the sample results to construct a 95% confidence interval for the population mean. What hypothesis-testing conclusion would you draw based on the confidence interval result?

48. Refer to Exercise 47:

a. Compute the p-value and make the hypothesis-testing conclusion.

b. Repeat the hypothesis-testing procedure using the standardized test statistic.

49. In making bids on building projects, Sonneborn Builders, Inc. assumes construction workers are idle no more than 15% of the time. For a normal 8-hour shift, this means that the average idle time per worker should be 72 minutes or less per day. A sample of 30 construction workers found a mean idle time of 80 minutes per day. The sample standard deviation was 20 minutes.

a. What is the p-value associated with the sample result?

b. Using a .05 level of significance and the p-value, test the hypothesis $H_0{:}\ \mu \leq 72$. What is your conclusion?

c. Repeat the above test using the standardized test statistic.

50. Refer to Exercise 49:

a. What is the probability of making the Type II error when the population mean idle time is 80 minutes?

b. What is the probability of making the Type II error when the population mean idle time is 75 minutes?

c. What is the probability of making the Type II error when the population mean idle time is 70?

d. Sketch the operating characteristic curve for this problem.

e. Sketch the power curve for this problem.

51. A federal funding program is available to low-income neighborhoods. To qualify for the funding a neighborhood must have a mean household income of less than $7,000 per year. Neighborhoods with mean annual household incomes of $7,000 or more do not qualify. Funding decisions are based on a sample of residents in the neighborhood. A hypothesis test with a .02 level of significance is conducted. If the funding guidelines call for a maximum probability of .05 of not funding a neighborhood with a mean annual household income of $6,500, what sample size should be used in the funding decision study? Use $\sigma = \$2{,}000$ as a planning value.

52. The bath soap production process in Exercise 46 uses the hypothesis H_0: $\mu = 120$ and H_1: $\mu \neq 120$ in order to test whether or not the production process is meeting the standard output of 120 bars per batch. Use a .05 level of significance for the test and a planning value of 5 for the standard deviation.

a. If the mean output drops to 117 bars per batch, the firm would like a 98% chance of concluding that the standard production output is not being met. How large a sample should be selected?

b. What is the probability of accepting H_0 and concluding that the process is satisfactory for each of the following actual mean outputs: 117, 118, 119, 120, 121, 122, and 123 bars per batch?

c. Show the OC and power curves for this situation.

53. During the past year the Dumont Clothing Store recorded 72% charge purchases and 28% cash purchases. A sample of 200 recent purchases shows that 160 were charge purchases. Does this suggest a change in the paying practices of the Dumont customers? Test with $\alpha = .05$.

54. The manager of K-Mark Supermarkets estimates that at least 30% of the Saturday customers purchase the price-reduced special advertised in the Friday newspaper. Use $\alpha = .05$ and test the manager's estimate if a sample of 250 Saturday customers show that 60 purchased the advertised special.

55. The filling machine for a production operation must be adjusted if more than 8% of the items being produced are underfilled. A random sample of 80 items from the day's production contained 9 underfilled items. Does the sample evidence indicate that the filling machine should be adjusted? Use $\alpha = .02$.

56. A radio station in a major resort area announced that at least 90% of the hotels and motels would be full for the Memorial Day weekend. The station went on to advise listeners to make reservations in advance if they planned to be in the resort over the weekend. On Saturday night a sample of 58 hotels and motels showed 49 with a no-vacancy sign and 9 with vacancies. What is your reaction to the radio station's claim based on the sample evidence? Use $\alpha = .05$ in making this statistical test. What is the p-value for the sample results?

57. In Exercise 56, assume we are willing to live with a .20 probability of making a Type II error if in fact only 85% of the hotels and motels were full. How large a sample would have to be taken to meet this Type II error requirement?

Application

Harris Corporation*

HARRIS

Melbourne, Florida

Harris Corporation/RF Communications Division is a major manufacturer of point-to-point radio communications equipment. It is a horizontally integrated manufacturing company with a multi-plant facility. Harris' primary factory for catalog products specializes in medium to high volume assembly; factory operations include printed circuit assembly, final assembly and testing.

A HYPOTHESIS TESTING APPLICATION

One of the higher volume items at the catalog products factory involves an assembly called an RF deck. The assembly consists of sixteen electronic components soldered to a machined casting which forms the plated surface of the deck. During a manufacturing run a problem developed in the soldering process; the flow of solder onto the deck did not meet the criteria established. A manufacturing engineer was summoned to solve the problem.

A preliminary investigation conducted by the manufacturing engineer uncovered the following factors which could cause the soldering problem:

1. impure flux or solder;
2. improper temperature of the soldering iron;
3. operator training;
4. contaminated RF deck surface;
5. defective plating from supplier.

After considering these factors, the engineer made a preliminary determination that the soldering problem was most likely due to defective plating from the supplier. But, before drawing a final conclusion, the engineer decided to set up an experiment to test this hypothesis. The objective of the experiment was to generate data that could be used for a hypothesis test on a population proportion.

The question raised by the engineer was: Did the proportion of defective platings exceed that which could be expected when the supplier's operation conforms to design specifications? Letting p indicate the proportion of defective platings in the Harris inventory and p_0 indicate the proportion of defective platings when the supplier

*The authors are indebted to Richard A. Marshall of the Harris Corporation for providing this application.

A business executive uses a radio communication car telephone manufactured by the Harris Corporation.

operation conforms with design specifications, the following hypothesis test was conducted

$$H_0: p \leq p_0$$
$$H_1: p > p_0$$

Accepting H_0 indicates that the Harris inventory has a defective plating proportion less than or equal to the design specification. However, rejecting H_0 indicates that the Harris inventory has a defective plating proportion greater than the design specification.

Tests made on a sample of the Harris inventory showed a sample proportion defective of $\bar{p} = .15$. This proportion defective resulted in the rejection of H_0; the current inventory had a higher defective proportion than would exist if the supplier's operation conformed to design specifications.

As a result the manufacturing engineer concluded that the plating was defective and pressure was applied to purchasing management to have the plating supplier held responsible for both rejected parts and damages.

FURTHER ANALYSIS AND RECONSIDERATION

The conclusion that the plating supplier was responsible for the soldering problem resulted in additional meetings with management at various levels of the organization. Finally, a meeting was called with the plant manager and the materials manager to review the data and decide on a course of action.

At this meeting several questions were raised concerning the experimental process that resulted in the conclusion that defective plating was the underlying problem. For example, one of the defective pieces had only one of sixteen components that did not accept solder; further investigation of this piece showed that the plating was not at fault, but that the component itself was contaminated. Another defective piece examined was scorched and bleeded. In this case, further investigation revealed high soldering temperatures as the most likely cause of the defect.

After reflection on the experimental process, the plant manager and materials manager concluded that, although the hypothesis test correctly indicated an undesirably high proportion of defective platings, it *did not prove* defective plating was the cause of the soldering problem. Additional experimentation was requested in order to identify other possible causes. This further investigation ultimately led to the conclusion that the true underlying problem was shelf contamination, not plating. Management thus recognized that the supplier was not at fault and accordingly did not press a complaint.

CONCLUSION

The hypothesis test described in this application pointed to an undesirable level of defective platings in the Harris inventory. However, the statistical evidence did not prove the supplier was the cause of the soldering problem. In this case, the fact that management did not blindly accept "official looking" statistics prevented a wrong decision from being made.

10 Statistical Inference about Means and Proportions with Two Populations

What you will learn in this chapter:

- how to construct interval estimates and conduct hypothesis tests about the difference between the means of two populations
- when and how to use the *t* distribution to conduct inferences about the difference between the means of two populations
- the difference between independent and matched samples
- how to compute a pooled variance estimate
- how to construct interval estimates and conduct hypothesis tests about the difference between the proportions of two populations

Contents

In the preceding three chapters we have developed statistical methodology for interval estimation and hypothesis testing for population means and population proportions. However, the statistical procedures we have discussed thus far have considered only single-population situations. In this chapter we expand our discussion to cases where *two populations* are involved. Specifically, we will be selecting samples and performing statistical analyses that will enable us to draw conclusions about the difference between means and/or proportions for two populations.

10.1 ESTIMATION OF THE DIFFERENCE BETWEEN THE MEANS OF TWO POPULATIONS—INDEPENDENT SAMPLES

In many practical decision-making situations we are faced with two separate populations where the difference between the means of the two populations is of prime importance. We know from Chapters 7 and 8 that we can take a simple random sample from a single population and use the sample mean $\bar{x}$ as a point estimator of the population mean. In the two-population case we will select two independent simple random samples, one from population 1 and another from population 2. Let

μ_1 = mean of population 1,
μ_2 = mean of population 2,
$\bar{x}_1$ = sample mean of the simple random sample taken from population 1 (i.e., the point estimator of μ_1),
$\bar{x}_2$ = sample mean of the simple random sample taken from population 2 (i.e., the point estimator of μ_2).

The actual difference in the two population means is given by $\mu_1 - \mu_2$. The point estimator of $\mu_1 - \mu_2$ is as follows:

Point Estimator of the Difference Between the Means of Two Populations

$$\bar{x}_1 - \bar{x}_2. \qquad (10.1)$$

Thus we see that a point estimator of the difference between two population means is the difference between the sample means for the two independent simple random samples.

Sampling Distribution of $\bar{x}_1 - \bar{x}_2$

In the study of the difference between the means of two populations, $\bar{x}_1 - \bar{x}_2$ is the point estimator of interest. This point estimator, just like the point estimators discussed previously, has its own sampling distribution. If we can identify the sampling distribution of $\bar{x}_1 - \bar{x}_2$, we can use it to develop an interval estimate of the difference between the two population means in much the same way that we used the sampling distribution of $\bar{x}$ for interval estimation about a single population mean. The properties of the sampling distribution of $\bar{x}_1 - \bar{x}_2$ are as follows:

Sampling Distribution of $\bar{x}_1 - \bar{x}_2$

$$\text{Mean: } E(\bar{x}_1 - \bar{x}_2) = \mu_1 - \mu_2, \tag{10.2}$$

$$\text{Standard Deviation: } \sigma_{\bar{x}_1 - \bar{x}_2} = \sqrt{\frac{\sigma_1^2}{n_1} + \frac{\sigma_2^2}{n_2}}. \tag{10.3}$$

where

σ_1 = standard deviation of population 1,
σ_2 = standard deviation of population 2,
n_1 = sample size for the simple random sample selected from population 1,
n_2 = sample size for the simple random sample selected from population 2.

Distribution form: Provided that the sample sizes are both *large* ($n_1 \geq 30$ and $n_2 \geq 30$), the sampling distribution of $\bar{x}_1 - \bar{x}_2$ can be approximated by a normal probability distribution.

Equation (10.2) shows that the point estimator $\bar{x}_1 - \bar{x}_2$ is an unbiased estimator of the difference between the two population means. Figure 10.1 shows the sampling distribution of $\bar{x}_1 - \bar{x}_2$ and its relationship to the individual sampling distributions of $\bar{x}_1$ and $\bar{x}_2$.

Let us use this knowledge of the sampling distribution of $\bar{x}_1 - \bar{x}_2$ to develop an interval estimate of the difference between the means of two populations. We shall consider two cases, one where the sample sizes are large ($n_1 \geq 30$ and $n_2 \geq 30$) and the other where the populations are assumed to have normal distributions with equal variances. Let us consider the large-sample-size case first.

Interval Estimation of $\mu_1 - \mu_2$: Large-Sample Case

Greystone Department Stores, Inc. operates two stores in Buffalo, New York, one located in the inner city and one located in a suburban shopping center. Over a period of time the manager of the Buffalo area stores has noticed that products that sell extremely well in one store do not always sell well in the other. One plausible explanation for the differences in sales at the two stores is that there are differences in the customers who shop at the two locations. Customer differences may be noticeable in such dimensions as age, education, income, and so on. While we could elect to study any of these dimensions, let us assume that the manager has asked about the difference in the average or mean ages of the customers that shop the two stores. Figure 10.2 provides an illustration of this two population situation.

Let us suppose that Greystone conducts a survey of customers at each store. The customer age data collected from two independent samples provide the following results:

Store	*Number of Customers Sampled*	*Sample Mean Age*	*Sample Standard Deviation*
Inner city	36	$\bar{x}_1$ = 40 years	s_1 = 9 years
Suburban	49	$\bar{x}_2$ = 35 years	s_2 = 10 years

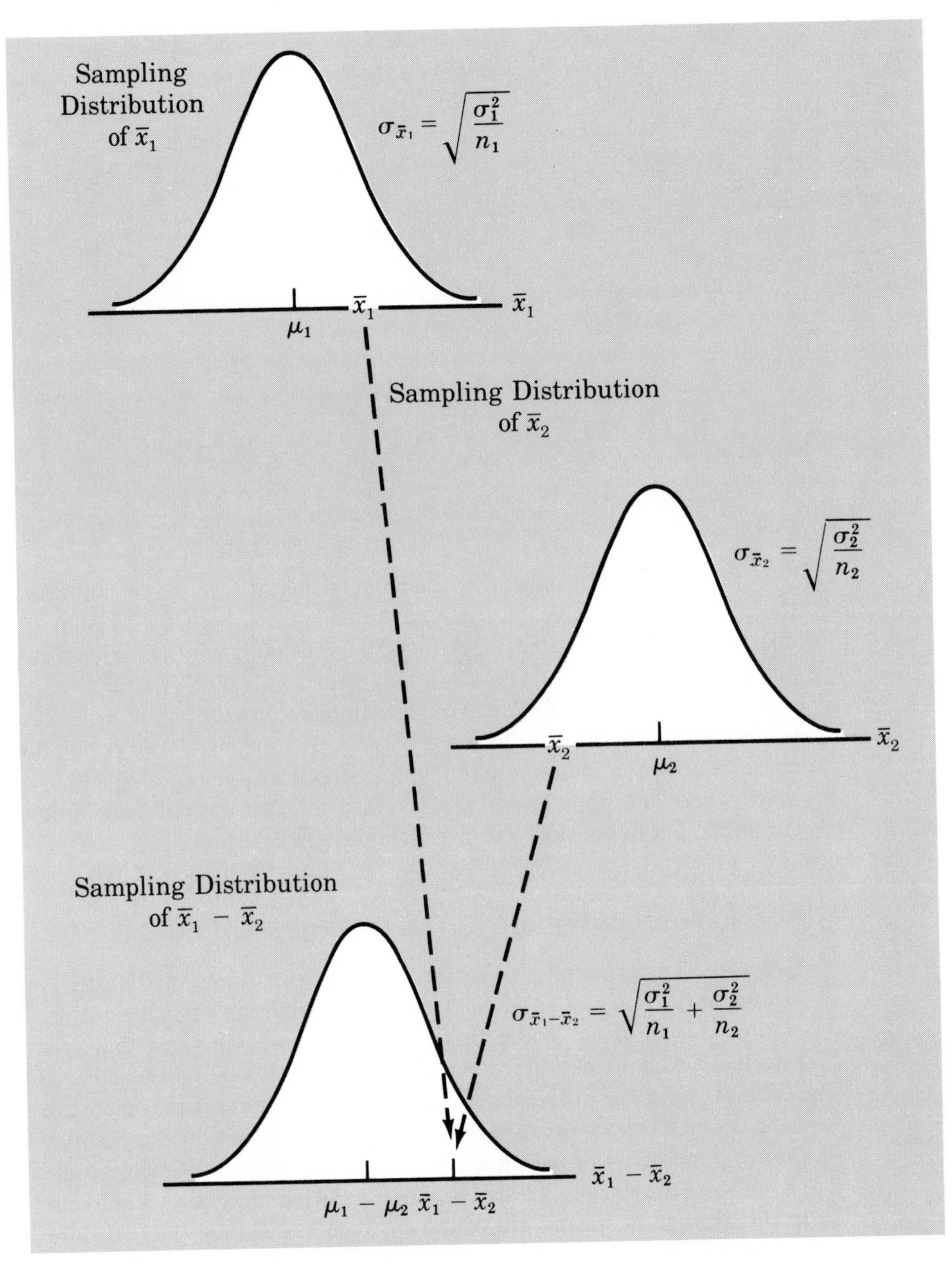

FIGURE 10.1 Sampling Distribution of $\bar{x}_1 - \bar{x}_2$ and Its Relationship to the Sampling Distributions of $\bar{x}_1$ and $\bar{x}_2$

Using (10.1), we find the point estimate of the difference between mean ages of the two populations to be $\bar{x}_1 - \bar{x}_2 = 40 - 35 = 5$ years. Specifically, we are led to believe that the customers at the inner city store have a mean age approximately 5 years greater than the mean age of the suburban store customers. However, as with all point estimates, we know that 5 years is only an approximate value of the difference

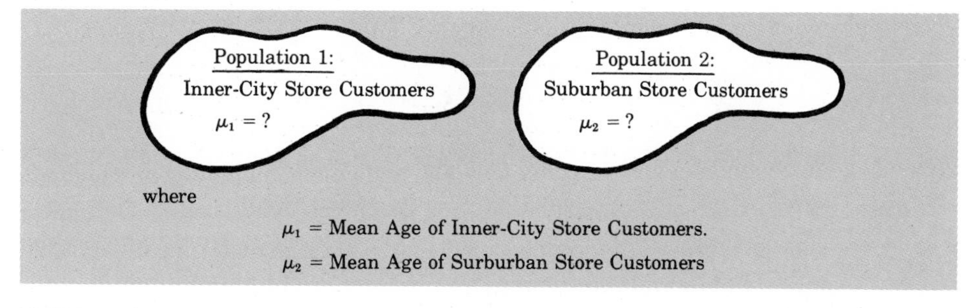

FIGURE 10.2 Two Customer Populations for the Greystone Department Store, Inc.

between mean ages of the two populations. Thus as in Chapter 8 we are interested in identifying a confidence interval for the difference between the means of the two populations.

In the large-sample-size case we know that the sampling distribution of $\bar{x}_1 - \bar{x}_2$ can be approximated by a normal probability distribution. With this approximation we can use the following expression to develop an interval estimate of the difference between the two population means:

Interval Estimate of the Difference Between the Means of Two Populations (Large-Sample Case with $n_1 \geq 30$ and $n_2 \geq 30$)

$$\bar{x}_1 - \bar{x}_2 \pm z_{\alpha/2}\sigma_{\bar{x}_1-\bar{x}_2}, \tag{10.4}$$

where $1 - \alpha$ is the confidence coefficient.

Let us use this procedure to develop a confidence interval estimate of the difference between the mean ages for the two populations in the Greystone Department Store study. Since the population standard deviations σ_1 and σ_2 are unknown, we cannot use (10.3) to calculate $\sigma_{\bar{x}_1-\bar{x}_2}$. However, we can use the sample standard deviations as estimates of the population standard deviations and estimate $\sigma_{\bar{x}_1-\bar{x}_2}$ as follows:

Estimate of $\sigma_{\bar{x}_1-\bar{x}_2}$

$$s_{\bar{x}_1-\bar{x}_2} = \sqrt{\frac{s_1^2}{n_1} + \frac{s_2^2}{n_2}}. \tag{10.5}$$

With large sample sizes, $s_{\bar{x}_1-\bar{x}_2}$ can be accepted as a good estimate of $\sigma_{\bar{x}_1-\bar{x}_2}$.

Making the calculations to estimate $\sigma_{\bar{x}_1-\bar{x}_2}$ for the Greystone problem, we have

$$s_{\bar{x}_1-\bar{x}_2} = \sqrt{\frac{(9)^2}{36} + \frac{(10)^2}{49}} = \sqrt{4.29} = 2.07.$$

With this value as the estimate of $\sigma_{\bar{x}_1-\bar{x}_2}$ and with $z_{\alpha/2} = z_{.025} = 1.96$, (10.4) provides the following 95% confidence interval estimate:

$$5 \pm (1.96)(2.07)$$

or

$$5 \pm 4.06.$$

Thus at a 95% level of confidence the interval estimate for the difference in mean ages of the two Greystone populations is .94 years to 9.06 years.

Interval Estimation of $\mu_1 - \mu_2$: Two Normal Populations with Equal Variances

Let us consider the interval estimation procedure for the difference between the means of two populations when the populations have normal distributions with equal variances, i.e., $\sigma_1^2 = \sigma_2^2 = \sigma^2$. We will again be assuming that independent random samples are selected from the populations. In this case the sampling distribution of $\bar{x}_1 - \bar{x}_2$ is normal regardless of the sample sizes involved. The mean of the sampling distribution is $\mu_1 - \mu_2$. Because of the equal variances, equation (10.3) can be written

$$\sigma_{\bar{x}_1 - \bar{x}_2} = \sqrt{\frac{\sigma^2}{n_1} + \frac{\sigma^2}{n_2}} = \sqrt{\sigma^2\left(\frac{1}{n_1} + \frac{1}{n_2}\right)}. \qquad (10.6)$$

The sampling distribution of $\bar{x}_1 - \bar{x}_2$ is shown in Figure 10.3.

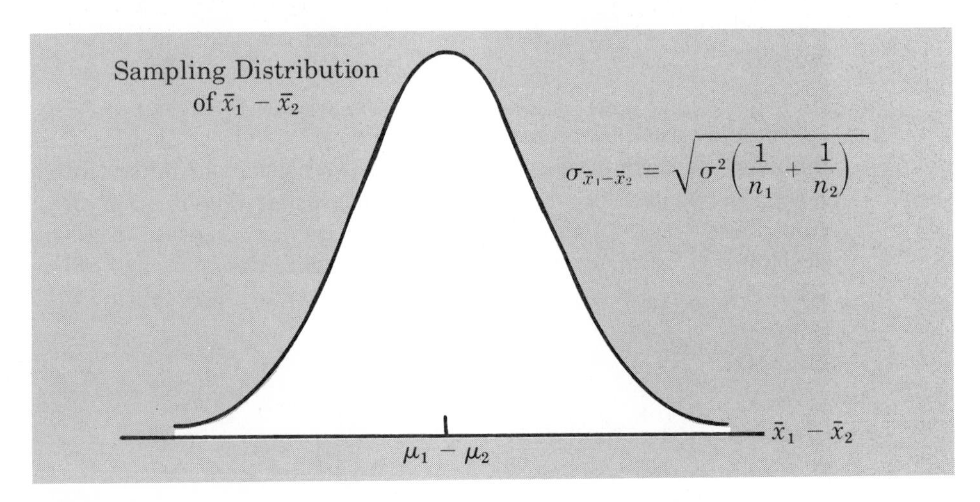

FIGURE 10.3 Sampling Distribution of $\bar{x}_1 - \bar{x}_2$ When Populations Are Normal with Equal Variances σ^2

If the variance, σ^2, is known, then (10.6) can be used to compute $\sigma_{\bar{x}_1 - \bar{x}_2}$ and (10.4) can be used to develop the interval estimate of the difference between the means of the two populations. However, if σ^2 is unknown, the two sample variances, s_1^2 and s_2^2, can be combined to compute the following estimate of σ^2:

$$s^2 = \frac{(n_1 - 1)s_1^2 + (n_2 - 1)s_2^2}{n_1 + n_2 - 2}. \qquad (10.7)$$

The process of combining the results of the two independent samples to provide one estimate of the population variance is referred to as *pooling*, and s^2 is referred to as the *pooled estimator* of σ^2.

With s^2 as the estimator of σ^2 and with (10.6), the following estimate of the standard deviation of $\bar{x}_1 - \bar{x}_2$ can be obtained:

Estimate of $\sigma_{\bar{x}_1 - \bar{x}_2}$

$$s_{\bar{x}_1 - \bar{x}_2} = \sqrt{s^2\left(\frac{1}{n_1} + \frac{1}{n_2}\right)}. \qquad (10.8)$$

Recall that in Chapter 8 we pointed out that whenever a population standard deviation σ is unknown and whenever the population possesses a normal distribution the t distribution can be used to develop interval estimates of a population mean. When σ is unknown and when normal populations exist for a two-population case, the t distribution can be used to compute a confidence interval for the difference between the two population means. This procedure is as follows:

Interval Estimate of the Difference Between the Means of Two Populations (Normal Populations With Equal Variances Estimated by s^2)

$$\bar{x}_1 - \bar{x}_2 \pm t_{\alpha/2} s_{\bar{x}_1 - \bar{x}_2}, \qquad (10.9)$$

where the t value is based on a t distribution with $n_1 + n_2 - 2$ degrees of freedom and where $1 - \alpha$ is the confidence coefficient.

Let us demonstrate the interval estimation procedure for the sampling study conducted by the Clearview National Bank. Independent random samples of checking account balances for customers at two Clearview branch banks show the following results:

Branch Bank	*Number of Checking Accounts*	*Sample Mean Balance*	*Sample Standard Deviation*
Cherry Grove	12	$\bar{x}_1 = \$1,000$	$s_1 = \$150$
Beechmont	10	$\bar{x}_2 = \$\ \ 920$	$s_2 = \$120$

Let us use the above data to develop a 90% confidence interval estimate of the difference between the mean checking account balances for the two branch banks. With (10.7), the pooled estimate of the population variance becomes

$$s^2 = \frac{(11)(150)^2 + (9)(120)^2}{12 + 10 - 2} = 18{,}855.$$

The corresponding estimate of the standard deviation of $\bar{x}_1 - \bar{x}_2$ is

$$s_{\bar{x}_1 - \bar{x}_2} = \sqrt{18{,}855\left(\frac{1}{12} + \frac{1}{10}\right)} = 58.79.$$

The appropriate t distribution for the interval estimation procedure has $n_1 + n_2 - 2 =$ 20 degrees of freedom. With $\alpha = .10$, the $t_{\alpha/2}$ value for this distribution is $t_{.05} = 1.725$. Thus the interval estimate becomes

$$\bar{x}_1 - \bar{x}_2 \pm t_{.05} s_{\bar{x}_1 - \bar{x}_2}$$
$$1000 - 920 \pm (1.725)(58.79)$$
$$80 \pm 101.4.$$

At a 90% level of confidence the interval estimate for the difference in mean account balances of the two branch banks is −\$21.40 to \$181.40. The fact that the interval includes a negative range of values indicates that the actual difference in the two means, $\mu_1 - \mu_2$, may be negative. Thus μ_2 could actually be larger than μ_1, indicating that the mean balance for the population could be greater for the Beechmont branch even though the results show a greater sample mean balance at the Cherry Grove branch. The fact that the confidence interval contains the value 0 can be interpreted as indicating that we do not have sufficient evidence to conclude that the population mean account balances differ at the two branches.

As a final comment, let us note that in developing interval estimates of the difference between two population means, (10.4) can be applied whenever both sample sizes are large. Thus although (10.9) applies for any sample sizes as long as the populations are normal and as long as the population variances are unknown but equal, (10.9) is not needed for the large-sample case. This is true because (10.4) with $s_{\bar{x}_1 - \bar{x}_2}$ substituted for $\sigma_{\bar{x}_1 - \bar{x}_2}$ can be used instead. As a result, common practice is to use (10.9) only when (10.4) cannot be applied. This occurs whenever one or both sample sizes are small. Statistical procedures to be presented in Chapters 11 and 12 are helpful in testing whether or not the assumptions of normal populations and equal variances are appropriate.

EXERCISES

1. A college admissions board is interested in estimating the difference between the mean grade point averages of students from two high schools. Independent simple random samples of students at the two high schools provide the following results:

Mt. Washington	*Country Day*
$n_1 = 46$	$n_2 = 33$
$\bar{x}_1 = 3.02$	$\bar{x}_2 = 2.72$
$s_1 = .38$	$s_2 = .45$

a. What is the point estimate of the difference between the means of the two populations?

b. Develop an interval estimate of the difference between the two population means with a confidence coefficient of .90.

c. Answer part b using a .95 confidence coefficient.

2. The Butler County Bank and Trust Company is interested in estimating the difference between the mean credit card balances at two of its branch banks. Independent samples of credit card customers provide the following results:

Branch 1	*Branch 2*
$n_1 = 32$	$n_2 = 36$
$\bar{x}_1 = \$500$	$\bar{x}_2 = \$375$
$s_1 = \$150$	$s_2 = \$130$

a. Develop a point estimate of the difference between the mean balances at the two branches.
b. Develop an interval estimate of the difference between mean balances. Use a confidence coefficient of .99.

3. An urban planning group is interested in estimating the difference between mean household income for two neighborhoods in a large metropolitan area. Independent samples of households in the neighborhoods provide the following results:

Neighborhood 1	*Neighborhood 2*
$n_1 = 8$	$n_2 = 12$
$\bar{x}_1 = \$15{,}700$	$\bar{x}_2 = \$14{,}500$
$s_1 = \$700$	$s_2 = \$850$

a. Develop a point estimate of the difference between mean incomes in the two neighborhoods.
b. Develop an interval estimate of the difference between mean incomes in the two neighborhoods. Show the results for confidence coefficients of .90 and .95.
c. What assumptions were made in order to compute the interval estimates in part b above?

4. Production quantities for two assembly line workers are shown below. Each data value indicates the amount produced during a randomly selected 1 hour period.

Worker 1	*Worker 2*
20	22
18	18
21	20
22	23
20	24

a. Develop a point estimate of the difference between the mean hourly production rates of the two workers. Which worker appears to have the higher mean production rate?
b. Develop a 90% confidence interval estimate for the difference between the mean production rates of the two workers. Consider the confidence interval estimate. Does the result provide conclusive evidence that the worker having the higher sample mean production rate is actually the worker with the overall higher production rate? Explain.

10.2 HYPOTHESIS TESTS ABOUT THE DIFFERENCE BETWEEN THE MEANS OF TWO POPULATIONS—INDEPENDENT SAMPLES

As part of a study to evaluate any differences in educational quality of two training centers, a company administers a standardized examination to individuals who were trained at the two centers. The examination scores are a major factor in assessing any quality differences at the centers.

Let

μ_1 = the mean examination score for the population of individuals trained at center A,
μ_2 = the mean examination score for the population of individuals trained at center B.

The hypotheses of interest are as follows:

$$H_0: \mu_1 - \mu_2 = 0,$$
$$H_1: \mu_1 - \mu_2 \neq 0.$$

The null hypothesis shows that we wish to test the assumption that there is no difference between the mean examination scores of individuals trained at the two centers. If the sample results cast doubt on this assumption, we will conclude H_1. This latter conclusion suggests that a quality differential exists for the two centers and indicates that a followup study investigating the reasons for the differential may be warranted.

Following the hypothesis-testing procedure from Chapter 9, we will make the assumption that H_0 is true. Using the difference between the sample means as the point estimator of the difference between the population means, we consider the sampling distribution of $\bar{x}_1 - \bar{x}_2$ when H_0 is true. Assuming the large-sample case, this distribution is as shown in Figure 10.4.

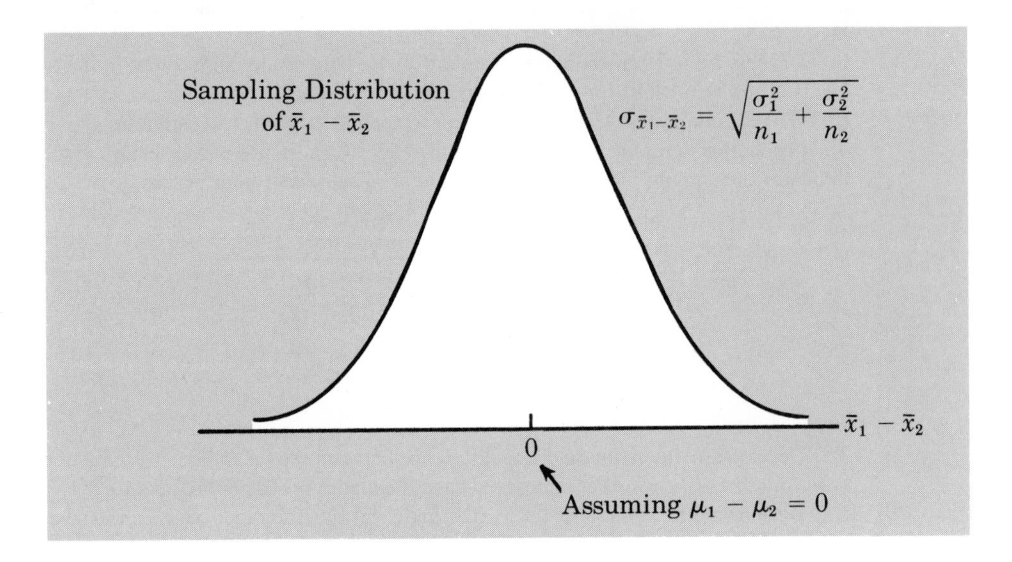

FIGURE 10.4 Sampling Distribution of $\bar{x}_1 - \bar{x}_2$ with $\mu_1 - \mu_2 = 0$

For $\alpha = .05$ and thus $z_{\alpha/2} = 1.96$, the critical values for the test are computed as shown in Figure 10.5. The decision rule expressed in terms of the value of the point estimator, $\bar{x}_1 - \bar{x}_2$, and the critical values are as follows:

$$\text{Accept } H_0 \text{ if } c_1 \leq \bar{x}_1 - \bar{x}_2 \leq c_2,$$
$$\text{Reject } H_0 \text{ if } \bar{x}_1 - \bar{x}_2 < c_1 \text{ or if } \bar{x}_1 - \bar{x}_2 > c_2.$$

Let us assume that independent random samples of individuals trained at the two centers provide the following examination score results:

Training Center A	*Training Center B*
$n_1 = 30$	$n_2 = 40$
$\bar{x}_1 = 82.5$	$\bar{x}_2 = 79$
$s_1 = 8$	$s_2 = 10$

Using the above sample standard deviations and (10.5), we compute an estimate of

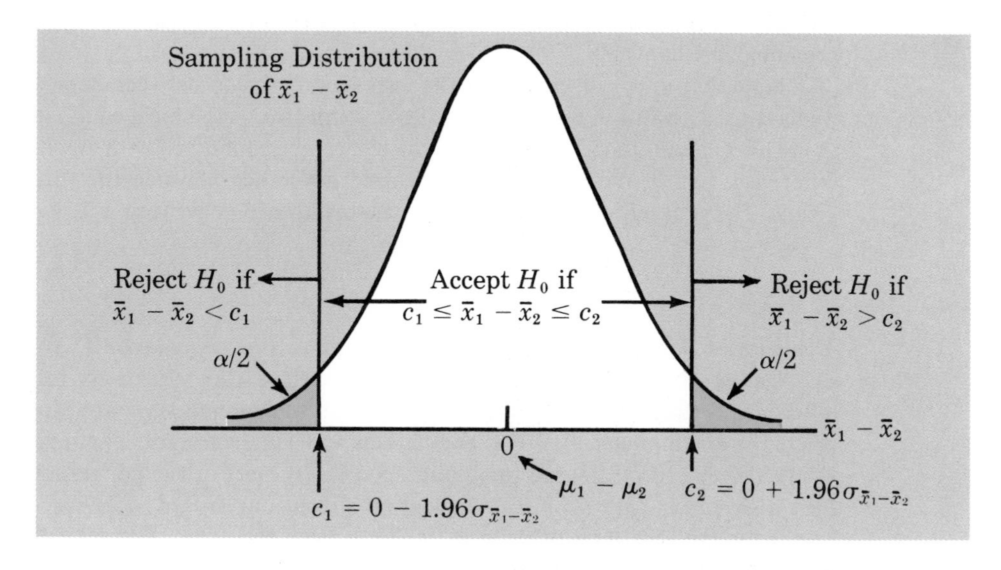

FIGURE 10.5 Critical Values for the Hypothesis Test H_0: $\mu_1 - \mu_2 = 0$ with $\alpha = .05$

$\sigma_{\bar{x}_1-\bar{x}_2}$ as follows:

$$s_{\bar{x}_1-\bar{x}_2} = \sqrt{\frac{s_1^2}{n_1} + \frac{s_2^2}{n_2}} = \sqrt{\frac{(8)^2}{30} + \frac{(10)^2}{40}} = 2.15.$$

Using this value in the computation of the critical values, we obtain

$$c_1 = 0 - 1.96(2.15) = -4.21,$$
$$c_2 = 0 + 1.96(2.15) = 4.21.$$

A value of $\bar{x}_1 - \bar{x}_2$ in the interval -4.21 to $+4.21$ leads to the acceptance of H_0, while a value of $\bar{x}_1 - \bar{x}_2$ outside this interval does not support H_0 and leads to its rejection in favor of H_1.

Let us return to the sample results for the examination scores. We find $\bar{x}_1 - \bar{x}_2 = 82.5 - 79 = 3.5$. Following the decision rule, we reach the conclusion of accepting H_0 on the basis that the sample evidence is not sufficient to conclude otherwise. Thus the sample results should lead the firm to suspect that a "similar" educational quality exists at the two centers.

In the above hypothesis test we were interested in determining if a difference existed between the means of the two populations. Since we did not have a prior belief about one mean being specifically greater than or less than the other, the hypotheses H_0: $\mu_1 - \mu_2 = 0$ and H_1: $\mu_1 - \mu_2 \neq 0$ were appropriate. In other hypothesis tests about the difference between the means of two populations, we may want to test a belief about one of the means being greater than or perhaps less than the other mean. In this case a one-tailed testing procedure would be appropriate. The form of a one-tailed test about the difference between two population means would be one of the following:

$$H_0: \mu_1 - \mu_2 \le 0 \qquad H_0: \mu_1 - \mu_2 \ge 0$$
$$H_1: \mu_1 - \mu_2 > 0 \qquad H_1: \mu_1 - \mu_2 < 0$$

These hypotheses are tested with use of the sampling distribution of $\bar{x}_1 - \bar{x}_2$ with critical values determined in a manner identical to the one-tailed test decision rules presented in Chapter 9.

Note that in our discussion of the tests on differences between population means we have used a value of 0 in the hypotheses. While this is the most common value used in the tests on differences, realize that the test can be conducted with any hypothesized difference. Let D_0 indicate the hypothesized difference between the two population means. The general form of a two-tailed test could then be written

$$H_0: \mu_1 - \mu_2 = D_0,$$
$$H_1: \mu_1 - \mu_2 \neq D_0.$$

The value of D_0 depends upon the particular application being studied.

Our discussion in this section has emphasized the hypothesis testing of the difference between two population means when the sample sizes are large. In cases where the assumptions of normal populations and equal variances are appropriate and where s^2 [see (10.7)] is used to obtain a pooled estimate of the population variances, the t distribution with $n_1 + n_2 - 2$ degrees of freedom is used to develop the critical values for the test. The procedures for both one- and two-tailed tests for this case parallel the previously discussed hypothesis-testing methodology.

EXERCISES

5. The Greystone Department Store study in Section 10.1 provided the following data on customer ages from independent random samples taken at two store locations:

Inner City Store	*Suburban Store*
$n_1 = 36$	$n_2 = 49$
$\bar{x}_1 = 40$ years	$\bar{x}_2 = 35$ years
$s_1 = 9$ years	$s_2 = 10$ years

For $\alpha = .05$, test the hypothesis $H_0: \mu_1 - \mu_2 = 0$ against the alternative hypothesis $H_1: \mu_1 - \mu_2 \neq 0$. What is your conclusion about the mean ages of the populations of customers at the two stores?

6. A firm is studying the delivery times for two raw material suppliers. The firm is basically satisfied with its current supplier A and will stay with this supplier provided that the mean delivery times are the same as or less than those of supplier B. However, if the firm finds that the mean delivery times from supplier B are less than those of supplier A, it will begin making raw material purchases from supplier B.

a. What are the null and alternative hypotheses for this situation?

b. Assume that independent samples show the following delivery time characteristics for the two suppliers:

Supplier A	*Supplier B*
$n_1 = 50$	$n_2 = 30$
$\bar{x}_1 = 14$ days	$\bar{x}_2 = 12.5$ days
$s_1 = 3$ days	$s_2 = 2$ days

Show the sampling distribution of $\bar{x}_1 - \bar{x}_2$ for the hypothesis test.

c. For $\alpha = .05$, what is the critical value and decision rule for the test?

d. What is your conclusion for the hypotheses from part a above? What action do you recommend in terms of supplier selection?

7. In a wage discrimination case involving male and female employees, independent samples of male and female employees with 5 years or more experience show the following hourly wage results:

Male Employees	*Female Employees*
$n_1 = 44$	$n_2 = 32$
$\bar{x}_1 = \$6.25$	$\bar{x}_2 = \$5.70$
$s_1 = \$1.00$	$s_2 = \$.80$

The null hypothesis is stated such that male employees have a mean hourly wage less than or equal to that of the female employees. Rejection of H_0 leads to the conclusion that male employees have a mean hourly wage exceeding the female employee wages. Test the hypothesis with $\alpha = .01$. Does wage discrimination appear to exist in this case?

8. A production line is designed on the assumption that the difference in mean assembly times for two operations is 5 minutes. Independent tests for the two assembly operations show the following results:

Operation A	*Operation B*
$n_1 = 100$	$n_2 = 50$
$\bar{x}_1 = 14.8$ minutes	$\bar{x}_2 = 10.4$ minutes
$s_1 = .8$ minutes	$s_2 = .6$ minutes

For $\alpha = .02$, test the hypothesis that the difference between the mean assembly times is 5 minutes.

9. Salary surveys of marketing and management majors show the following starting annual salary data:

Marketing Majors	*Management Majors*
$n_1 = 14$	$n_2 = 16$
$\bar{x}_1 = \$19{,}800$	$\bar{x}_2 = \$19{,}300$
$s_1 = \$\ 1{,}000$	$s_2 = \$\ 1{,}400$

Consider the test of the hypothesis that the mean annual salaries are the same for both majors.

a. What assumptions must be made in order to test the hypothesis?

b. Assume that these assumptions are appropriate. What is the pooled estimate of the population variance?

c. For $\alpha = .05$, can you conclude that a difference exists in the mean annual salary for the two majors?

10.3 INFERENCES ABOUT THE DIFFERENCE BETWEEN THE MEANS OF TWO POPULATIONS—MATCHED SAMPLES

Suppose that a manufacturing company has two methods available for employees to perform a certain production task. In order to maximize production output, the company would like to identify the method with the smallest mean completion time per unit. If a difference between mean completion times exists, the company will implement the method with the smaller mean completion time. If no difference between means can be detected, the choice between the two production methods will be based on a criterion other than completion time. The hypotheses to be tested are stated as follows:

Hypothesis	*Conclusion*
$H_0: \mu_1 - \mu_2 = 0$	No difference exists between the mean completion times of the two methods
$H_1: \mu_1 - \mu_2 \neq 0$	A difference exists between the mean completion times; select the method with the smaller mean completion time

In designing the sampling procedure that will be used to collect production time data and test the above hypotheses, we consider two alternative designs. One is based on *independent samples,* and the other is based on *matched samples*. The designs are described as follows:

Independent-sample design—A random sample of workers is selected and uses method 1. A second and independent random sample of workers is selected and uses method 2. The test of the difference between means is based on the procedures of Section 10.2.

Matched-sample design—One random sample of workers is selected with the workers first using one method and then using the other method. The order of the two methods is assigned randomly to the workers, with some workers performing method 1 first and others performing method 2 first. Each worker provides a pair of data values, one value for method 1 and another value for method 2.

Our interest in the matched-sample design is that since both production methods are tested under similar conditions (i.e., same workers), this design often leads to a smaller sampling error than the independent sample design. The primary reason for this is that each worker in a matched sample design provides data first under one method and then under the other method. Thus variation between workers is eliminated as a source of the sampling error. This variation between workers cannot be eliminated when the independent-sample design is used.

Let us demonstrate the analysis of a matched-sample design by assuming that this method is used to test the difference between the two production methods. A random sample of six workers is used. The data on completion times for the six workers are shown in Table 10.1. Note that each worker provides a pair of data values, one for each

TABLE 10.1 Task Completion Times for a Matched Sample of Six Workers

Worker	*Completion Time, Method 1 (minutes)*	*Completion Time, Method 2 (minutes)*	*Difference in Completion Times (d_i)*
1	6.0	5.4	.6
2	5.0	5.2	−.2
3	7.0	6.5	.5
4	6.2	5.9	.3
5	6.0	6.0	.0
6	6.4	5.8	.6

production method. Also note that the last column contains the difference (d_i) for each pair of values or for each worker in the sample.

The key to the analysis of the matched-sample design is to realize that we consider only the column of differences in the two methods. As a result we have six data values (.6, −.2, .5, .3, .0, and .6) that will be used in the analysis of the difference between means of the two production methods.

Let μ_d = the mean of the *difference* values for the population of workers. With this notation the null and alternative hypotheses are rewritten as follows:

Hypothesis	*Conclusion*
$H_0\colon \mu_d = 0$	No difference exists in the mean completion time of the two methods
$H_1\colon \mu_d \neq 0$	A difference exists in the mean completion time; select the method with the smaller mean completion time

The d notation is a reminder that the population involves difference data. The sample mean and sample standard deviation for the difference values in Table 10.1 are as follows:

$$\bar{d} = \frac{\sum_{i=1}^{n} d_i}{n} = \frac{1.8}{6} = .30,$$

$$s_d = \sqrt{\frac{\sum_{i=1}^{n} (d_i - \bar{d})^2}{n-1}} = \sqrt{\frac{.56}{5}} = \sqrt{.112} = .335.$$

In Chapter 9 we found that if the population can be assumed normal or near normal the t distribution with $n - 1$ degrees of freedom can be used to test the null hypothesis about a population mean. We make the assumption of a normal distribution of difference values. The critical values for the two-tailed test are

$$c_1 = 0 - t_{\alpha/2}\sqrt{\frac{s_d^2}{n}},$$

$$c_2 = 0 + t_{\alpha/2}\sqrt{\frac{s_d^2}{n}}.$$

Thus with $\alpha = .05$ and with $n - 1 = 5$ degrees of freedom ($t_{.025} = 2.571$), we obtain

$$c_1 = 0 - (2.571)\sqrt{(.335)^2/6} = -.35,$$
$$c_2 = 0 + (2.571)\sqrt{(.335)^2/6} = +.35.$$

The decision rule for the two-tailed test is expressed

Accept H_0 if $-.35 \le \bar{d} \le +.35$,
Reject H_0 if $\bar{d} < -.35$ or if $\bar{d} > +.35$.

With $\bar{d} = .30$ for the data in Table 10.1, the decision rule leads us to accept H_0. The sample data do not provide sufficient evidence to reject the assumption that there is no difference between the mean completion times of the two methods. Unless the firm wants to take a larger sample and study the issue further, it should select a production method on criterion other than completion time.

With the above sample results for an interval estimation of the difference between the two population means we would use the single-population methodology of Chapter 8 to obtain the following:

$$0.3 \pm t_{\alpha/2}\sqrt{\frac{s_d^2}{n}},$$

$$0.3 \pm 2.571\sqrt{(.335)^2/6}$$

$$0.3 \pm .35.$$

Thus the 95% confidence interval estimate of the difference in the means of the two production methods is −.05 to .65 minutes.

In the above example, workers performed the production task using first one method and then the other method. This is an example of a matched-sample design, where each sampled item (worker) provides a pair of data values. Although this is often the procedure used in the matched-samples analysis, it is possible to use different but "similar" items to provide the pair of data values. In this sense, a worker at one location could be matched with a "similar" worker at another location (similarity based on age, education, sex, experience, etc.). The pairs of workers would provide the difference data that could be used in the matched-sample analysis.

Since a matched-sample procedure for inferences about two population means generally provides a better precision than the two-independent-samples approach, it is the recommended design. However, in some applications the matching cannot be achieved, or perhaps the time and cost associated with matching is excessive. In these cases the independent-sample design should be used.

The example presented in this section employed a sample size of six workers. As such, the small-sample case existed, and the t distribution was used in both the test of hypothesis and interval estimation computations. If the sample size is large ($n \geq 30$), the use of the t distribution is unnecessary, and the statistical inferences can be based on the z values of the standard normal probability distribution.

EXERCISES

10. A manufacturer produces both a deluxe and a standard model automatic sander designed for home use. Selling prices obtained from a sample of retail outlets are as follows:

Retail Outlet	*Price, Deluxe Model*	*Price, Standard Model*
1	$39	$27
2	39	28
3	45	35
4	38	30
5	40	30
6	39	34
7	35	29

The manufacturer's suggested retail prices for the two models show a $10 differential in prices.

Use a .05 level of significance and test to see if the mean difference between prices of the two models is at the $10 value. What is the 95% confidence interval estimate of the difference between mean prices for the two models?

11. A market research firm used a sample of individuals to rate the potential to purchase for a particular product before and after they saw a new television commercial about the product. The potential to purchase ratings were based on a 0 to 10 scale, with higher values indicating a higher potential to purchase. The null hypothesis stated that the mean rating "after" would be less than or equal to the mean rating "before." Rejection of this hypothesis would provide the conclusion that the commercial improved the mean potential to purchase rating. Use $\alpha = .05$ and the following data to test the hypothesis and comment on the value of the commercial:

Individual	*Purchase Rating Before Seeing Commercial*	*Purchase Rating After Seeing Commercial*
1	5	6
2	4	6
3	7	7
4	3	4
5	5	3
6	8	9
7	5	7
8	6	6

12. A company attempts to evaluate the potential for a new bonus plan by selecting a random sample of five salespersons to use the bonus plan for a trial period. The weekly sales volumes before and after implementing the bonus plan are shown below:

Salesperson	*Weekly Sales Before*	*Weekly Sales After*
1	15	18
2	12	14
3	18	19
4	15	18
5	16	18

a. Use $\alpha = .05$ and test to see if it can be concluded that the bonus plan will result in an increase in the mean weekly sales.

b. Provide a 90% confidence interval estimate for the mean *increase* in weekly sales that can be expected if a new bonus plan is implemented.

10.4 INFERENCES ABOUT THE DIFFERENCE BETWEEN THE PROPORTIONS OF TWO POPULATIONS

A tax preparation firm is interested in comparing the quality of work at two of its regional offices. By randomly selecting samples of tax returns prepared at each office and having the sample returns verified for accuracy, the firm will be able to estimate the proportion of erroneous returns prepared at each office. Of particular interest in this section will be the difference between the proportions of erroneous returns existing at the two offices.

Let

p_1 = proportion of erroneous returns for population 1 (office 1),
p_2 = proportion of erroneous returns for population 2 (office 2),
$\bar{p}_1$ = sample proportion for a simple random sample taken from population 1 (i.e., the point estimator of p_1),
$\bar{p}_2$ = sample proportion for a simple random sample taken from population 2 (i.e., the point estimator of p_2).

The actual difference in two population proportions is given by $p_1 - p_2$. The point estimator of $p_1 - p_2$ is as follows:

Point Estimator of the Difference Between the Proportions of Two Populations

$$\bar{p}_1 - \bar{p}_2. \tag{10.10}$$

That is, the point estimator of the difference between two population proportions is the difference between the sample proportions of two independent simple random samples.

Sampling Distribution of $\bar{p}_1 - \bar{p}_2$

In the study of the difference between two population proportions, $\bar{p}_1 - \bar{p}_2$ is the point estimator of interest. As we have seen in several previous cases, the sampling distribution of the point estimator is a key factor in developing confidence interval estimates and in testing hypotheses about the parameters of interest. The properties of the sampling distribution of $\bar{p}_1 - \bar{p}_2$ are as follows:

Sampling Distribution of $\bar{p}_1 - \bar{p}_2$

Mean: $$E(\bar{p}_1 - \bar{p}_2) = p_1 - p_2, \tag{10.11}$$

Standard deviation: $$\sigma_{\bar{p}_1 - \bar{p}_2} = \sqrt{\frac{p_1(1 - p_1)}{n_1} + \frac{p_2(1 - p_2)}{n_2}}. \tag{10.12}$$

where

n_1 = sample size for the simple random sample selected from population 1,
n_2 = sample size for the simple random sample selected from population 2.

Distribution Form: Provided that the sample sizes are large [that is, n_1p_1, $n_1(1 - p_1)$, n_2p_2, and $n_2(1 - p_2)$ must all be greater than or equal to 5], the sampling distribution of $\bar{p}_1 - \bar{p}_2$ can be approximated by a normal probability distribution.

Equation (10.11) shows that the point estimator $\bar{p}_1 - \bar{p}_2$ is an unbiased estimator of the difference between the two population proportions. Figure 10.6 shows the sampling distribution of $\bar{p}_1 - \bar{p}_2$.

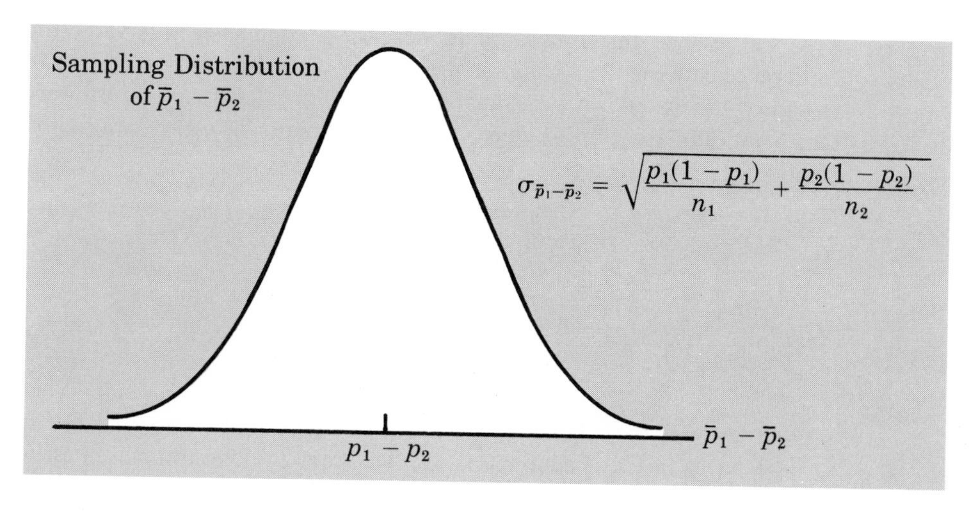

FIGURE 10.6 **Sampling Distribution of $\bar{p}_1 - \bar{p}_2$ (Large-Sample Case)**

Interval Estimation of $p_1 - p_2$

Let us assume that independent simple random samples of tax returns from the two offices show the following:

Office 1	*Office 2*
$n_1 = 250$	$n_2 = 300$
Number of returns with errors = 35	Number of returns with errors = 27

The sample proportions for the two offices are as follows:

$$\bar{p}_1 = \frac{35}{250} = .14,$$

$$\bar{p}_2 = \frac{27}{300} = .09.$$

With (10.10), the point estimate of the difference between the proportion of erroneous tax returns for the two populations is $\bar{p}_1 - \bar{p}_2 = .14 - .09 = .05$. Specifically, we are led to believe that office 1 possesses a 5% greater error rate than office 2. However, as with all point estimates, we know that the .05 difference is only an approximate value of the difference between the two population proportions. The following expression can be used to develop an interval estimate of the difference between the proportions of the two populations:

Interval Estimate of the Difference Between the Proportions of Two Populations [Large-Sample Case with n_1p_1, $n_1(1 - p_1)$, n_2p_2, and $n_2(1 - p_2)$ All Greater Than or Equal to 5]

$$\bar{p}_1 - \bar{p}_2 \pm z_{\alpha/2}\sigma_{\bar{p}_1-\bar{p}_2}, \tag{10.13}$$

where $1 - \alpha$ is the confidence coefficient.

Let us use this procedure to develop a confidence interval estimate of the difference between the population proportions of return errors at the two offices. Since p_1 and p_2 are unknown, we cannot use (10.12) to calculate $\sigma_{\bar{p}_1-\bar{p}_2}$. However, using $\bar{p}_1$, the point estimator of p_1, and $\bar{p}_2$, the point estimator of p_2, we can estimate $\sigma_{\bar{p}_1-\bar{p}_2}$ as follows:

Estimate of $\sigma_{\bar{p}_1-\bar{p}_2}$

$$s_{\bar{p}_1-\bar{p}_2} = \sqrt{\frac{\bar{p}_1(1-\bar{p}_1)}{n_1-1} + \frac{\bar{p}_2(1-\bar{p}_2)}{n_2-1}}. \qquad (10.14)$$

As we saw with the sampling distribution of a population proportion in Chapter 8, the $n - 1$ terms in the denominators are necessary to obtain unbiased estimates of the variance of the sampling distribution. In the large-sample-size case, $s_{\bar{p}_1-\bar{p}_2}$ can be accepted as a good estimate of $\sigma_{\bar{p}_1-\bar{p}_2}$ and used in (10.13) to obtain an interval estimate of $p_1 - p_2$.

Let us make these calculations. We have

$$s_{\bar{p}_1-\bar{p}_2} = \sqrt{\frac{.14(.86)}{249} + \frac{.09(.91)}{299}} = .0275.$$

With a 90% confidence interval, $z_{\alpha/2} = z_{.05} = 1.645$. Expression (10.13) provides the following confidence interval estimate:

$$(.14 - .09) \pm 1.645(.0275)$$
$$.05 \pm .045.$$

Thus the 90% confidence interval estimate of the difference in error rates at the two offices is .005 to .095.

Hypothesis Tests About $p_1 - p_2$

As an example of hypothesis tests concerning the difference between the proportions of two populations, let us consider the data collected in the preceding example and assume that the firm had been attempting to determine whether or not a difference exists in the error proportions at the two offices. Let us illustrate the statistical analysis we could use to test the following hypotheses:

$$H_0: p_1 - p_2 = 0,$$
$$H_1: p_1 - p_2 \neq 0.$$

Figure 10.7 shows the sampling distribution of $\bar{p}_1 - \bar{p}_2$ based on the assumption that there is no difference between the two population proportions. That is, $p_1 - p_2 = 0$. Also, note that the shaded regions in Figure 10.7 provide the rejection region for the above two-tailed test.

We may be tempted to use (10.14), the estimate of the standard error of the difference between proportions, $s_{\bar{p}_1-\bar{p}_2}$, in place of $\sigma_{\bar{p}_1-\bar{p}_2}$ and proceed with the calculation of the critical values c_1 and c_2. However, we generally adjust the formula for $s_{\bar{p}_1-\bar{p}_2}$ for the *special case* of a hypothesis test of the form $H_0: p_1 - p_2 = 0$. That is, with a 0 in the null hypothesis, (10.14) is modified to reflect the fact that when we

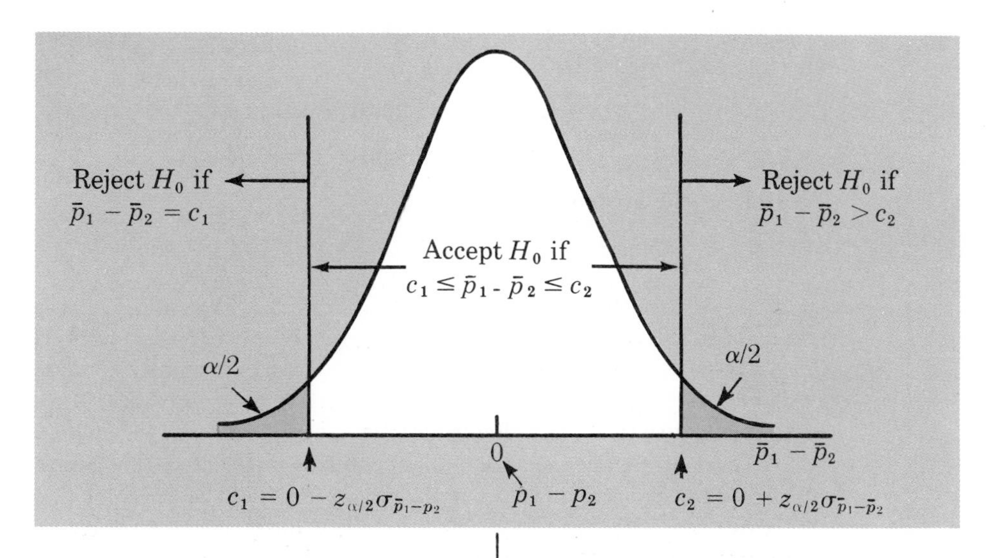

FIGURE 10.7 Sampling Distribution of $\bar{p}_1 - \bar{p}_2$ (Large-Sample Case) with the Assumption That $p_1 - p_2 = 0$ and the Hypothesis H_0: $p_1 - p_2 = 0$

assume the null hypothesis to be true we are assuming $p_1 = p_2$. Whenever this is the case, there is no need to use the individual values for $\bar{p}_1$ and $\bar{p}_2$ in the formula for $s_{\bar{p}_1 - \bar{p}_2}$. Only one value, $\bar{p}$, is needed, since it can be used as an estimate for both p_1 and p_2. Thus we will combine or *pool* the two sample results to provide one estimate of the population proportion. The pooled estimate is as follows:

$$\bar{p} = \frac{n_1 \bar{p}_1 + n_2 \bar{p}_2}{n_1 + n_2}. \qquad (10.15)$$

Since $\bar{p}$ can be used as an estimate of both p_1 and p_2, (10.14) is revised to

$$s_{\bar{p}_1 - \bar{p}_2} = \sqrt{\bar{p}(1 - \bar{p})\left(\frac{1}{n_1 - 1} + \frac{1}{n_2 - 1}\right)}. \qquad (10.16)$$

Remember, however, that (10.15) and (10.16) are only used whenever the null hypothesis for a statistical test about the difference between two population proportions is H_0: $p_1 - p_2 = 0$. In all other cases the formula for $s_{\bar{p}_1 - \bar{p}_2}$ is as previously shown in (10.14).

We can now proceed with our calculations as follows:

$$\bar{p} = \frac{250(.14) + 300(.09)}{550} = \frac{62}{550} = .113,$$

$$s_{\bar{p}_1 - \bar{p}_2} = \sqrt{(.113)(.887)\left(\frac{1}{249} + \frac{1}{299}\right)} = .027.$$

Note here that the values of $s_{\bar{p}_1 - \bar{p}_2}$ obtained with (10.14) and (10.16) were approximately equal. This case exists whenever the two sample sizes are the same or nearly equal. In general, however, with unequal sample sizes the results from the two equations tend to differ. Thus, it is recommended that (10.16) for $s_{\bar{p}_1 - \bar{p}_2}$ be used whenever the null hypothesis is H_0: $p_1 - p_2 = 0$.

Let us test the hypothesis with $\alpha = .10$. We have $z_{.05} = 1.645$, and the critical values shown in Figure 10.7 are as follows:

$$c_1 = 0 - 1.645(.027) = -.044,$$
$$c_2 = 0 + 1.645(.027) = +.044.$$

The decision rule becomes

$$\text{Accept } H_0 \text{ if } -.044 \leq \bar{p}_1 - \bar{p}_2 \leq +.044,$$
$$\text{Reject } H_0 \text{ if } \bar{p}_1 - \bar{p}_2 < -.044 \text{ or if } \bar{p}_1 - \bar{p}_2 > +.044.$$

A value of $\bar{p}_1 - \bar{p}_2$ in the interval $-.044$ to $+.044$ leads to the acceptance of H_0. A value of $\bar{p}_1 - \bar{p}_2$ outside this interval does not support H_0 and leads to its rejection in favor of H_1.

Let us return to the sample results. We have $\bar{p}_1 - \bar{p}_2 = .14 - .09 = .05$. Thus at the .10 level of significance the null hypothesis is rejected. The sample evidence indicates that there is a difference in the error proportions at the two offices.

As we saw with the hypothesis tests about differences between two population means, one-tailed tests can also be developed for the difference between two population proportions. The one-tailed rejection regions are established in a manner similar to the one-tailed hypothesis testing procedures for a single-population proportion.

EXERCISES

13. A sample of 400 items produced by supplier A contained 70 defective items. A sample of 300 items produced by supplier B contained 40 defective items. Compute a 90% confidence interval estimate for the difference in the proportion defective from the two suppliers.

14. During the primary elections of 1980 a particular presidential candidate showed the following preelection voter support in Wisconsin and Illinois:

State	*Voters Surveyed*	*Voters Favoring the Candidate*
Wisconsin	500	270
Illinois	360	162

Compute a 95% confidence interval estimate for the difference between the proportion of voters favoring the candidate in the two states.

15. Two loan officers at the North Ridge National Bank show the following data for defaults on loans that they have approved (the data are based on samples of loans granted over the past 5 years):

	Loans Reviewed in the Sample	*Defaulted Loans*
Loan Officer A	60	9
Loan Officer B	80	6

Use $\alpha = .05$ and test the hypothesis that the default rates are the same for the two loan officers.

16. A survey firm conducts door-to-door surveys on a variety of issues. Some individuals cooperate with the interviewer and complete the interview questionnaire, while others do not.

The following sample data are available (showing the response data for men and women):

	Sample Size	*Number Cooperating with the Survey*
Men	200	110
Women	300	210

a. Use $\alpha = .05$ and test the hypothesis that the response rate is the same for both men and women.
b. Compute the 95% confidence interval estimate for the difference between the proportion of men and the proportion of women that cooperate with the survey.
17. In a test of the quality of two television commercials, each commercial was shown in a separate test area six times over a 1 week period. The following week a telephone survey was conducted to identify individuals who had seen the commercials. The individuals who had seen the commercials were asked to state the primary message in the commercial. The following results were recorded:

	Number Reporting Having Seen the Commercial	*Number Recalling Primary Message*
Commercial A	150	63
Commercial B	200	60

a. Use $\alpha = .05$ and test the hypothesis that there is no difference in the recall proportions for the two commercials.
b. Compute a 95% confidence interval for the difference between the recall proportions for the two populations.

Summary

In this chapter we have discussed procedures for interval estimation and hypothesis testing involving two populations. Specifically, we showed how to make inferences about the differences between the means of two populations when independent simple random samples are selected. Two cases were considered, one where the sample sizes were large and one where the populations were normal with equal variances. The z values from the standard normal probability distribution are used for inferences about the difference between two population means when the sample sizes are large. The t distribution is used for the inferences when the populations are assumed normal with equal variances. This permits inferences in the small-sample case.

Inferences about the difference between the means of two populations were discussed for the matched-sample design. In the matched-sample design each data value from one sample is matched with a data value in the other sample. The difference in the pair of data values is then used in the statistical analysis. The matched-sample design is generally preferred over the independent-sample design because the matched-sample procedure eliminates element variability, thus tending to reduce the sampling error and to improve the precision of the estimate.

Finally, interval estimation and hypothesis testing about the difference between two population proportions were discussed. Statistical procedures for analyzing the difference between two population proportions are similar to the procedures for analyzing the difference between two population means.

Glossary

Pooled variance estimate—An estimate of the variance of a population based on the combination of two (or more) sample results. The pooled variance estimate is appropriate whenever the variances of two (or more) populations are assumed equal. For the two-population case, the pooled estimate of the variance is computed as follows:

$$s^2 = \frac{(n_1 - 1)s_1^2 + (n_2 - 1)s_2^2}{n_1 + n_2 - 2}.$$

Independent samples—Samples selected from two (or more) populations where the elements making up one sample are chosen independently of the elements making up the other sample(s).

Matched samples—Samples where each data value in one sample is matched with a corresponding data value in the other sample.

Key Formulas

Mean of $\bar{x}_1 - \bar{x}_2$

$$E(\bar{x}_1 - \bar{x}_2) = \mu_1 - \mu_2 \tag{10.2}$$

Standard Deviation of $\bar{x}_1 - \bar{x}_2$

$$\sigma_{\bar{x}_1 - \bar{x}_2} = \sqrt{\frac{\sigma_1^2}{n_1} + \frac{\sigma_2^2}{n_2}} \tag{10.3}$$

Interval Estimate of the Difference Between the Means of Two Populations (Large-Sample Case with $n_1 \geq 30$ and $n_2 \geq 30$)

$$\bar{x}_1 - \bar{x}_2 \pm z_{\alpha/2}\sigma_{\bar{x}_1 - \bar{x}_2} \tag{10.4}$$

Estimate of $\sigma_{\bar{x}_1 - \bar{x}_2}$

$$s_{\bar{x}_1 - \bar{x}_2} = \sqrt{\frac{s_1^2}{n_1} + \frac{s_2^2}{n_2}} \tag{10.5}$$

Standard Deviation of $\bar{x}_1 - \bar{x}_2$ When $\sigma_1^2 = \sigma_2^2 = \sigma^2$

$$\sigma_{\bar{x}_1 - \bar{x}_2} = \sqrt{\frac{\sigma^2}{n_1} + \frac{\sigma^2}{n_2}} = \sqrt{\sigma^2\left(\frac{1}{n_1} + \frac{1}{n_2}\right)} \tag{10.6}$$

Pooled Estimator of σ^2

$$s^2 = \frac{(n_1 - 1)s_1^2 + (n_2 - 1)s_2^2}{n_1 + n_2 - 2} \tag{10.7}$$

Estimate of $\sigma_{\bar{x}_1 - \bar{x}_2}$ When $\sigma_1^2 = \sigma_2^2 = \sigma^2$

$$s_{\bar{x}_1 - \bar{x}_2} = \sqrt{s^2\left(\frac{1}{n_1} + \frac{1}{n_2}\right)} \tag{10.8}$$

Interval Estimate of the Difference Between the Means of Two Populations (Normal Populations with Equal Variances Estimated by s^2)

$$\bar{x}_1 - \bar{x}_2 \pm t_{\alpha/2} s_{\bar{x}_1 - \bar{x}_2} \tag{10.9}$$

Sample Mean—Matched Samples

$$\bar{d} = \frac{\sum_{i=1}^{n} d_i}{n}$$

Sample Standard Deviation—Matched Samples

$$s_d = \sqrt{\frac{\sum_{i=1}^{n} (d_i - \bar{d})^2}{n - 1}}$$

Mean of $\bar{p}_1 - \bar{p}_2$

$$E(\bar{p}_1 - \bar{p}_2) = p_1 - p_2 \tag{10.11}$$

Standard Deviation of $\bar{p}_1 - \bar{p}_2$

$$\sigma_{\bar{p}_1 - \bar{p}_2} = \sqrt{\frac{p_1(1 - p_1)}{n_1} + \frac{p_2(1 - p_2)}{n_2}} \tag{10.12}$$

Interval Estimate of the Difference Between the Proportions of Two Populations [Large-Sample Case with $n_1 p_1$, $n_1(1 - p_1)$, $n_2 p_2$, and $n_2(1 - p_2)$ All Greater Than or Equal to 5]

$$\bar{p}_1 - \bar{p}_2 \pm z_{\alpha/2} \sigma_{\bar{p}_1 - \bar{p}_2} \tag{10.13}$$

Estimate of $\sigma_{\bar{p}_1 - \bar{p}_2}$

$$s_{\bar{p}_1 - \bar{p}_2} = \sqrt{\frac{\bar{p}_1(1 - \bar{p}_1)}{n_1 - 1} + \frac{\bar{p}_2(1 - \bar{p}_2)}{n_2 - 1}} \tag{10.14}$$

Pooled Estimate of the Population Proportion

$$\bar{p} = \frac{n_1 \bar{p}_1 + n_2 \bar{p}_2}{n_1 + n_2} \tag{10.15}$$

Estimate of $\sigma_{\bar{p}_1 - \bar{p}_2}$ When $p_1 = p_2$

$$s_{\bar{p}_1 - \bar{p}_2} = \sqrt{\bar{p}(1 - \bar{p})\left(\frac{1}{n_1 - 1} + \frac{1}{n_2 - 1}\right)} \tag{10.16}$$

Supplementary Exercises

18. Starting annual salaries for individuals with master's and bachelor's degrees in business were collected with use of two independent random samples. Use the data shown below and provide a 90% confidence interval estimate of the increase in starting salary that can be expected upon completion of the master's degree:

Master's Degree	*Bachelor's Degree*
$n_1 = 60$	$n_2 = 80$
$\bar{x}_1 = \$23{,}000$	$\bar{x}_2 = \$21{,}000$
$s_1 = \$\ 2{,}500$	$s_2 = \$\ 2{,}000$

19. Samples of dinner and luncheon receipts at a major downtown restaurant show the following results:

Dinner Receipts	*Luncheon Receipts*
$n_1 = 70$	$n_2 = 55$
$\bar{x}_1 = \$32.65$	$\bar{x}_2 = \$12.80$
$s_1 = \$\ 7.20$	$s_2 = \$\ 3.60$

Provide 90% and 98% confidence interval estimates of the difference between mean receipt amounts for the two meals.

20. A realtor is interested in estimating the difference between the mean selling prices of new homes in two sections of the city. Assume that the standard deviations of the selling prices are approximately $12,000 for both areas. How large a sample should be taken in each area to have a 95% probability that the sampling error for the difference between mean prices will be $5,000 or less? Use the same sample size for both sections of the city.

21. Safegate Foods, Inc. is redesigning the check-out lanes in its supermarkets throughout the country. Two designs have been suggested. Tests on customer check-out times have been collected at two stores where the two new systems have been installed. The sample data are as follows:

Times for Check-Out System A	*Times for Check-Out System B*
$n_1 = 120$	$n_2 = 100$
$\bar{x}_1 = 4.1$ minutes	$\bar{x}_2 = 3.3$ minutes
$s_1 = 2.2$ minutes	$\bar{s}_2 = 1.5$ minutes

Test at the .05 level of significance to determine if there is a difference in the mean check-out times for the two systems. Which system is preferred?

22. Samples of final examination scores for two statistics classes with different instructors showed the following results:

Instructor A's Class	*Instructor B's Class*
$n_1 = 12$	$n_2 = 15$
$\bar{x}_1 = 72$	$\bar{x}_2 = 78$
$s_1 = 8$	$s_2 = 10$

With $\alpha = .05$, test whether or not these data are sufficient to conclude that the mean grades differ for the two classes.

23. In a study of job attitudes and job satisfaction, a sample of 50 men and 50 women were asked to rate their overall job satisfaction on a 1 to 10 scale. A high rating indicates a higher degree of job satisfaction. Using the sample results shown below, does there appear to be a significant difference in the level of job satisfaction of men and women? Use $\alpha = .05$.

Men	*Women*
$\bar{x}_1 = 7.2$	$\bar{x}_2 = 6.4$
$s_1^2 = 2.8$	$s_2^2 = 1.8$

24. Figure Perfect, Inc. is a women's figure salon that specializes in weight reduction programs. Weights for a sample of clients before and after a 6 week introductory program are as follows:

Client	*Weight Before*	*Weight After*
1	140	132
2	160	158
3	210	195
4	148	152
5	190	180
6	170	164

Use $\alpha = .05$ and test to determine if the introductory program provides a weight loss.

25. A cable television firm is considering submitting bids for rights to operate in two regions of the state of Florida. Surveys of the two regions provide the following data on customer acceptance of the cable television service:

Region I	*Region II*
$n_1 = 500$	$n_2 = 800$
Number indicating likely to purchase = 175	Number indicating likely to purchase = 360

Develop a 99% confidence interval estimate of the difference between population proportions of likely to purchase customers in the two regions.

26. A large automobile insurance company selected samples of single and married male policyholders and recorded the number who had made an insurance claim over the past 3 year period:

Single Policyholders	*Married Policyholders*
$n_1 = 400$	$n_2 = 900$
Number making claims = 76	Number making claims = 90

a. Test with $\alpha = .05$ to determine if the claim rates differ between single and married male policyholders.

b. Provide a 95% confidence interval estimate of the difference between the claim proportions for the two populations.

27. A political opinion survey shows that of 200 Republicans surveyed 80 opposed the building of power plants using fission processes ("nuclear energy"). Similar results for a sample of 300 Democrats showed that 150 opposed nuclear energy. Do these results indicate that there is a significant difference in the attitudes of Republicans and Democrats toward nuclear energy? Use $\alpha = .05$.

Application

PHARMACEUTICAL DIVISION
PENNWALT
HEALTH PRODUCTS
CHEMICALS ■ EQUIPMENT

Pennwalt Corporation*

Rochester, New York

Pennwalt Corporation was founded in 1850 as a chemical company for the purpose of manufacturing caustic soda. The company steadily expanded its products to a variety of chemical and other markets, such as fruit sizing, dental products, agricultural needs, food processing, and pharmaceuticals. Today, Pennwalt is an international organization that employs over 9,000 people, with revenues of approximately one billion dollars. The techniques discussed in this application pertain to operations in the Pharmaceutical Division and involve statistical analyses that will enable us to draw conclusions about the differences between means.

One of the major responsibilities of the Pharmaceutical Division involves testing new drugs. This process consists of three stages: (1) initial preclinical testing; (2) testing for longterm usage and safety; and (3) actual clinical testing. At each successive stage the chances that the drug survives the rigorous testing decreases; however, the cost of testing increases geometrically. Thus it is important to weed out unlikely candidates in the early stages of the testing process. The statistical techniques that are discussed in this chapter play an important role in helping both to identify new drugs that should be rejected and to ensure that potential good candidates are not incorrectly rejected.

INITIAL PRECLINICAL TESTING

Once it has been shown that a new drug is chemically stable, it is sent to the pharmacology department for testing of efficacy (i.e., its capacity to produce the desired effect) in experimental animal models and to check for any untoward (i.e., unfavorable) effects. As part of this process, a statistician is asked to design an experiment that can be used to test for significant effect; this design must specify the number of observations to be collected, the methods of statistical analyses, and a statement of the statistical requirements that must be met. The efficacy of this new drug is checked by comparing its average efficacy with that of a standard drug if one is available, whereas the untoward effects are usually evaluated by the type of the effect and its severity. Based upon its intended use, the new drug may be tested in various subdisciplines (e.g., central nervous, cardiovascular, immunology and/or biochemical). In each discipline the chemical will be repeatedly tested in different settings;

*The authors are indebted to Dr. M. C. Trivedi, Pennwalt Corporation, Pharmaceutical Division, for providing this application.

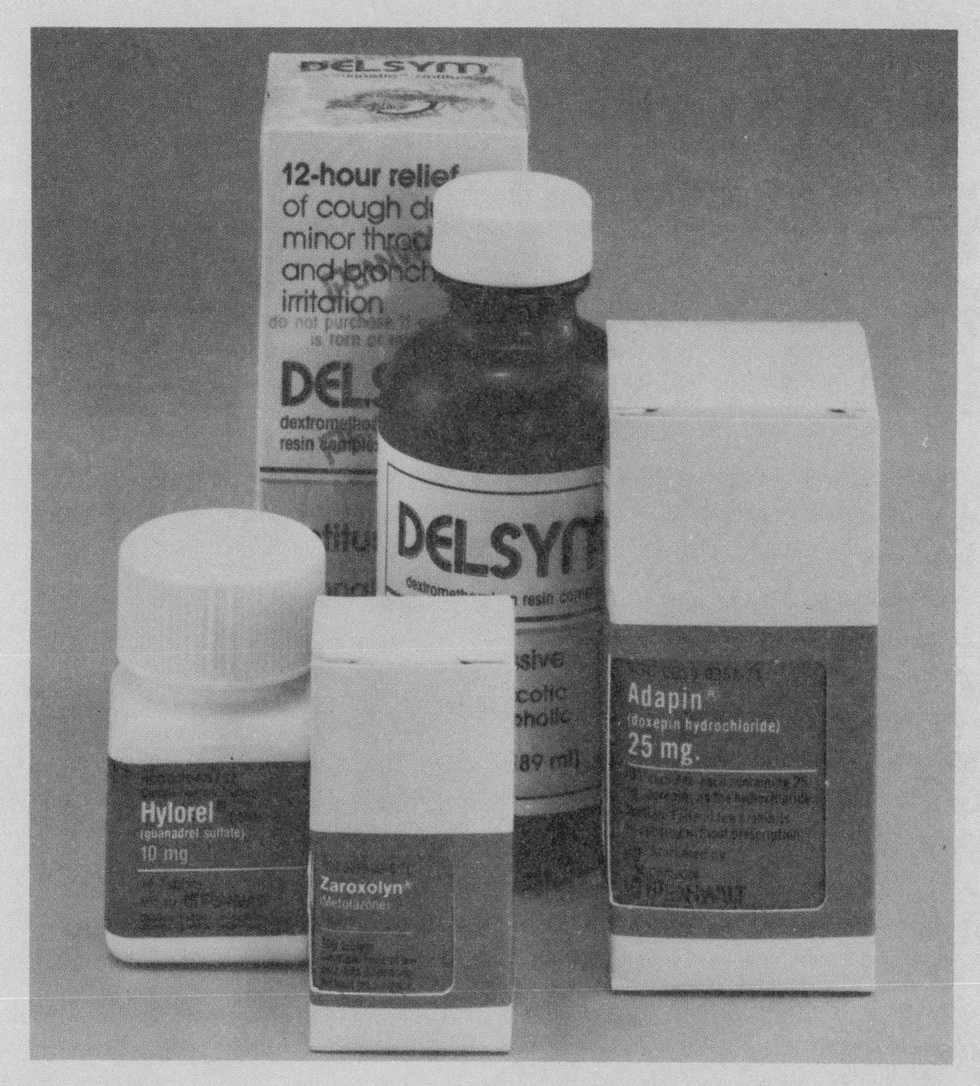

Some of the drug products manufactured by the pharmaceutical division of Pennwalt Corporation

typically, a total of more than 100 different tests are performed. Each test involves the comparison of the new drug with a standard drug in its category or with a placebo. Thus the actual statistical analysis involves testing for the difference between the means of two populations. The new drug can be rejected at any level of testing because of lack of efficacy or undesirable effects.

TESTING FOR LONGTERM USAGE AND SAFETY

The next stage of testing is in the Drug Safety Evaluation section. The new drug is tested here for its cumulative longterm effect. With use of at least two different animal species and with application of repeated doses, the new drug is tested for 15 days, 30 days, 90 days, and/or longer. For each experiment a protocol is prepared which is

application is made to the Food and Drug Administration (FDA) for permission to test the new drug in humans. This is known as an Investigational New Drug (IND) application. If the application is not rejected clinical testing may begin.

CLINICAL TESTING

The third stage of testing involves hospitals, physicians, nurses, and either volunteers or patients, depending upon the phase of testing. Each volunteer or patient is informed of the experimental nature of the drug, and a signed consent form is obtained before the person is admitted into the study. The first phase involves volunteers who are intensively monitored by the participating physician according to the protocol to detect any untoward effect. The second phase involves a selected set of patients to evaluate efficacy of the drug compared to a placebo or to a standard drug, if available. In drug testing, efficacy is defined in the form of one or more quantitative parameters that can be objectively monitored when a person is treated with the medication. For example, if the drug is an antihypertensive agent, the parameter of interest is blood pressure. A reduction in blood pressure implies a positive effect of the drug. Two major parameters of blood pressure, diastolic and systolic, are usually monitored. Each parameter may be monitored in three different positions: supine, sitting, and standing. This process may continue for a short period (e.g., 30 days) or for a long period (e.g., 4 to 5 years or beyond).

In addition to the blood pressure in this example, a battery of other efficacy and safety parameters, such as heart rate, body weight, and blood and urine chemistries, are monitored to detect any untoward effects. The person's sex may be a contributing factor. A person may have hypertension complicated by other diseases, or subjects may be in different phases of the life cycle (e.g., childhood, youth, pregnancy, old age); thus the drug has to be tested for each combination in order to determine its true effect in each case and in order to make sure that untoward effects, if any, are identified. During this phase the drug is tested in several hundred subjects. The statistical analysis involves comparing the average efficacy of the new drug with that of the standard drug and/or a placebo.

Another phase of clinical experiments involves evaluating the onset of drug effect and its duration. In these studies the new drug is compared with a standard drug, if applicable, and the average time to onset of the effect for the new drug is compared with that for the standard drug. Similarly, duration of effect is also compared (or studied by itself when there is no standard drug on the market).

If the new drug meets the requirements of the FDA, a New Drug Application (NDA) is filed with the FDA. Although the guidelines call for approval or disapproval within 6 months, the FDA usually takes longer to respond.

Besides being tested to meet the requirements of the FDA, the new drug is also tested to meet the needs of the marketing division. That is, the new drug may be good enough as a "me, too" or even a little bit better than the standard drug. However, the marketing division may not be able to sell it if it is not substantially better than the standard drug on the market. In this testing, one-sided statistical hypotheses are used when comparing average efficacies of the drugs.

11 Inferences About Population Variances

What you will learn in this chapter:

- situations where variance can be a critical part of the decision-making process
- how the chi-square distribution is used to make inferences about a population variance
- how the *F* distribution is used to make inferences about the variances of two populations

Contents

In the previous four chapters we have focused our attention on methods of statistical inference involving means and proportions. In this chapter we want to expand our discussion to situations involving inferences about variances. As an example of a case where a variance can provide important decision-making information, consider the production process of filling containers with a liquid detergent product. The filling mechanism for the process is adjusted so that the mean filling quantity is 16 ounces per container. While a mean of 16 ounces is desired, the variance of the fillings is also critical. That is, even with the filling mechanism properly adjusted for the mean of 16 ounces, we cannot expect every container to have exactly 16 ounces. By selecting a sample of containers we can compute a sample variance for the filling quantities. This value will serve as an estimate of the variance for the population of containers being filled by the production process. If the sample variance is modest, the production process will be continued. However, if the sample variance is excessive, overfilling and underfilling may exist even though the mean may be correct at 16 ounces. In this case the filling mechanism will be readjusted in an attempt to reduce the filling variance for the containers.

In the following section we consider inferences about the variance of a single population. Later we will discuss procedures that can be used to make inferences about the variances of two different populations.

11.1 INFERENCES ABOUT A POPULATION VARIANCE

In earlier chapters we used the sample variance

$$s^2 = \frac{\Sigma(x_i - \bar{x})^2}{n - 1} \tag{11.1}$$

as the point estimator of the population variance σ^2. In using the sample variance as a basis for making inferences about a population variance, we find the sampling distribution of the quantity $(n - 1)s^2/\sigma^2$ very helpful. This sampling distribution is described as follows:

Sampling Distribution of $(n - 1)s^2/\sigma^2$

Whenever a random sample of size n is selected from a *normal population,* the quantity

$$\frac{(n - 1)s^2}{\sigma^2} \tag{11.2}$$

has a *chi-square distribution* with $n - 1$ degrees of freedom, where s^2 is the sample variance and σ^2 is the population variance.

A graph of the sampling distribution of $(n - 1)s^2/\sigma^2$ is shown in Figure 11.1.

In the previous chapters we have shown that knowledge of a sampling distribution is essential for computing interval estimates and conducting hypothesis tests about a population parameter. Thus far the sampling distribution of interest has always been the sampling distribution of the point estimator. However, for inferences about a population variance we find it more convenient to work with the sampling distribution of $(n - 1)s^2/\sigma^2$ rather than the sampling distribution of the point estimator s^2. The

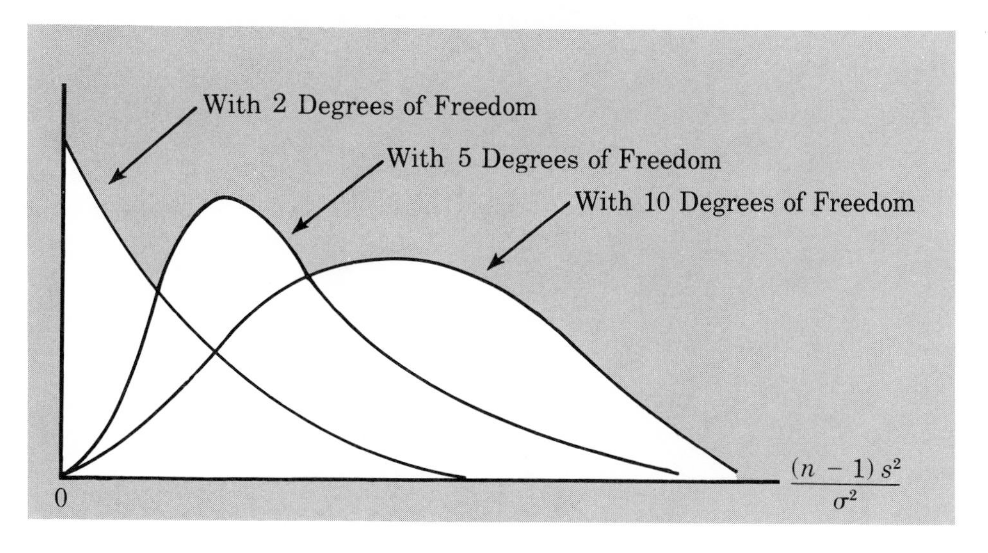

FIGURE 11.1 Examples of the Sampling Distribution of $(n - 1)s^2/\sigma^2$ (a Chi-Square Distribution) with 2, 5, and 10 Degrees of Freedom

reason for this is that with normal populations the sampling distribution of $(n - 1)s^2/\sigma^2$ is known to have a chi-square distribution with $n - 1$ degrees of freedom. Since tables of areas or probabilities are readily available for the chi-square distribution (see Table 3 of Appendix B), it becomes relatively easy to use the chi-square distribution to make interval estimates and hypothesis tests about the value of a population variance.

Interval Estimation of σ^2

Let us show how the chi-square distribution can be used to develop a confidence interval estimate of a population variance σ^2. As an illustration of this process, let us assume that we are interested in estimating the population variance for the production filling process mentioned at the beginning of this chapter. A sample of 20 containers is taken, and the sample variance for the filling quantities is found to be $s^2 = .0025$. However, we know we cannot expect the variance of a sample of 20 containers to provide the exact value of the variance for the population of containers produced by the production process. Thus our interest will be in developing an interval estimate for the population variance.

We will use the notation χ^2_α to denote the value for the chi-square distribution that provides an area or probability of α to the *right* of the stated χ^2_α value. For example, as noted in Figure 11.2, the chi-square distribution with 19 degrees of freedom is shown with $\chi^2_{.025} = 32.85$ indicating the upper 2.5% of the chi-square values and $\chi^2_{.975} = 8.91$ (97.5% of the chi-square values are to the right of $\chi^2_{.975}$) indicating the lower 2.5% of the chi-square values. Refer to Table 3 of Appendix B and check to see that the chi-square values with 19 degrees of freedom (19th row of the table) are correctly given by $\chi^2_{.025} = 32.85$ and $\chi^2_{.975} = 8.91$.

From the graph in Figure 11.2 we see that 95% of the chi-square (χ^2) values are between $\chi^2_{.975}$ and $\chi^2_{.025}$. That is, there is a .95 probability of having a χ^2 value such that

$$\chi^2_{.975} \leq \chi^2 \leq \chi^2_{.025}.$$

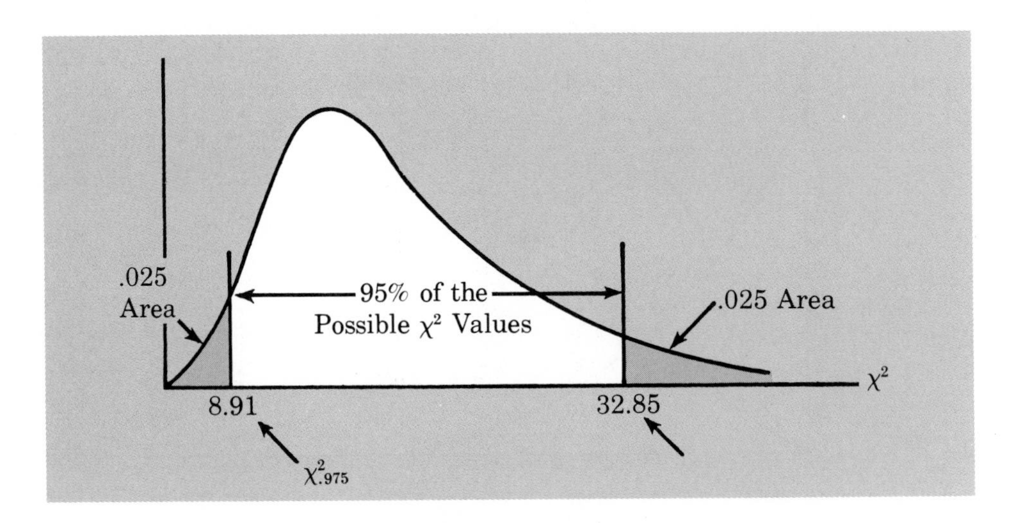

FIGURE 11.2 Example of a Chi-Square Distribution with 19 Degrees of Freedom

However, since we stated in (11.2) that $(n - 1)s^2/\sigma^2$ follows a chi-square distribution, we can substitute $(n - 1)s^2/\sigma^2$ for the above χ^2 and write

$$\chi^2_{.975} \leq \frac{(n - 1)s^2}{\sigma^2} \leq \chi^2_{.025} \tag{11.3}$$

In effect the above expression is providing an interval estimate for the quantity $(n - 1)s^2/\sigma^2$ in that 95% of all values for $(n - 1)s^2/\sigma^2$ will be in the interval $\chi^2_{.975}$ to $\chi^2_{.025}$. We now want to do some algebraic manipulations with (11.3) in order to establish an interval estimate for the population variance σ^2. Working with the leftmost inequality in (11.3), we have

$$\chi^2_{.975} \leq \frac{(n - 1)s^2}{\sigma^2}.$$

Thus

$$\sigma^2\chi^2_{.975} \leq (n - 1)s^2$$

or

$$\sigma^2 \leq \frac{(n - 1)s^2}{\chi^2_{.975}}. \tag{11.4}$$

Performance of similar algebraic manipulations with the rightmost inequality in (11.3) provides

$$\frac{(n - 1)s^2}{\chi^2_{.025}} \leq \sigma^2. \tag{11.5}$$

Finally, the results of (11.4) and (11.5) are combined to provide

$$\frac{(n - 1)s^2}{\chi^2_{.025}} \leq \sigma^2 \leq \frac{(n - 1)s^2}{\chi^2_{.975}}. \tag{11.6}$$

Since (11.3) will be true for 95% of the $(n - 1)s^2/\sigma^2$ values, (11.6) provides a 95% confidence interval estimate for the population variance σ^2.

Let us return to the problem of providing an interval estimate of the population variance of filling quantities. Recall that the sample of 20 containers provided a sample variance $s^2 = .0025$. With a sample of 20, we have 19 degrees of freedom. As shown in Figure 11.2 we have already found $\chi^2_{.975} = 8.91$ and $\chi^2_{.025} = 32.85$. Use of these values in (11.6) provides the following interval estimate for the population variance:

$$\frac{(19)(.0025)}{32.85} \leq \sigma^2 \leq \frac{(19)(.0025)}{8.91}$$

or

$$.0014 \leq \sigma^2 \leq .0053.$$

We take the square root of the above terms and find the following 95% confidence interval estimate for the population standard deviation:

$$.037 \leq \sigma \leq .073.$$

Thus we have illustrated the process of using the chi-square distribution to establish an interval estimate of a population variance and a population standard deviation as well. Note specifically that since $\chi^2_{.975}$ and $\chi^2_{.025}$ were used, the interval estimate has a .95 confidence coefficient. Extending (11.6) to the general case of any confidence coefficient, we have the following interval estimate for a population variance:

Interval Estimate of a Population Variance

$$\frac{(n-1)s^2}{\chi^2_{\alpha/2}} \leq \sigma^2 \leq \frac{(n-1)s^2}{\chi^2_{(1-\alpha/2)}}, \tag{11.7}$$

where the χ^2 values are based on a chi-square distribution with $n - 1$ degrees of freedom and where $1 - \alpha$ is the confidence coefficient.

Hypothesis Testing

Let us now consider an example and the statistical methodology necessary to test hypotheses concerning the value of a population variance. The St. Louis Metro Bus Company has recently made a concerted effort to promote an image of reliability by encouraging its drivers to maintain consistent schedules. As a standard policy the company expects arrival times at various bus stops to have low variability. In terms of the variance of arrival times, the company standard specifies an arrival time variance of 4 or less with arrival times measured in minutes. Periodically, the company collects arrival time data at various bus stops in order to determine if the variability guideline is being maintained. The sample results are used to make the following hypothesis test:

$$H_0: \sigma^2 \leq 4,$$
$$H_1: \sigma^2 > 4.$$

In assuming H_0 true, we are assuming that the variance for arrival times is within the company guidelines. We will reject H_0 only if the sample evidence clearly indicates that the guidelines are not being maintained. In this sense, rejection of H_0 suggests that followup steps are necessary in order to reduce the arrival time variability.

Assume for the moment that a random sample of ten bus arrivals will be taken at a particular downtown intersection. If the population of arrival times has a normal distribution, we know from (11.1) that the quantity $(n - 1)s^2/\sigma^2$ has a chi-square distribution with $n - 1$ degrees of freedom. With the null hypothesis assumed true at $\sigma^2 = 4$ and with a sample size of $n = 10$, we can conclude that

$$\frac{(n-1)s^2}{\sigma^2} = \frac{9s^2}{4}$$

has a chi-square distribution with $n - 1 = 9$ degrees of freedom. Thus once the sample data are obtained and the sample variance s^2 computed the following equation will provide an observed chi-square (χ^2) value

$$\chi^2 = \frac{9s^2}{4}. \tag{11.8}$$

The chi-square distribution showing the rejection region for the one-tailed test is shown in Figure 11.3. Note that we will reject H_0 only if the sample variance s^2 leads to a large χ^2 value [see (11.8)]. With $\alpha = .05$, Table 3 of Appendix B shows that with 9

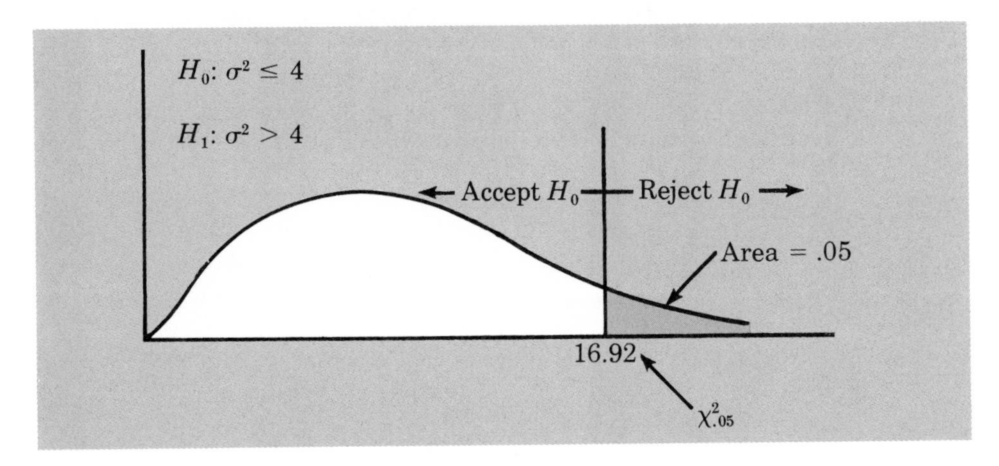

FIGURE 11.3 Chi-Square Distribution with 9 Degrees of Freedom

degrees of freedom $\chi^2_{.05} = 16.92$. With this as the critical value for the test, the decision rule is stated as follows:

$$\text{Accept } H_0 \text{ if } \chi^2 \leq 16.92,$$
$$\text{Reject } H_0 \text{ if } \chi^2 > 16.92.$$

Let us assume that the sample of arrival times for ten buses shows a sample variance of $s^2 = 4.8$. Is this sample evidence sufficient to reject H_0 and conclude that the buses are not meeting the company's arrival time variance guideline? With $s^2 = 4.8$ and (11.8) we obtain the following χ^2 value:

$$\chi^2 = \frac{9(4.8)}{4} = 10.80.$$

Refer to the decision rule for the test. We see that $\chi^2 = 10.80$ leads us to accept H_0.

That is, the sample variance of $s^2 = 4.8$ from the sample of ten bus arrivals is insufficient evidence to conclude that the bus arrival time variance is not meeting the company standard.

One-tailed tests concerning the value of a population variance as just demonstrated are in practice perhaps the most frequently encountered tests about population variances. That is, in situations involving arrival times, production times, filling weights, part dimensions, and so on, low variances generally are desired, while large variances tend to be unacceptable. With a statement about the maximum allowable variance, we frequently test the null hypothesis that the variance is less than or equal to this value against the alternative hypothesis that the variance is greater than this value. We now outline the decision rule for making this one-tailed test about a population variance:

Decision Rule for a One-Tailed Test About a Population Variance Where H_0: $\sigma^2 \leq \sigma_0^2$ and H_1: $\sigma^2 > \sigma_0^2$

$$\text{Accept } H_0 \text{ if } \chi^2 = \frac{(n-1)s^2}{\sigma_0^2} \leq \chi_\alpha^2,$$

$$\text{Reject } H_0 \text{ if } \chi^2 = \frac{(n-1)s^2}{\sigma_0^2} > \chi_\alpha^2,$$

where σ_0^2 is the hypothesized value for the population variance, α is the level of significance for the test, and where the χ_α^2 value is based on a chi-square distribution with $n - 1$ degrees of freedom.

However, just as we saw with population means and population proportions, other forms of the hypotheses are testable. The one-tailed test for H_0: $\sigma^2 \geq \sigma_0^2$ is similar to the test shown above with the exception that the one-tailed rejection region occurs in the lower tail at a critical value of $\chi_{1-\alpha}^2$. The two-tailed test with H_0: $\sigma^2 = \sigma_0^2$, as with other two-tailed tests, places an area of $\alpha/2$ in each tail in order to establish the two critical values for the test. The decision rule for conducting a two-tailed test about the value of a population variance is summarized as follows:

Decision Rule for a Two-Tailed Test About a Population Variance Where H_0: $\sigma^2 = \sigma_0^2$ and H_1: $\sigma^2 \neq \sigma_0^2$

Let

$$\chi^2 = \frac{(n-1)s^2}{\sigma_0^2};$$

$$\text{Accept } H_0 \text{ if } \chi_{(1-\alpha/2)}^2 \leq \chi^2 \leq \chi_{\alpha/2}^2,$$

$$\text{Reject } H_0 \text{ if } \chi^2 < \chi_{(1-\alpha/2)}^2 \text{ or if } \chi^2 > \chi_{\alpha/2}^2,$$

where σ_0^2 is the hypothesized value for the population variance, α is the level of significance for the test, and where the $\chi_{(1-\alpha/2)}^2$ and $\chi_{\alpha/2}^2$ values are based on a chi-square distribution with $n - 1$ degrees of freedom.

EXERCISES

1. In the St. Louis Metro Bus Company example, the sample of ten bus arrivals showed a sample variance of $s^2 = 4.8$.

a. Provide a 95% confidence interval estimate of the variance for the population of arrival times.

b. Assume that the sample variance of $s^2 = 4.8$ had been obtained from a sample of 25 bus arrivals. Provide a 95% confidence interval estimate of the variance for the population of arrival times.

c. What effect does a larger sample size have on the interval estimate of a population variance? Does this seem reasonable?

2. The variance in drug weights is very critical in the pharmaceutical industry. For a specific drug, with weights measured in grams, a sample of 18 units provided a sample variance of $s^2 = .36$.

a. Construct a 90% confidence interval estimate for the population variance for the weights of this drug.

b. Construct a 90% confidence interval estimate for the population standard deviation.

3. A sample of cans of soups produced by Carle Foods shows the following weights, measured in ounces:

12.2	11.9	12.0	12.2
11.7	11.6	11.9	12.0
12.1	12.3	11.8	11.9

Provide 95% confidence interval estimates for both the variance and the standard deviation of the population.

4. A certain part must be machined to very close tolerances or it is not acceptable to customers. Production specifications call for a maximum standard deviation in the lengths of the parts of .02 inches. The sample variance for 30 parts turns out to be $s^2 = .0005$. Using $\alpha = .05$, test to see if the production specifications are being violated.

5. City Trucking, Inc. claims consistent delivery times for its routine customer deliveries. A sample of 22 truck deliveries shows a sample variance of 1.5. Test to determine if the company can justifiably claim that the standard deviation in its delivery times is 1 hour or less. Use $\alpha = .10$.

6. Historical scores for an automobile driver's license examination have a mean of 80 and a variance of 60. A new examination has been designed. As part of the evaluation of the new examination, a sample of 41 drivers complete the new driver's test. The test scores for the sample have a variance of $s^2 = 35$. Test with $\alpha = .05$ to determine if the variance in scores for the population of new examinations is equal to the historical variance of 60.

11.2 INFERENCES ABOUT THE VARIANCES OF TWO POPULATIONS

In some statistical applications it is desirable to compare the variances of two populations. For instance, we might want to compare the variability in product quality resulting from two different production processes, the variability in assembly times for two assembly methods, or the variability in temperatures for two heating devices. In addition, recall that in Chapter 10 we developed a pooled variance estimate based on the assumption that two populations had equal variances. Thus we might want to compare the variances of two populations to determine if the equal variance assumption and, thus pooling can be justified.

In making comparisons about the variances of two populations we will be using data collected from two independent random samples, one from population 1 and another from population 2. In using the two sample variances, s_1^2 and s_2^2, as a basis for making inferences about the two population variances, σ_1^2 and σ_2^2, we find the sampling distribution of the ratio of the two sample variances, s_1^2/s_2^2, very helpful. The sampling distribution of this ratio is described as follows:

Sampling Distribution of s_1^2/s_2^2

Whenever independent random samples of sizes n_1 and n_2 are selected from normal populations with equal variances, the ratio

$$\frac{s_1^2}{s_2^2} \qquad (11.9)$$

has an F distribution with $n_1 - 1$ degrees of freedom for the numerator and $n_2 - 1$ degrees of freedom for the denominator, where s_1^2 is the sample variance for the random sample of n_1 items from population 1 and s_2^2 is the sample variance for the random sample of n_2 items from population 2.

In using the F distribution, we need to specify two sources of degrees of freedom. One is based upon the $n_1 - 1$ degrees of freedom associated with the *numerator* sample variance, and one is based on the $n_2 - 1$ degrees of freedom associated with the *denominator* sample variance. Thus whenever we consider a specific F distribution we must know both the numerator and denominator degrees of freedom. A graph of the F distribution with 20 degrees of freedom for both the numerator and denominator is shown in Figure 11.4.

We will use the notation F_α to denote the value for the F distribution that provides an area or probability of α to the *right* of the stated F_α value. For example, as noted in Figure 11.4, $F_{.05}$ denotes the upper 5% of the F values for an F distribution with 20

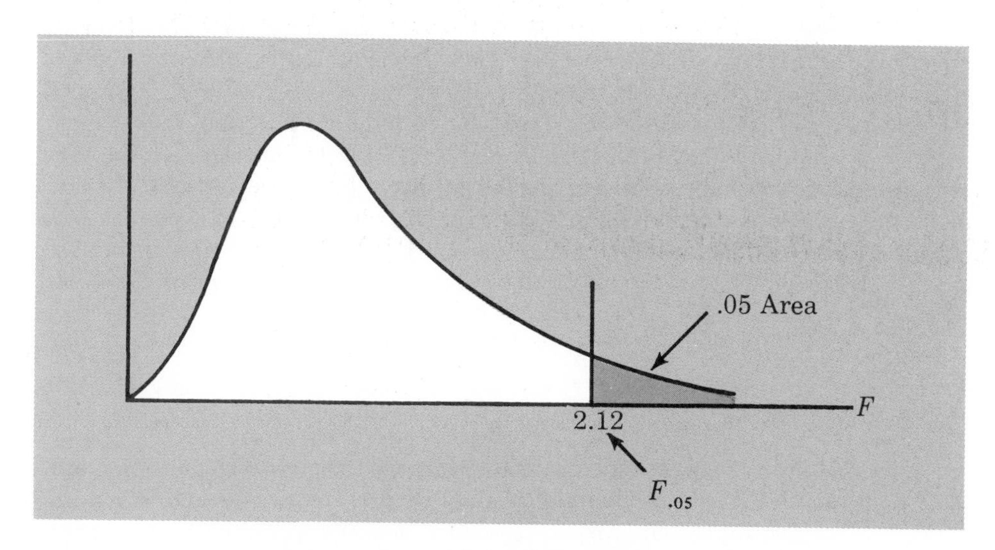

FIGURE 11.4 *F* **Distribution with 20 Degrees of Freedom for the Numerator and 20 Degrees of Freedom for the Denominator**

degrees of freedom for the numerator and 20 degrees of freedom for the denominator. Table 4 of Appendix B shows that for this particular F distribution $F_{.05} = 2.12$. Let us show how the F distribution can be used for a hypothesis test concerning the variances of two populations.

Dullus County Schools is renewing its school bus service contract for the coming year and must select one of the two bus companies, the Milbank Company or the Gulf Park Company. We will be interested in using the variance of the arrival or pickup/delivery times as a primary measure of the quality of the bus service. Low variance values will indicate the more consistent and higher-quality service. With equality of variances for the populations of arrival times associated with the two services, Dullus School administrators will select the company offering the better financial terms. However, if sample data on bus arrival times for the two companies indicate that a significant difference exists between the variances, the administrators may want to give special consideration to the company with the better or lower-variance service. The appropriate hypotheses and their associated conclusions and actions are as follows:

Hypothesis	*Conclusion and Action*
$H_0: \sigma_1^2 = \sigma_2^2$	Equal quality of service; base service selection decision on financial terms
$H_1: \sigma_1^2 \neq \sigma_2^2$	Unequal quality of service; give special consideration to the low-variance service

Assume that the above hypothesis test will be conducted with $\alpha = .10$. Furthermore, assume that we obtain samples of arrival times from school systems currently using the two school bus services. A sample of 25 arrival times is available for the Milbank service (population 1) and a sample of 16 arrival times is available for the Gulf Park service (population 2). The graph of the F distribution with $n_1 - 1 = 24$ degrees of freedom and $n_2 - 1 = 15$ degrees of freedom is shown in Figure 11.5. Note that the two-tailed rejection regions are indicated by the critical values at $F_{.95}$ and $F_{.05}$.

Let us assume that the two samples of bus arrival times show sample variances of $s_1^2 = 48$ for the Milbank service and $s_2^2 = 20$ for the Gulf Park service. What conclusion is now appropriate concerning the quality of the two bus services? Assume that the two populations of arrival times have normal distributions, and assume that H_0 is true with $\sigma_1^2 = \sigma_2^2$. The F distribution now can be used to reach a decision. Specifically, we compute $F = s_1^2/s_2^2$ and use the "accept" and "reject" regions shown in Figure 11.5. Thus we find

$$F = \frac{s_1^2}{s_2^2} = \frac{48}{20} = 2.40.$$

Using Table 4 of Appendix B to determine the critical value for the test, we find the upper-tail critical value with 24 and 15 degrees of freedom to be $F_{.05} = 2.29$. While the appendix does not provide $F_{.95}$ values, note that the determination of this lower-tail critical value is no longer necessary. We can already observe that $F = 2.40$ exceeds

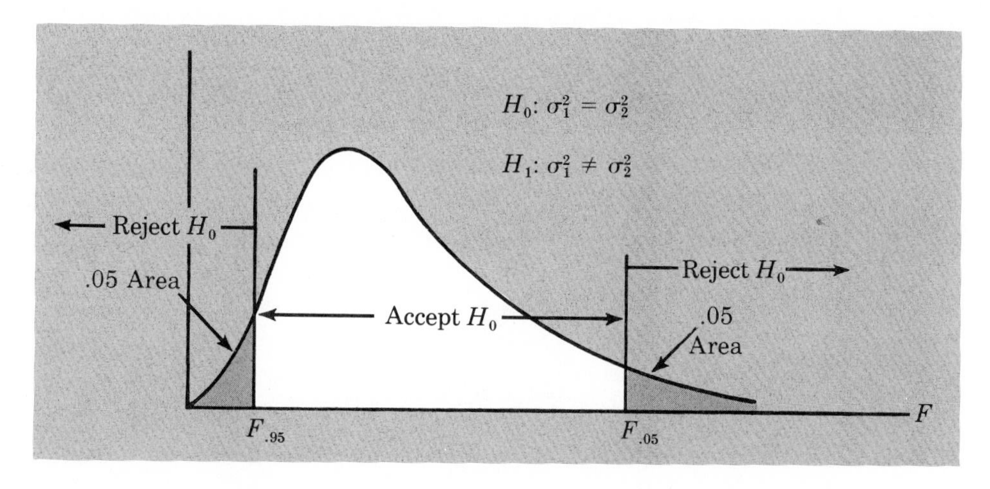

FIGURE 11.5 ***F* Distribution with 24 Degrees of Freedom for the Numerator and 15 Degrees of Freedom for the Denominator**

$F_{.05}$ = 2.29. Thus at the .10 level of significance H_0 is rejected. This leads us to the conclusion that the two bus services differ in terms of pickup/delivery time variances. Specifically, it is recommended that the Dullus School administrators give special consideration to the better, or lower-variance, service offered by the Gulf Park Company.

In the above statistical computations, you might feel that we were lucky in carrying out the test because the lower-tail critical value, $F_{.95}$, which could not be found in the table, was not even necessary. Thus we were able to draw the appropriate conclusion without knowing the $F_{.95}$ value. While lower-tail F values can be determined, it is quite common to find only upper-tail F values in statistical references simply because F tables would require several pages if very many F_α values were provided. Thus in order to limit the size of the appendices many statistical texts present only the upper-tail values, and then for only a relatively few α values. However, in hypothesis-testing situations involving H_0: $\sigma_1^2 = \sigma_2^2$, the lack of availability of the lower-tail critical values is not as restrictive as you might think. We can get around this apparent problem simply by denoting the population with the *larger sample variance* as population 1. That is, which population we denote population 1 and which we denote 2 is arbitrary. By labeling the population with the larger sample variance as population 1, we guarantee that $F = s_1^2/s_2^2$ will be greater than or equal to 1. Thus if a rejection of H_0 is to occur, it can occur *only* in the *upper tail.* While the lower-tail critical value still exists and could be correctly labeled $F_{.95}$, we do not need to know its value simply because the convention of using the population generating the largest sample variance as population 1 always places the ratio s_1^2/s_2^2 in the upper-tail direction. In the Dullus Schools example, population 1, the Milbank bus service, possessed the largest sample variance. Thus we proceeded directly with the test. If the Gulf Park bus service had provided the largest sample variance, we simply would have denoted Gulf Park as population 1 and followed the same statistical testing procedure. A summary of the two-tailed test for the equality of two population variances with this procedure is as follows:

Decision Rule for a Two-Tailed Test About Variances of Two Populations Where H_0: $\sigma_1^2 = \sigma_2^2$ and H_1: $\sigma_1^2 \neq \sigma_2^2$

Denote the population providing the *largest sample variance* as population 1. Let

$$F = s_1^2/s_2^2;$$

$$\text{Accept } H_0 \text{ if } F \leq F_{\alpha/2},$$

$$\text{Reject } H_0 \text{ if } F > F_{\alpha/2},$$

where α is the level of significance for the test and where the $F_{\alpha/2}$ value is based on an F distribution with $n_1 - 1$ degrees of freedom for the numerator and $n_2 - 1$ degrees of freedom for the denominator.

One-tailed tests involving two population variances are also possible. Again the F distribution is used, with the one-tailed rejection region enabling us to conclude whether or not one population variance is significantly greater or significantly less than the other. Only upper-tail F values are needed. For any one-tailed test we set up the null hypothesis so that the rejection region is in the upper tail. This can be accomplished by a judicious choice of which population is labeled population 1. For instance, suppose that we want to test as the null hypothesis that the variance of population B is less than or equal to the variance of population A. We would label population B population 1 in the general procedure which follows:

Decision Rule for a One-Tailed Test About Variances of Two Populations Where H_0: $\sigma_1^2 \leq \sigma_2^2$ and H_1: $\sigma_1^2 > \sigma_2^2$

Let

$$F = s_1^2/s_2^2;$$

$$\text{Accept } H_0 \text{ if } F \leq F_{\alpha},$$

$$\text{Reject } H_0 \text{ if } F > F_{\alpha},$$

where α is the level of significance for the test and where the F_α value is based on an F distribution with $n_1 - 1$ degrees of freedom for the numerator and $n_2 - 1$ degrees of freedom for the denominator.

EXERCISES

7. In Chapter 10 we discussed the problem of estimating the difference between the mean checking account balances at two branches of the Clearview National Bank. The data collected from two independent random samples are as follows:

Branch Bank	*Number of Checking Accounts*	*Sample Mean Balance*	*Sample Standard Deviation*
Cherry Grove	12	$\bar{x}_1$ = \$1,000	s_1 = \$150
Beechmont	10	$\bar{x}_2$ = \$ 920	s_2 = \$120

In using the t distribution to estimate the difference between means we made the assumption that the variances of the two populations were equal. This assumption was the basis for developing a pooled variance estimate. Use $\alpha = .10$ and conduct a test for the equality of the population variances. Do your results justify the use of a pooled variance estimate?

8. Two new assembly methods are tested with the following variances in assembly times:

Method	*Sample Size*	*Sample Variance*
A	31	$s_1^2 = 25$
B	25	$s_2^2 = 12$

Use $\alpha = .10$ and test for equality of the two population variances.

9. Independent random samples of parts manufactured by two suppliers show the following results:

Supplier	*Sample Size*	*Sample Variance of Part Sizes*
Durham Electric	41	$s_1^2 = 3.8$
Raleigh Electronics	31	$s_2^2 = 2.0$

The firm making the supplier selection decision is prepared to use the Durham supplier unless the test results show that the Raleigh supplier provides significantly better or significantly lower variance in part sizes. Use $\alpha = .05$ and conduct the statistical test that will help the firm select a supplier. Which supplier do you recommend?

10. The following sample data have been collected from two independent random samples:

Population	*Sample Size*	*Sample Mean*	*Sample Variance*
A	$n_A = 25$	$\bar{x}_A = 40$	$s_A^2 = 5$
B	$n_B = 21$	$\bar{x}_B = 50$	$s_B^2 = 11$

In a test for the difference between the two population means, the statistical analyst is considering using a pooled estimate of the population variance based on the assumption the variances of the two populations are equal. Is pooling appropriate in this case? Use $\alpha = .10$ for your test.

Summary

In this chapter we have presented statistical procedures that can be used to make inferences about population variances. In the process we have introduced two new probability distributions: the chi-square distribution and the F distribution. The chi-square distribution can be used as the basis for interval estimation and hypothesis tests concerning the variance of a normal population. In particular, we showed that with random samples of size n selected from a normal population, the quantity $(n - 1)s^2/\sigma^2$ has a chi-square distribution with $n - 1$ degrees of freedom.

We illustrated the use of the F distribution in making hypothesis tests concerning the variances of two normal populations. In particular, we showed that with independent random samples of sizes n_1 and n_2 selected from two normal populations with equal variances, $\sigma_1^2 = \sigma_2^2$, the sampling distribution of the ratio of the two sample variances, s_1^2/s_2^2, has an F distribution with $n_1 - 1$ degrees of freedom for the numerator and $n_2 - 1$ degrees of freedom for the denominator.

Key Formulas

Interval Estimate of a Population Variance

$$\frac{(n-1)s^2}{\chi^2_{\alpha/2}} \leq \sigma^2 \leq \frac{(n-1)s^2}{\chi^2_{(1-\alpha/2)}} \qquad (11.7)$$

Sampling Distribution of s_1^2/s_2^2 When $\sigma_1^2 = \sigma_2^2$

$$F = \frac{s_1^2}{s_2^2} \qquad (11.9)$$

Supplementary Exercises

11. Because of staffing decisions, management of the Gibson-Marimont Hotel is interested in the variability for the number of rooms occupied per day during a particular season of the year. A sample of 20 days of operation shows a sample mean of 290 rooms occupied per day and a sample standard deviation of 30 rooms.
a. What is the point estimate of the population variance?
b. Provide a 90% confidence interval estimate for the population variance.
c. Provide a 90% confidence interval estimate for the population standard deviation.
12. A random sample of 30 days of sales for United Mufflers, Inc. shows a sample mean of 22.5 mufflers sold per day, with a sample standard deviation of 6. Provide 95% confidence interval estimates of both the population variance and the population standard deviation for the muffler sales data.
13. Historical delivery times for Buffalo Trucking, Inc. have had a mean of 3 hours and a standard deviation of .5 hours. A sample of 22 deliveries over the past month provides a sample mean of 3.1 hours and a sample standard deviation of .75 hours.
a. Use a test of hypothesis about a population variance, and test to determine if the sample results support the historical delivery variability of $\sigma^2 = (.5)^2 = .25$. Use $\alpha = .05$.
b. Provide a 95% confidence interval estimate for both the population variance and the population standard deviation.
14. Part variability is very critical in the manufacturing of ball bearings. Large variances in the size of the ball bearings cause bearing failure and rapid wearout. Production standards call for a maximum variance of .0001 when the bearing sizes are measured in inches. A sample of 15 bearings shows a sample standard deviation of .014 inches.
a. Test with $\alpha = .10$ to determine if the sample bearings were taken from a population having a variance of .0001 or less.
b. Provide a 90% confidence interval estimate for the variance of the ball bearings in the population.
15. The filling variance for boxes of cereal is designed to be .02 or less. A sample of 41 boxes of cereal shows a sample standard deviation of .16 ounces. Test with $\alpha = .05$ to determine if the variance in the cereal box fillings is exceeding the standard.
16. A sample standard deviation for the number of passengers taking a particular airline flight is 8. A 95% confidence interval estimate of the standard deviation is 5.86 to 12.62.

a. Was a sample size of 10 or 15 used in the above statistical analysis?
b. If the sample standard deviation of $s = 8$ had been based on a sample of 25 flights, what change would you expect in the confidence interval for the population standard deviation? Compute a 95% confidence interval for σ if a sample of size 25 had been used.

17. Assume that the ball bearings referred to in Exercise 14 are produced on two different shifts. Sample results from the shifts are as follows:

First Shift	*Second Shift*
$n_1 = 22$	$n_2 = 25$
$s_1 = .12$	$s_2 = .09$

Are the sample results sufficient to conclude that the variances for the ball bearings differ for the two shifts? Test with $\alpha = .10$.

18. A firm gives a mechanical aptitude test to all job applicants. A sample of 20 male applicants shows a sample variance of 80 for the test scores. A sample of 16 female applicants shows a sample variance of 220. Test with $\alpha = .05$ to determine if the test score variances differ for male and female job applicants. If a difference in variances exists, which group has the higher variance in mechanical aptitude?

19. The accounting department analyzes the variance of the weekly unit costs reported by two production departments. A sample of 16 cost reports for each of the two departments shows cost variances of 2.3 and 5.4, respectively. Is this sample sufficient to conclude that the two production departments differ in terms of unit cost variances? Use $\alpha = .10$.

20. In using the t distribution to estimate the difference between two population means an analyst is interested in computing a pooled estimate of the variance of the populations. Pooling is justified only if it appears reasonable to assume that the two populations have equal variances. Use the following data to determine if pooling is appropriate for this situation (test with $\alpha = .10$):

Sample 1 80, 72, 75, 90, 78, 75, 72, 85.
Sample 2 50, 48, 45, 60, 65, 66, 70, 54.

If pooling is appropriate, what is the pooled estimate of the population variances?

Application

U.S. General Accounting Office*

Washington, D.C.

The United States General Accounting Office (GAO) is an independent, nonpolitical audit organization in the Legislative branch of the Federal government. GAO was created by the Budget and Accounting Act of 1921 and has three basic purposes:

To assist Congress, its committees, and its members carry out their legislative and oversight responsibilities, consistent with its role as an independent, nonpolitical agency.

To audit and evaluate the programs, activities, and financial operations of Federal departments and agencies, and to make recommendations toward more efficient and effective operations.

To carry out financial control and other functions with respect to Federal government programs and operations including accounting, legal, and claims settlement work.

GAO evaluators, the main occupation in GAO, determine the effectiveness of existing or proposed Federal programs and the efficiency, economy, legality, and effectiveness with which Federal agencies carry out their responsibilities. These evaluations culminate in reports to the Congress and to the heads of Federal departments and agencies. Such reports typically include recommendations to Congress concerning the need for enabling or remedial legislation and suggestions to agencies concerning the need for changes in programs or operations to improve their economy, efficiency, and effectiveness.

GAO evaluators analyze policies and practices, and the use of resources within and among Federal programs; identify problem areas and deficiencies in meeting program goals; develop and analyze alternative solutions to problems of program execution; and develop and recommend changes to enable the programs to better conform to Congressional goals and legislative intent. To effectively carry out their duties evaluators must be proficient in interviewing, data processing, records review, operations research, legislative research and statistical analysis techniques.

*The authors are indebted to Mr. Art Foreman and Mr. Dale Ledman of the U.S. General Accounting Office for providing this application.

STATISTICAL ANALYSIS AT GAO

GAO evaluators perform statistical analysis themselves and conduct audits of the validity of statistical analyses conducted by other governmental agencies and private organizations. Sampling procedures are frequently employed to collect data for studies. GAO evaluators will use probability samples wherever possible but must frequently work with whatever data is available and as a result must at times rely on judgment samples. Care must be exercised to ensure that the sample selected is representative of the population to which inferences are being made. Otherwise legislators and others might be led to draw invalid conclusions.

Regression and correlation analysis is another area in which GAO becomes involved. For instance a regression model might be developed to predict demand for hospital intensive care units. Such a model can have a great impact on government policies for medicare and medicaid reimbursement. GAO conducts careful evaluations of such models to ensure the validity of the results.

Other areas of statistical analysis have involved nonparametric studies of the consistency of parole determinations, probabilistic analysis of the disposition of disability applications, and time series analyses. In the following we describe an application where an hypothesis test concerning a population variance was used.

AN AUDIT OF A SEWAGE TREATMENT FACILITY

A few years ago a program was established by the Department of Interior to clean up the nation's rivers and lakes. As a part of the program, Federal grants were made to small cities scattered throughout the United States. Congress asked GAO to determine how effectively the program was operating. To make this determination GAO examined records and conducted site visits at several plants.

One objective of these audits was to ensure that the effluent (treated sewage) at the plants met certain standards. Among other things, the following characteristics of the effluents were examined:

1. Oxygen content.
2. PH level.
3. Amount of suspended solids.
4. Amount of soluble solids.

This application overviews an interesting finding that resulted from the audit of one particular plant.

A requirement of the program grants was that a variety of tests be taken and recorded daily with the records periodically sent to the state engineering department. GAO's investigation of the plant in question began with an examination of a sample of the records submitted to the state. Initially GAO auditors were concerned with determining whether or not various characteristics of the effluent were within acceptable limits. In this case the measurements were within acceptable limits, but the auditors noted that there was very little variability in the data. This led to further investigation along another line.

The PH level of water is 7. A certain variance is normal and expected for samples taken from different sources at different times. The apparent low variance in the sample data caused the auditors to conduct the following hypothesis test concerning the variance in PH level for the population:

$$H_0: \sigma^2 \geq \sigma_0^2$$
$$H_1: \sigma^2 < \sigma_0^2$$

where σ_0^2 is the population variance in PH level found at other properly functioning plants. The hypothesis test led to rejection of H_0 and the conclusion that the variance in PH level at the plant in question was significantly less than normal.

The auditors then made a plant visit to examine the measuring equipment and discuss their findings with the plant operator. They found that the measuring equipment was not even being used because the plant operator did not know how to operate it. Instead, the operator was told by an engineer what an acceptable PH level was and had simply recorded numbers within the acceptable range each day without actually conducting the required tests.

BENEFITS OF STUDY

This particular case caused GAO evaluators to investigate further the ability of sewage treatment plant operators to use the effluent measuring equipment as well as other equipment and chemicals used to treat the sewage. It was found that even though the plants and equipment purchased by the grant were modern and of high quality, many of the operators did not have adequate training to operate the equipment. This particular study and followup investigations led to a GAO recommendation that states receiving funds from the program be required to establish training programs for plant operators.

U.S. General Accounting Office Headquarters in Washington D.C.

12 Tests of Goodness of Fit and Independence

What you will learn in this chapter:

- what a goodness of fit test is
- what a contingency table is
- how to use the chi-square distribution to conduct tests of goodness of fit and independence
- how to conduct a goodness of fit test when the population is hypothesized to have either a multinomial, Poisson, or normal probability distribution

Contents

In Chapter 11 we introduced the chi-square distribution and illustrated how it could be used in estimation and hypothesis tests about a population variance. In this chapter we introduce two additional hypothesis-testing procedures, both based on the use of the chi-square distribution. As with other hypothesis-testing procedures, these tests compare sample results with those that are expected when the null hypothesis is true. The acceptance or rejection of the null hypothesis is based upon how "close" the sample or observed results are to the expected results.

In the following section we introduce a goodness of fit test involving a multinomial population. Later we discuss the test for independence using contingency tables and then show goodness of fit tests for Poisson and normal probability distributions.

12.1 GOODNESS OF FIT TEST—A MULTINOMIAL POPULATION

In this section we consider the case where each element of a population is assigned to one and only one of several classes or categories. Such a population is a *multinomial population.* For example, consider the market analysis being conducted by the J. Scott and Associates market research firm. The study involves a market share evaluation. Over the past year market shares have stabilized, with 30% for company A, 50% for company B, and 20% for company C. Company C, J. Scott's client, has developed a "new and improved" product that will replace its current entry in the market. Company management has asked J. Scott and Associates to determine if the new product will provide a shift in the market shares of the three competitors.

In this case the population of interest is a multinomial population, since each customer is classified as buying from company A, company B, or company C. Thus we have a multinomial population with three classifications or categories. Let us define the following notation:

$$p_A = \text{market share for company A},$$
$$p_B = \text{market share for company B},$$
$$p_C = \text{market share for company C}.$$

Based upon the assumption that company C's new product will not alter the market shares, the null and alternative hypotheses would be stated as follows:

$$H_0\colon p_A = .30,\ p_B = .50,\ \text{and } p_C = .20,$$
$$H_1\colon \text{The population proportions are not}$$
$$p_A = .30,\ p_B = .50,\ \text{and } p_C = .20.$$

If the sample results lead to the rejection of H_0, the firm will have evidence that the introduction of the new product has had an impact on the market.

Let us assume that the market research firm will use a consumer panel of 200 customers for the study. Each individual will be asked to specify a purchase preference among the three alternatives: company A's product, company B's product, and company C's new product. The 200 purchase preference responses are summarized below:

Company A's Product	*Company B's Product*	*Company C's New Product*
48	98	54

We now want to demonstrate a goodness of fit test that will determine if the sample of 200 customer purchase preferences supports the null hypothesis for the three market shares. The goodness of fit test is based on a comparison of the sample of *observed* results such as those shown above with the *expected* results under the assumption that the null hypothesis is true. Thus our next step is to compute expected purchase preferences for the 200 customers under the assumption that $p_A = .30$, $p_B = .50$, and $p_C = .20$. Doing this provides the expected results shown below:

Company A's Product	*Company B's Product*	*Company C's New Product*
200(.30) = 60	200(.50) = 100	200(.20) = 40

Thus we see that the expected frequency for each category is found by multiplying the sample size of 200 by the hypothesized proportion for the category.

The goodness of fit test now focuses on the differences between the observed frequencies and the expected frequencies. Large differences between observed and expected frequencies cast doubt on the assumption that the hypothesized proportions or market shares are correct. However, small differences between observed and expected frequencies will not provide sufficient evidence to reject the null hypothesis. In the case where observed and expected frequencies are equal, the sample data provide a perfect fit of the hypothesized distribution.

Whether the differences between the observed and expected frequencies are "large" or "small" is a question answered with the aid of the following calculations. Let

f_i = observed frequency for category i,
e_i = expected frequency for category i based on the assumption that the null hypothesis is true,
k = the number of categories.

If the null hypothesis is true and if the sample size is large, then the quantity

$$\chi^2 = \sum_{i=1}^{k} \frac{(f_i - e_i)^2}{e_i} \qquad (12.1)$$

has chi-square distribution. For the case where the null hypothesis involves proportions for k categories of a multinomial population, the appropriate chi-square distribution has $k - 1$ degrees of freedom. We will assume that the requirement of a *large* sample size is satisfied whenever the expected frequency for each category is 5 or more. The large sample rule is easily satisfied for our market-share study, and thus we can proceed with the computation of the chi-square (χ^2) value in (12.1) as follows:

$$\chi^2 = \frac{(48 - 60)^2}{60} + \frac{(98 - 100)^2}{100} + \frac{(54 - 40)^2}{40}$$

$$= \frac{144}{60} + \frac{4}{100} + \frac{196}{40} = 2.40 + .04 + 4.90 = 7.34$$

Suppose that we test the null hypothesis that the multinomial population has the proportions of $p_A = .30$, $p_B = .50$, and $P_C = .20$ at the $\alpha = .05$ level of significance. Since we will reject the null hypothesis only if the differences between the observed and expected frequencies and thus the computed χ^2 value are *large*, we will place a rejection area of .05 in the upper tail of chi-square distribution. Checking the chi-square distribution table (Table 3 of Appendix B), we find that with $k - 1 = 3 - 1 = 2$ degrees of freedom $\chi^2_{.05} = 5.99$. Thus as with similar one-tailed tests we will reject H_0 if the computed chi-square value exceeds the critical value of 5.99.

We check our previous calculations and find a computed $\chi^2 = 7.34$. Since this value is larger than the critical value of 5.99, we reject the null hypothesis. In rejecting H_0 we are concluding that the introduction of the new product by company C will alter the current market share structure. While the goodness of fit test itself permits no further conclusions, we can informally compare the observed and expected frequencies to obtain an idea of how the market share structure has changed.

Considering company C, we find that the observed frequency of 54 is larger than the expected frequency of 40. Since the expected frequency was based on current market shares, the larger observed frequency suggests that the new product will have a positive effect on company C's market share. Informal comparisons of the observed and expected frequencies for the other two companies indicate that company C's gain in market share will tend to hurt company A more than company B.

As illustrated in the above example, the goodness of fit test uses the chi-square distribution to determine if a hypothesized probability distribution for a population provides a good fit. Acceptance or rejection of the hypothesized population distribution is based upon differences between observed frequencies in a sample and the expected frequencies based on the assumed population distribution. Let us outline the general steps that can be used to conduct a goodness of fit test for any hypothesized population distribution:

1. Formulate a null hypothesis indicating a hypothesized distribution for the population.
2. Establish k categories for the population.
3. Use a simple random sample of n items and record the observed frequencies for each of k classes or categories.
4. Use the assumption that the null hypothesis is true and determine the probability or proportion associated with each category.
5. Multiply the category proportions in step 4 by the sample size to determine the expected frequencies for each category.
6. Use the observed and expected frequencies and (12.1) to compute a χ^2 value for the test.
7. Decision rule:

$$\text{Accept } H_0 \text{ if } \chi^2 \leq \chi^2_{\alpha},$$
$$\text{Reject } H_0 \text{ if } \chi^2 > \chi^2_{\alpha},$$

where α is the level of significance for the test.

EXERCISES

1. During the first 13 weeks of the television season, the Saturday evening 8:00 P.M. to 9:00 P.M. audience proportions were recorded as ABC 29%, CBS 28%, NBC 25%, and independents 18%. A sample of 300 homes 2 weeks after a Saturday night schedule revision showed the following viewing audience data: ABC 95 homes, CBS 70 homes, NBC 89 homes, and independents 46 homes. Test with $\alpha = .05$ to determine if the viewing audience proportions have changed.

2. A new container design has been adopted by a manufacturer. Color preferences indicated in a sample of 150 individuals are as follows:

Red	*Blue*	*Green*
40	64	46

Test using $\alpha = .10$ to see if the color preferences are the same. Hint: Formulate the null hypothesis as H_0: $p_1 = p_2 = p_3 = 1/3$.

3. Grade distribution guidelines for a statistics course at a major university are as follows: 10% A, 30% B, 40% C, 15% D, and 5% F. A sample of 120 statistics grades at the end of a semester showed 18 A's, 30 B's, 40 C's, 22 D's, and 10 F's. Use $\alpha = .05$ and test to see if the actual grades are within the grade distribution guidelines.

4. Consumer panel preferences for three proposed store displays are as follows:

Display A	*Display B*	*Display C*
43	53	39

Use $\alpha = .05$ and test to see if there is no preference among the three display designs

12.2 TEST OF INDEPENDENCE—CONTINGENCY TABLES

Another important application of the chi-square distribution involves using sample data to test for the independence of two variables. Let us illustrate the test of independence by considering the study conducted by the Alber's Brewery of Tucson, Arizona. Alber's manufactures and distributes three types of beers: a low-calorie light beer, a regular beer, and a dark beer. In an analysis of the market segments for the three beers, the firm's market research group has raised the question of whether or not preferences for the three beers differ among male and female beer drinkers. If beer preference is independent of the sex of the beer drinker, one advertising campaign will be initiated for all of Alber's beers. However, if beer preference depends upon the sex of the beer drinker, the firm will tailor its promotions toward different target markets.

A test of independence addresses the question of whether or not the beer preference (light, regular, or dark) is independent of the sex of the beer drinker (male, female). The hypotheses for this test of independence are as follows:

H_0: Beer preference is independent of the sex of the beer drinker,

H_1: Beer preference is not independent of the sex of the beer drinker.

Table 12.1 can be used to describe the situation being studied. By identification of the population as all male and female beer drinkers, a sample can be selected and each individual asked to state his or her preference for the three Alber's beers. Every

TABLE 12.1 Contingency Table—Beer Preference and Sex of Beer Drinker

		Beer Preference	
Sex	*Light*	*Regular*	*Dark*
Male	(cell 1)	(cell 2)	(cell 3)
Female	(cell 4)	(cell 5)	(cell 6)

individual in the sample will be classified in one of the six cells in the table. For example, an individual may be a male preferring regular beer (cell 2), a female preferring light beer (cell 4), a female preferring dark beer (cell 6), etc. Since we have listed all possible combinations of beer preference and sex, or, in other words, listed all possible contingencies, Table 12.1 is called a *contingency table.* The test of independence makes use of the contingency table format and for this reason is sometimes referred to as a *contingency table test.*

Let us assume that a simple random sample of 150 beer drinkers has been selected. After they have taste tested each beer the individuals in the sample are asked to state their preference or first choice. The contingency table in Table 12.2 summarizes the responses to the study. As we see in the contingency table, the data for the test of independence are collected in terms of counts or frequencies for each cell or category. Thus of the 150 individuals in the sample 20 were men who favored light beer, 40 were men who favored regular beer, 20 were men who favored dark beer, and so on.

Note that the data in Table 12.2 contain the sample or observed frequencies for each of six classes or categories. If we can determine the expected frequencies under the assumption of independence between beer preference and sex of the beer drinker, we can use the chi-square distribution, just as we did in the previous section, to determine whether or not there is a significant difference between observed and expected frequencies.

TABLE 12.2 Sample Results of Beer Preferences for Male and Female Beer Drinkers (Observed Frequencies)

	Beer Preference			
Sex	*Light*	*Regular*	*Dark*	*Total*
Male	20	40	20	80
Female	30	30	10	70
Total	50	70	30	150

Expected frequencies for the cells of the contingency table are based on the following rationale: First, we assume that the null hypothesis of independence between beer preference and sex of the beer drinker is true. Then, we note that the sample of 150 beer drinkers showed a total of 50 preferring light beer, 70 preferring regular beer, and 30 preferring dark beer. In terms of fractions we conclude that $50/150 = 1/3$ of the beer drinkers prefer light beer, $70/150 = 7/15$ prefer regular beer, and $30/150 = 1/5$ prefer dark beer. If the *independence* assumption is valid, we argue that these same fractions must be applicable to both male and female beer drinkers. Thus under the assumption of independence we would expect the 80 male beer drinkers to show that $(1/3)80 = 26.67$ prefer light beer, $(7/15)80 = 37.33$ prefer regular beer, and $(1/5)80 = 16$ prefer dark beer. Application of these same fractions to the 70 female beer drinkers provides the expected frequencies as shown in Table 12.3.

Let e_{ij} stand for the expected frequency for the contingency table category in row i and column j. With this notation let us reconsider the expected frequency calculation for males (row $i = 1$) who prefer regular beer (column $j = 2$)—that is, expected

TABLE 12.3 Expected Frequencies if Beer Preference Is Independent of the Sex of the Beer Drinker

Sex	*Light*	*Beer Preference* *Regular*	*Dark*	*Total*
Male	26.67	37.33	16.00	80
Female	23.33	32.67	14.00	70
Total	50	70	30	150

frequency e_{12}. Following our previous argument for the computation of expected frequencies, we showed that

$$e_{12} = (7/15)80 = 37.33$$

Writing this slightly differently, we find

$$e_{12} = (7/15)80 = (70/150)80 = \frac{(80)(70)}{150} = 37.33$$

Note that the 80 in the above expression is the total number of males (row 1), the 70 is the total number preferring regular beer (column 2), and the 150 is the total sample size. Thus we see that

$$e_{12} = \frac{(\text{Row 1 total})(\text{Column 2 total})}{\text{Sample size}}$$

Generalization of the above expression shows that the following formula provides the expected frequencies for a contingency table in the test for independence.

Expected Frequencies for Contingency Tables Under the Assumption of Independence

$$e_{ij} = \frac{(\text{Row } i \text{ total})(\text{Column } j \text{ total})}{\text{Sample size}}. \tag{12.2}$$

With the above formula for male beer drinkers preferring dark beer we find an expected frequency of $e_{13} = (80)(30)/150 = 16.00$, as shown previously in Table 12.3. Use (12.2) to verify that it can provide the other expected frequencies shown in Table 12.3.

The test procedure for comparing the observed frequencies of Table 12.2 with the expected frequencies of Table 12.3 is similar to the goodness of fit calculations made in the previous section. Specifically, the χ^2 value based on the observed and expected frequencies is computed as follows:

$$\chi^2 = \sum_i \sum_j \frac{(f_{ij} - e_{ij})^2}{e_{ij}} \tag{12.3}$$

where

f_{ij} = observed frequency for contingency table category in row i and column j,

e_{ij} = expected frequency for contingency table category in row i and column j.

The double summation in (12.3) is used to indicate that the calculation must be made for the combination of all rows with all columns.

Before we apply the above formula we must check to see that the expected frequencies in each cell are at least 5. This is the same check that we used in the previous section to determine whether or not the sample size was large enough for the χ^2 distribution assumption to be made. Since all expected frequencies in Table 12.3 are at least 5, we can conclude that the sample size is adequate. The resulting χ^2 value is found as follows:

$$\chi^2 = \frac{(20 - 26.67)^2}{26.67} + \frac{(40 - 37.33)^2}{37.33} + \cdots + \frac{(10 - 14.00)^2}{14.00}$$

$$= 1.67 + .07 + \cdots + 1.14 = 6.13.$$

The number of degrees of freedom for the appropriate χ^2 distribution is computed by multiplying the *number of rows minus 1* times the *number of columns minus 1*. With two rows and three columns, we have $(2 - 1)(3 - 1) = (1)(2) = 2$ degrees of freedom for the test of independence of beer preference and the sex of the beer drinker. With $\alpha = .05$ for the level of significance of the test, Table 3 of Appendix B shows an upper-tail χ^2 value of $\chi^2_{.05} = 5.99$. Note here that again we are using the upper-tail value because we will reject the null hypothesis only if the differences in observed and expected frequencies provide a large χ^2 value. In our example $\chi^2 = 6.13$ is greater than the critical value of $\chi^2_{.05} = 5.99$. Thus we reject the null hypothesis of independence and conclude that the preference for the beers is not independent of the sex of the beer drinkers.

Although the test for independence allows only the above conclusion, again we can informally compare the observed and expected frequencies in order to obtain an idea of how the dependence between the beer preference and sex of the beer drinker comes about. Refer to Table 12.2 and 12.3. We see that male beer drinkers have higher observed than expected frequencies for both regular and dark beers, while female beer drinkers have a higher observed than expected frequency for only the light beer. These observations give us an insight into the beer preference differences between the male and female beer drinkers.

EXERCISES

5. The number of units sold by three salespersons over a 3 month period are shown below:

	Product		
Salesperson	*A*	*B*	*C*
Troutman	14	12	4
Kempton	21	16	8
McChristian	15	5	10

Use $\alpha = .05$ and test for the independence of salesperson and type of product sold.

6. Starting positions for business and engineering gradutes are classified by industry as shown below:

	Industry			
Degree Major	*Oil*	*Chemical*	*Electrical*	*Computer*
Business	30	15	15	40
Engineering	30	30	20	20

Use $\alpha = .01$ and test for independence of degree major and industry type.

7. A sport preference poll shows the following data for men and women:

	Favorite Sport		
Sex	*Baseball*	*Basketball*	*Football*
Men	19	15	24
Women	16	18	16

Use $\alpha = .05$ and test for similar sport preferences by men and women. What is your conclusion?

8. Three suppliers provide the following data on defective parts:

	Part Quality		
Supplier	*Good*	*Minor Defect*	*Major Defect*
A	90	3	7
B	170	18	7
C	135	6	9

Use $\alpha = .05$ and test for independence between the supplier and the part quality. What does the result of your analysis tell the purchasing department?

9. A study of educational levels of voters and their political party affiliations showed the following results:

	Party Affiliation		
Educational Level	*Democratic*	*Republican*	*Independent*
Did Not Complete High School	40	20	10
High School Degree	30	35	15
College Degree	30	45	25

Use $\alpha = .01$ and test to see if party affiliation is independent of the educational level of the voters.

12.3 GOODNESS OF FIT TEST—POISSON AND NORMAL DISTRIBUTIONS

In Section 12.1 we introduced the goodness of fit test involving a multinomial population. In general the goodness of fit test can be used with any hypothesized probability distribution. In this section we illustrate the goodness of fit test procedure for cases where the population is hypothesized to have a Poisson or a normal distribution. As we shall see, the goodness of fit test and the use of the chi-square distribution for these tests follow the same general procedure used for the goodness of fit test in Section 12.1.

A Poisson Distribution

Let us illustrate the goodness of fit test for the case where the hypothesized population distribution is a Poisson distribution. As an example consider the arrivals of customers at Dubek's Food Market in Tallahassee, Florida. Dubek's management makes staffing decisions involving the number of clerks, the number of check-out lanes, and so on based on the anticipated arrivals of customers at the store. Because of some recent staffing problems Dubek's management has asked a local consulting firm to assist with the scheduling of clerks for the check-out lanes. The general objective of the consulting firm's work is to provide enough clerks to achieve a good level of service while maintaining reasonable payroll cost.

After reviewing the check-out lane operation the consulting firm makes a recommendation for a clerk scheduling procedure. The procedure, based on a mathematical analysis of waiting lines, is applicable only in situations where the number of customers arriving during a specified time period follows the Poisson probability distribution. Thus before the scheduling process is implemented data on customer arrivals must be collected and a statistical test conducted to see if an assumption of a Poisson distribution for arrivals appears reasonable.

We define the arrivals at the store in terms of the *number of customers* entering the store during 5 minute intervals. Thus the following null and alternative hypotheses are appropriate for the Dubek study:

H_0: The number of customers entering the store during 5 minute intervals has a Poisson probability distribution,

H_1: The number of customers entering the store during 5 minute intervals does not have a Poisson distribution.

If a sample of customer arrivals supports the acceptance of H_0, Dubek will be justified in implementing the consulting firm's scheduling procedure. However, if sample results lead to the rejection of H_0, the assumption of the Poisson distribution for the arrivals cannot be made and other scheduling procedures will have to be considered.

In order to test the assumption of a Poisson distribution for the number of arrivals during weekday morning hours, a store employee randomly selects a sample of 128 five minute intervals during weekday mornings over a 3 week period. For each 5 minute interval in the sample a store employee records the number of customer arrivals. In summarizing the data the employee determines the number of 5 minute intervals having no arrivals, the number of 5 minute intervals having one arrival, the number of 5 minute intervals having two arrivals, and so on. These data are summarized in Table

12.4. The number of customers arriving are the category descriptions, with the observed frequencies recorded in the column showing the number of 5 minute intervals having the corresponding number of customer arrivals.

Table 12.4 provides the observed frequencies for the ten categories. We now want to use a goodness of fit test to determine whether or not the sample of 128 time periods supports the hypothesized Poisson probability distribution. In order to conduct the goodness of fit test, we need to consider the expected frequency for each of the ten categories under the assumption that the Poisson distribution of arrivals is true. That is, we need to compute the expected number of time periods that no customers, one customer, two customers, and so on would arrive if, in fact, the customer arrivals have a Poisson distribution.

TABLE 12.4 Observed Frequencies of Dubek Customer Arrivals for a Sample of 128 5 Minute Time Periods

Category Description: Number of Customers Arriving		*Observed Frequency*
0		2
1		8
2		10
3		12
4		18
5		22
6		22
7		16
8		12
9		6
	Total sample	128

The Poisson probability function, which was first introduced in Chapter 6, is as follows:

$$f(x) = \frac{\mu^x e^{-\mu}}{x!}. \tag{12.4}$$

Let us apply this probability function to the Dubek problem. In (12.4) μ represents the mean or expected number of customers arriving per 5 minute period, and x is the random variable indicating the number of customers arriving during a 5 minute period. In this case x may be equal to 0, 1, 2, and so on. Finally, $f(x)$ is the probability that x customers will arrive in a 5-minute interval.

Before we use (12.4) to compute Poisson probabilities we must obtain an estimate of the mean (μ) number of customer arrivals during a 5 minute time period. The sample mean for the data in Table 12.4 provides this estimate. With no customers arriving in two periods, one customer arriving in eight periods, and so on, the total number of customers that arrived during the sample of 128 periods is given by $0(2) + 1(8) + 2(10) + \cdots + 9(6) = 640$. The 640 customer arrivals over the sample of 128 periods provides a mean arrival rate of $640/128 = 5$ customers per 5 minute period.

With this value for the mean of the Poisson probability distribution an estimate of the Poisson probability function for Dubek's Food Market becomes

$$f(x) = \frac{5^x e^{-5}}{x!}. \qquad (12.5)$$

Assume that the Poisson probabilty distribution is appropriate for the Dubek customer arrivals. The above probability function then can be evaluated for different values of x in order to determine the probability associated with each category of arrivals. These probabilities, which can also be found in Table 7 of Appendix B, are shown in Table 12.5. For example, the probability of zero customers arriving during a 5 minute interval is $f(0) = .0067$, the probability of one customer arriving during a 5 minute interval is $f(1) = .0337$, and so on. As we saw in Section 12.1, the expected frequencies for the categories are found by multiplying the probabilities by the sample size. For example, the expected number of periods with zero arrivals is given by $(.0067)(128) = .8576$, the expected number of periods with one arrival is given by $(.0337)(128) = 4.3136$, and so on.

Before we make the usual chi-square calculations to compare the observed and expected frequencies, note that in Table 12.5 four of the categories have an expected frequency less than 5. This condition violates the requirements necessary for use of the chi-square distribution. However, expected category frequencies less than 5 cause no difficulty, since categories can be combined to satisfy the "at least 5" expected frequency requirement. In particular, we will combine zero and one into a single category and then combine nine with "ten or more" into another single category. Thus the rule of a minimum expected frequency of 5 in each category is satisfied.

TABLE 12.5 Expected Frequencies of Dubek Customer Arrivals, Assuming a Poisson Probability Distribution with $\mu = 5$

Category Description: Number of Customers Arriving (x)	*Poisson Probability,* $f(x)$		*Expected Number of 5 Minute Time Periods with x Arrivals,* $128f(x)$
0	.0067		.8576
1	.0337		4.3136
2	.0842		10.7776
3	.1404		17.9712
4	.1755		22.4640
5	.1755		22.4640
6	.1462		18.7136
7	.1044		13.3632
8	.0653		8.3584
9	.0363		4.6464
10 or more	.0318		4.0704
		Total	128.0000

Let us redefine the categories for the observed results in Table 12.4 accordingly. We can proceed to compare the observed frequencies with the expected frequencies. This comparison is summarized in Table 12.6.

TABLE 12.6 Comparison of the Observed and Expected Frequencies for the Dubek Customer Arrivals

Category Description: Number of Customers Arriving	*Observed Frequency* (f_i)	*Expected Frequency* (e_i)	*Difference* ($f_i - e_i$)
0 or 1	10	5.1712	4.8288
2	10	10.7776	−.7776
3	12	17.9712	−5.9712
4	18	22.4640	−4.4640
5	22	22.4640	−.4640
6	22	18.7136	3.2864
7	16	13.3632	2.6368
8	12	8.3584	3.6416
9 or more	6	8.7168	−2.7168
Total	128	128.0000	

As in Section 12.1 the goodness of fit test focuses on the differences in observed and expected frequencies, $f_i - e_i$. Obviously, large differences cast doubt on the assumption that the customer arrivals have a Poisson distribution. However, small differences would tend to support the conclusion that the Poisson distribution provides a good fit.

Using the observed and expected frequencies shown in Table 12.6, we again compute the chi-square value,

$$\chi^2 = \sum_{i=1}^{k} \frac{(f_i - e_i)^2}{e_i}.$$

Doing so, we have

$$\chi^2 = \frac{(4.8288)^2}{5.1712} + \frac{(-.7776)^2}{10.7776} + \cdots + \frac{(-2.7168)^2}{8.7168}$$

$$= 4.5091 + .0561 + \cdots + .8468 = 10.9767.$$

We need now determine the appropriate degrees of freedom associated with this goodness of fit test. In general, the chi-square distribution for a goodness of fit test has $k - p - 1$ degrees of freedom, where k is the number of categories and where p is the number of population parameters estimated from the sample data. For the Poisson distribution goodness of fit test we are considering, Table 12.6 shows $k = 9$ categories. Since the sample data were used to estimate the mean of the Poisson distribution, $p = 1$. Thus $k - p - 1 = 9 - 1 - 1 = 7$ degrees of freedom exist for the chi-square distribution of interest in the Dubek study.

Suppose that we test the hypothesis that the probability distribution for the customer arrivals is Poisson with a .05 level of significance ($\alpha = .05$). This is an upper-tail test, since we will reject the null hypothesis only if the difference in the observed and expected frequencies—and thus the χ^2 value—becomes large. We check the χ^2 values in Table 3 of Appendix B and find that with 7 degrees of freedom $\chi^2_{.05} = 14.07$. As with similar one-tailed tests, we will reject the null hypothesis only if the computed value for χ^2 exceeds the χ^2_α value.

Checking our previous calculations for the Dubek study, we find a computed $\chi^2 = 10.9767$. Since this value is less than the critical value of 14.07, we cannot reject the null hypothesis. Thus for the Dubek study the assumption of a Poisson probability distribution for weekday morning customer arrivals has been supported. With this statistical finding, Dubek can proceed with the consulting firm's scheduling procedure for weekday mornings.

A Normal Distribution

The goodness of fit test for a normal probability distribution is also based on the use of the chi-square distribution. It is very similar to the procedure we have just discussed for the Poisson distribution. In particular, observed frequencies for several categories of sample data will be compared to expected frequencies under the assumption that the population has a normal distribution. Since the normal probability distribution is continuous, we must modify the way that the categories are defined and how the expected frequencies are computed. Let us demonstrate the goodness of fit test for a normal probability distribution by considering the job applicant test data for Chemline, Inc. shown in Table 12.7.

TABLE 12.7 Chemline Employee Aptitude Test Scores for 50 Randomly Chosen Job Applicants

71	66	61	65	54
93	60	86	70	70
73	73	55	63	56
62	76	54	82	79
76	68	53	58	85
80	56	61	61	64
65	62	90	69	76
79	77	54	64	74
65	65	61	56	63
80	56	71	79	84

Chemline hires approximately 400 new employees annually for its four plants located throughout the United States. Standardized tests are given by the personnel department, with performance on the test being a major factor in the employee hiring decision. With numerous tests being given annually, the personnel director has asked if a normal distribution could be applied to the population of test scores. If such a distribution can be applied, use of the distribution would be most helpful in evaluating specific test scores. That is, scores in the upper 20%, lower 40%, etc. could quickly be identified. Thus we would like to test the null hypothesis that the population of aptitude test scores follows a normal probability distribution.

Let us first use the data in Table 12.7 to develop estimates of the mean and standard deviation of the normal distribution that will be considered in the null hypothesis. We use as point estimators the sample mean $\bar{x}$ and the sample standard deviation s:

$$\bar{x} = \frac{\Sigma x_i}{n} = \frac{3{,}421}{50} = 68.42,$$

$$s = \sqrt{\frac{\Sigma (x_i - \bar{x})^2}{49}} = 10.41.$$

Using these values, we state the following hypotheses about the distribution of the job applicant test scores:

H_0: The population of test scores has a normal distribution with mean 68.42 and standard deviation 10.41,

H_1: The population of test scores does not have a normal distribution with mean 68.42 and standard deviation 10.41.

The hypothesized normal distribution is shown in Figure 12.1.

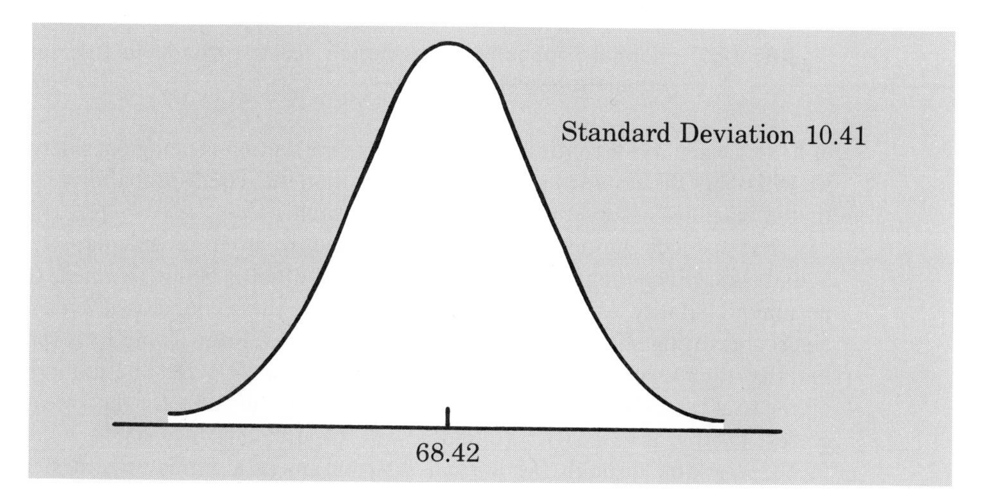

FIGURE 12.1 Hypothesized Normal Distribution of Test Scores for the Chemline Job Applicants

Now let us consider a way of defining the categories for a goodness of fit test involving a normal population. For the discrete probability distribution in the Poisson distribution test, the categories were readily defined in terms of the number of customers arriving, such as 0, 1, 2, and so on. However, with a continuous probability distribution, such as the normal, we will have to come up with a different procedure for defining the categories. Actually, we will need to define the categories in terms of *intervals* of test scores.

Recall the rule of thumb for an expected frequency of at least 5 in each interval or category. We will have to define the categories of test scores such that the expected frequencies will be at least 5 for each category. With a sample size of 50, one way of doing this is to divide the normal distribution into ten equal-probability intervals (see Figure 12.2). With a sample size of 50, we would expect five outcomes in each interval or category, and the rule of thumb for expected frequencies would be satisfied. This is the procedure we will follow for determining the number of categories for the goodness

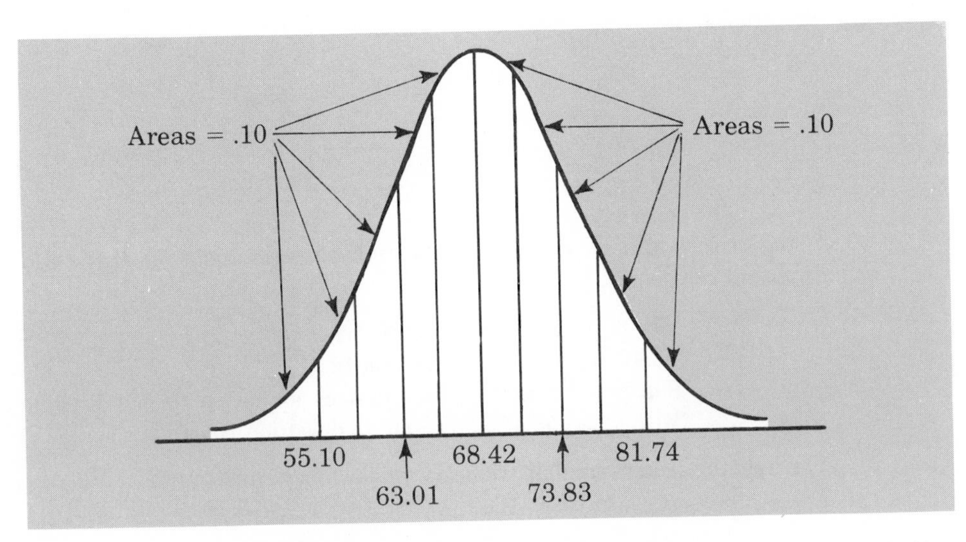

FIGURE 12.2 Normal Probability Distribution for the Chemline Example with Ten Equal-Probability Intervals

of fit test whenever a continuous probability distribution is being considered. Namely, we will break the assumed population distribution into equal-probability intervals such that at least five observations are expected in each.

Let us look more closely at the procedure used to calculate the category boundaries. Since the normal probability distribution is being assumed, the standard normal probability tables can be used to determine these boundaries. First consider the test score cutting off the lowest 10% of the test scores. From Table 1 of Appendix B we find that the z value for this test score is approximately -1.28. Therefore the test score of $x = 68.42 - 1.28(10.41) = 55.10$ provides this cutoff value for the lowest 10% of the scores. For the lowest 20%, we find $z = -.84$, and thus $x = 68.42 - .84(10.41) = 59.68$. Working through the normal distribution in a similar manner provides the following test score values:

Lower 30%:	$68.42 - .52(10.41) = 63.01$
Lower 40%:	$68.42 - .25(10.41) = 65.82$
Mid-score:	$68.42 + 0(10.41) = 68.42$
Upper 40%:	$68.42 + .25(10.41) = 71.02$
Upper 30%:	$68.42 + .52(10.41) = 73.83$
Upper 20%:	$68.42 + .84(10.41) = 77.16$
Upper 10%:	$68.42 + 1.28(10.41) = 81.74$

Some of these cutoff or interval boundary points have been identified on the graph in Figure 12.2.

With the categories or intervals of test scores now defined and with the known expected frequencies of five per category, we can return to the sample data of Table 12.7 and determine the observed frequencies for the categories. Doing so provides the results in Table 12.8. Also note that Table 12.8 contains a column of differences between the observed and expected frequencies.

With the results in Table 12.8, the goodness of fit calculations proceed exactly as before. Namely, we compare the observed and expected results by computing a χ^2 value as follows:

$$\chi^2 = \sum_{i=1}^{k} \frac{(f_i - e_i)^2}{e_i} = \frac{0^2}{5} + \frac{0^2}{5} + \frac{4^2}{5} + \cdots + \frac{1^2}{5} = 7.2.$$

TABLE 12.8 Observed and Expected Category Frequencies for Chemline Job Applicant Test Scores

Test Score Interval	*Observed Frequency* (f_i)	*Expected Frequency* (e_i)	*Difference* $(f_i - e_i)$
Less than 55.10	5	5	0
55.10 to 59.68	5	5	0
59.68 to 63.01	9	5	4
63.01 to 65.82	6	5	1
65.82 to 68.42	2	5	−3
68.42 to 71.02	5	5	0
71.02 to 73.83	2	5	−3
73.83 to 77.16	5	5	0
77.16 to 81.74	5	5	0
Over 81.74	6	5	1
Total	50	50	

To determine whether or not the computed χ^2 value of 7.2 is large, we need to refer to the appropriate chi-square probability distribution tables. Using the rule for computing the number of degrees of freedom for the goodness of fit test, we have $k - p - 1 = 10 - 2 - 1 = 7$, where there are $k = 10$ categories and $p = 2$ parameters (mean and standard deviation) estimated from the sample data. Using a .10 level of significance for this hypothesis test, we have $\chi^2_{.10} = 12.017$ for the upper-tail rejection region. With $7.2 < 12.017$ we conclude that the null hypothesis cannot be rejected. Thus the hypothesis that the probability distribution for the Chemline job applicant test scores is normal cannot be rejected.

EXERCISES

10. The number of automobile accidents occurring per day in a particular city is believed to have a Poisson distribution. A sample of 80 days during the past year shows the following data:

Number of Accidents	*Observed Frequency (days)*
0	34
1	25
2	11
3	7
4	3

Do these data support the belief that the number of accidents per day has a Poisson distribution? Use $\alpha = .05$.

11. The number of incoming phone calls occurring at a company switchboard during 1 minute intervals is believed to have a Poisson distribution. Use $\alpha = .10$ and the following data to test the assumption that the incoming phone calls have a Poisson distribution:

Number of Incoming Phone Calls during a 1 Minute Interval	*Observed Frequency*
0	15
1	31
2	20
3	15
4	13
5	4
6	2
	Total 100 one minute intervals

12. The weekly demand for a product is believed to be normally distributed. Use a goodness of fit test and the data in the following sample to test this assumption:

18, 25, 26, 27, 26, 25, 20, 22, 23, 25, 25, 28, 22, 27, 20, 19, 31, 26, 27, 25, 24, 21, 29, 28, 22, 24, 26, 25, 25, 24.

Use $\alpha = .10$. The sample mean is 24.5, and the sample standard deviation is 3.

13. Use $\alpha = .01$ and conduct a goodness of fit test to see if the following sample appears to have been selected from a normal population:

55	86	94	58	55
95	55	52	66	95
90	65	87	50	56
55	57	98	58	82
92	62	59	88	65

After you complete the goodness of fit calculations, construct a histogram of the data. Does the histogram representation support the conclusion reached with the goodness of fit test? (Note: $\bar{x} = 71$ and s is approximately 17)

Summary

In this chapter we introduced the goodness of fit test and the test of independence procedures, both of which are based on the use of the chi-square distribution. The purpose of the goodness of fit test is to determine whether a hypothesized probability distribution can be accepted as a distribution for a particular population of interest. The computations for conducting the goodness of fit test involve comparing observed frequencies from a sample with expected frequencies when the hypothesized probability distribution is assumed true. A chi-square distribution is used to determine if the differences in observed and expected frequencies are sufficient to reject the hypothesized probability distribution. We illustrated the goodness of fit test for assumed multinomial, Poisson, and normal probability distributions. For statistical procedures that are based on the assumption of a population with a normal probability distribution, the goodness of fit test may be applied in order to determine whether or not an assumed normal probability distribution is appropriate.

A test of independence for two variables is a straightforward extension of the methodology employed in the goodness of fit test for a multinomial population. A contingency table is used to determine the observed and expected frequencies. Then a chi-square value is computed. Large chi-square values, caused by large differences between observed and expected frequencies, lead to the rejection of the null hypothesis of independence.

Glossary

Goodness of fit test—A statistical test conducted to determine whether to accept or reject a hypothesized probability distribution for a population.

Contingency table—A table used to summarize observed and expected frequencies for a test of independence of population characteristics.

Key Formulas

Goodness of Fit Test

$$\chi^2 = \sum_{i=1}^{k} \frac{(f_i - e_i)^2}{e_i} \tag{12.1}$$

Expected Frequencies for Contingency Tables Under the Assumption of Independence

$$e_{ij} = \frac{(\text{Row } i \text{ total})(\text{Column } j \text{ total})}{\text{Sample size}} \tag{12.2}$$

Contingency Table Test

$$\chi^2 = \sum_i \sum_j \frac{(f_{ij} - e_{ij})^2}{e_{ij}} \tag{12.3}$$

Supplementary Exercises

14. In setting sales quotas, the marketing manager makes the assumption that order potentials are the same in each of four sales territories. A sample of 200 sales shows the following number of orders from each region:

Sales Territories			
I	*II*	*III*	*IV*
60	45	59	36

Do these data support the manager's assumptions? Use $\alpha = .05$.

15. A sample of parts provided the following contingency table data concerning part quality and production shift:

Shift	*Number Good*	*Number Defective*
First	368	32
Second	285	15
Third	176	24

Use $\alpha = .05$ and test the assumption that part quality is independent of the production shift. What is your conclusion?

16. A lending institution shows the following data regarding loan approvals by four different loan officers:

Loan Officer	*Loan Approval Decision* Approved	Rejected
Miller	24	16
McMahon	17	13
Games	35	15
Runk	11	9

Use $\alpha = .05$ and test to determine if the loan approval decision is independent of the loan officer reviewing the loan application.

17. A random sample of final examination grades for a college course are shown below:

55, 85, 72, 99, 48, 71, 88, 70, 59, 98, 80, 74, 93, 85, 74, 82, 90, 71, 83, 60, 95, 77, 84, 73, 63, 72, 95, 79, 51, 85, 76, 81, 78, 65, 75, 87, 86, 70, 80, 64.

Use $\alpha = .05$ and test to determine if a normal distribution may be assumed for the population of grades.

18. A salesperson makes four calls per day. A sample of 100 days shows the following frequencies of sales volumes:

Number of Sales	*Observed Frequency (days)*
0	30
1	32
2	25
3	10
4	3
	100

Use $\alpha = .05$ and test to determine if the population is a binomial distribution with a probability of purchase $p = .30$. Recall that in Chapter 6 the binomial probabilities were given by

$$f(x) = \frac{n!}{x!(n-x)!} p^x (1-p)^{n-x}.$$

For this exercise $n = 4$, $p = .30$, and $x = 0, 1, 2, 3$, and 4.

Application

United Way*

Rochester, New York

The United Way of Greater Rochester is a nonprofit fund-raising and social planning organization dedicated to improving the quality of life of residents in the six counties it serves. The annual United Way/Red Cross campaign, conducted each spring, helps support more than 140 human service agencies. These agencies meet a wide variety of human needs—physical, mental, and social—and serve people of all ages, backgrounds, and economic means.

The United Way relies on thousands of dedicated volunteers, many of whom serve year-round. Because of this volunteer involvement, the United Way is able to hold its operating costs to less than nine cents of every dollar raised—a remarkably low level.

MARKET RESEARCH STUDY

The United Way of Greater Rochester was interested in determining Rochester community perceptions of charities. Although a national survey of attitudes had been conducted by the United Way of America, it was believed that some of the general conclusions drawn from the national survey might not be applicable to the local United Way organizations. It was suggested that the United Way of Rochester conduct its own market research study to evaluate community perceptions for the purpose of recommending appropriate adjustments.

The initial research began with focus group interviews; subject groups included professional, service, and general worker categories. Conclusions drawn from these interviews included the following:

1. The more people are informed, the better they feel about giving to charities;
2. There exists a misconception by many individuals regarding how much charities allocate to administrative funds;
3. There is a great mistrust of large charities, including the feeling that many of their activities are dishonest.

After the completion of the focus group interviews, questionnaire development was started. The bases for selecting questions were the focus group interviews and the

*The authors are indebted to Dr. Philip R. Tyler, Marketing Consultant to the United Way of Greater Rochester, New York, for providing this application.

specific needs and interests that were set as guidelines by United Way personnel. These guidelines were stated as goals, which included increasing giving among groups now contributing, improving the effectiveness of public relations, and increasing the amount of money raised by identifying new target markets.

A draft of the questionnaire was discussed with United Way personnel. Minor adjustments were made in wording, and redundant questions were eliminated. The instrument was then tested for effectiveness on a small test group. The final questionnaire was then distributed to 440 individuals at 18 different organizations. From this group 323 completed questionnaires were obtained.

STATISTICAL ANALYSIS

SPSS (Statistical Package for the Social Sciences) is an integrated system of computer programs for statistical analysis. SPSS was selected to analyze the data because it is easy to use and provides a wide range of options for both analysis and output. The specific analysis of the data utilized the following SPSS procedures:

1. Procedure FREQUENCIES—This procedure produces one-way frequency distribution tables for specified variables. Using this procedure, the data were summarized in both graphical and tabular form.
2. Procedure CROSSTABS—Essentially this is a joint frequency distribution of two or more variables, with appropriate statistical analysis. For example, the chi-square test for contingency tables presented in this chapter was used to test for the independence of two variables.

As an illustration of the use of the chi-square test for independence in this study, consider the following question asked of respondents:

Of the funds collected, what percentage do you feel goes to United Way Administrative Expenses:

( ) Up to 10%
( ) 11%–20%
( ) 21% and over*

Each respondent was also asked to indicate their occupation according to the following classification:

( ) Production line/assemblers
( ) Maintenance/warehouse
( ) Craftsmen and foremen
( ) Clerical worker
( ) Sales worker
( ) Managers and administrators
( ) Professional and technical workers
( ) Other ___________

*This category was originally 21%–30%, 31%–50%, and 51% and over; these three classes were combined during the statistical analysis in order to obtain expected frequencies appropriate for the chi-square test.

In this case a test of independence addresses the question of whether the perception of United Way administrative expenses is independent of the occupation of the respondent. The hypotheses for the test of independence are as follows:

H_0: Perception of United Way administrative expenses is independent of the occupation of the respondent,

H_1: Perception of United Way administrative expenses is not independent of the occupation of the respondent.

The contingency table in Table 12A.1 summarizes the responses to the study and provides the necessary output statistics. We see that 290 useful responses were obtained. Note that there are (8-1)(3-1) = 14 degrees of freedom. With $\alpha = .05$ for the level of significance, Table 3 of Appendix B shows an upper-tail χ^2 value of $\chi^2_{.05} = 23.6848$. Since $\chi^2 = 49.7684$ is greater than the critical value of $\chi^2_{.05} = 23.6848$, we reject the null hypothesis of independence and conclude that perception of United Way administrative expenses is not independent of the respondent's occupation.

TABLE 12A.1 Perception of United Way Administrative Expenses by Occupation

A145 Occupation by A098 Question 21
A098

A145			TO 10% 1	11–20% 2	21% 3	ROW TOTAL
PROD LINE	COUNT	1.	3	3	11	17
	ROW PCT		17.65	17.65	64.71	5.86
	COL PCT		2.86	3.61	10.78	
	TOT PCT		1.03	1.03	3.79	
MAINT-WARE HSE	COUNT	2.	3	7	5	15
	ROW PCT		20.00	46.67	33.33	5.17
	COL PCT		2.86	8.43	4.90	
	TOT PCT		1.03	2.41	1.72	
CRAFTS-FOREMEN	COUNT	3.	7	5	3	15
	ROW PCT		46.67	33.33	20.00	5.17
	COL PCT		6.67	6.02	2.94	
	TOT PCT		2.41	1.72	1.03	
CLERICAL	COUNT	4.	15	27	35	77
	ROW PCT		19.48	35.06	45.45	26.55
	COL PCT		14.29	32.53	34.31	
	TOT PCT		5.17	9.31	12.07	
SALES	COUNT	5.	3	8	8	19
	ROW PCT		15.79	42.11	42.11	6.55
	COL PCT		2.86	9.64	7.84	
	TOT PCT		1.03	2.76	2.76	
MANAGERS	COUNT	6.	37	13	7	57
	ROW PCT		64.91	22.81	12.28	19.66
	COL PCT		35.24	15.66	6.86	
	TOT PCT		12.76	4.48	2.41	

TABLE 12A.1 Continued

A145 Occupation by A098 Question 21
A098

PRO-TECH	COUNT	7.	23	9	21	53
	ROW PCT		43.40	16.98	39.62	18.28
	COL PCT		21.90	10.84	20.59	
	TOT PCT		7.93	3.10	7.24	
OTHER	COUNT	8.	14	11	12	37
	ROW PCT		37.84	29.73	32.43	12.76
	COL PCT		13.33	13.25	11.76	
	TOT PCT		4.83	3.79	4.14	
	COLUMN		105	83	102	290
	TOTAL		36.21	28.62	35.17	100.00

CHI SQUARE = 49.76840 WITH 14 DEG OF FREEDOM SIGNIFICANCE = .00001

We can informally compare the observed and expected frequencies in order to obtain an idea of how the dependence between perception of administrative expenses and occupation comes about. As was pointed out above, actual administrative costs are less than 9%. Thus 36% of the respondents have an accurate perception of United Way administrative expenses. Note that 35% have a *very* inaccurate perception (21% and above); In this group, production-line employees, clerical, sales, and professional-

Poster Child for the annual United Way campaign

technical employees have more inaccurate perceptions than other groups. Certainly, other types of general observations such as this can be made.

MAIN FINDINGS

The study described and the resulting statistical analysis enabled conclusions to be drawn regarding the perceptions of Rochester people on several important issues. Some of the major findings include the following:

1. Perceptions of the United Way vary quite dramatically from company to company.
2. The United Way has the highest percentage of contributors of the charities studied. Percentage contribution varies significantly by occupation and education.
3. The perception of the funds used for administrative expenses varies significantly by occupation and education.
4. The most important considerations in deciding the size of a gift are financial constraints.
5. In-work solicitation is by far the most preferred method of collection.
6. Only 11.4% of the respondents do not feel well enough informed to make a proper United Way giving decision. This varies significantly by occupation.
7. Perceptions about the United Way in general vary significantly by occupation.

The general conclusions developed in this study were instrumental in defining adjustments to the United Way of Greater Rochester. The result has been improved communications and campaign efforts.

13 Experimental Design and Analysis of Variance

What you will learn in this chapter:

- how to use analysis of variance to test for the equality of the means of three or more populations
- how the F distribution is used in analysis of variance
- what an analysis of variance table is and how it is constructed
- the assumptions of analysis of variance
- how to make computations and conclusions for completely randomized, randomized block, and factorial experimental designs

Contents

In Chapter 10 we discussed how to test whether or not the means of two populations are equal. Recall that the test involved the selection of an independent random sample from each of the populations. In this chapter we present a statistical procedure for determining whether or not the means of more than two populations are equal. This technique is called the analysis of variance (ANOVA) procedure. We also discuss the process of designing an experiment that results in the collection of data; this process is referred to as *experimental design.*

In order to introduce the concepts of experimental design and analysis of variance, let us consider the situation facing Dr. Edward A. Johnson, Dean of the College of Business at a major Texas university.

13.1 THE DR. JOHNSON GMAT PROBLEM

The Graduate Management Admissions Test (GMAT) is a standardized test used by graduate schools of business to evaluate an applicant's ability to successfully pursue a graduate program in that field. Scores on the GMAT range from 200 to 800, with higher scores implying a higher aptitude.

In an attempt to improve the performance of his undergraduate students on the exam, Dr. Johnson is considering the use of a GMAT preparation program. The following three preparation programs have been proposed:

1. A 3 hour review session covering the types of questions generally asked on the GMAT.
2. A 1 day program covering relevant exam material, along with the taking and grading of a sample exam.
3. An intensive 10 week course involving the identification of each student's weaknesses and the setting up of individualized programs for improvement.

Before making a final decision as to the preparation program to adopt, Dr. Johnson has requested that further study be conducted in order to determine how the proposed programs affect GMAT scores received by the students.

Experimental Design Considerations

In the Dr. Johnson GMAT study or experiment the GMAT preparation program is referred to as a *factor.* Since there are three preparation programs, or *levels,* corresponding to this factor, we say that there are three *treatments* associated with the experiment: one treatment corresponds to the 3 hour review, another to the 1 day program, and the third to the 10 week course. In general, a treatment corresponds to a level of a factor.*

The three GMAT preparation programs or treatments define the three populations of interest in this experiment. One population corresponds to all students who take the 3 hour review session, another corresponds to those who take the 1 day program, and the third to those who take the 10 week course. Note that for each population the random variable of interest is the GMAT score, and our primary statistical objective for this experiment is to determine whether or not the mean

*The term treatment was originally used in experimental design because many of the applications were in agriculture, where the treatments often corresponded to different types of fertilizers applied to selected agricultural plots. Today the term is used in a more general context.

GMAT score is the same for all three populations. In experimental design terminology the random variable of interest is referred to as the *dependent variable,* the *response variable,* or simply the *response.*

Now that we have defined the response, factor, and treatments for the GMAT experiment, let us turn to the method of assigning the treatments. To begin with, let us assume that a random sample of three students has been selected from the population of seniors considering attending graduate school. In experimental design terminology, the students are referred to as the *experimental units.*

The experimental design that we will use for the Dr. Johnson GMAT experiment is referred to as a *completely randomized design.* This type of design requires that each of the three preparation programs or treatments be randomly assigned to one of the experimental units or students. For example, the 3 hour review session might be randomly assigned to the second student, the 1 day program to the first student, and the 10 week course to the third student. The concept of randomization as illustrated in this example is an important principle of all experimental designs.

Note that our experiment as described above would only result in one measurement or GMAT score for each treatment. In other words we have a sample size of 1 corresponding to each treatment. Thus to obtain additional data for each preparation program we must repeat or *replicate* our basic experimental process. For example, suppose that instead of selecting just 3 students at random we had selected 15 and then randomly assigned *each* of the three treatments to 5 of the students. Since each program is assigned 5 students, we say that five replicates have been obtained. The concept of *replication* is another important principle of experimental design. Figure 13.1 shows the completely randomized design for the Dr. Johnson GMAT study.

Data Collection

Once we are satisfied with the experimental design, we proceed by carrying out the study and collecting the data. In this case the students would prepare for the GMAT exam under their assigned preparation program. The students would then take the exam and have their GMAT scores recorded. Let us assume that this has been done and that the GMAT scores for the 15 students in the study are as shown in Table 13.1. Using these data we calculate the sample mean GMAT score for each of the three preparation programs, as shown below:

Type of Program	*Mean GMAT Score*
3 hour review	509
1 day program	526
10 week course	552

From these data it appears that the 10 week course may result in higher GMAT scores than either of the other methods. However, before we make any final recommendations we must remember that each of the above means is based upon the test results of just five students. Thus we are looking at three sample means drawn from the three populations representing all students who might participate in these programs prior to taking the GMAT. The real issue, then, is whether or not the three sample means observed are different enough for us to conclude that the means of the populations

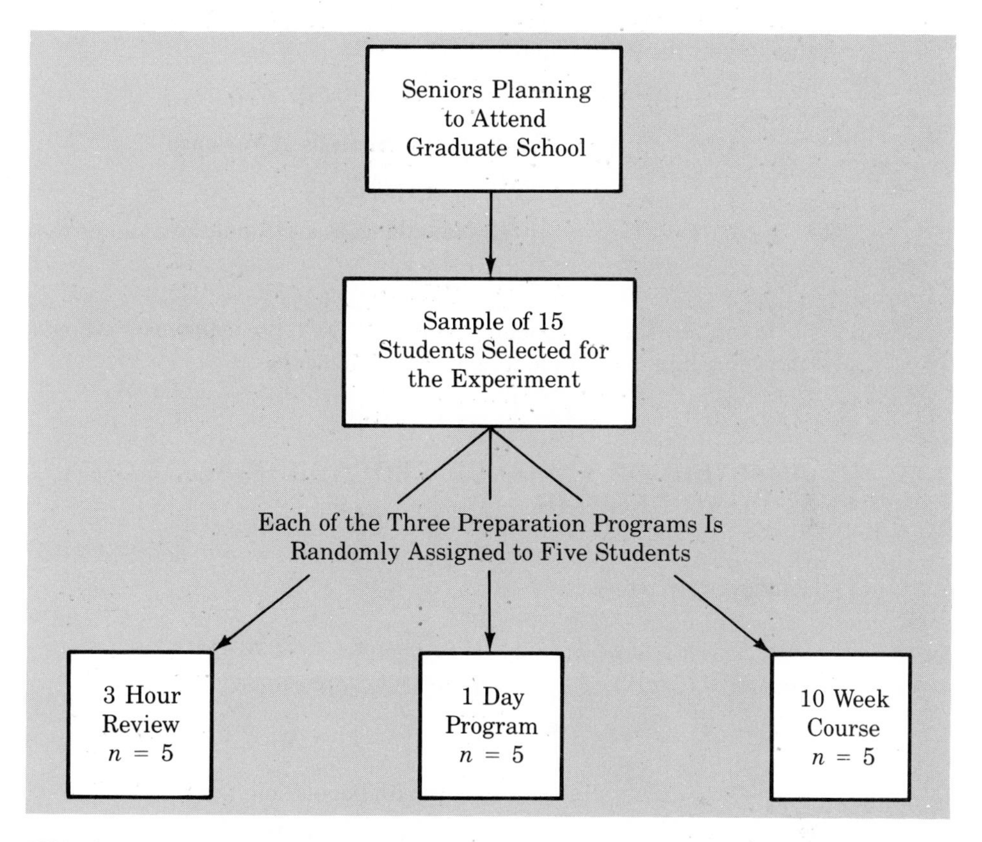

FIGURE 13.1 Completely Randomized Design for Evaluating the GMAT Preparation Programs

corresponding to the three preparation programs are different. To fully develop this question in statistical terms we introduce the following notation:

μ_1 = mean GMAT score for the population of all students who take the 3 hour review session,

μ_2 = mean GMAT score for the population of all students who take the 1 day program,

μ_3 = mean GMAT score for the population of all students who take the 10 week course.

TABLE 13.1 GMAT Scores for 15 Students

		Observation				
Treatment	3 Hour Review	491	579	451	521	503
	1 Day Program	588	502	550	520	470
	10 Week Course	533	628	501	537	561

We use $\bar{x}_1$, $\bar{x}_2$, and $\bar{x}_3$ to denote the corresponding sample means. The experimental results for the three programs yielded $\bar{x}_1 = 509$, $\bar{x}_2 = 526$, and $\bar{x}_3 = 552$. Although we will never know the actual values of μ_1, μ_2, and μ_3, what we want to do is use the sample values to test the following hypotheses:

Hypotheses for Analysis of Variance

$$H_0: \mu_1 = \mu_2 = \mu_3,$$
$$H_1: \text{Not all means are equal.}$$

In the next section we show how the ANOVA procedure can be used to determine if there is a difference in the three population means.

13.2 THE ANALYSIS OF VARIANCE PROCEDURE FOR COMPLETELY RANDOMIZED DESIGNS

The analysis of variance (ANOVA) procedure that we will describe is designed to test the following hypotheses:

$$H_0: \mu_1 = \mu_2 = \cdot \cdot \cdot = \mu_k,$$
$$H_1: \text{Not all } \mu_i \text{ are equal,}$$

where

μ_i = mean of the ith population

k = number of populations or treatments.

We assume that a simple random sample of size n has been selected from each of the k populations and that the sample means $\bar{x}_1, \bar{x}_2, \ldots, \bar{x}_k$ have been computed. In general, we will refer to the sample mean corresponding to the ith population as $\bar{x}_i$ and the overall sample mean as $\bar{\bar{x}}$. That is,

$$\bar{x}_i = \frac{\sum_j x_{ij}}{n}, \tag{13.1}$$

$$\bar{\bar{x}} = \frac{\sum_i \sum_j x_{ij}}{n_T}, \tag{13.2}$$

where

x_{ij} = jth observation corresponding to the ith treatment,

n_T = total sample size for the experiment.

Note that since the sample size is n for each of the k treatments, $n_T = kn$. For the Dr. Johnson GMAT experiment, where $k = 3$ and $n = 5$, $n_T = (3)(5) = 15$.

Assumptions for Analysis of Variance

The ANOVA procedure is based upon the following two assumptions:

1. The variable of interest for each population has a normal probability distribution. Example: In the GMAT situation this assumption would require that the variable of interest, GMAT score, be normally distributed for each of the three programs under study.
2. The variance associated with the variable must be the same for each population. Example: In the GMAT situation this assumption would require that the variance of GMAT scores be the same for students participating in each of the three programs. We denote the common population variance of the variable of interest as σ^2.

The logic behind the ANOVA procedure is based upon being able to develop two independent estimates of σ^2. One estimate is based upon the differences *between* the treatment means and the overall sample mean, and the other estimate is based upon the differences of observations *within* each treatment from the corresponding treatment mean. By comparing these two estimates of σ^2 we will be able to answer the question about whether or not the population means are equal.

Between-Treatments Estimate of Population Variance

The procedure for developing an estimate of σ^2 based upon the differences between the treatment means and the overall sample mean depends upon the assumption that the null hypothesis is true and that the two assumptions listed above are valid. In this case, if we let μ denote the common population mean (that is, $\mu_1 = \mu_2 = \cdots = \mu_k = \mu$), then all of the sample observations would represent data values drawn from the same normal probability distribution with mean μ and variance σ^2. If we let $\bar{x}$ denote the mean of a simple random sample of size n selected from this probability distribution, then the sampling distribution of $\bar{x}$ is normally distributed with mean μ and variance $\sigma_{\bar{x}}^2 = \sigma^2/n$. Figure 13.2 illustrates such a sampling distribution. Thus, under the null hypothesis, we can think of each $\bar{x}_i$ as a value drawn at random from this sampling distribution.

An estimate of the mean μ of this sampling distribution can be obtained by computing $\bar{\bar{x}}$, the overall sample mean. For the GMAT scores in Table 13.1 we use all the data to find

$$\sum_i \sum_j x_{ij} = (491 + 579 + 451 + \cdots + 537 + 561) = 7{,}935.$$

Then using (13.2) with $n_T = 15$ we compute an overall sample mean

$$\bar{\bar{x}} = \frac{7{,}935}{15} = 529$$

as the estimate of μ.

To estimate the variance of the sampling distribution of $\bar{x}$ (that is, $\sigma_{\bar{x}}^2$), we can use the variance of the individual sample means about the overall sample mean $\bar{\bar{x}}$. For the

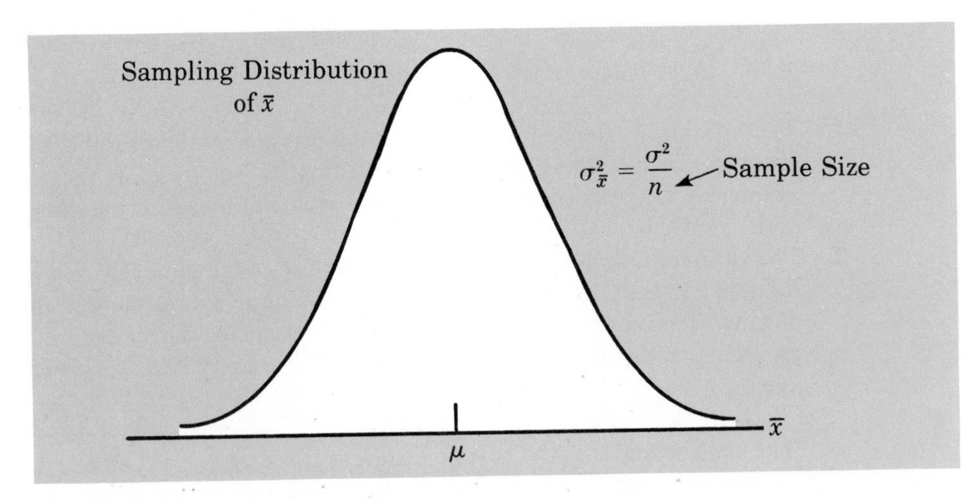

FIGURE 13.2 Sampling Distribution of $\bar{x}$ Given the Null Hypothesis $\mu_1 = \mu_2 \cdots = \mu_k = \mu$ and a Sample of Size n

case where each sample is the same size, the estimated variance, denoted $s_{\bar{x}}^2$, is

$$s_{\bar{x}}^2 = \frac{\sum_i^k (\bar{x}_i - \bar{\bar{x}})^2}{k - 1}, \tag{13.3}$$

where $\bar{x}_i$ = mean of the ith sample.

Since our previous computations of the three sample means showed $\bar{x}_1 = 509$, $\bar{x}_2 = 526$, and $\bar{x}_3 = 552$, we have

$$\begin{aligned}\sum_i (\bar{x}_i - \bar{\bar{x}})^2 &= (509–529)^2 + (526–529)^2 + (552–529)^2 \\ &= 400 + 9 + 529 = 938.\end{aligned}$$

With $k = 3$ populations or treatments, we have $k - 1 = 2$. Thus with (13.3) the estimate of the variance of the sampling distribution becomes

$$s_{\bar{x}}^2 = \frac{938}{2} = 469.$$

Since we know $\sigma_{\bar{x}}^2 = \sigma^2/n$, solving for σ^2 gives

$$\sigma^2 = n\,\sigma_{\bar{x}}^2, \tag{13.4}$$

where n is the sample size involved in computing $\bar{x}$. Hence an estimate of the population variance σ^2 can be obtained from (13.4) by multiplying n by the estimate of $\sigma_{\bar{x}}^2$:

$$\text{Estimate of } \sigma^2 = n\,(\text{Estimate of } \sigma_{\bar{x}}^2) = ns_{\bar{x}}^2. \tag{13.5}$$

With all samples of size $n = 5$ and with $s_{\bar{x}}^2 = 469$, for the GMAT problem we have

$$\text{Estimate of } \sigma^2 = 5\,(469) = 2{,}345.$$

This estimate of σ^2 is given the name *mean square between treatments* and is denoted MSTR.

From (13.3) and (13.5) we see that MSTR can be written

$$\text{MSTR} = \frac{n \sum_i (\bar{x}_i - \bar{\bar{x}})^2}{k - 1}. \tag{13.6}$$

The numerator of (13.6) is called the *sum of squares between treatments* and is denoted SSTR. The denominator, $k - 1$, represents the *degrees of freedom* corresponding to SSTR. Thus

$$\text{MSTR} = \frac{\text{SSTR}}{k - 1}. \tag{13.7}$$

Using our previous calculations we have the following values for SSTR, degrees of freedom, and MSTR for the GMAT problem:

$$\text{SSTR} = n \sum_i (\bar{x}_i - \bar{\bar{x}})^2 = 5\,(938) = 4{,}690, \qquad k - 1 = 3 - 1 = 2,$$

and thus

$$\text{MSTR} = \frac{\text{SSTR}}{k - 1} = \frac{4{,}690}{2} = 2{,}345.$$

Within-Treatments Estimate of Population Variance

We now develop a second estimate of σ^2 that is not based on the null hypothesis assumption that the population means are equal. It is instead based upon the variation of the sample observations "within" each treatment and is called the mean square due to error, denoted MSE. This estimate is also referred to as the *mean square within treatments.*

If each of the k samples is a simple random sample, then each of the k sample variances provides an estimate of σ^2. For each sample the sample variance is computed in the usual fashion:

$$\text{Variance of sample } i = s_i^2 = \frac{\sum_j (x_{ij} - \bar{x}_i)^2}{n - 1}. \tag{13.8}$$

For example, in considering the GMAT scores for the five students taking the 3 hour review, we have $n = 5$ and $\bar{x}_1 = 509$. Using (13.8) the variance estimate from this first sample becomes

$$s_1^2 = \frac{(491\text{–}509)^2 + (579\text{–}509)^2 + (451\text{–}509)^2 + (521\text{–}509)^2 + (503\text{–}509)^2}{5 - 1}$$

$$= \frac{324 + 4{,}900 + 3{,}364 + 144 + 36}{4} = \frac{8{,}768}{4} = 2{,}192.$$

The variance estimates for the two remaining samples can be similarily computed:

$$s_2^2 = \frac{\sum_j (x_{2j} - \bar{x}_2)^2}{n - 1} = \frac{8,168}{4} = 2,042,$$

$$s_3^2 = \frac{\sum_j (x_{3j} - \bar{x}_3)^2}{n - 1} = \frac{9,044}{4} = 2,261.$$

Rather than working with three separate estimates of the population variance we combine or *pool* the results to obtain MSE. To do so we must first compute the sum of the numerators and the sum of the denominators of the three within-treatment variance estimates. Dividing the numerator total by the denominator total provides MSE, the within-treatment estimate of the population variance. Thus for the Dr. Johnson GMAT data we obtain

$$\text{MSE} = \frac{8,768 + 8,168 + 9,044}{4 + 4 + 4} = \frac{25,980}{12} = 2,165.$$

Applying the procedure described above in a general context for the case of k treatments allows us to write MSE as follows:

$$\begin{aligned}\text{MSE} &= \frac{\sum_j (x_{1j} - \bar{x}_1)^2 + \sum_j (x_{2j} - \bar{x}_2)^2 + \cdots + \sum_j (x_{kj} - \bar{x}_k)^2}{(n-1) \quad + \quad (n-1) \quad + \cdots + \quad (n-1)} \\ &= \frac{\sum_i \sum_j (x_{ij} - \bar{x}_i)^2}{kn - k} \\ &= \frac{\sum_i \sum_j (x_{ij} - \bar{x}_i)^2}{n_T - k}. \end{aligned} \tag{13.9}$$

The numerator in the MSE computation is given the name *sum of squares within* or *sum of squares due to error* and is denoted SSE. Thus from the above result we see that SSE = 25,980 for the GMAT problem. The denominator of MSE is referred to as the *degrees of freedom* associated with the within-treatment variance estimate. In the above case the MSE value is based on 12 degrees of freedom, since $n_T = 15$, $k = 3$, and $n_T - k = 15 - 3 = 12$.

In general, we can compute MSE as follows:

$$\text{MSE} = \frac{\text{SSE}}{n_T - k}, \tag{13.10}$$

where the sum of squares due to error is given by

$$\text{SSE} = \sum_i \sum_j (x_{ij} - \bar{x}_i)^2 \tag{13.11}$$

Comparing the Variance Estimates: The *F* Test

We have now developed two estimates of σ^2. The first estimate (MSTR) is based upon the variation between treatment means, and our second estimate (MSE) is based upon the variation within each treatment. Recall that in order to compute MSTR we had to assume that the null hypothesis was true (that is, $\mu_1 = \mu_2 = \cdots = \mu_k$). However, in computing MSE this assumption was not required. In fact, regardless of whether or not the means of the k populations are equal, MSE will always provide an unbiased estimate of σ^2.

It can be shown that MSTR provides an unbiased estimate of σ^2 when H_0 is true. However, if the means of the k populations are not equal (that is, H_1 is true), MSTR is not an unbiased estimate of σ^2. In fact, in this case MSTR overestimates σ^2. This is the key to the analysis of variance procedure: that is, we test the hypotheses

$$H_0\colon \mu_1 = \mu_2 = \cdots = \mu_k,$$
$$H_1\colon \text{Not all } \mu_i \text{ are equal,}$$

by comparing the two estimates of the population variance MSTR and MSE. If MSTR and MSE are approximately equal, such that the ratio MSTR/MSE is near 1, we accept H_0 and conclude the means are equal. However, if MSTR is much larger than MSE, such that the ratio MSTR/MSE is much larger than 1, we reject H_0 and conclude that the means are not all equal.

To obtain a better intuitive feel for why larger values of MSTR are obtained when the null hypothesis is false, recall the formula for computing MSTR:

$$\text{MSTR} = \frac{\text{SSTR}}{k-1} = \frac{n\Sigma(\bar{x}_i - \bar{\bar{x}})^2}{k-1}. \tag{13.12}$$

Note that the numerator (SSTR) is based on the dispersion of the k sample means, $\bar{x}_i$'s, around the overall sample mean $\bar{\bar{x}}$. Now consider the two situations shown in Figure 13.3. In part A we have the sampling distribution of $\bar{x}$ under the assumption that H_0 is true with $\mu_1 = \mu_2 = \mu_3 = \mu$. In this case each sample comes from the same population, and there is only one sampling distribution. Although the three sample means $\bar{x}_1$, $\bar{x}_2$, and $\bar{x}_3$ are not the same, they are "close" to one another. In this case MSTR as computed in (13.12) will provide an unbiased estimate of σ^2.

The second diagram, part B of Figure 13.3, shows a situation where H_0 is false and the three means μ_1, μ_2, and μ_3 are not the same. What happens to the value of SSTR in this case? Note that since the sample means $\bar{x}_1$, $\bar{x}_2$, and $\bar{x}_3$ are coming from distributions with different μ's, they have different sampling distributions and show a much larger dispersion than when H_0 is true. In this case the value of SSTR in (13.12) will be larger, causing the value of MSTR to overestimate the population variance σ^2.

Let us assume for the moment that the null hypothesis is true and that $\mu_1 = \mu_2 = \cdots = \mu_k$. In this case it can be shown that MSTR and MSE provide two independent unbiased estimates of σ^2. Recall from Chapter 11 that the sampling distribution of the ratio of two independent estimates of σ^2 for a normal population follows an F probability distribution. Thus if the null hypothesis is true and the ANOVA assumptions are valid, the sampling distribution of MSTR/MSE is an F distribution with numerator degrees of freedom equal to $k - 1$ and denominator degrees of freedom equal to $n_T - k$.

On the other hand, if the means of the k populations are not all equal, the value of

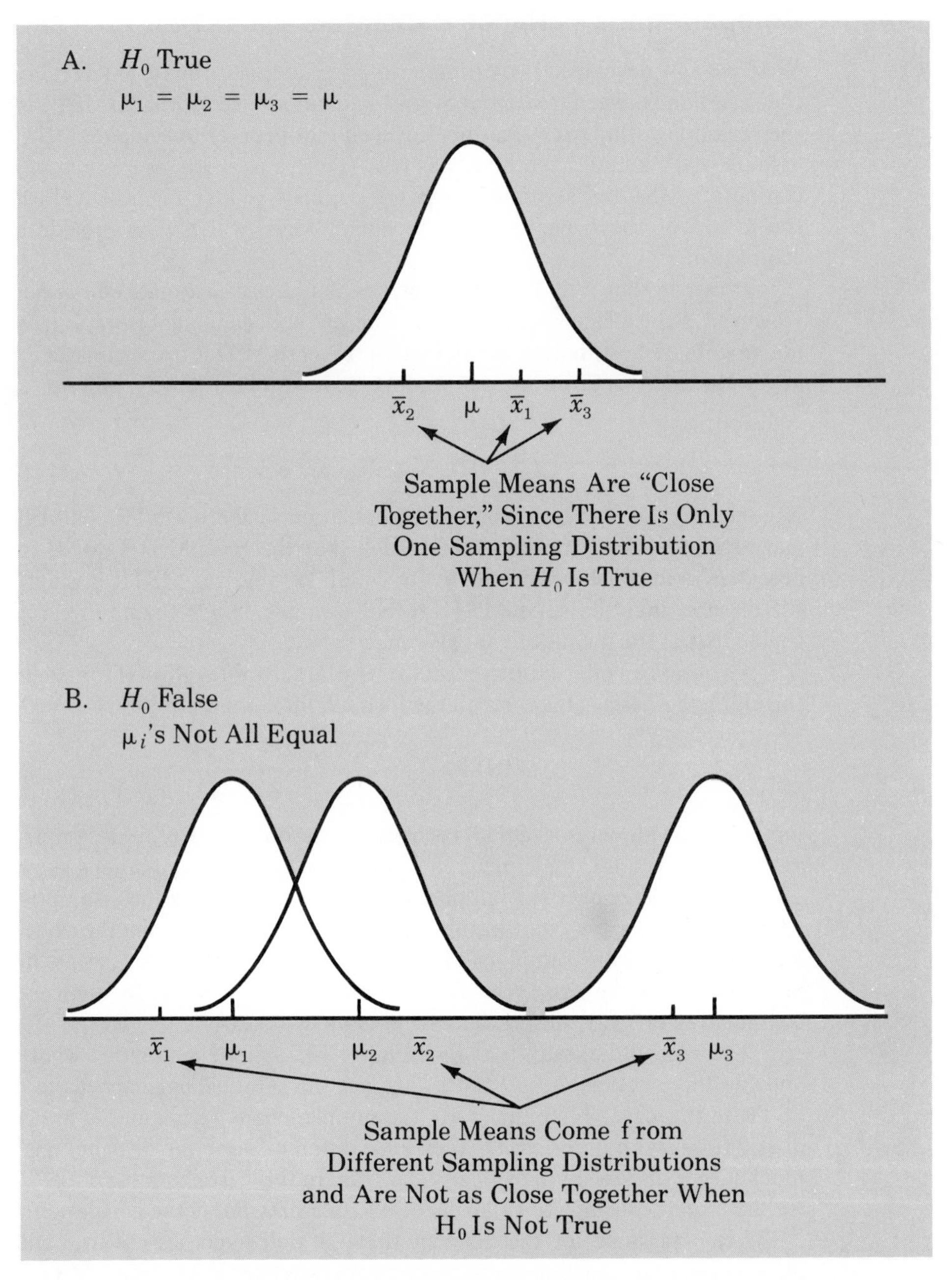

FIGURE 13.3 **Examples of the Sampling Distributions of $\bar{x}$ for the Cases of H_0 True (Part A) and H_0 False (Part B)**

MSTR/MSE will be inflated because MSTR overestimates σ^2. Hence we will reject H_0 if the resulting value of MSTR/MSE appears to be too "large" to have been selected at random from an F distribution with degrees of freedom $k - 1$ in the numerator and $n_T - k$ in the denominator. The value of F that will cause us to reject H_0 depends upon α, the level of significance. Once α is selected a critical value of F can be determined. Figure 13.4 shows the sampling distribution of MSTR/MSE and the

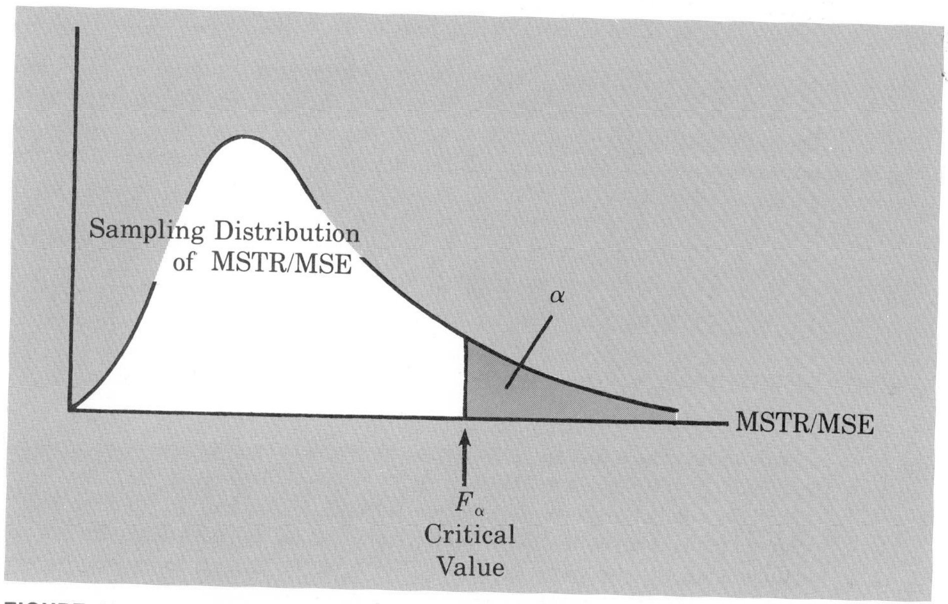

FIGURE 13.4 Sampling Distribution of MSTR/MSE; The Critical Value for Rejecting the Null Hypothesis of Equality of Means Is F_α

rejection region associated with a level of significance equal to α. Note that F_α denotes the critical value.

Let us finish the ANOVA procedure for our GMAT problem. Assume that Dr. Johnson was willing to accept a Type I error of $\alpha = .05$. From Table 4 of Appendix B we can determine the critical value by locating the value corresponding to numerator degrees of freedom equal to $k - 1 = 3 - 1 = 2$ and denominator degrees of freedom equal to $n_T - k = 15 - 3 = 12$. Thus we obtain the value $F_{.05} = 3.89$. Hence the appropriate decision rule for our GMAT problem is written

$$\text{If MSTR/MSE} \leq 3.89 \text{ accept } H_0,$$
$$\text{If MSTR/MSE} > 3.89 \text{ reject } H_0,$$

where

$$H_0\colon \mu_1 = \mu_2 = \mu_3,$$
$$H_1\colon \text{Not all } \mu_i \text{ are equal.}$$

Since MSTR/MSE $= 2{,}345/2{,}165 = 1.08$ is less than the critical value $F_{.05} = 3.89$, there is not sufficient statistical evidence to reject the null hypothesis that the means of the three programs are the same.

Although the three sample means observed were different, there is not sufficient evidence to conclude that a statistically significant difference exists in the three population means at the .05 level. Actually, this is not too surprising considering the relatively few observations that were available and the size of the variance that is associated with GMAT scores actually observed. If Dr. Johnson believed that additional investigation was still warranted, any further experiments should seek to increase the number of observations available in each sample.

EXERCISES

1. The Jacobs Chemical Company wants to estimate the mean time (minutes) required to mix a batch of material on machines produced by three different manufacturers. In order to limit the cost of testing, four batches of material were mixed on machines produced by each of the three manufacturers. The time needed to mix the material was recorded. The times in minutes are shown below:

Manufacturer 1	*Manufacturer 2*	*Manufacturer 3*
20	28	20
26	26	19
24	31	23
22	27	22

Use these data and test to see if the mean time needed to mix a batch of material is the same for each manufacturer. Use $\alpha = .05$.

2. Four different paints are advertised as having the same drying time. In order to check the manufacturers' claims, five paint samples were tested for each make of paint. The time in minutes until the paint was dry enough for a second coat to be applied was recorded. The following data were obtained:

Paint 1	*Paint 2*	*Paint 3*	*Paint 4*
128	144	133	150
137	133	143	142
135	142	137	135
124	146	136	140
141	130	131	153

At the $\alpha = .05$ level of significance test to see if the mean drying time is the same for each type of paint.

3. Three different brands of radial tires were compared for wear characteristics. For each brand of tire ten tires were randomly selected and subjected to standard wear-testing procedures. The average mileage obtained for each brand of tire and the sample standard deviations are shown below:

	Brand A	*Brand B*	*Brand C*
Average mileage	36,400	38,200	33,100
Sample standard deviation	1,650	1,800	1,500

Use these data and test to see if the mean mileage for all three brands of tires is the same. Use $\alpha = .05$.

4. Three top-of-the-line intermediate sized automobiles manufactured in the United States have been test driven and compared on a variety of criteria by a well known automotive magazine. In the area of gasoline mileage performance, five automobiles of each brand were each test driven 500 miles; the miles per gallon data obtained are shown below:

	Miles per Gallon Data				
Automobile A	19	21	20	19	21
Automobile B	19	20	22	21	23
Automobile C	24	26	23	25	27

Use the analysis of variance procedure with $\alpha = .05$ to determine if there is a significant difference in the mean miles per gallon for the three types of automobiles.

13.3 THE ANOVA TABLE AND OTHER CONSIDERATIONS

A convenient way to summarize the computations and results of the analysis of variance procedure involves the development of an analysis of variance (ANOVA) table. Before presenting the ANOVA table for the Dr. Johnson GMAT study, however, let us perform some more calculations with the data collected (see Table 13.1). Treating the entire data set as one sample of 15 observations, we have already found the overall sample mean to be $\bar{\bar{x}} = 529$. Using the entire data set as one sample, let us now compute an overall sample variance:

$$s^2 = \frac{\sum_i \sum_j (x_{ij} - \bar{\bar{x}})^2}{n_T - 1}. \qquad (13.13)$$

The numerator of (13.13) is called the total sum of squares about the mean (SST), and the denominator represents the degrees of freedom associated with this total sum of squares. Check for yourself that the 15 values in Table 13.1 provide the following value for SST:

$$\text{SST} = \sum_i \sum_j (x_{ij} - \bar{\bar{x}})^2$$
$$= (491\text{-}529)^2 + (579\text{-}529)^2 + \cdots + (537\text{-}529)^2 + (561\text{-}529)^2 = 30{,}670.$$

We can now develop the ANOVA table for a completely randomized design. Table 13.2 shows a general ANOVA table, and Table 13.3 shows the ANOVA table for the Dr. Johnson GMAT experiment.

TABLE 13.2 ANOVA Table for a Completely Randomized Design

Source of Variation	*Sum of Squares*	*Degrees of Freedom*	*Mean Square*	*F*
Between Treatments	SSTR	$k - 1$	$\text{MSTR} = \dfrac{\text{SSTR}}{k-1}$	$\dfrac{\text{MSTR}}{\text{MSE}}$
Error (within treatments)	SSE	$n_T - k$	$\text{MSE} = \dfrac{\text{SSE}}{n_T - k}$	
Total	SST	$n_T - 1$		

Note that the rows of the table provide information concerning the two sources of variation: "between treatments," which refers to the between-group variation, and "error," which refers to the within-group variation. The "Sum of Squares" column and the "Degrees of Freedom" column provide the corresponding values as defined in the previous section. The column labeled "Mean Square" is simply the sum of squares divided by the corresponding degrees of freedom. Recall that in Section 13.2 we showed that when we divided the sum of squares by the corresponding degrees of freedom we obtained two estimates of the population variance (MSTR and MSE). The ANOVA table simply summarizes these values in the "Mean Square" column.

TABLE 13.3 ANOVA Table for the Dr. Johnson GMAT Experiment

Source of Variation	*Sum of Squares*	*Degrees of Freedom*	*Mean Square*	*F*
Between treatments	4,690	2	$\frac{4{,}690}{2} = 2{,}345$	$\frac{2{,}345}{2{,}165} = 1.08$
Error (within treatments)	25,980	12	$\frac{25{,}980}{12} = 2{,}165$	
Total	30,670	14		

Finally, the last column in the table contains the F value corresponding to MSTR/MSE. Since the variance estimates are found in the "Mean Square" column, the F value is computed by dividing the mean square treatments (MSTR) by the mean square error (MSE). That is, F = MSTR/MSE.

What observation can you make from the "Sum of Squares" columns in Tables 13.2 and 13.3? Note in particular that the following condition holds:

$$\text{SST} = \text{SSTR} + \text{SSE} \tag{13.14}$$

or

$$\sum_i \sum_j (x_{ij} - \bar{\bar{x}})^2 = n \sum_i (\bar{x}_i - \bar{\bar{x}})^2 + \sum_i \sum_j (x_{ij} - \bar{x}_i)^2. \tag{13.15}$$

Thus we see that SST can be partitioned into two sums—one called the sum of squares between treatments and one called the sum of squares due to error. This is known as *partitioning* the sum of squares.

Note also that the degrees of freedom corresponding to SST ($n_T - 1 = 14$) can be partitioned into the degrees of freedom corresponding to SSTR ($k - 1 = 2$) and the degrees of freedom corresponding to SSE ($n_T - k = 12$). The analysis of variance procedure can be viewed as the process of partitioning the sum of squares and degrees of freedom into their corresponding sources: treatments and error. Dividing the sum of squares by the appropriate degrees of freedom provides the variance estimates and the F value used to test the hypothesis of equal population or treatment means.

Unbalanced Designs

The formula presented previously for computing SSTR applies only to *balanced designs.* Balanced designs are ones in which the sample size is the same for each treatment. Any experimental design for which the sample size is not the same for each treatment is said to be *unbalanced.* Although we would prefer a balanced design,* in some cases we must work with unequal sample sizes. For unbalanced designs the analysis of variance procedure described above may be used with the following modification of the formula for SSTR:

$$\text{SSTR} = \sum_i n_i (\bar{x}_i - \bar{\bar{x}})^2, \tag{13.16}$$

*With a balanced design the F statistic is less sensitive to small departures from the assumption that σ^2 is the same for each population, and choosing equal sample sizes maximizes the power of the test.

where

$$n_i = \text{sample size for the } i\text{th treatment.}$$

In such cases

$$n_T = n_1 + n_2 + \cdots + n_k$$
$$= \sum_i n_i.$$

Note that (13.16) yields the same formula we used earlier when the sample size is the same for each treatment, i.e. $n_i = n$.

Easing the Computational Burden

The computational aspect of the analysis variance procedure is devoted primarily to computing the appropriate sums of squares. Once the sums of squares are determined, the other computations in the ANOVA table are rather straightforward. The following step-by-step procedure is designed to ease the burden in computing the appropriate sums of squares. The formulas shown below can be applied to both balanced and unbalanced designs.

Computational Steps for Computing the Sums of Squares for a Completely Randomized Design

x_{ij} = value of the jth observation under treatment i,
T_i = sum of all observations in treatment i,
T = sum of all observations,
n_i = sample size for the ith treatment,
n_T = total sample size for the experiment.

Step 1: Compute the sum of squares about the mean (SST):

$$\text{SST} = \sum_i \sum_j (x_{ij} - \bar{\bar{x}})^2 = \sum_i \sum_j x_{ij}^2 - \frac{T^2}{n_T}. \qquad (13.17)$$

Step 2: Compute the sum of squares due to treatments (SSTR):

$$\text{SSTR} = \sum_i n_i (\bar{x}_i - \bar{\bar{x}})^2 = \sum_i \frac{T_i^2}{n_i} - \frac{T^2}{n_T}. \qquad (13.18)$$

Step 3: Compute the sum of squares due to error (SSE):

$$\text{SSE} = \text{SST} - \text{SSTR}. \qquad (13.19)$$

Using this computational procedure with the GMAT data in Table 13.1 we obtain the following results:

Step 1: $\quad \text{SST} = 4{,}228{,}285 - \dfrac{(7{,}935)^2}{15} = 30{,}670.$

Step 2: $$\text{SSTR} = \frac{(2{,}545)^2}{5} + \frac{(2{,}630)^2}{5} + \frac{(2{,}760)^2}{5} - \frac{(7{,}935)^2}{15} = 4{,}690.$$

Step 3: $$\text{SSE} = \text{SST} - \text{SSTR} = 30{,}670 - 4{,}690 = 25{,}980.$$

Computer Results for Analysis of Variance

As you can see, the computation of the sums of squares as required by the analysis of variance procedure can be quite a job even when a hand calculator is used in conjunction with the step-by-step procedure. Furthermore, the difficulty increases as the sample size and/or the number of treatments increases. Because of the computational burden, computer software packages are often used for ANOVA computations. Shown in Figure 13.5 is the output obtained from the MINITAB computer package for the GMAT problem.

ANALYSIS OF VARIANCE

DUE TO	DF	SS	MS=SS/DF	F-RATIO
FACTOR	2	4690.	2345.	1.08
ERROR	12	25980.	2165.	
TOTAL	14	30670.		

LEVEL	N	MEAN
1	5	509.0
2	5	526.0
3	5	552.0

FIGURE 13.5 Computer Output for GMAT Problem

We see that the computer output contains the familiar ANOVA table format. It should prove easy for you to interpret. Comparing Figure 13.5 with Table 13.3, we see that the same information is available, although some of the headings are a little different. "DUE TO" is the heading used for the source of variation column, "FACTOR" identifies the between-treatments row, and the sum of squares and degrees of freedom columns are interchanged. Note also that below the ANOVA table the computer output contains all of the sample means and respective sample sizes. The "F-RATIO" column contains the F statistic used for the hypothesis test.

EXERCISES

5. Solve Exercise 1 again, this time using the computational step-by-step procedure described in this section. Summarize your computations and results by setting up the ANOVA table.

6. Solve Exercise 4 again by using the computational step-by-step procedure. Set up the ANOVA table for this problem.

7. Three different methods for assembling a product were proposed by an industrial engineer. To investigate the number of units assembled correctly using each method, 30 employees were randomly selected and randomly assigned to the three proposed methods, such that ten workers were associated with each method. The number of units assembled correctly was recorded, and the analysis of variance procedure was applied to the resulting data set. The following results were obtained: SST = 10,800, SSTR = 4,560.

a. Set up the ANOVA table for this problem.

b. Using $\alpha = .05$, test for any significant difference in the means for the three assembly methods.

8. In an experiment designed to test the breaking strength of four types of cables, the following results were obtained: SST = 85.05, SSTR = 61.64, $n_T = 24$. Set up the ANOVA table and test for any significant difference in the mean breaking strength of the four cables. Use $\alpha = .05$.

9. To test for any significant difference in the time between breakdowns for four machines, the following data were obtained:

	Time (hours)					
Machine 1	6.4	7.8	5.3	7.4	8.4	7.3
Machine 2	8.7	7.4	9.4	10.1	9.2	9.8
Machine 3	11.1	10.3	9.7	10.3	9.2	8.8
Machine 4	9.9	12.8	12.1	10.8	11.3	11.5

Use the computational step-by-step procedure to develop the ANOVA table for this problem. At the $\alpha = .05$ level of significance, is there any difference in the mean time between breakdowns among the four machines?

13.4 RANDOMIZED BLOCK DESIGN

Thus far we have considered the completely randomized experimental design. Recall that in order to test for a difference in means we computed an F value using the ratio

$$F = \frac{\text{MSTR}}{\text{MSE}}. \tag{13.20}$$

A problem can arise whenever differences due to extraneous factors (ones not considered) cause the MSE term in this ratio to become large. In such cases the F value (13.20) can become small, signaling no difference between treatment means when in fact such a difference exists.

In this section we present an experimental design referred to as a *randomized block design.* The purpose of this design is to control some of the extraneous sources of variation by removing such variation from the MSE term. This design tends to provide a better estimate of the true error variance, and leads to a more powerful hypothesis test in terms of the ability to detect differences between treatment means. To illustrate, let us consider a stress study for air traffic controllers.

Air Traffic Controllers Stress Test

A study directed at measuring the fatigue and stress on air traffic controllers has resulted in proposals for modification and redesign of the controller's work station. After consideration of several designs for the work station, three specific alternatives have been selected as having the best potential for reducing controller stress. The key question is, "To what extent do the three alternatives differ in terms of their effect on controller stress?" To answer this question we need to design an experiment that will provide measurements of air traffic controller stress under each alternative.

In a completely randomized design a random sample of controllers would be assigned to each work station alternative. However, it is believed that controllers differ substantially in terms of their ability to handle stressful situations. What is high stress to one controller might be only moderate or even low stress to another. Thus when considering the within-group source of variation (MSE) we must realize that this variation includes both random error and error due to individual controller differences.

In fact, for this study management expected controller variability to be a major contributor to the MSE term.

One way to separate the effect of the individual differences is to use the randomized block design. This design will identify the variability stemming from individual controller differences and remove it from the MSE term. The randomized block design calls for a single sample of controllers. Each controller in the sample is tested using each of the three work station alternatives. In experimental design terminology, the work station is the *factor of interest,* and the controllers are referred to as the *blocks.* The three treatments or populations associated with the work station factor correspond to the three work station alternatives. For simplicity, we will refer to these alternatives as System A, System B, and System C.

The *randomized* aspect of the randomized block design refers to the fact that the order in which the treatments (systems) are assigned to the controllers is chosen randomly. If every controller were to test the three systems in the same order, any observed difference in systems might be due to the order of the test rather than to true differences in the systems.

To provide the necessary data, the three types of work stations were installed at the Cleveland Control Center in Oberlin, Ohio. Six controllers were selected at random and assigned to operate each of the systems. A followup interview and a medical examination of each controller participating in the study provided a measure of the stress for each controller on each system. The data are shown in Table 13.4.

A summary of the stress data collected is shown in Table 13.5. In this table we have included column totals (blocks) and row totals (treatments) as well as some other sums which will be helpful in making the sum of squares computations for the ANOVA procedure. Since lower stress values are viewed as better, the sample data available would seem to favor System B with its mean stress rating of 13. However, the usual question remains: do the sample results justify the conclusion that the mean stress levels for the three systems differ? That is, are the differences statistically significant? An analysis of variance computation similar to the one performed for the completely randomized design can be used to answer this statistical question.

The Analysis of Variance Procedure for a Randomized Block Design

The ANOVA procedure for the randomized block design requires us to partition the sum of squares total (SST) into three groups: sum of squares between treatments, sum of squares due to blocks, and sum of squares due to error. The formula for this partitioning is as follows:

$$\underset{\substack{\uparrow \\ \text{Sum of squares} \\ \text{total}}}{\text{SST}} = \underset{\substack{\uparrow \\ \text{treatments}}}{\text{SSTR}} + \underset{\substack{\uparrow \\ \text{blocks}}}{\text{SSB}} + \underset{\substack{\uparrow \\ \text{error}}}{\text{SSE}} \tag{13.21}$$

This sum of squares partition is summarized in the ANOVA table for the randomized block design as shown in Table 13.6. The notation used in this table is as follows.

k = the number of treatments,

b = the number of blocks,

n_T = the total sample size $n_T = kb$.

TABLE 13.4 A Randomized Block Design for the Air Traffic Controllers Stress Test

		Block 1 (Controller 1)	Block 2 (Controller 2)	Block 3 (Controller 3)	Block 4 (Controller 4)	Block 5 (Controller 5)	Block 6 (Controller 6)
Treatments	System A	15	14	10	13	16	13
	System B	15	14	11	12	13	13
	System C	18	14	15	17	16	13

Stress Value (arrow pointing to 18)

TABLE 13.5 Summary of Stress Data for the Air Traffic Controllers Stress Test

		Blocks							
		Controller 1	Controller 2	Controller 3	Controller 4	Controller 5	Controller 6	Row or Treatment Totals	Treatment Means (x_i)
Treatments	System A	15	14	10	13	16	13	81	13.5
	System B	15	14	11	12	13	13	78	13
	System C	18	14	15	17	16	13	93	15.5
Column or block totals		48	42	36	42	45	39	252 ← Overall Sum $\sum_i \sum_j x_{ij}$	

Sum of squares $\sum_i \sum_j x_{ij}^2 = (15)^2 + (14)^2 + (10)^2 + \cdots + (16)^2 + (13)^2 = 3{,}598$

Note that the ANOVA table in Table 13.6 also shows how the $n_T - 1$ total degrees of freedom are partitioned such that $k - 1$ go to treatments, $b - 1$ go to blocks, and $(k - 1)(b - 1)$ go to the error term. The mean square column shows the sum of squares divided by the degrees of freedom, and $F = \text{MSTR}/\text{MSE}$ is the F ratio used to test for a significant difference among the treatment means. The primary contribution of the randomized block design is that by including blocks we have removed the controller differences from the MSE term and obtained a more powerful test for the stress differences in the three work station alternatives.

TABLE 13.6 ANOVA Table for the Randomized Block Design with *k* Treatments and *b* Blocks

Source of Variation	*Sum of Squares*	*Degrees of Freedom*	*Mean Square*	*F*
Between treatments	SSTR	$k - 1$	$\text{MSTR} = \dfrac{\text{SSTR}}{k - 1}$	$\dfrac{\text{MSTR}}{\text{MSE}}$
Between blocks	SSB	$b - 1$	$\text{MSB} = \dfrac{\text{SSB}}{b - 1}$	
Error	SSE	$(k - 1)(b - 1)$	$\text{MSE} = \dfrac{\text{SSE}}{(k - 1)(b - 1)}$	
Total	SST	$n_T - 1$		

Computations and Conclusions

As we saw in the completely randomized design, the primary difficulty was computing the appropriate sum of squares values. The steps for easing this computation for a randomized block design are shown below.

Computational Steps for Computing Sums of Squares for a Randomized Block Design

In addition to k, b, and n_T as previously defined, we use the following notation:

x_{ij} = value of the observation under treatment i in block j,
$T_{i\cdot}$ = the total of all observations in treatment i,
$T_{\cdot j}$ = the total of all observations in block j,
T = the total of all observations,
$\bar{x}_{i\cdot}$ = sample mean of the ith treatment,
$\bar{x}_{\cdot j}$ = sample mean for the jth block,
$\bar{\bar{x}}$ = overall sample mean.

Step 1: Compute the total sum of squares (SST):

$$\text{SST} = \sum_i \sum_j (x_{ij} - \bar{\bar{x}})^2 = \sum_i \sum_j x_{ij}^2 - \frac{T^2}{n_T}. \tag{13.22}$$

Step 2: Compute the sum of squares between treatments (SSTR):

$$\text{SSTR} = \sum_i b\,(\bar{x}_{i.} - \bar{\bar{x}})^2 = \frac{\sum_i T_{i.}^2}{b} - \frac{T^2}{n_T}. \tag{13.23}$$

Step 3: Compute the sum of squares due to blocks (SSB):

$$\text{SSB} = \sum_j k\,(\bar{x}_{.j} - \bar{\bar{x}})^2 = \frac{\sum_j T_{.j}^2}{k} - \frac{T^2}{n_T}. \tag{13.24}$$

Step 4: Compute the sum of squares due to error (SSE):

$$\text{SSE} = \text{SST} - \text{SSTR} - \text{SSB}. \tag{13.25}$$

For the air traffic controller data in Table 13.5, these steps lead to the following sum of squares:

Step 1: $\text{SST} = 3{,}598 - \dfrac{(252)^2}{18} = 70,$

Step 2: $\text{SSTR} = \dfrac{(81)^2 + (78)^2 + (93)^2}{6} - \dfrac{(252)^2}{18} = 21,$

Step 3: $\text{SSB} = \dfrac{(48)^2 + (42)^2 + \cdots + (39)^2}{3} - \dfrac{(252)^2}{18} = 30,$

Step 4: $\text{SSE} = 70 - 21 - 30 = 19.$

These sums of squares divided by their degrees of freedom provide the corresponding mean square values shown in Table 13.7. The F ratio used to test for

TABLE 13.7 ANOVA Table for the Air Traffic Controller Stress Test

Source of Variation	*Sum of Squares*	*Degrees of Freedom*	*Mean Square*	*F*
Between treatments	21	2	10.5	$\dfrac{10.5}{1.9} = 5.53$*
Between blocks	30	5	6.0	
Error	19	10	1.9	
Total	70	17		

*Significant at $\alpha = .05$; critical $F = 4.10$

differences between treatment means is MSTR/MSE = 10.5/1.9 = 5.53. Checking the F values in Table 4 of Appendix B, we find that the critical F value at $\alpha = .05$ (2 numerator degrees of freedom and 10 denominator degrees of freedom) is 4.10. With $F = 5.53$, we reject the null hypothesis H_0: $\mu_1 = \mu_2 = \mu_3$ and conclude that the work station designs differ in terms of their mean stress effects on air traffic controllers.

Before leaving this section let us make some general comments about the randomized block design. The blocking as described in this section is referred to as a *complete* block design, the word "complete" indicating that each block is subjected to all k treatments. That is, all controllers (blocks) were tested in all three systems (treatments). Experimental designs employing blocking where some but not all treatments are applied to each block are referred to as *incomplete* block designs. A discussion of incomplete block designs is outside the scope of this text.

In addition, note that in the air traffic controller stress test each controller in the study was required to use all three systems. While this guarantees a complete block design, in some cases blocking is carried out with "similar" experimental units in each block. For example, assume that in a pretest of air traffic controllers the population of controllers was divided into groups ranging from extremely high stress individuals to extremely low stress individuals. The blocking could still have been accomplished by having three controllers from each of the stress classifications participate in the study. Each block would then be formed from three controllers in the same stress class. The randomized aspect of the block design would be conducted by randomly assigning the three controllers in each block to the three systems.

Finally, note that the ANOVA table shown in Table 13.7 provides an F value to test for treatment effects but *not* for blocks. The reason is that the experiment was designed to test a single factor—work station design. The blocking based on individual stress differences was conducted in order to remove this variation from the MSE term. However, the study was not designed to test specifically for individual differences in stress.

Some analysts compute $F = \text{MSB}/\text{MSE}$ and use this statistic to test for significance of the blocks. Then they use the result as a guide to whether or not this type of blocking would be desired in future experiments. However, regardless of whether or not the block effect is significant, the test for the single-factor significance is valid as computed. If individual stress difference is to be a factor in the study, a different experimental design should be used. A test of significance on blocks should not be performed to attempt to draw such a conclusion about a second factor.

EXERCISES

10. An automobile dealer conducted a test to determine if the time needed to complete a minor engine tuneup depends upon whether a computerized engine analyzer or an electronic analyzer is used. Because tuneup time varies among compact, intermediate, and full-sized cars, the three types of cars were used as blocks in the experiment. The data obtained are shown below:

		Car		
		Compact	*Intermediate*	*Full-size*
Analyzer	*Computerized*	50	55	63
	Electronic	42	44	46
		↑ Time (minutes)		

Using $\alpha = .05$, test for any significant differences.

11. An important factor in selecting software for word processing and data base management systems is the time required to learn how to use a particular system. In order to evaluate three file management systems, a firm designed a test involving five different word processing operators. Since operator variability was believed to be a significant factor, each of the five operators was trained on each of the three file management systems. The data obtained are shown below.

	Operator				
	1	*2*	*3*	*4*	*5*
System A	16	19	14	13	18
System B	16	17	13	12	17
System C	24	22	19	18	22

↑ Time (hours)

Using $\alpha = .05$, test to see if there is any difference in training time for the three software packages.

12. The following data were obtained for a randomized block design involving five treatments and three blocks: SST = 430, SSTR = 310, SSB = 85. Set up the ANOVA table and test for any significant differences. Use $\alpha = .05$.

13. Five different auditing procedures were compared with respect to total audit time. To control for possible variation due to the person conducting the audit, four accountants were selected randomly and treated as blocks in the experiment. The following values were obtained using the ANOVA procedure: SST = 100, SSTR = 45, SSB = 36. Using $\alpha = .05$, test to see if there is any significant difference in total audit time stemming from the auditing procedure used.

13.5 TESTS ON INDIVIDUAL TREATMENT MEANS

Suppose that in carrying out an analysis of variance procedure we reject the null hypothesis and conclude that the population or treatment means are not all the same. Sometimes we may be satisfied with this conclusion, but in other cases we will want to go a step further and determine where the differences occur. The purpose of this section is to show how to conduct statistical comparisons between pairs of treatment means.

Recall that in the Dr. Johnson GMAT study (Section 13.1 to 13.3) we accepted the null hypothesis and concluded that the three GMAT preparation programs led to the same mean test scores. Further tests comparing specific programs were unwarranted because we believed that no differences in means existed. However, in the air traffic controllers stress test we rejected the null hypothesis and concluded the three work station systems differed in terms of their effects on controller stress. In this case the followup question is, "We believe that the systems differ; where do the differences occur?"

Let us show the details of a procedure that could be used to test for the equality of two treatment means. For example, let us test to see if there is a significant difference between the means of Systems A and B. From the data in Table 13.5 we found the treatment means to be $\bar{x}_1 = 13.5$ for System A and $\bar{x}_2 = 13$ for System B; thus System B shows the better (lower) stress level. But is the sample information sufficient to justify the conclusion that a difference in stress level exists between System A and System B?

A simple test for a difference between two treatment means is based on the use of the t distribution (as presented for the two population case in Chapter 10). The mean square error (MSE) is used as an unbiased estimate of the population variance σ^2. A confidence interval estimate of the difference in the two population means (μ_1 and μ_2)

is given by the following expression:

Interval Estimate for $\mu_1 - \mu_2$

$$\bar{x}_1 - \bar{x}_2 \pm t_{\alpha/2}\sqrt{\text{MSE}\left(\frac{1}{n_1} + \frac{1}{n_2}\right)}, \tag{13.26}$$

where

$\bar{x}_1$ = sample mean for the first treatment,

$\bar{x}_2$ = sample mean for the second treatment,

$t_{\alpha/2}$ = t value for the test, where the number of degrees of freedom for the t statistic is given by the degrees of freedom for error (see ANOVA table "Degrees of Freedom" column),

MSE = mean square error (see ANOVA table),

n_1 = sample size for the first treatment,

n_2 = sample size for the second treatment.

If the confidence interval in (13.26) includes the value 0, we have to conclude that there is no significant difference between the treatment means. In this case we accept the hypothesis that no difference exists between the treatment means. However, if the confidence interval does not include the value 0, we conclude that there is a difference.

Let us return to the air traffic controllers stress test and compare the stress levels for System A and System B, at the .05 level of significance. The following data are needed:

$\bar{x}_1 = 13.5$ (System A), $\bar{x}_2 = 13$ (System B),

With $\alpha = .05$, $t_{.025} = 2.228$ (note: 10 degrees of freedom for error in ANOVA Table 13.7),

MSE = 1.9 (Table 13.7),

$n_1 = n_2 = 6$ (sample size for each treatment is six, since the study involved six blocks).

Using these data and (13.26) we have

$$13.5 - 13 \pm 2.228\sqrt{1.9(1/6 + 1/6)}$$
$$.5 \pm 2.228(.7958)$$
$$.5 \pm 1.77.$$

Thus a 95% confidence interval for the difference between System A and System B is given by -1.27 to 2.27. Since this interval includes 0, we are unable to reject the hypothesis that the systems provide the same level of stress.

At this point, analysis of variance has told us that the stress levels for the three systems are not all the same. However, the above result tells us that Systems A and B appear to have the same level of stress. With $\bar{x}_3 = 15.5$, System C appears to have the highest level of stress. Thus we should feel comfortable in concluding that the difference in population means is due to System C.

A word of caution is needed at this point: The above test of an individual difference should be applied only if we reject the null hypothesis of equal population means. That is, while it may appear natural to use (13.26) to compare all possible pairs of treatment means (System A versus System B, System A versus System C, System B versus System C), statistical problems can occur with this sequential approach. If a null hypothesis is true (that is, the two population means are equal) and a test is conducted using (13.26) with a Type I error probability of $\alpha = .10$, there is a probability of .10 of rejecting the null hypothesis when it is really true; hence the probability of making a correct decision is .90. If two tests are conducted in this manner, the probability that a correct decision is made on both tests is $(.9)(.9) = .81$. Thus the probability that *at least* one of the tests would result in rejecting a true null hypothesis is $1 - .81 = .19$.* Thus we see that the total Type I error probability using a sequential testing procedure to test two hypotheses at the .10 significance level is .19 and not .10. For three tests at the .10 level of significance the Type I error probability is $1 - (.90)(.90)(.90) = 1 - .729 = .271$. The probability of making a Type I error increases rapidly as the number of multiple tests increases.

A simple procedure to adjust (approximately) for this increasing probability of making the Type I error is to reduce the α level for each separate test. The aim is to reduce to a satisfactory level the α level for all tests taken together. With an α level of significance desired and m tests to be made, we would use α/m as the probability of making a Type I error on any one test. Then the overall probability of making a Type I error on any one of the tests will approximately equal α.

Because of the difficulty of the Type I error increasing when making multiple tests of difference between individual means, a variety of specialized tests have been developed. Often these tests carry the name of the developer. The better known tests for multiple comparisons include the Duncan multiple range test, the Newman-Kuels test, Tukey's test, and Scheffe's method. References in the Bibliography provide details for these methods. It is recommended that these tests be considered whenever multiple comparisons among treatment means are expected to be a major concern in the study.

EXERCISES

14. Refer to Exercise 1. Use the procedure described in this section to test for the equality of the mean for manufacturers 1 and 3. Use $\alpha = .05$. What conclusion can you make after carrying out this test?

15. In Exercise 2, does it make sense to use the procedure described in this section to test for the equality of the mean drying time for paints 1 and 4? Explain.

16. Refer to Exercise 4. Use the procedure described in this section to test for the equality of the mean mileage for companies A and B. What general conclusion can you make after carrying out this test? Use $\alpha = .05$.

13.6 FACTORIAL EXPERIMENTS

The experimental designs we have considered thus far enable statistical conclusions to be drawn about one factor. However, in many experiments we want to draw conclusions about more than one variable or factor. *Factorial experiments* and their

*This assumes that the two tests are independent and hence the joint probability of the two events can be obtained simply by multiplying the individual probabilities. In fact the two tests are not independent (MSE is used in each test), and hence the error involved is even greater than that shown.

corresponding analysis of variance computations are valuable designs when simultaneous conclusions are required about two or more factors. The term "factorial" is used because the experimental conditions include all possible combinations of the factors involved. For example, if there are a levels of factor A and b levels of factor B, the experiment will involve collecting data on ab treatment combinations. In this section we will show the analysis of a two-factor factorial experiment. This basic approach can be extended to experiments involving more than two factors.

As an illustration of a two-factor factorial experiment, let us return to the Dr. Johnson GMAT study (Sections 13.1 to 13.3). One factor in this study is the GMAT preparation program, which has three treatments: 3 hour review, 1 day program, and 10 week course. Let us assume that Dr. Johnson, Dean of the Business School, has talked to the deans of the College of Engineering and the College of Arts and Sciences and that these deans are interested in the GMAT preparation programs because a number of their students are also considering attending graduate business school. Thus, also of interest in this experiment is whether or not a student's undergraduate major affects the GMAT score and a second factor, undergraduate major, is defined. This factor also has three treatments: business, engineering, and arts and sciences. The factorial design for this experiment with three treatments corresponding to factor A, the preparation program, and three treatments corresponding to factor B, the undergraduate major, will have a total of $3 \times 3 = 9$ treatment combinations. These treatment combinations or experimental conditions are summarized in Table 13.8.

TABLE 13.8 Nine Treatment Combinations for the Two-Factor GMAT Experiment

		Factor B: Undergraduate Major		
	Treatments →	*Business*	*Engineering*	*Arts and Sciences*
Factor A:	*3 hour review*	1	2	3
Preparation	*1 day program*	4	5	6
Program	*10 week course*	7	8	9

Let us assume that a sample of two students will be selected corresponding to each of the nine treatment combinations shown in Table 13.8: two business majors will take the 3 hour review, two will take the 1 day program, and two will take the 10 week course. In addition, two engineering majors and two arts and sciences majors will take each of the three preparation programs. In experimental design terminology the sample size of two for each treatment combination indicates that we have two *replications* of the experiment. Additional replications and an increased sample size could easily be made, but we elected not to do so in order to minimize the computational aspects of this illustration.

This experimental design requires that six students who plan to attend graduate school be randomly selected from *each* of the three undergraduate majors. Then two students from each major should be assigned randomly to each preparation program, resulting in a total of 18 students being used in the study.

Let us assume that the students have been randomly selected, have participated in the preparation program, and have taken the GMAT. The scores obtained are shown in Table 13.9.

The analysis of variance computations using the data in Table 13.9 will provide

TABLE 13.9 GMAT Scores for the Two Factor Experiment

		Factor B: Undergraduate Major		
		Business	*Engineering*	*Arts and Sciences*
Factor A: Preparation Program	*3 hour review*	500 580	540 460	480 400
	1 day program	460 540	560 620	420 480
	10 week course	560 600	600 580	480 410

answers to the following questions:

Main effect (factor A): Do the preparation programs differ in terms of effect on GMAT scores?

Main effect (factor B): Do the undergraduate majors differ in terms of their ability to perform on the GMAT?

Interaction effect (factors A and B): Do students in some majors do better on one type of preparation program while others do better on a different type of preparation program?

The term "interaction" refers to a new effect that we can now study because we have used a factorial experiment. If the interaction effect has a significant impact on the GMAT scores, it will mean that the type of preparation program preferred depends on the undergraduate major. Let us proceed with the analysis of variance computations.

The Analysis of Variance Procedure

The analysis of variance procedure for the two factor factorial experiment is similar to the completely randomized experiment and the randomized block experiment in that we once again partition the sum of squares and the degrees of freedom into their respective sources. The formula for partitioning the sum of squares for the two factor factorial experiments is as follows:

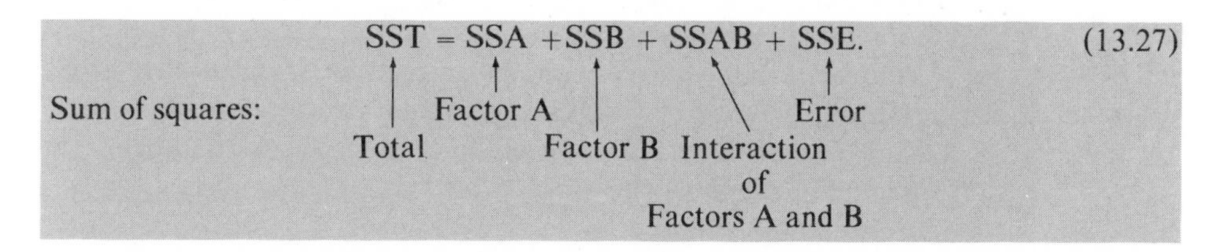

The partitioning of the sum of squares and degrees of freedom is summarized in Table 13.10. The following notation is used:

a = number of levels of factor A,

b = number of levels of factor B,

r = number of replications,

n_T = total number of observations taken in the experiment;

$$n_T = abr.$$

TABLE 13.10 ANOVA Table for the Two Factor Factorial Experiment with r Replications

Source of Variation	*Sum of Squares*	*Degrees of Freedom*	*Mean Square*	*F*
Factor A treatments	SSA	$a - 1$	$\text{MSA} = \dfrac{\text{SSA}}{a-1}$	$\dfrac{\text{MSA}}{\text{MSE}}$
Factor B treatments	SSB	$b - 1$	$\text{MSB} = \dfrac{\text{SSB}}{b-1}$	$\dfrac{\text{MSB}}{\text{MSE}}$
Interaction	SSAB	$(a-1)(b-1)$	$\text{MSAB} = \dfrac{\text{SSAB}}{(a-1)(b-1)}$	$\dfrac{\text{MSAB}}{\text{MSE}}$
Error	SSE	$ab(r-1)$	$\text{MSE} = \dfrac{\text{SSE}}{ab(r-1)}$	
Total	SST	$n_T - 1$		

Computations and Conclusions

The step-by-step computational procedure for computing the sum of squares for a two factor factorial design is shown below.

Computational Steps for Computing Sums of Squares for a Two-Factor Factorial Design

In addition to a, b, r, and n_T as defined above, we use the following notation:

x_{ijk} = observation corresponding to the kth replicate taken from treatment i of factor A and treatment j of factor B,
$T_{i.}$ = total of all observations in treatment i (factor A),
$T_{.j}$ = total of all observations in treatment j (factor B),
T_{ij} = total of all observations in the combination of treatment i (factor A) and treatment j (factor B),
T = total of all observations,
$\bar{x}_{i.}$ = sample mean for the observations in treatment i (factor A),
$\bar{x}_{.j}$ = sample mean for the observations in treatment j (factor B),
$\bar{x}_{ij}$ = sample mean for the observations in combination of treatment i (factor A) and treatment j (factor B),
$\bar{\bar{x}}$ = overall sample mean of all n_T observations

Step 1: Compute the sum of squares total, SST:

$$\text{SST} = \sum_i \sum_j \sum_k (x_{ijk} - \bar{\bar{x}})^2 = \sum_i \sum_j \sum_k x_{ijk}^2 - \frac{T^2}{n_T}. \tag{13.28}$$

Step 2: Compute the sum of squares for factor A:

$$SSA = br \sum_i (\bar{x}_{i.} - \bar{\bar{x}})^2 = \frac{\sum_i T_{i.}^2}{br} - \frac{T^2}{n_T}. \tag{13.29}$$

Step 3: Compute the sum of squares for factor B:

$$SSB = ar \sum_j (\bar{x}_{.j} - \bar{\bar{x}})^2 = \frac{\sum_j T_{.j}^2}{ar} - \frac{T^2}{n_T}. \tag{13.30}$$

Step 4: Compute the sum of squares for the interaction:

$$SSAB = r \sum_i \sum_j (\bar{x}_{ij} - \bar{x}_{i.} - \bar{x}_{.j} + \bar{\bar{x}})^2$$

$$= \frac{\sum_i \sum_j T_{ij}^2}{r} - \frac{T^2}{n_T} - SSA - SSB. \tag{13.31}$$

Step 5: Compute the sum of squares due to error:

$$SSE = SST - SSA - SSB - SSAB. \tag{13.32}$$

Table 13.11 shows the data collected in the experiment, along with the various sums that will help us with the sum of squares computations. Using (13.28) to (13.32)

TABLE 13.11 GMAT Scores for the Two-Factor Experiment

	Treatment Combination Totals T_{ij}	*Factor B: Undergraduate Major* Business	Engineering	Arts and Sciences	*Row Totals* $T_{i.}$	*Row Means*
Factor A: Preparation Program	3 hour review	500 580 1,080	540 460 1,000	480 400 880	2,960	493.3
	1 day program	460 540 1,000	560 620 1,180	420 480 900	3,080	513.3
	10 week course	560 600 1,160	600 580 1,180	480 410 890	3,230	538.3
	Column totals $T_{.j}$	3,240	3,360	2,670	9,270 (Overall Total T)	
	Column means	540	560	445		

Sum of Squares $\sum_i \sum_j \sum_k x_{ijk}^2 = (500)^2 + (580)^2 + \cdots + (410)^2 = 4{,}856{,}500$

we have the following sum of squares for the GMAT two-factor factorial experiment:

Step 1: $\text{SST} = 4{,}856{,}500 - \dfrac{(9{,}270)^2}{18} = 82{,}450.$

Step 2: $\text{SSA} = \dfrac{(2{,}960)^2 + (3{,}080)^2 + (3{,}230)^2}{6} - \dfrac{(9{,}270)^2}{18} = 6{,}100.$

Step 3: $\text{SSB} = \dfrac{(3{,}240)^2 + (3{,}360)^2 + (2{,}670)^2}{6} - \dfrac{(9{,}270)^2}{18} = 45{,}300.$

Step 4: $\text{SSAB} = \dfrac{(1{,}080)^2 + (1{,}000)^2 + \cdots + (890)^2}{2} - \dfrac{(9{,}270)^2}{18}$

$- 6{,}100 - 45{,}300 = 11{,}200.$

Step 5: $\text{SSE} = 82{,}450 - 6{,}100 - 45{,}300 - 11{,}200 = 19{,}850.$

These sums of squares divided by their corresponding degrees of freedom, as shown in Table 13.12, provide the appropriate mean square values for testing the two main

TABLE 13.12 ANOVA Table for the Two-Factor GMAT Study

Source of Variation	*Sum of Squares*	*Degrees of Freedom*	*Mean Square*	*F*
Factor A treatments	6,100	2	3,050	3,050/2,206 = 1.38
Factor B treatments	45,300	2	22,650	22,650/2,206 = 10.27*
Interaction (AB)	11,200	4	2,800	2,800/2,206 = 1.27
Error	19,850	9	2,206	
Total	82,450	17		

*Significant at $\alpha = .05$; critical $F = 4.26$

effects (preparation program and undergraduate major) and the interaction effect. The F ratio used to test for differences among preparation programs is 1.38. The critical F value at $\alpha = .05$ (with 2 numerator degrees of freedom and 9 denominator degrees of freedom) is 4.26. With $F = 1.38$, we cannot reject the null hypothesis and must conclude that there is no difference in the preparation provided by the three preparation programs. However, for the undergraduate major effect, $F = 10.27$ exceeds the critical F value of 4.26. Thus the analysis of variance results allow us to conclude that there is a difference in GMAT test scores among the three undergraduate majors; that is, the three undergraduate majors do not provide the same preparation for performance on the GMAT. Finally, the interaction F value of $F = 1.27$ (critical F value $= 3.63$ at $\alpha = .05$) means that we cannot identify a significant interaction effect. Thus there is no reason to believe that the three preparation programs differ in their ability to prepare the different majors for the GMAT.

Undergraduate major was found to be a significant factor. Checking the calculations in Table 13.11 we see that the sample means are as follows: business majors $\bar{x}_1 = 540$, engineering majors $\bar{x}_2 = 560$, and arts and sciences majors $\bar{x}_3 = 445$.

Tests on individual treatment means can be conducted; yet after reviewing the three sample means we would anticipate no difference in preparation for business and engineering graduates. However, the arts and sciences majors appear to be significantly less prepared for the GMAT than students in the other majors. Perhaps this observation will lead the Dean of the School of Arts and Sciences to consider other options for assisting these majors in preparing for graduate management admission tests.

Because of the computational effort involved in any modest to large-size factorial experiment, the computer usually plays an important role in making and summarizing the analysis of variance computations. The computer printout for the analysis of variance of the GMAT two-factor factorial experiment is shown in Figure 13.6. Note that "C2" identifies factor A, "C3" identifies factor B, and "C2*C3" identifies the interaction row.

```
ANALYSIS OF VARIANCE

DUE TO          DF        SS     MS=SS/DF
C2               2     6100.        3050.
C3               2    45300.       22650.
C2 * C3          4    11200.        2800.
ERROR            9    19850.        2206.
TOTAL           17    82450.
```

FIGURE 13-6 Computer Output for Two-Factor Design

EXERCISES

17. A mail order catalog firm designed a factorial experiment to test the effect of the size of a magazine advertisement and the advertisement design on the number of catalog requests received (1,000's). Three advertising designs and two different-size advertisements were considered. The data obtained are shown below:

	Size of Advertisement	
	Small	*Large*
Design A	8 12	12 8
Design B	22 14	26 30
Design C	10 18	18 14

Use the ANOVA procedure for factorial designs to test for any significant effects due to type of design, size of advertisement, or interaction. Use $\alpha = .05$.

18. An amusement park has been studying methods for decreasing the waiting time on rides by loading and unloading riders more efficiently. Two alternative loading/unloading methods have been proposed. To account for potential differences due to the type of ride and the possible interaction between the method of loading and unloading and the type of ride, a factorial experiment was designed. Using the data shown below, test for any significant effect due to the loading and unloading method, the type of ride, and interaction. Use $\alpha = .05$.

		Type of Ride	
	Roller Coaster	*Screaming Demon*	*Log Flume*
Method 1	41	52	50
	43	44	46
Method 2	49	50	48
	51	46	44

↑ Waiting Time (minutes)

19. The calculations for a factorial experiment involving four levels of factor A, three levels of factor B, and three replications resulted in the following data: SST = 280, SSA = 26, SSB = 23, SSAB = 175. Set up the ANOVA table and test for any significant main effect and any interaction effect. Use $\alpha = .05$.

Summary

In this chapter we have shown several experimental designs that can be employed with an analysis of variance procedure to lead to conclusions about differences in means of several populations or treatments. Specifically, we introduced the single-factor completely randomized, the randomized block, and the two-factor factorial experimental designs. The completely randomized design and the randomized block designs were used to draw conclusions about differences in the means of a single factor. The primary purpose of the blocking in the randomized block design was to remove extraneous sources of variation from the error term. This blocking provided a better estimate of the error variance and a better test to determine whether or not the population or treatment means of the single factor differed significantly.

Although the various experimental designs required different formulas and computations, we showed that the basis for the statistical test is the development of independent estimates of the population variance σ^2. In the single factor case one estimate, MSTR, is based upon the variation between the treatments. This value will provide an unbiased estimate of σ^2 only if the means $\mu_1, \mu_2, \ldots, \mu_k$ are all equal. A second estimate, MSE, is based upon the variation of the observations within each sample and will always provide an unbiased estimate of σ^2. By computing F = MSTR/MSE and using the F distribution, we developed a decision rule for determining whether to accept or reject the hypothesis that the treatment means are equal. In all the experimental designs considered, the partitioning of the sum of squares and degrees of freedom into their various sources enabled us to compute the appropriate values for making the analysis of variance calculations and tests.

Whenever an analysis of variance conclusion results in the rejection of the equal-means hypothesis, we may want to consider testing for a difference between the individual treatment means. We showed how the t distribution test could be used to compare two treatment means. However, we warned against indiscriminate use of this testing procedure because of the increasing probability of making a Type I error. By simultaneously making several tests for individual differences, the probability of erroneously claiming that a difference exists increases. Several specialized tests are available if multiple comparisons of the treatment means are to be considered.

Glossary

Analysis of variance (ANOVA) procedure—A statistical approach for determining whether or not the means of several different populations are equal.

Factor—Another word for the variable of interest in an ANOVA procedure.

Treatment—Different levels of a factor.

Single-factor experiment—An experiment involving only one factor with k populations or treatments.

Experimental units—The objects of interest in the experiment.

Completely randomized design—An experimental design where the experimental units are randomly assigned to the treatments.

Mean square—The sum of squares divided by its corresponding degrees of freedom. This quantity is used in the F ratio to determine if significant differences in means exist or not.

ANOVA table—A table used to summarize the analysis of variance computations and results. It contains columns showing the source of variation, the degrees of freedom, the sum of squares, the mean squares, and the F values.

Partitioning—The process of allocating the total sum of squares and degrees of freedom into the various components.

Blocking—The process of using the same or similar experimental units for all treatments. The purpose of blocking is to remove a source of variation from the error term and hence provide a more powerful test for a difference in population or treatment means.

Randomized block design—An experimental design employing blocking. The experimental unit(s) within a block are randomly ordered for the treatments.

Factorial experiments—An experimental design that permits statistical conclusions about two or more factors. All levels of each factor are considered with all levels of the other factors in order to specify the experimental conditions for the experiment.

Replication—The number of times each experimental condition is repeated in an experiment. It is the sample size associated with each treatment combination.

Main effect—The response produced by the different factors in a factorial design.

Interaction—The response produced when the levels of one factor interact with the levels of another factor in influencing the response variable.

Key Formulas

Completely Randomized Designs

The Sum of Squares About the Mean

$$\text{SST} = \sum_i \sum_j (\bar{x}_{ij} - \bar{\bar{x}})^2 = \sum_i \sum_j x_{ij}^2 - \frac{T^2}{n_T} \quad (13.17)$$

The Sum of Squares Due to Treatments

$$\text{SSTR} = \sum_i n_i(\bar{x}_i - \bar{\bar{x}})^2 = \sum_i \frac{T_i^2}{n_i} - \frac{T^2}{n_T} \quad (13.18)$$

The Sum of Squares Due to Error

$$\text{SSE} = \text{SST} - \text{SSTR} \tag{13.19}$$

The *F* Value

$$F = \frac{\text{MSTR}}{\text{MSE}} \tag{13.20}$$

Randomized Block Designs

The Total Sum of Squares

$$\text{SST} = \sum_i \sum_j (x_{ij} - \bar{\bar{x}})^2 = \sum_i \sum_j x_{ij}^2 - \frac{T^2}{n_T} \tag{13.22}$$

The Sum of Squares Between Treatments

$$\text{SSTR} = \sum_i b(\bar{x}_{i.} - \bar{\bar{x}})^2 = \frac{\sum_i T_{i.}^2}{b} - \frac{T^2}{n_T} \tag{13.23}$$

The Sum of Squares Due to Blocks

$$\text{SSB} = \sum_j k(\bar{x}_{.j} - \bar{\bar{x}})^2 = \frac{\sum_j T_{.j}^2}{k} - \frac{T^2}{n_T} \tag{13.24}$$

The Sum of Squares Due to Error

$$\text{SSE} = \text{SST} - \text{SSTR} - \text{SSB} \tag{13.25}$$

Factorial Experiments

The Total Sum of Squares

$$\text{SST} = \sum_i \sum_j \sum_k (x_{ijk} - \bar{\bar{x}})^2 = \sum_i \sum_j \sum_k x_{ijk}^2 - \frac{T^2}{n_T} \tag{13.28}$$

The Sum of Squares for Factor A

$$\text{SSA} = br \sum_i (\bar{x}_{i.} - \bar{\bar{x}})^2 = \frac{\sum_i T_{i.}^2}{br} - \frac{T^2}{n_T} \tag{13.29}$$

The Sum of Squares for Factor B

$$\text{SSB} = ar \sum_j (\bar{x}_{.j} - \bar{\bar{x}})^2 = \frac{\sum_j T_{.j}^2}{ar} - \frac{T^2}{n_T} \tag{13.30}$$

The Sum of Squares for the Interaction

$$\text{SSAB} = r\sum_i\sum_j(\bar{x}_{ij} - \bar{x}_{i.} - \bar{x}_{.j} + \bar{\bar{x}})^2$$

$$= \frac{\sum_i\sum_j T_{ij}^2}{r} - \frac{T^2}{n_T} - \text{SSA} - \text{SSB} \tag{13.31}$$

The Sum of Squares Due to Error

$$\text{SSE} = \text{SST} - \text{SSA} - \text{SSB} - \text{SSAB} \tag{13.32}$$

Supplementary Exercises

20. In your own words explain what the ANOVA procedure is used for.

21. What has to be true in order for MSTR to provide a good estimate of σ^2? Explain.

22. Why do we assume that the populations sampled all have the same variance when we apply the ANOVA procedure?

23. Explain why MSTR and MSE provide two independent estimates of σ^2.

24. Explain why MSTR provides an inflated estimate of σ^2 when the population means are not the same.

25. A simple random sample of the asking price ($1,000's) of four houses currently for sale in each of two residential areas resulted in the following data:

Area 1	*Area 2*
92	90
89	102
98	96
105	88

a. Use the procedure developed in Chapter 10 and test if the mean asking price is the same in both areas. Use $\alpha = .05$.

b. Use the ANOVA procedure to test if the mean asking price is the same. Compare your analysis with part a. Use $\alpha = .05$.

26. Suppose that in Exercise 25 data were collected for another residential area. The asking prices for the simple random sample from the third area were as follows: $81,000, $86,000, $75,000, and $90,000. Is the mean asking price for all three areas the same? Use $\alpha = .05$.

27. An analysis of the number of units sold by ten salespersons in each of four sales territories resulted in the following data:

	Sales Territory			
	1	2	3	4
Number of salespersons	10	10	10	10
Average number sold ($\bar{x}$)	130	120	132	114
Sample variance (s^2)	72	64	69	67

Test at the $\alpha = .05$ level if there is any significant difference in the mean number of units sold in the four sales territories.

28. Suppose that in Exercise 27 the number of salespersons in each territory was as follows: $n_1 = 10$, $n_2 = 12$, $n_3 = 10$, and $n_4 = 15$. Using the same data for $\bar{x}$ and s^2 as given in Exercise 27, test at the $\alpha = .05$ level if there is any significant difference in the mean number of units sold in the four sales territories.

29. In Chapter 3 we discussed various situations involving the monthly starting salaries for recent business school graduates. Use the data provided in Table 3.1 and test whether or not the mean monthly starting salaries for accounting, finance, management, and marketing graduates are the same. Use $\alpha = .05$.

30. Three different assembly methods have been proposed for a new product. In order to determine which assembly method results in the greatest number of parts produced per hour, 30 workers were randomly selected and assigned to use one of the proposed methods. The number of units produced by each worker is given below:

Method A	97	73	93	100	73	91	100	86	92	95
Method B	93	100	93	55	77	91	85	73	90	83
Method C	99	94	87	66	59	75	84	72	88	86

Use these data and test to see if the mean number of parts produced with each method is the same. Use $\alpha = .05$.

31. In order to test to see if there is any significant difference in the mean number of units produced per week by each of three production methods, the following data were collected:

Method 1	*Method 2*	*Method 3*
58	52	48
64	63	57
55	65	59
66	58	47
67	62	49

At the $\alpha = .05$ level of significance is there any difference in the means for the three methods?

32. Pappashales Restaurant is considering introducing a new specialty sandwich. For a determination of the effect of sandwich price on sales, the new sandwich was test marketed at three prices in selected company restaurants. The following data, in terms of the number of sandwiches sold per day, were obtained:

$1.49	*$1.79*	*$1.99*
925	910	860
850	845	935
930	905	820
955	860	845

At the $\alpha = .05$ level of significance is there any difference in the mean number of sandwiches sold per day for the three prices? What should management of Pappashales do?

33. Hargreaves Automotive Parts Inc. would like to compare the mileage for four different types of brake linings. Thirty linings of each type were produced and placed on a fleet of rental cars. The number of miles that each brake lining lasted until it no longer met the required federal safety standard was recorded, and an average value was computed for each type of lining. The

following data were obtained:

	Sample Size	*Sample Mean*	*Standard Deviation*
Type A	30	32,000	1,450
Type B	30	27,500	1,525
Type C	30	34,200	1,650
Type D	30	30,300	1,400

Use these data and test to see if the corresponding population means are equal. Use =.05.

34. A manufacturer of batteries for electronic toys and calculators is considering three new battery designs. An attempt was made to determine if the mean lifetime in hours is the same for each of the three designs. The following data were collected:

Design A	*Design B*	*Design C*
78	112	115
98	99	101
88	101	100
96	116	120

Test to see if the population means are equal. Use $\alpha = .05$.

35. Refer to Exercise 34. Use the procedure described in Section 13.5 to test for the equality of the mean for design A and design B. What conclusion can you make after carrying out the above test? Use $\alpha = .05$.

36. In Exercise 32, would it make sense to do a test on individual treatment means as described in Section 13.5? Explain.

37. A research firm tests the miles per gallon characteristics of three brands of gasoline. Because of different gasoline performance characteristics in different brands of automobiles, five brands of automobiles are selected and treated as blocks in the experiment. That is, each brand of automobile is tested with each type of gasoline. The results of the experiment are shown below:

		Blocks: Automobiles				
		A	*B*	*C*	*D*	*E*
	I	18	24	30	22	20
Gasoline Brands	*II*	21	26	29	25	23
	III	20	27	34	24	24

Miles per gallon

With $\alpha = .05$, is there a significant difference in the mean miles per gallon characteristics of the three brands of gasoline?

38. Analyze the experimental data provided in Exercise 37 using the ANOVA procedure for completely randomized designs. Compare your findings with those obtained in Exercise 37. What is the advantage of attempting to remove the block effect?

39. The following data were obtained for a randomized block design involving three treatments and four blocks: SST = 148, SSTR = 84, SSB = 50. Set up the ANOVA table and test for any significant differences. Use $\alpha = .05$.

40. A factorial experiment was designed to test for any significant differences in the time needed to perform English to foreign language translations using two computerized language translators. Since the type of language translated was also considered a significant factor,

translations were made using both systems for three different languages: Spanish, French, and German. Use the data shown below:

		Language	
	Spanish	*French*	*German*
System 1	8	10	12
	12	14	16
System 2	6	14	16
	10	16	22

↑ Translation Time (hours)

Test for any significant differences due to language translator, type of language, and interaction. Use $\alpha = .05$.

41. A manufacturing company designed a factorial experiment in order to determine if the number of defective parts produced by two machines differed. Since it was believed that the number of defective parts produced also depends upon whether or not raw material needed by the machine was loaded manually or using an automatic feed system, use the following data to test for any significant effect due to machine, loading system, and interaction. Use $\alpha = .05$.

	Loading System	
	Manual	*Automatic*
Machine 1	30	30
	34	26
Machine 2	20	24
	22	28

↗ number of defective parts

Burke
Marketing Services, Inc.

Burke Marketing Services Inc.*

Cincinnati, Ohio

For half a century Burke Marketing Services has been a leader in solving marketing problems; during this time, Burke has become the preeminent, most experienced custom survey research firm in the industry. Burke writes more proposals, on more projects, every day than any other research company in the world. Supported by state-of-the-art technology, four Burke divisions offer a wide range of research capabilities, providing answers to nearly every marketing question. The divisions are BASES, Test Marketing Group, Burke Marketing Research, and Consulting and Analytical Services.

The BASES system translates key attitudinal measures into sales estimates at four critical decision points in the new product development process: when the product is just a concept; when a prototype has been developed; at the stage when final packaging and so on has been developed; and in the test market or rollout stage. In effect, BASES answers the toughest new product question of them all: "Does this product have the potential to meet the company's minimum business requirements?"

The Test Marketing Group consists of two operating units: AdTel and Market Audits. AdTel conducts advertising tests using one-of-a kind cable TV systems in six test markets across the nation. The unique AdTel construction allows a company to develop an alternative media plan and test it alongside its current plan. This service allows a company to evaluate new product viability, test new packaging, compare strategies and schedules, and test weights and promotions.

Market Audits provides consumer products manufacturers with a complete record of their product's performance from which they may make marketing decisions. Each year Market Audits conducts over 100 in-store tests, primarily in food, drug, and mass merchandiser retail outlets. Market Audits offers a full complement of in-store research: mini-market testing, controlled store tests, standard retail sales audits, distribution checks, and observation studies.

Burke Marketing Research provides a complete range of ad hoc research services—television and radio commercial tests, magazine and newspaper advertising tests, concept generation and evaluation studies, package evaluations, claim support, and tracking studies, to name just a few. The Consulting and Analytical Services Division of Burke provides consulting in matters of marketing problems, research design, and interpretation of results to a wide range of clients. Consulting and

*The authors are indebted to Dr. Ronald Tatham of Burke Marketing Services for providing this application.

Burke's in-store research provides valuable information for clients

Analytical Services provides analytic support to all Burke divisions as well as to Burke clients.

STATISTICAL ANALYSIS AT BURKE MARKETING SERVICES, INC.

A wide range of statistical and other analytical techniques are used to facilitate research design and data interpretation. A simple division of research into experimental and survey research facilitates discussion.

Experimental research: BASES, Test Marketing Group, and Burke Marketing Research all conduct experiments and use regression, analysis of variance, and analysis of covariance to estimate the effect of the experiment. This is about 30% of the research conducted by the company.

Survey research: The results of survey research are often examined and interpretation facilitated through the use of such techniques as multivariate exploratory methods, regression, analysis of variance, and statistical tests of differences.

Almost all research at Burke involves statistical analysis of some type or the provision of statistics (standard errors, etc.) to the client to facilitate later analysis.

AN EXPERIMENTAL DESIGN APPLICATION

A major manufacturer of children's dry cereal continually strives for formula improvements that offer a better tasting product. To maintain confidentiality we will refer to the manufacturer as the Anon Company. Burke's Consulting and Analytical

Services Division was retained by Anon Company to evaluate potential new formulations designed to improve the taste of the cereal. The four key ingredients that Anon Company thought would enhance taste were

1. Ratio of wheat to corn in the cereal flake (2 levels),
2. Types of sweetness (3 types),
3. Flavor bits (present or absent), and
4. Manufacturing cooking time (short, long).

The sweeteners were sugar, honey, and an artificial sweetener additive. The flavor bits were small particles with a fruit taste that could be included in the cereal. The cooking time influences the "crunchiness" of the cereal.

Using all combinations of the four ingredients would lead to $2 \times 3 \times 2 \times 2 = 24$ different cereal formulations. If the 24 cereal formulations were tested independently, the sample sizes would have to be very large to provide sufficient power to measure the effects of the various ingredients on taste perception. In order to use fewer respondents, thus saving time and money, a variation of the factorial experiment was used; it is called a fractional factorial design. Only 9 of the 24 cereal formulations were used, but they were selected in such a way that the best possible measure of the independent effect of the four ingredients could be obtained.

The nine cereal formulations were then placed in blocks of size 3. This meant that each child in the sample would taste test three different cereals (the order of tasting was randomized for each child). There were 18 blocks, each involving a different combination of three of the nine cereals in the test. Table 13A.1 shows this experimental design. Each block of three cereals was then tested by six children,

TABLE 13A.1 Experimental Design for Cereal Taste Test. There Are 18 Blocks Each Containing a Different Set of Three Cereals

	Cereal Formulations								
Blocks	*C1*	*C2*	*C3*	*C4*	*C5*	*C6*	*C7*	*C8*	*C9*
1	x			x			x		
2	x			x				x	
3	x				x				x
4	x				x		x		
5	x					x		x	
6	x					x			x
7		x		x			x		
8		x		x				x	
9		x			x				x
10		x			x		x		
11		x				x		x	
12		x				x			x
13			x	x			x		
14			x	x				x	
15			x		x				x
16			x		x		x		
17			x			x		x	
18			x			x			x

resulting in an overall sample size of $6 \times 18 = 108$ children. To maintain control over the sampling Burke prepackaged the cereals into 108 packages of three cereals each. This ensured that the random sample of 108 children would have the desired characteristics.

RESULTS

An analysis of the sample results indicated the following:

The flake composition and sweetener type were *very* influential in taste evaluation.
The flavor bits actually *detracted* from the perception of the taste of the cereal.
The cooking time had *no* impact on the taste.

Note that managerial considerations such as time, cost, and control over the experimenters influenced the choice of the experimental design. These are often as important as statistical considerations.

14 Simple Linear Regression and Correlation

What you will learn in this chapter:

- how to use regression analysis to develop an equation that estimates how two variables are related
- what the least squares method is
- how to compute and interpret the coefficient of determination
- how to use the t and F distributions to test for significant relationships between variables
- what is meant by residual analysis
- how to use a regression equation for estimation and prediction
- how to compute and interpret the sample correlation coefficient

Contents

In day-to-day decision-making situations, conclusions and recommendations are often based upon the relationship between two or more variables. For example, after considering the relationship between advertising expenditures and subsequent sales volume, a marketing manager might attempt to predict sales volume from a known level of advertising expenditures. In another case a public utility might use the relationship between temperature and electricity use to predict demand for electricity. Although in some instances, a manager may rely on intuition as to how two variables are related, a more objective approach is to collect data on the two variables and then use statistical procedures to determine how the variables are related.

Regression analysis is a statistical technique that can be used to develop a mathematical equation showing how variables are related. In regression terminology the variable which is being predicted by the mathematical equation is called the *dependent* variable. The variable or variables being used to predict the value of the dependent variable are called the *independent* variables. In our sales-advertising expenditures example, the marketing manager's desire to predict sales volume would suggest making sales volume the dependent variable for the analysis. Advertising expenditure would be the independent variable used to predict the sales volume. Common statistical notation is to use y to denote the dependent variable and x to denote the independent variable.

In this chapter we consider the simplest type of regression analysis: situations involving one independent and one dependent variable for which the relationship between the variables is approximated by a straight line. This is called *simple linear regression* analysis. When there are two or more independent variables involved we have multiple regression; this type of analysis is covered in Chapter 15. In Chapter 15 we also show that in some cases the methods of simple linear regression can be used to analyze curvilinear as well as straight line relationships.

Another topic discussed in this chapter is correlation. In correlation analysis we are not concerned with identifying a mathematical equation relating an independent and dependent variable; we are concerned only with determining the extent to which the variables are linearly related. Correlation analysis is a procedure for making this determination and, if such a relationship exists, providing a measure of the relative strength of the relationship.

We caution the reader before beginning this chapter that neither regression nor correlation analysis can be interpreted as establishing cause–effect relationships. Regression and correlation analyses can indicate only how or to what extent the variables are associated with each other. Any conclusions about a cause and effect relationship must be based on the *judgment of the analyst*.

14.1 THE REGRESSION MODEL

An important concept that must be understood before we consider the computational aspects of regression analysis involves the distinction between a *deterministic model* and a *probabilistic model*. In a deterministic model the relationship between the dependent variable y and the independent variable x is such that if we specify the value of the independent variable, the value of the dependent variable can be determined *exactly*. For example, if a major oil company leases a service station under a contractual agreement of \$500 per month plus 10% of the gross sales, the relationship between the dealer's monthly payment (y) and the gross sales value (x) can be expressed as

$$y = 500 + .10x.$$

With this relationship a June gross sales of \$6,000 would provide a monthly payment of $y = 500 + .10\ (6{,}000) = \$1{,}100$, and a July gross sales of \$7,200 would provide a monthly payment of $y = 500 + .10\ (7{,}200) = \$1{,}220$. This type of relationship is deterministic: once the gross sales value (x) is specified the monthly payment (y) is determined exactly. Figure 14.1 shows graphically the relationship between gross sales and monthly payment.

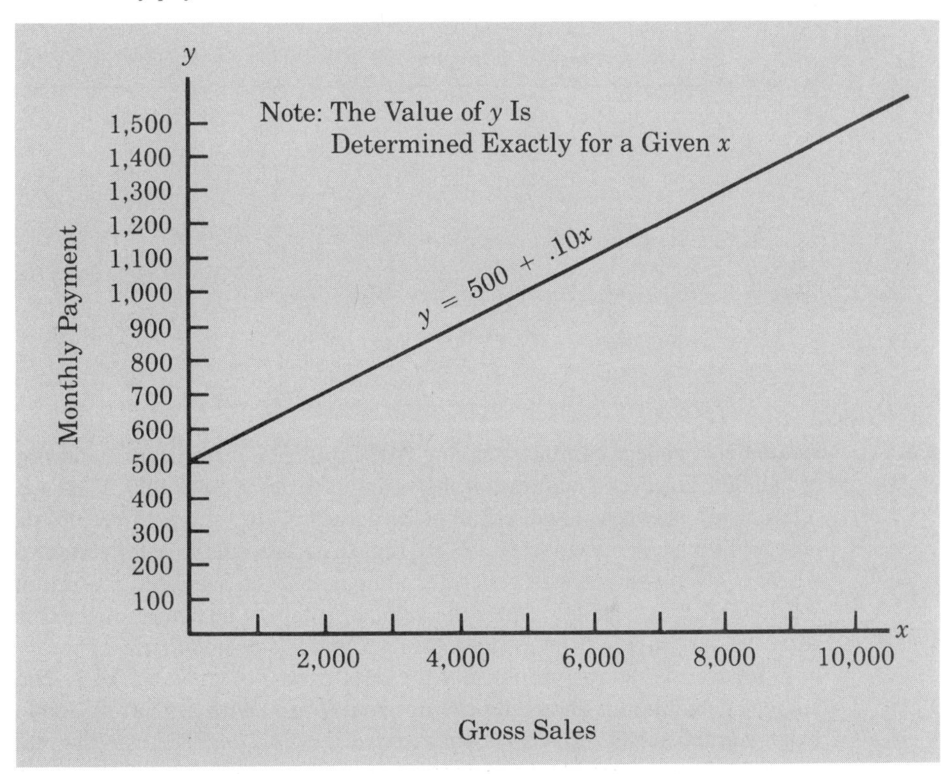

FIGURE 14.1 Illustration of a Deterministic Relationship

To illustrate a relationship between variables that is probabilistic rather than deterministic, let us consider the problem currently being faced by the management of Armand's Pizza Parlors, a chain of Italian-food restaurants located in a five state area. One of the most successful type of location for Armand's has been near a college campus.

Prior to opening a new restaurant Armand's management requires an estimate of yearly sales revenues. Such an estimate is used in planning the appropriate restaurant capacity, making initial staffing decisions, and deciding whether or not the potential revenues justify the cost of the operation.

Suppose management believes that the size of the student population on the nearby campus is related to the annual sales revenues. On an intuitive basis, management believes restaurants located near larger campuses generate more revenue than those near small campuses. To evaluate the relationship between student population (x) and annual sales (y), Armand's collected data from a sample of ten of its restaurants located near college campuses. These data are summarized in Table 14.1. We see that restaurant 1, with $x = 2$ and $y = 58$, is located near a campus with 2,000 students and has annual sales of $58,000; restaurant 2 is located near a campus with 6,000 students and has annual sales of $105,000; and so on.

Using the data in Table 14.1, consider restaurants 3 and 4, both of which are

TABLE 14.1 Data on Student Population and Annual Sales for Ten Armand's Restaurants

Restaurant	x = *Student Population* (1,000's)	y = *Annual Sales* ($1,000's)
1	2	58
2	6	105
3	8	88
4	8	118
5	12	117
6	16	137
7	20	157
8	20	169
9	22	149
10	26	202

located near college campuses having 8,000 students. Restaurant 3 shows annual sales of $88,000; however, restaurant 4 shows annual sales of $118,000. Thus we see that the relationship between y and x cannot be deterministic, since different values of y are observed for the same value of x. Note that this is also the case for restaurants 7 and 8, where a given campus size generates annual sales of $157,000 for restaurant 7 and $169,000 for restaurant 8. Since the value of y cannot be determined exactly from the value of x, we say that the model relating x and y is *probabilistic*.

Figure 14.2 shows graphically the data presented in Table 14.1. The size of the student population is shown on the horizontal axis, with annual sales on the vertical axis. A graph such as this is known as a *scatter diagram*. The usual practice is to plot the independent variable on the horizontal axis and the dependent variable on the vertical axis. The advantage of a scatter diagram is that it provides an overview of the data and enables us to draw preliminary conclusions about a possible relationship between the variables.

What preliminary conclusions can we draw from Figure 14.2? It appears that low sales volumes are associated with small student populations and higher sales volumes are associated with larger student populations. It also appears that the relationship between the two variables can be approximated by a straight line. In Figure 14.3 we have drawn a straight line through the data that appears to provide a good linear approximation of the relationship between the variables. However, observe that the relationship is not perfect. Indeed, few if any of the data items fall exactly on the line.

Since we are unable to guarantee a single value of y for each value of x, the underlying relationship can be explained only with a probabilistic model. Based on the observation that the relationship between student population and annual sales might be approximated by a straight line, we now make the assumption that the following probabilistic model is a realistic representation of the relationship between the two

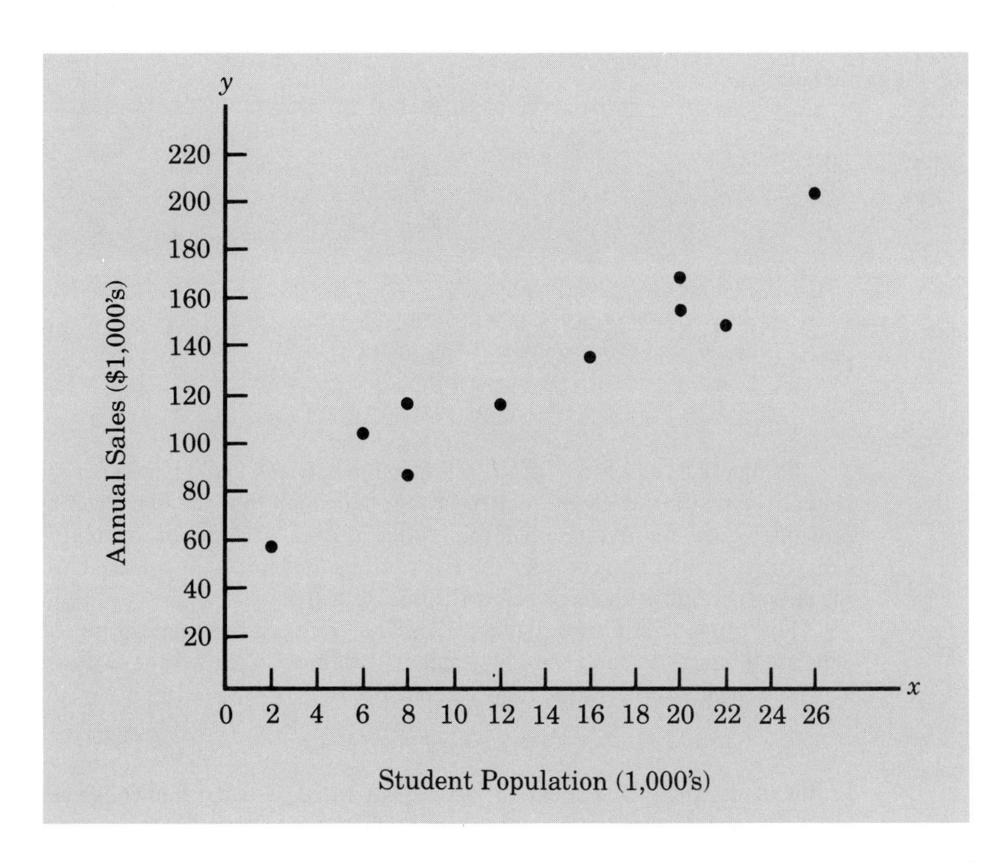

FIGURE 14.2 Scatter Diagram of Annual Sales versus Student Population

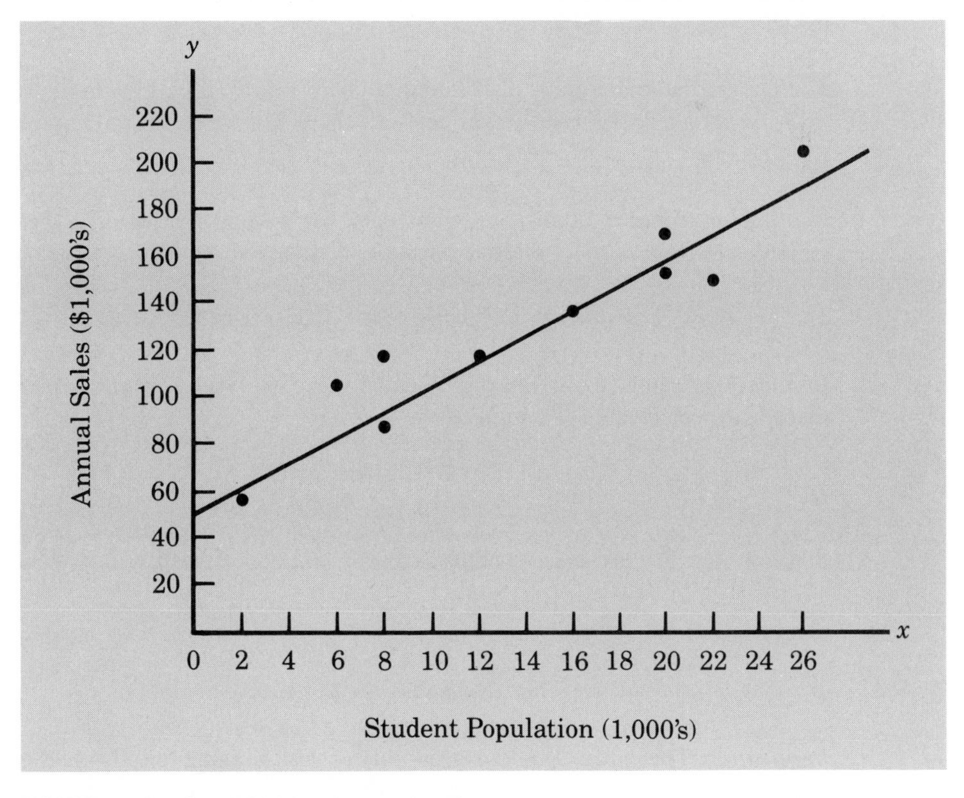

FIGURE 14.3 Straight-Line Approximation

variables:

$$y = \beta_0 + \beta_1 x + \epsilon, \tag{14.1}$$

where

β_0 = y axis intercept of the line given by $\beta_0 + \beta_1 x$,
β_1 = the slope of the line given by $\beta_0 + \beta_1 x$
ϵ = the error or deviation of the actual y value from the line given by $\beta_0 + \beta_1 x$.

Using (14.1) as a model of the relationship between x and y, we are saying that we believe the two variables are related in such a fashion that the line given by $\beta_0 + \beta_1 x$ provides a good approximation of the y value at each x. However, to identify the exact value of y we must also consider the error term ϵ (epsilon) that corresponds to how far the actual y value is above or below the line $\beta_0 + \beta_1 x$.

The probabilistic model that we believe reflects the relationship between the dependent variable y and the independent variable x is called the *regression model.* Thus for Armand's Pizza the regression model is

$$y = \beta_0 + \beta_1 x + \epsilon. \tag{14.2}$$

In this model the independent variable x is treated as being known, since we will be using the model to predict y given information about x. We refer to β_0 (the y intercept) and β_1 (the slope) as the *parameters* of the model.

The following assumptions are made about the error term ϵ in the regression model $y = \beta_0 + \beta_1 x + \epsilon$:

1. The error ϵ is a normally distributed random variable taking on positive or negative values to reflect the deviation between the y value and the value given by $\beta_0 + \beta_1 x$.

Implication. Since β_0 and β_1 are constants, for a known value of x the dependent variable $y = \beta_0 + \beta_1 x + \epsilon$ is also a normally distributed random variable.

2. The error ϵ has a mean or expected value of 0; that is, $E(\epsilon) = 0$.

Implication. Using (14.2) and the expected value operations from Chapter 5 we note that the expected value of y is given by

$$\begin{aligned} E(y) &= E(\beta_0 + \beta_1 x + \epsilon) \\ &= E(\beta_0) + E(\beta_1 x) + E(\epsilon). \end{aligned}$$

Since β_0 and $\beta_1 x$ are known values, $E(\beta_0) = \beta_0$ and $E(\beta_1 x) = \beta_1 x$. Thus since we assume that $E(\epsilon) = 0$,

$$E(y) = \beta_0 + \beta_1 x. \tag{14.3}$$

3. The variance of ϵ is denoted σ^2 and is the same for all values of x.

Implication. This means that the variability of y is the same for all values of x.

4. The values of the error are independent.

Implication. The size of the error for a particular value of x is not related to the size of the error for any other value of x.

Equation (14.3), shown above as an implication of Assumption 2, is referred to as the *regression equation.* In this equation $E(y)$ represents the average of all possible values of y that could occur at a stated value of x.

Let us return to the Armand's Pizza Parlor illustration. Consider any campus with 10,000 students ($x = 10$). The value of $E(y)$ in (14.3) represents the average annual sales that would occur for all restaurants located near college campuses with 10,000 students. Figure 14.4 provides a graphical interpretation of the model assump-

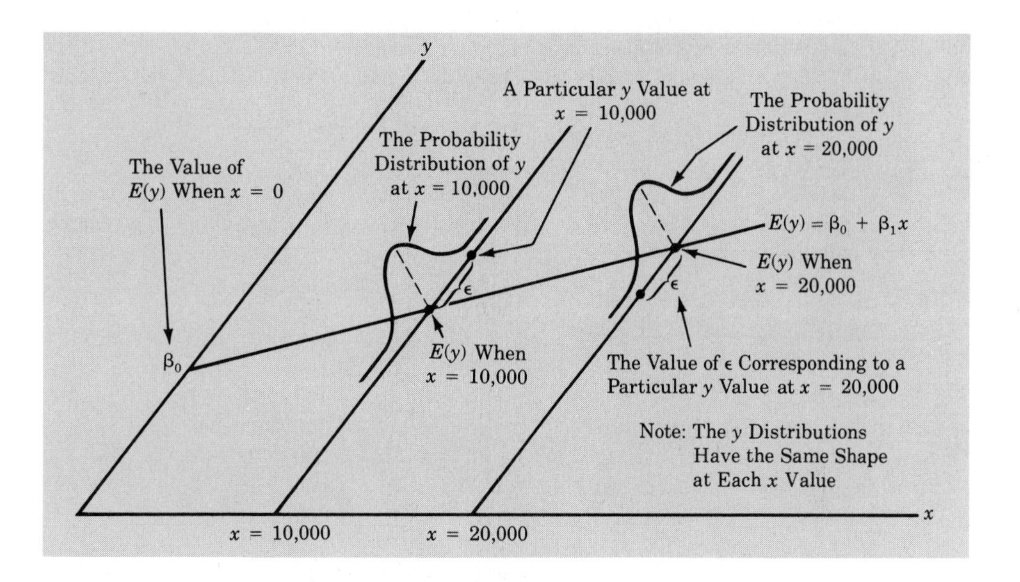

FIGURE 14.4 Illustration of Model Assumptions

tions, and their implications. As shown in Figure 14.4 the value of $E(y)$ changes according to the specific value of x considered. However, note that regardless of the x value the probability distribution of the errors (ϵ) and hence the probability distribution of y about the regression line form a normal distribution, each with the same shape and hence the same variance. The specific value of the error ϵ at any particular point depends upon whether the actual value of y is greater than or less than $E(y)$.

At this point we must keep in mind that we are also making an assumption or hypothesis about the form of the regression model and the associated regression equation. That is, we have assumed that a straight line represented by $\beta_0 + \beta_1 x$ is the basis for the relationship between the variables. We must not lose sight of the fact that some other model, such as $\beta_0 + \beta_1 x^2$, may turn out to be a better model for the underlying relationship. After using the sample data to estimate the parameters of the regression model we will want to conduct further analysis to determine whether or not the specific model that we have assumed appears to be valid.

EXERCISES

1. Consider the two models of the relationship between the variables x and y below where ϵ is an error term:

$$y = 6 + 4x \qquad \text{(Model 1)},$$

$$y = 6 + 4x + \epsilon \qquad \text{(Model 2)}.$$

a. Suppose that $x = 3$ and $\epsilon = 1$. Find the value of y for both models.
b. Suppose that $x = 3$ and $\epsilon = 0$. Find the value of y for both models.
c. State whether each of the above models is deterministic or probabilistic.

2. In simple linear regression it is assumed that a probabilistic model of the form $y = \beta_0 + \beta_1 x + \epsilon$ represents the relationship between the dependent variable y and the independent variable x. It is also assumed that ϵ is a normally distributed random variable with a mean of 0 and variance of σ^2 (see Assumptions 2 and 3).
a. Write the regression equation for $\beta_0 = 2$ and $\beta_1 = 3$.
b. Draw a graph of the regression equation in part a. Put $E(y)$ on the vertical axis and x on the horizontal axis.
c. Write the regression equation for $\beta_0 = 0$ and $\beta_1 = 2$.
d. Graph the regression equation in part c.

3. The following data were collected regarding the monthly starting salaries and the grade point averages (GPA) for undergraduate students who had obtained a degree in business administration:

GPA	*Monthly Salary*
2.6	$1,100
3.4	1,400
3.6	1,800
3.2	1,300
3.5	1,600
2.9	1,200

a. Develop a scatter diagram for these data with GPA on the horizontal axis.
b. What does the scatter diagram developed in part a indicate about the relationship between the two variables?
c. Draw a straight line through the data to approximate a linear relationship between GPA and salary.
d. Do the deviations about the line seem to satisfy the assumptions for the error term in the model of (14.1)?

4. Eddie's Restaurants collected the following data on the relationship between advertising and sales at a sample of five restaurants:

Advertising Expenditures ($1,000's)	*Sales ($1,000's)*
1.0	19.0
4.0	44.0
6.0	40.0
10.0	52.0
14.0	53.0

a. Develop a scatter diagram for these data with advertising expenditures on the horizontal axis and sales on the vertical axis.
b. Draw a graph of the straight line $E(y) = 24 + 2.5x$ on the scatter diagram. Does the line appear to provide a good approximation to the relationship between advertising expenditures and sales?
c. Draw a graph of the curve $E(y) = 60x/(2 + x)$. [Hint: Find values of $E(y)$ for $x = 1, 4, 6,$ 10, and 14 and draw a smooth curve through the points.] Does this curve provide a good approximation to the relationship between advertising expenditures and sales?
d. Does the curvilinear approximation of part c provide a better fit to the data than the linear approximation of part b?
e. For the advertising expenditures in the sample data compute the errors obtained by using the equations in both parts b and c to predict sales. For which equation is the sum of absolute values of the errors greatest?

5. Tyler Realty collected the following data regarding the selling price of new homes and the size of the homes measured in terms of square footage of living space:

Square Footage	*Selling Price*
2,500	$124,000
2,400	$108,000
1,800	$ 92,000
3,000	$146,000
2,300	$110,000

a. Develop a scatter diagram for these data with square footage on the horizontal axis.
b. Try to approximate the relationship between square footage and selling price by drawing a straight line through the data.
c. Does there appear to be a linear relationship? If so, do the assumptions about the error term seem reasonable?

6. The owner of a local grocery store varied the price of a 1 pound loaf of bread for six consecutive weeks. The following data show the price per loaf and the number of loaves sold that week:

Price	*Loaves Sold*
$.60	220
$.62	200
$.58	280
$.60	250
$.64	190
$.62	240

a. Develop a scatter diagram for the above data with price on the horizontal axis.
b. What does the scatter diagram developed in part a indicate about the relationship between price and the number of loaves sold?

14.2 ESTIMATING β_0 AND β_1: THE LEAST SQUARES METHOD

In Section 14.1 we hypothesized the following regression model relating x, the size of the student population, and y, the annual sales: $y = \beta_0 + \beta_1 x + \epsilon$. Moreover, we saw that the assumption that the mean or expected value of ϵ is 0 leads to a regression equation given by $E(y) = \beta_0 + \beta_1 x$ [see equation (14.3)]. Recall that β_0 and β_1 are unknown constants referred to as the parameters of the regression model. In this section we will see how a technique known as the least squares method can be used to develop estimates of β_0 and β_1 based upon the data available.

Using the least squares method to develop estimates of β_0 and β_1 leads to the following *estimated regression equation:*

$$\hat{y} = b_0 + b_1 x, \tag{14.4}$$

where

$$\hat{y} = \text{estimate of } E(y),$$
$$b_0 = \text{estimate of } \beta_0,$$
$$b_1 = \text{estimate of } \beta_1.$$

Note that the estimated regression equation is similar to the true regression equation with $\hat{y}$ substituted for $E(y)$, b_0 substituted for β_0, and b_1 substituted for β_1.

Let us now define

y_i = ith data value (observation) for the dependent variable,
x_i = ith data value (observation) for the independent variable,
$\hat{y}_i = b_0 + b_1 x_i$ = estimated value of the dependent variable corresponding to the ith data value for the independent variable.

The least squares method chooses the estimated regression equation that makes the sum of squares of the differences between the observed values of the dependent variable (y_i) and the estimated values of the dependent variable ($\hat{y}_i$) a minimum. In other words, the least squares method is a procedure for determining the values of b_0 and b_1 (and hence $\hat{y}_i$) such that these values minimize

$$\sum_{i=1}^{n} (y_i - \hat{y}_i)^2. \tag{14.5}$$

Using differential calculus it can be shown (see the appendix to this chapter) that the values of b_0 and b_1 that minimize (14.5) can be found by using (14.6) and (14.7):

$$b_1 = \frac{\Sigma (x_i - \bar{x})(y_i - \bar{y})}{\Sigma (x_i - \bar{x})^2} = \frac{\Sigma x_i y_i - (\Sigma x_i \Sigma y_i)/n}{\Sigma x_i^2 - (\Sigma x_i)^2/n} \tag{14.6}$$

$$b_0 = \bar{y} - b_1 \bar{x}, \tag{14.7}$$

where

x_i = value of the independent variable for the ith observation,
y_i = value of the dependent variable for the ith observation,
$\bar{x}$ = mean value for the independent variable,
$\bar{y}$ = mean value for the dependent variable,
n = total number of observations.

The summations above are taken over all n values of x and y. Note that in (14.6) we show two formulas for b_1. The second is more convenient when using a calculator to compute b_1.

Some of the calculations necessary to develop the least squares estimated regression equation for the Armand's Pizza problem are shown in Table 14.2. In our

TABLE 14.2 Calculations Necessary to Develop the Least Squares Estimated Regression Equation for Armand's Pizza

Restaurant (i)	x_i	y_i	$x_i y_i$	x_i^2
1	2	58	116	4
2	6	105	630	36
3	8	88	704	64
4	8	118	944	64
5	12	117	1,404	144
6	16	137	2,192	256
7	20	157	3,140	400
8	20	169	3,380	400
9	22	149	3,278	484
10	26	202	5,252	676
Totals	140	1,300	21,040	2,528
	Σx_i	Σy_i	$\Sigma x_i y_i$	Σx_i^2

example there are ten restaurants or observations; hence $n = 10$. Using (14.6) and (14.7), we can now compute the slope and intercept of the estimated regression equation for Armand's Restaurants. The calculation of the slope (b_1) proceeds as follows:

$$\begin{aligned} b_1 &= \frac{\Sigma x_i y_i - (\Sigma x_i \Sigma y_i)/n}{\Sigma x_i^2 - (\Sigma x_i)^2/n} \\ &= \frac{21{,}040 - (140)(1300)/10}{2528 - (140)^2/10} \\ &= \frac{2840}{568} \\ &= 5. \end{aligned}$$

The calculation of the y intercept (b_0) is as follows:

$$\bar{x} = \frac{\Sigma x_i}{n} = \frac{140}{10} = 14,$$

$$\bar{y} = \frac{\Sigma y_i}{n} = \frac{1300}{10} = 130,$$

$$\begin{aligned} b_0 &= \bar{y} - b_1 \bar{x} \\ &= 130 - 5(14) \\ &= 60. \end{aligned}$$

Thus the estimated regression equation found by using the method of least squares is

$$\hat{y} = 60 + 5x.$$

In Figure 14.5 we show the graph of this equation.

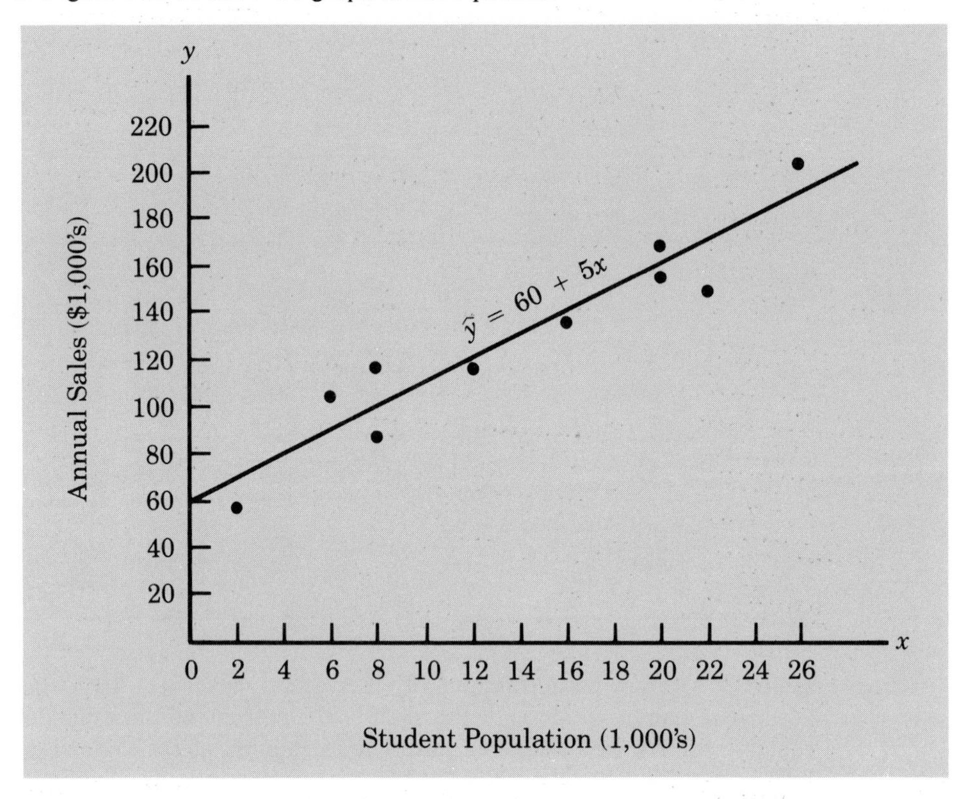

FIGURE 14.5 **Graph of the Estimated Regression Equation for Armand's Pizza: $\hat{y} = 60 + 5x$**

The slope of the estimated regression equation ($b_1 = 5$) is positive, implying that as student population increases, annual sales increase. In fact we can conclude (since sales are measured in \$1,000's and student population in 1,000's) that an increase in the student population of 1,000 is associated with an increase of \$5,000 in expected annual sales; that is, sales are expected to increase by \$5.00 per student.

If we believe that the least squares estimated regression equation adequately describes the relationship between x and y, then it would seem reasonable to use the estimated regression equation to predict the value of y for a given value of x. For example, if we wanted to predict annual sales for a restaurant location near a campus with 16,000 students we would compute

$$\begin{aligned} \hat{y} &= 60 + 5(16) \\ &= 140. \end{aligned}$$

Hence we would expect sales of \$140,000 per year. In the following sections we will discuss methods for assessing the appropriateness of using the estimated regression equation for estimation and prediction.

Rounding Inaccuracies

Although it was not necessary to do any rounding in computing b_0 and b_1 for the Armand's Pizza problem, we recommend that when you are solving such problems using a calculator you carry as many significant digits as possible. To see why, suppose that in computing b_1 we find that $b_1 = 1{,}022/1{,}960 = .5214$ (to four decimal places of accuracy). If $\bar{x} = 4{,}000$ and $\bar{y} = 2{,}100$, then we obtain $b_0 = 2{,}100 - .5214(4{,}000) = 14.4$. The estimated regression equation would then be

$$\hat{y} = 14.4 + .5214x.$$

If we had rounded the calculation of b_1 to two decimal places we would have obtained $b_1 = .52$ and $b_0 = 2{,}100 - .52(4{,}000) = 20$. The estimated regression equation would then have been

$$\hat{y} = 20 + .52x.$$

Compare the two estimated regression equations. We see that the values of b_1 are the same to two decimal places, but the values of b_0 differ by 5.60. Thus we see that preliminary rounding of b_1 produces a rounding error in the estimate of b_0. For the remainder of this chapter we will follow the practice of computing values of b_0 and b_1 to at least four decimal places. Rounding of these values to fewer places will be done only after the calculations have been completed.

EXERCISES

7. Given below are five observations taken for two variables, x and y:

Observation	x_i	y_i
1	2	25
2	3	25
3	5	20
4	1	30
5	8	16

a. Develop a scatter diagram for this data.
b. Graph the line $\hat{y} = 30 - 2x$ on the scatter diagram. Does it appear to provide a good fit to the data?
c. Use the method of least squares to compute an estimated regression equation for the data.
d. Compute $\Sigma\,(y_i - \hat{y}_i)^2$ for the lines in parts b and c. Which has the smallest sum of squared differences? Why?

8. Shown below are the data from Exercise 3 concerning GPA and monthly starting salaries for business administration students:

GPA	*Salary*
2.6	\$1,100
3.4	1,400
3.6	1,800
3.2	1,300
3.5	1,600
2.9	1,200

a. Use the least squares method to develop the estimated regression equation.
b. Predict the monthly starting salary for a student with a 3.0 grade point average and for a student with a 3.5 grade point average.

9. Shown below are the data from Exercise 5 concerning square footage and selling price for new homes:

Square Footage (1,000's)	*Selling Price ($1,000's)*
2.5	124
2.4	108
1.8	92
3.0	146
2.3	110

a. Develop an estimated regression equation using the least squares method.
b. Predict the selling price for a home with 2,700 square feet.

10. The data from Exercise 6 concerning price per loaf of bread and number of loaves sold are repeated below:

Price	*Loaves Sold*
$.60	220
$.62	200
$.58	280
$.60	250
$.64	190
$.62	240

a. Develop an estimated regression equation that can be used to predict the number of loaves sold given the price.
b. Predict the number of loaves sold at a price of $.63.
c. Use the estimated regression equation to predict the number of loaves sold at a price of $.64. How close does this predicted value come to the number of loaves the grocer actually sold at a price of $.64?

11. Shown below are some data that a sales manager has collected concerning annual sales and years of experience:

Salesperson	*Years of Experience*	*Annual Sales ($1,000's)*
1	1	80
2	3	97
3	4	92
4	4	102
5	6	103
6	8	111
7	10	119
8	10	123
9	11	117
10	13	136

a. Develop a scatter diagram for these data.
b. Graph the line $\hat{y} = 78 + 4.2x$ on the scatter diagram.

c. For the line in part b compute $\hat{y}_i$ for each data value.

d. For the line in part b compute $\Sigma (y_i - \hat{y}_i)^2$. This is the quantity that is minimized by the least squares line.

12. Develop the least squares estimated regression equation for the data in Exercise 11.

a. Compute $\Sigma (y_i - \hat{y}_i)^2$ for the least squares line.

b. Compare the value of $\Sigma (y_i - \hat{y}_i)^2$ for the least squares line with that obtained using the line $\hat{y} = 78 + 4.2x$ (see Exercise 11d). Which is smaller?

14.3 THE COEFFICIENT OF DETERMINATION

We have introduced the least squares method as a technique for finding b_0 and b_1 by minimizing the sum of squares of the differences between the observed values of the dependent variable (y_i) and the predicted values of the dependent variable ($\hat{y}_i$). Note that the differences between y_i and $\hat{y}_i$ actually represent the errors in using $\hat{y}_i$ to estimate y_i. That is, the error for the ith observation is $y_i - \hat{y}_i$. This difference is also referred to as the ith *residual.* Thus the resulting sum of squares is referred to as the sum of squares due to error, or the residual sum of squares. For convenience, we use SSE to represent this quantity.* Hence

$$\text{Sum of squares due to error} = \text{SSE} = \Sigma (y_i - \hat{y}_i)^2. \qquad (14.8)$$

Table 14.3 shows the calculations required to compute SSE for the Armand's Pizza

TABLE 14.3 Calculation of SSE for the Armand's Pizza Parlors Problem

Restaurant (i)	x_i = *Student Population* (1,000's)	y_i = *Annual Sales* ($1,000's)	$\hat{y}_i = 60 + 5x_i$	$y_i - \hat{y}_i$	$(y_i - \hat{y}_i)^2$
1	2	58	70	−12	144
2	6	105	90	15	225
3	8	88	100	−12	144
4	8	118	100	18	324
5	12	117	120	−3	9
6	16	137	140	−3	9
7	20	157	160	−3	9
8	20	169	160	9	81
9	22	149	170	−21	441
10	26	202	190	12	144
				SSE	1,530

Parlors problem. Thus SSE = 1,530 is a measure of the error involved in using the estimated regression equation $\hat{y} = 60 + 5x$ to predict y_i.

Now suppose that we were asked to develop an estimate of annual sales without using the size of the student population. We could not use the estimated regression equation and would have to use the value of the sample mean, $\bar{y} = 130$, as the best estimate of annual sales. In Table 14.4 we show the errors that would result from using

*SSE was the criterion minimized in using the least squares procedure to develop an estimated regression equation.

TABLE 14.4 Computation of Total Sum of Squares About the Mean ($\bar{y} = 130$)

Restaurant (i)	x_i = *Student Population* *(1,000's)*	y_i = *Annual Sales* *($1,000's)*	$(y_i - \bar{y})$	$(y_i - \bar{y})^2$
1	2	58	−72	5,184
2	6	105	−25	625
3	8	88	−42	1,764
4	8	118	−12	144
5	12	117	−13	169
6	16	137	7	49
7	20	157	27	729
8	20	169	39	1,521
9	22	149	19	361
10	26	202	72	5,184
				15,730
				SST

$\bar{y}$ to estimate annual sales at the ten Armand's Pizza Parlors. The corresponding sum of squares, denoted SST, is referred to as the sum of squares about the mean. Hence

$$\text{Sum of squares about the mean} = \text{SST} = \Sigma\,(y_i - \bar{y})^2. \tag{14.9}$$

For the Armand's Pizza Parlors problem SST = 15,730.

In Figure 14.6 we show the least squares regression line $\hat{y} = 60 + 5x$ and the line corresponding to $y = \bar{y} = 130$. Note in general that the points cluster more closely around the estimated regression line than they do about the line $\bar{y} = 130$. For example, for the tenth restaurant we see that the error is much larger when $\bar{y}$ is used as an estimate of y_{10} than when $\hat{y}_{10}$ is used. We can think of SST as a measure of how well the observations cluster about the $\bar{y}$ line and SSE as a measure of how well the observations cluster about the $\hat{y}$ line.

Based upon the discussion above we conclude that SST is a measure of error when regression analysis is not used and SSE is a measure of error when regression analysis is used. Thus the difference between SST and SSE must be the sum of squares "explained by" the estimated regression equation. This sum of squares is commonly called the sum of squares due to regression, denoted SSR. The sum of squares due to regression can be written as

$$\text{SSR} = \Sigma\,(\hat{y}_i - \bar{y})^2.$$

The relationship among SSE, SST, and SSR is given by

$$\text{SST} = \text{SSR} + \text{SSE}. \tag{14.10}$$

Because of (14.10) SST is usually referred to as the total sum of squares.

Substituting for SST, SSR and SSE in (14.10) we obtain

$$\Sigma\,(y_i - \bar{y})^2 = \Sigma\,(\hat{y}_i - \bar{y})^2 + \Sigma\,(y_i - \hat{y}_i)^2.$$

Using (14.10), then, we can conclude that the sum of squares explained by the

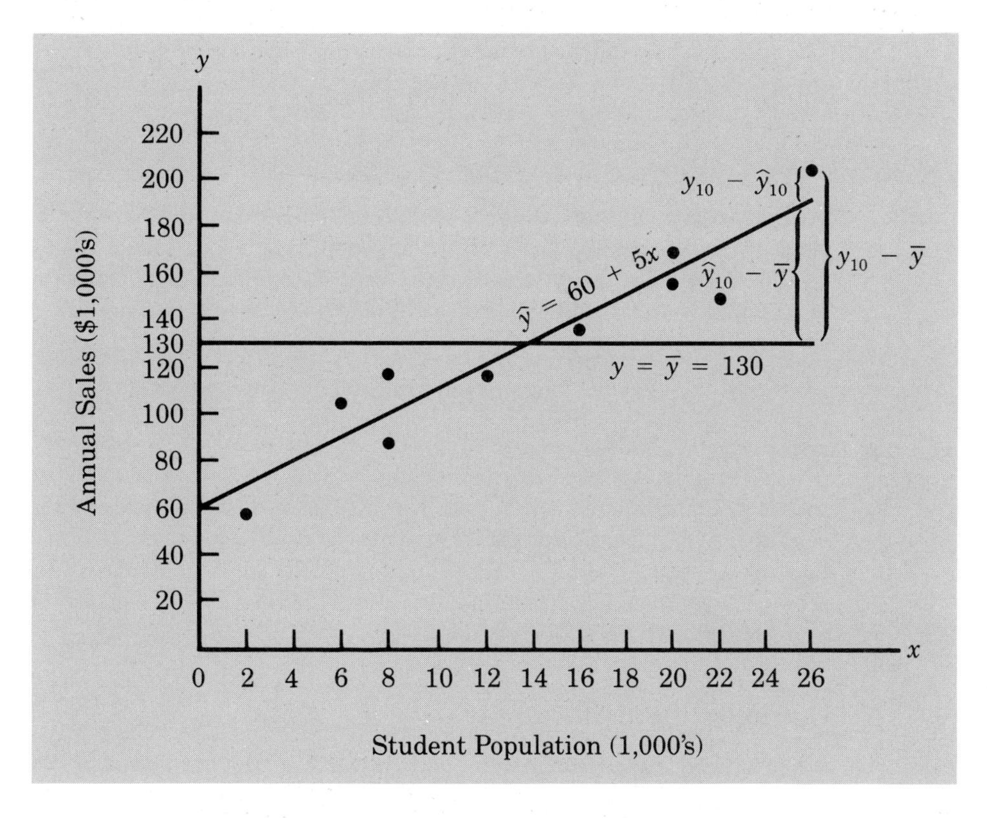

FIGURE 14.6 Deviations About the Estimated Regression Line and the Line $y = \bar{y}$ for the Armand's Pizza Problem

regression relationship for the Armand's Pizza Parlor problem is

$$\text{SSR} = \text{SST} - \text{SSE} = 15{,}730 - 1{,}530 = 14{,}200.$$

Now let us see how these sums of squares can be used to provide a measure of the strength of the regression relationship. We would have the strongest possible relationship if every observation happened to lie on the least squares fitted line—the line would pass through each point, and we would have SSE = 0. Hence for a perfect fit SST must equal SSR, and thus the ratio SSR/SST = 1. On the other hand, a poorer fit to the observed data results in a larger SSE. Since SST = SSE + SSR, however, the largest SSE (and hence worst fit) would occur when SSE = SST. In this case SSR = 0, and the estimated regression line does not help predict y. Thus the worst possible fit corresponds to SSR = 0 and the ratio SSR/SST = 0.

If we were to use the ratio SSR/SST to evaluate the strength of the regression relationship, we would have a measure that could take on values between 0 and 1, with values closer to 1 implying a stronger relationship.

The fraction SSR/SST is called the *coefficient of determination* and is represented by the symbol r^2:

Coefficient of Determination

$$r^2 = \frac{\text{SSR}}{\text{SST}}. \qquad (14.11)$$

The value of the coefficient of determination for our example is

$$r^2 = \frac{\text{SSR}}{\text{SST}} = \frac{14{,}200}{15{,}730} = .90.$$

To better interpret r^2, we can think of SST as the measure of how good $\bar{y}$ is as a predictor of annual sales volume. After developing the estimated regression equation, we compute SSE as the measure of the goodness of $\hat{y}$ as a predictor of annual sales volume. Thus SSR, the difference between SST and SSE, really measures the portion of SST that is explained by the estimated regression equation. Thus we can think of r^2 as

$$r^2 = \frac{\text{Sum of squares explained by regression}}{\text{Total sum of squares (before regression)}}.$$

When it is expressed as a percentage, r^2 can be interpreted as the percentage of the total sum of squares (SST) that can be explained using the estimated regression equation. Statisticians often use r^2 as a measure of the goodness of fit of a regression line to the data. Thus for our sample problem we conclude that the estimated regression equation has accounted for 90% of the total sum of squares. We should be very pleased with such a large r^2 value.

Computational Efficiencies

When using a calculator to compute the value of the coefficient of determination, computational efficiencies can be realized by computing SSR directly (instead of computing SSR = SST − SSE). It can be shown that

$$\text{SSR} = \frac{[\Sigma\, x_i y_i - (\Sigma\, x_i\, \Sigma\, y_i)/n]^2}{\Sigma\, x_i^2 - (\Sigma\, x_i)^2/n}. \tag{14.12}$$

In addition, we need not compute SST using the expression $\Sigma\,(y_i - \bar{y})^2$; this expression can be algebraically expanded to provide

$$\text{SST} = \Sigma\, y_i^2 - (\Sigma\, y_i)^2/n. \tag{14.13}$$

For the Armand's Pizza Parlors problem, part of the calculations needed to compute SSR and SST using the above formulas are shown in Table 14.5. Using the values in this table along with (14.12) and (14.13) we can compute SSR and SST as follows:

$$\begin{aligned}\text{SSR} &= \frac{[21{,}040 - (140 \times 1{,}300)/10]^2}{2{,}528 - (140)^2/10}\\ &= \frac{8{,}065{,}600}{568}\\ &= 14{,}200.\\ \text{SST} &= 184{,}730 - (1300)^2/10\\ &= 15{,}730.\end{aligned}$$

Note that since these are equivalent formulas we get the same values for SSR and SST that we obtained previously. Thus as before

$$r^2 = \frac{\text{SSR}}{\text{SST}} = \frac{14{,}200}{15{,}730} = .90.$$

TABLE 14.5 Calculations Used in Computing r^2

Restaurant (i)	x_i	y_i	x_iy_i	x_i^2	y_i^2
1	2	58	116	4	3,364
2	6	105	630	36	11,025
3	8	88	704	64	7,744
4	8	118	944	64	13,924
5	12	117	1,404	144	13,689
6	16	137	2,192	256	18,769
7	20	157	3,140	400	24,649
8	20	169	3,380	400	28,561
9	22	149	3,278	484	22,201
10	26	202	5,252	676	40,804
Totals	140	1,300	21,040	2,528	184,730
	Σx_i	Σy_i	Σx_iy_i	Σx_i^2	Σy_i^2

EXERCISES

13. Refer again to the data in Exercise 7.

a. Compute SSE $= \Sigma (y_i - \hat{y}_i)^2$.

b. Compute SST $= \Sigma (y_i - \bar{y})^2$.

c. Compute SSR.

d. Compute the coefficient of determination, r^2.

14. Refer again to the data in Exercise 8.

a. Use (14.12) to compute SSR.

b. Use (14.13) to compute SST.

c. What percentage of the total sum of squares is accounted for by the regression model?

15. Refer again to the data in Exercise 9.

a. Compute SSE, SST, and SSR using (14.8), (14.9), and (14.10).

b. Now compute SSR and SST using (14.12) and (14.13). Do you get the same results as in part a?

c. Compute the coefficient of determination. Comment on the strength of the regression relationship.

16. A medical laboratory estimates the amount of protein in liver samples through the use of a regression model. A spectrometer emitting light shines through a substance containing the sample, and the amount of light absorbed is used to estimate the amount of protein in the sample. A new regression formula is developed daily because of differing amounts of dye in the solution. On one day six samples with known protein concentrations gave the following absorbance readings:

Absorbance Reading (x)	*mgs of Protein* (y)
0.509	0
0.756	20
1.020	40
1.400	80
1.570	100
1.790	127

a. Use these data to develop an estimated regression equation relating the light absorbance reading to milligrams of protein present in the sample.

b. Compute r^2. Would you feel comfortable using this regression model to estimate the amount of protein in a sample?
c. In a sample just received the light absorbance reading was 0.941. Estimate the amount of protein in the sample.
17. There is a similarity between (14.6) and (14.12). State a formula for computing SSR once b_1 has been computed. Note that once b_1 has been calculated little additional work is necessary to compute SSR.
18. The data from Exercise 7 are repeated below:

Observation	x_i	y_i
1	2	25
2	3	25
3	5	20
4	1	30
5	8	16

a. Compute $\bar{x}$ and $\bar{y}$.
b. Substitute the values of $\bar{x}$ and $\bar{y}$ for x and $\hat{y}$ in the estimated regression equation. Do these values satisfy the equation?
c. Will the least squares line always pass through the point corresponding to $(\bar{x}, \bar{y})$? Why or why not?

14.4 TESTING FOR SIGNIFICANCE

In the previous section we saw how the coefficient of determination (r^2) could be used as a measure of the strength of the regression relationship. Larger values of r^2 indicated a stronger relationship, smaller values a weaker one. However, this measure did not allow us to conclude that the relationship was or was not statistically significant. In order to draw conclusions concerning statistical significance we must take the sample size into consideration. It makes intuitive sense that a regression relationship involving 20 sample points would be more statistically significant than one involving 5 sample points even with the same r^2 value. In this section we show how to conduct significance tests that will allow us to draw conclusions about the existence of a regression relationship.

An Estimate of σ^2

In the previous section we used the sum of squares due to error, SSE, as a measure of the variability of the actual observations about the estimated regression line. This quantity is also used to develop an estimate of σ^2, the variance of ϵ, and consequently the variance of the actual y values about the regression line. Recall from our earlier definition of sample variance that we divided the sum of the squared deviations about the sample mean by $n - 1$ to obtain an unbiased estimate of the population variance. We used $n - 1$ instead of n because 1 degree of freedom was lost when the sample mean was used to compute the sum of the squared deviations about the mean. One degree of freedom was lost because 1 parameter used in computing the sum of squares, the population mean, had to be estimated from the sample data. In regression analysis we must estimate the parameters β_0 and β_1 to compute SSE; that is, we use the estimates b_0 and b_1—obtained from the sample data—to compute the sum of squares due to error. To see how this is done, note that SSE can be written

$$\text{SSE} = \Sigma\,(y_i - \hat{y}_i)^2 = \Sigma\,(y_i - b_0 - b_1 x_i)^2. \tag{14.14}$$

For this reason two degrees of freedom are lost; hence we must divide SSE by $n - 2$ to obtain an unbiased estimate of σ^2. The estimate obtained is called the mean square due to error, denoted MSE:

$$\text{Estimate of } \sigma^2 = \text{MSE} = \frac{\text{SSE}}{n - 2}. \tag{14.15}$$

From the data in Table 14.3 we see that SSE = 1,530; thus for the Armand's Pizza example we have

$$\text{MSE} = \frac{1{,}530}{8} = 191.25.$$

In the discussion that follows we will see that this unbiased estimate of σ^2 is used in computing the F statistic in the test for the significance of the regression equation.

F Test

Recall that the underlying regression equation is assumed to be $E(y) = \beta_0 + \beta_1 x$. If there really exists a relationship of this form between x and y, β_1 would have to be different from 0. Thus a conclusion regarding the significance of the relationship can be tested using the following hypotheses:

$$H_0\colon \beta_1 = 0,$$
$$H_1\colon \beta_1 \neq 0.$$

The logic behind the use of the F test for determining whether or not the relationship between x and y is statistically significant is based upon our being able to develop two independent estimates of σ^2. We have just seen that MSE provides an estimate of σ^2. If the null hypothesis H_0: $\beta_1 = 0$ is true, the mean square due to regression (denoted MSR) provides another *independent* estimate of σ^2.

To compute MSR we first note that for any sum of squares the mean square is the sum of squares divided by its degrees of freedom. The number of degrees of freedom for the sum of squares due to regression, SSR, is always equal to the number of independent variables. Since in this chapter we are concerned only with models involving one independent variable, the number of regression degrees of freedom is 1. Using DF as an abbreviation for degrees of freedom we can write

$$\text{Mean square due to regression} = \text{MSR} = \frac{\text{SSR}}{\text{Regression DF}}$$

$$\text{MSR} = \frac{\text{SSR}}{\text{Number of independent variables}}. \tag{14.16}$$

For the Armand Pizza Parlor problem, we find that MSR = SSR/1 = 14,200.

If the null hypothesis (H_0:$\beta_1 = 0$) is true, MSR and MSE are two independent estimates of σ^2. In this case the sampling distribution of MSR/MSE follows an F distribution with numerator degrees of freedom equal to 1 and denominator degrees of freedom equal to $n - 2$. The test concerning the significance of the regression relationship is based on the following F statistic:

$$F = \frac{\text{MSR}}{\text{MSE}}. \tag{14.17}$$

Given any sample size, the numerator of the F statistic will increase as more of the variability in y is explained by the regression model and decrease as less is explained. Similarly, the denominator will increase if there is more variability about the estimated regression line and decrease if there is less variability. Thus one would intuitively expect large values of $F = \text{MSR}/\text{MSE}$ to cast doubt on the null hypotheses and lead us to the conclusion that there is a significant relationship between x and y. Indeed, this is correct; small values of F lead to acceptance of H_0, and large values lead to rejection.

Let us now conduct the F test for our Armand's Pizza Parlor problem. Assume that the maximum acceptable Type I error is $\alpha = .01$. From Table 4 of Appendix B we can determine the critical F value by locating the value corresponding to numerator degrees of freedom equal to 1 (the number of independent variables) and denominator degrees of freedom equal to $n - 2 = 10 - 2 = 8$. Thus we obtain $F = 11.26$. Hence the appropriate decision rule for our Armand's Pizza problem is written

$$\text{Accept } H_0 \text{ if MSR/MSE} \leq 11.26,$$

$$\text{Reject } H_0 \text{ if MSR/MSE} > 11.26.$$

Since $\text{MSR/MSE} = 14{,}200/191.25 = 74.25$ is greater than the critical value ($F_{.01} = 11.26$), we can reject H_0 and conclude that there is a significant statistical relationship between annual sales and the size of the student population.

We caution here that rejection of H_0 does not yet permit us to conclude that the relationship between x and y is *linear*. However, it is valid to conclude that x and y are related and that a linear approximation explains a significant amount of the variability in y over the range of x values observed in our sample. To illustrate this qualification we call your attention to Figure 14.7, where an F test (on $\beta_1 = 0$) yielded the conclusion that x and y were related. The figure shows that the actual relationship is nonlinear. In the graph we see that the linear approximation is very good for the values of x used in developing the least squares line, but it is very bad for larger values of x.

Given a significant relationship, we should feel confident in using the regression equation for predictions corresponding to x values within the range of the x values for our sample. For the Armand's Pizza Parlor problem this corresponds to values of x between 2 and 26. But unless there are reasons to believe the model is valid beyond this range, predictions outside the range of the independent variable should be made with caution. For the Armand's Pizza problem, since the regression relationship has been found significant at the .01 level, we should feel confident using it to predict sales whenever the student population is between 2,000 and 26,000.

t Test

The F test has been used to test the null hypothesis H_0: $\beta_1 = 0$. A t test also exists for testing this null hypothesis. In regression models with only one independent variable the t test and the F test yield the same results. But with more than one independent

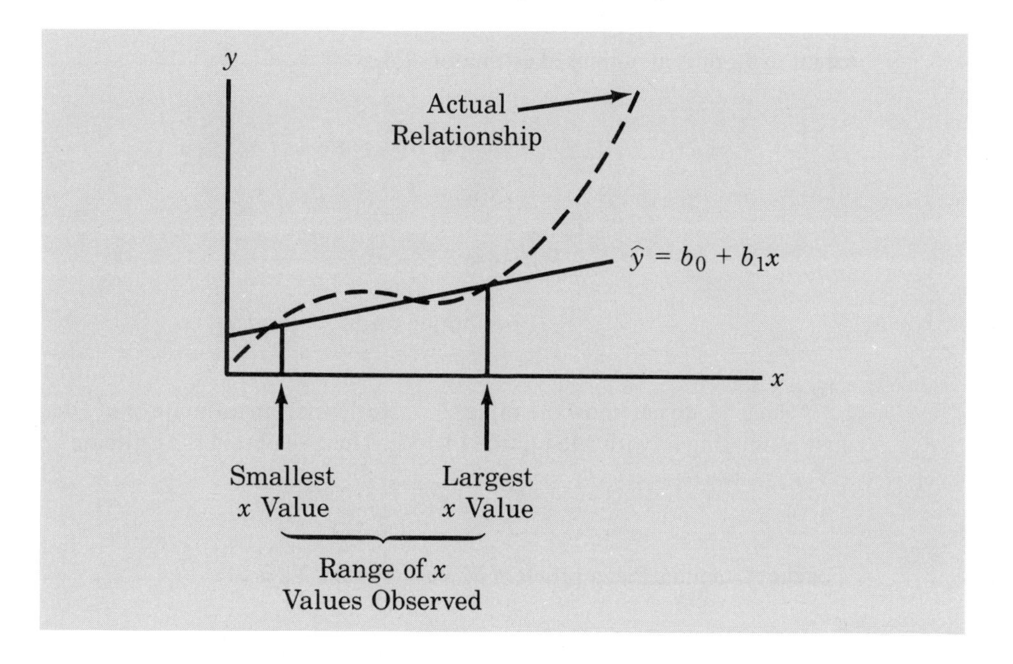

FIGURE 14.7 Example of a Linear Approximation of a Nonlinear Relationship

variable only the F test can be used to test for a significant relationship between a dependent variable and a set of independent variables. The t test is then used to determine whether or not the coefficient of a particular independent variable is zero. Here we will introduce the t test and show that it leads to the same conclusion as the F test. In the next chapter it will be used to test the significance of individual parameters in a multiple regression model.

The hypotheses we will be testing are the same as before:

$$H_0\colon \beta_1 = 0,$$
$$H_1\colon \beta_1 \neq 0.$$

Before presenting the t test we need to consider the properties of b_1, the least squares estimator of β_1. First, let us consider what would have happened if we had used a different random sample for the same regression study. For example, suppose that in our regression study relating annual sales to student population we had used the sales records of ten different restaurants. A regression analysis of this new data set, or sample, probably would result in an estimated regression equation similar to our previous estimated regression equation $\hat{y} = 60 + 5x$. However, it is doubtful that we would obtain an intercept of exactly 60 and a slope of exactly 5. Thus b_0 and b_1 computed using the least squares formulas are themselves random variables whose values depend upon which data items are included in the sample. Recall the discussion of sampling distributions in Chapter 7 and note that b_0 and b_1 must have their own sampling distributions.

Using the least squares formula for b_1 and the assumptions regarding ϵ, the sampling distribution of b_1 is defined below. Note that since the expected value of b_1 is equal to β_1, b_1 is an unbiased estimator of β_1:

Sampling Distribution of b_1

$$\text{Mean: } E(b_1) = \beta_1, \tag{14.18}$$

$$\text{Variance: } \sigma_{b_1}^2 = \sigma^2 \frac{1}{\Sigma x_i^2 - n\bar{x}^2}, \tag{14.19}$$

Distribution form: Normal.

Since we do not know the value of σ^2, we develop an estimate of $\sigma_{b_1}^2$, denoted $s_{b_1}^2$, by first estimating σ^2 with MSE [see (14.15)]. Thus we obtain the estimate

$$s_{b_1}^2 = \text{MSE} \frac{1}{\Sigma x_i^2 - n\bar{x}^2}. \tag{14.20}$$

For the Armand's Pizza problem MSE = 191.25. Thus

$$\begin{aligned} s_{b_1}^2 &= 191.25 \frac{1}{2{,}528 - 10(14)^2} \\ &= 191.25 \frac{1}{568} \\ &= .3367. \end{aligned}$$

Hence

$$s_{b_1} = \sqrt{.3367} = .5803.$$

The t test regarding β_1 is based on the fact that

$$\frac{b_1 - \beta_1}{s_{b_1}}$$

follows a t distribution with $n - 2$ degrees of freedom. If the null hypothesis is true, then $\beta_1 = 0$ and we find that b_1/s_{b_1} follows a t distribution with $n - 2$ degrees of freedom. Using b_1/s_{b_1} as a standardized test statistic we would use the following decision rule to test H_0: $\beta_1 = 0$ versus H_1: $\beta_1 \neq 0$:

$$\text{Accept } H_0 \text{ if } -t_{\alpha/2} \leq \frac{b_1}{s_{b_1}} \leq t_{\alpha/2},$$

$$\text{Reject } H_0 \text{ if } \left| \frac{b_1}{s_{b_1}} \right| > t_{\alpha/2}.$$

For the Armand's Pizza problem we have $b_1 = 5$ and $s_{b_1} = .5803$. Thus we have $b_1/s_{b_1} = 5/.5803 = 8.6162$. From Table 2 of Appendix B we find that the t value corresponding to $\alpha = .01$ and 8 degrees of freedom is $t_{.005} = 3.355$. Since $b_1/s_{b_1} = 8.6162 > 3.355$, we reject H_0 and conclude at the .01 level of significance that β_1 is not equal to zero.

Finally, we note in closing this section that the reason that the t test and the F test give the same results is that $F = t^2$ and the critical value for the F test is the square of the critical value for the t test. Hence whenever one of the tests leads to rejection of the null hypothesis the other will also.

EXERCISES

19. The data from Exercise 7 are repeated below:

Observation	x_i	y_i
1	2	25
2	3	25
3	5	20
4	1	30
5	8	16

a. Compute SSR, SST, and SSE.
b. Compute MSE and MSR.
c. Use the F test to test the hypotheses

$$H_0: \beta_1 = 0,$$

$$H_1: \beta_1 \neq 0,$$

at the $\alpha = .05$ level of significance.

20. Given below are five observations collected in a regression study on two variables:

Observation	x_i	y_i
1	2	2
2	4	3
3	5	2
4	7	6
5	8	4

a. Develop the estimated regression equation for these data.
b. Compute MSE and MSR.
c. Use the F test to test the hypotheses

$$H_0: \beta_1 = 0,$$

$$H_1: \beta_1 \neq 0,$$

at the $\alpha = .05$ level of significance.

21. Refer to Exercise 8, where an estimated regression equation relating GPA to monthly starting salaries was developed. SSR and SST were computed in Exercise 14.

a. Compute MSE as an estimate of σ^2.
b. Compute $s_{b_1}^2$ and s_{b_1}.
c. Use the t test to determine if grade point average and salary are related at the $\alpha = .05$ level of significance.

22. Refer to Exercise 9, where an estimated regression equation relating square footage to selling prices of new homes was developed. Test whether or not selling price and square footage are related at the $\alpha = .10$ level of significance.

23. Refer to Exercise 10, where an estimated regression equation relating price per loaf and number of loaves of bread sold was developed. Test whether or not price and the number of loaves sold are related at the $\alpha = .05$ level of significance.

24. Refer to Exercise 12, where an estimated regression equation relating years of experience and annual sales was developed.

a. Use an F test and an $\alpha = .05$ level of significance to determine whether or not annual sales and years of experience are related.

b. Use a t test and an $\alpha = .05$ level of significance to determine whether or not annual sales and years of experience are related.
c. Square the values of $t = b_1/s_{b_1}$ and $t_{.025}$ found in part b and compare these with the corresponding F values found in part a. Comment on whether or not you feel the F test and t test could yield different conclusions.
25. Refer to Exercise 16, where an estimated regression equation relating light absorbance readings and milligrams of protein present in a liver sample was developed. Test whether or not the absorbance readings and amount of protein present are related at the $\alpha = .01$ level of significance.

14.5 RESIDUAL ANALYSIS

In the previous section we showed how statistical tests can be used to determine whether or not the regression relationship is significant. These tests are only valid—and hence should only be used—when the model assumptions regarding the error are appropriate. These assumptions were presented in Section 14.1 and are restated below:

1. The error ϵ is a normally distributed random variable.
2. $E(\epsilon) = 0$.
3. The variance of ϵ, denoted σ^2, is the same for all values of x.
4. The values of the error are independent.

These assumptions regarding ϵ are important because statisticians have used them as the basis for developing the tests introduced in the previous section. Residual analysis is helpful in making a judgment as to whether or not these assumptions appear reasonable in a particular application. In addition, residual analysis can often provide insight as to whether or not a different type of model—for instance, one involving a different functional form might better describe the underlying relationship.

Recall from our earlier discussions that a residual is simply the difference between an observed value of y and the predicted value using the estimated regression equation; that is, the ith residual is given by $y_i - \hat{y}_i$. In checking the validity of the model assumptions, statisticians usually develop one or more residual plots. One of the most common is a graph with each residual plotted against the predicted value $\hat{y}_i$. Shown in Figure 14.8 are the residual plots for three different regression analyses.

Let us see how these plots can be used to determine if the model assumptions are valid. If the variance of ϵ is constant (i.e., the same for all values of x) and the model we have proposed is adequate, the residual plot against $\hat{y}_i$ should give an overall impression of a horizontal band of points. Part A of Figure 14.8 shows the type of pattern to be expected in this case. On the other hand, if the variance of ϵ is not constant—for example, the variability about the regression line is greater for larger values of $\hat{y}_i$—then we would observe a pattern such as that of part B of Figure 14.8. Finally, if we observe a residual pattern such as that of Part C of Figure 14.8 we would conclude that the assumption of a linear relationship between x and y is not appropriate.

Let us now look at a residual plot for the Armand's Pizza problem (the residuals were computed in Table 14.3). A plot of the residuals is shown in Figure 14.9. Note that the largest residual is 18, the smallest -21. Also, the residuals appear to be randomly distributed, with no particular patterns such as those shown in parts B and C

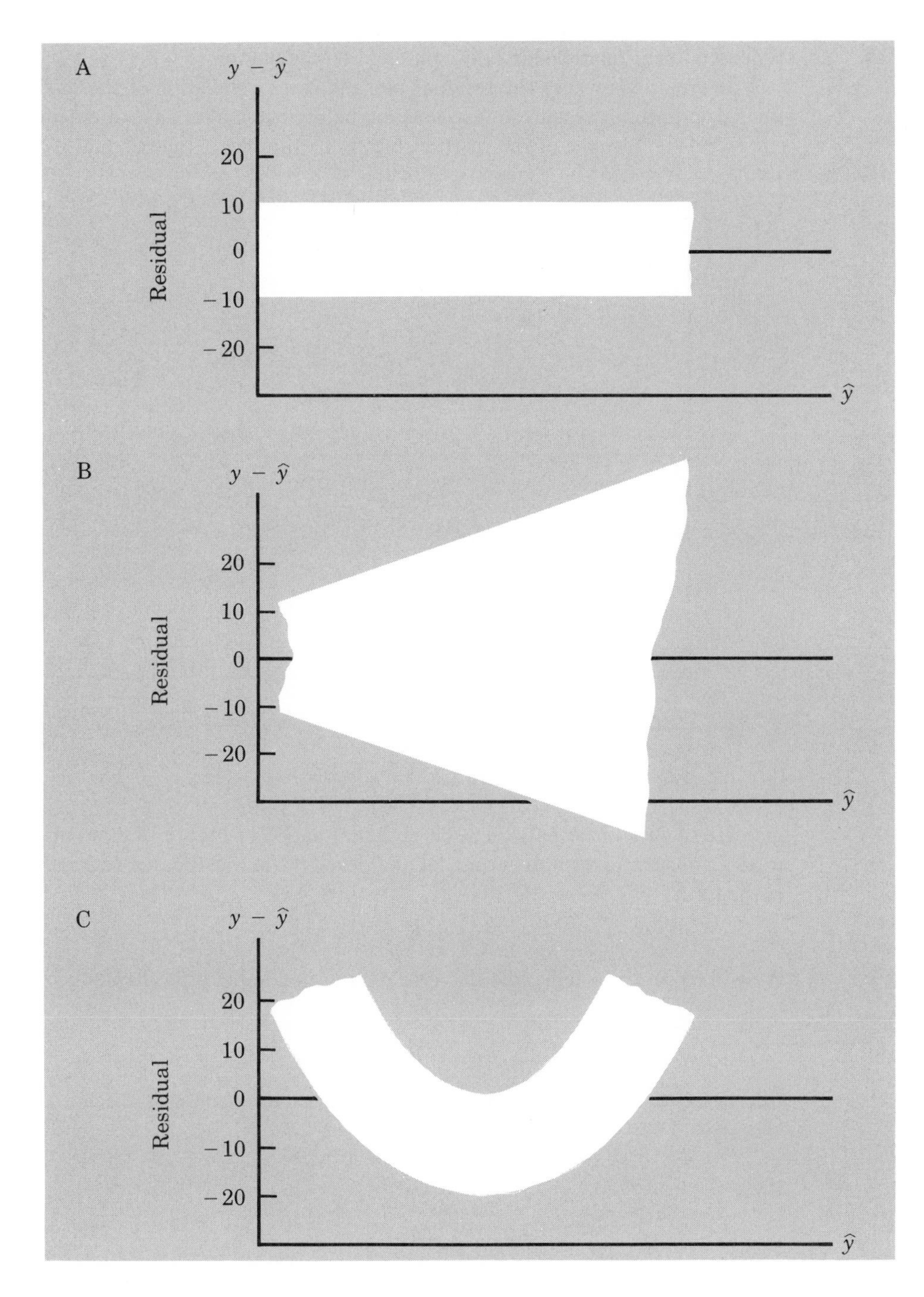

FIGURE 14.8 Residual Plots from Three Regression Studies

of Figure 14.9. We thus conclude that the model assumptions are satisfied and that our proposed linear relationship between x and y is adequate.

In situations where the residual plot indicates a violation of the model assumptions, appropriate measures must be taken before one makes any statements about the statistical significance of the relationship. For example, if the residual plot against $\hat{y}$

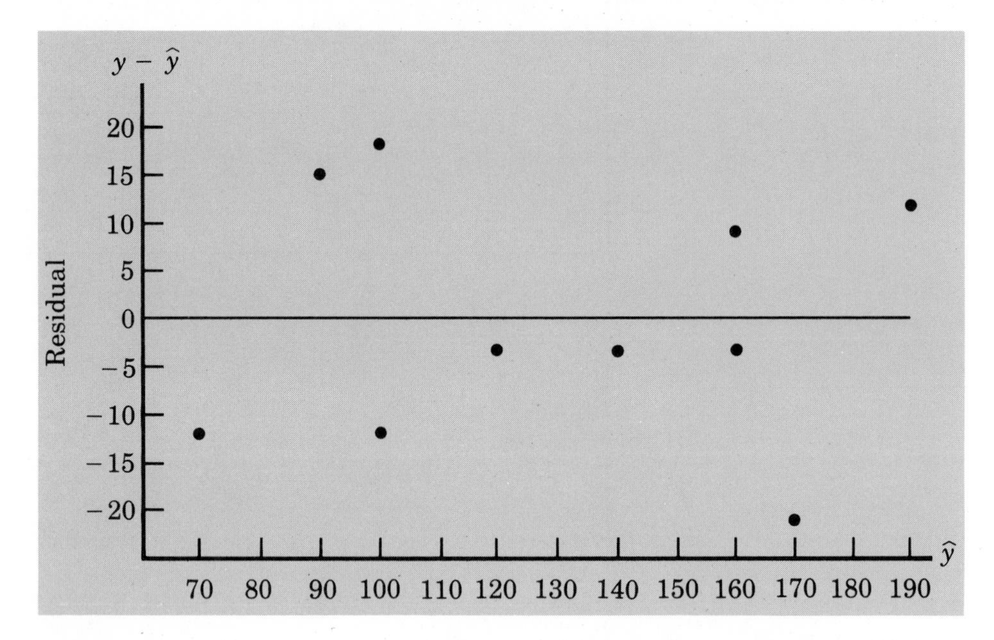

FIGURE 14.9 Residual Plot for Armand's Pizza Parlor Regression Study

indicated a curvilinear pattern, such as that of part C of Figure 14.8, we might attempt to add a quadratic term to our model and develop a new estimated regression equation of the form

$$\hat{y} = b_0 + b_1x + b_2x^2.$$

We will discuss how to handle this type of situation in the next chapter.

EXERCISES

26. In Exercise 4 data concerning advertising expenditures and sales at Eddie's Restaurants were given. These data are repeated below:

Advertising Expenditures (\$1,000's)	*Sales (\$1,000's)*
1.0	19.0
4.0	44.0
6.0	40.0
10.0	52.0
14.0	53.0

a. Let x equal advertising expenditures (\$1,000's) and y equal sales (\$1,000's). Use the method of least squares to develop a straight line approximation to the relationship between the two variables.

b. Test whether or not sales and advertising expenditures are related at the $\alpha = .10$ level of significance.

c. Prepare a residual plot of $y - \hat{y}$ versus $\hat{y}$ using part a to obtain the values of $\hat{y}$.

d. What conclusions can you draw from residual analysis? Should this model be used, or should we look for a better one?

27. The following data were used in a regression study:

Observation	x_i	y_i
1	2	4
2	3	5
3	4	4
4	5	6
5	7	4
6	7	6
7	7	9
8	8	5
9	9	11

a. Develop an estimated regression equation for this data.

b. Construct a plot of the residuals. Do the assumptions concerning the error terms seem to be satisfied?

28. Refer to Exercise 12, where an estimated regression equation relating years of experience and annual sales was developed.

a. Compute the residuals and construct a residual plot for this problem.

b. Do the assumptions concerning the error terms seem reasonable in light of the residual plot?

14.6 USING THE ESTIMATED REGRESSION EQUATION FOR PREDICTION

For the Armand's Pizza Parlor problem we concluded that annual sales (y) and the size of the student population (x) are related. Moreover, the estimated regression equation $\hat{y} = 60 + 5x$ appears to adequately describe the relationship between x and y. Now we can begin to develop interval estimates of annual sales for a given student population.

There are two types of interval estimates to consider. The first is an interval estimate of the mean value of y for a particular value of x. For instance, we might want an interval estimate of the *expected* annual sales for an Armand's Pizza Parlor located near any campus with a student population of 10,000. The expected annual sales for campuses with 10,000 students represents the average of the annual sales for all restaurants located near college campuses with 10,000 students.

The second type of interval estimate that we will consider is appropriate in situations where we want to predict an individual value of y corresponding to a given value of x. For instance, we might be interested in developing a prediction interval for the annual sales of a restaurant site near Talbot College, a school with 10,000 students. In this case our interest is in predicting the annual sales for one particular site, as opposed to predicting the average sales for all restaurants located near campuses with 10,000 students.

Estimating the Mean Value of y for a Given Value of x

Suppose we wanted to develop an estimate of the mean or expected value of annual sales for all restaurant sites near campuses with 10,000 students. First, recall that for the Armand's Pizza problem we hypothesized that the expected value of annual sales is given by

$$E(y) = \beta_0 + \beta_1 x.$$

Thus when the student population size is 10,000, $x = 10$ and hence $E(y) = \beta_0 + \beta_1(10)$.

The least squares method was used to develop estimates of β_0 and β_1 and hence an estimate of the regression equation; for the Armand's Pizza problem the estimated regression equation was found to be $\hat{y} = 60 + 5x$. Thus for restaurants near a campus with 10,000 students we would obtain $\hat{y} = 60 + 5(10) = 110$. The corresponding estimate of expected annual sales would be \$110,000.

In general, the point estimate of $E(y)$ for a particular value of x is the corresponding value of $\hat{y}$ given by the estimated regression equation. We denote the particular value of x by x_p, the mean value of y at x_p by $E(y_p)$, and the estimate of $E(y_p)$ by $\hat{y}_p = b_0 + b_1 x_p$.

Since b_0 and b_1 are only estimates of β_0 and β_1, we cannot expect that our estimated value $\hat{y}_p$ will exactly equal $E(y_p)$. For instance, in our example above we do not expect that the mean annual sales for all sites with 10,000 students to exactly equal \$110,000, our estimated value. If we want to make an inference about how close $\hat{y}_p$ is to the true mean value $E(y_p)$, however, we will have to consider the variability that exists when we develop estimates based on the estimated regression equation. Statisticians have developed the following estimate of the variance of $\hat{y}_p$:

$$\text{Estimated variance of } \hat{y}_p = s^2_{\hat{y}_p} = \text{MSE}\left(\frac{1}{n} + \frac{(x_p - \bar{x})^2}{\Sigma x_i^2 - n\bar{x}^2}\right). \qquad (14.21)$$

For the Armand's Pizza Parlor problem the estimated variance of $\hat{y}_p$ for restaurant sites near campus populations with 10,000 students is

$$\begin{aligned} s^2_{\hat{y}_p} &= 191.25\left(\frac{1}{10} + \frac{(10 - 14)^2}{2{,}528 - 10(14)^2}\right) \\ &= 191.25(.1282) \\ &= 24.52. \end{aligned}$$

Thus

$$s_{\hat{y}_p} = \sqrt{s^2_{\hat{y}_p}} = \sqrt{24.52} = 4.95.$$

The confidence interval estimate of $E(y_p)$ is as follows:

Interval Estimate of $E(y_p)$

$$\hat{y}_p \pm t_{\alpha/2}\, s_{\hat{y}_p}, \qquad (14.22)$$

where the confidence coefficient is $1 - \alpha$ and the t value has $n - 2$ degrees of freedom.

Thus to develop a 95% confidence interval estimate of the expected annual sales for restaurant sites with campus populations of 10,000 students, we need to find the t value from Table 2 of Appendix B corresponding to $n - 2 = 10 - 2 = 8$ degrees of freedom and $\alpha = .05$. Doing so, we find $t_{.025} = 2.306$. Hence the resulting confidence interval is

$$[b_0 + b_1(10)] \pm 2.306\, s_{\hat{y}_p}$$
$$[60 + 5(10)] \pm 2.306\,(4.95)$$
$$110 \pm 11.45.$$

In terms of dollars the 95% confidence interval estimate is \$110,000 ± \$11,415. Thus we obtain \$98,585 to \$121,415 as an interval estimate of the expected sales volume for all restaurant sites near campuses with 10,000 students.

Note that the estimated variance of $\hat{y}_p$ [see (14.21)] is smallest when the given value of x_p is $\bar{x}$. In this case $x_p = \bar{x}$ and (14.21) becomes

$$s^2_{\hat{y}_p} = \text{MSE}\left(\frac{1}{n} + \frac{(\bar{x} - \bar{x})^2}{\Sigma x_i^2 - n\bar{x}^2}\right),$$
$$= \frac{\text{MSE}}{n}$$

which implies that we can expect to make our best predictions at the mean of the independent variable.

Predicting an Individual Value of y for a Given Value of x

In the preceding discussion we showed how to develop a confidence interval estimate of the expected annual sales for all restaurant sites near campuses with 10,000 students. Now we turn to the problem of developing an interval estimate for an individual value of y corresponding to a particular value of x. Let us develop a prediction interval for the annual sales of a restaurant site near Talbot College, a school with 10,000 students.

Recall that the underlying regression model is assumed to be

$$y = \beta_0 + \beta_1 x + \epsilon.$$

Letting y_p and ϵ_p denote the values of y and ϵ corresponding to x_p, a particular value of x, we can write the regression model:

$$y_p = \beta_0 + \beta_1 x_p + \epsilon_p. \tag{14.23}$$

Since $\beta_0 + \beta_1 x_p$ is the mean value of y at x_p, which we denoted $E(y_p)$, we can rewrite (14.24) as follows:

$$y_p = E(y_p) + \epsilon_p. \tag{14.24}$$

Assume for the moment that we actually knew the value of $E(y_p)$. Since ϵ_p is the value of the unknown random variable ϵ, which has a mean of 0, it would seem reasonable to use $E(y_p)$ as our estimate of y_p. However, since in practice we do not know the value of $E(y_p)$, the best estimate of y_p must be given by our best estimate of $E(y_p)$. Thus in general the point estimate for an individual y_p is $\hat{y}_p = b_0 + b_1 x_p$, the same as that for the mean value of y_p given x_p. Hence for the Armand's Pizza problem our point estimate of annual sales for Talbot College is $\hat{y}_p = 60 + 5(10) = 110$. Note that this is the same as our point estimate of the mean annual sales for all restaurants near campuses with 10,000 students. In order to develop an interval estimate of sales for the individual Talbot College site we must first determine the variance associated with the estimate.

Since $y_p = E(y_p) + \epsilon_p$, the variance associated with our estimate of y_p is made up of the sum of the following two components:

1. The variance associated with using $\hat{y}_p$ to estimate $E(y_p)$, which we previously found to be $s^2_{\hat{y}_p}$.
2. The variance of individual y values about the mean $E(y_p)$, an estimate of which is given by MSE.

Statisticians have shown that the variance of the estimate of an individual value of y_p, which we denote s^2_{ind}, is given by

$$s^2_{\text{ind}} = \text{MSE} + s^2_{\hat{y}_p}$$

$$s^2_{\text{ind}} = \text{MSE} + \text{MSE}\left(\frac{1}{n} + \frac{(x_p - \overline{x})^2}{\Sigma x_i^2 - n\overline{x}^2}\right)$$

$$s^2_{\text{ind}} = \text{MSE}\left(1 + \frac{1}{n} + \frac{(x_p - \overline{x})^2}{\Sigma x_i^2 - n\overline{x}^2}\right). \tag{14.25}$$

For the Armand's Pizza problem, the value of (14.25) corresponding to the prediction of annual sales for a particular restaurant site near a campus with 10,000 students is computed as follows:

$$s^2_{\text{ind}} = 191.25\left(1 + \frac{1}{10} + \frac{(10 - 14)^2}{2{,}528 - 10(14)^2}\right)$$

$$= 191.25\ (1.1282)$$

$$= 215.77.$$

Thus

$$s_{\text{ind}} = \sqrt{s^2_{\text{ind}}} = \sqrt{215.77} = 14.69.$$

The confidence interval estimate* of y_p is given by (14.26):

Interval Estimate of y_p

$$\hat{y}_p \pm t_{\alpha/2}\, s_{\text{ind}}, \tag{14.26}$$

where the confidence coefficient is $1 - \alpha$ and the t value has $n - 2$ degrees of freedom.

Thus a 95% confidence interval for sales of an individual restaurant near a campus with 10,000 students is

$$[b_0 + b_1(10)] \pm t_{\alpha/2}(14.69)$$

$$[60 + 5(10)] \pm 2.306\ (14.69)$$

$$110 \pm 33.875.$$

*The interval estimate for $E(y_p)$ is called a confidence interval because $E(y_p)$ is a parameter. Some authors prefer to refer to the interval estimate for an individual value of y_p as a prediction interval because y_p is a random variable. In this text we refer to both interval estimates as confidence intervals.

In terms of dollars, then, the 95% confidence interval for annual sales for an individual restaurant located near a campus with 10,000 students is $76,125 to $143,875. We note that this prediction interval is greater in width than the confidence interval for mean sales of all sites near campuses with 10,000 students ($98,585 to 121,415). This difference simply reflects the fact that we are able to predict the mean annual sales volume with more precision than we can the annual sales for any particular site.

EXERCISES

29. As an extension of Exercise 8 develop a 95% confidence interval for estimating the mean starting salary for students with 3.0 grade point averages.

30. As an extension of Exercise 9 develop a 90% confidence interval for predicting the mean selling price for homes with 2,200 square feet of living space.

31. For Armand's Pizza develop a 95% confidence interval for sales at a site near Mills College, where the student population is 8,000 students.

32. As an extension of Exercise 8 develop a 95% confidence interval for estimating the starting salary of Joe Heller, who has a grade point average of 3.0.

33. As an extension of Exercise 9 develop a 95% confidence interval for the selling price of a home on Highland Terrace with 2,800 square feet.

34. For Armand's Pizza develop a 90% confidence interval for sales at sites near campuses with 15,000 students.

35. State in your own words why a smaller confidence interval is obtained when a mean value is predicted than when an individual value is predicted.

14.7 COMPUTER SOLUTION OF REGRESSION PROBLEMS

As we have seen, the computational aspects associated with regression analysis can be quite time-consuming. In this section we will discuss how the computational burden can be simplified by using a computer software package. The general procedure followed in using computer packages is for the user to input the data (x and y values for the sample) together with some instructions concerning the types of analyses that are required. The software package performs the analysis and prints the results in an output report. Before discussing the details of this approach, we will discuss the use of the analysis of variance (ANOVA) table as a device for summarizing the calculations performed in regression analysis. The ANOVA table is an important component of the output report produced by most software packages.

The ANOVA Table

In Chapter 13 we saw how the ANOVA table could provide a convenient summary of the computational aspects of analysis of variance. In regression analysis a similar table can be developed. Table 14.6 shows the general form of the ANOVA table for two-variable regression studies and Table 14.7 shows the ANOVA table for the Armand's Pizza Parlor problem. It can be seen that the relationship that holds among the sum of squares (that is, SST = SSR + SSE) also holds for the degrees of freedom. That is

$$\text{Total DF} = \text{Regression DF} + \text{Error DF}.$$

TABLE 14.6 General Form of the ANOVA Table for Two-Variable Regression Analysis

Source of Variation	*Sum of Squares*	*Degrees of Freedom*	*Mean Square*
Regression	SSR	1	$MSR = \frac{SSR}{1}$
Error	SSE	$n - 2$	$MSE = \frac{SSE}{n-2}$
Total (about the mean)	SST	$n - 1$	

TABLE 14.7 ANOVA Table for the Armand's Pizza Parlor Problem

Source of Variation	*Sum of Squares*	*Degrees of Freedom*	*Mean Square*
Regression	14,200	1	$\frac{14{,}200}{1} = 14{,}200$
Error	1,530	8	$\frac{1{,}530}{8} = 191.25$
Total (about the mean)	15,730	9	

Computer Output

The computer output for most software packages appears in a format similar to that of Tables 14.6 and 14.7. In Figure 14.10 we show the computer output resulting from running the Armand's Pizza problem on MINITAB, one of the more widely available software packages. The interpretation of the specific output sections is as follows:

1. The computer package prints the estimated regression equation as Y = 60.0 + 5.00X1. Thus X1 is the independent variable (which we denoted x) and Y represents the estimated value of the dependent variable (which we denoted $\hat{y}$).
2. A table is printed which shows the value of b_0 and b_1, the standard deviation of each coefficient, and the t value obtained by dividing each coefficient value by its standard deviation. Thus to test H_0: $\beta_1 = 0$ versus H_1: $\beta_1 \neq 0$, we could compare 8.62 (the "T-RATIO" corresponding to the X1 coefficient) to the table value. This is the procedure we described in the last part of Section 14.4.
3. The computer package prints the estimate of σ, $s = 13.83$, as well as information regarding the strength of the relationship. Note that "R-SQUARED = 90.3 PERCENT" is the coefficient of determination, which we denoted r^2. The following line of output showing "R-SQUARED = 89.1 PERCENT, ADJUSTED FOR D.F." will be discussed in Chapter 15.
4. The ANOVA table (as described earlier in this section) is printed below the heading "ANALYSIS OF VARIANCE." Note that MSR is given as 14,200 and MSE as 191.3. Using these values to compute F = MSR/MSE we would obtain $F = 74.23$.

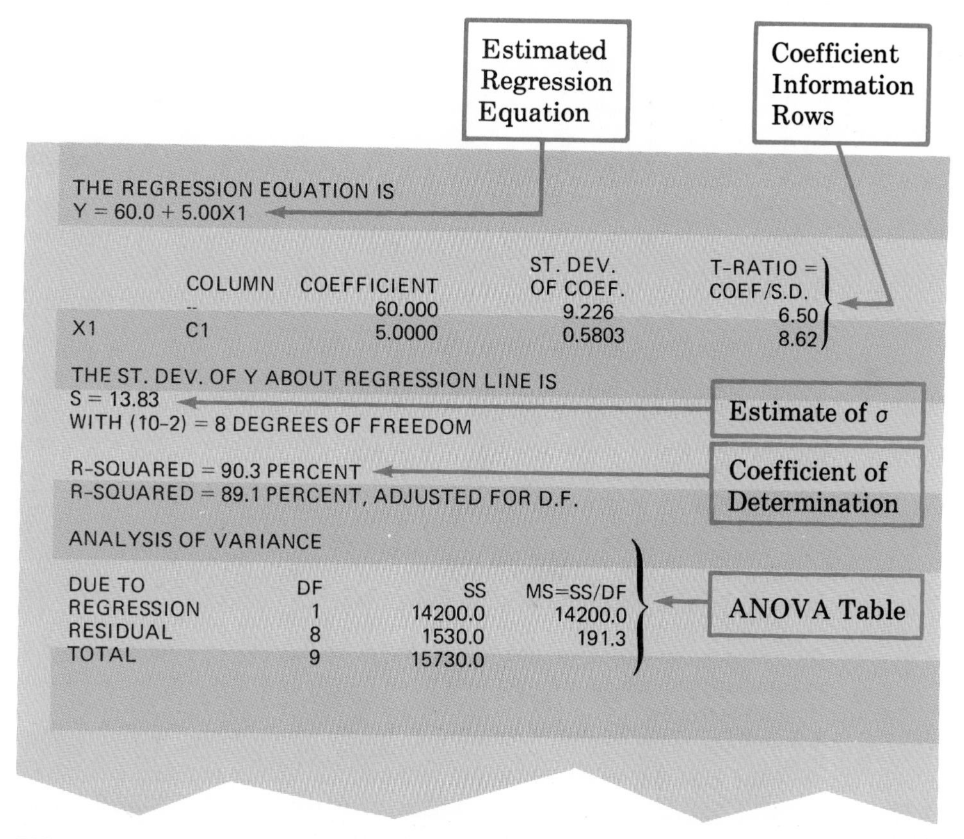

FIGURE 14.10 Armand's Pizza Parlor Solution Using the MINITAB Statistical Computing System

As you can see, the computer output from MINITAB is fairly easy to interpret given our current background in regression analysis. Many other computer packages, most with rather cryptic names such as SPSS, SAS, BMDP, and so on, are available for solving regression problems. After a brief period of familiarity with the control language associated with each of these packages, computer output such as that shown in Figure 14.10 can be obtained easily. When large data sets are involved, computer packages provide the only practical means for solving regression problems.

EXERCISES

36. The commercial division of a real estate firm is conducting a regression analysis of the relationship between x, annual gross rents (\$1,000's), and y, selling price (\$1,000's) for apartment buildings. Data have been collected on a number of properties recently sold, and the output at the top of p. 440 has been obtained in a computer run:

a. How many apartment buildings were in the sample?

b. Write the estimated regression equation.

c. What is the value of s_{b_1}?

d. Compute the F statistic and test the significance of the relationship at an $\alpha = .05$ level of significance.

e. Estimate the selling price of an apartment building with gross annual rents of \$50,000.

```
THE REGRESSION EQUATION IS
Y = 20.0 + 7.21X1

                                          ST. DEV.          T-RATIO =
          COLUMN    COEFFICIENT           OF COEF.          COEF/S.D.
            --            20.000           3.2213              6.21
X1          C1            7.2100           1.3626              5.29

THE ST. DEV. OF Y ABOUT REGRESSION LINE IS
S = 38.54
WITH (9-2) = 7 DEGREES OF FREEDOM

R-SQUARED = 80.0 PERCENT

ANALYSIS OF VARIANCE

DUE TO          DF         SS      MS=SS/DF
REGRESSION       1    41587.3       41587.3
RESIDUAL         7    10396.8        1485.3
TOTAL            8    51984.1
```

37. Shown below is a portion of the computer output for a regression analysis relating maintenance expense (dollars per month) to usage (hours per week) of a particular brand of computer terminal:

```
THE REGRESSION EQUATION IS
Y = 6.1092 + .8951X1

                                          ST. DEV.          T-RATIO =
          COLUMN    COEFFICIENT           OF COEF.          COEF/S.D.
            --            6.1092           0.9361              6.53
X1          C1            0.8951           0.1490              6.01

THE ST. DEV. OF Y ABOUT REGRESSION LINE IS
S = 6.61
WITH (10-2) = 8 DEGREES OF FREEDOM

R-SQUARED = 81.9 PERCENT

ANALYSIS OF VARIANCE

DUE TO          DF         SS      MS=SS/DF
REGRESSION       1    1575.76       1575.76
RESIDUAL         8     349.14         43.64
TOTAL            9    1924.90
```

a. Write the estimated regression equation.

b. Test to see (use a t test) if monthly maintenance expense is related to usage at the .01 level of significance.

c. Use the estimated regression equation to predict monthly maintenance expense for any terminal that is used 25 hours per week.

38. A regression model relating x, number of salespersons at a branch office, to y, annual sales at the office ($1,000's), has been developed. Shown below is the computer output from a regression analysis of the data:

a. Write the estimated regression equation.

```
THE REGRESSION EQUATION IS
Y = 80.0 + 50.00X1

                                          ST. DEV.        T-RATIO =
        COLUMN    COEFFICIENT             OF COEF.        COEF/S.D.
          --             80.0              11.333             7.06
X1        C1             50.0               5.482             9.12

THE ST. DEV. OF Y ABOUT REGRESSION LINE IS
S = 9.06
WITH (30-2) = 28 DEGREES OF FREEDOM

R-SQUARED = 74.8 PERCENT

ANALYSIS OF VARIANCE

DUE TO          DF         SS      MS=SS/DF
REGRESSION       1     6828.6        6828.6
RESIDUAL        28     2298.8          82.1
TOTAL           29     9127.4
```

b. How many branch offices were involved in the study?

c. Compute the F statistic and test the significance of the relationship at an $\alpha = .05$ level of significance.

d. Predict the annual sales at the Memphis branch office. This branch has 12 salespersons.

14.8 CORRELATION ANALYSIS

As we indicated in the introduction to this chapter, there are some situations in which the decision maker is not as concerned with the equation that relates two variables as in measuring the extent to which the two variables are related. In such cases a statistical technique referred to as correlation analysis can be used to determine the strength of the relationship between the two variables.* The output of a correlation study is a number referred to as the correlation coefficient. Because of the way in which it is defined, values of the correlation coefficient are always between -1 and $+1$. A value of $+1$ indicates that x and y are perfectly related in a positive linear sense. That is, all the points lie on a straight line that has a positive slope. A value of -1 indicates that x and y are perfectly related in a negative linear sense. That is, all the points lie on a straight line that has a negative slope. Values of the correlation coefficient close to zero indicate that x and y are not linearly related.

To begin with, let us see how the sample correlation coefficient can be computed as a byproduct of a regression analysis.

Determining the Sample Correlation Coefficient from the Regression Analysis Results

In this discussion we will assume that the least squares estimated regression equation is $\hat{y} = b_0 + b_1x$. In such cases the sample correlation coefficient can be computed using

*In correlation analysis it is assumed that x and y are both random variables and are both normally distributed.

one of the following formulas:

Sample Correlation Coefficient

$$r = \pm \sqrt{\text{Coefficient of determination}}, \tag{14.27}$$

$$r = b_1 \left(\frac{s_x}{s_y} \right), \tag{14.28}$$

where

$$s_x = \sqrt{\Sigma (x_i - \bar{x})^2/(n-1)} \quad \text{(Sample standard deviation for } x\text{)},$$

$$s_y = \sqrt{\Sigma (y_i - \bar{y})^2/(n-1)} \quad \text{(Sample standard deviation for } y\text{)}.$$

With (14.27) it is not clear what the proper sign of the correlation coefficient should be. However, from (14.28) we see that the sign of the sample correlation coefficient must be the same as the sign of b_1, the slope of the estimated regression equation. For the Armand's Pizza Parlor problem

$$\begin{aligned} r &= \pm \sqrt{0.90} \\ &= +0.95. \end{aligned}$$

Note that the sign of the sample correlation coefficient in this case is given by the positive root, since the slope ($b_1 = 5$) is positive.

The sample correlation coefficient is really a point estimator of the population correlation coefficient, a measure of the actual linear relationship between x and y in the population.

Let us use the Greek letter ρ (rho) to denote the population correlation coefficient. A statistical test for the significance of a linear relationship between x and y can be performed by testing the following hypotheses:

$$\begin{aligned} H_0&: \rho = 0, \\ H_1&: \rho \neq 0. \end{aligned}$$

It can be shown that these hypotheses are equivalent to the hypotheses regarding the significance of β_1, the slope of the regression equation. Recall that the appropriate hypotheses in this case are

$$\begin{aligned} H_0&: \beta_1 = 0, \\ H_1&: \beta_1 \neq 0. \end{aligned}$$

Since we earlier rejected the null hypothesis H_0: $\beta_1 = 0$ (see Section 14.4) we can reject the null hypothesis H_0: $\rho = 0$ in this case and conclude that x and y are linearly related.

Determining the Sample Correlation Coefficient without Performing a Regression Analysis

If the decision maker is concerned not about the form of the relationship between x and y but rather whether there is a relationship or not, the sample correlation coefficient can be computed without performing a regression study. In such cases the following formulas can be used:

Sample Correlation Coefficient

$$r = \left(\frac{1}{n-1}\right)\frac{\Sigma (x_i - \bar{x})(y_i - \bar{y})}{s_x s_y}, \tag{14.29}$$

$$r = \frac{\Sigma x_i y_i - (\Sigma x_i \Sigma y_i)/n}{\sqrt{\Sigma x_i^2 - (\Sigma x_i)^2/n}\sqrt{\Sigma y_i^2 - (\Sigma y_i)^2/n}} \tag{14.30}$$

For the Armand's Pizza problem the calculations required for use of (14.30) are shown in Table 14.8 and below:

$$\begin{aligned} r &= \frac{21{,}040 - (140)(1300)/10}{\sqrt{2528 - (140)^2/10}\sqrt{184{,}730 - (1300)^2/10}} \\ &= \frac{2{,}840}{\sqrt{568}\sqrt{15{,}730}} \\ &= .95. \end{aligned}$$

Note that, as expected, this value agrees with that obtained by computing r as the square root of the coefficient of determination.

TABLE 14.8 Calculations Required to Compute the Sample Correlation Coefficient

Restaurant (i)	x_i	y_i	$x_i y_i$	x_i^2	y_i^2
1	2	58	116	4	3,364
2	6	105	630	36	11,025
3	8	88	704	64	7,744
4	8	118	944	64	13,924
5	12	117	1,404	144	13,689
6	16	137	2,192	256	18,769
7	20	157	3,140	400	24,649
8	20	169	3,380	400	28,561
9	22	149	3,278	484	22,201
10	26	202	5,252	676	40,804
	140	1,300	21,040	2,528	184,730
	Σx_i	Σy_i	$\Sigma x_i y_i$	Σx_i^2	Σy_i^2

To test for a significant relationship without performing a regression study, statisticians have developed a test procedure for testing the following hypotheses regarding ρ, the population correlation coefficient:

$$H_0: \rho = 0,$$
$$H_1: \rho \neq 0.$$

It can be shown that if H_0 is true, then the value of

$$r\sqrt{\frac{n-2}{1-r^2}} \tag{14.31}$$

has a t distribution with $n - 2$ degrees of freedom. For $\alpha = .05$ and $n - 2 = 10 - 2 = 8$ degrees of freedom, we see that the appropriate t value from Table 2 of Appendix B is 2.306. Thus if the value of (14.31) exceeds 2.306 or is less than -2.306 we must reject the null hypothesis H_0: $\rho = 0$.

For our current problem, the value of (14.31) is

$$.95\sqrt{\frac{8}{1-.90}} = 8.5.$$

Since 8.5 exceeds the t value of 2.306, we reject H_0 and hence conclude that x and y are linearly related. We note that this test yields the same result as the test on β_1 in Section 14.4.

A Note Regarding the Use of the Sample Correlation Coefficient

The sample correlation coefficient is a frequently used measure of association. However, care must be used in interpreting values of r. The single most important point to keep in mind is that r is a measure of *linear association*. Thus $r = 0$ does not mean that x and y are not related, but that they are not *linearly* related. Figure 14.11 depicts a case where x and y obviously are related but the sample correlation coefficient is 0. The data and calculations that verify that $r = 0$ in this case are shown in Table 14.9 and

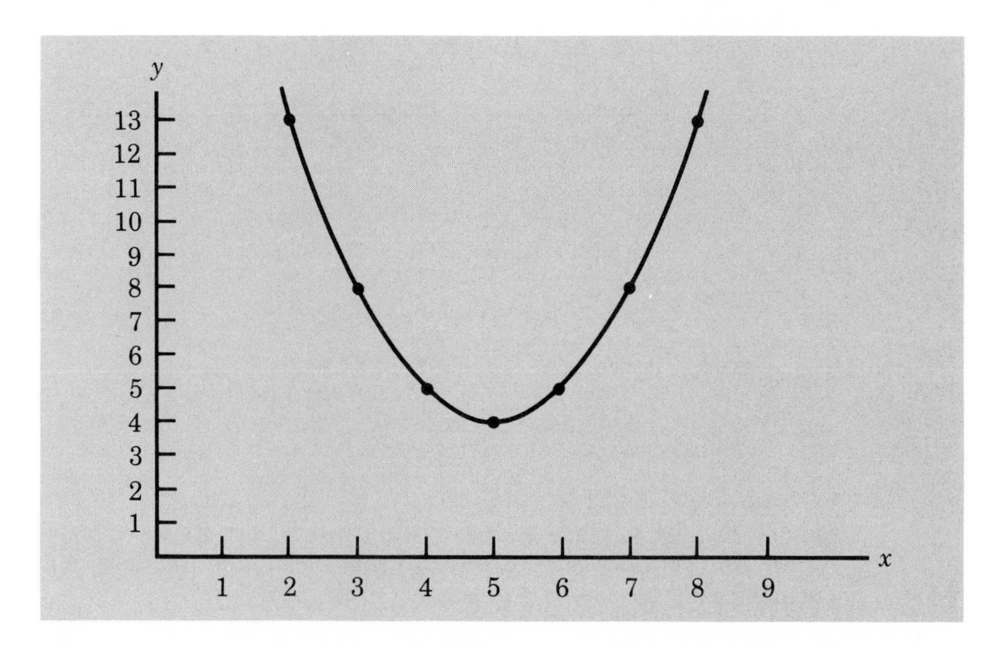

FIGURE 14.11 Illustration That $r = 0$ Does Not Imply That x and y Are Not Related

TABLE 14.9 Computations for Example of Figure 14.11 Showing That $r = 0$

i	x_i	y_i	x_i^2	y_i^2	x_iy_i
1	2	13	4	169	26
2	3	8	9	64	24
3	4	5	16	25	20
4	5	4	25	16	20
5	6	5	36	25	30
6	7	8	49	64	56
7	8	13	64	169	104
Totals	35	56	203	532	280

below:

$$r = \frac{(280) - (35)(56)/7}{\sqrt{203 - (35)^2/7}\,\sqrt{532 - (56)^2/7}}$$

$$= \frac{280 - 280}{\sqrt{28}\,\sqrt{84}}$$

$$= 0.$$

EXERCISES

39. Refer to Exercise 36 and compute the sample correlation coefficient between annual gross rent and selling price of apartment buildings.

40. The following estimated regression equation has been developed to estimate the relationship between x, the number of units produced per week, and y, the total weekly cost of production ($):

$$\hat{y} = 60 + 3.2x.$$

The standard deviation of weekly production is 10 units, and the standard deviation of weekly cost is $35.00. Compute the sample correlation coefficient r.

41. Eight observations on two random variables are given below:

x_i	y_i
2	11
9	4
6	6
8	5
4	9
7	4
5	9
6	7

a. Compute r.
b. Test the hypotheses

$$H_0: \rho = 0,$$
$$H_1: \rho \neq 0,$$

at the $\alpha = .01$ level of significance.

42. Use the data in Exercise 3 to compute the sample correlation coefficient between grade point average and salary.

Summary

In this chapter we introduced the topics of regression and correlation analysis. We discussed how regression analysis can be used to develop an equation showing how variables are related and how correlation analysis can be used to determine the strength of the relationship between variables.

Before concluding our discussion, however, we would like to emphasize a potential misinterpretation of these studies. In the introduction to this chapter we indicated that it is important to realize that regression and correlation analyses can indicate only how or to what extent the variables are associated with each other. These techniques cannot be interpreted directly as showing cause and effect relationships. For instance, suppose that a regression study of sales volumes and salesperson's grocery expenses indicated that high grocery expenses are associated with high sales volumes. Although these variables are related, it is doubtful that just by increasing grocery expenses we can cause an increase in sales volumes. On the other hand, it may seem reasonable that a relationship between advertising expenditures and sales volume is truly a cause and effect relationship. That is, it seems reasonable to suppose that increasing advertising expenditures could cause an increase in sales volume. Thus we have two associations or relationships—grocery expenses versus sales volume and advertising expenditures versus sales volume. One of these probably is not causal, the other probably is. In both cases the regression analysis has indicated only an association. Any conclusions about a cause and effect relationship must be based on the judgment of the analyst.

Glossary

Note: The definitions here are all stated with the understanding that simple linear regression and correlation is being considered.

Independent variable—The variable that is doing the predicting or explaining. It is denoted by x in the regression equation.

Dependent variable—The variable that is being predicted or explained. It is denoted by y in the regression equation.

Deterministic model—A relationship between an independent variable and a dependent variable whereby specifying the value of the independent variable allows one to compute exactly the value of the dependent variable.

Probabilistic model—A relationship between an independent variable and a dependent variable in which specifying the value of the independent variable is not sufficient to allow determination of the value of the dependent variable.

Simple linear regression—The simplest kind of regression, involving only two variables that are related approximately by a straight line.

Regression equation—The mathematical equation relating the independent variable to the expected value of the dependent variable; that is, $E(y) = \beta_0 + \beta_1 x$.

Estimated regression equation—The estimate of the regression equation obtained by the least squares method; that is, $\hat{y} = b_0 + b_1 x$.

Scatter diagram—A graph of the available data in which the independent variable appears on the horizontal axis and the dependent variable appears on the vertical axis.

Residual—The difference between the actual value of the dependent variable and the value predicted using the estimated regression equation; i.e., $y_i - \hat{y}_i$.

Least squares method—The approach used to develop the estimated regression equation which minimizes the sum of squares of the vertical distances from the points to the least squares fitted line.

Coefficient of determination (r^2)—A measure of the variation explained by the estimated regression equation. It is a measure of how well the estimated regression equation fits the data.

Sample correlation coefficient (r)—A statistical measure of the linear association between two variables.

Key Formulas

Regression Model

$$y = \beta_0 + \beta_1 x + \epsilon \tag{14.1}$$

Regression Equation

$$E(y) = \beta_0 + \beta_1 x \tag{14.3}$$

Estimated Regression Equation

$$\hat{y} = b_0 + b_1 x \tag{14.4}$$

Least Squares Estimates

$$b_1 = \frac{\Sigma(x_i - \overline{x})(y_i - \overline{y})}{\Sigma(x_i - \overline{x})^2} = \frac{\Sigma x_i y_i - (\Sigma x_i \Sigma y_i)/n}{\Sigma x_i^2 - (\Sigma x_i)^2/n} \tag{14.6}$$

$$b_0 = \overline{y} - b_1 \overline{x} \tag{14.7}$$

Sum of Squares Due to Error

$$\text{SSE} = \Sigma(y_i - \hat{y}_i)^2. \tag{14.8}$$

Sum of Squares About the Mean

$$\text{SST} = \Sigma(y_i - \overline{y})^2 \tag{14.9}$$

Relationship Among SST, SSR, and SSE

$$\text{SST} = \text{SSR} + \text{SSE} \tag{14.10}$$

Coefficient of Determination

$$r^2 = \frac{\text{SSR}}{\text{SST}} \tag{14.11}$$

Formula for SSR

$$\text{SSR} = \frac{[\Sigma x_i y_i - (\Sigma x_i \Sigma y_i)/n]^2}{\Sigma x_i^2 - (\Sigma x_i)^2/n}. \tag{14.12}$$

Formula for SST

$$\text{SST} = \Sigma y_i^2 - (\Sigma y_i)^2/n \tag{14.13}$$

Estimate of σ^2

$$\text{MSE} = \frac{\text{SSE}}{n - 2} \tag{14.15}$$

The F Statistic

$$F = \frac{\text{MSR}}{\text{MSE}} \tag{14.17}$$

Mean Square Due to Regression

$$\text{MSR} = \frac{\text{SSR}}{\text{Regression DF}} = \frac{\text{SSR}}{\text{no. of independent variables}} \tag{14.16}$$

Variance of b_1

$$\sigma_{b_1}^2 = \sigma^2\left(\frac{1}{\Sigma x_i^2 - n\bar{x}^2}\right) \tag{14.19}$$

Estimated Variance of b_1

$$s_{b_1}^2 = \text{MSE}\left(\frac{1}{\Sigma x_i^2 - n\bar{x}^2}\right) \tag{14.20}$$

Estimated Variance of $\hat{y}_p$

$$s_{\hat{y}_p}^2 = \text{MSE}\left(\frac{1}{n} + \frac{(x_p - \bar{x})^2}{\Sigma x_i^2 - n\bar{x}^2}\right) \tag{14.21}$$

Interval Estimate of $E(y_p)$

$$\hat{y}_p \pm t_{\alpha/2} s_{\hat{y}_p} \tag{14.22}$$

Estimated Variance When Predicting an Individual Value

$$s_{\text{ind}}^2 = \text{MSE}\left(1 + \frac{1}{n} + \frac{(x_p - \bar{x})^2}{\Sigma x_i^2 - n\bar{x}^2}\right) \tag{14.25}$$

Interval Estimate of y_p

$$\hat{y}_p \pm t_{\alpha/2} s_{\text{ind}} \tag{14.26}$$

Determining the Sample Correlation Coefficient from the Regression Analysis

$$r = \pm\sqrt{\text{Coefficient determination}} \tag{14.27}$$

or

$$r = b_1\left(\frac{s_x}{s_y}\right) \tag{14.28}$$

where

$$s_x = \sqrt{\Sigma(x_i - \bar{x})^2/(n-1)}$$
$$s_y = \sqrt{\Sigma(y_i - \bar{y})^2/(n-1)}$$

Determining the Sample Correlation Coefficient without Performing a Regression Analysis

$$r = \left(\frac{1}{n-1}\right)\frac{\Sigma(x_i - \bar{x})(y_i - \bar{y})}{s_x s_y} \tag{14.29}$$

or

$$r = \frac{\Sigma x_i y_i - (\Sigma x_i \Sigma y_i)/n}{\sqrt{\Sigma x_i^2 - (\Sigma x_i)^2/n}\sqrt{\Sigma y_i^2 - (\Sigma y_i)^2/n}} \tag{14.30}$$

Supplementary Exercises

43. What is the difference between regression analysis and correlation analysis?

44. Does a high value of r^2 imply that two variables are causally related? Explain.

45. In your own words explain the difference between an interval estimate of the mean value of y for a given x and an interval estimate for an individual value of y for a given x.

46. How do we measure how closely the actual data points are to the estimated regression line? That is, how do we measure the goodness of fit of the regression line?

47. What is the purpose of testing whether or not $\beta_1 = 0$?

48. What is the purpose of residual analysis?

49. In a manufacturing process the assembly line speed (feet/minute) was thought to affect the number of defective parts found during the inspection process. To test this theory, management devised a situation where the same batch of parts was inspected visually at a variety of line speeds. The following data were collected:

Line Speed	*Number of Defective Parts Found*
20	21
20	19
40	15
30	16
60	14
40	17

a. Develop the estimated regression equation that relates line speed to the number of defective parts found.

b. At the $\alpha = .05$ level of significance determine whether or not line speed and number of defective parts found are related.
c. Did the estimated regression equation provide a good fit to the data?
d. Develop a 95% confidence interval to predict the mean number of defective parts for a line speed of 50 feet per minute.

50. The PJH&D Company is in the process of deciding whether or not to purchase a maintenance contract for its new word processing system. They feel that maintenance expense should be related to usage and have collected the following information on weekly usage (hours) and annual maintenance expense:

Weekly Usage (Hours)	*Annual Maintenance Expense ($100's)*
13	17.0
10	22.0
20	30.0
28	37.0
32	47.0
17	30.5
24	32.5
31	39.0
40	51.5
38	40.0

a. Develop the estimated regression equation that relates annual maintenance expense, in hundreds of dollars, to weekly usage.
b. Test the significance of the relationship in part a at the $\alpha = .05$ level significance.
c. Develop a residual plot of $y - \hat{y}$ versus $\hat{y}$. Are you satisfied that the model is adequate?
d. PJH&D expects to operate the word processor 30 hours per week. Develop a 95% prediction interval for the company's annual maintenance expense.
e. If the maintenance contract costs $3,000 per year, would you recommend purchasing it? Why or why not?

51. Management wishes to investigate the relationship between the number of unauthorized days that an employee is absent per year and the distance (miles) between home and work for employees. A sample of ten employees was chosen, and the following data were collected:

Distance to Work (miles)	*Number of Days Absent*
1	8
3	5
4	8
6	7
8	6
10	3
12	5
14	2
14	4
18	2

a. Develop a scatter diagram for the above data. Does a linear relationship appear reasonable? Explain.
b. Develop the least squares estimated regression line.
c. Is there a significant relationship between the two variables? Use $\alpha = .05$.

d. Did the estimated regression line provide a good fit? Explain.
e. Use the estimated regression line developed in part b to develop a 95% confidence interval estimate of the expected number of days absent for employees living 5 miles from the company.
52. The management of a chain of fast-food restaurants would like to investigate the relationship between the daily sales volume of a company restaurant and the number of competitor restaurants within a 1 mile radius of the firm's restaurant. The following data have been collected:

Number of Competitors Within 1 Mile	*Sales ($)*
1	3,600
1	3,300
2	3,100
3	2,900
3	2,700
4	2,500
5	2,300
5	2,000

a. Develop the least squares estimated regression line that relates daily sales volume to the number of competitor restaurants within a 1 mile radius.
b. Is there a significant relationship between the two variables? Use $\alpha = .05$.
c. Did the estimated regression line provide a good fit? Explain.
d. Use the estimated regression line developed in part a to develop a 95% interval estimate of the daily sales volume for a particular company restaurant that has four competitors within a 1 mile radius.
53. The Regional Transit Authority for a major metropolitan area would like to determine if there is any relationship between the age of a bus and the annual maintenance cost. A sample of ten buses resulted in the following data:

Age of Bus (years)	*Maintenance Cost ($)*
1	350
2	370
2	480
2	520
2	590
3	550
4	750
4	800
5	790
5	950

a. Compute the sample correlation coefficient for the above data.
b. Using the sample correlation coefficient, test to see if the two variables are significantly related. Use $\alpha = .10$.
54. Reconsider the Regional Transit Authority problem presented in Exercise 53.
a. Develop the least squares estimated regression line.
b. Test to see if the two variables are significantly related at $\alpha = .05$.
c. Did the least squares line provide a good fit to the observed data? Explain.
d. Develop a 90% confidence interval estimate of the maintenance cost for a specific bus that is 4 years old.

55. The computer analysis teacher at Givens College is interested in the relation between time spent using the computer system and total points earned in the course. Data collected on ten students who took the course last quarter are given below:

Hours Using Computer System	*Total Points Earned*
45	40
30	35
90	75
60	65
105	90
65	50
90	90
80	80
55	45
75	65

a. Compute the sample correlation coefficient for the above data.
b. Use the sample correlation coefficient and test to see if there is a significant relationship at the $\alpha = .05$ level.

56. Reconsider the Givens College data in Exercise 55.
a. Develop an estimated regression line relating total points earned to hours spent using the computer system.
b. Test at the $\alpha = .05$ level the significance of the model.
c. Develop a residual plot of $y - \hat{y}$ versus $\hat{y}$. Do the model assumptions seem unreasonable?
d. Predict the total points earned by Mark Sweeney. He spent 95 hours on the computer system.
e. Develop a 90% confidence interval for the total points earned by Mark Sweeney.

APPENDIX TO CHAPTER 14: CALCULUS BASED DERIVATION OF LEAST SQUARES FORMULAS

As mentioned in the chapter, the least squares method is a procedure for determining the values of b_0 and b_1 that minimize the sum of squared deviations of the actual y values from the estimated regression line. The sum of squared deviations is given by

$$\Sigma(y_i - \hat{y}_i)^2.$$

Substituting $\hat{y}_i = b_0 + b_1 x_i$, we get

$$\Sigma(y_i - b_0 - b_1 x_i)^2 \tag{14A.1}$$

as the expression that must be minimized.

To minimize (14A.1) we must take the partial derivatives with respect to b_0 and b_1, set them equal to zero, and solve. Doing so we get

$$\frac{\partial\,\Sigma(y_i - b_0 - b_1 x_i)^2}{\partial b_0} = -2\Sigma(y_i - b_0 - b_1 x_i) = 0, \tag{14A.2}$$

$$\frac{\partial\,\Sigma(y_i - b_0 - b_1 x_i)^2}{\partial b_1} = -2\Sigma x_i(y_i - b_0 - b_1 x_i) = 0. \tag{14A.3}$$

Dividing (14A.2) by 2 and summing each term individually yields

$$-\Sigma y_i + \Sigma b_0 + \Sigma b_1 x_i = 0.$$

Bringing Σy_i to the other side of the equal sign and noting that $\Sigma b_0 = nb_0$, we obtain

$$nb_0 + (\Sigma x_i)b_1 = \Sigma y_i. \tag{14A.4}$$

Similar algebraic simplification applied to (14A.3) yields

$$(\Sigma x_i)b_0 + (\Sigma x_i^2)b_1 = \Sigma x_i y_i. \tag{14A.5}$$

(14A.4) and (14A.5) are known as the *normal equations*. Solving (14A.4) for b_0 yields

$$b_0 = \frac{\Sigma y_i}{n} - b_1 \frac{\Sigma x_i}{n}. \tag{14A.6}$$

Using (14A.6) to substitute for b_0 in (14A.5) provides

$$\frac{\Sigma x_i \Sigma y_i}{n} - \frac{(\Sigma x_i)^2}{n} b_1 + (\Sigma x_i^2)b_1 = \Sigma x_i y_i. \tag{14A.7}$$

Rearranging (14A.7), we obtain

$$b_1 = \frac{\Sigma x_i y_i - (\Sigma x_i \Sigma y_i)/n}{\Sigma x_i^2 - (\Sigma x_i)^2/n} \tag{14A.8}$$

Since $\bar{y} = \Sigma y_i/n$ and $\bar{x} = \Sigma x_i/n$, we can rewrite (14A.6):

$$b_0 = \bar{y} - b_1\bar{x}. \tag{14A.9}$$

Equations (14A.8) and (14A.9) are the formulas we used in the chapter to compute the coefficients in the estimated regression equation.

Application

Monsanto

Monsanto Company*

St. Louis, Missouri

Monsanto Company traces its roots to an investment of $5,000 and a dusty warehouse on the Mississippi riverfront, where in 1901 John F. Queeny began manufacturing saccharin. From the maiden name of his wife, Olga Monsanto, came the name of Queeny's new company, and from his small investment and meager beginnings came what is today the nation's fourth largest chemical company.

Today more than a thousand products are made by Monsanto. Industrial chemical products include phenol, acetic acid, potassium hydroxide, maleic anhydride, and adipic acid. Monsanto also produces a long and widely varied list of petrochemicals, rubber chemicals, plasticizers, manmade fibers, agricultural products, nutrition chemicals, polymers, detergent intermediates, semiconductor materials, process control equipment, and instruments. AstroTurf® synthetic playing surface and Lasso® and Roundup® herbicides are some of the familiar products made by Monsanto.

World headquarters are located in St. Louis, Missouri. Monsanto is truly a worldwide corporation, with 146 manufacturing facilities, 20 laboratories or technical centers, and marketing operations in 65 countries. Approximately 52,800 people are employed worldwide, with about 38,000 employees in the United States.

REGRESSION ANALYSIS AT MONSANTO

Monsanto utilizes regression analysis in many facets of its operations. Much of Monsanto's statistical analysis is done using prepared computer packages because of the complexity of many problems and the power and speed of computer analysis techniques. The following example, taken from product applications research in the Nutrition Chemicals Division, demonstrates the basic approach and utility of regression analysis.

Monsanto, through its Nutrition Chemicals Division, manufactures and markets a methionine supplement for use in poultry, swine, and cattle feeds. Poultry growers in particular work with very high volumes and low profit margins and have invested large amounts in the accurate definition of nutritional requirements for poultry. Optimal feed compositions result in more rapid growth and in a higher final body weight for a given feed intake. Feed efficiency, which relates gain in body weight to amount of feed

*The authors are indebted to James R. Ryland and Robert M. Schisla, Senior Research Specialists, Monsanto Nutrition Chemical Division, St. Louis, Missouri, for providing this application.

Chicks thrive on feed produced with Monsanto's Methionine supplement

consumed, is monitored closely over growth cycles. The success of poultry growers in improving feed efficiency is reflected in the low cost of poultry products relative to other meat products.

Monsanto conducts research at its own research farms and through several major universities in the United States to develop the growth response to methionine supplementation. The data shown in the scattergram (Figure 14A.1) give the results of one such experiment. Body weight after a fixed period of growth is shown as a function of the amount of methionine added to the feed, expressed as a total feed content of sulfur-containing amino acids.

With the data collected in this experiment a linear regression analysis was performed with supplemental methionine as the independent variable and body weight as the dependent variable. The estimated regression equation developed is as follows:

$$y = 0.21 + 0.42x,$$

where

y = body weight in kilograms,

x = percentage of sulfur-containing amino acids in the feed.

The coefficient of determination, r^2, was 0.78, indicating a reasonably good fit. The F value of 81.5 pointed to a significant relationship between body weight and the percentage of supplemental methionine used in feed.

The use of linear models such as this is widespread in the poultry industry, but the dangers of using such a model to predict outside the range of the values of x are well recognized. An objective of the research was to find the relationship between methionine and body weight such that feed could be developed with the optimal levels

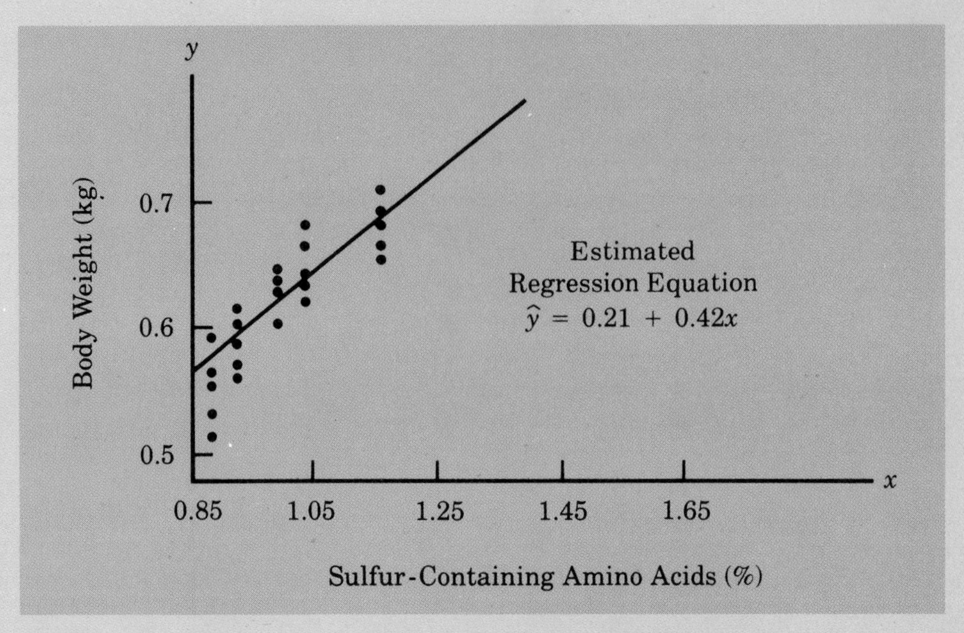

FIGURE 14A.1 Body Weight as a Function of Sulfur-Containing Amino Acids (Sample of 25)

of supplemental methionine to thereby provide the maximum possible body weight. Note that the regression model shows a positive relationship between x and y. Thus one might argue that increasing the supplemental methionine will continue to increase the body weight. However, this cannot be expected to continue without limit. In fact, further research has shown that as methionine supplementation is increased, body

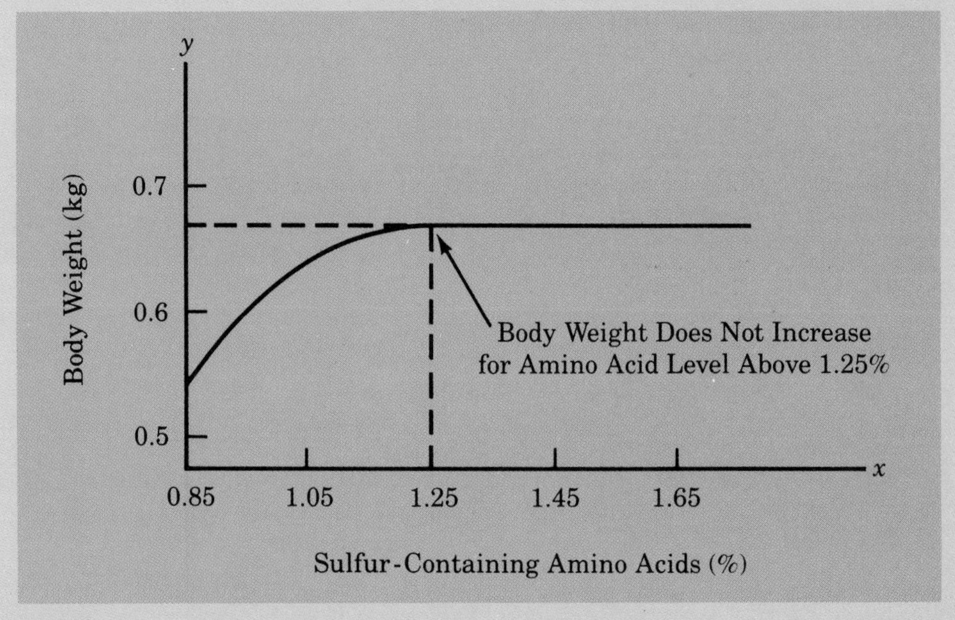

FIGURE 14A.2 Nonlinear Regression Model for Body Weight as a Function of Sulfur-Containing Amino Acids

weight eventually levels off and may even decline as methionine is increased beyond nutritional requirements. The real point of interest—the level of supplemental methionine which provides peak growth performance—cannot be found with the above straight line relationship. Residual analysis here showed that a curvilinear relationship might provide a better model.

Experimentation was continued with the values of x (sulfur-containing amino acid percentage) increased up to 1.8%. Body weight began to level off as anticipated. Since the relationship was clearly nonlinear, Monsanto researchers and collaborating researchers in universities worked to obtain the best nonlinear regression model for the complete set of data. Using an advanced form of regression analysis and a computer package which uses the least squares method to develop a nonlinear regression model, the following estimated regression equation was developed:

$$\hat{y} = 0.55 + 0.12(1 - 3{,}410e^{-9.57x})$$

where

y = body weight in kilograms,

x = percentage sulfur-containing amino acids.

The F value again showed a significant relationship, and the coefficient of determination, r^2, of 0.88 showed a good fit for the complete set of data. A graph of the above equation is shown in Figure 14A.2. Of particular interest is the optimal body weight, which occurs at a sulfur-containing amino acid level of 1.25%.

Thus regression analysis was essential in enabling the poultry growers to obtain optimum weight gain per feed dollar. This application is typical of the important role regression analysis has played at Monsanto as the company works to develop quality products for its customers.

15 Multiple Regression

What you will learn in this chapter:

- what multiple regression analysis is
- the important role computer packages play in performing multiple regression analysis
- how to use the t and F distributions to test for significant relationships in multiple regression analysis
- the concept of multicollinearity
- the use of dummy variables in regression analysis
- what is meant by curvilinear regression functions and a general linear model
- how multiple regression can be used for analysis of variance

Contents

In Chapter 14 we discussed how regression analysis can be used to develop a mathematical equation representing the relationship between two variables. Recall that the variable being predicted or explained by the mathematical equation is called the dependent variable; the variable being used to predict or explain the value of the dependent variable is called the independent variable. In this chapter we continue our study of regression analysis by considering situations which involve two or more independent variables and by showing how the techniques of regression can be adapted to model curvilinear relationships. The study of regression models involving more than one independent variable is called multiple regression analysis.

15.1 THE MULTIPLE REGRESSION MODEL

Consider a situation involving the sale of a new product (y) in a certain region. Suppose that we believe sales are related to the population size (x_1) and the average disposable income (x_2) of people in the region by the following regression model:

$$y = \beta_0 + \beta_1 x_1 + \beta_2 x_2 + \epsilon. \tag{15.1}$$

The relationship shown in (15.1) is a multiple regression model involving two independent variables. Note that if $\beta_2 = 0$, then x_2 is not related to y and hence our multiple regression model reduces to the one-independent-variable model discussed in Chapter 14; that is, $y = \beta_0 + \beta_1 x_1 + \epsilon$.

The multiple regression model of (15.1) can be extended to the case of p independent variables simply by adding more terms. Equation (15.2) shows the general case:

Multiple Regression Model

$$y = \beta_0 + \beta_1 x_1 + \beta_2 x_2 + \cdots + \beta_p x_p + \epsilon. \tag{15.2}$$

Note that if $\beta_3, \beta_4, \ldots, \beta_p$ all equal zero, (15.2) reduces to the two-independent-variable multiple regression model of (15.1).

The same assumptions we made for the error term ϵ in Chapter 14 also apply in multiple regression analysis:

1. ϵ is a normally distributed random variable taking on positive or negative values to reflect the error or deviation between the y value and the value given by $\beta_0 + \beta_1 x_1 + \beta_2 x_2 + \cdots + \beta_p x_p$.
2. The error has a mean or expected value of 0; that is, $E(\epsilon) = 0$.
3. The error has variance σ^2 which remains the same regardless of the values of the independent variables $x_1, x_2, \ldots, x_p$.
4. The values of the error are independent.

Using Assumption 2 we can take the expected value of both sides of (15.2) to obtain the multiple regression equation shown in (15.3):

Multiple Regression Equation

$$E(y) = \beta_0 + \beta_1 x_1 + \cdots + \beta_p x_p. \tag{15.3}$$

This equation relates the values of the independent variables to the corresponding average or expected value of the dependent variable.

To obtain more insight into the form of the relationship given by (15.3), consider for the moment the following two-independent-variable regression equation:

$$E(y) = \beta_0 + \beta_1 x_1 + \beta_2 x_2. \tag{15.4}$$

The graph of this equation is a plane in three dimensional space. Figure 15.1 shows such a graph with x_1 and x_2 on the horizontal axis and y on the vertical axis. Note that ϵ is shown as the difference between the actual y value and the expected value of y, $E(y)$, when $x_1 = x_1^*$ and $x_2 = x_2^*$.

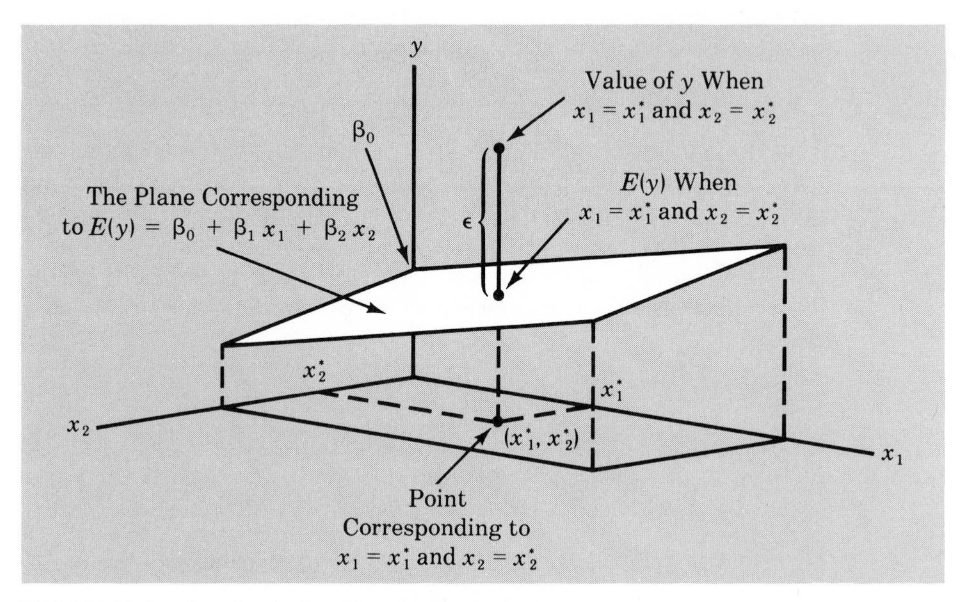

FIGURE 15.1 Graph of the Regression Equation for Multiple Regression Analysis Involving Two Independent Variables

In regression analysis the term response variable is often used in place of the term dependent variable. Furthermore, since the multiple regression equation generates a plane or surface, its graph is referred to as a response surface.

In the previous chapter the least squares method was used to develop estimates of β_0 and β_1 for the simple linear regression model. In multiple regression analysis the least squares method is used in an analogous manner to develop estimates of the parameters $\beta_0, \beta_1, \beta_2, \ldots, \beta_p$. These estimates are denoted $b_0, b_1, b_2, \ldots, b_p$; the corresponding estimated regression equation is written as follows:

Estimated Regression Equation

$$\hat{y} = b_0 + b_1 x_1 + \cdots + b_p x_p. \tag{15.5}$$

At this point you should begin to see the similarity between the concepts of multiple regression analysis and those of the previous chapter. We have just extended the concepts of simple linear regression to the case involving more than one independent variable. In the next section we will begin to apply these concepts to the problem facing the Butler Trucking Company.

15.2 THE BUTLER TRUCKING PROBLEM

Butler Trucking is an independent trucking company located in southern California. A major portion of Butler's business involves deliveries throughout its local area.

To develop better work schedules management would like to develop an estimated regression equation that will help predict total daily travel time for its drivers. Initially management felt that travel time should be closely related to miles traveled. A random sample of 10 days of operation was taken; the data obtained are presented in Table 15.1, and the corresponding scatter diagram is shown in Figure 15.2.

TABLE 15.1 Preliminary Data for the Butler Trucking Problem

Day	*Miles Traveled*	*Travel Time (hours)*
1	100	9.3
2	50	4.8
3	100	8.9
4	100	5.8
5	50	4.2
6	80	6.8
7	75	6.6
8	80	5.9
9	90	7.6
10	90	6.1

The scatter diagram indicates that the number of miles traveled (x_1) and the travel time (y) appear to be positively related; as x_1 increases, y increases. After observing the scatter diagram, management hypothesized the following regression model:

$$y = \beta_0 + \beta_1 x_1 + \epsilon.$$

Note that this is nothing more than the simple linear regression model with x_1 replacing x. As a result of this notational change we use x_{1i} to denote the ith observation for the independent variable x_1. Table 15.2 shows the application of the least squares formulas with these notational differences from Chapter 14 to compute b_0 and b_1. After rounding, the estimated regression equation relating travel time to miles traveled is given by $\hat{y} = 1.13 + .067x_1$.

Table 15.3 shows the computation of the residuals and SSE for the estimated regression equation $\hat{y} = 1.13 + .067x_1$; we see that SSE = 9.5669. Figure 15.3 shows a plot of the residuals versus $\hat{y}$. This residual plot appears to indicate that the assumption of constant variance (Assumption 3) is not appropriate; the variability of the y values about the estimated regression line is increasing as $\hat{y}$ gets larger. Although not shown, the value of SST = 24.0 and hence SSR = SST − SSE = 24.0 − 9.5669 = 14.4331.

TABLE 15.2 Least Squares Calculations for the Model Involving One Independent Variable

Day (i)	x_{1i} = *Miles Traveled*	y_i = *Travel Time (hours)*	$x_{1i}y_i$	x_{1i}^2
1	100	9.3	930	10,000
2	50	4.8	240	2,500
3	100	8.9	890	10,000
4	100	5.8	580	10,000
5	50	4.2	210	2,500
6	80	6.8	544	6,400
7	75	6.6	495	5,625
8	80	5.9	472	6,400
9	90	7.6	684	8,100
10	90	6.1	549	8,100
	815	66.0	5,594	69,625
	Σx_{1i}	Σy_i	$\Sigma x_{1i}y_i$	Σx_{1i}^2

$$\bar{x}_1 = \frac{815}{10} = 81.5 \qquad \bar{y} = \frac{66}{10} = 6.6$$

$$b_1 = \frac{\Sigma x_{1i}y_i - (\Sigma x_{1i}\,\Sigma y_i)n}{\Sigma x_{1i}^2 - (\Sigma x_{1i})^2/n} = \frac{5{,}594 - (815)(66)/10}{69{,}625 - (815)^2/10} = \frac{215}{3{,}202.5} = .0671$$

$$b_0 = \bar{y} - b_1\bar{x}_1 = 6.6 - (.0671)(81.5) = 1.1314$$

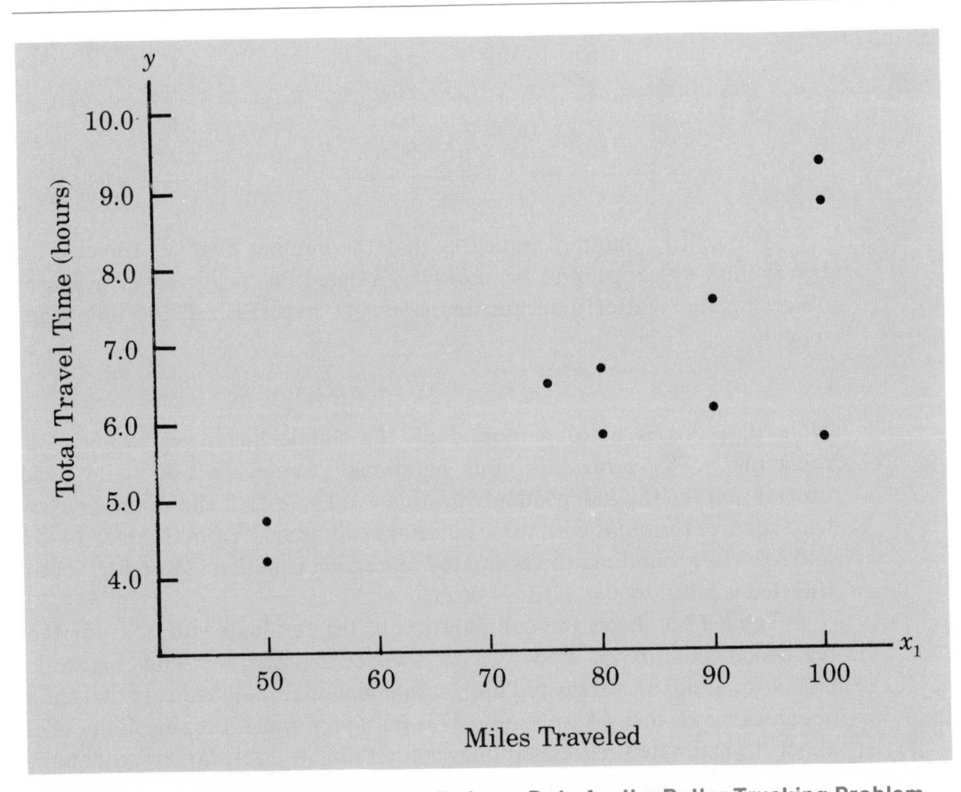

FIGURE 15.2 Scatter Diagram of Preliminary Data for the Butler Trucking Problem

TABLE 15.3 Computation of Residuals and SSE for Butler Trucking Using $\hat{y} = 1.13 + .067x_1$

Miles Traveled (x_{1i})	*Travel Time* (y_i)	*Predicted Travel Time* $(\hat{y}_i)$	*Residual* $(y_i - \hat{y}_i)$	$(y_i - \hat{y}_i)^2$
100	9.3	7.830	1.470	2.1609
50	4.8	4.480	0.320	0.1024
100	8.9	7.830	1.070	1.1449
100	5.8	7.830	−2.030	4.1209
50	4.2	4.480	−0.280	0.0784
80	6.8	6.490	0.310	0.0961
75	6.6	6.155	0.445	0.1980
80	5.9	6.490	−0.590	0.3481
90	7.6	7.160	0.440	0.1936
90	6.1	7.160	−1.060	1.1236
			SSE	9.5669

Thus the coefficient of determination is

$$r^2 = \frac{\text{SSR}}{\text{SST}} = \frac{14.4331}{24.0} = .60.$$

Since $r^2 = .60$, we conclude that 60% of the variability in travel time is explained

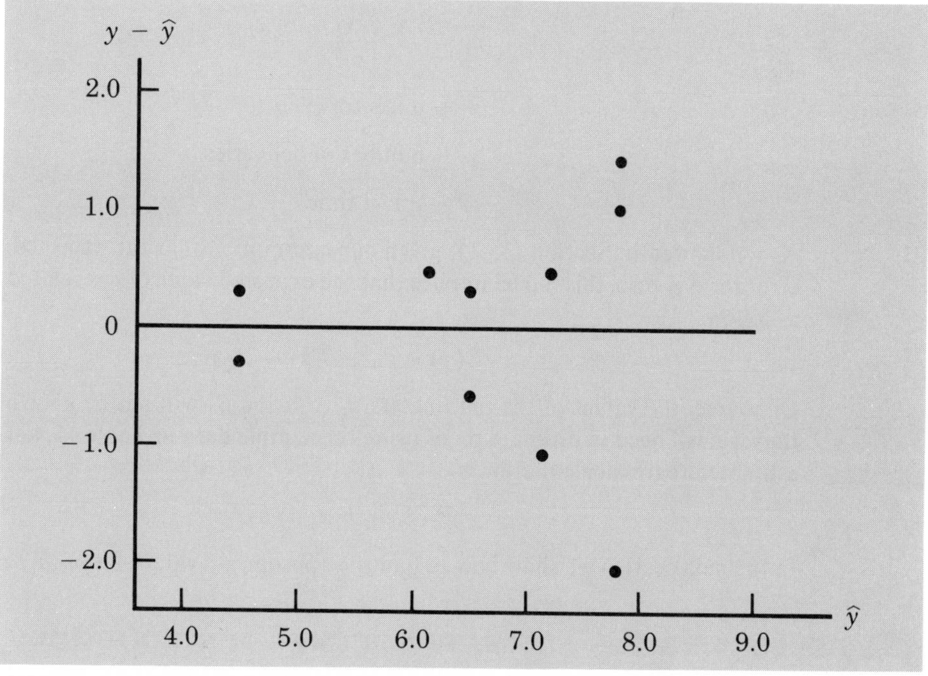

FIGURE 15.3 Residual Plot for the Butler Trucking Problem Corresponding to $\hat{y} = 1.13 + .067x_1$

by the relationship with miles traveled. However, note that 40% of the variance is still unexplained, and note the possible problem we observed with the residual plot; we should doubt the adequacy of our existing model.

Looking for a possible alternative, management suggested that perhaps the number of deliveries made could also be used to help predict travel time and hence improve the regression model. The data, with the addition of the number of deliveries, are shown in Table 15.4, where x_{2i} denotes the number of deliveries on day i.

TABLE 15.4 Data for the Butler Trucking Problem with Miles Traveled (x_1) and Number of Deliveries (x_2) as the Independent Variables

Day	x_{1i} = *Miles Traveled*	x_{2i} = *Number of Deliveries*	y_i = *Travel Time (hours)*
1	100	4	9.3
2	50	3	4.8
3	100	4	8.9
4	100	2	5.8
5	50	2	4.2
6	80	1	6.8
7	75	3	6.6
8	80	2	5.9
9	90	3	7.6
10	90	2	6.1

With the number of deliveries included as a second independent variable, the following multiple regression model for Butler Trucking is obtained:

$$y = \beta_0 + \beta_1 x_1 + \beta_2 x_2 + \epsilon, \tag{15.6}$$

where

$$x_1 = \text{miles traveled},$$

$$x_2 = \text{number of deliveries},$$

$$y = \text{travel time}.$$

As we showed in Section (15.1), given our assumption that the expected value of the error term is zero, this model implies that the expected value of y is related to x_1 and x_2 as follows:

$$E(y) = \beta_0 + \beta_1 x_1 + \beta_2 x_2. \tag{15.7}$$

Of course, the values of the parameters β_0, β_1, and β_2 will not be known in practice, thus we will need to estimate them using the sample data in Table 15.4. The resulting estimated regression equation is

$$\hat{y} = b_0 + b_1 x_1 + b_2 x_2. \tag{15.8}$$

In the next section we show how to find the appropriate values for b_0, b_1, and b_2.

15.3 DEVELOPING THE ESTIMATED REGRESSION EQUATION

In Chapter 14 we presented formulas for estimating b_0 and b_1 for the regression model $y = \beta_0 + \beta_1 x + \epsilon$. In the general multiple regression case the usual presentation of formulas for computing the coefficients of the estimated regression equation involves the use of matrix algebra and is beyond the scope of this text. However, for the special case of two independent variables we can show what is involved. In this section we will concern ourselves with finding b_0, b_1, and b_2 for the two-independent-variable Butler Trucking problem. In subsequent sections we will show how to obtain the estimated regression equations using computer software packages.

In the previous section we showed that by including the effect of number of deliveries we obtain the following estimated regression equation:

$$\hat{y} = b_0 + b_1x_1 + b_2x_2,$$

where

$$x_1 = \text{miles traveled,}$$
$$x_2 = \text{number of deliveries.}$$

Using this notation, the predicted value for the ith observation is

$$\hat{y}_i = b_0 + b_1x_{1i} + b_2x_{2i},$$

where

$$x_{1i} = i\text{th value of } x_1,$$
$$x_{2i} = i\text{th value of } x_2,$$
$$\hat{y}_i = \text{predicted value of travel time (hours) when } x_1 = x_{1i} \text{ and } x_2 = x_{2i}.$$

For the case of two independent variables the residuals are defined as follows:

$$\text{Residual for } i\text{th observation} = y_i - \hat{y}_i = y_i - (b_0 + b_1x_{1i} + b_2x_{2i}) \qquad (15.9)$$

The least squares method determines the values of b_0, b_1, and b_2 that minimize the sum of squared residuals. Thus we must choose b_0, b_1, and b_2 to satisfy the following criterion:

$$\min \Sigma\, (y_i - b_0 - b_1x_{1i} - b_2x_{2i})^2. \qquad (15.10)$$

Using calculus it can be shown (see chapter appendix) that the values of b_0, b_1, and b_2 that minimize (15.10) must satisfy the three equations below, called the *normal equations:*

Normal Equations—Two Independent Variables

$$nb_0 + (\Sigma\, x_{1i})b_1 + (\Sigma\, x_{2i})b_2 = \Sigma\, y_i, \qquad (15.11)$$

$$(\Sigma\, x_{1i})b_0 + (\Sigma\, x_{1i}^2)b_1 + (\Sigma\, x_{1i}x_{2i})b_2 = \Sigma\, x_{1i}y_i, \qquad (15.12)$$

$$(\Sigma\, x_{2i})b_0 + (\Sigma\, x_{1i}x_{2i})b_1 + (\Sigma\, x_{2i}^2)b_2 = \Sigma\, x_{2i}y_i. \qquad (15.13)$$

In order to apply the normal equations we must first use the data to find the values of the coefficients of b_0, b_1, b_2 and the values for the right hand sides of these equations. The necessary data for Butler Trucking are contained in Table 15.5.

Using the information in Table 15.5 we can substitute into the normal equations (15.11) to (15.13) to obtain the following normal equations for the Butler Trucking

TABLE 15.5 Calculation of Coefficients for Normal Equations

y_i	x_{1i}	x_{2i}	x_{1i}^2	x_{2i}^2	$x_{1i}x_{2i}$	$x_{1i}y_i$	$x_{2i}y_i$
9.3	100	4	10,000	16	400	930	37.2
4.8	50	3	2,500	9	150	240	14.4
8.9	100	4	10,000	16	400	890	35.6
5.8	100	2	10,000	4	200	580	11.6
4.2	50	2	2,500	4	100	210	8.4
6.8	80	1	6,400	1	80	544	6.8
6.6	75	3	5,625	9	225	495	19.8
5.9	80	2	6,400	4	160	472	11.8
7.6	90	3	8,100	9	270	684	22.8
6.1	90	2	8,100	4	180	549	12.2
66.0	815	26	69,625	76	2,165	5,594	180.6

problem:

$$10b_0 + 815b_1 + 26b_2 = 66.0, \tag{15.14}$$

$$815b_0 + 69{,}625b_1 + 2{,}165b_2 = 5{,}594.0, \tag{15.15}$$

$$26b_0 + 2{,}165b_1 + 76b_2 = 180.6. \tag{15.16}$$

Since the least squares estimates b_0, b_1, and b_2 must satisfy these three equations simultaneously, in order to obtain values for b_0, b_1, and b_2 we will have to solve this system of three simultaneous linear equations in three variables. The solution* is given by $b_0 = .0367$, $b_1 = .0562$, and $b_2 = .7639$. Thus the estimated regression equation for Butler Trucking is

$$\hat{y} = .0367 + .0562x_1 + .7639x_2\,. \tag{15.17}$$

Note on Interpretation of Coefficients

One observation can be made at this time concerning the relationship between the estimated regression equation with only miles traveled as an independent variable and the one which includes the number of deliveries as a second independent variable. The value of b_1 is not the same in both cases. In simple linear regression we interpret b_1 as the amount of change in y for a 1 unit change in the independent variable. In multiple regression analysis this interpretation must be modified somewhat. That is, in multiple regression analysis we interpret each regression coefficient as follows: b_i represents the change in y corresponding to a 1 unit change in x_i when all other independent variables are held constant. For example, in the Butler Trucking problem involving two independent variables $b_1 = .0562$. Thus .0562 hours is the expected increase in travel time corresponding to an increase of 1 mile in the distance traveled when the number of deliveries is held constant. Similarly, since $b_2 = .7639$, the expected increase in travel time corresponding to an increase of one delivery when the number of miles traveled is held constant is .7639 hours.

*In the chapter appendix we show in detail how the solution is obtained.

Computer Solution

It can be shown that for multiple regression problems involving p independent variables there are $p + 1$ normal equations that must be solved simultaneously for the estimated coefficients $b_0, b_1, b_2, \ldots, b_p$. The computational effort involved requires more sophisticated solution procedures than we have used in the solution of (15.14) to (15.16). Fortunately, computer software packages can be used to obtain these solutions with very little effort on the part of the user.

In Figure 15.4 we show the computer output from the MINITAB computer package for the version of the Butler Trucking problem involving the two independent variables, miles traveled and number of deliveries. Note that in the column labeled "COEFFICIENT" the values are the same as we obtained (except for rounding) for b_0, b_1, and b_2. We will discuss the remainder of the output in the following sections.

```
THE REGRESSION EQUATION IS
Y = 0.0367 + 0.0562X1 + 0.764X2

                                          ST. DEV.       T-RATIO =
        COLUMN        COEFFICIENT         OF COEF.       COEF/S.D.
        --                  0.037         1.326               0.03
X1      MILES               0.05616       0.01564             3.59
X2      DELIVERIES          0.7639        0.3053              2.50

THE ST. DEV. OF Y ABOUT REGRESSION LINE IS
S = 0.8494
WITH (10-3) = 7 DEGREES OF FREEDOM

R-SQUARED = 79.0 PERCENT
R-SQUARED = 72.9 PERCENT, ADJUSTED FOR D.F.

ANALYSIS OF VARIANCE

DUE TO          DF        SS            MS=SS/DF
REGRESSION       2        18.9499         9.4749
RESIDUAL         7         5.0501         0.7214
TOTAL            9        24.0000
```

FIGURE 15.4 Output Data for Multiple Regression Problem Solution Using MINITAB

EXERCISES

1. A shoe store has developed the following estimated regression equation relating sales to inventory investment and advertising expenditures:

$$\hat{y} = 25 + 10x_1 + 8x_2,$$

where x_1 = inventory investment ($1,000's),

x_2 = advertising expenditures ($1,000's),

y = sales ($1,000's).

a. Estimate sales if there is a $15,000 investment in inventory and an advertising budget of $10,000.

b. Interpret the parameters (b_1 and b_2) in this estimated regression equation.

2. The owner of TAI Movie Theaters, Inc. would like to investigate the effect of television advertising on weekly gross revenue for special promotion films. The following historical data were developed:

Weekly Gross Revenue ($1,000's)	*Television Advertising ($1,000's)*
96	5.0
90	2.0
95	4.0
92	2.5
95	3.0
94	3.5
94	2.5
94	3.0

a. Using these data, develop an estimated regression equation relating weekly gross revenue to television advertising expenditure.
b. Estimate the weekly gross revenue in a week in which $3,500 is spent on television advertising.
3. As an extension of Exercise 1 consider the possibility of incorporating the effect of newspaper advertising as well as television advertising on weekly gross revenue. The following data were developed from historical records:

Weekly Gross Revenue ($1,000's)	*Newspaper Advertising ($1,000's)*	*Television Advertising ($1,000's)*
96	1.5	5.0
90	2.0	2.0
95	1.5	4.0
92	2.5	2.5
95	3.3	3.0
94	2.3	3.5
94	4.2	2.5
94	2.5	3.0

Let

$$x_1 = \text{newspaper advertising (\$1,000's)},$$
$$x_2 = \text{television advertising (\$1,000's)},$$
$$y = \text{weekly gross revenue (\$1,000's)}.$$

a. Write the normal equations that must be solved to find b_0, b_1, and b_2 for the estimated regression equation. (That is, insert the numerical values, where appropriate, for Σx_{1i}, Σx_{1i}^2, $\Sigma x_{1i}y_i$, etc.)
b. Solve the normal equations above to find an estimated regression equation relating weekly gross revenue to newspaper and television advertising.
c. Is the coefficient for television advertising expenditures the same in Exercise 2a and 3b? Interpret this coefficient in each case.
4. Heller Company believes that the quantity of lawnmowers sold depends on the price of its mower and the price of a competitor's mower. Let

$$y = \text{quantity sold (1,000's)},$$
$$x_1 = \text{price of competitor's mower (dollars)},$$
$$x_2 = \text{price of Heller's mower (dollars)}.$$

Management would like an estimated regression equation that relates quantity sold to the price of the Heller Mower and the competitor's mower. The following data are available concerning prices in ten different cities.

Competitor's Price (x_1)	*Heller's Price* (x_2)	*Quantity Sold* (y)
120	100	102
140	110	100
190	90	120
130	150	77
155	210	46
175	150	93
125	250	26
145	270	69
180	300	65
150	250	85

a. Write the normal equations.
b. Solve the normal equations to obtain the values of the parameters in the estimated regression equation.
c. Predict the quantity sold in a city where Heller prices its mower at $160 and the competitor prices its mower at $170.
d. Interpret b_1 and b_2.

15.4 TEST FOR A SIGNIFICANT RELATIONSHIP

The regression equation that we have assumed for the Butler Trucking problem involving two independent variables is

$$E(y) = \beta_0 + \beta_1 x_1 + \beta_2 x_2.$$

Therefore the appropriate test for determining whether or not there is a significant relationship among x_1, x_2, and y is as follows:

H_0: $\beta_1 = \beta_2 = 0$,

H_1: At least one of the two coefficients is not equal to zero.

If we reject H_0, we conclude that there is a significant relationship among x_1, x_2, and y and that the estimated regression equation is useful in predicting y.

The test used to determine if there is a significant relationship in the multiple regression case is an F test very similar to the one introduced in Chapter 14 for models involving one independent variable. First, we will show how it is used by applying it in the Butler Trucking problem. Then we will generalize the application of the test to models involving p independent variables.

The ANOVA Table for the Butler Trucking Problem

Recall that in Chapter 14 we discussed the use of the F test for determining whether or not the relationship between x and y is statistically significant. The test statistic used is F = MSR/MSE, where MSR denotes the mean square due to regression and MSE denotes the mean square due to error. Before we try to apply this test for the multiple regression case let us review the calculations needed to compute SST, SSR, and SSE.

In Chapter 14 we stated that the relationship among SST, SSR, and SSE is SST = SSR + SSE. This equation also holds for multiple regression analysis. We will first compute SST $= \Sigma\,(y_i - \bar{y})^2$ and SSE $= \Sigma\,(y_i - \hat{y}_i)^2$. Then we obtain SSR by subtraction; that is, SSR = SST − SSE. Table 15.6 shows the computation of SST and

TABLE 15.6 Computation of Sum of Squares for Butler Trucking ($\hat{y} = .0367 + .0562x_1 + .7639x_2$)

y_i	$y_i - \bar{y}$	$(y_i - \bar{y})^2$	$\hat{y}_i$	$y_i - \hat{y}_i$	$(y_i - \hat{y}_i)^2$
9.3	2.7	7.29	8.7084	0.5916	0.3500
4.8	−1.8	3.24	5.1364	−0.3364	0.1132
8.9	2.3	5.29	8.7084	0.1916	0.0367
5.8	−0.8	0.64	7.1807	−1.3807	1.9063
4.2	−2.4	5.76	4.3725	−0.1725	0.0298
6.8	0.2	0.04	5.2936	1.5064	2.2692
6.6	0.0	0.00	6.5405	0.0595	0.0035
5.9	−0.7	0.49	6.0574	−0.1574	0.0248
7.6	1.0	1.00	7.3829	0.2171	0.0471
6.1	−0.5	0.25	6.6191	−0.5191	0.2695
66.0		24.00			5.0501
$\bar{y} = 66/10 = 6.6$		SST			SSE

SSE for the Butler Trucking problem corresponding to the estimated regression equation $\hat{y} = .0367 + .0562x_1 + .7639x_2$. We see tht SST = 24.00 and SSE = 5.0501. Using these two values we obtain SSR = 24.00 − 5.0501 = 18.9499.

In the general multiple regression situation involving p independent variables, the numbers of degrees of freedom for SSR, SSE, and SST are as follows:

Sum of Squares	*Degrees of Freedom (DF)*
SSR	p
SSE	$n - p - 1$
SST	$n - 1$

Thus for the Butler Trucking problem SST has 10 − 1 = 9 degrees of freedom. In addition, since we have $p = 2$ independent variables, SSR has 2 degrees of freedom. Finally, since $n - p - 1 = 10 - 2 - 1 = 7$, SSE has 7 degrees of freedom. Note that—just as before—the numbers of degrees of freedom for SSR and SSE add to $n - 1$, the degrees of freedom for SST.

Finally, recall that MSR and MSE are calculated by dividing the appropriate sum of squares by its degrees of freedom. That is,

$$\text{MSR} = \frac{\text{SSR}}{\text{Regression DF}} = \frac{\text{SSR}}{p},$$

$$\text{MSE} = \frac{\text{SSE}}{\text{Error DF}} = \frac{\text{SSE}}{n - p - 1}.$$

For the Butler trucking problem then we see that

$$\text{MSR} = \frac{18.9499}{2} = 9.475,$$

$$\text{MSE} = \frac{5.0501}{7} = .7214.$$

The sums of squares, degrees of freedom, and the corresponding mean squares are conveniently summarized in the ANOVA table shown in Table 15.7. Note that the same table was provided by the computer output shown in Figure 15.4.

TABLE 15.7 ANOVA Table for the Butler Trucking Problem with Two Independent Variables

Source of Variation	*Sum of Squares*	*Degrees of Freedom*	*Mean Square*
Regression	SSR = 18.9499	2	$\text{MSR} = \frac{18.9499}{2} = 9.475$
Error	SSE = 5.0501	7	$\text{MSE} = \frac{5.0501}{7} = .7214$
Total (about the mean)	SST = 24.00	9	

The *F* Test for the Butler Trucking Problem

It can be shown that if the null hypothesis (H_0: $\beta_1 = \beta_2 = 0$) is true and the four underlying regression model assumptions are valid, then the sampling distribution of MSR/MSE follows an *F* distribution. The number of numerator degrees of freedom is equal to the degrees of freedom associated with the sum of squares due to regression, and the denominator degrees of freedom is equal to the degrees of freedom associated with the sum of squares due to error. Recall that the sum of squares due to regression (SSR) measures the amount of the variabilty in y explained by the regression model. Thus we would expect large values of F = MSR/MSE to cast doubt on the null hypothesis (no relationship between the dependent and independent variables). Indeed, this is true; small values of *F* lead to the acceptance of H_0, and large values lead to the rejection of H_0.

Let us return to the Butler Trucking problem and test the significance of the multiple regression model. First, we compute MSR/MSE:

$$\frac{\text{MSR}}{\text{MSE}} = \frac{18.9499/2}{5.0501/7}$$

$$= 13.133.$$

To test the hypothesis that $\beta_1 = \beta_2 = 0$ we must determine whether or not 13.133 is a value that appears likely when random sampling from an *F* distribution with 2 numerator degrees of freedom and 7 denominator degrees of freedom.

Suppose that we are willing to tolerate a maximum Type I error of $\alpha = .05$. The critical value from the *F* distribution table (Table 4 of Appendix B) is 4.74. That is, if we are sampling randomly from an *F* distribution with numerator and denominator degrees of freedom equal to 2 and 7, respectively (as we would be if H_0 were true), then

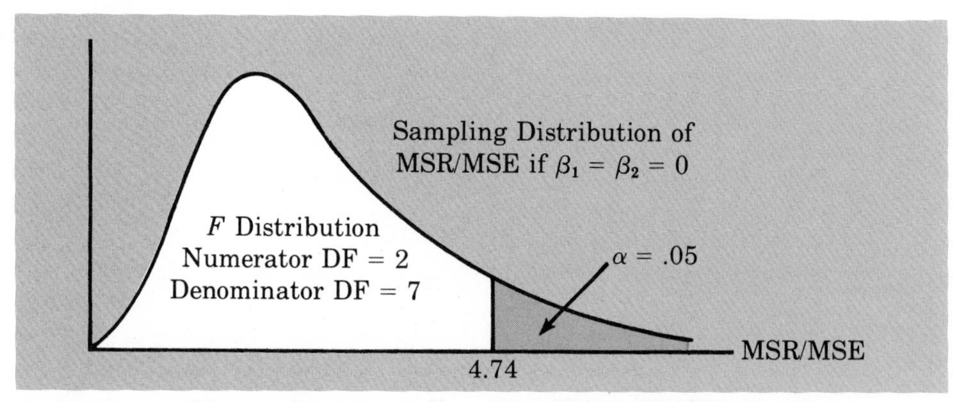

FIGURE 15.5 Determination of the Critical Value and Rejection Region for a Test for a Significant Regression Relation. Probability of a Type I Error is $\alpha = .05$

only 5% of the time would we get a value larger than 4.74. Figure 15.5 illustrates the determination of the critical region.

We can use the above analysis to formulate a decision rule for determining whether to accept or reject H_0 at the .05 significance level as follows:

$$\text{Accept } H_0 \text{ if } \frac{\text{MSR}}{\text{MSE}} \leq 4.74,$$

$$\text{Reject } H_0 \text{ if } \frac{\text{MSR}}{\text{MSE}} > 4.74.$$

Since the value of MSR/MSE was 13.133, we can reject H_0. Hence we conclude that there is a significant relationship between total travel time and the two independent variables. Thus the estimated regression equation should be useful in predicting y for values of the independent variables within the range of those included in the sample.

The General ANOVA Table and *F* Test

Now that we know how the F test can be applied for a multiple regression model with two independent variables, let us generalize our test to the case involving a model with p independent variables. The appropriate hypothesis test to determine if there is a significant relationship is as follows:

H_0: $\beta_1 = \beta_2 = \cdots = \beta_p = 0,$

H_1: at least one of the p coefficients is not equal to zero.

Again, if we reject H_0 we can conclude that there is a significant relationship and that the estimated regression equation is useful for predicting or explaining the dependent variable y.

The general form of the ANOVA table for the multiple regression case involving p independent variables is shown in Table 15.8. The only change from the two-variable case is the degrees of freedom corresponding to SSR and SSE. Here the sum of squares due to regression has p degrees of freedom corresponding to the p independent

TABLE 15.8 ANOVA Table for a Multiple Regression Model Involving p Independent Variables

Source	Sum of Squares	Degrees of Freedom	Mean Square
Regression	SSR	p	$\text{MSR} = \dfrac{\text{SSR}}{p}$
Error	SSE	$n - p - 1$	$\text{MSE} = \dfrac{\text{SSE}}{n - p - 1}$
Total (about the mean)	SST	$n - 1$	

variables; hence

$$\text{MSR} = \frac{\text{SSR}}{p}. \tag{15.18}$$

In addition, the sum of squares due to error has $n - p - 1$ degrees of freedom; thus

$$\text{MSE} = \frac{\text{SSE}}{n - p - 1}. \tag{15.19}$$

Hence the F statistic for the case of p independent variables is computed as follows:

$$F = \frac{\text{MSR}}{\text{MSE}} = \frac{\text{SSR}/p}{\text{SSE}/(n - p - 1)}. \tag{15.20}$$

When looking up the critical value from the F distribution table the numerator degrees of freedom are p and the denominator degrees of freedom are $n - p - 1$.

t Test for Significance of Individual Parameters

If after using the F test we conclude that the multiple regression relationship is significant (that is, we conclude that at least one of the $\beta_i \neq 0$), it is often of interest to conduct tests to see if the individual parameters β_i are significant. The t test is a statistical method for testing the significance of the individual parameters.

The hypothesis test we wish to conduct is the same for the coefficient of each independent variable. It is stated as follows:

$$H_0\colon \quad \beta_i = 0,$$
$$H_1\colon \quad \beta_i \neq 0.$$

Recall that in Chapter 14 we learned how to conduct such a test for the case where there is only one independent variable. The hypotheses were:

$$H_0\colon \quad \beta_1 = 0,$$
$$H_1\colon \quad \beta_1 \neq 0.$$

To test this hypothesis we computed the sample statistic b_1/s_{b_1}, where b_1 was the least squares estimate of β_1 and s_{b_1} was an estimate of the standard deviation of the sampling distribution of b_1. We learned that the sampling distribution of b_1/s_{b_1} follows a t distribution with $n - 2$ degrees of freedom. Thus to conduct the hypothesis test we chose a value of α, found a rejection region determined by $t_{\alpha/2}$, computed b_1/s_{b_1} and rejected H_0 if

$$b_1/s_{b_1} > t_{\alpha/2}$$

or

$$b_1/s_{b_1} < -t_{\alpha/2}.$$

The procedure for testing individual parameters in the multiple regression case is essentially the same. The only differences are in the number of degrees of freedom for the appropriate t distribution and in the formula for computing s_{b_i}. The number of degrees of freedom is the same as for the sum of squares due to error. Thus we use $n - p - 1$ degrees of freedom, where p is the number of independent variables. (Note that for the case of one independent variable this reduces to the $n - 2$ degrees of freedom used in Chapter 14.) The formula for s_{b_i} is more involved, and we do not present it here; however, s_{b_i} is calculated and printed by most computer software packages for multiple regression analysis.

Let us return now to the Butler Trucking problem to test the significance of the parameters β_1 and β_2. Note that in the MINITAB printout (Figure 15.4) the values of b_1, b_2, s_{b_1}, and s_{b_2} were given as

$$b_1 = .05616, \qquad s_{b_1} = .01564,$$

$$b_2 = .7639 \qquad s_{b_2} = .3053.$$

Therefore for the parameters β_1 and β_2 we obtain

$$\frac{b_1}{s_{b_1}} = \frac{.05616}{.01564} = 3.59$$

$$\frac{b_2}{s_{b_2}} = \frac{.7639}{.3053} = 2.50.$$

Note that both of these values were provided by the MINITAB output of Figure 15.4 under the column labeled "T-RATIO = COEF/S.D." Using $\alpha = .05$ and $10 - 2 - 1 = 7$ degrees of freedom we can find the appropriate t value for our hypothesis tests in Table 2 of Appendix B. We obtain

$$t_{.025} = 2.365.$$

Now, since $b_1/s_{b_1} = 3.59 > 2.365$, we reject the hypothesis that $\beta_1 = 0$. Furthermore, since $b_2/s_{b_2} = 2.50 > 2.365$, we reject the hypothesis that $\beta_2 = 0$.

Multicollinearity

We have used the term independent variable in regression analysis to refer to any variable being used to predict or explain the value of the dependent variable. The term does not mean, however, that the independent variables themselves are independent in any statistical sense. Quite the contrary, most independent variables in a multiple regression problem are correlated to some degree with one another. For example, in the Butler Trucking problem involving the two independent variables x_1 (miles traveled) and x_2 (number of deliveries) we could treat the miles traveled as the dependent variable and the number of deliveries as the independent variable in order to determine if these two variables are themselves related. Using (14.30), then, we could compute the sample correlation coefficient r to determine the extent to which these variables are related. We did so and obtained $r = .28$. Thus there is some degree of linear association between the two independent variables. In multiple regression analysis we use the term *multicollinearity* to refer to the correlation among the independent variables.

To provide a better perspective of the potential problems of multicollinearity, let us consider a modification of the Butler Trucking problem. Instead of x_2 being the number of deliveries, let x_2 denote the number of gallons of gasoline consumed. Clearly, x_1 (the miles traveled) and x_2 are related; that is, we know that the number of gallons of gasoline used depends upon the number of miles traveled. Thus we would conclude logically that x_1 and x_2 are highly correlated independent variables.

Assume that we obtain the equation $\hat{y} = b_0 + b_1x_1 + b_2x_2$ and find that the F test shows that the regression is significant. Then suppose that we were to conduct a t test on β_1 to determine if $\beta_1 \neq 0$, and we cannot reject H_0: $\beta_1 = 0$. Does this mean that travel time is not related to miles traveled? Not necessarily. What it probably means is that with x_2 already in the model, x_1 does not contribute a significant addition toward determining the value of y. This would seem to make sense in our example, since if we know the amount of gasoline consumed we do not gain much information useful in predicting y by knowing the miles traveled. Similarly, a t test might lead us to conclude $\beta_2 = 0$ on the grounds that with x_1 in the model knowledge of the amount of gasoline consumed does not add much.

To summarize, the difficulty caused by multicollinearity in conducting t tests for the significance of individual parameters is that it is possible to conclude that none of the individual parameters are significantly different from zero when an F test on the overall multiple regression equation indicates a significant relationship. This problem is avoided when there is very little correlation among the independent variables.

Ordinarily multicollinearity does not affect the way in which we perform our regression analysis or interpret the output from a study. However, when multicollinearity is severe—that is, when two or more of the independent variables are highly correlated with one another—we can run into difficulties interpreting the results of t tests on the individual parameters. In addition to the type of problem illustrated above, severe cases of multicollinearity have been shown to result in least squares estimates that even have the wrong sign. That is, in simulated studies where researchers created the underlying regression model and then applied the least squares technique to develop estimates of β_0, β_1, β_2, and so on, it has been shown that under conditions of multicollinearity the least squares estimates can even have a sign opposite to that of the parameter being estimated. For example, β_2 might actually be $+10$ and b_2, its estimate, might turn out to be -2. Thus little faith can be placed in the individual coefficients themselves if multicollinearity is present to a high degree.

Statisticians have developed several tests for determining whether or not multicollinearity is high enough to cause these types of problems. One simple test, referred to as the "rule of thumb" test, says that multicollinearity is a potential problem if the absolute value of the sample correlation coefficient exceeds .7 for any two of the independent variables. The other types of tests are more advanced and beyond the scope of this text.

If possible, every attempt should be made to avoid including independent variables that are highly correlated. In practice, however, it is rarely possible to adhere to this policy strictly. Thus the decision maker should be warned that when there is reason to believe that substantial multicollinearity is present it is difficult to separate out the effect of the individual independent variables on the dependent variable.

EXERCISES

5. In Exercise 1 the following estimated regression equation for relating sales to inventory investment and advertising expenditures was given:

$$\hat{y} = 25 + 10x_1 + 8x_2.$$

The data used to develop the model came from a survey of ten stores. In addition to the estimated regression equation, it was found as a result of a computer run that SST = 16,000 and SSR = 12,000.

a. Compute SSE, MSE, and MSR.

b. Use an F test and an $\alpha = .05$ level of significance to determine if there is a relationship among the variables.

6. Refer to Exercise 3.

a. Use a level of significance of $\alpha = .01$ to test the hypotheses

$$H_0: \beta_1 = \beta_2 = 0,$$

$$H_1: \beta_1 \text{ or } \beta_2 \text{ is not equal to zero.}$$

b. Use a level of significance of $\alpha = .05$ to test the significance of β_1. (Note: $s_{b_1} = .3207$). Should x_1 be dropped from the model?

c. Use a level of significance of $\alpha = .05$ to test the significance of β_2. (Note: $s_{b_2} = .3041$). Should x_2 be dropped from the model?

7. Refer to Exercise 4 involving the Heller Company. Test the significance of the overall model at $\alpha = .05$.

8. Shown below is a partial computer printout from a multiple regression problem involving two independent variables.

```
THE REGRESSION EQUATION IS
Y = 11.61 + 2.16X1 + 4.80X2

                                          ST. DEV.      T-RATIO =
         COLUMN        COEFFICIENT        OF COEF.      COEF/S.D.
           --              11.61            3.07
X1       VAR1               2.16            0.69
X2       VAR2               4.80            1.03

ANALYSIS OF VARIANCE

DUE TO              DF        SS      MS=SS/DF
REGRESSION                   90.3
RESIDUAL            12
TOTAL                       108.6
```

a. Find the appropriate values for regression and total degrees of freedom.

b. Find SSE, MSR, MSE, and the t ratios for the coefficients.

c. Compute F and test at the $\alpha = .01$ level whether a significant relationship exists or not.

d. Use the t test and $\alpha = .05$ to determine if either of the independent variables should be dropped from the model.

9. The following estimated regression equation was developed for a model involving two independent variables:

$$\hat{y} = 40.7 + 8.63x_1 + 2.71x_2$$

After dropping x_2 from the model the least squares method was used again to obtain an estimated regression equation involving only x_1 as an independent variable:

$$\hat{y} = 42.0 + 9.01x_1.$$

a. Give an interpretation of the coefficient of x_1 in both models.

b. Could multicollinearity explain why the coefficient of x_1 differs in the two models? If so, how?

15.5 DETERMINING MODEL ACCEPTABILITY

The tests that we discussed in the previous section should be used only if we believe that the model assumptions presented in Section 15.1 regarding ϵ are appropriate. These assumptions regarding ϵ are important because statisticians have used these assumptions as the basis for developing the tests introduced in Section 15.4. Residual analysis will allow us to make a judgment as to whether or not these assumptions appear to be satisfied. In addition, residual analysis can often provide insight as to whether or not a different type of model—for example, one involving a different functional form—might better describe the observed relationship.

Residual Analysis

Residual analysis in multiple regression is similar to that done in regression involving one independent variable. As we indicated in our discussion of residual analysis for models involving one independent variable, one of the most common residual plots that can be used to examine the assumptions regarding ϵ is a plot of the residuals against $\hat{y}_i$.* Figure 15.6 shows this type of residual plot for the Butler Trucking problem

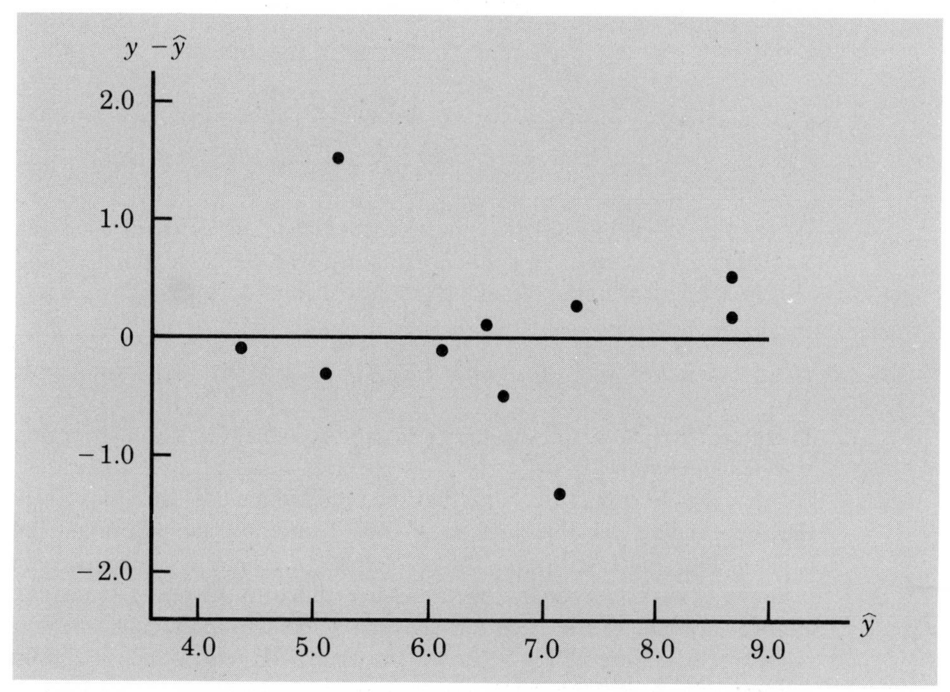

FIGURE 15.6 Residuals Plotted Against $\hat{y}$ for Model with Miles Traveled and Number of Deliveries as Independent Variables

*In multiple regression analysis it is also common to examine plots of the residuals versus each of the independent variables. Some of the references in the bibliography discuss this.

corresponding to the estimated regression equation $\hat{y} = .0367 + .0562x_1 + .7639x_2$. Comparing this plot with the general types of plots illustrated in Figure 14.8, we conclude that the plot of Figure 15.6 does not show any residual patterns that would lead us to question the assumptions of the model. Therefore we would conclude that the model assumptions appear to be satisfied and hence that the statistical conclusions developed in the previous section are justified.

Strength of Relationship

At this point we have concluded that the assumptions for our model appear reasonable and that the estimated regression equation $\hat{y} = .0367 + .0562x_1 + .7639x_2$ is statistically significant. That is, there is a significant relationship among x_1, x_2, and y and the estimated regression equation is useful in predicting y. Now we would like to compute a measure of the strength of the relationship, or the goodness of fit of the regression equation to the data.

In Chapter 14 we used the coefficient of determination (r^2) to evaluate the strength of the regression relationship. Recall that r^2 was computed as

$$r^2 = \frac{\text{SSR}}{\text{SST}}.$$

In multiple regression analysis we compute a similar quantity, called the multiple coefficient of determination:

Multiple Coefficient of Determination

$$R^2 = \frac{\text{SSR}}{\text{SST}}. \tag{15.21}$$

When multiplied by 100 the multiple coefficient of determination represents the percentage of variability in y that is explained by the estimated regression equation.

In the case of Butler Trucking (refer to Table 15.7) we find

$$R^2 = \frac{18.9499}{24.0000} = .7896.$$

Therefore 78.96% of the variability in y is explained by the relationship with miles traveled and number of deliveries.

Refer to Section 15.1. Note that the regression model with only miles traveled as the independent variable had $r^2 = .60$. Therefore the percentage of variability explained has increased from 60% to 78.96%. In general it is always true that R^2 will increase as more independent variables are added to the regression equation because adding variables to the equation causes the prediction errors to be smaller, hence reducing SSE. Since SST = SSR + SSE, when SSE gets smaller SSR must get larger, causing R^2 = SSR/SST to increase.

Many analysts recommend adjusting R^2 for the number of independent variables in order to avoid overestimating the impact of adding an independent variable on the amount of explained variability. This so-called *adjusted multiple coefficient of determination* is computed as follows:

Adjusted Multiple Coefficient of Determination

$$R_a^2 = 1 - (1 - R^2)\frac{n-1}{n-p-1}. \tag{15.22}$$

For the Butler Trucking problem we obtain

$$\begin{aligned} R_a^2 &= 1 - (1 - .7896)\frac{10-1}{10-2-1} \\ &= 1 - (.2104)(1.2857) \\ &= .7295. \end{aligned}$$

Note that both the value of R^2 and the value of R_a^2 are provided by the MINITAB output shown in Figure 15.4.

EXERCISES

10. Refer to the computer output from a regression run shown in Exercise 8.

a. Compute R^2.

b. Compute R_a^2.

c. Does the model appear to explain a large amount of the variability in the data?

11. Refer to Exercise 5.

a. For the estimated regression equation given, compute R^2.

b. Compute R_a^2.

c. Does the model appear to explain a large amount of the variability in the data?

15.6 THE USE OF QUALITATIVE VARIABLES

So far the variables that we have used to build a model to predict total travel time have been quantitative variables; that is, variables that are measured in terms of numerical values (e.g., miles traveled and number of deliveries). Frequently, however, we will wish to use variables that are not measured in terms of numerical values. We refer to such variables as *qualitative variables*. For instance, suppose that we were interested in predicting sales for a product which was available in either bottles or cans. Clearly the independent variable "container type" could influence the dependent variable "sales"—but container type is a qualitative, not a quantitative variable. The distinguishing feature of qualitative variables is that there is no natural measure of "how much"; these variables are used to refer to attributes that are either present or not present.

Let us see how qualitative variables might be used in the context of the Butler Trucking problem. Suppose that management felt that the type of truck should also be considered in attempting to predict total travel time. Butler Trucking has only two types of trucks: pickups and vans. Thus truck type is an example of a qualitative variable. Table 15.9 shows the expanded data set for the Butler Trucking Company with the addition of truck type as a third independent variable. To incorporate the effect of truck type into a model to predict total travel time we define the following

TABLE 15.9 Data for the Butler Trucking Problem, Including Truck Type

Day (i)	Miles traveled (x_{1i})	Number of Deliveries (x_{2i})	Truck Type (x_{3i})		Travel Time (hours) y_i
1	100	4	Van	1	9.3
2	50	3	Pickup	0	4.8
3	100	4	Van	1	8.9
4	100	2	Pickup	0	5.8
5	50	2	Pickup	0	4.2
6	80	1	Van	1	6.8
7	75	3	Van	1	6.6
8	80	2	Pickup	0	5.9
9	90	3	Pickup	0	7.6
10	90	2	Van	1	6.1

variable:

$$x_3 = \begin{cases} 0 & \text{if the truck is a pickup,} \\ 1 & \text{if the truck is a van.} \end{cases}$$

When preparing the data, whenever an observation involves a pickup truck we will set $x_3 = 0$; whenever an observation involves a van we will set $x_3 = 1$. In regression analysis this type of qualitative variable is commonly referred to as a *dummy* or *indicator* variable.

Adding this dummy variable to our previous regression function for predicting travel time results in the following regression equation:

$$E(y) = \beta_0 + \beta_1 x_1 + \beta_2 x_2 + \beta_3 x_3. \tag{15.23}$$

We can see that when $x_3 = 0$, and hence the truck is a pickup, our regression equation reduces to

$$\begin{aligned} E(y) &= \beta_0 + \beta_1 x_1 + \beta_2 x_2 + \beta_3(0) \\ &= \beta_0 + \beta_1 x_1 + \beta_2 x_2. \end{aligned} \tag{15.24}$$

However, if we want to predict y when a van is used, $x_3 = 1$ and our regression equation becomes

$$\begin{aligned} E(y) &= \beta_0 + \beta_1 x_1 + \beta_2 x_2 + \beta_3(1) \\ &= \beta_0 + \beta_1 x_1 + \beta_2 x_2 + \beta_3. \end{aligned} \tag{15.25}$$

If we subtract (15.24), the expected travel time for a pickup, from (15.25), the expected travel time for a van, we obtain

$$\underbrace{(\beta_0 + \beta_1 x_1 + \beta_2 x_2 + \beta_3)}_{\text{Expected travel time for a van}} - \underbrace{(\beta_0 + \beta_1 x_1 + \beta_2 x_2)}_{\text{Expected travel time for a pickup}} = \beta_3.$$

Thus β_3 can be interpreted as the difference in the expected travel time between a van and a pickup truck.

When we fit an estimated regression equation using the least squares method and incorporate the possible effect of truck type we obtain the following estimated regression equation:

$$\hat{y} = b_0 + b_1x_1 + b_2x_2 + b_3x_3.$$

As usual, b_3 turns out to be the least squares estimate of β_3 and hence our best estimate of the effect of truck type.

Computer Solution

In Figure 15.7 we show the computer output obtained using the MINITAB statistical computer software system. The estimated regression equation is

$$\hat{y} = .522 + .0464x_1 + .7102x_2 + .900x_3.$$

Thus we see that $b_3 = .900$. Hence the best estimate of the difference in the expected travel time when a van is used instead of a pickup truck is .9 hours, or 54 minutes.

```
THE REGRESSION EQUATION IS
Y = 0.522 + 0.0464X1 + 0.710X2
    + 0.900X3

                                              ST. DEV.        T-RATIO =
       COLUMN          COEFFICIENT            OF COEF.        COEF/S.D.
       --                    0.522               1.210             0.43
X1     MILES               0.04640             0.01500             3.09
X2     DELIVERIES           0.7102              0.2725             2.61
X3     TRUCK                0.9000              0.5281             1.70

THE ST. DEV. OF Y ABOUT REGRESSION LINE IS
S = 0.7531
WITH (10-4) = 6 DEGREES OF FREEDOM

R-SQUARED = 85.8 PERCENT
R-SQUARED = 78.7 PERCENT, ADJUSTED FOR D.F.

ANALYSIS OF VARIANCE

DUE TO          DF          SS     MS=SS/DF
REGRESSION       3     20.5969       6.8656
RESIDUAL         6      3.4031       0.5672
TOTAL            9     24.0000
```

FIGURE 15.7 **MINITAB Output for Model Involving Miles Traveled, Number of Deliveries, and Truck Type**

To test for the significance of x_3 given that x_1 and x_2 are in the model, the appropriate hypotheses are

$$H_0\colon\ \beta_3 = 0,$$
$$H_1\colon\ \beta_3 \neq 0.$$

Using $\alpha = .05$ and $n - p - 1 = 10 - 3 - 1 = 6$ degrees of freedom we find that the appropriate t value (see Table 2 of Appendix B) is

$$t_{.025} = 2.447.$$

From the computer output (Figure 15.7) we see that $b_3/s_{b_3} = 1.70$. Since $1.70 < 2.447$, we cannot reject H_0 and must conclude that truck type is not a significant factor in predicting travel time once the effects of miles traveled and number of deliveries have been accounted for. Note that we concluded not that truck type is not significant but that it is not significant once the effect of x_1 and x_2 have been accounted for. This is an important point that will be discussed fully in the next section.

EXERCISES

12. The following regression model has been proposed to predict sales at a fast-food outlet:

$$y = \beta_0 + \beta_1 x_1 + \beta_2 x_2 + \beta_3 x_3 + \epsilon,$$

where

$$x_1 = \text{number of competitors within 1 mile,}$$

$$x_2 = \text{population within 1 mile (1,000's),}$$

$$x_3 = \begin{cases} 1 & \text{if drive-up window present,} \\ 0 & \text{otherwise,} \end{cases}$$

$$y = \text{sales (\$1,000's).}$$

The following estimated regression equation was developed after 20 outlets were surveyed:

$$\hat{y} = 10.1 - 4.2x_1 + 6.8x_2 + 15.3x_3.$$

a. What is the expected amount of sales attributable to the drive-up window?
b. Predict sales for a store with two competitors, a population of 8,000 within 1 mile, and no drive-up window.
c. Predict sales for a store with one competitor, a population within 1 mile of 3,000, and a drive-up window.
13. In order to investigate the relationship among the service time to repair a machine and (1) the number of months since the machine was serviced and (2) whether a mechanical failure or an electrical failure had occurred, the following data were obtained:

Repair Time (hours)	*Time Since Previous Service Call (months)*	*Type of Failure*
2.9	2	Electrical
3.0	6	Mechanical
4.8	8	Electrical
1.8	3	Mechanical
2.9	2	Electrical
4.9	7	Electrical
4.2	9	Mechanical
4.8	8	Mechanical
4.4	4	Electrical
4.5	6	Electrical

Ignore for now the type of failure associated with the machine. Develop a simple linear regression equation to predict the repair time given the number of months since the previous service call.
14. Does the equation that you developed in Exercise 13 provide a good fit for the observed data? Explain.
15. This problem is an extension of the situation described in Exercise 13. In an attempt to incorporate the possible effect of the type of failure the following dummy variable was added to

the function:

$$x_2 = \begin{cases} 1 & \text{if the failure was electrical,} \\ 0 & \text{if the failure was mechanical.} \end{cases}$$

With the addition of this variable, the following regression equation was proposed:

$$E(y) = \beta_0 + \beta_1 x_1 + \beta_2 x_2,$$

where

x_1 = number of months since the previous service call,

y = repair time (hours).

What is the interpretation of β_2 in this regression equation?

16. A computer package was used to estimate the parameters of the regression equation proposed in Exercise 15. The output from the package is shown below:

SOURCE	SUM OF SQUARES	DF	F
REGRESSION	9.009	2	21.3570
ERROR	1.4792	7	

COEFFICIENT OF DETERMINATION	.8592
STANDARD ERROR OF ESTIMATE	.4590

VARIABLE	REGRESSION COEFFICIENT	STANDARD DEVIATION-COEFFICIENT
1	.3876	.0626
2	1.2627	.3141
INTERCEPT	.9305	

Write the equation that should be used to predict the repair time given this output.

17. At the $\alpha = .05$ level of significance test whether the estimated regression equation developed in Exercise 16 represents a significant relationship between the independent variables and the dependent variable.

18. Does the estimated regression equation developed in Exercise 16 provide a better fit than the equation developed in Exercise 13? Explain.

19. Use the estimated regression equation developed in Exercise 16 to determine on the average how much longer it takes to service a machine involving an electrical failure than one with a mechanical failure.

15.7 DETERMINING WHEN TO ADD OR DELETE VARIABLES

In Section 15.4 we discussed the use of an F test for determining the significance of the estimated regression equation. In this section we will show how an F test can also be used to determine whether or not it is advantageous to add one variable—or a group of variables—to a multiple regression model. This test is based on a determination of the amount of reduction in the error sum of squares resulting from adding one or more independent variables to the model. We will first illustrate the use of the test by determining for Butler Trucking whether or not it is advantageous to add x_2, the number of deliveries, to the original regression model. This model included only the effect of x_1, the miles traveled.

With miles traveled as the only independent variable the least squares procedure provided the following estimated regression equation:

$$\hat{y} = 1.13 + .067x_1.$$

Table 15.3 showed the computation of the residuals and SSE for this equation. Referring to Table 15.3, we see that with this one independent variable SSE = 9.5669.

When x_2, the number of deliveries, was added as a second independent variable we obtained the following estimated regression equation:

$$\hat{y} = .0367 + .0562x_1 + .7639x_2.$$

Table 15.6 showed the computation of the error sum of squares for this model, SSE = 5.0501. Clearly, adding x_2 resulted in a reduction of SSE. The question we want to answer is, "Does adding the variable x_2 lead to a *significant* reduction in SSE?"

For our work in this section we will use the notation SSE (x_1) to denote the error sum of squares when x_1 is the only independent variable, SSE(x_1,x_2) the error sum of squares when x_1 and x_2 are both independent variables, and so on. Hence the reduction in SSE resulting from adding x_2 to the model involving just x_1 is

$$\text{SSE}(x_1) - \text{SSE}(x_1,x_2) = 9.5669 - 5.0501 = 4.5168.$$

An F test is conducted to determine whether or not this reduction is significant.

The numerator of the F statistic is the reduction in SSE divided by the number of variables added to the original model. Here only one variable, x_2, has been added; thus the numerator is

$$\frac{\text{SSE}(x_1) - \text{SSE}(x_1,x_2)}{1} = 4.5168.$$

The numerator is a measure of the reduction in SSE per variable added to the model. The denominator of the F statistic is the mean square error for the model that includes all of the variables. In our example this corresponds to the model containing both x_1 and x_2; thus $p = 2$ and hence

$$\text{MSE} = \frac{\text{SSE}(x_1,x_2)}{n - p - 1} = \frac{5.0501}{7} = .7214.$$

The following F statistic provides the basis for testing whether or not the addition of x_2 is statistically significant:

$$F = \frac{\dfrac{\text{SSE}(x_1) - \text{SSE}(x_1,x_2)}{1}}{\dfrac{\text{SSE}(x_1,x_2)}{10 - 2 - 1}}. \qquad (15.26)$$

The numerator degrees of freedom for this F test equal the number of variables added to the model, and the denominator degrees of freedom equal $10 - 2 - 1$.

For the Butler Trucking problem, we obtain

$$F = \frac{\dfrac{4.5168}{1}}{\dfrac{5.0501}{7}} = \frac{4.5168}{.7214} = 6.26.$$

Refer to Table 4 of Appendix B. We find that for a level of significance of $\alpha = .05$,

$$F_{.05} = 5.59.$$

Since

$$F = 6.26 > F_{.05} = 5.59,$$

we reject the null hypothesis that x_2 is not statistically significant; in other words, adding x_2 to the model involving only x_1 results in a significant reduction in the error sum of squares.

When we want to test for the significance of adding only one additional independent variable to an existing model, the result found with the F test just described could also be obtained by using the t test for the significance of an individual parameter (described in Section 15.4). Indeed, the F statistic we just computed is the square of the t statistic used to test the hypothesis that an individual parameter is zero.

Since the t test is equivalent to the F test when only one variable is being added to the model, we can now further clarify the proper use of the t test on the individual parameters. If an individual parameter is not significant, the corresponding variable can be dropped from the model. However, no more than one variable can ever be dropped from a model on the basis of a t test; if one variable is dropped, a second variable that was not significant initially might become significant.

We now turn briefly to a consideration of whether or not the addition of more than one variable—as a set—results in a significant reduction in the error sum of squares.

The General Case

Consider the following multiple regression model involving q independent variables:

$$y = \beta_0 + \beta_1 x_1 + \beta_2 x_2 + \cdots + \beta_q x_q + \epsilon. \tag{15.27}$$

If we add variables $x_{q+1}, x_{q+2}, \ldots, x_p$ to this model, we obtain a model involving p independent variables:

$$y = \beta_0 + \beta_1 x_1 + \beta_2 x_2 + \cdots + \beta_q x_q + \beta_{q+1} x_{q+1} + \beta_{q+2} x_{q+2} + \cdots + \beta_p x_p + \epsilon. \tag{15.28}$$

To test whether or not the addition of $x_{q+1}, x_{q+2}, \ldots, x_p$ is statistically significant, the null and alternative hypotheses can be stated as follows:

$$H_0\colon\ \beta_{q+1} = \beta_{q+2} = \cdots = \beta_p = 0$$

H_1: At least one of the coefficients is not equal to zero.

The following F statistic provides the basis for testing whether or not the additional variables are statistically significant:

$$F = \frac{\dfrac{\text{SSE}(x_1, x_2, \cdots, x_q) - \text{SSE}(x_1, x_2, \cdots, x_q, x_{q+1}, \cdots, x_p)}{p - q}}{\dfrac{\text{SSE}(x_1, x_2, \cdots, x_q, x_{q+1}, \cdots, x_p)}{n - p - 1}} \tag{15.29}$$

Note that for the special case where $q = 1$ and $p = 2$ (15.29) reduces to (15.26).

EXERCISES

20. In a regression analysis involving 27 observations the following estimated regression equation was developed:

$$\hat{y} = 16.3 + 2.3x_1 + 12.1x_2 - 5.8x_3 .$$

Also, the following standard errors were obtained:

$$s_{b_1} = .53, \qquad s_{b_2} = 8.15, \qquad s_{b_3} = 1.30.$$

At an $\alpha = .05$ level of significance conduct the following hypothesis tests:

a. H_0: $\beta_1 = 0$ versus H_1: $\beta_1 \neq 0$.
b. H_0: $\beta_2 = 0$ versus H_1: $\beta_2 \neq 0$.
c. H_0: $\beta_3 = 0$ versus H_1: $\beta_3 \neq 0$.
d. Can any of the variables be dropped from the model? Why or why not?

21. In a regression analysis involving 30 observations the following estimated regression equation was obtained:

$$\hat{y} = 17.6 + 3.8x_1 - 2.3x_2 + 7.6x_3 + 2.7x_4.$$

For this model SST = 1805 and SSR = 1760.

a. Compute R^2.
b. Compute R_a^2.
c. At $\alpha = .05$ test the significance of the relationship among the variables.

22. Refer again to Exercise 21. Variables x_1 and x_4 were dropped from the model, and the following estimated regression equation was obtained:

$$\hat{y} = 11.1 - 3.6x_2 + 8.1x_3.$$

For this model SST = 1805 and SSR = 1705.

a. Compute SSE(x_1,x_2,x_3,x_4).
b. Compute SSE(x_2,x_3)
c. Use an F test and an $\alpha = .05$ level of significance to determine if x_1 and x_4 contribute significantly to the model.

15.8 ESTIMATION AND PREDICTION

Estimating the mean value of y and predicting an individual value of y in multiple regression is similar to that for the case of regression analysis involving one independent variable. First, recall that in Chapter 14 we showed that the point estimate of the expected value of y for a given value of x was the same as the point estimate of an individual value of y. In both cases we used $\hat{y} = b_0 + b_1x$ as the point estimate.

In multiple regression we use the same procedure. That is, we substitute the given values of $x_1, x_2, \ldots, x_p$ into the estimated regression equation and use the corresponding value of $\hat{y}$ as the point estimate. For example, suppose that for the Butler Trucking problem we wanted to use the estimated regression equation involving x_1 (miles traveled) and x_2 (number of deliveries) to do the following:

1. Estimate the mean value of travel time for all trucks that travel 50 miles and make two deliveries;
2. Predict the travel time for *one specific* truck that travels 50 miles and makes two deliveries.

Using the estimated regression equation $\hat{y} = .0367 + .0562x_1 + .7639x_2$ with $x_1 = 50$ and $x_2 = 2$ we obtain the following value of $\hat{y}$:

$$\hat{y} = .0367 + .0562(50) + .7639(2) = 4.3745.$$

Hence our point estimate of travel time in both cases is approximately 4.4 hours.

To develop interval estimates for the mean value of y and for an individual value of y we use a procedure similar to that for the case of regression analysis involving one independent variable. The formulas required, however, are beyond the scope of the text. But, computer packages for multiple regression analysis will often provide confidence intervals once the values of $x_1, x_2 \ldots, x_p$ are specified by the user. In Table 15.10 we show 95% confidence interval estimates for the Butler Trucking problem for

TABLE 15.10 95% Confidence Interval Estimates for the Butler Trucking Problem

Value of x_1	Value of x_2	*Expected Value of y*		*Individual Value of y*	
		Lower Limit	*Upper Limit*	*Lower Limit*	*Upper Limit*
50	2	3.0841	5.6649	1.9869	6.7621
50	3	3.7127	6.5642	2.6750	7.6018
80	1	3.9907	6.7097	2.9006	7.6926
80	2	5.2984	6.8226	3.9120	8.2091
100	2	6.0774	8.2916	4.8908	9.4782
100	4	7.4853	9.9394	6.3584	11.0662

selected values of x_1 and x_2. Note that the interval estimate for an individual value of y is wider than the interval estimate for the expected value of y. This simply reflects the fact that for given values of x_1 and x_2 we can predict the mean travel time for all trucks with more precision than we can the travel time for one specific truck.

EXERCISES

23. Refer to the estimated regression equation in Exercise 22. Let $x_2 = 10$ and $x_3 = 20$. Estimate the value of y.

24. The following estimated regression equation has been developed to predict annual sales for account executives:

$$\hat{y} = 160 + 8x_1 + 15x_2,$$

where

$$\hat{y} = \text{sales (\$1,000's)},$$
$$x_1 = \text{years of experience},$$
$$x_2 = \begin{cases} 1 & \text{if account executive attended training program,} \\ 0 & \text{otherwise.} \end{cases}$$

a. Estimate expected annual sales for an employee with 3 years of experience who did not attend the training program.

b. Estimate annual sales for a given employee with 2 years of experience who did attend the training program.

c. What is the expected increase in sales as a result of attendance at the training program?

15.9 CURVILINEAR REGRESSION FUNCTIONS

As an illustration of a situation involving a regression equation not representing a straight line relationship, consider the problem being analyzed by Nugent Industries, a manufacturer of hardware and building supplies. Management has been investigating sales of one product, with particular attention given to the relationship between each customer's annual order size and the number of sales calls made per year. A random sample of seven customer accounts was obtained; the data are presented in Table 15.11 and Figure 15.8.

TABLE 15.11 Annual Order Size and Number of Sales Calls for Nugent Industries

Number of Sales Calls (x)	*Annual Sales* (\$1,000's) (y)
2	12
3	17
4	16
5	24
6	26
7	34
8	46

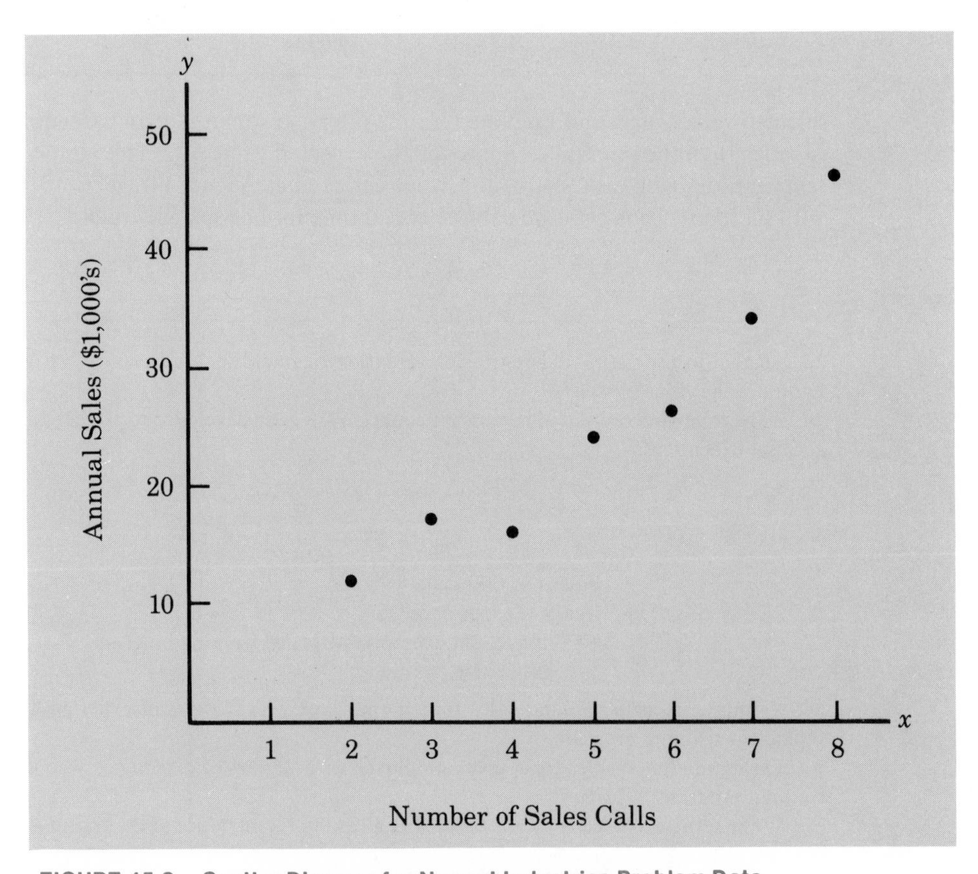

FIGURE 15.8 Scatter Diagram for Nugent Industries Problem Data

From Figure 15.8 it would seem reasonable to hypothesize that the underlying relationship between x and y might be approximated by a straight line. Using the methods of simple linear regression, the following regression equation and corresponding coefficient of determination were obtained:

$$\hat{y} = -1.07 + 5.21x,$$
$$r^2 = .91.$$

An F test for significance was conducted; the conclusion ($\alpha = .05$) was that x and y are related.

The list of residuals and the corresponding residual plot of $(y_i - \hat{y}_i)$ versus $\hat{y}_i$ are shown in Table 15.12 and Figure 15.9, respectively. Analyzing Table 15.12, we

TABLE 15.12 Residuals for the Estimated Regression Equation $\hat{y} = -1.07 + 5.21x$

Customer (i)	x_i	y_i	$\hat{y}_i$	$y_i - \hat{y}_i$
1	2	12	9.35	2.65
2	3	17	14.56	2.44
3	4	16	19.77	−3.77
4	5	24	24.98	−0.98
5	6	26	30.19	−4.19
6	7	34	35.40	−1.40
7	8	46	40.61	5.39

conclude that there is something disturbing about a "run" of four negative residuals for values of x equal to 3, 4, 5, and 6; that is, the fitted model is consistently overpredicting over this range of values of the independent variable. Moreover, the shape of the residual plot does not appear to have the points randomly scattered above and below the line as $\hat{y}$ increases; it looks more like that in Figure 14.8, part c. Thus although the estimated regression equation is statistically significant and the value of r^2 is very high, we should begin to doubt whether or not we have hypothesized the correct underlying model. If we look more carefully at the scatter diagram in Figure 15.8, we conclude that perhaps a curvilinear relationship would be more appropriate.

Suppose that we hypothesize that y is related to x by the following regression model:

$$y = \beta_0 + \beta_1 x^2 + \epsilon.$$

At first glance this appears to be a very different type of model than we have previously considered. However, if we substitute z for x^2, we can rewrite the above regression model as

$$y = \beta_0 + \beta_1 z + \epsilon.$$

Since we could have used the symbol z instead of x to denote the independent variable, this form of the model suggests that the solution procedure used to obtain estimates of β_0 and β_1 for the model $y = \beta_0 + \beta_1 x + \epsilon$ can be used. Rewriting the formulas for the

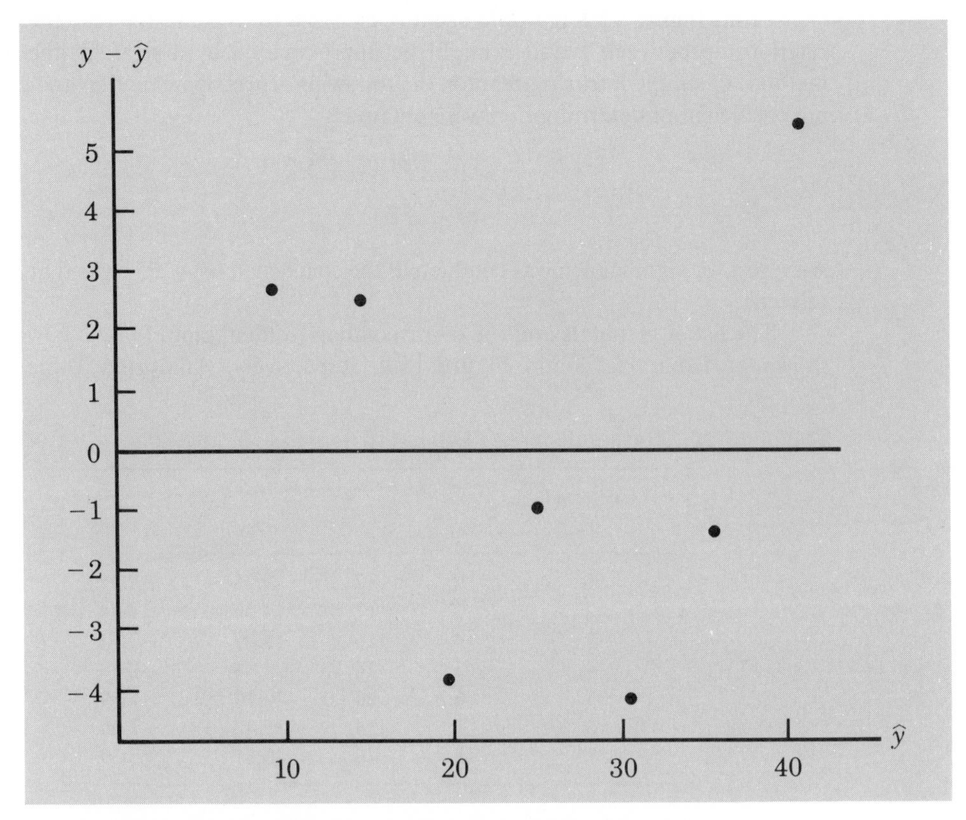

FIGURE 15.9 Residual Plot for the Estimated Regression Equation $\hat{y} = -1.07 + 5.21x$

least squares estimates b_0 and b_1 after substituting z for x, we obtain

$$b_1 = \frac{\Sigma z_i y_i - (\Sigma z_i \Sigma y_i)/n}{\Sigma z_i^2 - (\Sigma z_i)^2/n}, \tag{15.30}$$

$$b_0 = \bar{y} - b_1 \bar{z}. \tag{15.31}$$

Thus for the Nugent Industries problem we have $z_i = x_i^2$. The only difference in computing the coefficients is that we must use the values of x^2 instead of x in our formulas for computing b_0 and b_1.

The calculations of b_0 and b_1 for this problem are summarized in Table 15.13. Rounding to two decimal places and substituting x^2 for z leads to the estimated regression equation

$$\hat{y} = 9.64 + .53x^2.$$

This equation represents a curvilinear (quadratic) relationship between x and y.

A residual plot of $y_i - \hat{y}_i$ versus $\hat{y}_i$ corresponding to the estimated regression equation $\hat{y} = 9.64 + .53x^2$ is shown in Figure 15.10. The unusual pattern observed previously now appears to have been removed. It turns out that the relationship between y and x^2 is statistically significant and that the value of r^2 for this equation is .97. On the basis of this analysis, we recommend that $\hat{y} = 9.64 + .53x^2$ be used for developing predictions of annual sales given the number of sales calls per year.

TABLE 15.13 Calculations for the Estimated Regression Equation $\hat{y} = b_0 + b_1x^2$

Customer (i)	x_i	$z_i = x_i^2$	y_i	z_iy_i	z_i^2
1	2	4	12	48	16
2	3	9	17	153	81
3	4	16	16	256	256
4	5	25	24	600	625
5	6	36	26	936	1,296
6	7	49	34	1,666	2,401
7	8	64	46	2,944	4,096
Totals	35	203	175	6,603	8,771

$$\bar{z} = \frac{\Sigma z_i}{n} = \frac{203}{7} = 29$$

$$\bar{y} = \frac{\Sigma y_i}{n} = \frac{175}{7} = 25$$

$$b_1 = \frac{\Sigma z_iy_i - (\Sigma z_i \Sigma y_i)/n}{\Sigma z_i^2 - (\Sigma z_i)^2/n} = \frac{(6{,}603) - (203)(175)/7}{(8{,}771) - (203)^2/7} = .5298$$

$$b_0 = \bar{y} - b_1\bar{z} = 25 - .5298(29) = 9.6358$$

$$\hat{y} = 9.6358 + .5298z$$

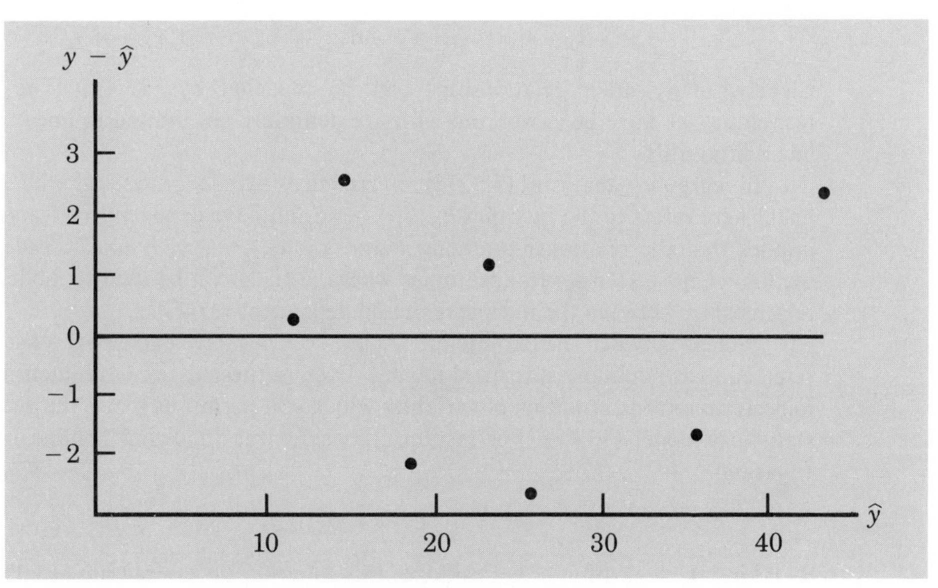

FIGURE 15.10 Residual Plot for the Estimated Regression Equation $\hat{y} = 9.64 + .53x^2$

However, we caution that this model should not be used to make predictions outside the range of x values observed. Obviously one would expect diminishing returns per sales call beyond some point.

A General Linear Model

The multiple regression model with p independent variables can be generalized to handle curvilinear functions involving some or all of the independent variables. A general regression model involving p independent variables can be written

$$y = \beta_0 + \beta_1 z_1 + \beta_2 z_2 + \cdots + \beta_p z_p + \epsilon. \tag{15.32}$$

In (15.32) each $z_j, j = 1, 2, \ldots, p$ is a function of some other variables $x_1, x_2, \ldots, x_k$. The simplest case occurs when $p = k = 1$ and $z_1 = x_1$. Then equation (15.32) becomes

$$y = \beta_0 + \beta_1 x_1 + \epsilon.$$

This is the simple linear regression model introduced in Chapter 14 with x_1 instead of x. In the case where $p = k$ and $z_j = x_j$, (15.32) becomes

$$y = \beta_0 + \beta_1 x_1 + \beta_2 x_2 + \cdots + \beta_p x_p + \epsilon.$$

This is the multiple regression model introduced earlier in this chapter.

More complex types of relationships can be modeled easily with (15.32). For example, if $p = 1$ and $z_1 = x_1^2$ we have the model used earlier in this section (with x_1 replacing x) for Nugent Industries. As another example, if $p = 2$, $z_1 = x_1$, and $z_2 = x_1^2$, we obtain

$$y = \beta_0 + \beta_1 x_1 + \beta_2 x_1^2 + \epsilon.$$

Similarly, if $p = 5$, $z_1 = x_1$, $z_2 = x_2$, $z_3 = x_1^2$, $z_4 = x_2^2$, and $z_5 = x_1 x_2$, then we obtain

$$y = \beta_0 + \beta_1 x_1 + \beta_2 x_2 + \beta_3 x_1^2 + \beta_4 x_2^2 + \beta_5 x_1 x_2 + \epsilon.$$

Clearly, many other relationships can be modeled by (15.32). The regression techniques we have been working with are definitely not limited to linear or straight line relationships.

In regression analysis (15.32) is referred to as a *linear statistical model.* The term linear here refers to the fact that $\beta_0, \beta_1, \ldots, \beta_p$ all have exponents of 1 and in no way implies that the relationship among y and $x_1, x_2, \ldots, x_p$ is linear. Indeed, in this section we have seen several examples where (15.32) can be used to model nonlinear relationships between the independent and dependent variables.

Models in which the parameters ($\beta_0, \beta_1, \ldots, \beta_p$) have exponents other than 1 are referred to as nonlinear statistical models. Even in these cases it is sometimes possible to perform a transformation of variables which will permit us to use the general linear statistical model (15.32). For instance, consider the following nonlinear regression equation:

$$E(y) = \beta_0 \beta_1^x. \tag{15.33}$$

By taking the logarithm of both sides of this equation we can rewrite (15.33) as

$$\log E(y) = \log \beta_0 + x \log \beta_1. \tag{15.34}$$

Now if we let $y' = \log E(y)$, $\beta_0' = \log \beta_0$, and $\beta_1' = \log \beta_1$, we can rewrite (15.31) as

$$y' = \beta_0' + \beta_1' x.$$

It is clear that the formulas for simple linear regression can now be used to develop estimates of β_0' and β_1'. Denoting the estimates as b_0' and b_1' leads to the following estimated regression equation:

$$\hat{y}' = b_0' + b_1' x. \tag{15.35}$$

To obtain predictions of the original dependent variable y given a value of x, we would first substitute the value of x into (15.35) and compute $\hat{y}'$. The antilog of $\hat{y}'$ would be our prediction of y, or the expected value of y.

In closing this section we should make it clear that there are many nonlinear models which cannot be transformed into an equivalent linear model. However, such models have had limited use in business and economic applications. Furthermore, the mathematical background needed for study of such models is beyond the scope of this text.

EXERCISES

25. The highway department is doing a study on the relationship between traffic flow and speed. The following model has been hypothesized:

$$y = \beta_0 + \beta_1 x + \epsilon$$

where

$$y = \text{traffic flow in vehicles per hour,}$$

$$x = \text{vehicle speed in miles per hour.}$$

The following data have been collected during rush hour for six highways leading out of the city:

Traffic Flow (y)	Vehicle Speed (x)
1,256	35
1,329	40
1,226	30
1,335	45
1,349	50
1,124	25

a. Develop an estimated regression equation for these data.
b. Test at an $\alpha = .01$ level the significance of the relationship.

26. In working further with the problem of Exercise 25 statisticians suggested the use of the following curvilinear estimated regression equation:

$$\hat{y} = b_0 + b_1 x + b_2 x^2$$

a. Use the data of Exercise 25 to estimate the parameters of this estimated regression equation.
b. Test at an $\alpha = .01$ level the significance of the relationship.
c. Estimate the traffic flow in vehicles per hour at speeds of 38 miles per hour.

27. The following estimated regression equation has been developed for the relationship between y, sales ($1,000's), and x, store size (sq. ft. × 10,000):

$$\hat{y} = 150 + 100x - 10x^2.$$

Ten stores were included in the sample. Values of SST = 168,000 and SSR = 140,000 were obtained.

a. Compute R^2 and R_a^2.
b. Test at $\alpha = .05$ the significance of the relationship.

15.10 MULTIPLE REGRESSION APPROACH TO ANALYSIS OF VARIANCE

In Section 15.6 we discussed the use of dummy variables in multiple regression analysis. In this section we show how the use of dummy variables in a multiple regression equation can provide another approach to solving analysis of variance problems. We will demonstrate the multiple regression approach to analysis of variance by applying it to the GMAT problem introduced in Chapter 13.

Recall that the objective of the GMAT study was to determine if the three preparation programs (3 hour review session, 1 day program, 10 week course) were different in terms of their effect on GMAT test scores. Sample observations were available from the population of students who took the 3 hour review session, the population of students who took the 1 day program, and the population of students who took the 10 week course.

We begin the regression approach to this problem by defining two dummy variables that will be used to indicate the population from which each sample observation was selected. Since there are three populations in the GMAT problem, we need two dummy variables. In general, if there are k populations we need to define $k - 1$ dummy variables. For the GMAT illustration we define x_1 and x_2 as shown in Table 15.14.

TABLE 15.14 GMAT Problem Dummy Variables

x_1	x_2	*These Values Are Used Whenever*
0	0	Observation is associated with the 3 hour review session
1	0	Observation is associated with the 1 day program
0	1	Observation is associated with the 10 week course

We can use the dummy variables x_1 and x_2 to relate the GMAT score of each student (y) to the type of GMAT preparation program:

$$\begin{aligned} E(y) &= \text{expected value of the GMAT score} \\ &= \beta_0 + \beta_1 x_1 + \beta_2 x_2. \end{aligned}$$

Thus if we are interested in the expected value of the GMAT score for a student who completed the 3 hour review session, our procedure for assigning numerical values to the dummy variables x_1 and x_2 would result in setting $x_1 = x_2 = 0$. Now our multiple regression equation reduces to

$$E(y) = \beta_0 + \beta_1(0) + \beta_2(0) = \beta_0.$$

Thus we can interpret β_0 as the expected value of the GMAT score for students who complete the 3 hour review session.

Next let us consider the forms of the multiple regression equation for each of the other programs. For the 1 day program $x_1 = 1$ and $x_2 = 0$, and

$$E(y) = \beta_0 + \beta_1(1) + \beta_2(0) = \beta_0 + \beta_1.$$

For the 10 week program $x_1 = 0$ and $x_2 = 1$, and

$$E(y) = \beta_0 + \beta_1(0) + \beta_2(1) = \beta_0 + \beta_2.$$

We see that $\beta_0 + \beta_1$ represents the expected value of the GMAT score for students who complete the 1 day program; $\beta_0 + \beta_2$ represents the expected value of the GMAT score for students who complete the 10 week course.

We now want to estimate the coefficients β_0, β_1, and β_2 and hence develop estimates of the expected value of the GMAT score for each program. The sample consisting of 15 observations of x_1, x_2, and y was entered into a standard multiple regression computer software package. The actual input data and the output from the multiple regression package are shown in Table 15.15 and Figure 15.11, respectively.

Refer to Figure 15.11. We see that the estimates of β_0, β_1, and β_2 are $b_0 = 509$, $b_1 = 17$, and $b_2 = 43$. Thus our best estimate of the expected value of the GMAT score

TABLE 15.15 Input Data for the GMAT Problem

Observations Correspond to	x_1	x_2	y
	0	0	491
	0	0	579
3 hour review	0	0	451
	0	0	521
	0	0	503
	1	0	588
	1	0	502
1 day program	1	0	550
	1	0	520
	1	0	470
	0	1	533
	0	1	628
10 week course	0	1	502
	0	1	537
	0	1	561

```
THE REGRESSION EQUATION IS
Y = 509. + 17.0X1 + 43.0X2

                                      ST. DEV.       T-RATIO =
        COLUMN   COEFFICIENT          OF COEF.       COEF/S.D.
          --          509.00             20.81           24.46
X1        C1           17.00             29.43            0.58
X2        C2           43.00             29.43            1.46

THE ST. DEV. OF Y ABOUT REGRESSION LINE IS
S = 46.58
WITH (15-3) = 12 DEGREES OF FREEDOM

R-SQUARED = 15.3 PERCENT
R-SQUARED =  1.2 PERCENT, ADJUSTED FOR D.F.

ANALYSIS OF VARIANCE

DUE TO              DF          SS     MS=SS/DF
REGRESSION           2        4690         2345
RESIDUAL            12       25980         2165
TOTAL               14       30670
```

FIGURE 15.11 Multiple Regression Output for the GMAT Problem

for each type of program is as follows:

Type of Program	*Estimate of E(y)*
3 hour review	$b_0 = 509$
1 day program	$b_0 + b_1 = 509 + 17 = 526$
10 week course	$b_0 + b_2 = 509 + 43 = 552$

Note that the best estimate of the expected value of the GMAT score for each program obtained from the regression analysis is the same as the sample means found earlier when applying the ANOVA procedure. That is, $\bar{x}_1 = 509$, $\bar{x}_2 = 526$, and $\bar{x}_3 = 552$.

Now let us see how we can use the output from the multiple regression package in order to perform the ANOVA test on the difference in the means for the three programs. First, we observe that if there is no difference in the means, then

$$E(y) \text{ for the 1 day program } - E(y) \text{ for the 3 hour review} = 0,$$
$$E(y) \text{ for the 10 week course } - E(y) \text{ for the 3 hour review} = 0.$$

Since β_0 equals $E(y)$ for the 3 hour review and $\beta_0 + \beta_1$ equals $E(y)$ for the 1 day program, the first difference above is equal to $(\beta_0 + \beta_1) - \beta_0 = \beta_1$. Moreover, since $\beta_0 + \beta_2$ equals $E(y)$ for the 10 week course, the second difference above is equal to $(\beta_0 + \beta_2) - \beta_0 = \beta_2$. Hence we would conclude that there is no difference in the three means if $\beta_1 = 0$ and $\beta_2 = 0$. Thus the null hypothesis for a test for difference of means can be stated as

$$H_0\colon \beta_1 = \beta_2 = 0.$$

Recall that in Section 15.4 we showed that in order to test this type of null hypothesis about the significance of the regression relationship we must compare the value of MSR/MSE to the critical value from an F distribution with numerator and denominator degrees of freedom equal to the degrees of freedom for the regression sum of squares and the error sum of squares, respectively. In our current problem the regression sum of squares has 2 degrees of freedom and the error sum of squares has 12 degrees of freedom. Thus the values for MSR and MSE are

$$\text{MSR} = \frac{\text{SSR}}{2} = \frac{4{,}690}{2} = 2{,}345,$$

$$\text{MSE} = \frac{\text{SSE}}{12} = \frac{25{,}980}{12} = 2{,}165.$$

Hence the computed F value is

$$F = \frac{\text{MSR}}{\text{MSE}} = \frac{2{,}345}{2{,}165} = 1.0831.$$

At the $\alpha = .05$ level of significance the critical value of F with 2 numerator degrees of freedom and 12 denominator degrees of freedom is 3.89. Since the observed value of F is less than the critical value of 3.89, we cannot reject the null hypothesis $H_0: \beta_1 = \beta_2 = 0$. Hence we cannot conclude that the means for the three preparation programs are different.

EXERCISES

28. The Jacobs Chemical Company wants to estimate the mean time (minutes) required to mix a batch of material on machines produced by three different manufacturers. In order to limit the cost of testing, four batches of material were mixed on machines produced by each of the three manufacturers. The times needed to mix the material were recorded and are shown below:

Manufacturer 1	*Manufacturer 2*	*Manufacturer 3*
20	28	20
26	26	19
24	31	23
22	27	22

a. Write a multiple regression equation that can be used to analyze the data.
b. What are the best estimates of the coefficients in your regression equation?
c. In terms of the regression equation coefficients, what hypotheses do we have to test to see if the mean time to mix a batch of material is the same for each manufacturer?
d. What is the value of MSR/MSE?
e. For the $\alpha = .05$ level of significance what conclusion should be drawn?

29. Four different paints are advertised as having the same drying time. In order to check the manufacturers' claims, five paint samples were tested for each brand of paint. The time in minutes until the paint was dry enough for a second coat to be applied was recorded for each sample. The following data were obtained:

Paint 1	Paint 2	Paint 3	Paint 4
128	144	133	150
137	133	143	142
135	142	137	135
124	146	136	140
141	130	131	153

The following dummy variables were used in a multiple regression analysis of the problem.

x_1	x_2	x_3	
0	0	0	Paint 1
1	0	0	Paint 2
0	1	0	Paint 3
0	0	1	Paint 4

A computer run with the above data resulted in the following estimated regression equation:

$$\hat{y} = 133 + 6x_1 + 3x_2 + 11x_3$$

a. For an $\alpha = .05$ level of significance, test for a difference in mean drying times of the paints.

b. What is your estimate of mean drying time for paint 2? How is it obtained from the computer output?

Summary

In this chapter we have shown how extensions of the concepts of simple linear regression can be used to develop an estimated regression equation for predicting y that involves several independent variables. We noted that our interpretation of the coefficients had to be modified somewhat for this case. That is, we interpreted b_i as an estimate of the change in the dependent variable y that would result from a 1 unit change in independent variable x_i when the other independent variables do not change.

In multiple regression it is necessary to conduct an F test based on $F = \text{MSR}/\text{MSE}$ to determine if the overall regression relationship is significant. We saw that the role of the t test was to determine whether or not each of the independent variables provided a significant addition to the model given the presence of the other independent variables. We also cautioned that the t test should never be used to drop more than one independent variable from a multiple regression model since if one variable is dropped, a second variable that was not significant initially might become significant.

In Section 15.9 we saw that the general linear model for regression analysis could also be used in cases where there is a nonlinear relationship among the variables. This capability was used to develop and analyze a model in which a customer's annual order size was related in a nonlinear fashion to the number of sales calls. The versatility of regression analysis was further illustrated in Section 15.10, where we showed that it could be used to solve analysis of variance problems.

A key part of any multiple regression study is the use of a computer software package for carrying out the computational work. Many excellent packages exist and, after a short learning period, can be used to develop the estimated regression equation, conduct the appropriate significance tests, and prepare residual plots. We illustrated the use of such a statistical package for the Butler Trucking problem.

Glossary

Multiple regression model—A regression model in which more than one independent variable is used to predict the dependent variable.

Multicollinearity—A term used to describe the case then the independent variables in a multiple regression model are correlated.

Qualitative variable—A variable that cannot be measured.

Dummy variable—A variable that takes on the values of 0 or 1 and is used to incorporate the effects of qualitative variables in a regression model.

Key Formulas

Multiple Regression Model

$$y = \beta_0 + \beta_1 x_1 + \beta_2 x_2 + \cdots + \beta_p x_p + \epsilon \tag{15.2}$$

Multiple Regression Equation

$$E(y) = \beta_0 + \beta_1 x_1 + \cdots + \beta_p x_p \tag{15.3}$$

Estimated Regression Equation

$$\hat{y} = b_0 + b_1 x_1 + \cdots + b_p x_p \tag{15.5}$$

Mean Square Due to Regression

$$\text{MSR} = \frac{\text{SSR}}{p} \tag{15.18}$$

Mean Square Due to Error

$$\text{MSE} = \frac{\text{SSE}}{n - p - 1} \tag{15.19}$$

The *F* Statistic

$$F = \frac{\text{MSR}}{\text{MSE}} = \frac{\text{SSR}/p}{\text{SSE}/(n - p - 1)} \tag{15.20}$$

Multiple Coefficient of Determination

$$R^2 = \frac{\text{SSR}}{\text{SST}} \tag{15.21}$$

Adjusted Multiple Coefficient of Determination

$$R_a^2 = 1 - (1 - R^2)\left(\frac{n-1}{n-p-1}\right) \tag{15.22}$$

***F* Statistic for Determining When to Add or Delete x_2**

$$F = \frac{\dfrac{\text{SSE}(x_1) - \text{SSE}(x_1, x_2)}{1}}{\dfrac{\text{SSE}(x_1, x_2)}{n-2-1}} \tag{15.26}$$

General Linear Statistical Model

$$y = \beta_0 + \beta_1 z_1 + \beta_2 z_2 + \cdot\cdot\cdot + \beta_p z_p + \epsilon \tag{15.32}$$

Supplementary Exercises

30. In a regression analysis involving 18 observations and four independent variables it was determined that SSR = 18,051.63 and SSE = 1,014.3.

a. Determine R^2 and R_a^2.

b. Test the significance of the relationship at the $\alpha = .01$ level of significance.

31. The following estimated regression equation involving three independent variables has been developed:

$$\hat{y} = 18.31 + 8.12x_1 + 17.9x_2 - 3.6x_3.$$

Computer output indicates that $s_{b_1} = 2.1$, $s_{b_2} = 9.72$, and $s_{b_3} = .71$. There were 15 observations in the study.

a. Test $H_0\colon \beta_1 = 0$ at $\alpha = .05$

b. Test $H_0\colon \beta_2 = 0$ at $\alpha = .05$

c. Test $H_0\colon \beta_3 = 0$ at $\alpha = .05$

d. Would you recommend dropping any of the independent variables from the model?

32. Shown below is a partial computer output from a regression analysis:

```
THE REGRESSION EQUATION IS
Y = 8.103 + 7.602X1 + 3.111X2

                                        ST. DEV.         T-RATIO =
        COLUMN    COEFFICIENT           OF COEF.         COEF/S.D.
        --               8.103             2.667
X1      VAR1             7.602             2.105
X2      VAR2             3.111             0.613

THE ST. DEV. OF Y ABOUT REGRESSION LINE IS
S = 3.35
WITH (15-3) = 12 DEGREES OF FREEDOM

R-SQUARED = 92.3 PERCENT
R-SQUARED =       PERCENT, ADJUSTED FOR D.F.

ANALYSIS OF VARIANCE

DUE TO              DF          SS        MS=SS/DF
REGRESSION                    1612
RESIDUAL
TOTAL
```

a. Compute the appropriate t ratios.
b. Test for the significance of β_1 and β_2 at $\alpha = .05$.
c. Compute the entries in the DF, SS, and MS = SS/DF columns.
d. Compute R_a^2.

33. Bauman Construction Company makes bids on a variety of projects. In an effort to estimate the bid to be made by one of its competitors Bauman has obtained data on 15 previous bids and developed the following estimated regression equation:

$$\hat{y} = 80 + 45x_1 - 3x_2,$$

where

$$\hat{y} = \text{competitor's bid (\$1,000's)},$$
$$x_1 = \text{square feet (1,000's)}$$
$$x_2 = \text{local index of construction activity.}$$

a. Estimate the competitor's bid on a project involving 50,000 square feet and an index of construction activity of 70.
b. If SSR = 19,780 and SST = 21,533, test at $\alpha = .01$ the significance of the relationship.

34. In order to evaluate the effectiveness of a special energy-saving package in new home construction, a builders' association developed the following data regarding annual energy costs, the square footage of the residence, and whether or not the energy-saving package had been installed:

Square Footage (1,000 sq. ft.)	*Insulation Package*	*Annual Energy Costs ($100's)*
1.5	Standard	9
2.0	Standard	12
2.2	Energy saving	14
1.8	Standard	11
2.4	Standard	12
2.8	Energy saving	19
2.0	Standard	12
2.4	Energy saving	14
1.9	Energy saving	11
3.0	Energy saving	18
2.1	Standard	13
2.8	Energy saving	17

Use just the data involving the square footage and the annual energy cost to develop an estimated regression equation to predict annual energy cost given the square footage.

35. Did the equation developed in Exercise 34 provide a good fit to the observed data? Explain.

36. This question is an extension of Exercise 34. In order to incorporate the effect of using the special energy-saving package the builders defined the following dummy variable:

$$x_2 = \begin{cases} 1 & \text{if the energy-saving package was used,} \\ 0 & \text{if the standard package was used.} \end{cases}$$

With this variable the following regression equation was proposed:

$$E(y) = \beta_0 + \beta_1 x_1 + \beta_2 x_2$$

where

$$x_1 = \text{square footage (1,000 sq. ft.)},$$
$$y = \text{annual cost (\$100's).}$$

In the regression equation proposed above, what is the interpretation of β_2?

37. This question is an extension of Exercise 36. Using a computer software package the least squares estimated regression equation $\hat{y} = b_0 + b_1x_1 + b_2x_2$ was developed:

```
THE REGRESSION EQUATION IS
Y = .1398 + 5.7764X1 + .8230X2

                                         ST. DEV.          T-RATIO =
          COLUMN   COEFFICIENT           OF COEF.          COEF/S.D.
            --         0.1398
X1        SQFT         5.7764              .9206               6.275
X2        ENPK         0.8230              .7971               1.032

THE ST. DEV. OF Y ABOUT REGRESSION LINE IS
S = 1.0663
WITH (12-3) = 9 DEGREES OF FREEDOM

R-SQUARED = 90.07 PERCENT
R-SQUARED = 87.86 PERCENT, ADJUSTED FOR D.F.

ANALYSIS OF VARIANCE

DUE TO            DF          SS      MS=SS/DF
REGRESSION         2     92.7671        46.384
RESIDUAL           9     10.2329         1.137
TOTAL             11    103.0000
```

a. Is the energy-saving package worth including in the construction of a new home?

b. What equation would you recommend that management use in order to predict total energy cost?

APPENDIX TO CHAPTER 15: CALCULUS BASED DERIVATION AND SOLUTION OF NORMAL EQUATIONS FOR REGRESSION WITH TWO INDEPENDENT VARIABLES

In order to show how the least squares method is applied in multiple regression we derive the normal equations for the two-independent-variables case. The least squares criterion calls for the minimization of the following expression for the residual sum of squares:

$$\text{SSE} = \Sigma\,(y_i - b_0 - b_1x_{1i} - b_2x_{2i})^2. \tag{15A.1}$$

To minimize (15A.1) we must take the partial derivatives of SSE with respect to b_0, b_1, and b_2. We can then set the partial derivatives equal to zero and solve for the estimated regression coefficients b_0, b_1, and b_2. Taking the partial derivatives and setting them equal to zero provides:

$$\frac{\partial \text{SSE}}{\partial b_0} = -2\,\Sigma\,(y_i - b_0 - b_1x_{1i} - b_2x_{2i}) = 0, \tag{15A.2}$$

$$\frac{\partial \text{SSE}}{\partial b_1} = -2\,\Sigma\,x_{1i}\,(y_i - b_0 - b_1x_{1i} - b_2x_{2i}) = 0, \tag{15A.3}$$

$$\frac{\partial \text{SSE}}{\partial b_2} = -2\,\Sigma\,x_{2i}\,(y_i - b_0 - b_1x_{1i} - b_2x_{2i}) = 0. \tag{15A.4}$$

Dividing (15A.2) by 2 and summing the terms individually yields

$$-\Sigma y_i + \Sigma b_0 + \Sigma b_1 x_{1i} + \Sigma b_2 x_{2i} = 0.$$

Bringing Σy_i to the right hand side of the equation and noting that $\Sigma b_0 = nb_0$, we obtain the normal equation given in this chapter as (15.11):

$$nb_0 + (\Sigma x_{1i})b_1 + (\Sigma x_{2i})b_2 = \Sigma y_i. \qquad (15.11)$$

Similar algebraic simplification applied to (15A.3) and (15A.4) leads to the normal equations given by (15.12) and (15.13):

$$(\Sigma x_{1i})b_0 + (\Sigma x_{1i}^2)b_1 + (\Sigma x_{1i}x_{2i})b_2 = \Sigma x_{1i}y_i, \qquad (15.12)$$

$$(\Sigma x_{2i})b_0 + (\Sigma x_{1i}x_{2i})b_1 + (\Sigma x_{2i}^2)b_2 = \Sigma x_{2i}y_i. \qquad (15.13)$$

Application of these procedures to a regression model involving p independent variables would lead to $p + 1$ normal equations of this type. However, matrix algebra is usually used for that type of derivation.

Solving the Normal Equations for the Butler Trucking Company

In the chapter we developed a regression model for Butler Trucking involving two independent variables, miles traveled and number of deliveries. Substituting the data for this problem (see Section 15.3) into (15.11) to (15.13) provided the following normal equations:

$$10b_0 + 815b_1 + 26b_2 = 66.0, \qquad (15.14)$$

$$815b_0 + 69{,}625b_1 + 2{,}165b_2 = 5{,}594.0, \qquad (15.15)$$

$$26b_0 + 2{,}165b_1 + 76b_2 = 180.6. \qquad (15.16)$$

By multiplying (15.14) by 81.5 and subtracting the result from (15.15) we can eliminate b_0 and obtain an equation involving b_1 and b_2 only:

$$\begin{array}{r} 815b_0 + 69{,}625.0b_1 + 2{,}165b_2 = 5{,}594.0 \\ -815b_0 - 66{,}422.5b_1 - 2{,}119b_2 = -5{,}379.0 \\ \hline 3{,}202.5b_1 + 46b_2 = 215.0\,. \end{array} \qquad (15A.5)$$

Now multiply (15.14) by 2.6 and subtract the result from (15.16). This manipulation yields a second equation involving b_1 and b_2 only:

$$\begin{array}{r} 26b_0 + 2{,}165b_1 + 76.0b_2 = 180.6 \\ -26b_0 - 2{,}119b_1 - 67.6b_2 = -171.6 \\ \hline 46b_1 + 8.4b_2 = 9.0\,. \end{array} \qquad (15A.6)$$

With equations (15A.5) and (15A.6) we can solve simultaneously for b_1 and b_2. Multiplying (15A.6) by 46/8.4 and subtracting the result from (15A.5) gives us an equation involving b_1 only:

$$\begin{array}{r} 3{,}202.5000b_1 + 46b_2 = 215.0000 \\ -\ 251.9048b_1 - 46b_2 = -49.2857 \\ \hline 2{,}950.5952b_1 = 165.7143\,. \end{array} \qquad (15A.7)$$

Using (15A.7) to solve for b_1 we get

$$b_1 = \frac{165.7143}{2{,}950.5952} = .056163.$$

Using this value for b_1 we can substitute into (15A.6) to solve for b_2:

$$\begin{aligned} 46(.056163) + 8.4b_2 &= 9, \\ 2.583498 + 8.4b_2 &= 9, \\ 8.4b_2 &= 6.416502, \\ b_2 &= .7638693. \end{aligned}$$

Now we can substitute the values obtained for b_1 and b_2 into (15.14), thus obtaining b_0:

$$\begin{aligned} 10b_0 + 815(.056163) + 26(.7638693) &= 66.0, \\ 10b_0 + 45.772845 \quad + 19.860601 \quad &= 66.0, \\ 10b_0 &= .366554, \\ b_0 &= .0366554. \end{aligned}$$

Rounding to four significant digits, we obtain the following estimated regression equation for Butler Trucking:

$$\hat{y} = .0367 + .0562x_1 + .7639x_2. \qquad (15A.8)$$

Application

Champion
Champion International Corporation

Champion International Corporation*

Stamford, Connecticut

Champion International Corporation is one of the largest forest products companies in the world, employing over 41,000 people in the United States, Canada, and Brazil. Champion manages over 3 million acres of timberlands in the United States. Its objective is to maximize the return of this timber base by converting trees into three basic product groups: (1) building materials, such as lumber and plywood; (2) white paper products, including printing and writing grades of white paper; and (3) brown paper products, such as linerboard and corrugated containers. Given the highly competitive markets within the forest products industry, survival dictates that Champion must maintain its position as a low cost producer of quality products. This requires an ambitious capital program to improve the timber base and to build additional modern, cost effective timber conversion facilities.

THE MANAGEMENT SCIENCE FUNCTION

The Management Science function at Champion International Corporation is organizationally structured within the Management Information Services Department and operates as an internal consulting service within the company. Approximately 40% of the project activity is involved with facility and production planning, 30% with physical distribution, 20% with process improvement, and 10% with capital budgeting. The primary techniques used are mathematical programming (e.g., linear programming), simulation, and statistical analyses. Of the statistical analyses performed, multivariate (multiple) regression and forecasting techniques, such as exponential smoothing, are used most frequently.

A MULTIPLE REGRESSION APPLICATION

A pulp mill is a facility in which wood chips and chemicals are processed to produce wood pulp. The wood pulp is then used at a paper mill to produce paper products. Wood is made up primarily of cellulose fibers held together by a substance called lignin. These cellulose fibers will eventually become wood pulp. The pulping process used at a pulp mill chemically dissolves this lignin binding and separates the cellulose fibers. To start, the wood chips and some chemicals are cooked in a pressure cooker called a digester which softens the chips and dissolves the lignin binding; the softened chips are piped into a blow tower, where they are smashed against a steel target to separate the

*The authors are indebted to Marian Williams and Bill Griggs of Champion International Corporation for providing this application.

cellulose fibers, as shown in Figure 15A.1. The pulp is then chemically washed in a pulp washer, from which it flows immediately into screens which filter out any undesirable material. Although cellulose is composed of white fibers, wood pulp, no

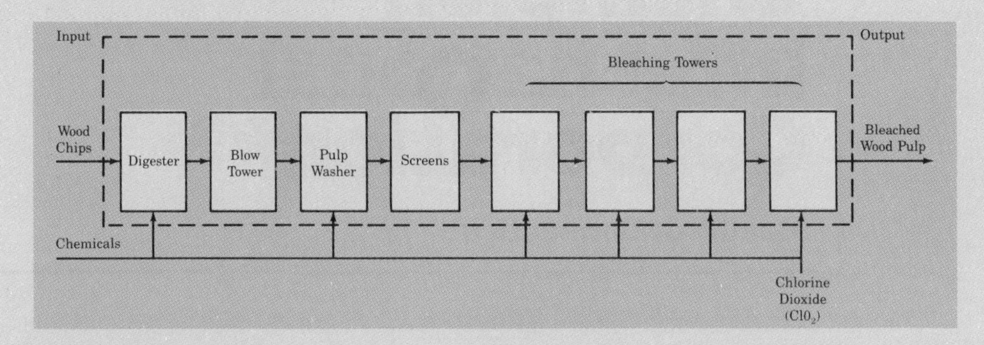

FIGURE 15A.1 The Champion Bleached Wood Pulping Process

matter how well cooked and washed, contains a fair amount of lignin compounds which gives the pulp a dark brownish color.* In order to produce white paper products the pulp must be bleached to break down these lignin compounds. Therefore the next step in the pulping process is bleaching. The pulp flows into a four stage bleaching process. In each tower in this process a different bleaching chemical is used to gradually bleach the pulp to a white color. The bleached pulp is now ready for further processing, such as the addition of dyes and other chemicals to prepare it for paper production.

A key bleaching agent used in the last tower of the bleaching process is chlorine dioxide (ClO_2). It can bleach pulp to a high white brightness, not obtainable with other bleaching agents, without significant loss of pulp fiber strength. Chlorine dioxide gas has a yellow green color with an irritating odor and is highly corrosive to most metals. Under certain conditions chlorine dioxide gas will decompose violently (i.e., explode). Owing to the combustible nature of chlorine dioxide gas, it is usually produced—at least at Champion pulp mill facilities—on site and is piped in solution form into the bleaching tower of the pulp mill.

One of Champion International's integrated pulp and paper mills

*John H. Ainsworth, *Paper—The Fifth Wonder,* Kaukaune, Wisconsin: Thomas Publishing Co., 1958.

One of Champion's pulp mill facilities presently has two ClO_2 generation processes—the R3 process and the sulfur dioxide (SO_2) process. The R3 process operates very efficiently, but the SO_2 process does not. A study was undertaken to look at process control and efficiency improvement of the SO_2 process. To begin the study, the acquisition of a basic understanding of the following areas was necessary:

SO_2 process. The SO_2 process requires the following four feed chemicals: (1) sulfuric acid (H_2SO_4), (2) sodium chlorate ($NaClO_3$), (3) sulfur dioxide gas (SO_2), and (4) air. These chemicals flow at metered rates into the ClO_2 generator, as shown in Figure 15A.2. ClO_2 gas is produced in the generator and flows into the ClO_2 absorber,

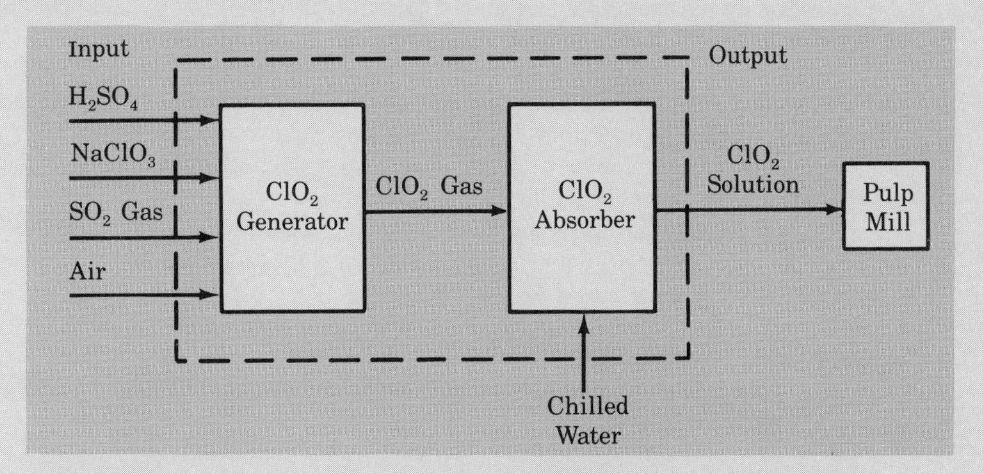

FIGURE 15A.2 A CLO_2 Generation Process at a Champion Pulp Mill Facility

where chilled water absorbs the ClO_2 gas to form a ClO_2 solution. This ClO_2 solution is piped into the pulp mill.

Process control. There is a control room for the generator which contains a dial for each feed chemical and also for chilled water flow. These dials regulate the amount of each chemical that is to be used to produce ClO_2 and are manually set and adjusted by a shift operator. Since the generator is run 24 hours a day, there are three operators—one each shift—who control the generator.

The demand for ClO_2 at the pulp mill dictates the amount produced. This causes the ClO_2 production rate to be variable. The chilled water flow rate to the absorber is measured in gallons per minute (gpm) and represents the production rate of ClO_2 solution. The pulp mill requires that the ClO_2 solution have a ClO_2 concentration between 15 and 16* grams per liter (gpl). If the chilled water flow is increased and the chemical feeds to the generator remain the same, then the concentration of the ClO_2 in the solution may fall out of the acceptable range. Consequently, any significant change in chilled water flow requires an adjustment of the chemical feeds that together produce ClO_2 gas.

It is the operator's task to maintain the appropriate ClO_2 production rate and concentration in the ClO_2 solution. Moreover, the operator must maintain concentrations of $NaClO_3$ and H_2SO_4 in the generator solution within acceptable ranges set by the chemical engineers. Controlling the concentration of these chemicals in the

*All numerical values have been modified to protect proprietary information.

generator solution improves the efficiency and the stability of the chemical reactions that produce ClO_2 gas.

Data recorded. Every hour, on the hour, the operator records on a log sheet the chemical feed rates, the chilled water flow rate, and some generator temperature data. Every 2 hours, on the hour, a sample of the generator solution and a sample of the ClO_2 solution are taken. Titrations are performed on the samples to determine concentrations of H_2SO_4 and $NaClO_3$ in the generator solution and the ClO_2 in the ClO_2 solution. The operator uses these results to adjust the feed rates if the concentrations are out of acceptable ranges and/or if a production rate change is necessary.

The basic problem in operating the generator revolves around how to set the chemical feed rates. The generator manufacturer provided suggested chemical feed rates only for operation near design capacity. Owing to the variable production rate environment in which the generator must run, ClO_2 production is usually much less than design capacity. The chemical feed rates were being set by the operators based on experience and the concentration tests. As a result the process was becoming over controlled by the operators. The chemical engineers at the mill requested that a set of control equations, one for each chemical feed, be developed from the operating log data to aid the operators in setting chemical feed rates.

The approach taken was to obtain a representative sample of the log data and apply multiple regression to develop an equation for each chemical feed. In order to do this, the following steps were taken:

1. Log sheets for 2½ months were obtained. A sample log sheet is presented in Table 15A.1. The data were entered into the computer, with each hour considered a separate observation.

TABLE 15A.1 Sample ClO_2 Manufacturing Log Sheet (Date: 6/27/83)

Time	*Chemical Feed*				*Concentration Test*			*Production Rate*
	Air (*cfm*)	*SO_2* (*cfm*)	*$NaClO_3$* (*gpm*)	*H_2SO_4* (*gpm*)	*$NaClO_3$* (*gpl*)	*H_2SO_4* (*gpl*)	*ClO_2* (*gpl*)	*Chilled Water Flow* (*gpm*)
Acceptable Range					40–45	600–620	15–16	
7 AM	400	51	6.1	3.6				300
8	400	51	6.1	3.6	42.0	610	14.9	300
9	400	56	6.5	3.7				350
10	400	56	6.5	3.7	45.0	615	15.5	350

2. The data were plotted so that engineers could look for unusual or nonrepresentative observations that should be eliminated.
3. A new data set was formed by only using those observations that contained concentration test data. Lag variables were created for chemical feeds and the previous concentration tests. There were over 400 observations in this data set.
4. Taking one chemical feed at a time, a stepwise regression was performed. For instance, an initial $NaClO_3$ multiple regression equation of the following form was

developed:

$y =$ (dependent variable) difference between the amount of ClO_2 produced in time period t and the amount produced in time period $t - 2$
$x_1 =$ (independent variable) average of the amount of $NaClO_3$ fed in time period $t - 1$ and time period $t - 2$
$x_2 =$ (independent variable) concentration test for ClO_2 in time period $t - 2$

$$\hat{y} = b_0 + b_1 x_1 + b_2 x_2,$$

We found $R^2 = .50$.

5. The residuals were plotted; they appeared random.

RESULTS

An SAS (Statistical Analysis System) computer program was developed to perform the above analysis. Four regression equations were developed and were programmed into a microcomputer at the mill. The operators key into the computer the concentration test data and the desired production rate. The computer then calculates the appropriate chemical feeds and displays them on a screen.

Since the operators have begun using the control equations, the generator efficiency has increased and the number of times that the concentrations fall within acceptable ranges has increased significantly.

16 Time Series Analysis and Forecasting

What you will learn in this chapter:

- the four components of a time series
- how to use moving averages and exponential smoothing to develop forecasts
- how to use regression analysis to identify the trend component of a time series
- how to forecast using the classical decomposition method
- how to identify seasonal adjustments for a time series

Contents

A critical aspect of managing any organization is planning for the future. Indeed, the long-run success of an organization is closely related to how well management is able to foresee the future and develop appropriate strategies. Good judgment, intuition, and an awareness of the state of the economy may give a manager a rough idea or "feeling" of what is likely to happen in the future. However, it is often difficult to convert this "feeling" into hard data such as next quarter's sales volume or next year's raw material cost per unit. The purpose of this chapter is to introduce several methods that can help predict many future aspects of a business operation.

Let us suppose for a moment that we have been asked to provide quarterly estimates of the sales volume for a particular product during the coming 1 year period. Production schedules, raw material purchasing plans, inventory policies, and sales quotas will all be affected by the quarterly estimates we provide. Consequently, poor estimates may result in poor planning and hence result in increased costs for the firm. How should we go about providing the quarterly sales volume estimates?

We will certainly want to review the actual sales data for the product in past periods. Suppose that we have actual sales data for each quarter over the past 3 years. From these historical data we can identify the general level of sales and determine whether or not there is any trend such as an increase or decrease in sales volume over time. A further review of the data might reveal a seasonal pattern, such as peak sales occurring in the third quarter of each year and sales volume bottoming out during the first quarter. By reviewing historical data over time we are in a better position to understand the pattern of past sales and hence better able to predict future sales for the product.

The historical sales data referred to above form what is called a *time series.* Specifically, a time series is a set of observations measured at successive points in time or over successive periods of time. In this chapter we will introduce several procedures that can be used to analyze time series data. The objective of this analysis will be to provide good *forecasts* or predictions of future values of the time series.

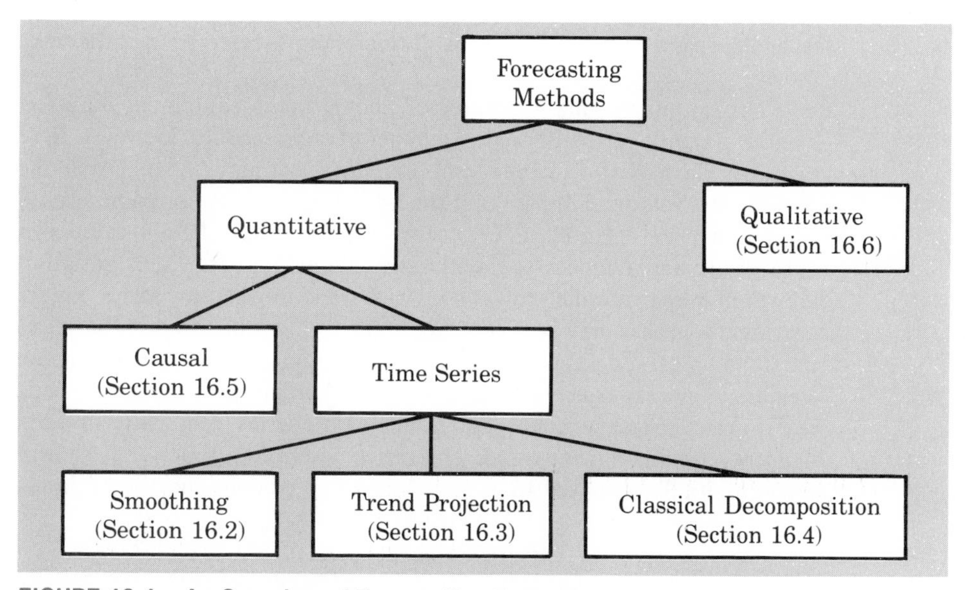

FIGURE 16.1 An Overview of Forecasting Methods

Forecasting methods can be classified as quantitative or qualitative. Quantitative forecasting methods are based on an analysis of historical data concerning a time series and possibly other related time series. If the historical data used are restricted to past values of the series that we are trying to forecast, the forecasting procedure is called a time series method. In this chapter we discuss three time series methods: smoothing (moving averages and exponential smoothing), trend projection, and classical decomposition. If the historical data used in a quantitative forecasting method involve other time series that are believed to be related to the time series we are trying to forecast, we say that we are using a causal method. We discuss the use of multiple regression analysis as a causal forecasting method. Qualitative forecasting methods generally utilize the judgment of experts to make forecasts. An advantage of these procedures is that they can be applied in situations where no historical data are available. We discuss some of these approaches in Section 16.6. Figure 16.1 provides an overview of the different types of forecasting methods.

16.1 THE COMPONENTS OF A TIME SERIES

In order to explain the pattern or behavior of the data in a time series it is often helpful to think of the time series as consisting of several components. The usual assumption is that four separate components—trend, cyclical, seasonal, and irregular—combine to make the time series take on specific values. Let us look more closely at each of these components of a time series.

Trend Component

In time series analysis the measurements may be taken every hour, day, week, month, or year or at any other regular interval.* Although time series data generally exhibit random fluctuations, the time series may still show gradual shifts or movements to relatively higher or lower values over a longer period of time. This gradual shifting of the time series, which is usually due to long-term factors such as changes in the population, changes in demographic characteristics of the population, changes in technology, and changes in consumer preferences, is referred to as the *trend* in the time series.

For example, a manufacturer of photographic equipment may see substantial month-to-month variability in the number of cameras sold. However, in reviewing the sales over the past 10 to 15 years this manufacturer may find a gradual increase in the annual sales volume. Suppose that the sales volume was approximately 1,800 cameras per month in 1973, 2,200 cameras per month in 1978, and 2,600 cameras per month in 1983. While actual month-to-month sales volumes may vary substantially, this gradual growth in sales over time shows an upward trend for the time series. Figure 16.2 shows a straight line that may be a good approximation of the trend in the sales data. While the trend for camera sales appears to be linear and increasing over time, sometimes the trend in a time series is better described by other patterns.

Figure 16.3 shows some other possible time series trend patterns. In part A of this figure we see a nonlinear trend. This curve describes a time series showing very little growth initially, followed by a period of rapid growth, and then a leveling off. This

*We restrict our attention here to time series where the values of the series are recorded at equal intervals. Treatment of cases where the observations are not made at equal intervals is beyond the scope of this text.

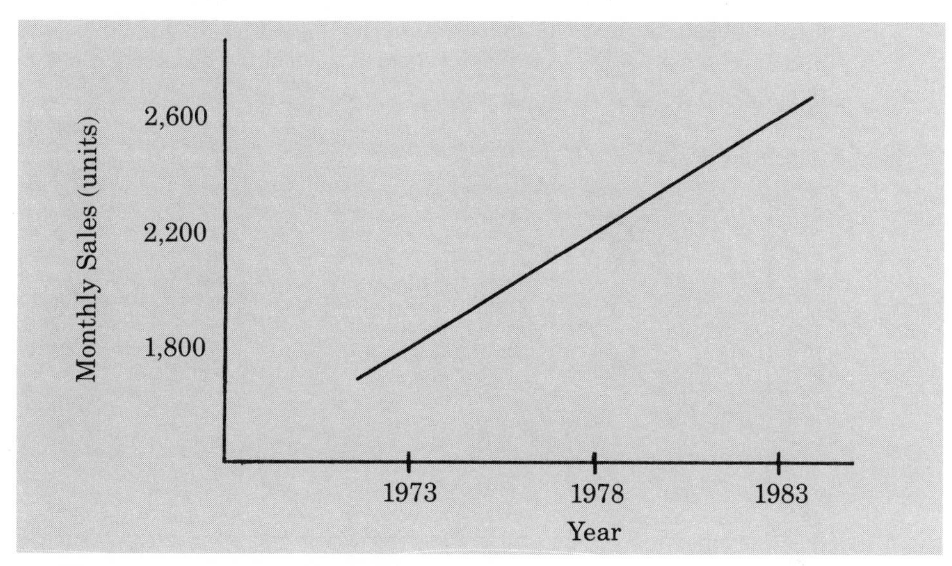

FIGURE 16.2 Linear Trend of Camera Sales

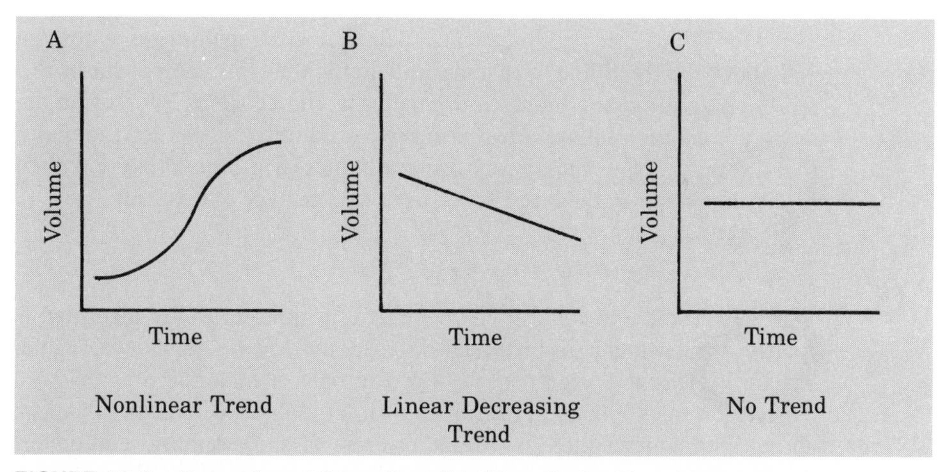

FIGURE 16.3 Examples of Some Possible Time Series Trend Patterns

might be a good approximation to sales for a product from introduction through a growth period and into a period of market saturation. The linear decreasing trend in part B of Figure 16.3 is useful for time series displaying a steady decrease over time. The horizontal line in part C of Figure 16.3 is used for a time series that does not show any consistent increase or decrease over time. It is actually the case of no trend.

Cyclical Component

While a time series may exhibit a gradual shifting or trend pattern over long periods of time, we cannot expect all future values of the time series to be exactly on the trend line. In fact, time series often show alternating sequences of points below and above the trend line. Any regular pattern of sequences of points above and below the trend line is attributable to the cyclical component of the time series. Figure 16.4 shows the graph of a time series with an obvious cyclical component. The observations are taken at intervals 1 year apart.

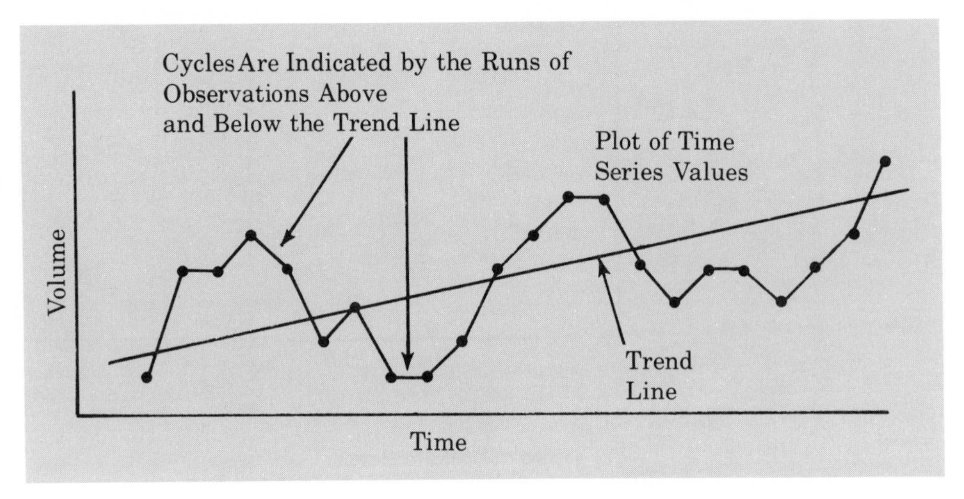

FIGURE 16.4 Trend and Cyclical Components of a Time Series. Data Points Are 1 Year Apart

Many time series exhibit cyclical behavior with regular runs of observations below and above the trend line. The general belief is that this component of the time series represents multiyear cyclical movements in the economy. For example, periods of moderate inflation followed by periods of rapid inflation can lead to many time series that alternate below and above a generally increasing trend line (e.g., housing costs). Many time series in the late 1970s displayed this type of behavior.

Seasonal Component

While the trend and cyclical components of a time series are identified by analyzing multiyear movements in historical data, many time series show a regular pattern of variability within 1 year periods. For example, a manufacturer of swimming pools expects low sales activity in the fall and winter months, with peak sales occurring in the spring and summer months. Manufacturers of snow removal equipment and heavy clothing, however, expect just the opposite yearly pattern. It should not be surprising that the component of the time series that represents the variability in the data due to seasonal influences is called the seasonal component. While we generally think of seasonal movement in a time series as occurring within 1 year, the seasonal component can also be used to represent any repeating pattern that is less than 1 year in duration. For example, daily traffic volume data show within-the-day "seasonal" behavior, with peak levels during rush hours, moderate flow during the rest of the day and early evening, and light flow from midnight to early morning.

Irregular Component

The irregular component of the time series is the residual or "catchall" factor that accounts for the deviation of the actual time series value from what we would expect if the trend, cyclical, and seasonal components completely explained the time series. It accounts for the random variability in the time series. The irregular component is caused by the short term, unanticipated, and nonrecurring factors that affect the time series. Since this component accounts for the random variability in the time series, it is unpredictable. We cannot attempt to predict its impact on the time series in advance.

16.2 FORECASTING USING SMOOTHING METHODS

In this section we discuss forecasting techniques that are appropriate for a fairly stable time series, one that exhibits no significant trend, cyclical, or seasonal effects. In such situations the objective of the forecasting method is to "smooth out" the irregular component of the time series through some type of averaging process. We begin with a consideration of the method known as moving averages.

Moving Averages

The *moving averages* method consists of computing an average of the *most recent n* data values in the time series. This average is then used as the forecast for the next period. Mathematically, the moving average calculation is made as follows:

Moving Average

$$\text{Moving average} = \frac{\Sigma\,(\text{most recent } n \text{ data values})}{n} \qquad (16.1)$$

The term "moving" average is based on the fact that as a new observation becomes available for the time series, it replaces the oldest observation in (16.1), and a new average is computed. As a result the average will change or "move" as new observations become available.

To illustrate the moving averages method, consider the 12 weeks of data presented in Table 16.1 and Figure 16.5. These data show the number of gallons of gasoline sold by a gasoline distributor in Bennington, Vermont over the past 12 weeks.

TABLE 16.1 Gasoline Sales Time Series

Week	*Sales (1,000's of Gallons)*
1	17
2	21
3	19
4	23
5	18
6	16
7	20
8	18
9	22
10	20
11	15
12	22

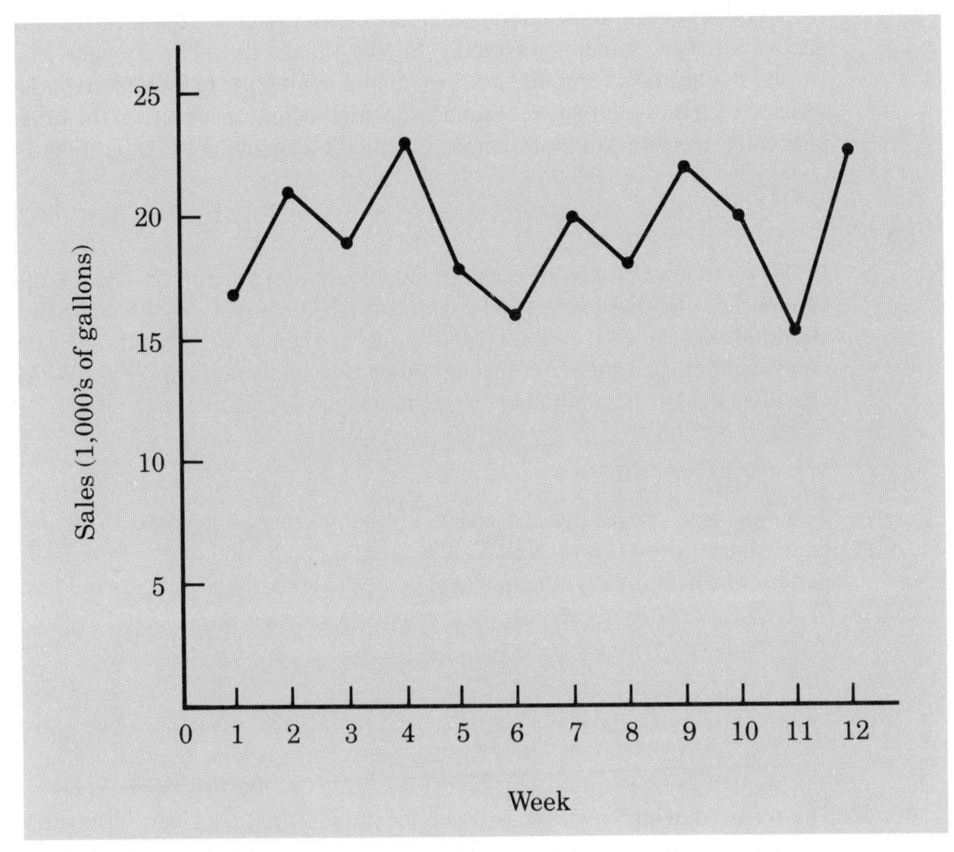

FIGURE 16.5 Graph of Gasoline Sales Time Series

In order to use moving averages to forecast the gasoline sales time series, we must first select the number of data values to be included in the moving average. As an example, let us compute forecasts based upon a 3 week moving average. The moving average calculation for the first 3 weeks of the gasoline sales time series is as follows:

$$\text{Moving average (weeks 1–3)} = \frac{17 + 21 + 19}{3} = 19.$$

This moving average value is then used as the forecast for week 4. Since the actual value observed in week 4 is 23, we see that the forecast error in week 4 is $23 - 19 = 4$. In general, the error associated with any forecast is the difference between the observed value of the time series and the forecast value.

The calculation for the second 3 week moving average is shown below:

Week

1	2	3	4	5	6	7
17	21	19	23	18	16	20

$$\text{Moving average (weeks 2–4)} = \frac{21 + 19 + 23}{3} = 21.$$

Hence the forecast for week 5 is 21. The error associated with this forecast is $18 - 21 = -3$. Thus we see that the forecast error can be positive or negative depending upon whether the forecast is too low or too high.

A complete summary of the 3 week moving average calculations for the gasoline sales time series is shown in Table 16.2 and Figure 16.6.

An important consideration in using any forecasting method is the accuracy of the forecast. Clearly, we would like the forecast errors to be small. The last two columns of Table 16.2, which contain the forecast errors and the forecast errors squared, can be used to develop measures of accuracy.

One measure of forecast accuracy you might think of using would be to simply sum the forecast errors over time. The problem with this measure is that if the errors are random (as they should be if our choice of forecasting method is appropriate), some errors will be positive and some errors will be negative, resulting in a sum near zero regardless of the size of the individual errors. Indeed, we see from Table 16.2 that the sum of forecast errors for the gasoline sales time series is zero. This difficulty can be avoided by either squaring or taking the absolute value of each of the individual forecast errors.

For the gasoline sales time series we can use the last column of Table 16.2 to compute the average of the sum of the squared errors. Doing so we obtain

$$\text{Average of the sum of squared errors} = \frac{92}{9} = 10.22.$$

This average of the sum of squared errors is commonly referred to as the *mean squared error* (MSE). The mean squared error is an often used measure of the accuracy of a forecasting method.

Another commonly used measure of forecast accuracy is the *mean absolute deviation* (MAD). This measure is simply the average of the sum of the absolute deviations for all the forecast errors. Using the errors given in Table 16.2, we obtain

$$\text{Mean absolute deviation (MAD)} = \frac{4 + 3 + 4 + 1 + 0 + 4 + 0 + 5 + 3}{9}$$
$$= 2.67.$$

One major difference between MSE and MAD is that the MSE measure is influenced much more by large forecast errors than by small errors (since for the MSE

TABLE 16.2 Summary of 3 Week Moving Average Calculations

Week	*Time Series Value*	*Moving Average Forecast*	*Forecast Error*	*(Error)^2*
1	17			
2	21			
3	19			
4	23	19	4	16
5	18	21	−3	9
6	16	20	−4	16
7	20	19	1	1
8	18	18	0	0
9	22	18	4	16
10	20	20	0	0
11	15	20	−5	25
12	22	19	3	9
		Totals	0	92

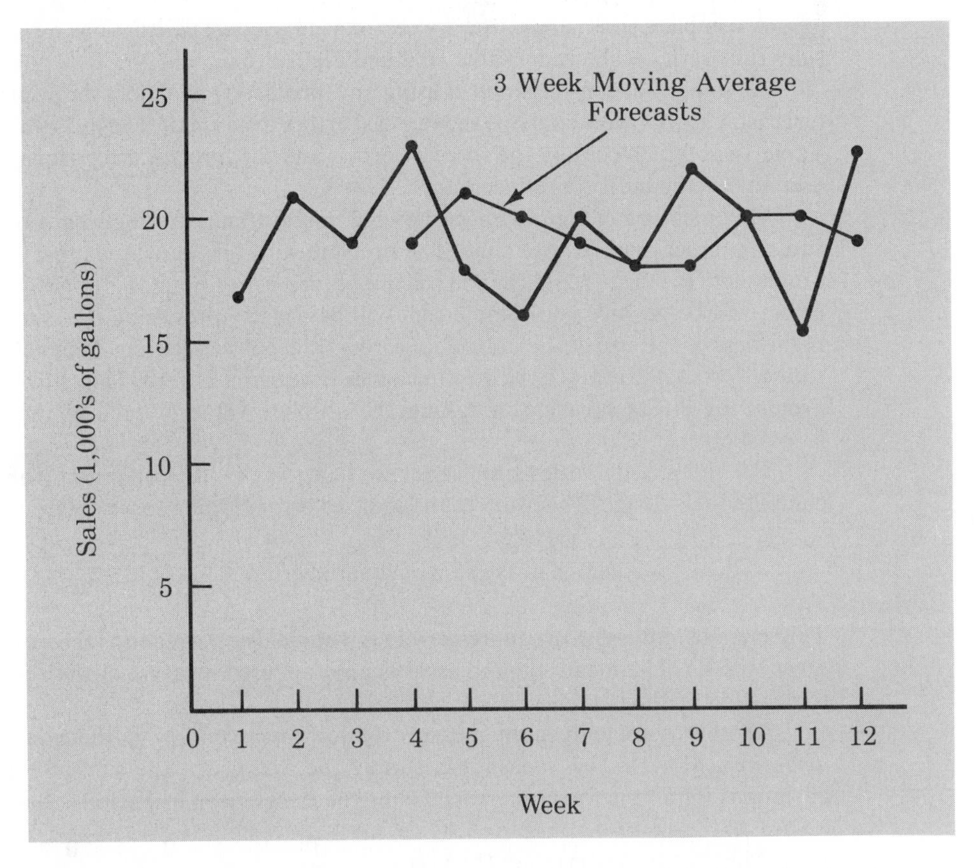

FIGURE 16.6 **Graph of Gasoline Sales Time Series and 3 Week Moving Average Forecasts**

measure the errors are squared). The selection of the best measure of forecasting accuracy is not a simple matter. Indeed, forecasting experts often disagree as to which measure should be used. In this chapter we will use the MSE measure.

As we indicated previously, in order to use the moving averages method we must first select the number of data values to be included in the moving average. It should not be too surprising that for a particular time series different length moving averages will differ in their ability to accurately forecast the time series. One possible approach to choosing the proper length is to use "trial and error" to identify the length that minimizes the MSE measure of forecast accuracy. Then, if we are willing to assume that the length which is best for the past will also be best for the future, we would forecast the next value in the time series using the number of data values that minimized the MSE for the historical time series. Exercise 1 at the end of the section will ask you to consider 4 week and 5 week moving averages for the gasoline sales data. A comparison of the mean square error for each will indicate the number of weeks of data you may want to include in the moving average calculation.

Weighted Moving Averages

In the moving averages method each observation in the moving average calculation receives the same weight. One possible variation, known as *weighted moving averages,* involves selecting different weights for each data value and then computing a weighted mean as the forecast. In most cases the most recent observation receives the most weight, and the weight decreases for older data values. For example, using the gasoline sales time series, let us illustrate the computation of a weighted 3 week moving average, where the most recent observation receives a weight three times as great as that given the oldest observation, and the next oldest observation receives a weight twice as great as the oldest. The weighted moving average forecast for week 4 would be computed as follows:

$$\text{Weighted moving average forecast for week 4} = \tfrac{3}{6}(19) + \tfrac{2}{6}(21) + \tfrac{1}{6}(17) = 19.33$$

Note that for the weighted moving average the sum of the weights is equal to one. This was also true for the simple moving average, where each weight was $\frac{1}{3}$. However, recall that the simple or unweighted moving average provided a forecast of 19. Exercise 2 at the end of the section asks you to calculate the remaining values for the 3 week weighted moving average and compare the forecast accuracy with what we have obtained for the unweighted moving average.

Although we have spent a fair amount of time describing moving averages, in practice the method is not frequently used as a forecasting technique. Instead, a special type of weighted moving average known as exponential smoothing is preferred. Nonetheless, as we will see in Section 16.4, moving averages play an important role in identifying the seasonal component of a time series.

Exponential Smoothing

Exponential smoothing is a forecasting technique that uses a smoothed value of the time series in one period to forecast the value of the time series in the next period. The basic exponential smoothing model is as follows:

Exponential Smoothing Model

$$F_{t+1} = \alpha Y_t + (1 - \alpha)F_t, \qquad (16.2)$$

where

F_{t+1} = the forecast of the time series for period $t + 1$,

Y_t = the actual value of the time series in period t,

F_t = the forecast of the time series for period t,

α = the smoothing constant ($0 \leq \alpha \leq 1$).

With (16.2) the forecast for any period is a weighted average of the previous actual values for the time series. To see this, let us suppose that we have three periods of data, Y_1, Y_2, and Y_3. Then the forecast for period 4 becomes

$$F_4 = \alpha Y_3 + (1 - \alpha)F_3.$$

The forecast value for period 4 is clearly a weighted average of Y_3 and F_3, with weights of α and $1 - \alpha$, respectively. But from (16.2) we also know that

$$F_3 = \alpha Y_2 + (1 - \alpha)F_2,$$
$$F_2 = \alpha Y_1 + (1 - \alpha)F_1.$$

Since there are no previous data values for the time series, the first forecast value is taken to be equal to Y_1. That is, $F_1 = Y_1$. Using this value for F_1, F_2 is written

$$F_2 = \alpha Y_1 + (1 - \alpha)Y_1 = Y_1.$$

Substituting $F_2 = Y_1$ in the expression for F_3, we have

$$F_3 = \alpha Y_2 + (1 - \alpha)Y_1.$$

Finally, substituting this expression for F_3 in the above expression for F_4, we obtain

$$\begin{aligned} F_4 &= \alpha Y_3 + (1 - \alpha)[\alpha Y_2 + (1 - \alpha)Y_1] \\ &= \alpha Y_3 + \alpha(1 - \alpha)Y_2 + (1 - \alpha)^2 Y_1. \end{aligned}$$

Hence we see that F_4 is a weighted average of the first three time series values. Note that the sum of the coefficients or weights for Y_1, Y_2, and Y_3 equals 1. A similar argument can be made to show that any forecast F_{t+1} is a weighted average of the previous t time series values.

An advantage of exponential smoothing is that it is a simple procedure that requires very little historical data for its use. Once the smoothing constant α has been selected, only two pieces of information are required in order to compute the forecast for the next period. From (16.2) we see that with a given α we can compute the forecast for period $t + 1$ simply by knowing the actual and forecast time series values for period t; that is, Y_t and F_t.

As an illustration of the exponential smoothing model, consider the gasoline sales time series presented previously in Table 16.2 and Figure 16.5. With no forecast available for period 1 we begin our calculations by letting F_1 equal the actual value of the time series in period 1. That is, with $Y_1 = 17$, we will assume $F_1 = 17$ simply to get the exponential smoothing computations started. Using a smoothing constant of $\alpha = .2$, the forecast for period 2 becomes

$$F_2 = .2Y_1 + (1 - .2)F_1 = .2(17) + .8(17) = 17.$$

Referring to the time series data in Table 16.2, we find an actual time series value in period 2 of $Y_2 = 21$. Thus period 2 has a forecast error of $21 - 17 = 4$.

Continuing with the exponential smoothing computations provides the following forecast for period 3:

$$F_3 = .2Y_2 + .8F_2 = .2(21) + .8(17) = 17.8.$$

Once the actual time series value in period 3, $Y_3 = 19$, is known, we can generate a forecast for period 4 as follows:

$$F_4 = .2Y_3 + .8F_3 = .2(19) + .8(17.8) = 18.04.$$

As stated previously, we note that each forecast is obtained from a simple calculation using the actual and forecast time series value from the previous period.

By continuing the exponential smoothing calculations we are able to determine the weekly forecast values and the corresponding weekly forecast errors, as shown in Table 16.3. For week 12, we have $Y_{12} = 22$ and $F_{12} = 18.48$. Can you use this information to generate a forecast for week 13 before the actual value of week 13 becomes known? Using the exponential smoothing model, we have

$$F_{13} = .2Y_{12} + .8F_{12} = .2(22) + .8(18.48) = 19.18.$$

Thus the exponential smoothing forecast of the amount sold in week 13 is 19.18, or 19,180 gallons of gasoline. With this forecast the firm can make plans and decisions

TABLE 16.3 **Summary of the Exponehtial Smoothing Forecasts and Forecast Errors for Gasoline Sales with Smoothing Constant $\alpha = .2$**

Week (t)	*Time Series Value* (Y_t)	*Exponential Smoothing Forecast* (F_t)	*Forecast error* ($Y_t - F_t$)
1	17	17.00	*
2	21	17.00	4.00
3	19	17.80	1.20
4	23	18.04	4.96
5	18	19.03	−1.03
6	16	18.83	−2.83
7	20	18.26	1.74
8	18	18.61	−.61
9	22	18.49	3.51
10	20	19.19	.81
11	15	19.35	−4.35
12	22	18.48	3.52

*Forecast error for week 1 is not considered because F_1 was assumed equal to Y_1 in order to begin the smoothing computations.

accordingly. The accuracy of the forecast will not be known until the firm conducts its business through week 13. Hopefully the exponential smoothing model has provided a good forecast for the unknown 13th week gasoline sales volume.

Figure 16.7 shows the plot of the actual and the forecast time series values. Note in particular how the forecasts "smooth out" the irregular fluctuations in the time series.

In the preceding calculations we used a smoothing constant of $\alpha = .2$. Any value of α between 0 and 1 is acceptable; however, some values will yield better forecasts than others. Some insight into choosing a good value for α can be obtained by rewriting the basic exponential smoothing model as follows:

$$\begin{aligned} F_{t+1} &= \alpha Y_t + (1 - \alpha)F_t \\ &= \alpha Y_t + F_t - \alpha F_t \\ &= \underbrace{F_t}_{\text{Forecast in Period } t} + \alpha\underbrace{(Y_t - F_t)}_{\text{Forecast Error in Period } t}. \end{aligned} \tag{16.3}$$

We see that the new forecast F_{t+1} is equal to the previous forecast F_t plus an adjustment, which is α times the most recent forecast error, $Y_t - F_t$. That is, the forecast in period $t + 1$ is obtained by adjusting the forecast in period t by a fraction of the forecast error. If the time series is very volatile and contains substantial random variability, a small value for the smoothing constant is preferred. The reason for this choice is that since much of the forecast error is due to random variability, we do not want to overreact and adjust the forecasts too quickly. For a fairly stable time series with relatively little random variability, larger values of the smoothing constant have the advantage of quickly adjusting the forecasts when forecasting errors occur and therefore allowing the forecast to react faster to changing conditions.

The criterion that we shall use to determine a desirable value for the smoothing

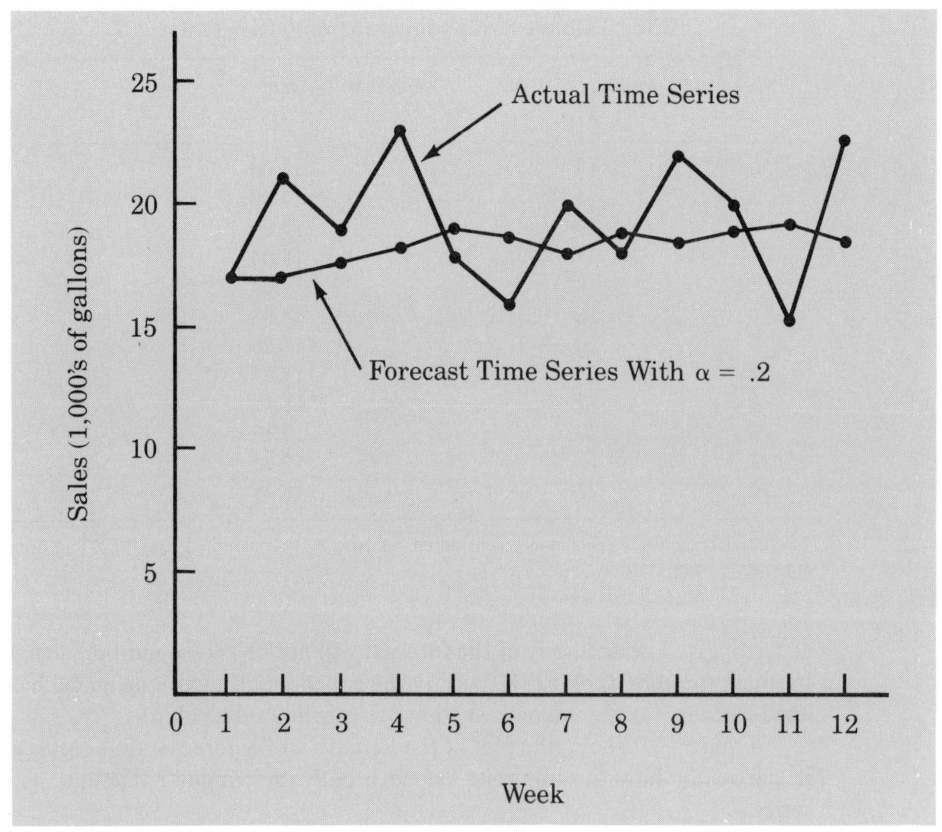

FIGURE 16.7 Graph of Actual and Forecast Gasoline Sales Time Series with Smoothing Constant $\alpha = .2$

constant α is the same as the criterion we proposed earlier for determining the number of periods of data to include in the moving averages calculation. That is, we choose the value of α that minimizes the mean square error (MSE).

A summary of the mean square error calculations for the exponential smoothing forecast of gasoline sales with $\alpha = .2$ is shown in Table 16.4. Note that there is one less squared error term than the number of time periods. This is because we had no past values with which to make a forecast for period 1. As a result we set $F_1 = Y_1$, and do not obtain a squared error in period 1.

Would a different value of α have provided better results in terms of a lower MSE value? Perhaps the most straightforward way to answer this question is to simply try another value for α. We will then compare its mean square error with the MSE value of 8.98 obtained by using a value of .2 for the smoothing constant.

The exponential smoothing results with $\alpha = .3$ are shown in Table 16.5. With MSE = 9.35, we see that for the current data set a smoothing constant of $\alpha = .3$ results in less forecast accuracy than a smoothing constant of $\alpha = .2$. Thus we would be inclined to prefer the original smoothing constant of .2. With a trial-and-error calculation with other values of α a "good" value for the smoothing constant can be found. This value can be used in the exponential smoothing model to provide forecasts for the future. At a later date, after a number of new time series observations have been obtained, it is good practice to analyze the newly collected time series data to see if the smoothing constant should be revised to provide better forecasting results.

TABLE 16.4 **Mean Square Error Computations for Forecasting Gasoline Sales with $\alpha = .2$**

Week (t)	Time Series Value (Y_t)	Forecast (F_t)	Forecast Error ($Y_t - F_t$)	Squared Error ($(Y_t - F_t)^2$)
1	17	17.00	—	—
2	21	17.00	4.00	16.00
3	19	17.80	1.20	1.44
4	23	18.04	4.96	24.60
5	18	19.03	−1.03	1.06
6	16	18.83	−2.83	8.01
7	20	18.26	1.74	3.03
8	18	18.61	− .61	.37
9	22	18.49	3.51	12.32
10	20	19.19	.81	.66
11	15	19.35	−4.35	18.92
12	22	18.48	3.52	12.39
			Total	98.80

$$\text{Mean Square Error (MSE)} = \frac{98.80}{11} = 8.98$$

TABLE 16.5 **Mean Square Error Computations for Forecasting Gasoline Sales with $\alpha = .3$**

Week (t)	Time Series Value (Y_t)	Forecast (F_t)	Forecast Error ($Y_t - F_t$)	Squared Error ($(Y_t - F_t)^2$)
1	17	17.00	—	—
2	21	17.00	4.00	16.00
3	19	18.20	.80	.64
4	23	18.44	4.56	20.79
5	18	19.81	−1.81	3.28
6	16	19.27	−3.27	10.69
7	20	18.29	1.71	2.92
8	18	18.80	− .80	.64
9	22	18.56	3.44	11.83
10	20	19.59	.41	.17
11	15	19.71	−4.71	22.18
12	22	18.30	3.70	13.69
				Total 102.83

$$\text{Mean Square Error (MSE)} = \frac{102.83}{11} = 9.35$$

EXERCISES

1. Refer to the gasoline sales time series data in Table 16.1.

a. Compute 4 week and 5 week moving averages for the time series.

b. Compute the mean square error (MSE) for the 4 week and 5 week moving average forecasts.

c. What appears to be the best number of weeks of past data to use in the moving average computation? Remember that the MSE for the 3 week moving average is 10.22.

2. Refer again to the gasoline sales time series data in Table 16.1.
a. Using a weight of ½ for the most recent observation, ⅓ for the second most recent, and ⅙ for third most recent, compute a 3 week weighted moving average for the time series.
b. Compute the mean square error for the weighted moving average in part a. Do you prefer this weighted moving average to the unweighted moving average? Remember that the MSE for the unweighted moving average is 10.22.
c. Suppose you are allowed to choose any weight as long as they sum to one. Could you always find a set of weights that would make the MSE smaller for a weighted moving average than an unweighted moving average? Why or why not?
3. Use the gasoline time series data from Table 16.1 to show the exponential smoothing forecasts using $\alpha = .1$. Using the mean squared error criterion, would you prefer a smoothing constant of $\alpha = .1$ or $\alpha = .2$ for the gasoline sales time series?
4. Using a smoothing constant of $\alpha = .2$, (16.2) shows that the forecast for the 13th week of the gasoline sales data from Table 16.1 is given by $F_{13} = .2Y_{12} + .8F_{12}$. However, the forecast for week 12 is given by $F_{12} = .2Y_{11} + .8F_{11}$. Thus we could combine these two results to show that the forecast for the 13th week can be written

$$F_{13} = .2Y_{12} + .8(.2Y_{11} + .8F_{11}) = .2Y_{12} + .16Y_{11} + .64F_{11}.$$

a. Making use of the fact that $F_{11} = .2Y_{10} + .8F_{10}$ (and similarly for F_{10} and F_9), continue to expand the expression for F_{13} until it is written in terms of the past data values Y_{12}, Y_{11}, Y_{10}, Y_9, Y_8, and the forecast for period 8.
b. Refer to the coefficients or weights for the past data Y_{12}, Y_{11}, Y_{10}, Y_9, and Y_8; what observation do you make about how exponential smoothing weights past data values in arriving at new forecasts? Compare this weighting pattern with the weighting pattern of the moving averages method.
5. The following time series shows the sales of a particular product over the past 12 months:

Month	*Sales*
1	105
2	135
3	120
4	105
5	90
6	120
7	145
8	140
9	100
10	80
11	100
12	110

Use $\alpha = .3$ to compute the exponential-smoothing values for the time series.
6. Analyze the forecasting errors for the time series in Exercise 5 by using a smoothing constant of .5. Does a smoothing constant of .3 or .5 appear to provide the better forecasts?

16.3 FORECASTING TIME SERIES USING TREND PROJECTION

In this section we will see how to forecast the values of a time series that exhibits a long term linear trend. Specifically, let us consider the time series data for bicycle sales of a particular manufacturer over the past 10 years, as shown in Table 16.6 and Figure 16.8. Note that 21,600 bicycles were sold in year 1, 22,900 were sold in year 2, and so

TABLE 16.6 Bicycle Sales Data

Year (t)	*Sales (1,000's)* (Y_t)
1	21.6
2	22.9
3	25.5
4	21.9
5	23.9
6	27.5
7	31.5
8	29.7
9	28.6
10	31.4

on; in year 10, the most recent year, 31,400 bicycles were sold. While the graph in Figure 16.8 shows some up-and-down movement over the past 10 years, the time series seems to have an overall increasing or upward trend in the number of bicycles sold.

One should not expect the trend component of a time series to follow each and every "up" and "down" movement. Rather, the trend component should reflect the gradual shifting—in our case, growth—of the time series values. After we view the time series data in Table 16.6 and the graph in Figure 16.8 we might agree that a linear trend as shown in Figure 16.9 has the potential of providing a reasonable description of the long-run movement in the series. Thus we can now concentrate on finding the linear function that best approximates the trend.

Using the television-set sales data to illustrate the calculations involved, we will

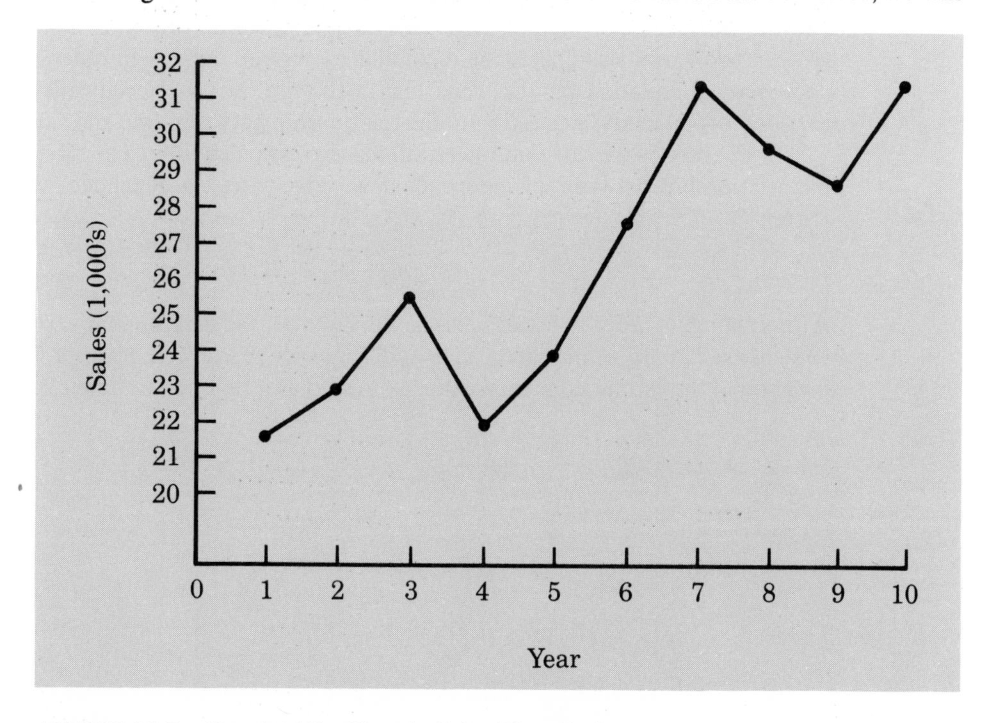

FIGURE 16.8 Graph of the Bicycle Sales Time Series

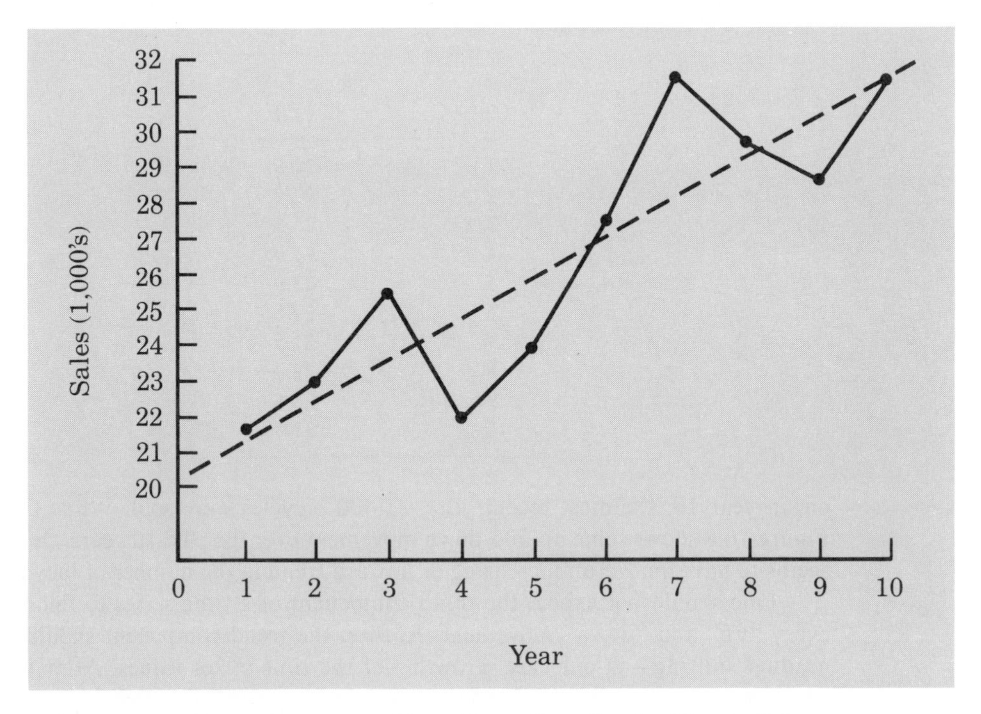

FIGURE 16.9 Trend Represented by a Linear Function for Bicycle Sales

now describe how regression analysis can be used to identify a linear trend for a time series. Recall in the discussion of simple linear regression in Chapter 14 we described how the least squares method was used to find the best straight line relationship between two variables. This is the methodology we will employ in order to develop a mathematical equation for the trend line in the time series. Specifically, we will be using regression analysis to estimate the relationship between time and sales volume.

In Chapter 14 we saw that the estimated regression equation describing a straight line relationship between an independent variable x and a dependent variable y is written

$$\hat{y} = b_0 + b_1 x. \tag{16.4}$$

In forecasting, in order to better focus on the fact that the independent variable is time, we shall use t in (16.4) instead of x; in addition, we will use T_t in place of $\hat{y}$. Thus for a linear trend the estimated sales volume expressed as a function of time can be written

$$T_t = b_0 + b_1 t, \tag{16.5}$$

where

T_t = forecast value (based upon trend) of the time series in period t,

b_0 = intercept of the trend line,

b_1 = slope of the trend line,

t = point in time.

In this linear trend relationship, we will let $t = 1$ for the time of the first observation on the time series data, $t = 2$ for the time of the second observation, and so on. Note that for our time series on television set sales $t = 1$ corresponds to the oldest time series value and $t = 10$ corresponds to the most recent year's data. The formulas for computing the estimated regression coefficients (b_1 and b_0) for (16.4) were presented in Chapter 14. They are shown again below, with t replacing x and Y_t replacing y_i:

$$b_1 = \frac{\Sigma\, tY_t - (\Sigma\, t\, \Sigma\, Y_t)/n}{\Sigma\, t^2 - (\Sigma\, t)^2/n} \tag{16.6}$$

$$b_0 = \overline{Y} - b_1\bar{t}, \tag{16.7}$$

where

Y_t = actual value of the time series in period t,

n = number of periods,

$\overline{Y}$ = average value of the time series; that is, $\overline{Y} = \Sigma\, Y_t/n$,

$\bar{t}$ = average value of t; that is, $\bar{t} = \Sigma\, t/n$.

Using these relationships for b_0 and b_1 and the television set sales data of Table 16.6, we have the following calculations:

	t	Y_t	tY_t	t^2
	1	21.6	21.6	1
	2	22.9	45.8	4
	3	25.5	76.5	9
	4	21.9	87.6	16
	5	23.9	119.5	25
	6	27.5	165.0	36
	7	31.5	220.5	49
	8	29.7	237.6	64
	9	28.6	257.4	81
	10	31.4	314.0	100
Totals	55	264.5	1,545.5	385

$$\bar{t} = \frac{55}{10} = 5.5 \text{ years},$$

$$\overline{Y} = \frac{264.5}{10} = 26.45 \text{ thousands},$$

$$b_1 = \frac{1545.5 - (55)(264.5)/10}{385 - (55)^2/10} = \frac{90.75}{82.50} = 1.10,$$

$$b_0 = 26.45 - 1.10(5.5) = 20.4.$$

Therefore

$$T_t = 20.4 + 1.1t \tag{16.8}$$

is the expression for the linear trend component of the bicycle sales time series.

Trend Projections

The slope of 1.1 indicates that over the past 10 years the firm has experienced an average growth in sales of around 1100 units per year. If we assume that the past 10-year trend in sales is a good indicator of the future, then (16.8) can be used to project the trend component of the time series. For example, substituting $t = 11$ into (16.8) yields next year's trend projection, T_{11}:

$$T_{11} = 20.4 + 1.1(11) = 32.5.$$

Thus using the trend component only we would forecast sales of 32,500 bicycles next year.

The use of a linear function to model the trend is common. However, as we discussed earlier, sometimes time series exhibit a nonlinear trend. Figure 16.10 shows

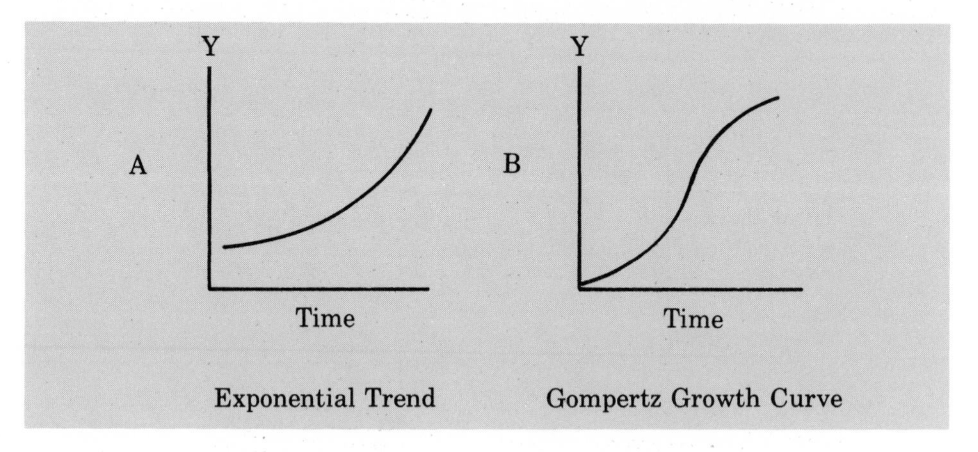

FIGURE 16.10 Some Possible Functional Forms for Nonlinear Trend Patterns

two common nonlinear trend functions. More advanced texts discuss in detail how to solve for the trend component when a nonlinear function is used and how to decide when to use such a function. For our purposes it is sufficient to note that the analyst should choose the function that provides the best fit to the data.

EXERCISES

7. The enrollment data for a state college for the past 6 years are shown below:

Year	*Enrollment*
1	20,500
2	20,200
3	19,500
4	19,000
5	19,100
6	18,800

Develop a linear trend expression. Comment on what is happening to enrollment at this institution.

8. Average attendance figures at home football games for a major university show the following pattern for the past 7 years:

Year	*Attendance*
1	28,000
2	30,000
3	31,500
4	30,400
5	30,500
6	32,200
7	30,800

Use a linear functional form and develop the trend expression (16.4) for this time series.

9. Automobile sales at B. J. Scott Motors, Inc., provided the following 10 year time series:

Year	*Sales*
1	400
2	390
3	320
4	340
5	270
6	260
7	300
8	320
9	340
10	370

Plot the time series and comment on the appropriateness of a linear trend. What type of functional form do you believe would be most appropriate for the trend pattern of this time series?

10. The president of a small manufacturing firm has been concerned about the continual growth in manufacturing costs over the past several years. Shown below is a time series of the cost per unit for the firm's leading product over the past 8 years:

Year	*Cost/Unit ($)*
1	20.00
2	24.50
3	28.20
4	27.50
5	26.60
6	30.00
7	31.00
8	36.00

a. Show a graph of this time series. Does a linear trend appear to exist?

b. Develop a linear trend expression for the above time series. What is the average cost increase that the firm has been realizing per year?

16.4 FORECASTING TIME SERIES USING CLASSICAL DECOMPOSITION

To use the classical decomposition approach to forecasting we must specify how the four time series components—trend (T), cyclical (C), seasonal (S), and irregular (I)—are combined into a model which describes the behavior of the time series. We will assume that the time series can best be described with a multiplicative time series model. This model assumes that if the trend, cyclical, seasonal, and irregular components can be identified and measured, multiplying the measures of the four components will provide the actual time series value, denoted by Y_t. Mathematically this multiplicative model is written as follows:

Multiplicative Model

$$Y_t = T_t \times C_t \times S_t \times I_t. \tag{16.9}$$

In this model T_t is the trend measured in units of the item being forecast. However, the C_t, S_t, and I_t components are all measured in relative terms, with values above 1.00 indicating a cyclical effect above the trend, a seasonal effect* above the normal or average level, or an irregular effect above the combined trend, cyclical, and seasonal components. Values below 1.00 for C_t, S_t, and I_t would show below-average levels for each component, respectively. To illustrate, suppose that we have a trend projection of 540 units and values of C_t, S_t, and I_t given by 1.10, .85, and 1.02, respectively. The value of the time series would be $540(1.10)(.85)(1.02) = 515$.

In this section we will illustrate the classical decomposition approach by working with the quarterly data presented in Table 16.7 and Figure 16.11. These data show the

TABLE 16.7 Quarterly Data for Television Set Sales

Year	*Quarter*	*Sales (1,000's)*
1	1	4.8
	2	4.1
	3	6.0
	4	6.5
2	1	5.8
	2	5.2
	3	6.8
	4	7.4
3	1	6.0
	2	5.6
	3	7.5
	4	7.8
4	1	6.3
	2	5.9
	3	8.0
	4	8.4

*We note that when the time series observations are made at yearly intervals the seasonal component takes on a value of 1 and need not be included in the model.

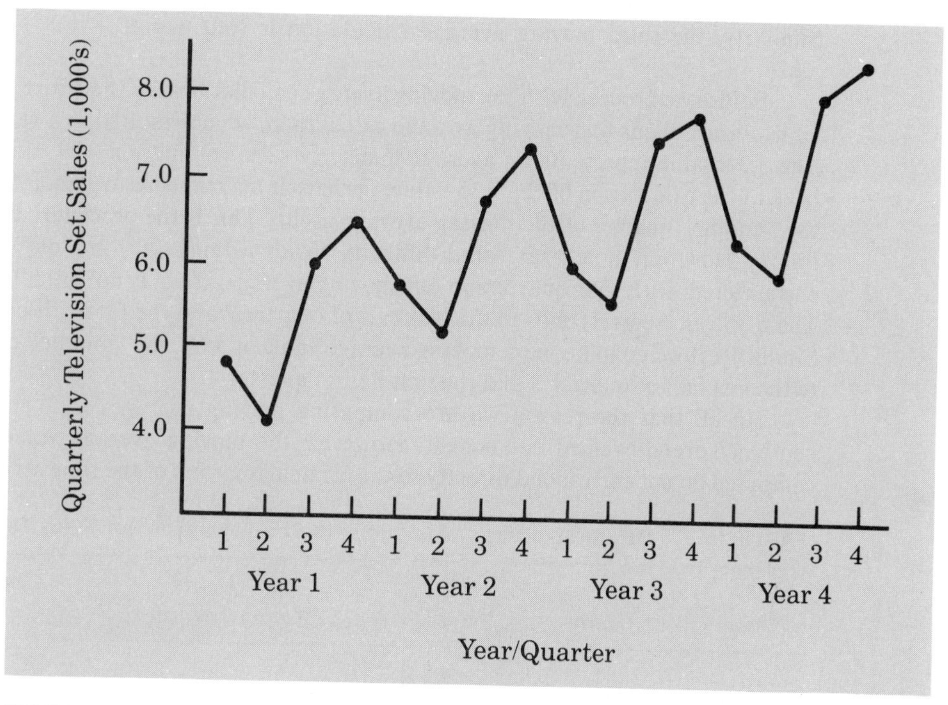

FIGURE 16.11 Graph of Quarterly Television Set Sales Time Series

television set sales (in thousands of units) for a particular manufacturer over the past 4 years. We begin our analysis by showing how to identify the seasonal component of the time series.

Calculating the Seasonal Factors

By looking at Figure 16.11 we can begin to identify a seasonal pattern for the television set sales. Specifically, we observe that sales are lowest in the second quarter of each year, followed by higher sales levels in quarters 3 and 4. The computational procedure used to identify each quarter's seasonal influence begins with the use of moving averages to measure the combined trend-cyclical ($T_t C_t$) component of the time series. That is, we eliminate the seasonal and irregular components, S_t and I_t.

In using moving averages to do this, we use 1 year of data in each calculation. Since we are working with a quarterly series we will use four data values in each moving average. The moving average calculation for the first four quarters of the television set sales data is as follows:

$$\text{First moving average} = \frac{4.8 + 4.1 + 6.0 + 6.5}{4} = \frac{21.4}{4} = 5.35.$$

Note that the moving average calculation for the first four quarters yields the average quarterly sales over the first year of the time series. Continuing the moving average calculation, we next add the 5.8 value for the first quarter of year 2 and drop the 4.8 for the first quarter of year 1. Thus the second moving average is

$$\text{Second moving average} = \frac{4.1 + 6.0 + 6.5 + 5.8}{4} = \frac{22.4}{4} = 5.6.$$

Similarly, the third moving average calculation is $(6.0 + 6.5 + 5.8 + 5.2)/4 = 5.875$.

Before we proceed with the moving average calculations for the entire time series, let us return to our first moving average calculation, which resulted in a value of 5.35. The 5.35 value represents an average quarterly sales volume for year 1. As we look back at the calculation of the 5.35 value, perhaps it makes sense to associate 5.35 with the "middle" quarter of the moving average group. This is the procedure that we will follow. However, note that some difficulty in identifying the "middle" quarter is encountered; with four quarters in our moving average, there is no "middle" quarter. The 5.35 value corresponds to the last half of quarter 2 and the first half of quarter 3. Similarly, if we go to the next moving average value of 5.60, the "middle" corresponds to the last half of quarter 3 and the first half of quarter 4.

Recall that the reason we are computing moving averages is to measure the combined trend-cyclical component. However, the moving average values we have computed do not correspond directly to the original quarters of the time series. We can

TABLE 16.8 Moving Average Calculations for the Television Set Sales Time Series

Year	*Quarter*	*Sales (1,000's)*	*Four-Quarter Moving Average*	*Centered Moving Average*
1	1	4.8		
	2	4.1		
			5.350	
	3	6.0		5.475
			5.600	
	4	6.5		5.738
			5.875	
2	1	5.8		5.975
			6.075	
	2	5.2		6.188
			6.300	
	3	6.8		6.325
			6.350	
	4	7.4		6.400
			6.450	
3	1	6.0		6.538
			6.625	
	2	5.6		6.675
			6.725	
	3	7.5		6.763
			6.800	
	4	7.8		6.838
			6.875	
4	1	6.3		6.938
			7.000	
	2	5.9		7.075
			7.150	
	3	8.0		
	4	8.4		

resolve this difficulty by using the midpoints between successive moving average values. For example, since 5.35 corresponds to the first half of quarter 3 and 5.60 corresponds to the last half of quarter 3, we will use $(5.35 + 5.60)/2 = 5.475$ as the moving average value for quarter 3. Similarly, we associate a moving average value of $(5.60 + 5.875)/2 = 5.738$ with quarter 4. A complete summary of the moving average calculations for the television set sales data is shown in Table 16.8.

Note that if the number of data points in a moving average calculation is an odd number, the middle point will correspond to one of the points in the time series. Then we would not have to adjust the moving average values to correspond to a particular data point, as we did in the calculations in Table 16.8.

Let us pause for a moment to consider what the moving averages in Table 16.8 tell us about this time series. A plot of the actual time series values and their corresponding moving averages is shown in Figure 16.12. Note particularly how the moving average values tend to "smooth out" the fluctuations in the time series. Since the moving average values are for four quarters of data, they do not include the fluctuations due to seasonal influences. Furthermore, since the irregular movements tend to average out to zero over a number of periods, we have also eliminated the fluctuations due to the irregular influences.

With the seasonal and irregular influences removed, or averaged out, the centered moving average values identify the combined trend-cyclical component of the time series. Thus for the television set sales data the smoothed values in Figure 16.12 measure the trend-cyclical component over the most recent 4 years of data.

With quarterly time series data, the data values observed are due to the combined trend, cyclical, seasonal, and irregular components of the series. Again using the

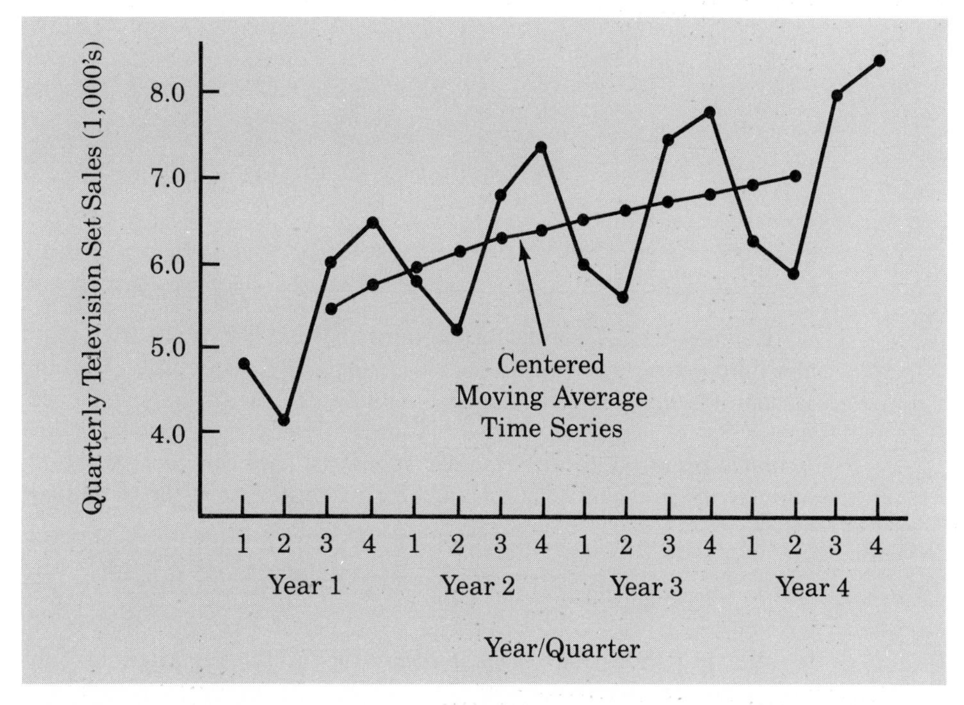

FIGURE 16.12 Graph of Quarterly Television Set Sales Time Series and Moving Averages

notation of the multiplicative model, we have

$$Y_t = T_t \times C_t \times S_t \times I_t. \tag{16.10}$$

For the 4 years of quarterly data (see Table 16.8) the actual time series values (Y_t) and the combined trend-cyclical (T_tC_t) component (see the centered moving average results) are known. Thus with (16.10) we can solve for and identify the combined seasonal-irregular component as follows:

$$S_tI_t = \frac{Y_t}{T_tC_t}. \tag{16.11}$$

By dividing each time series observation Y_t by the corresponding moving average value T_tC_t we can identify the seasonal-irregular effect in the time series. The resulting S_tI_t values are summarized in Table 16.9.

TABLE 16.9 Seasonal-Irregular Factors for the Television Set Sales Time Series

Year	*Quarter*	*Quarterly Sales* (Y_t)	**Four-Quarter Moving Average** (T_tC_t)	**Seasonal-Irregular Component** ($S_tI_t = Y_t/T_tC_t$)
1	1	4.8		
	2	4.1		
	3	6.0	5.475	1.096
	4	6.5	5.738	1.133
2	1	5.8	5.975	.971
	2	5.2	6.188	.840
	3	6.8	6.325	1.075
	4	7.4	6.400	1.156
3	1	6.0	6.538	.918
	2	5.6	6.675	.839
	3	7.5	6.763	1.109
	4	7.8	6.838	1.141
4	1	6.3	6.938	.908
	2	5.9	7.075	.834
	3	8.0		
	4	8.4		

Consider the S_tI_t results for the third quarter. The results from years 1, 2, and 3 show third-quarter values of 1.096, 1.075, and 1.109, respectively. Thus in all cases the seasonal-irregular component appears to have an above average influence in the third quarter. Since the year-to-year fluctuations in the seasonal-irregular component can be attributed primarily to the irregular component, we can average the S_tI_t values to eliminate the irregular influence and obtain an estimate of the third-quarter seasonal influence:

$$\text{Seasonal effect of third quarter} = \frac{1.096 + 1.075 + 1.109}{3} = 1.09.$$

We refer to 1.09 as the *seasonal factor* for the third quarter. In Table 16.10 we summarize the calculations involved in computing the seasonal factors for the television set sales time series. Thus, we see that the seasonal factors for all four

TABLE 16.10 Seasonal Component Calculations for the Television Set Sales Time Series

Quarter	*Seasonal-Irregular Component Values* (S_tI_t)	*Seasonal Factor* (S_t)
1	.971, .918, .908	.93
2	.840, .839, .834	.84
3	1.096, 1.075, 1.109	1.09
4	1.133, 1.156, 1.141	1.14

quarters are as follows: quarter 1, .93; quarter 2, .84; quarter 3, 1.09; and quarter 4, 1.14.

Interpretation of the values in Table 16.10 provides some observations about the "seasonal" component in television-set sales. The best sales quarter is the fourth quarter, with sales averaging 14% above the average quarterly value. The worst, or slowest, sales quarter is the second quarter, with its seasonal factor at .84, showing the sales average 16% below the average quarterly sales. The seasonal component corresponds nicely to the intuitive expectation that television viewing interest and thus television purchase patterns tend to peak in the fourth quarter, with its coming winter season and fewer outdoor activities. The low second-quarter sales reflect the reduced television interest resulting from the spring and presummer activities of the potential customers.

One final adjustment is sometimes necessary in obtaining the seasonal factors. The multiplicative model requires that the average seasonal factor equal 1.00; that is, the sum of the factors must equal 4.00. This is necessary if the seasonal effects are to even out over the year, as they must. The average of the seasonal factors in our example is equal to 1.00, and hence this type of adjustment is not necessary. In other cases a slight adjustment will be necessary. The adjustment can be made by simply multiplying each seasonal factor by the number of seasons divided by the sum of the unadjusted seasonal factors. For example, for quarterly data we would multiply each seasonal factor by 4/(sum of the unadjusted seasonal factors). Some of the problems at the end of the chapter will require this adjustment in order to obtain the appropriate seasonal factors.

Deseasonalizing to Identify Trend

To identify the trend for a time series containing seasonal effects we must first remove the effect of season from the original time series. This process is referred to as *deseasonalizing the time series.* Using the notation of the multiplicative model, we have

$$Y_t = T_t \times C_t \times S_t \times I_t.$$

Since we have just identified the seasonal factors for this model, we can solve for and identify the combined trend, cyclical, and irregular components as

$$T_tC_tI_t = \frac{Y_t}{S_t}.$$

TABLE 16.11 Deseasonalized Values for the Television Set Sales Time Series

Year	Quarter	Sales (1,000's) (Y_t)	Seasonal Factor (S_t)	Deseasonalized Sales ($Y_t/S_t = T_tC_tI_t$)
1	1	4.8	.93	5.16
	2	4.1	.84	4.88
	3	6.0	1.09	5.50
	4	6.5	1.14	5.70
2	1	5.8	.93	6.24
	2	5.2	.84	6.19
	3	6.8	1.09	6.24
	4	7.4	1.14	6.49
3	1	6.0	.93	6.45
	2	5.6	.84	6.67
	3	7.5	1.09	6.88
	4	7.8	1.14	6.84
4	1	6.3	.93	6.77
	2	5.9	.84	7.02
	3	8.0	1.09	7.34
	4	8.4	1.14	7.37

By dividing each time series observation by the corresponding seasonal factor we have removed the effect of season from the time series. The deseasonalized values are summarized in Table 16.11. A graph of the deseasonalized television set sales time series is shown in Figure 16.13.

Looking at Figure 16.13, we see that while the graph shows some up-and-down movement over the past 16 quarters, the time series seems to have an upward linear

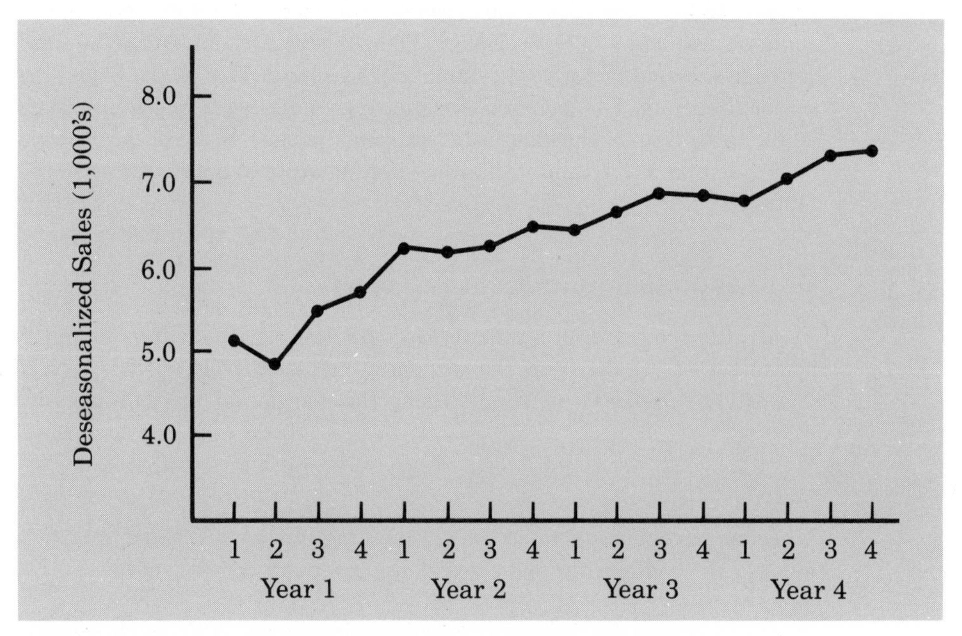

FIGURE 16.13 Deseasonalized Television Set Sales Time Series

trend. To identify this trend, we will use the same procedure we introduced for identifying trend when forecasting with annual data; in this case, since we have deseasonalized the data, quarterly sales values can be used. Thus for a linear trend the estimated sales volume expressed as a function of time can be written

$$T_t = b_0 + b_1 t,$$

where

T_t = trend value for television set sales in period t,

b_0 = intercept of the trend line,

b_1 = slope of the trend line.

As we did before, we will let $t = 1$ for the time of the first observation on the time series data, $t = 2$ for the time of the second observation, and so on. Thus for the deseasonalized television set sales time series $t = 1$ corresponds to the first deseasonalized quarterly sales value and $t = 16$ corresponds to the most recent deseasonalized quarterly sales value. The formulas for computing the value of b_0 and the value of b_1 are shown again

$$b_1 = \frac{\Sigma\, tY_t - (\Sigma\, t\, \Sigma\, Y_t)/n}{\Sigma\, t^2 - (\Sigma\, t)^2/n}$$

$$b_0 = \bar{Y} - b_1 \bar{t}.$$

Note, however, that Y_t now refers to the deseasonalized time series value at time t and not the actual value of the time series. Using the above relationships for b_0 and b_1 and the deseasonalized sales data of Table 16.11, we have the following calculations:

	t	Y_t *(Deseasonalized)*	tY_t	t^2
	1	5.16	5.16	1
	2	4.88	9.76	4
	3	5.50	16.50	9
	4	5.70	22.80	16
	5	6.24	31.20	25
	6	6.19	37.14	36
	7	6.24	43.68	49
	8	6.49	51.92	64
	9	6.45	58.05	81
	10	6.67	66.70	100
	11	6.88	75.68	121
	12	6.84	82.08	144
	13	6.77	88.01	169
	14	7.02	98.28	196
	15	7.34	110.10	225
	16	7.37	117.92	256
Totals	136	101.74	914.98	1,496

$$\bar{t} = \frac{136}{16} = 8.5,$$

$$\bar{Y} = \frac{101.74}{16} = 6.359,$$

$$b_1 = \frac{914.98 - (136)(101.74)/16}{1496 - (136)^2/16} = \frac{50.19}{340} = .148,$$

$$b_0 = 6.359 - .148\,(8.5) = 5.101.$$

Therefore

$$T_t = 5.101 + .148t \tag{16.12}$$

is the expression for the linear trend component of the time series.

The slope of .148 indicates that over the past 16 quarters, the firm has experienced an average deseasonalized growth in sales of around 148 sets per quarter. If we assume that the past 16-quarter trend in sales data is a reasonably good indicator of the future, then (16.12) can be used to project the trend component of the time series for future quarters. For example, substituting $t = 17$ into (16.12) yields next quarter's trend projection, T_{17}:

$$T_{17} = 5.101 + .148(17) = 7.617.$$

Using the trend component only, we would forecast sales of 7,617 television sets for the next quarter. In a similar fashion, if we use the trend component only, we would forecast sales of 7765, 7913, and 8061 television sets in quarters 18, 19, and 20, respectively.

Seasonal Adjustments

Now that we have a forecast of sales for each of the next four quarters based upon trend, we must adjust these forecasts to account for the effect of season. For example, since the seasonal factor for the first quarter of year 5($t = 17$) is .93, the quarterly forecast can be obtained by multiplying the forecast based upon trend ($T_{17} = 7{,}617$) times the seasonal factor (0.93). Thus the forecast for the next quarter is 7,617(.93) = 7,084. Table 16.12 shows the quarterly forecast for quarters 17, 18, 19, and 20. The quarterly forecasts show the high volume fourth quarter with a 9,190 unit forecast, while the low volume second quarter has a 6,523 unit forecast.

TABLE 16.12 Quarter-by-Quarter Short Range Forecasts for the Television Set Sales Time Series

Year	*Quarter*	*Trend Forecast*	*Seasonal Factor (see Table 16.10)*	*Quarterly Forecast*
5	1	7,617	.93	(7,617)(.93) = 7,084
	2	7,765	.84	(7,765)(.84) = 6,523
	3	7,913	1.09	(7,913)(1.09) = 8,625
	4	8,061	1.14	(8,061)(1.14) = 9,190

Although each of these forecasts could be adjusted further to account for a possible cyclical effect, such an analysis is beyond the scope of this text.

EXERCISES

11. The quarterly sales data for a college textbook over the past 3 years are as follows:

Quarter	*Year 1*	*Year 2*	*Year 3*
1	1690	1800	1850
2	940	900	1100
3	2625	2900	2930
4	2500	2360	2615

a. Show the four-quarter moving average values for this time series. Plot both the original time series and the moving averages on the same graph.
b. Compute seasonal factors for the four quarters.
c. When does the textbook publisher experience the largest seasonal effect? Does this appear reasonable? Explain.

12. Identify the monthly seasonal factors for the following 3 years of expenses for a 6 unit apartment house in southern Florida. Use a 12 month moving average calculation.

Month	*Year 1*	*Year 2*	*Year 3*
January	170	180	195
February	180	205	210
March	205	215	230
April	230	245	280
May	240	265	290
June	315	330	390
July	360	400	420
August	290	335	330
September	240	260	290
October	240	270	295
November	230	255	280
December	195	220	250

16.5 FORECASTING TIME SERIES USING REGRESSION MODELS

In our discussion of regression analysis in Chapters 14 and 15 we showed how one or more independent variables could be used to predict the value of a single dependent variable. In particular we saw that in many cases an estimated regression equation can be developed that would enable the use of known values of the independent variables to provide good predictions of the value of the dependent variable. According to the logic and methodology of regression analysis the time series value that we would like to forecast can be viewed as the dependent variable. Thus if we can identify a good set of related independent or predictor variables, we may be able to develop an estimated regression equation for predicting or forecasting the time series.

The approach we used in Section 16.3 to fit a linear trend line to the bicycle sales time series is a special case of regression analysis. In that example two variables—bicycle sales and time—were shown to be linearly related.* The inherent complexity of most real-world problems necessitates the consideration of more than one variable to predict the variable of interest. The statistical technique known as multiple regression analysis can be used in such situations.

Recall that in order to develop an estimated regression equation we need a sample of observations for the dependent variable and all independent variables. In time series analysis the n periods of time series data provide a sample of n observations on each variable that can be used in the analysis. For a function involving k independent variables we use the following notation:

$$Y_t = \text{actual value of the time series in period } t,$$
$$x_{1t} = \text{value of independent variable 1 in period } t,$$
$$x_{2t} = \text{value of independent variable 2 in period } t,$$
$$\vdots$$
$$x_{kt} = \text{value of independent variable } k \text{ in period } t.$$

The n periods of data necessary to develop the estimated regression equation would appear as follows:

				Value of Independent Variables				
Period	*Time Series Value* (Y_t)	x_{1t}	x_{2t}	x_{3t}	·	·	·	x_{kt}
1	Y_1	x_{11}	x_{21}	x_{31}	·	·	·	x_{k1}
2	Y_2	x_{12}	x_{22}	x_{32}	·	·	·	k_{k2}
⋮	⋮	⋮	⋮	⋮	⋮	⋮	⋮	⋮
n	Y_n	x_{1n}	x_{2n}	x_{3n}	·	·	·	x_{kn}

As you might imagine, there are a number of possible choices for the independent variables in a forecasting model. One possible choice for an independent variable is simply time expressed in terms of the numbers of the periods in the time series. This is the choice we made in Section 16.3 when we estimated the trend of the time series using a linear function of the independent variable time. Letting

$$x_{1t} = t,$$

we obtain an estimated regression equation of the form

$$\hat{Y}_t = b_0 + b_1 t,$$

where $\hat{Y}_t$ is the estimate of the time series value Y_t and where b_0 and b_1 are the estimated regression coefficients. In a more complex model additional terms could be

*In a purely technical sense the number of bicycles sold is not thought of as being related to time; instead, time is used as a surrogate for variables that the number of bicycles sold is actually related to but that are either unknown or too difficult or too costly to measure.

added corresponding to time raised to other powers. For example, if

$$x_{2t} = t^2$$

and

$$x_{3t} = t^3,$$

the estimated regression equation would then become

$$\begin{aligned}\hat{Y}_t &= b_0 + b_1 x_{1t} + b_2 x_{2t} + b_3 x_{3t} \\ &= b_0 + b_1 t + b_2 t^2 + b_3 t^3.\end{aligned}$$

Note that this model provides a forecast of a time series with nonlinear characteristics over time.

Other regression-based forecasting models employ a mixture of economic and demographic independent variables. For example, in forecasting the sale of refrigerators we might select independent variables such as the following:

x_{1t} = price in period t,
x_{2t} = total industry sales in period $t - 1$,
x_{3t} = number of building permits for new houses in period $t - 1$,
x_{4t} = population forecast for period t,
x_{5t} = advertising budget for period t.

According to the usual multiple regression procedure, an estimated regression equation with five independent variables would be used to develop forecasts.

Whether or not a regression approach provides a good forecast depends largely on how well we are able to identify and obtain data for independent variables that are closely related to the time series. Generally, during the development of an estimated regression equation we will want to consider many possible sets of independent variables. Part of the regression analysis procedure should focus on the selection of the set of independent variables that provides the best forecasting model.

In the chapter introduction we stated that *causal forecasting models* utilized time series related to the one being forecast in an effort to better explain the cause of a time series behavior. Regression analysis is the tool most often used in developing these causal models. The related time series become the independent variables, and the time series being forecast is the dependent variable.

Another type of regression based forecasting model occurs whenever the independent variables are all previous values of the same time series. For example, if the time series values are denoted $Y_1, Y_2, \ldots, Y_n$, then with a dependent variable Y_t we might try to find an estimated regression equation relating Y_t to the most recent time series values Y_{t-1}, Y_{t-2}, and so on. With the three most recent periods as independent variables, the estimated regression equation would be

$$\hat{Y}_t = b_0 + b_1 Y_{t-1} + b_2 Y_{t-2} + b_3 Y_{t-3}.$$

Regression models where the independent variables are previous values of the time series are referred to as *autogressive models*.

Finally, another regression based forecasting approach is one that incorporates a mixture of the independent variables previously discussed. For example, we might select a combination of time variables, some economic/demographic variables, and some previous values of the time series variable itself.

16.6 QUALITATIVE APPROACHES TO FORECASTING

In the previous sections we have discussed several types of quantitative forecasting methods. Since each of these techniques requires historical data on the variable of interest, in situations where no historical data are available these techniques cannot be applied. Furthermore, even when historical data are available a significant change in environmental conditions affecting the time series may make the use of past data questionable in predicting future values of the time series. For example, a government imposed gas rationing program would cause one to question the validity of a gas sales forecast based on past data. Qualitative forecasting techniques offer an alternative in these, and other, cases.

One of the most commonly used qualitative forecasting methods is the *Delphi approach.* This technique, originally developed by a research group at the Rand Corporation, attempts to obtain forecasts through "group consensus." In the usual application of this technique the members of a panel of experts—all of whom are physically separated from and unknown to each other—are asked to respond to a series of questionnaires. The responses from the first questionnaire are tabulated and used to prepare a second questionnaire which contains information and opinions of the whole group. Each respondent is then asked to reconsider and possibly revise his or her previous response in light of the group information that has been provided. This basic process continues until the coordinator feels that some degree of consensus has been reached. Note that the goal of the Delphi approach is not to produce a single answer as output but to produce instead a relatively narrow spread of opinions within which the "majority" of experts concur.

The qualitative procedure referred to as *scenario writing* consists of developing a conceptual scenario of the future based upon a well defined set of assumptions. Thus by starting with a different set of assumptions many different future scenarios can be presented. The job of the decision maker is to decide which scenario is most likely to occur in the future and then to make decisions accordingly.

Subjective or intuitive qualitative approaches are based upon the ability of the human mind to process a variety of information that is, in most cases, difficult to quantify. These techniques are often used in group work, wherein a committee or panel seeks to develop new ideas or solve complex problems through a series of "brainstorming sessions." In such sessions individuals are freed from the usual group restrictions of peer pressure and criticism, since any idea or opinion can be presented without regard to its relevancy and, even more importantly, without fear of criticism.

Summary

The purpose of this chapter has been to provide an introduction to the basic methods of time series analysis and forecasting. First, we showed that in order to explain the behavior of a time series, it is often helpful to think of the time series as consisting of four separate components: trend, cyclical, seasonal, and irregular. By isolating these components and measuring their apparent effect, it is possible to forecast future values of the time series.

We discussed how smoothing methods can be used to forecast a time series that exhibits no significant trend, seasonal, or cyclical effect. The moving averages approach consists of computing an average of past data values and then using this average as the forecast for the next period. The exponential smoothing method is a more preferred technique which uses a weighted average of past time series values to compute a forecast.

When the time series exhibits only a long term trend, we showed how regression analysis could be used to make trend projections. When both trend and seasonal influences are significant, we showed how classical multiplicative decomposition could be used to isolate the effects of the two factors and prepare better forecasts. Finally, regression analysis was described as a procedure for developing so-called causal forecasting models. A causal forecasting model is one that relates the time series value (dependent variable) to other independent variables that are believed to explain (cause) the time series behavior.

Qualitative forecasting methods were discussed as approaches that could be used when little or no historical data were available. These methods are also considered most appropriate when the past pattern of the time series is not expected to continue into the future.

It is important to realize that time series analysis and forecasting is a major field in its own right. In this chapter we have just scratched the surface of the field of time series and forecasting methodology.

Glossary

Time series—A set of observations measured at successive points in time or over successive periods of time.

Forecast—A projection or prediction of future values of a time series.

Multiplicative time series model—A model that assumes that the separate components of trend, cyclical, seasonal, and irregular effects can be multiplied together to identify the actual time series value $Y_t = T_t \times C_t \times S_t \times I_t$.

Trend—The long-run shift or movement in the time series observable over several periods of data.

Cyclical component—The component of the time series model that results in periodic above-trend and below-trend behavior of the time series lasting more than 1 year.

Seasonal component—The component of the time series model that shows a periodic pattern over 1 year or less.

Irregular component—The component of the time series model that reflects the random variation of the actual time series values beyond what can be explained by the trend, cyclical, and seasonal components.

Moving averages—A method of forecasting or smoothing a time series by averaging each successive group of data points. The moving averages method can be used to identify the combined trend/cyclical component of the time series.

Weighted moving averages—A method of forecasting or smoothing a time series by computing a weighted average of past data values. The sum of the weights must equal one.

Exponential smoothing—A forecasting technique that uses a weighted average of past time series values in order to arrive at smoothed time series values which can be used as forecasts.

Smoothing constant—A parameter of the exponential-smoothing model which provides the weight given to the most recent time series value in the calculation of the forecast value.

Mean square error (MSE)—One approach to measuring the accuracy of a forecasting model. This measure is the average of the sum of the squared difference between the forecast values and the actual time series values.

Mean absolute deviation (MAD)—A measure of forecast accuracy. MAD is the average of the sum of the absolute value of the forecast errors.

Deseasonalized time series—A time series that has had the effect of season removed by dividing each original time series observation by the corresponding seasonal factor.

Causal forecasting methods—Forecasting methods that relate a time series to other variables that are believed to explain or cause its behavior.

Autoregressive model—A time series model that uses a regression relationship based on past time series values to predict the future time series values.

Delphi approach—A qualitative forecasting method that obtains forecasts through "group consensus."

Scenario writing—A qualitative forecasting method which consists of developing a conceptual scenario of the future based upon a well defined set of assumptions.

Key Formulas

Moving Average

$$\text{Moving average} = \frac{\Sigma\ (\text{Most recent } n \text{ data values})}{n} \tag{16.1}$$

Exponential Smoothing Model

$$F_{t+1} = \alpha Y_t + (1 - \alpha)F_t, \tag{16.2}$$

or

$$\underset{\text{Forecast in Period } t}{F_{t+1} = F_t} + \alpha\underbrace{(Y_t - F_t)}_{\text{Forecast Error in Period } t}. \tag{16.3}$$

Linear Trend Relationship

$$T_t = b_0 + b_1 t \tag{16.5}$$

Multiplicative Model

$$Y_t = T_t \times C_t \times S_t \times I_t. \tag{16.9}$$

Supplementary Exercises

13. The number of component parts used in a production process each week in the last 10 weeks showed the following:

Week	*Parts*	*Week*	*Parts*
1	200	6	210
2	350	7	280
3	250	8	350
4	360	9	290
5	250	10	320

Use a smoothing constant of .25 and develop the exponential smoothing values for this time series. Indicate your forecast for next week.

14. A chain of grocery stores experienced the following weekly demand (cases) for a particular brand of automatic-dishwater detergent:

Week	*Demand*
1	22
2	18
3	23
4	21
5	17
6	24
7	20
8	19
9	18
10	21

Use exponential smoothing with $\alpha = .2$ in order to develop a forecast for week 11.

15. United Dairies, Inc. supplies milk to several independent grocers throughout Dade County, Florida. Management of United Dairies would like to develop a forecast of the number of half-gallons of milk sold per week. Sales data for the past 12 weeks are as follows:

Week	*Sales (Units)*
1	2750
2	3100
3	3250
4	2800
5	2900
6	3050
7	3300
8	3100
9	2950
10	3000
11	3200
12	3150

Use the above 12 weeks of data and exponential smoothing with $\alpha = .4$ to develop a forecast of demand for the 13th week.

16. Canton Supplies, Inc. is a service firm that employs approximately 100 individuals. Because of the necessity of meeting monthly cash obligations, management of Canton Supplies would like to develop a forecast of monthly cash requirements. Owing to a recent change in operating policy, only the past 7 months of data were considered to be relevant. Use the historical data shown below to develop a forecast of cash requirements for each of the next 2 months using trend projection.

Month	1	2	3	4	5	6	7
Cash Required ($1,000's)	205	212	218	224	230	240	246

17. The Costello Music Company has been in business for 5 years. During this time the sale of electric organs has grown from 12 units in the first year to 76 units in the most recent year. Fred Costello, the firm's owner, would like to develop a forecast of organ sales for the coming year. The historical data are shown below:

Year	1	2	3	4	5
Sales	12	28	34	50	76

a. Show a graph of this time series. Does a linear trend appear to exist?
b. Develop a linear trend expression for the above time series. What is the average increase in sales that the firm has been realizing per year?

18. Hudson Marine has been an authorized dealer for C&D marine radios for the past 7 years. The number of radios sold each year is shown below:

Year	1	2	3	4	5	6	7
Number Sold	35	50	75	90	105	110	130

a. Show a graph of this time series. Does a linear trend appear to exist?
b. Develop a linear trend expression for the above time series.
c. Use the linear trend developed in part b and prepare a forecast for annual sales in year 8.

19. Refer to Exercise 18. Suppose that the quarterly sales values for the 7 years of historical data are as follows:

Year	*Quarter 1*	*Quarter 2*	*Quarter 3*	*Quarter 4*	*Total Sales*
1	6	15	10	4	35
2	10	18	15	7	50
3	14	26	23	12	75
4	19	28	25	18	90
5	22	34	28	21	105
6	24	36	30	20	110
7	28	40	35	27	130

a. Show the four-quarter moving average values for this time series. Plot both the original time series and the moving average series on the same graph.
b. Compute the seasonal factors for the four quarters.
c. When does Hudson Marine experience the largest seasonal effect? Does this seem reasonable? Explain.

20. Consider the Costello Music Company problem presented in Exercise 17. The quarterly sales data are shown below:

Year	*Quarter 1*	*Quarter 2*	*Quarter 3*	*Quarter 4*	*Total Yearly Sales*
1	4	2	1	5	12
2	6	4	4	14	28
3	10	3	5	16	34
4	12	9	7	22	50
5	18	10	13	35	76

a. Compute the seasonal factors for the four quarters.

b. When does Costello Music experience the largest seasonal effect? Does this appear reasonable? Explain.

21. Refer to the Hudson Marine data presented in Exercise 19.

a. Deseasonalize the data and use the deseasonalized time series to identify the trend.

b. Use the results of part a to develop a quarterly forecast for next year based upon trend.

c. Use the seasonal factors developed in Exercise 19 to adjust the forecasts developed in part b to account for the effect of season.

22. Consider the Costello Music Company time series presented in Exercise 20.

a. Deseasonalize the data and use the deseasonalized time series to identify the trend.

b. Use the results of part a to develop a quarterly forecast for next year based upon trend.

c. Use the seasonal factors developed in Exercise 20 to adjust the forecasts developed in part b to account for the effect of season.

Application

The Cincinnati Gas & Electric Company*

Cincinnati, Ohio

The Cincinnati Gas Light and Coke Company was chartered by the State of Ohio on April 3, 1837. Under this charter the company manufactured gas by distillation of coal and sold it for lighting purposes. During the last quarter of the 19th century the company successfully marketed gas for lighting, heating, and cooking and as fuel for gas engines.

In 1901 the Cincinnati Gas Light and Coke Company and the Cincinnati Electric Light Company merged to form the Cincinnati Gas & Electric Company (CG&E). This new company was able to shift from manufactured gas to natural gas and adopt the rapidly emerging technologies in generating and distributing electricity. CG&E operated as a subsidiary of the Columbia Gas Electric Company from 1909 until 1944.

Today CG&E is a privately owned public utility serving approximately 370,000 gas customers and 600,000 electric customers. The company's service area covers approximately 3,000 square miles in and around the Greater Cincinnati area. In 1981 the Company's revenues exceeded 1 billion dollars and its assets totaled approximately 2.5 billion dollars.

FORECASTING AT CG&E

As in any modern company, forecasting at CG&E is an integral part of operating and managing the business. Depending upon the decision to be made, the forecasting techniques used range from judgment and graphical trend projections to sophisticated multiple regression models.

Forecasting in the utility industry offers some unique perspectives as compared to other industries. Since there are no finished-goods or in-process inventories of electricity, this product must be generated to meet the instantaneous requirements of the customers. Electrical shortages are not just lost sales, but "brownouts" or "blackouts." This situation places an unusual burden on the utility forecaster. On the positive side, the demand for energy and the sale of energy is more predictable than for many other products. Also, unlike the situation in a multiproduct firm, a great amount of forecasting effort and expertise can be concentrated on the two products: gas and electricity.

*The authors are indebted to Dr. Richard Evans, The Cincinnati Gas & Electric Company, Cincinnati, Ohio, for providing this application.

FORECASTING ELECTRIC ENERGY AND PEAK LOADS

The two types of forecasts discussed in this section are the long range forecasts of electric peak load and electric energy. The largest observed electric demand for any given period, such as an hour, a day, a month, or a year, is defined as the peak load. The cumulative amount of energy generated and used over the period of an hour is referred to as electric energy.

Until the mid 1970s the seasonal pattern of both electric energy and electric peak load were very regular; the time series for both of these exhibited a fairly steady exponential growth. Business cycles had little noticeable effect on either. Perhaps the most serious shift in the behavior of these time series came from the increasing installation of air conditioning units in the Greater Cincinnati area. This fact caused an accelerated growth in the trend component and also in the relative magnitude of the summer peaks. Nevertheless, the two time series were very regular and generally quite predictable.

Trend projection was the most popular method used to forecast electric energy and electric peak load. The forecast accuracy was quite acceptable and even enviable when compared to forecast errors experienced in other industries.

A NEW ERA IN FORECASTING

In the mid-1970s a variety of actions by the government, the off-and-on energy shortages, and price signals to the consumer began to affect the consumption of electric energy. As a result the behavior of the peak load and electric energy time series became

Cincinnati Gas & Electric Company linesman works on electric high voltage transmission tower

more and more unpredictable. Hence a simple trend projection forecasting model was no longer adequate. As a result a special forecasting model—referred to as an econometric model—was developed by CG&E to better account for the behavior of these time series.

The purpose of the econometric model is to forecast the annual energy consumption by residential, commercial, and industrial classes of service. These forecasts are then used to develop forecasts of summer and winter peak loads. First, energy consumption in the industrial and commercial classes is forecast. For an assumed level of economic activity, the projection of electric energy is made along with a forecast of employment in the area. The employment forecast is converted to a forecast of adult population through the use of unemployment rates and labor force participation rates. Household forecasts are then developed through the use of demographic statistics on the average number of persons per household. The resulting forecast of households is used as an indicator of residential customers.

At this point a comparison is made with the demographic projections for the area population. The differences between the residential customers forecast and the population forecast are reconciled to produce the final forecast of residential customers. This forecast becomes the principal independent variable in forecasting residential electric energy.

Summer and winter peak loads are then forecast by applying class peak contribution factors to the energy forecasts. The contributions that each class makes toward the peak are summed to establish the peak forecast.

A number of economic and demographic time series are used in the construction of the above econometric model. Simply speaking, the entire forecasting system is a compilation of several statistically verified multiple regression equations.

IMPACT AND VALUE OF THE FORECASTS

The forecast of the annual electric peak load guides the timing decisions for constructing future generating units. The financial impact of these decisions is great. For example, the last generating unit built by the Company cost nearly 600 million dollars, and the interest rate on a recent first mortgage bond was 16%. At this rate, annual interest costs would be nearly 100 million dollars. Obviously, a timing decision which leads to having the unit available no sooner than necessary is crucial.

The energy forecasts are important in other ways also. For example, purchases of coal and nuclear fuel for the generating units are based on the forecast levels of energy needed. The revenue from the electric operations of the Company is determined from forecasted sales, which in turn enters into the planning of rate changes and external financing. These planning and decision-making processes are among the most important management activities in the company. It is imperative that the decision makers have the best forecast information available to assist them in arriving at these decisions.

17 Indexes and Index Numbers

What you will learn in this chapter:

- how index numbers help in understanding business and economic conditions
- how to compute and interpret price relatives and aggregate price indexes
- the difference between the Laspeyres and Paasche weighted aggregate price indexes
- the purpose and interpretation of the Consumer Price Index, the Producer Price Index, and the Dow Jones averages
- how to compute and interpret quantity indexes

Contents

Each month the United States Government publishes a variety of indexes which are designed to help individuals better understand current business and economic conditions. Perhaps the most widely known and cited of these indexes is the Consumer Price Index (CPI). As its name implies, the CPI is an indicator of what is happening to prices consumers are paying for items purchased. Specifically, the CPI measures changes in price over a period of time. With a given starting point or *base period* and its associated index of 100, the CPI can be used to compare current period consumer prices with those in the base period. For example, a CPI of 225 reflects the condition that consumer prices as a whole are running approximately 125% above the base period prices for the same items. Although relatively few individuals know exactly what a CPI of 225 means, they do know enough about the CPI to understand that an increase means higher prices.

The Consumer Price Index is perhaps the most widely known index. However, many other governmental and private-sector indexes are available to help us understand and measure how conditions in one period compare with conditions in other periods. The purpose of this chapter is to describe the most widely used types of indexes, specifically emphasizing how the indexes are computed and interpreted.

17.1 PRICE RELATIVES

The simplest form of a price index shows how the current price per unit for a given item compares to a base period price per unit for the same item. For example, Table 17.1 shows the cost of 1 gallon of regular gasoline for the years 1979 to 1983. To facilitate comparisons with other years the actual cost-per-gallon figure can be converted to a *price relative* index which expresses the unit price in each period as a percentage of the unit price in a base period.

TABLE 17.1 Cost per Gallon of Regular Gasoline

Year	*Cost ($)/Gallon*
1979	.93
1980	1.20
1981	1.35
1982	1.30
1983	1.15

For the gasoline prices in Table 17.1 and with 1979 as the base year, the price relatives for 1 gallon of regular gasoline in the years 1979 to 1983 can be calculated:

$$\text{Price relative in period } t = \frac{\text{Price in period } t}{\text{Base period price}}(100). \qquad (17.1)$$

These calculations are shown in Table 17.2. Note how easily the price in any one year can be compared with the price in the base year by knowing the price relative or price index. For example, the price relative of 145 in 1981 shows that the gasoline cost in 1981 was 45% above the 1979 base-year cost. Similarly, the 1983 price relative of 124 shows a 24% increase in gasoline cost in 1983 over the 1979 base-year cost.

Price relatives, such as the ones for gasoline, are extremely helpful in terms of understanding and interpreting changing economic and business conditions over time.

TABLE 17.2 Calculation of Price Relatives for the Cost of a Gallon of Gasoline for 1979–1983

Year	*Price Relatives (Base 1979 = 100)*
1979	($.93/$.93)100 = 100
1980	($1.20/$.93)100 = 129
1981	($1.35/$.93)100 = 145
1982	($1.30/$.93)100 = 140
1983	($1.15/$.93)100 = 124

The primary contribution of such index numbers is that they facilitate comparison and interpretation of price changes relative to prices of a base period.

17.2 AGGREGATE PRICE INDEXES

While price relatives can be used to identify price changes over time for individual items, we often are more interested in the general price change for a group of items taken as a whole. For example, if we want an index that measures the change in the overall cost of living over time, we will want the index to be based on the price changes for a variety of items, including food, housing, clothing, transportation, medical care, etc. An aggregate price index is developed for the specific purpose of measuring the combined change of a group of items.

Let us consider the development of an aggregate price index for a group of items falling under the heading of normal automotive operating expenses. For purposes of this illustration we limit the items included in the group to gasoline, oil, tire, and insurance expenses. We will not attempt to include other maintenance and repair expenses in this price index.

Table 17.3 provides the data for the four components of our automotive operating

TABLE 17.3 1979 and 1983 Prices for Automotive Operating Expenses

		Unit Price ($)	
Item	*Unit*	*1979* (P_0)	*1983* (P_t)
Gasoline	Gallon	.93	1.15
Oil	Quart	.95	1.75
Tires	Single radial	50.00	100.00
Insurance	Standard annual policy (adult over 25)	230.00	350.00

expense index for the years 1979 and 1983. With 1979 as the base period, an aggregate price index for the four components will give us a measure of the change in normal automotive operating expenses over the 1979–1983 period.

An unweighted aggregate index can be developed by simply summing the unit prices in the year of interest (e.g., 1983) and dividing this sum by the sum of the unit prices in the base year (1979). Let

$$P_{it} = \text{unit price for item } i \text{ in period } t,$$

$$P_{i0} = \text{unit price for item } i \text{ in the base period.}$$

An unweighted aggregate price index in period t, denoted I_t, is given by

$$I_t = \frac{\Sigma P_{it}}{\Sigma P_{i0}} (100), \tag{17.2}$$

where the sum is over all items in the group.

An unweighted aggregate index for normal automotive operating expenses in 1983 ($t = 1983$) is given by

$$I_{1983} = \frac{1.15 + 1.75 + 100.00 + 350.00}{.93 + .95 + 50.00 + 230.00} (100)$$

$$= \frac{452.90}{281.88} (100) = 161.$$

From the unweighted aggregate price index we might be tempted to conclude that the price of normal automotive operating expenses increased 61% over the period from 1979 to 1983. But, note that the unweighted aggregate approach to establishing a composite price index for automotive expenses is heavily influenced by the items with large per unit prices. Consequently, items with relatively low unit prices, such as gasoline and oil, are dominated by the high-unit-price items such as tires and insurance. In effect, then, the unweighted aggregate index for automotive operating expenses tends to measure only the price changes in the tire and insurance items.

The sensitivity of an unweighted index to one or more high priced items prevents this form of aggregate index from being widely used. A weighted aggregate price index provides a way around this difficulty.

The philosophy behind the weighted aggregate index is that each item in the group should be weighted according to its importance. In most cases the *quantity* of usage provides the best measure of importance. Thus one must obtain a measure of the quantity of usage for the various items in the group. Table 17.4 provides quantity of usage information for each item of automotive operating expense based on the assumption of operating a midsize automobile approximately 15,000 miles per year. The quantities listed show the expected annual usage for this type of driving situation.

TABLE 17.4 Quantity of Usage Information for Automotive Operating Expenses of Midsize Automobiles for 15,000 Miles Per Year

Item	*Unit*	*Annual Usage**
Gasoline	Gallon	1,000
Oil	Quart	15
Tires	Single radial	2
Insurance	Standard annual policy	1

*Tire usage was computed based on a 30,000-mile-life assumption or an average tire replacement rate of two per year.

Let Q_i = quantity for item i. The weighted aggregate price index in period t is given by

$$I_t = \frac{\Sigma P_{it} Q_i}{\Sigma P_{i0} Q_i} (100). \tag{17.3}$$

The above sums are over all items in the group.

Let $t = 1983$, and use the quantity weights in Table 17.4. We obtain a weighted aggregate price index for normal automotive operating expenses in 1983:

$$I_{1983} = \frac{1.15(1{,}000) + 1.75(15) + 100.00(2) + 350.00(1)}{.93(1{,}000) + .95(15) + 50.00(2) + 230.00(1)}(100)$$

$$= \frac{1{,}726.25}{1{,}274.25}(100) = 135.$$

From this weighted aggregate price index we would conclude that the price of normal automotive operating expenses has increased 35% over the period from 1979 to 1983.

Most individuals will agree that compared with the unweighted aggregate index the above weighted index provides a more accurate indication of the price change for automotive operating expenses over the 1979 to 1983 period. It is based on dividing total operating costs in 1983 by total operating costs in 1979. In general, the weighted aggregate index with quantities of usage as weights is the preferred method for establishing a price index for a group of items.

In the above weighted aggregate price index formula (17.3) note that the quantity term Q_i does not have a second subscript to indicate the time period. The reason for this is that the quantities Q_i are considered *fixed* and do not vary with time, as the prices do. The fixed weights or quantities are specified by the designer of the index at levels believed to be representative of typical usage. Once established, they are held constant or fixed for all periods of time the index is in use. Indexes for years other than 1983 require the gathering of new price data P_{it}, but the weighting quantities Q_i remain the same.

A special case of the fixed-weight aggregate index is when the quantities are determined from base-year usages. In this case we write $Q_i = Q_{i0}$, with the 0 subscript indicating base-year quantity weights. In this case (17.3) would become

$$I_t = \frac{\Sigma P_{it} Q_{i0}}{\Sigma P_{i0} Q_{i0}}(100). \tag{17.4}$$

Whenever the fixed quantity weights are determined from base-year usage, the weighted aggregate index is given the name *Laspeyres index*.

Another option exists for determining quantity weights. This option differs from the Laspeyres index in that the quantities are revised each period. A quantity Q_{it} has to be determined for each year that the index is computed. The weighted aggregate index in period t with these quantity weights is given by

$$I_t = \frac{\Sigma P_{it} Q_{it}}{\Sigma P_{i0} Q_{it}}(100). \tag{17.5}$$

Note that the same quantity weights are used for the base period (period 0) and for period t. However, the weights are based on usage in period t, not the base period. This weighted aggregate index is referred to as the *Paasche index*. It has the advantage of being based on current usage patterns. While use of current quantity weights has some appeal, this method of computing a weighted aggregate index has two disadvantages: the normal usage quantities Q_{it} must be redetermined each year, thus adding to the time and cost of data collection, and each year the index numbers for previous years must be recomputed to reflect the effect of the new quantity weights. Because of these disadvantages the Laspeyres index is more widely used.

EXERCISES

1. A large manufacturer purchases an identical component from three independent suppliers, each of which differs in terms of unit price and quantity supplied. The relevant data for 1981 and 1983 are shown below:

Supplier	*Quantity (1981)*	*Unit Price ($)* 1981	*Unit Price ($)* 1983
A	150	5.45	6.00
B	200	5.60	5.95
C	120	5.50	6.20

a. Compute the price relatives for each of the component suppliers separately. Compare the price increases by the various suppliers over the 2 year period.
b. Compute an unweighted aggregate price index for the component part in 1983.
c. Compute a 1983 weighted aggregate price index for the component part. What is the interpretation of this index for the manufacturing firm?

2. R&B Beverages, Inc. provides a complete line of beer, wine, and soft-drink products for distribution through retail outlets in central Iowa. Unit price data and quantities sold in cases are shown below for 1982 and 1983:

Item	*Quantity (1982) (cases)*	*Unit Prices ($)* 1982	*Unit Prices ($)* 1983
Beer	35,000	11.00	12.00
Wine	5,000	38.00	40.00
Soft Drink	60,000	6.85	7.00

Compute a weighted aggregate index for the R&B Beverage sales in 1983, with 1982 as the base period.

3. Under the LIFO inventory valuation method, a price index for inventory must be established for tax purposes. The quantity weights are based on year-ending inventory levels. Use the beginning-of-the-year price per unit as the base-period price and develop a weighted aggregate index for the total inventory value at the end of the year. What type of weighted aggregate price index must be developed for the LIFO inventory valuation?

Product	*Ending Inventory*	*Beginning Unit Price ($)*	*Ending Unit Price ($)*
A	500	.15	.19
B	50	1.60	1.80
C	100	4.50	4.20
D	40	12.00	13.20

17.3 COMPUTING AN AGGREGATE INDEX FROM PRICE RELATIVES

In Section 17.1 we defined the concept of a price relative and showed how a price relative could be computed with knowledge of the current-period unit price and the base-period unit price. We now want to show how aggregate price indexes like the ones developed in Section 17.2 can be computed directly from information about the price relative of each item in the group. Because of the limited use of unweighted indexes we restrict our attention to weighted aggregate price indexes. Let us return to the automotive operating expenses example of the preceding section. The necessary information for the four items is shown in Table 17.5.

Let w_i be the weight applied to the price relative for item i. The general expression for a weighted average of price relatives is given by

$$I_t = \frac{\sum \frac{P_{it}}{P_{i0}} w_i}{\Sigma w_i}(100). \tag{17.6}$$

The proper choice of weights in (17.6) will allow us to compute from the price relatives the weighted aggregate price indexes of the previous section. The proper choice of weights is given by multiplying the base-period price by the quantity of usage:

$$w_i = P_{i0}Q_i. \tag{17.7}$$

TABLE 17.5 Price Relatives for Automotive Operating Expense Items for 1979 and 1983

Item	*Unit Price ($)* 1979 ($P_0$)	1983 (P_t)	*Price Relative* (P_t/P_0) 100	*Annual Usage*
Gasoline	.93	1.15	124	1,000
Oil	.95	1.75	184	15
Tires	50.00	100.00	200	2
Insurance	230.00	350.00	152	1

Substitution of the value for w_i shown in (17.7) into (17.6) provides the following expression for a weighted price relatives index:

$$I_t = \frac{\sum \frac{P_{it}}{P_{i0}}(P_{i0}Q_i)(100)}{\Sigma P_{i0}Q_i}. \tag{17.8}$$

With the canceling of the P_{i0} terms in the numerator, the weighted price relatives index becomes

$$I_t = \frac{\Sigma P_{it}Q_i}{\Sigma P_{i0}Q_i}(100).$$

Thus we see that the weighted price relatives index with $w_i = P_{i0}Q_i$ provides a price index identical to the weighted aggregate index presented in Section 17.2 [see (17.2)].

With the weights given by (17.7), the weighted price relatives given by (17.8) are the same as the weighted aggregate indexes of the previous section. A choice of base-period quantities (that is, $Q_i = Q_{i0}$) in (17.7) leads to a Laspeyres index. A choice of current-period quantities (that is, $Q_i = Q_{it}$) in (17.7) leads to a Paasche index.

Let us return to our automotive operating expenses data in Table 17.5. We can use (17.6) to compute a weighted average of price relatives. The results obtained using the weights specified by (17.7) are presented in Table 17.6. The index number of 136 shows a 36% increase in automotive operating expenses. Except for rounding differ-

TABLE 17.6 Automotive Operating Expenses Index (1979–1983) Based on Weighted Price Relatives

Item	*Price Relatives*	*Base Price ($)*	*Quantity* Q_i	*Weight* $w_i = P_{i0}Q_j$	*Weighted Price Relatives* $(P_{it}/P_{i0})(100)w_i$
Gasoline	124	.93	1,000	930.00	115,320
Oil	184	.95	15	14.25	2,622
Tires	200	50.00	2	100.00	20,000
Insurance	152	230.00	1	230.00	34,960
			Totals	1,274.25	172,902

$$I_{1983} = \frac{172,902}{1,274.25} = 136$$

ences, this is the same as the increase identified by the weighted aggregate index computation in Section 17.2.

EXERCISES

4. The Mitchell Chemical Company produces a special industrial chemical that is a blend of three chemical ingredients. The beginning-year cost per pound, the ending-year cost per pound, and the blend proportions are shown below:

Ingredient	*Beginning Cost/Pound ($)*	*Ending Cost/Pound ($)*	*Quantity (pounds) per 100 Pounds of Product*
A	2.50	3.95	25
B	8.75	9.90	15
C	.99	.95	60

a. Compute the price relatives for the three ingredients.
b. Compute a weighted average of the price relatives in order to develop a 1 year price index for the product. What is your interpretation of this index value?

5. An investment portfolio consists of four stocks. The number of shares, purchase price, and current price are shown below:

Stock	*Purchase Price/Share ($)*	*Current Price/Share ($)*	*Number of Shares*
Holiday Trans	15.50	17.00	500
NY Electric	18.50	20.25	200
KY Gas	26.75	26.00	500
PQ Soaps	42.25	45.50	300

Construct a weighted average of price relatives as an index of the performance of the portfolio to date. Interpret this price index.

6. Compute the price relatives for the R&B Beverage Company products in Exercise 2. Use a weighted average of price relatives to show that this method provides the same index as the weighted aggregate method.

17.4 SOME IMPORTANT PRICE INDEXES

We have identified the procedures used to compute price indexes for single items or groups of items. Now let us consider some price indexes that are important measures of business and economic conditions. Specifically, we will consider the Consumer Price Index, the Producer Price Index, and the Dow Jones averages.

Consumer Price Index

Perhaps the most widely known and cited measure of change in general economic conditions is the Consumer Price Index (CPI). This index, published monthly by the U.S. Bureau of Labor Statistics, is the primary measure of the cost of living in the United States. The group of items used to develop the index consists of a *market basket* of 400 items including food, housing, clothing, transportation, and medical items. The CPI is a weighted aggregate index with fixed weights.* The weight applied to each item in the market basket derives from a usage survey of urban families throughout the United States.

In May, 1983 the CPI, computed with a 1967 base index of 100, was 297.1. This means that the cost of purchasing the market basket of goods and services had increased 197.1% since 1967. The 30-year time series of the CPI from 1950 to 1980 is shown in Figure 17.1. Note how the CPI measure reflects the sharp inflationary behavior of the economy in the late 1970s.

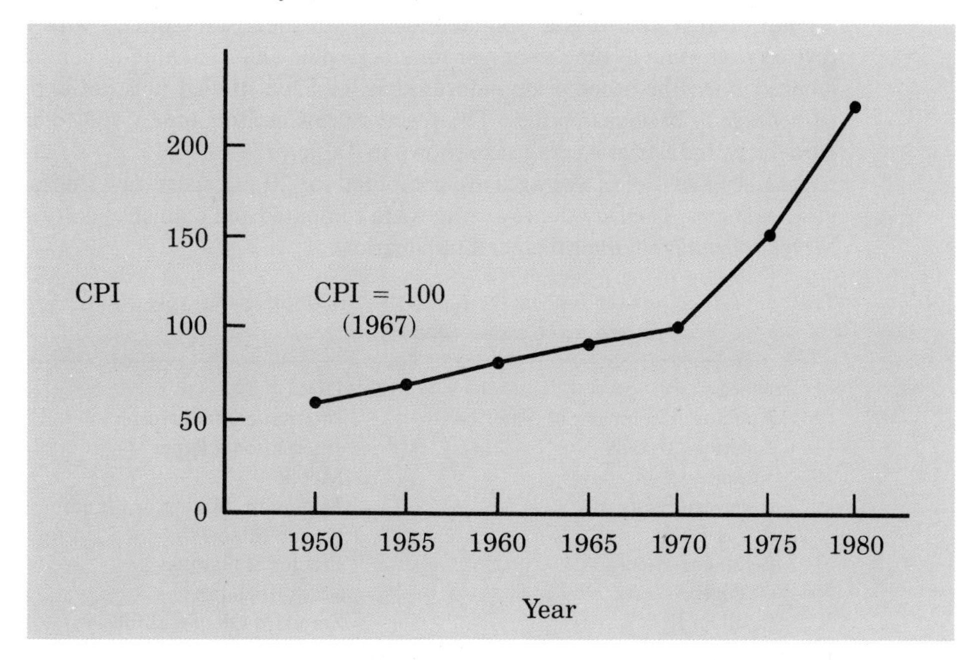

FIGURE 17.1 Consumer Price Index, 1950 to 1980, with Base Year 1967

*There are actually two Consumer Price Indexes. The Bureau of Labor Statistics publishes a Consumer Price Index for all urban consumers (CPI-U) and a revised Consumer Price Index for urban wage earners and clerical workers (CPI-W). The CPI-U is the one most widely quoted, and it is published regularly in *The Wall Street Journal*.

Producer Price Index

The Producer Price Index (PPI), also published monthly by the U.S. Bureau of Labor Statistics, measures the monthly changes in prices in primary markets in the United States. This index replaces the old Wholesale Price Index. The index is based on prices for "the first transaction of each product in nonretail markets. All commodities sold in commercial transactions in these markets are represented, including those imported for sale. The survey includes crude, manufactured, and processed goods at each level of processing and includes the output of industries classified as manufacturing, agriculture, forestry, fishing, mining, gas and electricity, and public utilities."* One of the common uses of this index is as a leading indicator of the future trend of consumer prices and the cost of living. An increase in the PPI reflects producer price increases that will eventually be passed on to the consumer through higher retail prices.

Weights for the various items in the index are based on the value of shipments. The index is a weighted average of price relatives calculated by the Laspeyres method. In May, 1983 the PPI, computed with a 1967 base index of 100, was 284.3.

Dow Jones Average

The Dow Jones averages are indexes which are designed to show price trends and movements on the New York Stock Exchange. The best known of the Dow Jones indexes is the Dow Jones Industrial Average, which is based on common stock prices of 30 industrial stocks. It is a weighted average of these stock prices, with the weights revised from time to time to adjust for stock splits and switching of companies in the index. Unlike the other price indexes that we have studied it is not expressed as a percentage of base-year prices. The specific firms used in June, 1983 to compute the Dow Jones Industrial Average are shown in Table 17.7.

Other Dow Jones averages are computed for 20 transportation stocks and for 15 utilities stocks. The Dow Jones averages are computed and published daily in *The Wall Street Journal* and other financial publications.

TABLE 17.7 Thirty Industrial Common Stocks Used in the Dow Jones Industrial Average Price Index (June 1983)

Allied	IBM
Aluminum Company of America	International Harvester
American Brands	International Paper
American Can	Merck
American Express	Minnesota Mining & Manufacturing
A.T.&T.	Owens-Illinois
Bethlehem Steel	Procter & Gamble
DuPont	Sears Roebuck
Eastman Kodak	Standard Oil of California
Exxon	Texaco
General Electric	Union Carbide
General Foods	United Technologies
General Motors	US Steel
Goodyear	Westinghouse Electric
Inco	Woolworth

*"Brief Description of Series," *Chartbook on Prices, Wages, and Productivity,* p. 2, September 1979.

17.5 DEFLATING A SERIES BY PRICE INDEXES

Many business and economic series reported over time, such as company sales, industry sales, and inventories, are measured in dollar amounts. These time series often show an increasing growth pattern over time which is generally interpreted as showing an increase in the physical volume associated with these activities. For example, a total dollar amount of inventory up by 10% is generally interpreted to mean that the physical inventory is 10% larger. Such interpretations can be very misleading whenever a time series is measured in terms of dollars, since the total dollar amount is a combination of both price and quantity change. Thus in periods where price changes are significant the changes in the dollar amounts may be very misleading unless we are able to adjust the time series to eliminate the price change effect.

For example, from 1976 to 1980 the total amount of spending in the construction industry increased approximately 75%. At first glance this figure suggests an excellent growth in construction activity. However, during this period of time construction prices were increasing just as fast as—or sometimes even faster than—this 75% rate. In fact, while total construction spending was increasing, construction activity was staying relatively constant or, as in the case of new housing starts, showing a decrease. Thus in order for us to correctly interpret construction spending activity over the 1976–1980 period it is necessary to adjust the total spending series by a price index in order to remove the price-increase effect from the time series. Whenever we remove the price-increase effect from the time series, we say we are *deflating the series.*

In the area of personal income and wages we often hear discussions concerning issues such as "real wages" or the "purchasing power" of wages. These concepts are based on the notion of deflating an hourly wage index. For example, Figure 17.2 shows the pattern of hourly wages of factory workers for the period 1976–1980. At first glance, we see the sharply increasing trend in wages, with excellent growth in wages from approximately \$4.90/hour to \$7.00/hour. Should the factory workers be pleased

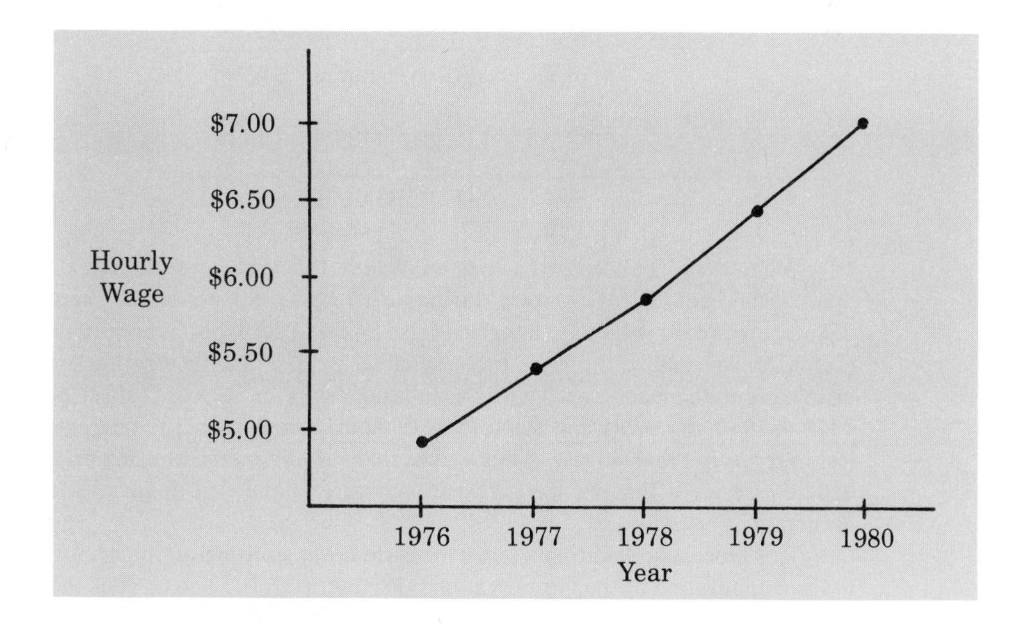

FIGURE 17.2 Hourly Wages of Factory Workers

with this growth in hourly wages? Perhaps yes; but on the other hand, if the cost of living has increased just as fast, maybe the answer is no. If we can compare the purchasing power of the $7.00/hour wage in 1980 with the purchasing power of the $4.90 wage in 1976, we will have a better idea of the relative improvement in wages.

Table 17.8 shows both the hourly wage rate and Consumer Price Index for the period 1976–1980. Note that we are here using 1976 as the base for the CPI. With these data we will show how the CPI can be used to deflate the index of hourly wages. In effect we shall be removing the consumer price increases from the hourly wage index in an attempt to measure the change in purchasing power of the wages. Thus we will be better able to determine what has happened to "real wages" over the time span.

TABLE 17.8 Hourly Wages of Factory Workers and Consumer Price Index, 1976–1980

Year	*Hourly Wage ($)*	*CPI (1976 Base)*
1976	4.90	100
1977	5.40	105
1978	5.85	113
1979	6.40	122
1980	7.00	138

The calculations used to deflate the hourly wage index are not difficult. The deflated series is found by dividing the hourly wage rate in each year by the corresponding value of the CPI. The deflated hourly wage index for factory workers is shown in Table 17.9. A graph showing both the actual wage rates and the deflated or "real wages" is shown in Figure 17.3.

TABLE 17.9 Deflated Series of Hourly Wages for Factory Workers

Year	*Deflated Hourly Wage*
1976	($4.90/100)(100) = $4.90
1977	($5.40/105)(100) = $5.14
1978	($5.85/113)(100) = $5.18
1979	($6.40/122)(100) = $5.25
1980	($7.00/138)(100) = $5.07

What does the deflated series of wages tell us about the "real wages" or "purchasing power" of workers during the 1976–1980 period? In terms of 1976 dollars, the hourly wage rate has risen from $4.90 to $5.07, or approximately 3.5%. In fact, after we remove the price-increase effect we see that factory workers are doing little more than keeping even with the inflationary price increases of the period. From 1979 to 1980, even with the $6.40 to $7.00 hourly wage increase, the factory workers lost in terms of "real wages." Thus we see that the advantage of using price indexes to deflate a series is that we have a clearer picture of the real dollar changes that are occurring.

This process of deflating a series measured over time has an important application in the computation of the Gross National Product (GNP). The GNP is the total value of all goods and services produced in a given country. Obviously, over time the GNP will show gains which are in part due to price increases if the GNP is not deflated by a price index. Thus in order to adjust the total value of goods and services to reflect

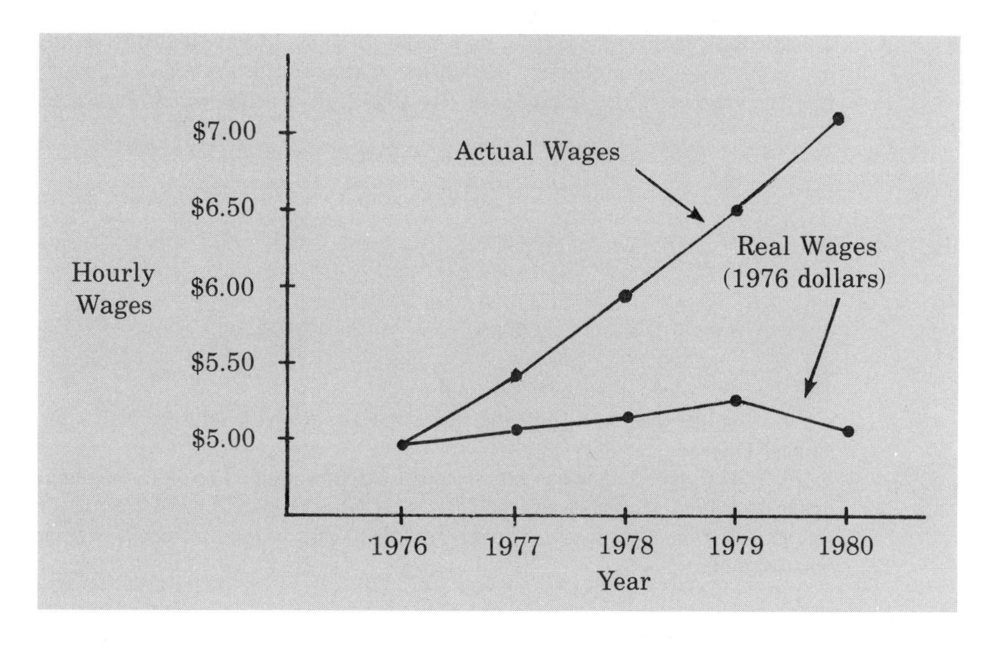

FIGURE 17.3 Actual and 1976 Constant-Dollar "Real Wages" of Factory Workers

actual changes in the volume of goods and services produced and sold, the GNP must be computed with a price index deflator. The process is similar to that previously discussed in the "real wages" computation.

EXERCISES

7. The United States Department of Commerce reported total personal income for the 5 years 1976 to 1980 as follows:

Year	*Total Personal Income (billions of dollars)*
1976	1,300
1977	1,440
1978	1,605
1979	1,800
1980	2,050

Use the Consumer Price Index information in Table 17.8 to deflate the personal income series. What has been the percent increase in "real personal income" from 1978 to 1980?

8. The United States Department of Commerce reported total inventories of all manufacturers for the 5 years 1976 to 1980 as follows:

Year	*Total Inventories (billions of dollars)*
1976	155
1977	163
1978	178
1979	198
1980	227

a. Use the Consumer Price Index information in Table 17.8 to deflate this series, and comment on the pattern of manufacturers' inventories in terms of constant dollars.
b. The Producers Price Index for 1976 to 1980 is given below, with 1976 as a base year:

Year	*PPI (1976 base)*
1976	100
1977	103
1978	110
1979	120
1980	135

Use the PPI to deflate the inventories series.
c. Do you feel that the CPI or the PPI is most appropriate to use as a deflator for inventory values? Discuss.
9. Dooley Retail Outlets has experienced the following total retail sales volumes for selected years since 1965. Also shown is the CPI with the index base of 1967. Deflate the sales volume figures based on 1967 constant dollars and comment on the firm's sales volumes in terms of deflated dollars.

Year	*Retail Sales ($)*	*CPI (1967 base)*
1965	380,000	95
1970	520,000	110
1975	700,000	156
1980	870,000	229

17.6 PRICE INDEXES: OTHER CONSIDERATIONS

In the previous sections we described several methods used to compute price indexes, we discussed the use of some important indexes, and we presented a procedure for using price indexes to deflate a time series. There are several other issues that must be considered in order to enhance our understanding of how price indexes are constructed and how they are used. Some of these considerations are discussed in this section.

Selection of Items

The primary purpose of a price index is to measure the price changes over time for a specified class of items, products, etc. Whenever the class in question has an extremely large number of items, it is clear that the index cannot be based on all items in the class. Rather a sample of representative items must be used. By collecting price and quantity information for the selected items we hope to obtain a good idea of the price behavior of all items that the index is representing. For example, in the Consumer Price Index the total number of items that might be considered in the population of normal purchase items for a consumer could run as high as 2,000 or more. However, the index is based on the price-quantity characteristics of just 400 items. The selection of the specific items in the index is not a trivial question. Surveys of user purchase patterns as well as good judgment go into the selection process. A simple random sample of all potential items for the index is not used to select the 400 items.

In addition to the initial selection process, the group of items in the index must be periodically reviewed and revised whenever actual purchase patterns change. Thus the issue of which items to include in an index is a key question that must be resolved before an index can be developed or revised.

Selection of a Base Period

Many indexes are established with a base-period value of 100 at some specific time. All future values of the index are then related to the base period. But what base period is appropriate for an index? This is not an easy question, and the answer must be based on the judgment of the developer of the index.

Many of the indexes established by the United States government as of 1983 use a 1967 base period. As a general guideline it is believed that the base period should not be too far from the current period. For example, a Consumer Price Index that used a 1945 base period would be difficult for most individuals to relate to because of unfamiliarity with conditions in 1945. Thus the base period for most indexes is adjusted periodically to a more recent period of time. For the CPI it is anticipated that the base period will be moved up into the 1970s in the next few years.

Quality Changes

The purpose of a price index is to attempt to measure changes in prices over time. Ideally, price data are collected for the same set of items at several times, and then the index is computed. A basic assumption is that the prices are identified for the same items each period. A problem is encountered when a product changes in quality from one period to the next. For example, a manufacturer may alter the quality of a product by using less expensive materials, fewer features, and so on from year to year. While the price may go up in following years, the price is for a lower quality product. In some instances this means that the price has actually gone up more than is represented by the list price for the item. However, it is very difficult, if not impossible, to adjust the index for decreases in the quality of an item.

On the other hand, a substantial quality improvement may cause an increase in price for basically the same product. Thus a portion of the price related to the quality improvement should be excluded from the index computation. Again, however, it is extremely difficult, if not impossible, to adjust the index for the price increase that is related to the higher quality factor.

While quality changes cause some concern in the development of price indexes, common practice is simply to ignore minor quality changes in developing a price index. Major quality changes must be handled on the basis that major changes in effect change the product description from period to period. If a product description is changed the index must be modified to account for it. For example, the product could be deleted from the index.

17.7 QUANTITY INDEXES

Although the previous sections have emphasized the important area of price indexes, there are other types of indexes. In particular, one other use of index numbers is to measure changes in quantity levels over time.

Recall that in the development of the weighted aggregate price index in Section 17.2 to compute an index number for period t we required data on unit prices at a base period (P_0) and period t (P_t). The formula for a weighted aggregate price index is restated below:

$$I_t = \frac{\Sigma P_{it} Q_i}{\Sigma P_{i0} Q_i}(100). \tag{17.3}$$

The numerator, $\Sigma P_{it} Q_i$, represents the total value of fixed quantities of the index items in period t. The denominator, $\Sigma P_{i0} Q_i$, represents the total value of the same fixed quantities of the index items in year 0.

The weighted aggregate quantity index is computed in a fashion quite similar to that for a weighted aggregate price index. Quantities for each item are measured in the base period and period t, with Q_{i0} and Q_{it}, respectively, representing these quantities for item i. The quantities are then weighted by a fixed price, the value added, etc. Note that the "value added" to a product is the sales value minus the cost of purchased inputs. The formula for computing a weighted aggregate quantity index for period t is

$$I_t = \frac{\Sigma Q_{it} w_i}{\Sigma Q_{i0} w_i}(100). \tag{17.9}$$

In some quantity indexes the weight for item i is taken to be the base-period price (P_{i0}), in which case the weighted aggregate quantity index becomes

$$I_t = \frac{\Sigma Q_{it} P_{i0}}{\Sigma Q_{i0} P_{i0}}(100). \tag{17.10}$$

Quantity indexes can also be computed on the basis of weighted quantity relatives. One formula for this version of a quantity index is as follows:

$$I_t = \frac{\sum \frac{Q_{it}}{Q_{i0}} (Q_{i0} P_i)(100)}{\Sigma Q_{i0} P_i}. \tag{17.11}$$

This formula is the quantity version of the weighted price relatives as developed in Section 17.3 [see (17.8)].

The Index of Industrial Production, developed by the Federal Reserve Board, is probably the best known quantity index. This index is reported monthly and has a base period of 1967. It is designed to measure changes in volume of production levels for a variety of manufacturing classifications in addition to mining and utilities. In May, 1983 this index was reported in *The Wall Street Journal* at 144.3

EXERCISES

10. A trucking firm handles four commodities for a particular distributor. Total shipments for the commodities in 1979 and 1983, as well as the 1979 prices, are shown below:

Commodity	*Shipments 1979*	*Shipments 1983*	*Price/Shipment 1979*
A	120	95	$1,200
B	86	75	$1,800
C	35	50	$2,000
D	60	70	$1,500

Develop a weighted aggregate quantity index with a 1979 base. Use base-period prices for weights. Comment on the growth or decline in quantities over this period.

11. An automobile dealer reports the following 1981 and 1983 sales for three models:

Model	*1981 Sales*	*1983 Sales*	*Mean Price per Sale (1983)*
Sedan	200	170	$6,200
Sport	100	80	$7,000
Wagon	75	60	$5,800

Develop a weighted aggregate quantity index using the 2 years of data.

Summary

Price and quantity indexes are important measures of changes in price and quantity levels within the business and economic environment. Price relatives are simply the ratio of the current unit price of an item to a base-period unit price, with a value of 100 indicating no difference in the current- and base-period prices. Aggregate price indexes are created as a composite measure of the overall change in prices for a given group of items or products. Usually the items in an aggregate price index are weighted by their quantity of usage. A weighted aggregate price index can also be computed by weighting the price relatives for the items in the index.

The Consumer Price Index and the Producer Price Index are both widely quoted indexes using 1967 as a base year. The Dow Jones Industrial Average is another widely quoted price index. It is a weighted average of the prices of 30 common stocks listed on the New York Stock Exchange. Unlike many other indexes, it is not stated as a percentage of some base-period value.

Often price indexes are used to deflate some other economic series reported over time. We saw how the CPI could be used to deflate hourly wages to obtain an index of "real wages." Issues such as selecting the items to be included in the index, selecting a base period for the index, and adjusting for changes in quality are important additional considerations in the development of an index number. Quantity indexes were briefly discussed, and the Index of Industrial Production was mentioned as an important quantity index.

Glossary

Price relative—A price index for a given item which is computed by dividing a current unit price by a base-period unit price and multiplying the result by 100.

Aggregate price index—A composite price index based on the prices of a group of items.

Weighted aggregate price index—A composite price index where the prices of the items in the composite are weighted by their relative importance.

Laspeyres index—A weighted aggregate price index where the weight for each item is its base-period quantity.

Paasche index—A weighted aggregate price index where the weight for each item is its current-period quantity.

Consumer Price Index—A monthly price index that uses the price changes in a market basket of consumer goods and services to measure the changes in consumer prices over time.

Producer Price Index—A monthly price index that is designed to measure changes in prices for goods sold in primary markets (i.e., first purchase of a commodity in nonretail markets).

Dow Jones averages—Aggregate price indexes reflecting the prices of stocks listed on the New York Stock Exchange.
Quantity index—An index that is designed to measure changes in quantities over time.
Index of Industrial Production—A quantity index which is designed to measure changes in the physical volume or production levels of industrial goods over time.

Key Formulas

Price Relative in Period *t*

$$\frac{\text{Price in Period } t}{\text{Base Period Price}}(100) \tag{17.1}$$

Unweighted Aggregate Price Index in Period *t*

$$I_t = \frac{\Sigma P_{it}}{\Sigma P_{i0}}(100) \tag{17.2}$$

Weighted Aggregate Price Index in Period *t*

$$I_t = \frac{\Sigma P_{it}Q_i}{\Sigma P_{i0}Q_i}(100) \tag{17.3}$$

Weighted Average of Price Relatives

$$I_t = \frac{\sum \frac{P_{it}}{P_{i0}} w_i}{\Sigma w_i}(100) \tag{17.6}$$

Weighting Factor for (17.6)

$$w_i = P_{i0}Q_i \tag{17.7}$$

Weighted Aggregate Quantity Index for Period *t*

$$I_t = \frac{\Sigma Q_{it}w_i}{\Sigma Q_{i0}w_i}(100) \tag{17.9}$$

Quantity Version of the Weighted Price Relatives

$$I_t = \frac{\sum \frac{Q_{it}}{Q_{i0}}(Q_{i0}P_i)(100)}{\Sigma Q_{i0}P_i} \tag{17.11}$$

Supplementary Exercises

12. Nickerson Manufacturing Company shows the following data on units shipped and quantities shipped for each of its four products:

Products	*Base-Period Quantities (1980)*	*Mean Shipping Cost per Unit ($)* 1980	1984
A	2,000	10.50	15.90
B	5,000	16.25	32.00
C	6,500	12.20	17.40
D	2,500	20.00	35.50

a. Compute the price relative for each product.
b. Compute a weighted aggregate price index that reflects the shipping cost change over the 4 year period.

Use the price data in Exercise 12 to compute a Paasche index for the shipping cost if 1984 quantities are 4,000, 3,000, 7,500, and 3,000 for each of the four products.

Boran Stockbrokers, Inc. selects four stocks for the purpose of developing its own index of stock market behavior. Costs per share for a 1981 base period, January 1983, and March 1983 are shown below. Base-year quantities have been set based on historical volumes for the four stocks.

			Cost per Share ($)		
Stock	*Industry*	*1981 Quantity*	*1981 Base*	*January 1983*	*March 1983*
A	Oil	100	31.50	32.75	32.50
B	Computer	150	65.00	59.00	57.50
C	Steel	75	40.00	42.00	39.50
D	Real Estate	50	18.00	16.50	13.75

Use the 1981 base period to compute the Boran index for January 1983 and March 1983. Comment on what the index tells you about what is happening in the stock market.

Compute the price relatives for the four stocks making up the Boran index in Exercise 14. Use the weighted aggregates of price relatives for the January 1983 and March 1983 Boran indexes.

Consider the price relatives and quantity information for the following grain production in Iowa:

Product	*1978 Quantities (millions of bushels)*	*Base Price per Unit ($)*	*1978 to 1980 Price Relatives*
Corn	500	2.50	103
Soybeans	300	6.25	100
Oats	100	3.00	98

What is the weighted aggregates price index for the Iowa grains?

17. Dairy product price and quantity data for the years 1970 and 1980 are shown below. Quantities are based on estimated annual usage for a family of two adults and two children.

Product	*1970 Quantities*	*1970 Price ($)*	*1980 Price ($)*
Milk (gallons)	125	.79	2.09
Eggs (dozens)	50	.49	.95
Butter (pounds)	50	.60	.91
Cheese (pounds)	25	.79	1.49

a. Compute the price relative for each product.
b. Compute a weighted aggregate price index for dairy products.

18. Starting faculty salaries (9 month basis) for assistant professors of business administration at a major Midwestern university are shown below:

Year	*Starting Salary ($)*	*CPI (1967 Base)*
1967	11,500	100
1970	14,000	110
1975	17,500	156
1980	23,000	229
1983	32,000	297.1

Use the CPI to deflate the salary data to 1967 constant dollars. Comment on the trend in salaries in higher education as indicated by these data.

19. A particular stock shows the following 5 year historical price per share (also shown is the Consumer Price Index with a 1976 base period):

Year	*Price per Share ($)*	*CPI (1976 Base)*
1976	51.00	100
1977	54.00	105
1978	58.00	113
1979	59.50	122
1980	59.00	138

Deflate the stock price series and comment on the investment aspects of this stock. What would the 1980 price per share be for this stock if it had kept pace with the cost of living as indicated by the Consumer Price Index?

20. A major manufacturing company has the following quantity and product value information for 1980 and 1984:

Product	*1980 Quantities*	*1984 Quantities*	*Value ($)*
A	800	1,200	30.00
B	600	500	20.00
C	200	500	25.00

Compute a weighted aggregate quantity index for the data. Comment on what this quantity index means.

of Labor Statistics Index of Wholesale Prices, and corrected for differences in employment trends between the state and the nation.

Construction. McGraw-Hill Information Systems Company's reports on construction contracts in New York State are subdivided into residential buildings, nonresidential buildings, and public works and utilities. The first two are taken on a square-foot basis, the third on dollars corrected for price by Engineering News Record's Construction Cost Index. Relative weights are assigned the first two based on differences in cost per square foot (averaged over 1967 to 1971), and final weights for the buildings total and public utilities are based on 1967 contract values. After weighting, price correction and summation, the results are converted into work completed (since contracts are only the intention to build) by spreading each month's contracts over the next 8 months in proportions based on seasonality of construction labor-hours. A time period of 8 months allows the trend for contracts to most resemble that for employment. Exceptions are made for very large contracts, which are separately spread over the estimated completion time of each. Adjustments are made for strikes or extreme weather.

Transportation, Communication, Public Utilities. This index is the sum of a number of segments, weighted by employed persons (census) in 1960 adjusted for employment trends to 1967 levels, which each segment indexed (1967 = 100) before weighting and combining.

1. Transportation: Motor transportation, railroad transportation, waterborne transportation, and air transportation make up the transportation segment of this index; each part is treated separately and weighted by estimated 1967 employment levels.
2. Communication: Commercial telephone stations in use in the New York Telephone Company system (which covers all but a few sections of the state) is taken as representing all communication industries.
3. Public Utilities: The weight for utilities is divided between electricity and gas in proportion to their respective employment in 1967 using these indicators: (1) production of electric energy in New York State from the United States Power Commission (kilowatts) and (2) gas sales in New York State from the American Gas Institute (therms).

Finance, Insurance, Real Estate. The trends for each of these three sections of the industry are computed separately and combined with weights based on employed persons in 1960 (census) adjusted for the employment trend to 1967 levels. Each part is indexed (1967 = 100) before weighting and combining.

1. Finance: This is subdivided into two parts, with weights based on the estimated number of employed persons in each in 1967. The Securities Markets part is measured by the market value of stocks and bonds traded on the New York Stock Exchange. The Financial Institutions part is measured by total deposits held in savings banks, savings and loan associations, and commercial banks in New York State.
2. Insurance: The total of premium payments to insurance companies licensed to do business in New York State (life, including accident and health; nonprofit hospital and medical; fire and marine; casualty; title insurance; fraternal benefit societies; and cooperative fire), about 85 percent of the national total, given annual U.S.

Department of Commerce indexes reflect business activity on Wall Street

benchmark data. These are from the annual reports of the New York State Insurance Department. Final estimates for months between benchmarks are straight line interpolations.

3. Real Estate: Annual benchmark data are estimates of occupied square feet of office and loft space, weighted by the relative assessed valuation per square foot. Totals of office space reported for Manhattan by the Real Estate Board of New York, plus an estimate of loft space based on projections of totals reported for earlier years, are inflated to a New York City total by ratios determined from assessed valuation of all space in each class, whether occupied or not. These totals are further inflated to a state estimate by the ratio of employment in real estate in the state to that in New York City.

VALUE TO NEW YORK STATE

The components of the Business Activity Index have been described in some detail to provide you with an appreciation for the application of practical statistical methods compatible with the availability and limitation of economic data to measure changes in regional economic activity.

The Business Activity Index is one of the principal gauges used to appraise the monthly business trends throughout New York and to assess the annual state of the State's economy. This type of statistical index assists in the determination of the need for additional development strategies and programs to improve the economic well-being of the state's population.

18 Nonparametric Methods

What you will learn in this chapter:

- the difference between parametric and nonparametric methods
- the advantages of nonparametric methods
- when nonparametric methods are applicable
- how to use the Wilcoxon and Mann-Whitney tests for differences between two populations
- how to use and interpret the sign test
- how to compute the Spearman rank correlation coefficient

Contents

When the methods of statistical inference are based upon the assumption that the population has a certain probability distribution, such as the normal, the resulting collection of statistical tests and procedures is referred to as *parametric methods*. In these cases we are usually concerned with making inferences about one or more of the assumed probability distribution's parameters. For example, the t distribution used in Chapters 8, 9, 10, 14, and 15, the χ^2 distribution used in Chapters 11 and 12, and the F distribution used in Chapters 11, 13, 14, and 15 are all concerned with the values of the parameters of an assumed normal probability distribution.

In this chapter we introduce several statistical procedures that do not require knowledge of the form of the probability distribution from which the measurements come. The methods of statistical inference we shall study here are called *nonparametric methods*. Since nonparametric methods do not require assumptions about the form of the population probability distribution they are often referred to as *distribution-free methods*.

From this discussion we see that one reason for using nonparametric methods is that in some situations there is insufficient knowledge about the form of the population distribution. Thus the assumptions necessary for use of parametric tests cannot be made.

A second reason for using nonparametric methods concerns data measurement. Nonparametric methods are often applied to rank order or preference data. For instance, a ranking of ten individuals from best to worst according to sales potential is an example of rank order data. Such data differ from the continuous data that we are more familiar with in the sense that the usual numerical measures (e.g., mean, standard deviation, etc.) are not applicable. Hence parametric procedures cannot be used with rank order data, whereas nonparametric procedures can be. Preference data are the type of data generated when people express preference for one product over another, one service over another, etc. Parametric procedures cannot be applied with these data, but nonparametric ones can.

In this chapter we present an introduction to some of the commonly used statistical procedures that can be classified as nonparametric or distribution-free methods. The emphasis will be on the type of problems that can be solved, how the statistical calculations are made, and how appropriate conclusions can be developed to assist management in the decision-making process.

18.1 WILCOXON RANK-SUM TEST

In this section we present a nonparametric test to determine if there are any differences between two populations. The nonparametric test is based upon independent random samples from each population. Recall that in Chapter 10 we conducted a parametric test to determine if there was any difference between the means of two populations. The hypotheses tested were

$$H_0: \mu_1 = \mu_2,$$

$$H_1: \mu_1 \neq \mu_2.$$

In the small-sample case, the parametric method used to test these hypotheses is based on two assumptions:

1. Both populations are normally distributed.
2. The population variances are equal.

The nonparametric method we will use here does not require either of these assumptions.

A nonparametric alternative for analyzing data based upon two independent samples was developed by Wilcoxon and is referred to as the *Wilcoxon rank-sum test.* The only assumption made about the population is that the data values are continuous; thus when we compare them for purposes of ranking, there will be few ties.* Instead of testing for the differences in the two populations means, the Wilcoxon rank-sum procedure tests whether or not the two samples come from identical populations. The hypotheses for the Wilcoxon rank-sum test are given below:

H_0: The two populations are identical,

H_1: The two populations are not identical.

We shall demonstrate the methodology of the Wilcoxon rank-sum test by using it to conduct a test on the populations of account balances at two branches of the Third National Bank. Data collected from two independent simple random samples, one from each branch, are shown in Table 18.1.

The first step in the Wilcoxon rank-sum procedure is to rank the *combined* data from the two samples from low to high. Using the combined set of 22 observations shown in Table 18.1, the lowest value of $750 (item 6 of sample 2) is ranked number 1. Continuing the ranking, we have

Account Balance	*Item*	*Rank*
750	6 of sample 2	1
800	5 of sample 2	2
805	7 of sample 1	3
850	2 of sample 2	4
.	.	.
.	.	.
.	.	.
1,195	4 of sample 1	21
1,200	3 of sample 1	22

Item 6 of sample 1 and item 4 of sample 2 both have the same account balance, $950. We could give one of these items a rank of 12 and the other a rank of 13, but this could lead to an erroneous conclusion. In order to avoid this difficulty the usual treatment for tied data values is to assign each value the rank equal to the average of the ranks associated with the tied items. Thus the tied observations of $950 are both assigned ranks of 12.5. Table 18.2 shows the entire data set with the rank of each observation.

The next step in the Wilcoxon rank-sum test is to sum the ranks for each sample. These sums are shown in Table 18.2. The test procedure can be based upon the sum of the ranks for either sample. In the following discussion we use the sum of the ranks for the sample from branch 1. We will denote this sum by the symbol W. Thus in our example $W = 169.5$.

Let us investigate for a moment the properties of this sum. First of all, what is the smallest value that W could assume in this example? Clearly, the smallest value for W

*The purpose of this assumption is to make ties theoretically impossible. In practice, however, we will usually have a few ties because of rounding or the fact that the continuity assumption is not strictly satisfied.

TABLE 18.1 Account Balances for Two Branches of the Third National Bank

Branch 1		*Branch 2*	
Sampled Account	*Account Balance ($)*	*Sampled Account*	*Account Balance ($)*
1	1,095	1	885
2	955	2	850
3	1,200	3	915
4	1,195	4	950
5	925	5	800
6	950	6	750
7	805	7	865
8	945	8	1,000
9	875	9	1,050
10	1,055	10	935
11	1,025		
12	975		

corresponds to the case where the 12 sample accounts from branch 1 have ranks 1, 2, 3, . . . , 12. The sum of the ranks in this case is $1 + 2 + \cdots + 12 = 78$. Thus the smallest possible value for W is 78.

Next, let us consider the maximum possible value for W. Clearly, the maximum value for W occurs if the ranks of the 12 sampled items are 11, 12, . . . , 22. Note that this occurs if the ranks for the ten items in sample 2 are 1, 2, . . . , 10. Thus the maximum value for W is equal to $11 + 12 + \cdots + 22 = 198$. What we have shown, then, is that for this example W must take on a value between 78 and 198.

Note that values of W close to 78 imply that branch 1 has smaller account balances. Values of W close to 198 imply that branch 1 has higher account balances. Thus if the null hypothesis of two identical populations is true we would expect the value of W to be somewhere close to the average of the above two values, or (78 +

TABLE 18.2 Combined Ranking of the Data in the Two Samples from the Third National Bank

Branch 1			*Branch 2*		
Sampled Account	*Account Balance ($)*	*Rank*	*Sampled Account*	*Account Balance ($)*	*Rank*
1	1,095	20	1	885	7
2	955	14	2	850	4
3	1,200	22	3	915	8
4	1,195	21	4	950	12.5
5	925	9	5	800	2
6	950	12.5	6	750	1
7	805	3	7	865	5
8	945	11	8	1,000	16
9	875	6	9	1,050	18
10	1,055	19	10	935	10
11	1,025	17		Sum of ranks	83.5
12	975	15			
	Sum of ranks	169.5			

198)/2 = 138. Using this insight and the fact that the actual sum of the ranks for sample 1 is $W = 169.5$, we might be tempted to conclude that branch 1 is carrying the higher account balances. However, when we realize that the data shown are for relatively small sample sizes, we may not be ready to draw a general conclusion about the differences in the two populations. The Wilcoxon rank-sum test provides a statistical procedure for determining if there is sufficient evidence to conclude that the populations are in fact different.

Let n_1 denote the size of the sample selected from the first population and n_2 denote the size of the sample selected from the second population. As shown above, W denotes the sum of the ranks in the first sample. It can be shown that if the two samples are selected from identical populations and if n_1 and n_2 each is 10 or greater, the sampling distribution of W can be approximated by a normal distribution with mean

$$\mu_W = \tfrac{1}{2}\, n_1(n_1 + n_2 + 1), \tag{18.1}$$

and standard deviation

$$\sigma_W = \sqrt{\tfrac{1}{12}\, n_1 n_2 (n_1 + n_2 + 1)}. \tag{18.2}$$

For the Third National Bank example,

$$\mu_W = \tfrac{1}{2}\, 12(12 + 10 + 1)$$
$$= 138,$$

$$\sigma_W = \sqrt{\tfrac{1}{12}\, 12(10)(12 + 10 + 1)}$$
$$= 15.17.$$

The sampling distribution of W is shown in Figure 18.1.

Following the usual hypothesis-testing procedure we develop a decision rule based on the sampling distribution of W as shown in Figure 18.1. With $\alpha = .05$ the decision rule will be to reject H_0 (identical populations) if the observed value of W is greater than

$$\mu_W + 1.96\sigma_W = 138 + 1.96(15.17) = 167.73$$

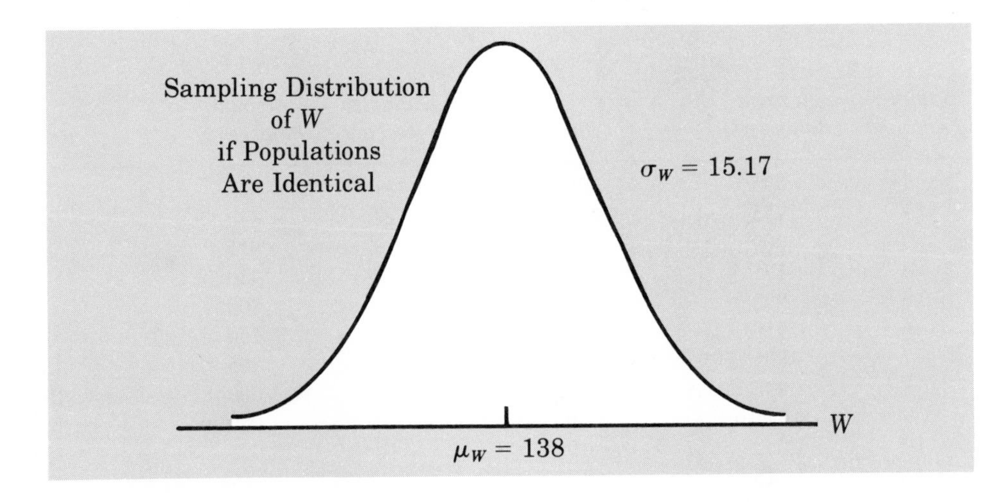

FIGURE 18.1 Sampling Distribution of *W* for the Third National Bank Example

or less than

$$\mu_W - 1.96\sigma_W = 138 - 1.96(15.17) = 108.27.$$

By comparing the actual value of $W = 169.5$ with the above critical values, we see that the decision rule calls for the rejection of H_0. Thus we conclude that the two populations are not identical. The probability distribution of account balances at branch 1 is not the same as that at branch 2.

In summary, the Wilcoxon rank-sum test follows the steps outlined below in order to determine if the two samples are from the same population.

1. Rank the two combined samples of data from lowest to highest, with tied values being assigned the mean of the tied rankings.
2. Compute W, the sum of the ranks for the first sample.
3. Make the test for significant difference in the two populations. That is, compare the observed value of W with the critical values based on the properties of the sampling distribution of W.

Exceptions arise if either sample size is less than ten or if numerous ties occur in the ranking of the data. In the first case, the small sample size(s) invalidates the use of the normal approximation as shown in Figure 18.1. However, a table of exact probabilities for W in the small-sample-size case is available in most books on nonparametric methods. This table can be used to determine if a significant difference exists in the two populations. If numerous tied ranks are observed, a slight modification in the formula for σ_W is required. The necessary modification can be found in most advanced nonparametric reference texts.

Mann-Whitney Test

A test equivalent to the Wilcoxon rank-sum test was proposed by Mann and Whitney and is referred to as the *Mann-Whitney test.* Instead of working directly with the sum of the ranks to carry out the test under the assumption of identical populations, Mann and Whitney proposed the use of a test statistic U whose value is related to W by the following expression:

$$\begin{aligned} U &= (\text{largest possible value of } W) - W \\ &= \left(n_1 n_2 + \frac{n_1(n_1 + 1)}{2}\right) - W. \end{aligned}$$

For the Third National Bank example the largest possible value of W was 198. Thus the value of U would be

$$U = 198 - 169.5 = 28.5.$$

It can be shown that if the two populations are identical the sampling distribution of U can be approximated for large n_1 and n_2 (that is, $n_1, n_2 \geq 10$) by a normal distribution with mean

$$\mu_U = \frac{1}{2} n_1 n_2 \tag{18.3}$$

and standard deviation

$$\sigma_U = \sqrt{1/12\, n_1 n_2 (n_1 + n_2 + 1)}. \tag{18.4}$$

Let us compare these results with those for the normal approximation to W in the Wilcoxon rank-sum test. We see that only the means differ ($\mu_W \neq \mu_U$ but $\sigma_W = \sigma_U$). For the Third National Bank example $\mu_U = 60$ and $\sigma_U = 15.17$. Thus at the $\alpha = .05$ level of significance the decision rule will be to reject H_0 (identical populations) if the observed value of U is greater than

$$\mu_U + 1.96\sigma_U = 60 + 1.96(15.17) = 89.73$$

or less than

$$\mu_U + 1.96\sigma_U = 60 - 1.96(15.17) = 30.27.$$

Since the value of U observed, $U = 28.5$, is less than 30.27, we reject H_0. Thus we see that the Mann-Whitney test procedure leads to the same conclusion as the Wilcoxon rank-sum test. Hence either test can be used, depending upon the preference of the user.

The nonparametric tests discussed in this section are used to determine whether or not two populations are identical. The parametric tests, such as the t test described in Chapter 10, test the equality of two population means. When we reject the hypothesis that the means are equal with the parametric methods, we conclude that the populations differ only in their means. When we reject the hypothesis that the populations are identical using one of the methods of this section, we cannot state how they differ. The populations could have different means, different variances, and/or different forms. Nonetheless, if we had assumed that the populations were the same in every way except for the means, a rejection of H_0 using a nonparametric method would have implied that the means differed. The major advantage of the nonparametric methods, however, is that they do not require any assumptions about the form of the probability distribution from which the measurements come.

EXERCISES

1. Mileage performance tests were conducted for two models of automobiles. Twelve automobiles of each model were randomly selected and a miles-per-gallon rating for each model was developed based upon 1,000 miles of highway driving. The data are shown below:

Model 1		*Model 2*	
Automobile	*mpg*	*Automobile*	*mpg*
1	20.6	1	21.3
2	19.9	2	17.6
3	18.6	3	17.4
4	18.9	4	18.5
5	18.8	5	19.7
6	20.2	6	21.1
7	21.0	7	17.3
8	20.5	8	18.8
9	19.8	9	17.8
10	19.8	10	16.9
11	19.2	11	18.0
12	20.5	12	20.1

Use $\alpha = .10$ and test for a significant difference in the populations of miles-per-gallon ratings for the two models.

2. Starting salaries were recorded for ten recent graduates at each of two community colleges:

Eastern College		*Western College*	
Student	*Monthly Salary ($)*	*Student*	*Monthly Salary ($)*
1	890	1	1,000
2	950	2	1,020
3	1,200	3	1,140
4	1,150	4	1,000
5	1,300	5	975
6	1,350	6	925
7	990	7	900
8	1,050	8	1,025
9	1,400	9	1,075
10	1,450	10	930

Use $\alpha = .10$ and test for the differences in the starting salaries from the two colleges.

3. Independent random samples of faculty at two colleges were taken, and annual salaries ($1,000s) were recorded. Use the Mann-Whitney test and a .10 level of significance to test if there is a difference between faculty salaries at the two colleges.

College 1	*College 2*
36	22
18	16
22	19
15	15
19	12
27	24
42	25
48	19
31	14
29	18
33	

18.2 WILCOXON SIGNED-RANK TEST

In the previous section we described two nonparametric statistical tests that could be used to determine if two independent samples were selected from identical populations. An important alternative to the two independent samples approach occurs whenever we pair or match items such that each observation from one sample has a corresponding or matched observation from the other sample. In some cases the same individual or same unit generates the observations or measurements for *both populations* and therefore provides the matched or paired observations. In other cases two separate but "similar" individuals or units are identified. One of the individuals generates a measurement for one population, and the second individual generates a measurement for the other population.

In Chapter 10 we described a parametric test of differences between population means based on two matched samples. That parametric test is often referred to as a paired difference test. The methodology of the parametric paired difference test required us to measure the actual observed difference for each pair of observations. Then, under the assumption that the population of differences between the pairs was normally distributed, a t test was used to test the null hypothesis of no difference in population means. If some question exists concerning the appropriateness of the assumption of normally distributed differences, or if it is possible only to rank order the differences from most similar to most dissimilar, a nonparametric statistical method becomes desirable. Wilcoxon provided the methodology for the nonparametric analysis of the paired differences or matched-sample analysis. The resulting test is referred to as the *Wilcoxon signed-rank test*. The Wilcoxon signed-rank test requires us to assume that we have taken a random sample of the paired differences and that the population of differences is continuous (so that few ties will occur) and symmetric.

To demonstrate the use of the Wilcoxon signed-rank test let us consider a manufacturing firm that is attempting to determine if a difference exists in two production methods. A sample of 11 workers was selected, and each worker completed the production task using each of the two production methods. The production method that each worker performed first was based upon a random assignment process. Thus each worker in the sample provides a pair of observations, as shown in Table 18.3. Table 18.3 also provides the difference in the completion times. A positive value

TABLE 18.3 Production Task Completion Times (Minutes)

Worker	*Method 1*	*Method 2*	*Difference*
1	10.2	9.5	.7
2	9.6	9.8	−.2
3	9.2	8.8	.4
4	10.6	10.1	.5
5	9.9	10.3	−.4
6	10.2	9.3	.9
7	10.6	10.5	.1
8	10.0	10.0	.0
9	11.2	10.6	.6
10	10.7	10.2	.5
11	10.6	9.8	.8

indicates that method 1 required more time, and a negative value indicates that method 2 required more time. From the data we see that most of the workers performed better using method 2; in fact, only two of the workers had better results using method 1. The statistical question is whether or not the data indicate that the methods are significantly different in terms of completion times. We note that this is equivalent to testing whether or not the populations of completion times are identical for the two methods. Thus the null and alternative hypotheses can be written

H_0: The two populations of task completion times are identical,

H_1: The two populations of task completion times are not identical.

The first step of the Wilcoxon signed-rank test requires that we rank the *absolute value* of the differences in the two methods. To do this we first discard any differences of zero and then rank the remaining absolute differences from lowest to highest. Tied differences are assigned average rank values, as in the Wilcoxon rank-sum test. The ranking of the absolute values of differences in the example problem is shown in Table 18.4. Note that the difference of 0 for worker 8 is discarded from the ranking; then the smallest absolute difference of .1 is ranked number 1, with the largest absolute difference of .9 ranked number 10. The absolute differences of .4 for workers 3 and 5 are assigned the average rank of 3.5, while the differences of .5 for workers 4 and 10 are assigned the average rank of 5.5.

TABLE 18.4 Ranking of Absolute Differences for the Production Task Completion Time Example

Worker	*Difference*	*Absolute Value of Difference*	*Rank*
1	.7	.7	8
2	−.2	.2	2
3	.4	.4	3.5
4	.5	.5	5.5
5	−.4	.4	3.5
6	.9	.9	10
7	.1	.1	1
8	0	0	—
9	.6	.6	7
10	.5	.5	5.5
11	.8	.8	9

Once the ranks of the absolute differences have been determined, the ranks are given the sign of the original difference in the data. For example, the .1 difference ranked number 1 is given the value of +1 because the actual difference was +.1 for worker 7, the .2 difference ranked number 2 is given a −2 signed rank value because the actual difference is −.2, etc. The complete list of signed ranks, together with their sum, is shown in Table 18.5.

TABLE 18.5 Signed Ranks for the Production Task Completion Time Example

Worker	*Difference*	*Rank of Absolute Difference*	*Signed Rank*
1	.7	8	+8
2	−.2	2	−2
3	.4	3.5	+3.5
4	.5	5.5	+5.5
5	−.4	3.5	−3.5
6	.9	10	+10
7	.1	1	+1
8	.0	—	—
9	.6	7	+7
10	.5	5.5	+5.5
11	.8	9	+9
		Sum of signed ranks	+44

Let us return to the original hypothesis of identical completion time populations for the two methods. We would expect the sum of the signed rank values to be approximately 0, since the positive ranks and negative ranks should be similar and cancel each other in the sum. Thus the test for significance under the Wilcoxon signed-rank test involves determining if the computed sum of signed ranks (+44 in our example) is significantly different from 0.*

Let T denote the sum of the signed rank values. It can be shown that if the two populations are identical and n is 10 or more, the sampling distribution of T can be approximated by a normal probability distribution with mean μ_T and standard deviation σ_T given by

$$\mu_T = 0, \tag{18.5}$$

$$\sigma_T = \sqrt{\frac{n(n+1)(2n+1)}{6}}. \tag{18.6}$$

For the production task example, we have $n = 10$, since we discarded the observation with 0 difference. Thus

$$\sigma_T = \sqrt{\frac{10(11)(21)}{6}} = 19.62.$$

The distribution of T under the identical population assumption is shown in Figure 18.2. Also shown are the critical values corresponding to $\alpha = .05$.

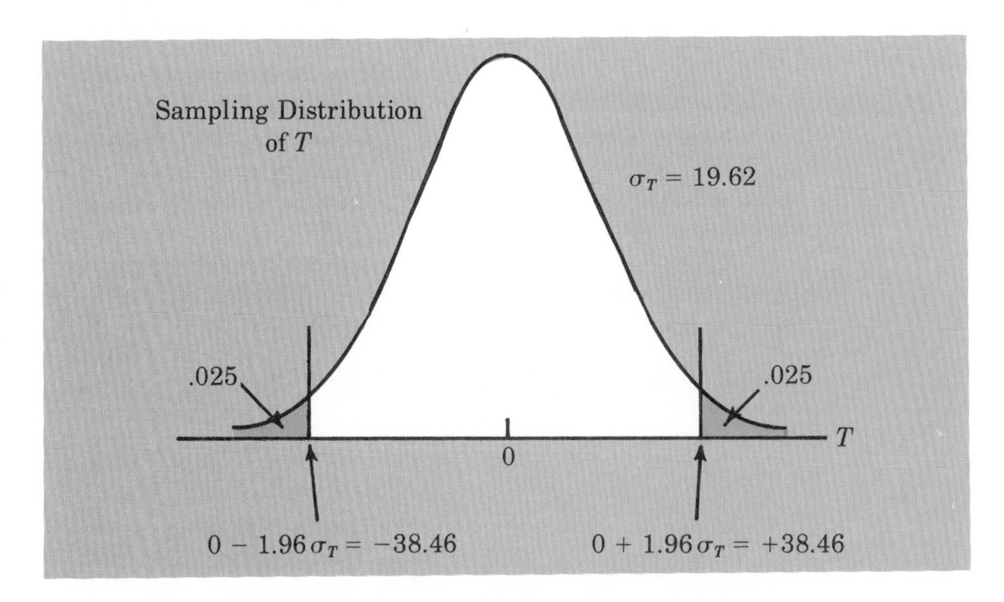

FIGURE 18.2 Sampling Distribution of the Wilcoxon *T* Value for the Production Task Completion Time Example

*This test is equivalent to a test of the null hypothesis that the median of the differences equals 0.

Compare the computed T value of 44 with the critical values of $+38.46$ and -38.46. We see that the null hypothesis that the populations are identical can be rejected since the computed T value is greater than 38.46. Thus the nonparametric Wilcoxon signed-rank test provides the conclusion that the two production methods differ. Although we have identified a difference, we are unable to conclude in what way the population distributions differ. If the firm had been primarily interested in the mean production times of the two methods and had felt that it was reasonable to assume a normal population of differences, then perhaps the parametric test of differences in population means should have been considered more carefully.

EXERCISES

4. Harding Investors, Inc. provides a 6 week training program for newly hired management trainees. As part of the normal evaluation procedure, the firm gives each trainee a pretest and posttest. Use a one-tailed test with $\alpha = .05$ and analyze the following data as part of the evaluation of the firm's management training program. What is your conclusion?

Trainee	*Pretest Score*	*Posttest Score*
1	45	65
2	60	70
3	65	63
4	60	67
5	52	60
6	62	58
7	57	70
8	70	65
9	72	80

5. Eight test market cities were selected as part of a market research study designed to evaluate the effectiveness of a particular advertising campaign. The sales dollars for each city were recorded for the week prior to the promotional program. Then the campaign was conducted for 2 weeks, with new sales data collected for the week immediately following the campaign. The resulting data with sales in thousands of dollars are shown below:

City	*Precampaign Sales*	*Postcampaign Sales*
Dayton	100	105
Cincinnati	120	140
Columbus	95	90
Cleveland	140	130
Indianapolis	80	82
Louisville	65	55
St. Louis	90	105
Pittsburgh	140	152

Use $\alpha = .05$. What conclusion would you draw concerning the value of the advertising program?

6. Twelve homemakers were asked to estimate the retail selling price of two models of refrigerators. The estimates of selling price provided by the homemakers are shown below:

Homemaker	*Model 1*	*Model 2*
1	$650	$ 900
2	760	720
3	740	690
4	700	850
5	590	920
6	620	800
7	700	890
8	690	920
9	900	1,000
10	500	690
11	610	700
12	720	700

Use these data and test at the .05 level of significance to determine if there is a difference in the homemaker's perception of selling price for the two models.

18.3 THE SIGN TEST FOR PAIRED COMPARISONS

The paired differences or matched-sample approach is a good experimental design for identifying differences in two populations. If quantitative measures, such as the task completion times of the example in the previous section, are obtainable, then, depending on the assumptions, either the parametric paired differences test based on matched samples (Section 10.3) or the nonparametric Wilcoxon signed-rank test may be appropriate methods to use. Recall that the parametric test requires the assumption of a normal population of differences in addition to quantitative measures of each difference. The Wilcoxon signed-rank test requires less precise measurements of the differences, but we still need the sign of each difference and a ranking of the absolute differences.

In some situations when we want to test for differences in populations, it is not possible to obtain the data in the form needed for the above two tests. For instance, a common market research application involves using a sample of potential customers to identify preferences for two brands of products such as coffee, beer, detergent, etc. In such cases potential customers are asked to state a preference for one of the two brands. Thus quantitative measures of differences are not available. Furthermore, it is not possible to rank order the absolute differences across individuals. A nonparametric method, called the sign test, has been designed to handle this type of situation. This method can be used to analyze data from pairs of observations where the data recorded are preferences.

As an example, consider the market research study conducted for Mueller Beverage Products of Milwaukee, Wisconsin. In the study 24 individuals were asked to express their preference for one of two beers: the individual's usual beer and Mueller's Old Brew. For the test each individual was provided with a glass of his or her usual beer and a glass of Mueller's Old Brew. The two glasses were not labeled, and thus the individuals had no way of knowing beforehand which of the two glasses was their usual beer. The brand that each individual tasted first was randomly selected. After tasting the beer in each glass, the individuals were asked to state a preference. If the test resulted in the individual stating a preference for Mueller's Old Brew, a "+" was recorded. On the other hand, if the individual stated a preference for his or her usual beer, a "−" was recorded. The data collected are shown in Table 18.6.

TABLE 18.6 Sign Test Data Collection for the Mueller Beer Study

Individual	*Brand Preferred*	*Value Recorded*
1	Old brew	+
2	Old brew	+
3	Usual brew	−
4	Old brew	+
5	Usual brew	−
6	Old brew	+
7	Usual brew	−
8	Old brew	+
9	Old brew	+
10	Usual brew	−
11	Old brew	+
12	Usual brew	−
13	Usual brew	−
14	Usual brew	−
15	Old brew	+
16	Usual brew	−
17	Old brew	+
18	Old brew	+
19	Old brew	+
20	Usual brew	−
21	Old brew	+
22	Old brew	+
23	Usual brew	−
24	Old brew	+

Let us now proceed with the sign test to determine if preferences for the two brands of beer do in fact differ. The null hypothesis being tested is that there is no difference in preferences or—in terms of the proportion preferring a particular brand—that the proportion of preferences is $p = .50$. In effect, then, the sign test is comparing the observed proportion ($14/24 = .58$) for Old Brew to the hypothesized proportion ($p = .50$). We state below the null and alternative hypotheses:

H_0: $p = .50$ No difference in preference exists,

H_1: $p \neq .50$ A difference in preference for one product over the other exists.

Under the assumption of no differences, $p = .50$ and the number of "+" values in the sample follows the binomial distribution. In Chapter 6 we stated that the normal distribution was a good approximation to the binomial if both $np \geq 5$ and $n(1 - p) \geq 5$. Thus since we assume $p = .50$ in this test, the normal distribution may be used to approximate the binomial distribution with samples of ten or more. Using this normal approximation, the mean for the number of "+" values is $\mu = 24(.50) = 12$ and $\sigma =$

$\sqrt{24(.50)(.50)} = 2.45$. We will choose a level of significance of $\alpha = .05$. The sampling distribution with the critical values is shown in Figure 18.3.

Finally, by comparing 14, the number of "+" signs observed in the study, with the critical values shown in Figure 18.3, we see that we are unable to reject H_0. Hence there is insufficient evidence to conclude that a difference in preference exists between

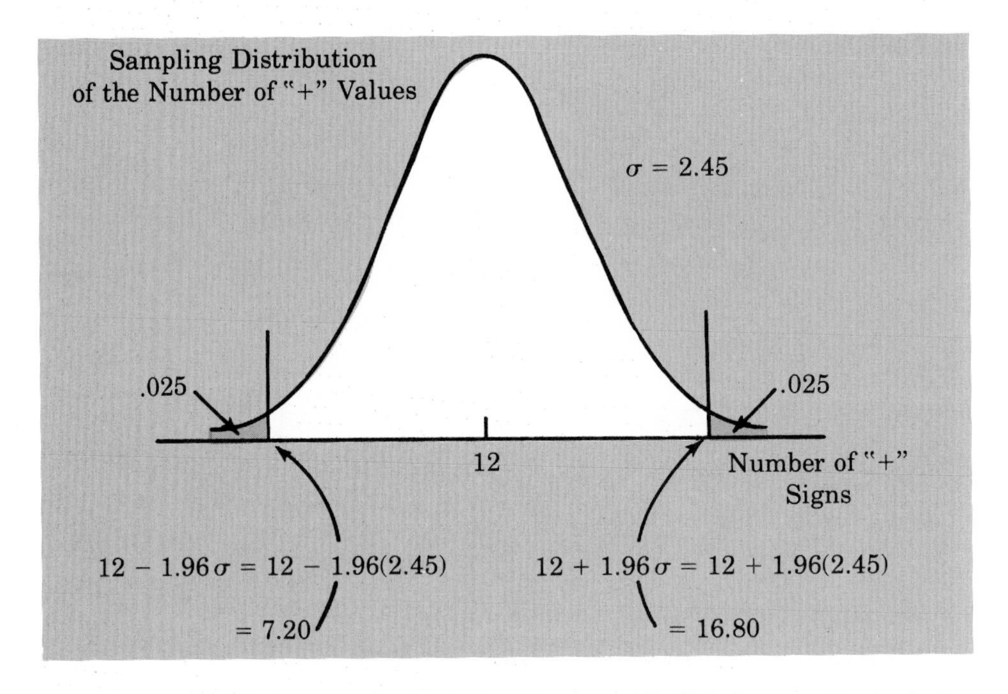

FIGURE 18.3 Sampling Distribution for the Number of "+" Ratings for Mueller's Old Brew Beer

Mueller's Old Brew and the individual's usual brand. While this is the conclusion supported by the sign test, it actually may be a positive indicator for the Mueller product. That is, the fact that we were not able to detect a significant difference between Mueller's Old Brew and an individual's usual brand is indicative of a possibility that individuals may switch to Old Brew on the basis of it being of similar quality to their own individual brand.

In this example the individuals were all able to express a preference. In some situations, however, one or more preferences may not be stated. In such cases the individual's response can be removed from the study and the analysis conducted with a smaller sample size.

The sign test is also applicable for situations encountered by the Wilcoxon signed-rank test. However, it is usually not recommended in these cases because the sign test does not take advantage of all the data available. For example, if paired observations were 10.5 and 9.8, the Wilcoxon signed-rank test would make use of the sign of the difference (+) and would rank order this difference according to its absolute value (.7). The sign test of this section is not as powerful because it ignores the rank-order information.

EXERCISES

7. The following data show the preferences indicated by ten individuals in taste tests involving two brands of coffee:

Individual	*Brand A Versus Brand B*	*Individual*	*Brand A Versus Brand B*
1	+	6	+
2	+	7	−
3	+	8	+
4	−	9	−
5	+	10	+

With $\alpha = .10$, test for a significant difference in the preferences for the two brands. A "+" indicates a preference for brand A over brand B.

8. Use the sign test and perform the statistical analysis that will help us determine whether or not the task completion times for two production methods discussed in the previous section differ. The data are repeated below:

Worker	*Method 1 (minutes)*	*Method 2 (minutes)*
1	10.2	9.5
2	9.6	9.8
3	9.2	8.8
4	10.6	10.1
5	9.9	10.3
6	10.2	9.3
7	10.6	10.5
8	10.0	10.0
9	11.2	10.6
10	10.7	10.2
11	10.6	9.8

Use a "+" if the difference is positive and a "−" if the difference is negative. Test with $\alpha = .05$. Do you prefer the sign test or the Wilcoxon signed-rank test for the analysis of the problem? Explain.

9. Mayfield Products, Inc. has collected data on preferences of 12 individuals concerning cleaning power of two brands of detergent. The individuals and their preferences are shown below (a "+" indicates a preference for brand A):

Individual	*Brand A Versus Brand B*	*Individual*	*Brand A Versus Brand B*
1	−	7	−
2	+	8	+
3	+	9	+
4	+	10	−
5	−	11	+
6	+	12	+

With $\alpha = .10$, test for a significant difference in the preference for the two brands.

18.4 RANK CORRELATION

Correlation was introduced in Chapter 14 as a measure of the linear association between two variables for which quantitative measures are available. In this section we consider measures of association between two variables when only rank-order data are available. The *Spearman rank-correlation coefficient, r_s*, has been developed for this purpose. The Spearman rank-correlation coefficient is just the ordinary sample correlation coefficient r applied to the rank order data. That is, we could compute r_s by treating the ranks as the values of the variables in the formula presented in Section 14.8 for the sample correlation coefficient. However, we shall give a simpler formula in this section.

To show how r_s is computed and to illustrate the use of rank correlation in a decision-making context, we consider a company that wants to determine if those individuals who were expected at the time of employment to be better salespersons actually turn out to have better sales records. To investigate this question the vice president in charge of personnel has carefully reviewed the original job interviews, academic records, and personal recommendations of ten members of the firm's sales force. With this review the vice president ranked the ten individuals in terms of their potential for success, basing the assessment solely upon the information available at the time of employment. Then a list of the number of units sold by each salesperson over the first 2 years was obtained. With these actual sales values a second ranking of the ten salespersons was carried out. Table 18.7 shows the relevant data and the two rankings.

The statistical question involves determining whether or not there is agreement between the ranking of potential at the time of employment and the ranking based upon the actual sales performance over the first 2 years. To help us answer this question, we will compute the Spearman rank-correlation coefficient, r_s.

To present the formula for computing the rank-correlation coefficient we must introduce the following notation:

n = total number of items or individuals being ranked,

x_i = rank of item or individual i according to the first criterion (e.g., sales potential),

y_i = rank of item or individual i according to the second criterion (e.g., actual sales performance):

$$d_i = x_i - y_i.$$

With these symbols, the formula for computing the Spearman rank-correlation coefficient is as follows:

$$r_s = 1 - \frac{6\Sigma d_i^2}{n(n^2 - 1)}. \qquad (18.7)$$

Table 18.8 shows the calculation of r_s for our illustration.

We see that the Spearman rank-correlation coefficient shows a positive relationship with .73 correlation between sales potential and sales performance. The Spearman rank-correlation coefficient ranges from -1.0 to $+1.0$, with an interpretation similar to that for the sample correlation coefficient presented in Chapter 14. It is a measure of the monotonicity of a relationship. Positive values near 1 indicate a monotonically

TABLE 18.7 Sales Potential and Actual 2 Year Sales in Units for Ten Salespersons

Salesperson	*Ranking of Potential*	*2 Year Sales (units)*	*Ranking According to 2 Year Sales*
A	2	400	1
B	4	360	3
C	7	300	5
D	1	295	6
E	6	280	7
F	3	350	4
G	10	200	10
H	9	260	8
I	8	220	9
J	5	385	2

TABLE 18.8 Computation of the Spearman Rank-Correlation Coefficient for Sales Potential and Sales Performance

Salesperson	x_i = *Ranking of Potential*	y_i = *Ranking of Sales Performance*	$d_i = x_i - y_i$	d_i^2
A	2	1	1	1
B	4	3	1	1
C	7	5	2	4
D	1	6	−5	25
E	6	7	−1	1
F	3	4	−1	1
G	10	10	0	0
H	9	8	1	1
I	8	9	−1	1
J	5	2	3	9
				$\Sigma d_i^2 = 44$

$$r_s = 1 - \frac{6\Sigma d_i^2}{n(n^2 - 1)} = 1 - \frac{6(44)}{10(100 - 1)} = .73$$

increasing relationship (increases in one variable are accompanied by increases in the other variable), whereas negative values near −1 indicate a monotonically decreasing relationship (decreases in one variable are accompanied by increases in the other variable).

We caution that r_s should not be interpreted as a measure of linear association between two variables; r_s is a measure of linear association between the *ranks* of the variables. For example, a value of r_s near 1 may indicate that two variables are related in a highly nonlinear (but monotonically increasing) fashion.

We now have found the Spearman rank-correlation coefficient. We may wish now to test whether or not the correlation found in the sample of ten salespersons justifies a conclusion that there is a nonzero correlation between the rankings based upon sales potential and the rankings based upon sales performance for all personnel in the sales force. It can be shown that if the two rankings are independent and if n is 10 or more, the sampling distribution of r_s can be approximated by a normal probability distribution with mean

$$\mu_{r_s} = 0$$

and standard deviation

$$\sigma_{r_s} = \sqrt{\frac{1}{(n-1)}}.$$

Following the usual hypothesis-testing procedure, we develop a decision rule based on the sampling distribution of r_s. With $\alpha = .05$, the decision rule will be to reject the null hypothesis that the two rankings are independent if the observed value of r_s is greater than

$$\mu_{r_s} + 1.96\sigma_{r_s} = 0 + 1.96\sqrt{1/9} = +.65$$

or less than

$$\mu_{r_s} - 1.96\sigma_{r_s} = 0 - 1.96\sqrt{1/9} = -.65$$

Thus since the Spearman rank-correlation coefficient shows a value of .73, we can reject the null hypothesis and conclude that a significant rank correlation exists between sales potential and sales performance for the population of the entire sales force.

EXERCISES

10. Consider the following two sets of rankings for six items:

Case 1			*Case 2*		
Item	*First Ranking*	*Second Ranking*	*Item*	*First Ranking*	*Second Ranking*
A	1	1	A	1	6
B	2	2	B	2	5
C	3	3	C	3	4
D	4	4	D	4	3
E	5	5	E	5	2
F	6	6	F	6	1

Note that in the first case the rankings are identical, while in the second case the rankings are exactly opposite. What value should you expect for the Spearman rank-correlation coefficient for each of these cases? Perform the calculation of the rank-correlation coefficient for each case.

11. In the baseball draft process eight players are ranked by a scout in terms of speed and then in terms of power hitting:

Player	*Speed Ranking*	*Power-Hitting Ranking*
A	1	8
B	2	5
C	3	6
D	4	7
E	5	2
F	6	3
G	7	4
H	8	1

Use the Spearman rank-correlation coefficient to measure the association between speed and power. Use $\alpha = .05$ and test for the significance of this correlation coefficient.

12. A sample of 15 students obtained the following rankings on midterm and final examinations in a statistics course:

Midterm Rank	*Final Rank*
1	4
2	7
3	1
4	3
5	8
6	2
7	5
8	12
9	6
10	9
11	14
12	15
13	11
14	10
15	13

Compute the Spearman rank-correlation coefficient for the data and test for a significant correlation with $\alpha = .10$.

13. A student organization surveyed both recent graduates and current students in an attempt to obtain information on the quality of teaching at a particular university. An analysis of the responses provided the following rankings for ten professors on the basis of teaching ability:

Professor	*Ranking by Current Students*	*Ranking by Recent Graduates*
1	4	6
2	6	8
3	8	5
4	3	1
5	1	2
6	2	3
7	5	7
8	10	9
9	7	4
10	9	10

Do the rankings given by the current students agree with the rankings given by the recent graduates? Use $\alpha = .10$ and test for a significant rank correlation.

Summary

Methods of statistical inference that are based upon an assumed population distribution, such as the normal, are called parametric methods. In many cases the assumptions concerning the population distribution are inappropriate. As a result, statistical methods that do not require assumptions about the population probability distribution were developed. These methods are called nonparametric or distribution-free methods. There are two major advantages in working with nonparametric methods. First, as we have just noted, fewer assumptions are required; hence the methods can be applied to a wider range of problems. Second, many of the nonparametric methods, in contrast to the parametric methods, require only rank-order or preference data.

The Wilcoxon rank-sum test provides a nonparametric procedure for identifying differences in two populations whenever two independent samples are selected. The Wilcoxon signed-rank test provides a similar methodology for cases when the samples involve matched pairs. The sign test provides a nonparametric procedure for identifying differences in two populations when the only data available are preferences between matched items. This test does not even require rank-order data. The Spearman rank-correlation coefficient provides a measure of association between two ranked sets of items.

Glossary

Parametric methods—The collection of statistical methods that require assumptions about the population distribution.

Nonparametric methods—A collection of statistical methods that generally requires very few, if any, assumptions about the population distribution. These methods can be applied when only rank-order or preference data are available.

Distribution-free methods—Another name for nonparametric statistical methods suggested by the lack of assumptions required concerning the population distribution.

Wilcoxon rank-sum test—A nonparametric statistical test for identifying differences between two populations based on the analysis of two independent samples.

Mann-Whitney test—A nonparametric statistical test for identifying differences between two populations based on the analysis of two independent samples. It is equivalent to the Wilcoxon rank-sum test.

Wilcoxon signed-rank test—A nonparametric statistical test for identifying differences between two populations based on the analysis of two matched or paired samples.

Sign test—A nonparametric statistical test for identifying differences between two populations based on the analysis of two matched or paired samples. Preference data are all that is required.

Spearman rank-correlation coefficient—A correlation measure based on rank-order data for two variables. It provides a measure of monotonicity of the relationship between the two variables.

Key Formulas

Wilcoxon Rank-Sum Test

$$\mu_W = \tfrac{1}{2}n_1(n_1 + n_2 + 1), \tag{18.1}$$

$$\sigma_W = \sqrt{\tfrac{1}{12}n_1 n_2(n_1 + n_2 + 1)} \tag{18.2}$$

Mann-Whitney Test

$$U = (\text{Largest Possible Value of } W) - W$$

$$= \left(n_1 n_2 + \frac{n_1(n_1 + 1)}{2}\right) - W \qquad \text{(Page 581)}$$

(Page 581)

$$\mu_U = \tfrac{1}{2} n_1, n_2 \qquad (18.3)$$

$$\sigma_U = \sqrt{\tfrac{1}{12} n_1 n_2 (n_1 + n_2 + 1)} \qquad (18.4)$$

Wilcoxon Signed-Rank Test

$$\mu_T = 0 \qquad (18.5)$$

$$\sigma_T = \sqrt{\frac{n(n + 1)(2n + 1)}{6}} \qquad (18.6)$$

Rank Correlation

$$r_s = 1 - \frac{6\Sigma d_i^2}{n(n^2 - 1)} \qquad (18.7)$$

Supplementary Exercises

14. A certain brand of microwave oven was priced at 10 stores in Dallas and 13 stores in San Antonio. The data are presented below:

Dallas	*San Antonio*
445	460
489	451
405	435
485	479
439	475
449	445
436	429
420	434
430	410
405	422
	425
	459
	430

Use a .05 level of significance and test whether or not prices for the microwave oven are the same in the two cities.

15. Independent random samples of houses in two different neighborhoods of a large city have been collected and data on assessed valuation recorded. These data are presented below:

Neighborhood 1	*Neighborhood 2*
18,000	16,500
16,000	20,500
12,000	23,000
20,000	17,500
19,000	22,000
17,000	21,000
16,500	21,500
19,000	19,500
15,500	17,000
16,000	23,500
17,500	21,000
18,000	22,000

Test at the $\alpha = .05$ level of significance whether or not assessed values are the same in each neighborhood.

16. The following data show product weights for items produced on two production lines:

Production Line 1	*Production Line 2*
13.6	13.7
13.8	14.1
14.0	14.2
13.9	14.0
13.4	14.6
13.2	13.5
13.3	14.4
13.6	14.8
12.9	14.5
14.4	14.3
	15.0
	14.9

Test for a difference between the product weights for the two lines. Use $\alpha = .10$.

17. In a coffee taste test 48 individuals stated a preference for one of two well-known brands. Results showed 28 favoring brand A, 16 favoring brand B, and 4 undecided. Use the sign test with $\alpha = .10$ and determine whether or not there is a significant difference in the preferences for the two brands of coffee.

18. In a television preference poll a sample of 180 individuals was asked to state a preference for one of the two shows aired at the same time on Friday evenings: 100 favored "Big Town Detective," 65 favored "The Friday Variety Special," and 15 were unable to state a preference for one over the other. Is there evidence of a significant difference in the preferences for the two shows? Use $\alpha = .05$ for the test.

19. Menu planning at the Hampshire House Restaurant involves the question of customer preferences for steak and seafood. A sample of 250 customers was asked to state a preference for the two menu items: 140 stated a preference for steak, and 110 stated a preference for seafood. Use $\alpha = .05$ and test for a difference in the preference for the two menu items.

20. A group of investment analysts ranked 12 companies, first with respect to book value and

then with respect to growth potential:

Company	*Ranking of Book Value*	*Ranking of Growth Potential*
1	12	2
2	2	9
3	8	6
4	1	11
5	9	4
6	7	5
7	3	12
8	11	1
9	4	7
10	5	10
11	6	8
12	10	3

For these data does a relationship exist between the companies' book value and growth potential? Use $\alpha = .05$.

21. In a poll of men and women viewers, preferences for the top ten shows led to the following rankings:

Television Show	*Ranking by Men*	*Ranking by Women*
1	1	5
2	5	10
3	8	6
4	7	4
5	2	7
6	3	2
7	10	9
8	4	8
9	6	1
10	9	3

Is there a relationship between the rankings provided for the two groups? Use $\alpha = .10$.

22. Two individuals provided the following preference rankings of seven soft-drink products:

Soft Drink	*Ranking by Individual 1*	*Ranking by Individual 2*
A	1	3
B	3	2
C	5	5
D	6	7
E	7	6
F	4	1
G	2	4

For $\alpha = .05$, is there a significant rank correlation for the two individuals?

Application

West Shell Realtors*

Cincinnati, Ohio

West Shell Realtors was founded in 1958 with one office and a sales staff of three people. The company's first-year sales were $900,000. In 1964 the Company began a long term expansion program, with new offices being added almost yearly. Since that time West Shell has grown to its current size, with 17 offices located in southwest Ohio, southeast Indiana, and northern Kentucky. With 450 associates, West Shell's sales in 1982 were approximately $230 million, with over 7,000 sales transactions.

As part of the company's expansion program, a 1968 merger with the Robson-Middendorf Company added a commercial and industrial division to the previously residentially focused company. Expansion in 1971 added property management to West Shell's range of services. Currently, West Shell is a recognized leader in the industry, with a motto that describes a full line of realty services for its customers.

West Shell is heavily involved with employee transfers and relocations for a wide variety of corporations. Currently, West Shell's relocation department accounts for approximately 47% of the Company's residential business. Selling houses for employees transferred to other locations and finding houses for employees transferred into the greater Cincinnati area are part of West Shell's total service for its customers.

STATISTICS IN REAL ESTATE

As you might expect, a real estate firm such as West Shell must monitor sales performance closely if it is to remain competitive. Monthly reports are generated for each of West Shell's 17 offices as well as for the total company. Statistical summaries of total sales dollars, number of units sold, mean selling price per unit, and so on are essential in keeping both office managers and the company's top management informed of progress and/or trouble spots in the organization. Monthly progress reports also show statistical summaries of the percentage over or under budget for each office in the Company.

In addition to monthly statistical summaries for ongoing operations, the Company uses statistical considerations to guide corporate plans and strategies. Managers must determine where to focus sales efforts from among the total area served by the Company. Statistical considerations, such as the ratio of total units sold to total units in

*The authors are indebted to Rodney Fightmaster, Vice President, West Shell Realtors, Cincinnati, Ohio, for providing this application.

a particular area provides information which guides the sales effort. Phone calls and canvassing efforts by the sales force are targeted in the high potential areas.

STATISTICAL ANALYSIS FOR OFFICE LOCATION

West Shell has implemented a strategy of planned expansion over the past 20 years. Each time an expansion plan calls for the establishment of a new sales office, the Company must address the question of the best place to locate the new office. Selling prices of homes, turnover rates, forecast sales volumes, and so on are the types of data that assist in evaluating and comparing alternative office location sites.

Residential sales are a large portion of West Shell's business

In one such instance the company had identified two areas as prime candidates for a new office location: Clifton and Roselawn. The statistical issues involved determining in what ways the two areas were alike and in what ways they differed. There were a variety of factors to be considered in comparing the two areas, but let us focus on one such factor: the selling price of homes in the two areas. Were the selling prices in the two areas similar or different? The actual sales price of units sold over a period of time could be viewed as a sample of sales for the area. In cases where the number of units sold were relatively small and where a normal distribution assumption for selling prices was inappropriate, nonparametric statistical methods could be employed to help answer the question concerning differences between the two areas.

For example, if a sample of 25 sales in the Clifton area showed a mean selling price of $55,250 and a sample of 18 sales in the Roselawn area showed a mean selling price of $50,375, the Wilcoxon rank-sum test or the Mann-Whitney test could be used to determine statistically whether or not the population of sales in two areas appeared identical or not. Using a .05 level of significance, the critical values for the Wilcoxon rank-sum test are 470 and 630. Suppose the value of W for the total sample of 43 recorded sales in the two areas was 595.2. In this case the nonparametric test leads to an acceptance of the hypothesis that the two populations are identical. The selection basis for the location of the new office should now focus on criteria other than unit selling price, since the areas are believed similar on this factor.

The real estate business has been and continues to be extremely competitive. At West Shell statistical data and statistical considerations play a meaningful role in helping the Company maintain a leadership role in the industry.

19 Decision Theory

What you will learn in this chapter:

- how to describe a decision problem in terms of decision alternatives, states of nature, and payoffs
- how to use payoff tables and decision trees to analyze decision problems
- how to use decision criteria such as maximin, minimax, minimax regret, expected monetary value, and expected opportunity loss to identify the best decision alternative
- how to determine the potential value of additional information in decision-making situations
- how to use a Bayesian approach to decision making

Contents

Decision theory or decision analysis can be used to determine optimal strategies when a decision maker is faced with several decision alternatives and an uncertain or risk filled pattern of future events. For example, a manufacturer of a new style or line of seasonal clothing would like to manufacture large quantities of the product if consumer acceptance and consequently demand for the product are going to be high. However, the manufacturer would like to produce much smaller quantities if consumer acceptance and demand for the product are going to be low. Unfortunately, seasonal clothing items require the manufacturer to make a production quantity decision before the demand is actually known. The actual consumer acceptance of the new product will not be determined until the items have been placed in the stores and the customers have had the opportunity to purchase them. The selection of the best production volume decision from among several production volume alternatives when the decision maker is faced with the uncertainty of future demand is a problem suited for a decision theory analysis. In this chapter we will introduce the concepts and procedures associated with the decision theory approach to problem solving.

We begin our study of decision theory by considering problem situations in which there is a reasonably small number of decision alternatives and possible future events. The concept of a payoff table is introduced to provide a structure for this type of decision situation and to illustrate the fundamentals involved in the decision theory approach to any situation. This analysis is then extended to show how additional information obtained through experimentation can be combined with the decision maker's preliminary information in order for an optimal decision strategy to be developed.

19.1 STRUCTURING THE DECISION SITUATION: PAYOFF TABLES

In order to illustrate the decision theory approach let us consider the case of Political Systems, Inc. (PSI), a newly formed computer service firm specializing in information services such as surveys, data analysis, and so on for individuals running for political office. PSI is in the final stages of selecting a computer system for its Midwest branch, located in Chicago. While the firm has decided on a computer manufacturer, it is currently attempting to determine the size of the computer system that would be the most economical to lease. We will use decision theory to help PSI make its computer-leasing decision.

The first step in the decision theory approach for a given problem situation is identification of the alternatives that may be considered by the decision maker. For PSI, the final decision will be to lease one of three computer systems which differ in size and capacity. The three decision alternatives, denoted d_1, d_2, and d_3, are as follows:

d_1 = lease the large computer system,

d_2 = lease the medium sized computer system,

d_3 = lease the small computer system.

Obviously, the determination of the best decision will depend upon what PSI management foresees as the possible market acceptance of their service and consequently the possible demand or load on the PSI computer system. Often the future events associated with a problem situation are uncertain. That is, while a decision maker may have an idea of the variety of possible future events, he or she will often be unsure as to which particular event will occur. Thus the second step in a decision theory appoach is to identify the future events that might occur. These future events, which are not under the control of the decision maker, are referred to as the *states of nature* for the problem. It is assumed that the list of possible states of nature includes everything that can happen and that the individual states of nature do not overlap. That is, the states of nature are defined so that one and only one of the listed states of nature will occur.

When asked about the states of nature for the PSI decision problem, management viewed the possible acceptance of their service as an either–or situation. That is, PSI management believed that the firm's overall level of acceptance in the marketplace would be one of two possibilities: high acceptance or low acceptance. Thus the PSI states of nature, denoted by s_1 and s_2, are as follows:

$$s_1 = \text{high customer acceptance of PSI services,}$$

$$s_2 = \text{low customer acceptance of PSI services.}$$

It is given that there are three decision alternatives and the two states of nature. Which computer system should PSI lease? In order to answer this question, we will need information on the profit associated with each combination of a decision alternative and a state of nature. For example, what profit would PSI experience if the firm decided to lease the large computer system (d_1) and market acceptance was high (s_1)? What profit would PSI experience if the firm decided to lease the large computer system (d_1) and market acceptance was low (s_2)? And so on.

In decision theory terminology we refer to the outcome resulting from making a certain decision and the occurrence of a particular state of nature as the *payoff*. Using the best information available, management of PSI has estimated the payoffs or profits for the PSI computer-leasing problem. These estimates are presented in Table 19.1. A

TABLE 19.1 Payoff Table for the PSI Computer-Leasing Problem

		States of Nature	
Decision Alternatives		*High Acceptance* (s_1)	*Low Acceptance* (s_2)
Lease a large system	(d_1)	$200,000	$ −20,000
Lease a medium sized system	(d_2)	$150,000	$ 20,000
Lease a small system	(d_3)	$100,000	$ 60,000

table of this form is referred to as a *payoff table*. In general, entries in a payoff table can be stated in terms of profits, costs, or any other measure of output that may be appropriate for the particular situation being analyzed. The notation we will use for the entries in the payoff table is $V(d_i,s_j)$, which denotes the payoff associated with decision alternative d_i and state of nature s_j. With this notation we see that $V(d_3,s_1) =$ $100,000.

The identification of the decision alternatives, the states of nature, and the determination of the payoff associated with each decision alternative and state of nature combination are the first three steps in the decision theory approach. The question we now turn to is: How can the decision maker best utilize the information presented in the payoff table to arrive at a decision? As we shall see, there are several criteria that may be used.

19.2 TYPES OF DECISION-MAKING SITUATIONS

Before discussing specific decision-making criteria, let us consider the types of decision-making situations that we may encounter. The classification scheme for decision-making situations is based upon the knowledge that the decision maker has about the states of nature. In this regard, there are two types of decision-making situations:

1. Decision making under certainty—The process of choosing a decision alternative when the state of nature is known.
2. Decision making under uncertainty—The process of choosing a decision alternative when the state of nature is not known.

In the case of decision making under certainty there will only be one state-of-nature column in the payoff table, and the optimal decision is the one corresponding to the best payoff in the column. For example, if PSI knew for certain that market acceptance of its service was going to be high, column s_2 could be removed from the payoff table. The optimal solution would be the large system (d_1), since d_1 provides the largest profit ($200,000) in the s_1 column.

Most situations in which the decision theory approach is applied involve decision making under uncertainty. In these cases the selection of the best alternative is more difficult. First the decision maker must select a criterion. Then he or she must determine which decision alternative is best under the chosen criterion.

19.3 CRITERIA FOR DECISION MAKING UNDER UNCERTAINTY WITHOUT USE OF PROBABILITIES

In some situations of decision making under uncertainty the decision maker may have very little confidence in his or her ability to assess the probabilities of the various states of nature. In such cases the decision maker might prefer to choose a decision criterion that does not require any knowledge of the probabilities of the states of nature. Three of the most popular criteria available for these cases are maximin (or minimax), maximax (or minimin), and minimax regret.

Because different criteria will sometimes lead to different decision recommendations it is important for the decision maker to know the criteria available. The decision maker can then select the specific criterion which, according to his or her judgment, is the most appropriate. We now discuss the above three criteria for decision making under uncertainty by showing how each could be used to solve the PSI computer-leasing problem.

Maximin

The *maximin* decision criterion is a pessimistic or conservative approach to arriving at a decision. In this approach the decision maker attempts to *max*imize his or her *min*imum possible profits; hence the term maximin. Using the information contained in the payoff table, the decision maker would first list the minimum payoff that is possible for each decision alternative. He or she would then select the decision from the new list that results in maximum payoff. Table 19.2 illustrates this process for the PSI problem.

TABLE 19.2 PSI Minimum Payoff for Each Decision Alternative

Decision Alternatives		*Minimum Payoff*	
Large system	(d_1)	$-20,000	
Medium system	(d_2)	$20,000	Maximum of the minimum payoff values
Small system	(d_3)	$60,000 ←	

Since $60,000, corresponding to the decision to lease a small system, yields the maximum of the minimum payoffs, the decision to lease a small system is recommended as the maximin decision. This decision criterion is considered conservative because it concentrates on the worst possible payoffs and then recommends the decision alternative that avoids the possibility of extremely "bad" payoffs. In using the maximin criterion PSI is guaranteed a profit of at least $60,000. While PSI may still make more, it cannot make *less* than the maximin criterion value of $60,000.

For problems in which costs are to be minimized, the conservative maximin approach is reversed in that the decision maker first lists the maximum cost for each decision alternative. The recommended decision then corresponds to the *mini*mun of the *maxi*mum costs. Thus this criterion, used for minimization problems, is referred to as *minimax*.

Maximax

While maximin offers a pessimistic decision criterion, maximax provides an optimistic criterion. With this criterion for maximization problems, the decision maker selects the decision that *maxi*mizes his or her *max*imum payoff; hence the name *maximax*. In applying this criterion, the decision maker first determines the maximum payoff possible for each decision alternative. The decision maker then identifies the decision that provides the overall maximum payoff. Table 19.3 shows the result of applying this criterion for the PSI problem.

TABLE 19.3 PSI Maximum Payoff for Each Decision Alternative

Decision Alternatives		*Maximum Payoff*	
Large system	(d_1)	$200,000 ←	Maximum of the maximum payoff values
Medium system	(d_2)	$150,000	
Small system	(d_3)	$100,000	

Since $200,000, corresponding to the decision to lease a large system, yields the maximum of the maximum payoffs, the decision to lease a large system is the recommended maximax decision. This decision criterion reflects an optimistic point of view because it simply recommends the decision alternative that provides the possibility of obtaining the best of all payoffs, $200,000. While the use of this criterion provides the opportunity for a large payoff, it also exposes the company to the possibility of a $20,000 loss. Hence we would definitely not recommend using this criterion if information was available indicating that state of nature s_2 was highly likely.

For minimization problems the maximax criterion reverses to the *mini*mum of the *mini*mum cost values, or the *minimin* criterion.

Minimax Regret

Suppose we make the decision to lease the small system d_3 and afterwards learn that market acceptance of the PSI service is high, s_1. Table 19.1 shows the resulting profit to be $100,000. However, now that we know that state of nature s_1 has occurred, we see that the large-system decision d_1 yielding a profit of $200,000 would have been the optimal decision. This difference between the optimal payoff ($200,000) and the payoff experienced ($100,000) is referred to as the *opportunity loss* or *regret* associated with our d_3 decision when state of nature s_1 occurs ($200,000 − $100,000 = $100,000). If we had made decision d_2 and state of nature s_1 had occurred, the opportunity loss or regret for this decision and state of nature would have been $200,000 − $150,000 = $50,000.

The general expression for opportunity loss or regret is as follows:

Opportunity Loss or Regret

$$R(d_i,s_j) = V^*(s_j) - V(d_i,s_j), \tag{19.1}$$

where

$R(d_i,s_j)$ = regret associated with decision alternative d_i and state of nature s_j,
$V^*(s_j)$ = best payoff value[1] under state of nature s_j.

For our d_3 decision and state of nature s_1, $V^*(s_1)$ = $200,000 and $V(d_3,s_1)$ = $100,000. Thus

$$R(d_3,s_1) = \$200{,}000 - \$100{,}000 = \$100{,}000.$$

From (19.1) we can compute the regret associated with all combinations of decision alternatives d_i and states of nature s_j. We simply replace each entry in the payoff table with the value found by subtracting the entry from the largest entry in its column. Table 19.4 shows the regret, or opportunity loss, table for the PSI problem.

The next step in applying the minimax regret criterion requires the decision

TABLE 19.4 Regret or Opportunity Loss Table for the PSI Problem

		States of Nature	
Decision Alternatives		*High Acceptance* (s_1)	Low Acceptance (s_2)
Large system	(d_1)	0	$80,000
Medium system	(d_2)	$50,000	$40,000
Small system	(d_3)	$100,000	0
		↖ Regret or opportunity loss	

[1]In cost minimization problems $V^*(s_j)$ will be the smallest entry in column j. Thus for minimization problems (19.1) must be changed to $R(d_i,s_j) = V(d_i,s_j) - V^*(s_j)$.

analyst to identify the maximum regret for each decision alternative. These data are shown in Table 19.5. The decision is made by selecting the alternative corresponding to the *mini*mum of the *max*imum regret values; hence the name *minimax regret*. For the PSI problem, the decision to lease a medium sized computer system, with a corresponding regret of $50,000, is the recommended minimax regret decision.

TABLE 19.5 PSI Maximum Regret or Opportunity Loss for Each Decision Alternative

Decision Alternatives		*Maximum Regret or Opportunity Loss*	
Large system	(d_1)	$ 80,000	
Medium system	(d_2)	$ 50,000 ←	Minimum of the maximum regret
Small system	(d_3)	$100,000	

Note that the three decision criteria discussed in this section have each led to different recommendations. This is not in itself bad. It simply reflects the difference in decision-making philosophies that underly the various criteria. Ultimately, the decision maker will have to choose the most appropriate criterion and then make the final decision accordingly. The major criticism of the criteria discussed in this section is they do not consider any information about the probabilities of the various states of nature. In the next section we discuss criteria that utilize probability information in selecting a decision alternative.

EXERCISES

1. Suppose that a decision maker faced with four decision alternatives and four states of nature develops the following profit payoff table:

	States of Nature			
Decisions	s_1	s_2	s_3	s_4
d_1	14	9	10	5
d_2	11	10	8	7
d_3	9	10	10	11
d_4	8	10	11	13

If the decision maker knows nothing about the chances or probability of occurrence of the four states of nature, what is the recommended decision under each of the following criteria:

a. Maximin?

b. Maximax?

c. Minimax regret?

Which decision criterion do you prefer? Explain. Is it important for the decision maker to establish the most appropriate decision criterion before analyzing the problem? Explain.

2. Assume that the payoff table in Exercise 1 provides *cost* rather than profit payoffs. What is the recommended decision under each of the following criteria:

a. Minimax?

b. Minimin?

c. Minimax regret?

3. McHuffter Condominiums, Inc. of Pensacola, Florida has recently purchased land near the Gulf and is attempting to determine the size of the condominium development it should build.

Three sizes of developments are being considered: small (d_1), medium (d_2), and large (d_3). At the same time an uncertain economy makes it difficult to ascertain the demand for the new condominiums. McHuffter's management realizes that a large development followed by a low demand could be very costly to the company. However, if McHuffter makes a conservative, small-development decision and then finds a high demand, the firm's profits will be lower than they might have been. With the three levels of demand—low, medium, and high—McHuffter's management has prepared the following payoff table giving profit in $1,000's:

		Demand	
Decision	*Low*	*Medium*	*High*
Small	400	400	400
Medium	100	600	600
Large	−300	300	900

If nothing is known about the demand probabilities, show the decision recommendations under the maximin, maximax, and minimax regret criteria.

19.4 CRITERIA FOR DECISION MAKING UNDER UNCERTAINTY WITH USE OF PROBABILITIES

In many situations good probability estimates can be developed for the states of nature. Two decision criteria which make use of these probability estimates in the selection of a decision alternative are expected monetary value and expected opportunity loss. Let us now see how these criteria can be applied when making decisions under uncertainty.

Expected Monetary Value

The *expected monetary value* criterion requires the analyst to compute the expected value for each decision alternative and then select the alternative yielding the best expected value. Let

$$P(s_j) = \text{probability of occurrence for state of nature } s_j,$$
$$N = \text{number of possible states of nature.}$$

Recall from our study of probability in Chapter 4 that since one and only one of the N states of nature can occur the associated probabilities must satisfy the following two conditions:

$$P(s_j) \geq 0 \qquad \text{for all states of nature } j, \tag{19.2}$$

$$\Sigma P(s_j) = P(s_1) + P(s_2) + \cdots + P(s_N) = 1. \tag{19.3}$$

The expected monetary value, denoted EMV, of a decision alternative d_i is given as follows:

Expected Monetary Value

$$\text{EMV}(d_i) = \Sigma P(s_j)V(d_i,s_j). \tag{19.4}$$

In words, the expected monetary value of a decision alternative is the sum of weighted payoffs for the alternative. The weight for a payoff is the probability of the associated state of nature and therefore the probability that the payoff occurs. Let us now return to the PSI problem to see how the expected monetary value criterion can be applied.

Suppose that PSI management believes that the high-acceptance state of nature, while very desirable, has only a .3 probability of occurrence, while the low-acceptance state of nature has a .7 probability. Thus $P(s_1) = .3$ and $P(s_2) = .7$. Using the payoff values $V(d_i,s_j)$ shown in Table 19.1 expected monetary values for the three decision alternatives can be calculated:

$$\begin{aligned}\text{EMV}(d_1) &= .3(200{,}000) + .7(-20{,}000) = \$46{,}000,\\ \text{EMV}(d_2) &= .3(150{,}000) + .7(\ 20{,}000) = \$59{,}000,\\ \text{EMV}(d_3) &= .3(100{,}000) + .7(\ 60{,}000) = \$72{,}000.\end{aligned}$$

Thus according to the expected monetary value criterion the small-system decision d_3, with an expected monetary value of \$72,000, is the recommended decision.

Note, however, that if the probabilities of the states of nature change, a different decision alternative might be selected. For example, if $P(s_1) = .6$ and $P(s_2) = .4$, we find the following expected monetary values:

$$\begin{aligned}\text{EMV}(d_1) &= .6(200{,}000) + .4(-20{,}000) = \$112{,}000,\\ \text{EMV}(d_2) &= .6(150{,}000) + .4(\ 20{,}000) = \$\ 98{,}000,\\ \text{EMV}(d_3) &= .6(100{,}000) + .4(\ 60{,}000) = \$\ 84{,}000.\end{aligned}$$

We now see that decision alternative d_1, with an expected monetary value of \$112,000, is the recommended decision with these probabilities.

Expected Opportunity Loss

In the previous section we defined the concept of an opportunity loss or regret associated with each decision alternative and state-of-nature combination. For PSI we developed the opportunity loss table shown in Table 19.4. The *expected opportunity loss* criterion uses the probabilities of the states of nature as weights for the opportunity loss values and computes the expected value of the opportunity loss (EOL) as follows:

Expected Opportunity Loss

$$\text{EOL}(d_i) = \Sigma P(s_j)R(d_i,s_j), \tag{19.5}$$

where $R(d_i,s_j)$ denotes the regret or opportunity loss for decision alternative d_i and state of nature s_j [see (19.1)].

Again with $P(s_1) = .3$ and $P(s_2) = .7$ for PSI and with the opportunity loss data of Table 19.4, the expected opportunity losses for the three decision alternatives become

$$\begin{aligned}\text{EOL}(d_1) &= .3(0) + .7(80{,}000) = \$56{,}000,\\ \text{EOL}(d_2) &= .3(\ 50{,}000) + .7(40{,}000) = \$43{,}000,\\ \text{EOL}(d_3) &= .3(100{,}000) + .7(0) = \$30{,}000.\end{aligned}$$

Since we would want to minimize the expected opportunity loss, the small-system decision d_3, with the smallest expected opportunity loss of $30,000, is recommended.

While expected opportunity loss offers an alternate criterion and approach to decision making under uncertainty, the optimal decision using the expected opportunity loss criterion will *always* be the same as the optimal decision using the expected monetary value criterion. Since the recommended decisions are identical, only one criterion need be applied in a given decision-making situation. In practice the expected monetary value has been the most widely used and accepted criterion for decision making under uncertainty. However, the expected opportunity loss provides the same result. As we will see in Section 19.6, the EOL associated with the best decision provides an indication of the value of collecting additional information about the probabilities for the states of nature.

EXERCISES

4. The payoff table presented in Exercise 1 is repeated below:

	States of Nature			
Decision	s_1	s_2	s_3	s_4
d_1	14	9	10	5
d_2	11	10	8	7
d_3	9	10	10	11
d_4	8	10	11	13

Suppose that the decision maker obtains some information that enables the following probability estimates: $P(s_1) = .5$, $P(s_2) = .2$, $P(s_3) = .2$, $P(s_4) = .1$.

a. Use the expected monetary value criterion to determine the optimal decision.

b. Now assume that the entries in the payoff table are costs. Use the expected monetary value criterion to determine the minimum cost solution.

c. Show that the expected opportunity loss criterion leads to the same decisions recommended by the expected monetary value criterion in parts a and b.

5. Hale's TV Productions is considering producing a pilot for a comedy series for a major TV network. While the network may reject the pilot and the series, it may also purchase the program for 1 or 2 years. Hale may decide to produce the pilot, but he also has an offer of $100,000 to transfer the rights for the series to a competitor. Hale's profits are summarized in the following payoff table, with profit in $1,000's:

	States of Nature		
Decision	*Reject*	*1 Year*	*2 Years*
Produce pilot (d_1)	−100	50	150
Sell to competitor (d_2)	100	100	100

If the probability estimates for the states of nature are $P(\text{reject}) = .2$, $P(1\text{ year}) = .3$, $P(2\text{ years}) = .5$, what should the company do?

6. Consider the McHuffter Condominium problem presented in Exercise 3. If $P(\text{low}) = .20$, $P(\text{medium}) = .35$, and $P(\text{high}) = .45$, what is the decision recommended under the expected monetary value criterion?

19.5 DECISION TREES

While decision problems involving a modest number of decision alternatives and a modest number of states of nature can be analyzed by use of payoff tables, they can also be analyzed by use of a graphical representation of the decision-making process. Recall that in Chapter 4 we showed how a graphical representation of a problem situation called a tree diagram could assist in the solution process. In this section we will present a similar diagram, called a decision tree, which can aid in the solution of decision theory problems.

Figure 19.1 shows a decision tree for the PSI computer-leasing problem. Note

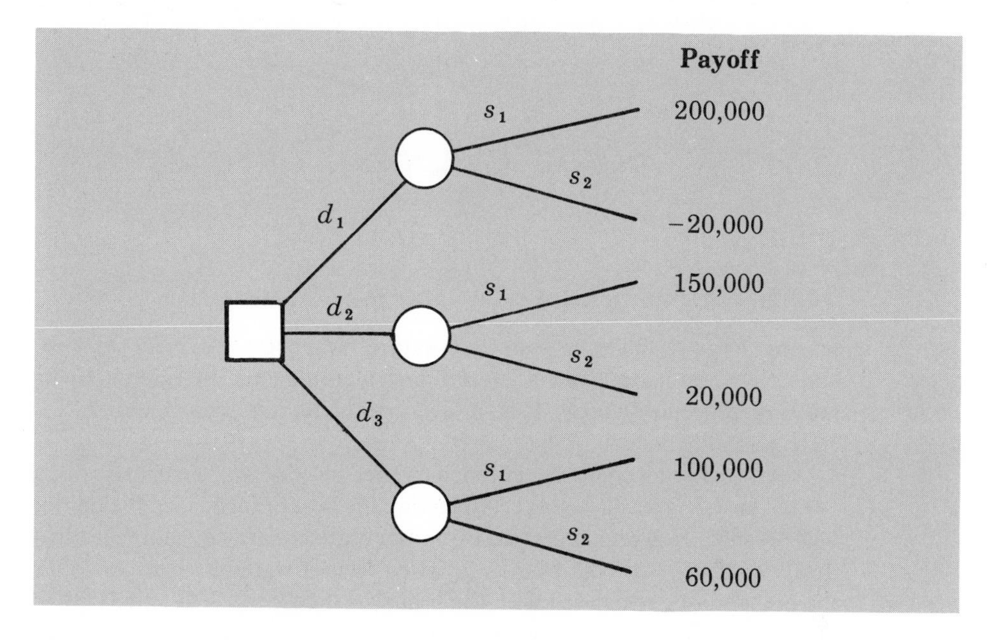

FIGURE 19.1 Decision Tree for the PSI Problem

that the decision tree, like the tree diagrams introduced in Chapter 4, shows the natural or logical progression that will occur in the decision-making process. First, the firm must make its decision (d_1,d_2, or d_3). Then, once the decision is in operation, the state of nature (s_1 or s_2) will occur. The number at each endpoint of the tree represents the payoff associated with a particular chain of events. For example, the topmost payoff of $200,000 arises whenever management makes the decision to purchase a large system (d_1) and market acceptance turns out to be high (s_1). The next-lower terminal point of $−20,000 is reached when management has made the decision to lease the large system (d_1) and the true state of nature turns out to be a low degree of market acceptance (s_2). Thus we see that each possible sequence of events for the PSI problem is represented in the decision tree.

According to the general terminology associated with decision trees, we will refer to the intersection or junction points of the tree as *nodes* and the arcs or connectors between the nodes as *branches*. Figure 19.2 shows the PSI decision tree with the nodes numbered 1 to 4 and the branches labeled as decision or state-of-nature branches. When the branches *leaving* a given node are decision branches, we refer to the node as a decision node. Decision nodes are denoted by squares. Similarly, when the branches

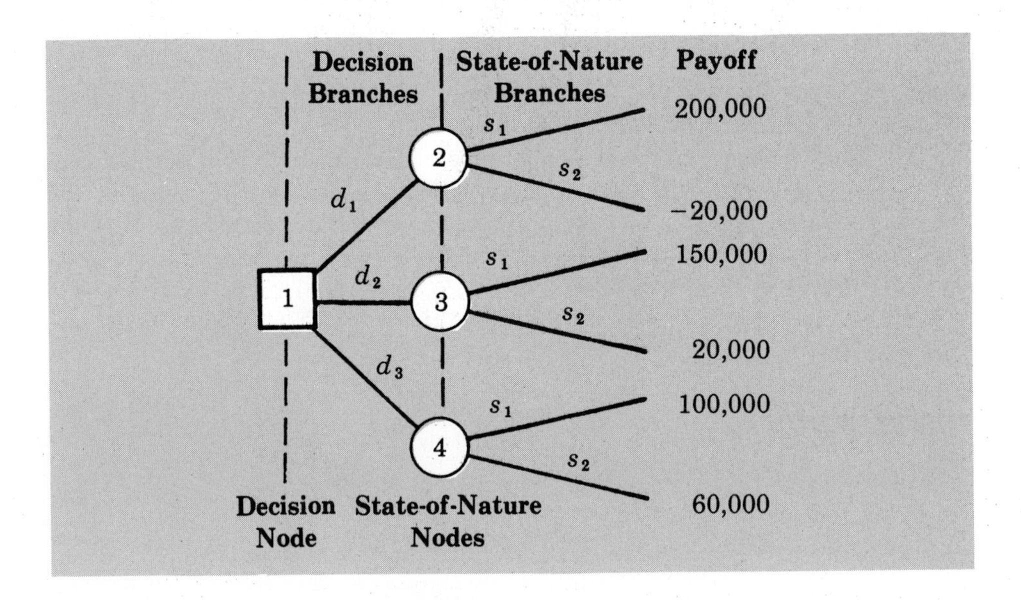

FIGURE 19.2 PSI Decision Tree with Node and Branch Labels

leaving a given node are state-of-nature branches, we refer to the node as a state-of-nature node. State-of-nature nodes are denoted by circles. Using this node-labeling procedure, node 1 is a decision node, whereas nodes 2, 3, and 4 are state-of-nature nodes.

At decision nodes the decision maker selects the particular decision branch (d_1,d_2, or d_3) that will be taken. Selecting the best branch is equivalent to making the best decision. However, the state-of-nature branches are not controlled by the decision maker. Thus the specific branch followed from a state-of-nature node depends upon the probabilities associated with the branches. For $P(s_1) = .3$ and $P(s_2) = .7$ we show the PSI decision tree with state-of-nature branch probabilities in Figure 19.3.

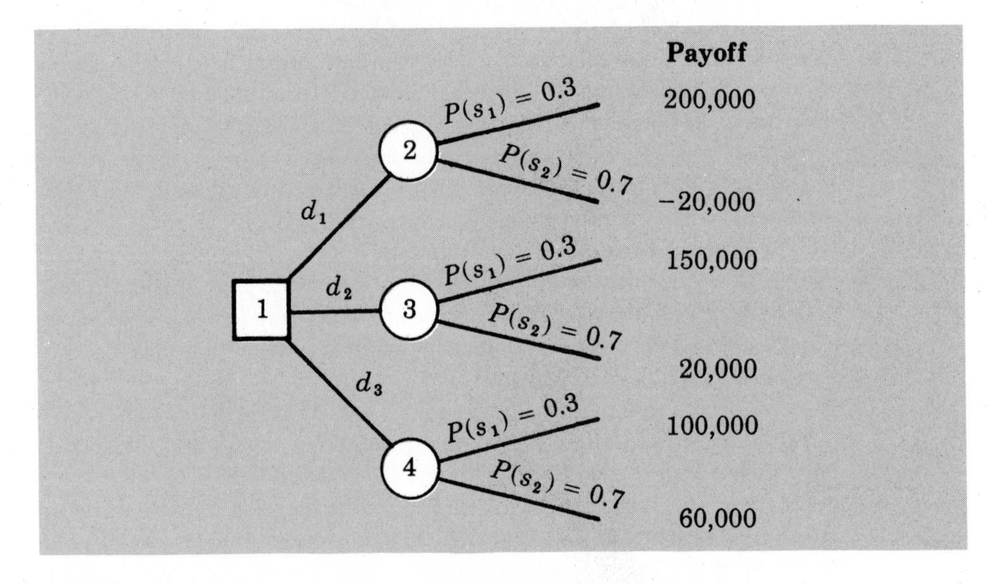

FIGURE 19.3 PSI Decision Tree with State-of-Nature Branch Probabilities

We will now use the branch probabilities and the expected monetary value criterion to arrive at the optimal decision for PSI. Working *backward* through the decision tree, we first compute the expected monetary value at each state-of-nature node. That is, at each state-of-nature node we weight the possible payoffs by their chance of occurrence. The expected monetary values for nodes 2, 3, and 4 are computed as follows:

$$\text{EMV(node 2)} = .3(200{,}000) + .7(-20{,}000) = 46{,}000,$$
$$\text{EMV(node 3)} = .3(150{,}000) + .7(\ \ 20{,}000) = 59{,}000,$$
$$\text{EMV(node 4)} = .3(100{,}000) + .7(\ \ 60{,}000) = 72{,}000.$$

We now continue backward through the tree to the decision node. Since the expected monetary values for nodes 2, 3, and 4 are known, the decision maker can view decision node 1 as shown in Figure 19.4. The decision maker controls the branch leaving a decision node and since we are trying to maximize expected profits, the best-decision branch at node 1 is d_3. Thus the decision tree analysis leads us to recommend d_3, with an expected monetary value of $72,000. Note that this is the same recommendation that was obtained using the expected monetary value criterion in conjunction with the payoff table.

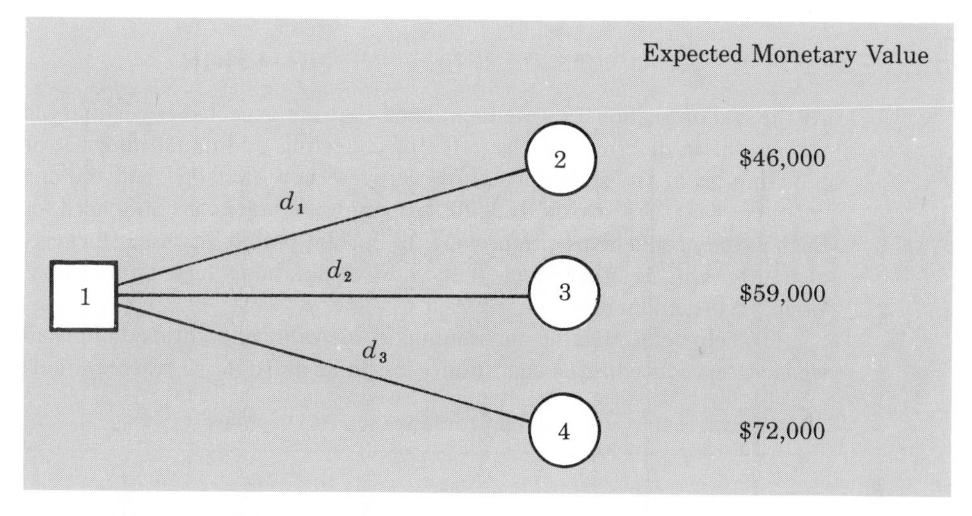

FIGURE 19.4 PSI Decision Node 1

We have seen how decision trees can be used to analyze a decision under uncertainty. While other decision problems may be substantially more complex than the PSI problem, if there are a reasonable number of decision alternatives and states of nature the decision tree approach outlined in this section can be used. First, the analyst must draw a decision tree consisting of decision and state-of-nature nodes and branches that describe the sequential nature of the problem. If the expected monetary value criterion is to be used, the next step is to determine the probabilities for each of the state-of-nature branches. The analyst then computes the expected monetary value at each state-of-nature node. The decision branch leading to the state-of-nature node with the best expected monetary value is then selected. The decision alternative associated with this branch is the best decision using the expected monetary value criterion.

EXERCISES

7. Construct a decision tree for the Hale's TV Productions problem (Exercise 5). What is the expected value at each state-of-nature node?
8. Construct a decision tree for the McHuffter Condominiums problem (Exercise 6). What is the expected value at each state-of-nature node? What is the optimal decision?
9. Martin's Service Station is considering investing in a heavy duty snowplow this fall. Martin has analyzed the situation carefully and feels that this would be a very profitable investment if the snowfall is heavy. A small profit could still be made if the snowfall is moderate, but Martin would lose money if the snowfall is light. Specifically, Martin forecasts a profit of $7,000 if snowfall is heavy and $2,000 if it is moderate, but a $9,000 loss if it is light. From the weather bureau's long range forecast Martin estimates *P*(heavy snowfall) = .4, *P*(moderate snowfall) = .3, and *P*(light snowfall) = .3. Prepare a decision tree for Martin's problem.

19.6 EXPECTED VALUE OF PERFECT INFORMATION

At the end of Section 19.4 we mentioned that the expected opportunity loss criterion was useful in determining the value of collecting additional information about the probabilities of the states of nature. Suppose now that PSI had the opportunity to conduct a market research study to thoroughly evaluate consumer need for its service. Such a study could help by improving the current probability assessments for the states of nature. On the other hand, if the cost of obtaining such information exceeds its value, PSI should not seek it.

To help determine the maximum possible value of additional information for PSI we have reproduced PSI's opportunity loss table as Table 19.6. Recall that the optimal

TABLE 19.6 Opportunity Loss Table for the PSI Problem

	States of Nature	
Decision Alternatives	*High Acceptance* (s_1)	*Low Acceptance* (s_2)
Large system (d_1)	0	$80,000
Medium system (d_2)	$ 50,000	$40,000
Small system (d_3)	$100,000	0

decision using either the expected monetary value or expected opportunity loss criterion was to lease the small system, d_3, when $P(s_1) = .3$ and $P(s_2) = .7$. Let us now concentrate on the losses associated with this decision.

We see that if state of nature s_1 occurs, d_3 will not have been the best decision and PSI will have an opportunity loss of $100,000 because d_1 was not selected. That is, given perfect information that s_1 was going to occur, PSI could increase its profit $100,000 by selecting d_1 instead of d_3. On the other hand, if state of nature s_2 occurs, d_3 will have been the best decision and the opportunity loss will be $0. Thus perfect information that s_2 was going to occur would be of no value to the company, since PSI would have made the optimal decision without it.

What is the expected value of this perfect information? With $P(s_1) = .3$ and

$P(s_2) = .7$ and the opportunity loss values, we see that 30% of the time PSI could save $100,000, while 70% of the time the savings would be $0. Thus the *expected value of perfect information* (EVPI) for PSI's problem is

$$\text{EVPI} = (.3)(\$100{,}000) + (.7)(\$0) = \$30{,}000.$$

In Table 19.7 we summarize what is involved in computing the expected value of perfect information. Note that EVPI is the same as the *expected opportunity loss of the optimal decision* (that is, d_3; see Section 19.4). Thus if we had used expected opportunity loss as a decision criterion, or if in our analysis we had computed the expected opportunity loss of the optimal decision, we would have already computed the expected value of perfect information.

TABLE 19.7 Perfect Information in the PSI Problem

Possible Information	*Action and Payoff If Decision Is Made Before Information Is Available*	*Action and Payoff If Decision Is Made After Information Is Available*	*Value of Perfect Information (Opportunity Loss of d_3)*	*Probability of Information*
High acceptance	Lease small system: $100,000	Lease large system: $200,000	$100,000	.3
Low acceptance	Lease small system $60,000	Lease small system: $60,000	$0	.7
		EVPI = .3(100,000) + .7(0) = $30,000		

Generally speaking, we would not expect a market research study to provide "perfect" information, but the information provided might be worth a good portion of the $30,000. In any case, PSI's management knows it should never pay more than $30,000 for any information, no matter how good. Provided that the market survey cost is reasonably small—say, $5,000 to $10,000—it appears economically desirable for PSI to consider the market research study.

Before leaving this section we note the general expression for computing the expected value of perfect information (EVPI) from a payoff table. Let

d^* = optimal decision for the problem prior to obtaining perfect information,
$P(s_j)$ = probability of state of nature s_j,
N = number of states of nature,
$R(d^*,s_j)$ = opportunity loss of regret value for decision d^* and state of nature s_j.

Then we have the following expression for EVPI:

Expected Value of Perfect Information

$$\text{EVPI} = \Sigma P(s_j)R(d^*,s_j). \tag{19.6}$$

EXERCISES

10. Consider the Hale's TV Productions problem (Exercise 5). What is the maximum that Hale should be willing to pay for inside information on what the network will do?
11. Consider the Martin's Service Station problem (Exercise 9). What is the expected value of perfect information?
12. Consider the McHuffter Condominium problem (Exercise 6). What is the expected value of perfect information?

19.7 DECISION THEORY WITH EXPERIMENTS

In decision theory situations involving decision making under uncertainty, we have seen how probability information about the states of nature affects the expected value calculations and thus possibly the decision recommendation. Frequently decision makers have preliminary or prior probability estimates for the states of nature which are initially the best probability values available. However, in order to make the best possible decision, the decision maker may want to seek additional information about the states of nature. This new information can be used to revise or update the prior probabilities so that the final decision is based upon more accurate probability estimates for the states of nature.

The seeking of additional information is most often accomplished through experiments designed to provide the most current data available about the states of nature. Raw material sampling, product testing, and test market research are examples of experiments that may enable a revision or updating of the state-of-nature probabilities. In the remaining sections of this chapter we will consider the PSI computer-leasing problem and show how new information can be used to revise the state-of-nature probabilities. We will then show how these revised probabilities can be used to develop an optimal decision strategy for PSI.

Recall that management had assigned a probability of .3 to the state of nature s_1 and a probability of .7 to the state of nature s_2. At this point we will refer to these initial probability estimates, $P(s_1)$ and $P(s_2)$, as the *prior* probabilities for the states of nature. Using these prior probabilities we found that the decision to lease the small system, d_3, was optimal, yielding an expected monetary value of $72,000. Applying the criterion of minimizing expected opportunity loss, we obtained the same decision recommendation and also learned that the expected opportunity loss (EOL) of the optimal decision d_3 was $30,000. In addition, we showed that since the expected value of perfect information was equal to the EOL of the optimal decision, the expected value of new information about the states of nature could potentially be worth as much as $30,000.

Suppose that PSI decides to consider hiring a market research firm to study the potential acceptance of the PSI service. The market research study will provide new information which can be combined with the prior probabilities through a Bayesian procedure (see Chapter 4) to obtain updated or revised probability estimates for the states of nature. These revised probabilities are called *posterior* probabilities. The complete process of revising probabilities is depicted in Figure 19.5.

We usually refer to the new information obtained through research or experimentation as an *indicator*. Since in many cases the experiment conducted to obtain the additional information will consist of taking a statistical sample, the new information is also often referred to as *sample information*.

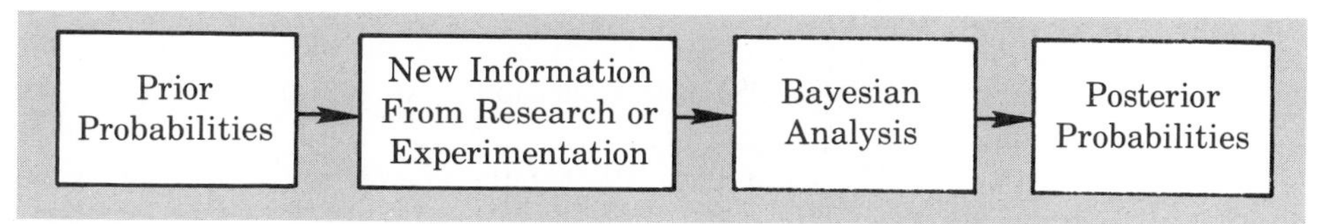

FIGURE 19.5 Probability Revision Based on New Information

Using the indicator terminology, we can denote the outcomes of the PSI marketing research study as follows:

I_1 = favorable market research report—i.e., in the market research study the individuals contacted generally express considerable interest in PSI's services;

I_2 = unfavorable market research report—i.e., in the market research study the individuals contacted generally express little interest in PSI's services.

If we are given one of these possible indicators, our objective is to provide improved estimates of the probabilities of the various states of nature according to the findings of the market research study. The end result of the Bayesian revision process depicted in Figure 19.5 is a set of posterior probabilities of the form $P(s_j \mid I_k)$, where $P(s_j \mid I_k)$ represents the conditional probability that state of nature s_j will occur given that the outcome of the market research study was indicator I_k.

To make effective use of this indicator information we must know something about the probability relationships between the indicators and the states of nature. For example, in the PSI problem, if the state of nature ultimately turns out to be high customer acceptance, what is the probability that the market research study will result in a favorable report? In this case we are asking about the conditional probability of indicator I_1 given state of nature s_1, written $P(I_1 \mid s_1)$. In order to carry out the analysis, we will need conditional probability relationships for all indicators given all states of nature—i.e., $P(I_1 \mid s_1)$, $P(I_1 \mid s_2)$, $P(I_2 \mid s_1)$, and $P(I_2 \mid s_2)$. Historical relative frequency data and/or subjective probability estimates are usually the primary source for these conditional probability values.

In the PSI case the past record of the marketing research company on similar studies has led to the following estimates of the relevant conditional probabilities:

	Market Research Report	
States of Nature	*Favorable* (I_1)	*Unfavorable* (I_2)
High acceptance (s_1)	$P(I_1 \mid s_1) = .8$	$P(I_2 \mid s_1) = .2$
Low acceptance (s_2)	$P(I_1 \mid s_2) = .1$	$P(I_2 \mid s_2) = .9$

Note that these probability estimates indicate that a great degree of confidence can be placed in the market research report. When the true state of nature is s_1, the market research report will be favorable 80% of the time and unfavorable only 20%. When the true state is s_2, the report will make the correct indication 90% of the time. Now let us see how this additional information can be incorporated into the decision-making process.

19.8 DEVELOPING A DECISION STRATEGY

A decision strategy is simply a policy or decision rule that is to be followed by the decision maker. In the PSI case a decision strategy would consist of a rule to follow according to the outcome of the market research study. The rule would recommend a particular decision based upon whether the market research report was favorable or unfavorable. We will employ a decision tree analysis to find the optimal decision strategy for PSI.

Figure 19.6 shows the decision tree for the PSI computer-leasing problem provided that a market research study is conducted. Note that as you move from left to right the tree shows the natural or logical order that will occur in the decision-making process. First, the firm will obtain the market research report indicator (I_1 or I_2); then, a decision (d_1,d_2, or d_3) will be made; finally, the state of nature (s_1 or s_2) will occur. The decision and the state of nature combine to provide the final profit or payoff.

Using decision tree terminology, we have now introduced an *indicator node,* node 1, and *indicator branches* I_1 and I_2. Since the branches emanating from indicator nodes are not under the control of the decision maker but are determined by chance, these nodes are depicted by a circle just like the state-of-nature nodes. We see that nodes 2 and 3 are decision nodes, while nodes 4, 5, 6, 7, 8, and 9 are state-of-nature nodes. For decision nodes the decision maker must select the specific branch d_1, d_2, or d_3 that will be taken. Selecting the best decision branch is equivalent to making the best decision. However, since the indicator and state-of-nature branches are not controlled by the decision maker, the specific branch leaving an indicator or a state-of-nature node will depend upon the probability associated with the branch. Thus before we can carry out an analysis of the decision tree and develop a decision strategy we must compute the probability of each indicator branch $P(I_k)$ and the probability of each state-of-nature branch. Note from the decision tree that the state-of-nature branches occur *after* the indicator branches. Thus when we attempt to compute state-of-nature branch probabilities we will need to consider which indicator was previously observed. That is, we will express state-of-nature probabilities in terms of the probability of state of nature s_j given that indicator I_k was observed. Thus all state-of-nature probabilities will be expressed in a $P(s_j \mid I_k)$ form.

Computing Branch Probabilities: The Decision Tree Approach

In Section 4.6 we introduced a method of computing probabilities known as Bayes' theorem. Recall that the purpose of Bayes' theorem was to enable us to revise initial estimates of the probability of events according to some additional information about the events. In this section we will show how Bayes' theorem can be used to revise the prior probabilities for the PSI problem with use of the market research indicator. The prior probabilities for the states of nature in the PSI problem were given as $P(s_1) = .3$ and $P(s_2) = .7$. In Section 19.7 we identified the relationship between the market research indicator and states of nature with the probabilities

$$P(I_1 \mid s_1) = .8, \qquad P(I_2 \mid s_1) = .2,$$
$$P(I_1 \mid s_2) = .1, \qquad P(I_2 \mid s_2) = .9.$$

However, in order to further develop the decision tree in Figure 19.6 we need indicator branch probabilities $P(I_k)$ and state-of-nature branch probabilities $P(s_j \mid I_k)$. The problem now facing us is how to use the given prior probability estimates $P(s_j)$ and the

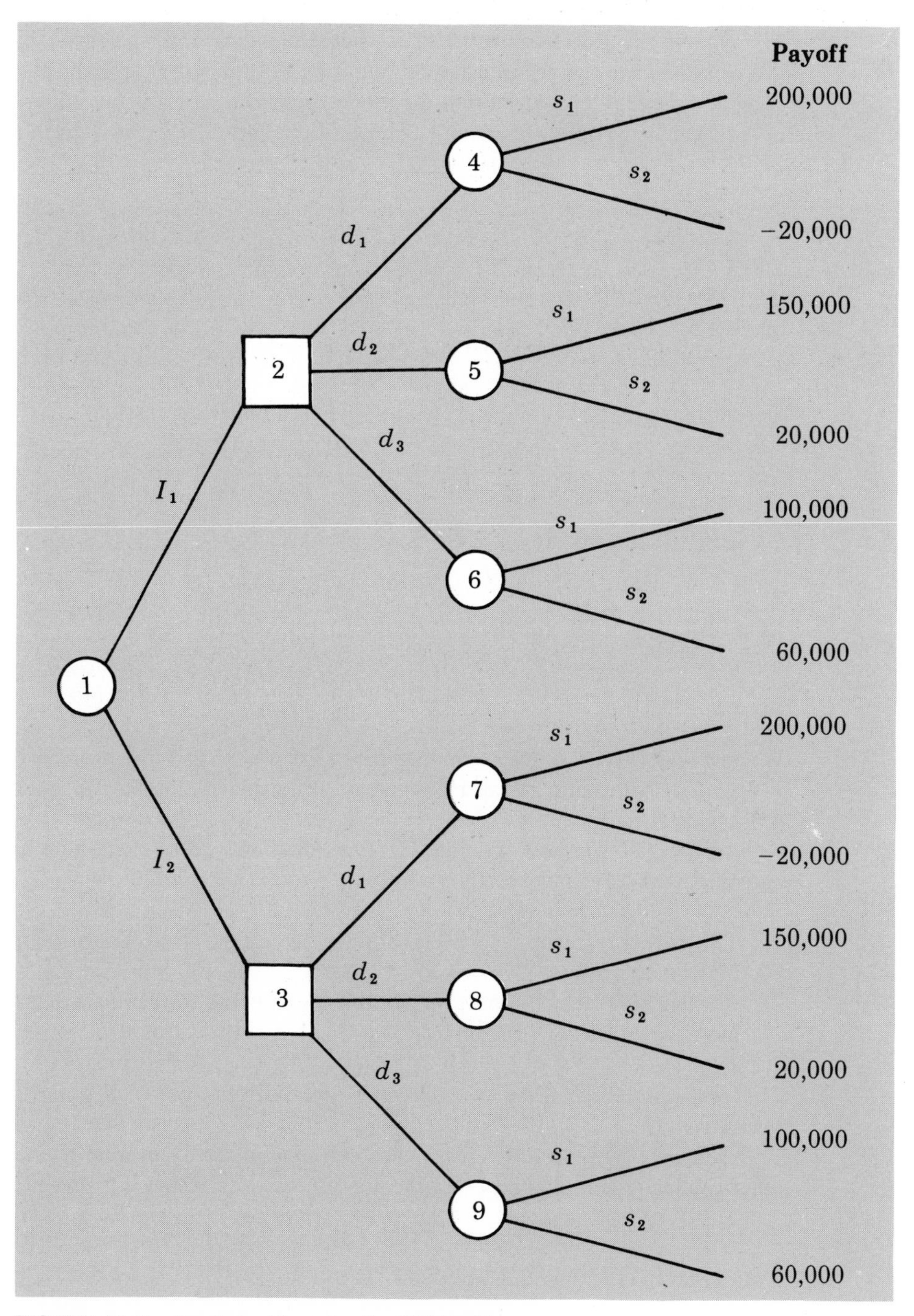

FIGURE 19.6 Decision Tree for the PSI Problem

given conditional probability estimates $P(I_k \mid s_j)$ to calculate the branch probabilities $P(I_k)$ and $P(s_j \mid I_k)$.

Let us concentrate on computing the state-of-nature branch probabilities $P(s_j \mid I_k)$. In the process we will also compute the indicator branch probabilities $P(I_k)$. We will illustrate the procedure by calculating $P(s_1 \mid I_1)$, the probability that market acceptance is high (s_1) given that the market research report is favorable (I_1). Shown in (19.7) is the basic expression for the conditional probability of s_1 given I_1:

$$P(s_1 \mid I_1) = \frac{P(I_1 \cap s_1)}{P(I_1)}, \tag{19.7}$$

where $P(I_1 \cap s_1)$ is referred to as the joint probability. The value of $P(I_1 \cap s_1)$ gives the probability that I_1, a favorable market research report, and s_1, a high consumer acceptance, both occur.

Unfortunately, we cannot use (19.7) directly to compute $P(s_1 \mid I_1)$, since both $P(I_1 \cap s_1)$ and $P(I_1)$ are unknown. However, substituting I_1 for s_1 in (19.7) and vice-versa provides the following conditional probability expression:

$$P(I_1 \mid s_1) = \frac{P(I_1 \cap s_1)}{P(s_1)}. \tag{19.8}$$

Using (19.8) and solving for $P(I_1 \cap s_1)$, we have

$$P(I_1 \cap s_1) = P(I_1 \mid s_1)P(s_1). \tag{19.9}$$

By substituting (19.9) for $P(I_1 \cap s_1)$ in (19.7), we have

$$P(s_1 \mid I_1) = \frac{P(I_1 \mid s_1)P(s_1)}{P(I_1)}. \tag{19.10}$$

We are now a step closer to finding $P(s_1 \mid I_1)$, since we have been given values of $P(I_1 \mid s_1) = .8$ and $P(s_1) = .3$. However, we still need a value for the indicator branch probability $P(I_1)$.

In order to see how we find $P(I_1)$, we first recognize that there are only two outcomes that correspond to I_1:

1. The market research report is favorable (I_1) and the state of nature turns out to be high acceptance (s_1), written ($I_1 \cap s_1$).
2. The market research report is favorable (I_1) and the state of nature turns out to be low acceptance (s_2), written ($I_1 \cap s_2$).

The probabilities of these two outcomes are written $P(I_1 \cap s_1)$ and $P(I_1 \cap s_2)$, respectively.

Since ($I_1 \cap s_1$) and ($I_1 \cap s_2$) are mutually exclusive events (if one occurs, the other cannot), the probability that the market research report is favorable is given by

$$P(I_1) = P(I_1 \cap s_1) + P(I_1 \cap s_2). \tag{19.11}$$

Recall that in (19.9) we showed that

$$P(I_1 \cap s_1) = P(I_1 \mid s_1)P(s_1).$$

Using this expression form with s_2 instead of s_1, we have

$$P(I_1 \cap s_2) = P(I_1 \mid s_2)P(s_2).$$

Thus (19.11) can now be revised as follows:

$$P(I_1) = P(I_1 \mid s_1)P(s_1) + P(I_1 \mid s_2)P(s_2). \tag{19.12}$$

Since $P(I_1 \mid s_1)$, $P(I_1 \mid s_2)$, $P(s_1)$, and $P(s_2)$ are known, (19.12) can be used to calculate $P(I_1)$. Then (19.10) can be used to calculate $P(s_1 \mid I_1)$.

Before we make these calculations let us examine the general form of (19.12) and (19.10). Specifically, we want to be able to calculate any indicator branch probability $P(I_k)$ and any state-of-nature branch probability $P(s_j \mid I_k)$. Assume that we have N states of nature, $s_1, s_2, \ldots, s_N$. Now (19.11) generalizes to

$$P(I_k) = P(I_k \cap s_1) + P(I_k \cap s_2) + \cdots + P(I_k \cap s_N). \tag{19.13}$$

We can use the relationship established by (19.9) to write

$$P(I_k) = P(I_k \mid s_1)P(s_1) + P(I_k \mid s_2)P(s_2) + \cdots + P(I_k \mid s_N)P(s_N) \tag{19.14}$$

or

$$P(I_k) = \Sigma P(I_k \mid s_j)P(s_j). \tag{19.15}$$

Thus (19.10) generalizes to

$$P(s_j \mid I_k) = \frac{P(I_k \mid s_j)P(s_j)}{P(I_k)}, \tag{19.16}$$

where $P(I_k)$ is computed from either (19.14) or (19.15). Equation (19.16) is known as *Bayes' rule* or *Bayes' theorem.* It provides a general expression for computing the conditional probability of state of nature s_j given that indicator I_k has occurred. Thus with prior probabilities $P(s_j)$ and conditional probabilities of the form $P(I_k \mid s_j)$, the Bayesian procedures defined in (19.15) and (19.16) can be used to compute indicator probabilities $P(I_k)$ and revised or posterior state-of-nature probabilities $P(s_j \mid I_k)$.

Let us return to the PSI decision. We have been given the prior probabilities $P(s_1) = .3$ and $P(s_2) = .7$ and the conditional probabilities $P(I_1 \mid s_1) = .8$, $P(I_2 \mid s_1) = .2$, $P(I_1 \mid s_2) = .1$, and $P(I_2 \mid s_2) = .9$. Moving from left to right through the decision tree (Figure 19.6), we can use (19.15) to first find the probabilities of the indicator branches:

$$\begin{aligned} P(I_1) &= P(I_1 \mid s_1)P(s_1) + P(I_1 \mid s_2)P(s_2) \\ &= (.8)(.3) + (.1)(.7) = .31, \\ P(I_2) &= P(I_2 \mid s_1)P(s_1) + P(I_2 \mid s_2)P(s_2) \\ &= (.2)(.3) + (.9)(.7) = .69. \end{aligned}$$

Note that after we obtained $P(I_1)$, we could have found $P(I_2)$ by using the fact that there are only two indicators, I_1 and I_2. Thus $P(I_1) + P(I_2) = 1$, and hence $P(I_2) = 1 - P(I_1) = 1 - .31 = .69$.

The above calculations tell us that with the given prior probabilities for the states of nature, $P(s_j)$, and the conditional probabilities, $P(I_k \mid s_j)$, the probability of a favorable market research report I_1 is .31 and the probability of an unfavorable report I_2 is .69.

Now we can use (19.16) to compute the revised or posterior probabilities for the state-of-nature branches, given an indicator of I_1 or I_2 from the market research study.

For example, if indicator I_1 occurs, we have

$$P(s_1 \mid I_1) = \frac{P(I_1 \mid s_1)P(s_1)}{P(I_1)} = \frac{(.8)(.3)}{(.31)} = .7742,$$

$$P(s_2 \mid I_1) = \frac{P(I_1 \mid s_2)P(s_2)}{P(I_1)} = \frac{(.1)(.7)}{(.31)} = .2258.$$

Similar calculations for the indicator I_2 will show $P(s_1 \mid I_2) = .0870$ and $P(s_2 \mid I_2) = .9130$.

These revised probabilities provide the probability estimates of the state-of-nature branches after the market research study is complete. For example, if the report is favorable, $P(s_1 \mid I_1) = .7742$ indicates that there is a .7742 probability that the market acceptance will be high (s_1). However, we realize that the final market acceptance of the PSI service may be low, even though the market research report is favorable. This probability is given by $P(s_2 \mid I_1) = .2258$. Note, however, that the state-of-nature probabilities given I_1 have been substantially revised from the prior values of $P(s_1) = .3$ and $P(s_2) = .7$. Similar interpretations can be made for the case when the market research report is unfavorable (I_2). Figure 19.7 shows the PSI decision tree after all indicator and state-of-nature branch probabilites have been computed.

Although the above procedure can be used to compute branch probabilities for decision theory problems when the decision tree approach is employed, the calculations can become quite cumbersome as the problem size grows larger. Thus in order to assist you in applying Bayes' rule to compute branch probabilities, we present a tabular procedure that will make it easier to carry out the computations, especially for large decision theory problems.

A Tabular Procedure for Computing Branch Probabilities

In Section 4.6 we introduced a tabular approach for the Bayes' theorem calculations. The procedure used for computing the probabilities of the indicator and state-of-nature branches can be carried out by utilizing a similar tabular approach. First, for each indicator I_k we form a table consisting of the following five column headings:

Column 1: States of nature s_j,
Column 2: Prior probabilities $P(s_j)$,
Column 3: Conditional probabilities $P(I_k \mid s_j)$,
Column 4: Joint probabilities $P(I_k \cap s_j)$,
Column 5: Posterior probabilities $P(s_j \mid I_k)$.

Then, the following procedure can be utilized to calculate $P(I_k)$ and the $P(s_j \mid I_k)$ values:

Step 1: In column 1 list the states of nature appropriate to the problem being analyzed.
Step 2: In column 2 enter the prior probability corresponding to each state of nature listed in column 1.
Step 3: In column 3 write the appropriate value of $P(I_k \mid s_j)$ for each state of nature specified in column 1.
Step 4: To compute each entry in column 4, multiply each entry in column 2 by the corresponding entry in column 3.

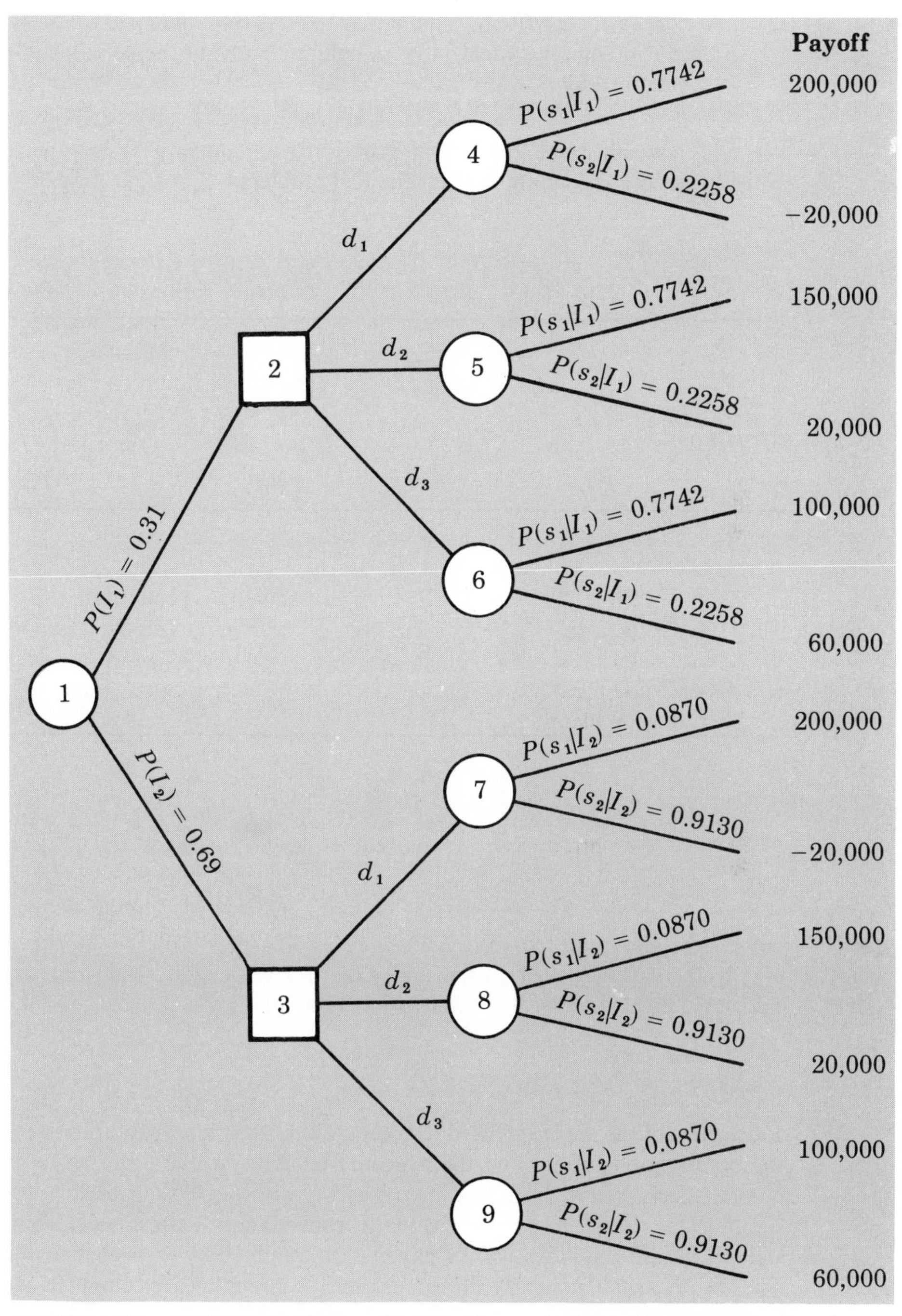

FIGURE 19.7 PSI Decision Tree with Indicator and State-of-Nature Branch Probabilities

Step 5: Add the entries in column 4. The sum is the value of $P(I_k)$. For convenience, write the sum under the last column entry.

Step 6: To compute each entry in column 5 divide the corresponding entry in column 4 by $P(I_k)$.

We will now illustrate the above procedure to compute $P(I_1)$ and the revised probabilities corresponding to I_1 for the PSI problem:

Steps 1, 2, and 3:

s_j	$P(s_j)$	$P(I_1 \mid s_j)$	$P(I_1 \cap s_j)$	$P(s_j \mid I_1)$
s_1	.3	.8		
s_2	.7	.1		

Steps 4 and 5:

s_j	$P(s_j)$	$P(I_1 \mid s_j)$	$P(I_1 \cap s_j)$	$P(s_j \mid I_1)$
s_1	.3	.8	.24	
s_2	.7	.1	.07	
			$P(I_1) = .31$	

Step 6:

s_j	$P(s_j)$	$P(I_1 \mid s_j)$	$P(I_1 \cap s_j)$	$P(s_j \mid I_1)$
s_1	.3	.8	.24	$\frac{.24}{.31}$ .7742
s_2	.7	.1	.07	$\frac{.07}{.31} = .2258$
			$P(I_1) = .31$	

Note that $P(I_1)$, $P(s_1 \mid I_1)$, and $P(s_2 \mid I_1)$ are exactly the same as the values calculated by applying (19.15) and (19.16) directly.

An Optimal Decision Strategy

Regardless of the approach used to compute the branch probabilities, we can now use the branch probabilities and the expected monetary value criterion to arrive at the optimal decision for PSI. Working *backward* through the decision tree, we first compute the expected monetary value at each state-of-nature node. That is, at each state-of-nature node the possible payoffs are weighted by their chance of occurrence. Thus the expected monetary values for nodes 4 through 9 are computed as follows:

$$\text{EMV(node 4)} = (.7742)(200{,}000) + (.2258)(-20{,}000) = 150{,}324,$$

$$\text{EMV(node 5)} = (.7742)(150{,}000) + (.2258)(20{,}000) = 120{,}646,$$

$$\text{EMV(node 6)} = (.7742)(100{,}000) + (.2258)(60{,}000) = 90{,}968,$$

$$\text{EMV(node 7)} = (.0870)(200{,}000) + (.9130)(-20{,}000) = -860,$$

$$\text{EMV(node 8)} = (.0870)(150{,}000) + (.9130)(\ \ 20{,}000) = \ \ 31{,}310,$$

$$\text{EMV(node 9)} = (.0870)(100{,}000) + (.9130)(\ \ 60{,}000) = \ \ 63{,}480.$$

We now continue backward through the decision tree to the decision nodes. Because the expected values for nodes 4, 5, and 6 are known, the decision maker can view decision node 2 as shown in Figure 19.8. Since the decision maker controls the branch

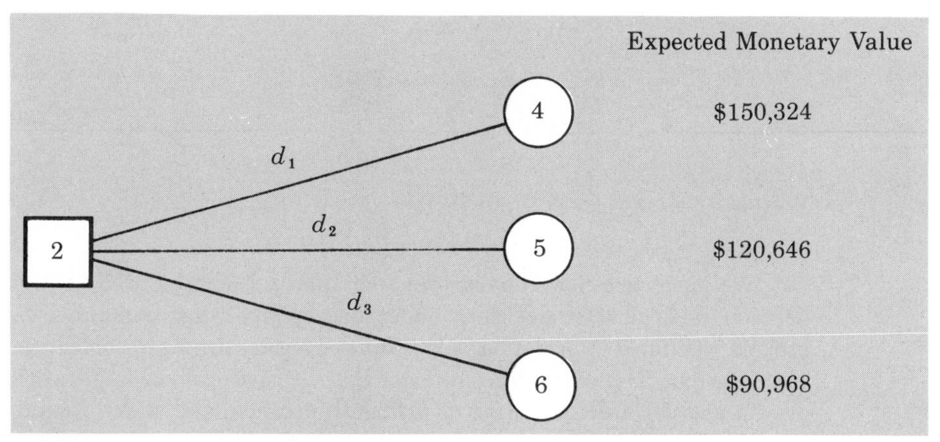

FIGURE 19.8 PSI Decision Node 2

leaving a decision node, and since we are trying to maximize expected profits, the optimal decision at node 2 is d_1. Since d_1 leads to an expected value of \$150,324, we say EMV(node 2) = \$150,324 if the optimal decision of d_1 is made.

A similar analysis of decision node 3 shows that the optimal decision branch at this node is d_3. Thus EMV(node 3) becomes \$63,480 provided that the optimal decision of d_3 is made.

As a final step, we can continue working backward to the indicator node and establish its expected value. The branches at node 1 are shown in Figure 19.9. Since node 1 has probability branches, we cannot select the best branch. Rather, we must compute the expected value over all possible branches. Thus we have

$$\text{EMV(Node 1)} = (.31)(\$150{,}324) + (.69)(\$63{,}480) = \$90{,}402.$$

The value of \$90,402 is viewed as the expected value of the optimal decision strategy when the market research study is used or as the expected monetary value using sample information (i.e., that provided by the market research report).

Note that the final decision has not yet been determined. We will need to know the results of the market research study before deciding whether to lease a large system (d_1) or a small system (d_3). The results of the decision theory analysis at this point, however, have provided us with the following optimal *decision strategy* if the market research study is conducted:

Decision Strategy

If	*Then*
Report favorable (I_1)	Lease large system (d_1)
Report unfavorable (I_2)	Lease small system (d_3)

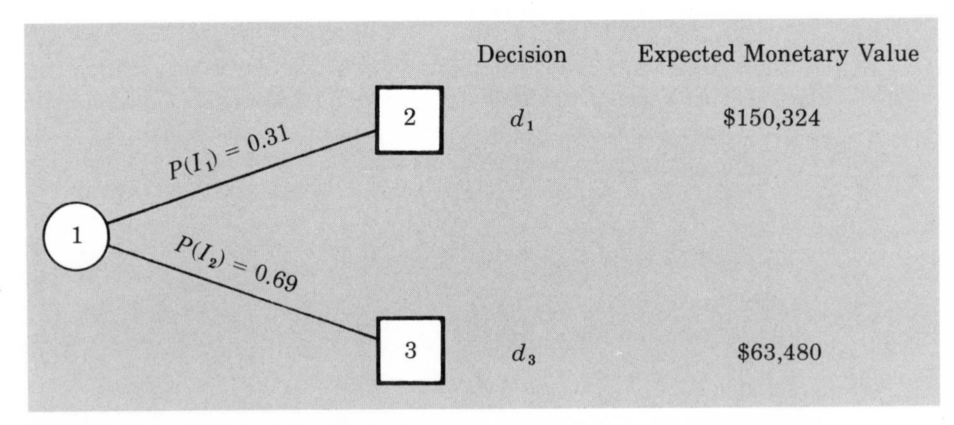

FIGURE 19.9 PSI Decision Node 1

We have seen how the decision tree approach can be used to develop optimal decision strategies for decisions under uncertainty when experiments are used to provide additional information. While other decision strategy problems may not be as simple as the PSI problem, the approach that we have outlined is still applicable. First, draw a decision tree consisting of indicator, decision, and state-of-nature nodes and branches such that the tree describes the specific decision-making process. Posterior probability calculations must be made in order to establish indicator and state-of-nature branch probabilities. Then by working backward through the tree, computing expected values at state-of-nature and indicator nodes, and selecting the best decision branch at decision nodes, the analyst can determine an optimal decision strategy and the associated expected value for the problem.

EXERCISES

13. Suppose that you are given a decision situation with three possible states of nature: s_1, s_2, and s_3. The prior probabilities are $P(s_1) = .2$, $P(s_2) = .5$, and $P(s_3) = .3$. Indicator information I is obtained, and it is known that $P(I \mid s_1) = .1$, $P(I \mid s_2) = .05$, and $P(I \mid s_3) = .2$. Compute the revised or posterior probabilities $P(s_1 \mid I)$, $P(s_2 \mid I)$, and $P(s_3 \mid I)$.

14. The payoff table for a decision problem with two states of nature and three decision alternatives is presented below:

	s_1	s_2
d_1	15	10
d_2	10	12
d_3	8	20

The prior probabilities for s_1 and s_2 are $P(s_1) = .8$ and $P(s_2) = .2$.

a. Use only the prior probabilities and the expected monetary value criterion to find the optimal decision.

b. Suppose that some indicator information I is obtained with $P(I \mid s_1) = .2$ and $P(I \mid s_2) = .75$. Find the posterior probabilities $P(s_1 \mid I)$ and $P(s_2 \mid I)$. Recommend a decision alternative based on these probabilities.

15. Consider the following decision tree representation of a decision theory problem with two indicators, two decision alternatives, and two states of nature:

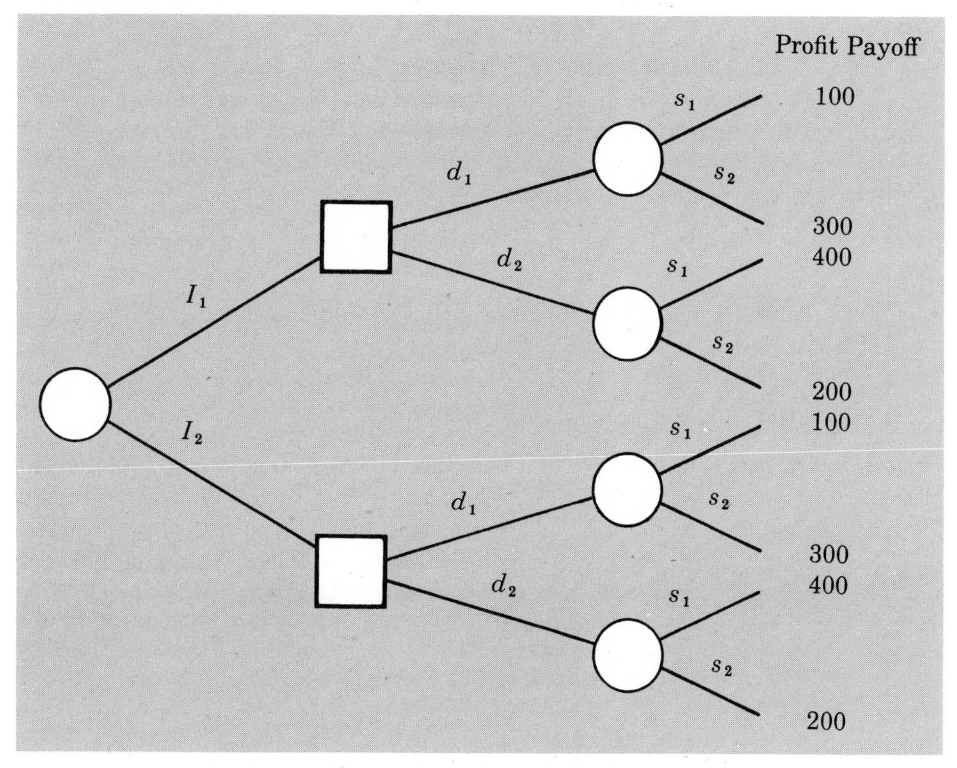

Assume that the following probability information is given:

$$P(s_1) = .4, \qquad P(I_1 \mid s_1) = .8, \qquad P(I_2 \mid s_1) = .2,$$
$$P(s_2) = .6, \qquad P(I_1 \mid s_2) = .4, \qquad P(I_2 \mid s_2) = .6.$$

a. What are the values for $P(I_1)$ and $P(I_2)$?

b. What are the values of $P(s_1 \mid I_1)$, $P(s_2 \mid I_1)$, $P(s_1 \mid I_2)$, and $P(s_2 \mid I_2)$?

c. Use the decision tree approach and determine the optimal decision strategy. What is the expected monetary value of your solution?

19.9 EXPECTED VALUE OF SAMPLE INFORMATION

In the PSI problem management now has a decision strategy of leasing the large computer system if the market research report is favorable and leasing the small computer system if the market research report is unfavorable. Since the additional information provided by the market research firm will result in an added cost for PSI in terms of the fee paid to the research firm, PSI management may question the value of this market research information.

The value of information is often measured by calculating what is referred to as the *expected value of sample information* (EVSI). For maximization problems[2]

$$\text{EVSI} = \begin{bmatrix}\text{Expected value of the}\\ \text{optimal decision } \textit{with}\\ \text{sample information}\end{bmatrix} - \begin{bmatrix}\text{Expected value of the}\\ \text{optimal decision } \textit{without}\\ \text{sample information}\end{bmatrix}. \qquad (19.17)$$

For PSI the market research information is considered the "sample" information. The decision tree calculations indicated that the expected value of the optimal decision with the market research information was \$90,402. The expected value of the optimal decision without the market research information was \$72,000. From (19.17) the expected value of the market research report is

$$\text{EVSI} = \$90{,}402 - \$72{,}000 = \$18{,}402.$$

Thus PSI should be willing to pay up to \$18,402 for the market research information.

Efficiency of Sample Information

In Section 19.6 we saw that the expected value of perfect information (EVPI) for the PSI problem was \$30,000. While we never expected the market research report to obtain perfect information, we can use an *efficiency* measure to express the value of the report. With perfect information having an efficiency rating of 100%, the efficiency rating (E) for sample information is computed as follows:

Efficiency of Sample Information

$$E = \frac{\text{EVSI}}{\text{EVPI}}(100). \qquad (19.18)$$

For our PSI example

$$E = \frac{18{,}402}{30{,}000}(100) = 61\%.$$

In other words, the information from the market research firm is 61% as "efficient" as perfect information.

Low efficiency ratings for information might lead the decision maker to look for other types of information. On the other hand, high efficiency ratings indicate that the information is almost as good as perfect information and that additional sources of information are probably not worthwhile.

[2]In minimization problems the expected value with sample information will be less than or equal to the expected value without sample information. Thus in minimization problems

$$\text{EVSI} = \begin{bmatrix}\text{Expected value of the}\\ \text{optimal decision } \textit{without}\\ \text{sample information}\end{bmatrix} - \begin{bmatrix}\text{Expected value of the}\\ \text{optimal decision } \textit{with}\\ \text{sample information}\end{bmatrix}.$$

EXERCISES

16. The payoff table and probability information for Exercise 15 are shown below:

	s_1	s_2
d_1	100	300
d_2	400	200

$$P(s_1) = .4, \quad P(I_1 \mid s_1) = .8, \quad P(I_2 \mid s_1) = .2,$$
$$P(s_2) = .6, \quad P(I_1 \mid s_2) = .4, \quad P(I_2 \mid s_2) = .6.$$

a. What is your decision without the indicator information?
b. What is the expected value of the indicator or sample information (EVSI)?
c. What is the expected value of perfect information (EVPI)?
d. What is the efficiency of the indicator information?

17. The payoff table for Hale's TV Productions (Exercise 5) is as follows (in $1,000's):

	States of Nature		
	s_1	s_2	s_3
Produce pilot (d_1)	−100	50	150
Sell to competitor (d_2)	100	100	100
Probability of states of nature	.2	.3	.5

For a consulting fee of $2,500 an agency will review the plans for the comedy series and indicate the overall chances of a favorable network reaction to the series. If the special agency review results in a favorable (I_1) or an unfavorable (I_2) evaluation, what should Hale's decision strategy be? Assume that Hale believes that the following conditional probabilities are realistic appraisals of the agency's evaluation accuracy:

$$P(I_1 \mid s_1) = .3, \quad P(I_2 \mid s_1) = .7,$$
$$P(I_1 \mid s_2) = .6, \quad P(I_2 \mid s_2) = .4,$$
$$P(I_1 \mid s_3) = .9, \quad P(I_2 \mid s_3) = .1.$$

a. Show the decision tree for this problem.
b. What are the recommended decision strategy and the expected value, assuming that the agency information is obtained?
c. What is the EVSI? Is the $2,500 consulting fee worth the information? What is the maximum that Hale should be willing to pay for the consulting information?

18. McHuffter Condominiums (Exercises 3 and 6) is conducting a survey which will help evaluate the demand for the new condominium development. McHuffter's payoff table (profit) is as follows:

		States of Nature	
Decision	*Low* (s_1)	*Medium* (s_2)	*High* (s_3)
Small (d_1)	400	400	400
Medium (d_2)	100	600	600
Large (d_3)	−300	300	900
Probability of states of nature	.20	.35	.45

The survey will result in three indicators of demand—weak (I_1), average (I_2), or strong (I_3). The conditional probabilities are as follows:

	$P(I_k \mid s_k)$		
	I_1	I_2	I_3
s_1	.6	.3	.1
s_2	.4	.4	.2
s_3	.1	.4	.5

a. What is McHuffter's optimal strategy?
b. What is the value of the survey information?
c. What are the EVPI and the efficiency of the survey information?

19.10 OTHER TOPICS IN DECISION THEORY

In our discussion of decision theory models we have considered only decision situations where there are a finite number of states of nature that can be listed. The next step would be to consider situations where the states of nature were so numerous that it would be impractical, if not impossible, to treat states of nature as a discrete random variable consisting of a finite number of values. For example, let us suppose that we are attempting to price a new product and are concerned with the potential sales volume we might experience at different prices. We might think of the states of nature as being all possible sales volumes from 0 to 200,000 units. Although there are a finite number of states of nature, no units sold, one unit sold, and so on, we recognize that attempting to deal with this large number of possible states of nature is extremely impractical. The solution procedure that is used in such circumstances is to treat the state of nature as a continuous random variable. For example, perhaps a reasonable approximation of the state of nature (that is, sales volume) is that sales are normally distributed with a mean of 100,000 units and a standard deviation of 25,000 units. Although decision theory techniques have been developed to handle such situations, we shall not attempt to present these procedures in this chapter.

Another area of decision theory is concerned with alternative measures of the payoffs in a decision theory situation. In our PSI example we used profit in dollars as the measure of the payoff. Then the criterion of best expected *monetary* value was used to select the best decision. While decision theory applications are often based on expected monetary value, perhaps there are other measures of payoff that should be used.

For example, let us consider a situation in which we have two alternative investments. Investment A yields a certain profit of \$50,000. Investment B yields a 50% chance of making \$100,002 but also a 50% chance of making nothing. Thus the expected value for B is

$$\begin{aligned} E(\text{B}) &= .5(100{,}002) + .5(0) \\ &= \$50{,}001. \end{aligned}$$

With the expected monetary value criterion as the measure of the payoff, we would select decision alternative B. However, many decision makers, if not most, would select alternative A; that is, some decision makers prefer the no-risk \$50,000 profit over the higher expected value but risky 50–50 chance of a \$100,002 profit. In decision theory terminology, if alternative A is preferred, we say that alternative A has a higher *utility,* where utility is a measure of the decision maker's preference considering monetary value as well as the risk involved. Ideally we would like to measure payoffs in terms of the decision maker's utility and select optimal decision strategies based on expected utility rather than expected monetary value. However, attempting to assess realistic utility functions for the decision maker can be very difficult, and we shall not attempt to present utility theory in this text.

Summary

In this chapter we have introduced the decision theory or decision analysis approach to decision making. We have discussed in detail the decision theory procedures designed to solve problems with a limited number of decision alternatives and a finite list of possible states of nature. The goal of the decision theory approach was to identify the best decision alternative given an uncertain or risk-filled pattern of future events (that is, states of nature).

After defining decision making under certainty and uncertainty, we discussed the decision criteria of maximin, maximax, and minimax regret for solving problems of decision making under uncertainty without use of probabilities. We then discussed the use of expected monetary value and expected opportunity loss as criteria for solving problems of decision making under uncertainty with use of probabilities. We also showed how additional information about the states of nature can be used to revise or update the probability estimates and develop an optimal decision strategy for the problem. The notions of expected value of sample information, expected value of perfect information, and efficiency were used to evaluate the contribution of the additional information.

Glossary

States of nature—The uncontrollable future events that can affect the outcome of a decision.

Payoff—The outcome measure such as profit, cost, etc. Each combination of a decision alternative and a state of nature has a specific payoff.

Payoff table—A tabular representation of the payoffs for a decision problem.

Decision making under certainty—The process of choosing a decision alternative when the state of nature is known.

Decision making under uncertainty—The process of choosing a decision alternative when the state of nature is not known.

Maximin—A maximization decision criterion for decisions under uncertainty that seeks to maximize the minimum payoff.

Minimax—A minimization decision criterion for decisions under uncertainty that seeks to minimize the maximum payoff.

Maximax—A maximization decision criterion for decisions under uncertainty that seeks to maximize the maximum payoff.

Minimin—A minimization decision criterion for decisions under uncertainty that seeks to minimize the minimum payoff.

Opportunity loss or regret—The amount of loss (lower profit or higher cost) due to not making the best decision for each state of nature.

Minimax regret—A maximization or minimization decision criterion for decisions under uncertainty that seeks to minimize the maximum regret.

Expected monetary value—A decision criterion for decisions under uncertainty. The expected monetary value weights the payoff for each decision by its probability of occurrence.

Expected opportunity loss—The expected value criterion applied to opportunity loss or regret values. Also a decision criterion for decisions under uncertainty. It yields the same optimal decision as the expected monetary value criterion.

Decision tree—A graphical representation of the decision-making situation from decision to state of nature to payoff.

Nodes—The intersection or junction points of the decision tree.

Branches—Lines or arcs connecting nodes of the decision tree.

Expected value of perfect information (EVPI)—The expected value of information that would tell the decision maker exactly which state of nature was going to occur (that is, perfect information). EVPI is equal to the expected opportunity loss of the best decision alternative when no additional information is available.

Indicators—Information about the states of nature obtained by experimentation. An indicator may be the result of a sample.

Prior probabilities—The probabilities of the states of nature prior to obtaining experimental information.

Posterior (revised) probabilities—The probabilities of the states of nature after use of Bayes' theorem to adjust the prior probabilities based upon given indicator information.

Bayesian revision—The process of adjusting prior probabilities to create the posterior probabilities based upon information obtained by experimentation.

Expected value of sample information (EVSI)—The difference between the expected value of an optimal strategy based on new information and the "best" expected value without any new information. It is a measure of the economic value of new information.

Efficiency—The ratio of EVSI to EVPI. Perfect information is 100% efficient.

Key Formulas

Opportunity Loss or Regret

$$R(d_i,s_j) = V^*(s_j) - V(d_i,s_j) \tag{19.1}$$

Expected Monetary Value

$$\text{EMV}(d_i) = \Sigma P(s_j)V(d_i,s_j). \tag{19.4}$$

Expected Opportunity Loss

$$\text{EOL}(d_i) = \Sigma P(s_j)R(d_i,s_j) \tag{19.5}$$

Expected Value of Perfect Information

$$\text{EVPI} = \Sigma P(s_j)R(d^*,s_j) \tag{19.6}$$

Calculation of Indicator Branch Probability

$$P(I_k) = \Sigma P(I_k \mid s_j)P(s_j) \tag{19.15}$$

Calculation of State-of-Nature Branch Probability

$$P(s_j \mid I_k) = \frac{P(I_k \mid s_j)P(s_j)}{P(I_k)} \tag{19.16}$$

Expected Value of Sample Information[2]

$$\text{EVSI} = \begin{bmatrix} \text{Expected value of the} \\ \text{optimal decision } \textit{with} \\ \text{sample information} \end{bmatrix} - \begin{bmatrix} \text{Expected value of the} \\ \text{optimal decision } \textit{without} \\ \text{sample information} \end{bmatrix} \tag{19.17}$$

Efficiency of Sample Information

$$E = \frac{\text{EVSI}}{\text{EVPI}}(100) \tag{19.18}$$

Supplementary Exercises

19. Refer again to the investment problem faced by Martin's Service Station (Exercise 9). Martin can purchase a blade to attach to his service truck that can also be used to plow driveways and parking lots. Since this truck must also be available to start cars, etc., Martin will not be able to generate as much revenue plowing snow if he elects this alternative. But he will keep his loss smaller if there is light snowfall. Under this alternative Martin forecasts a profit of $3,500 if snowfall is heavy and $1,000 if it is moderate, and a $1,500 loss if snowfall is light.

a. Prepare a new decision tree showing all three alternatives.

b. Under the expected monetary value criterion, what is the optimal decision?

c. Develop a table showing the opportunity loss for each decision/state of nature combination. Which decision minimizes expected opportunity loss?

d. What is the expected value of perfect information?

20. The Gorman Manufacturing Company must decide whether it should purchase a component part from a supplier or manufacture the component at its Milan, Michigan plant. If demand is high, it would be to Gorman's advantage to manufacture the component. However, if demand is low, Gorman's unit manufacturing cost will be high because of underutilization of equipment. The projected profit in thousands of dollars for Gorman's make-or-buy decision is shown below:

	Demand		
	Low	*Medium*	*High*
Manufacture component	−20	40	100
Purchase component	10	45	70

The states of nature have the following probabilities: $P(\text{low demand}) = .35$, $P(\text{medium demand}) = .35$, and $P(\text{high demand}) = .30$.

a. Use a decision tree to recommend a decision.

b. Use EVPI to determine whether Gorman should attempt to obtain a better estimate of demand.

21. In order to save on gasoline expenses, Rona and Jane agreed to form a carpool for traveling to and from work. After limiting the travel routes to two alternatives, Rona and Jane could not agree on the best way to travel to work. Jane preferred the expressway, since it was usually the

[2] In minimization problems the expected value with sample information will be less than or equal to the expected value without sample information. Thus in minimization problems

$$\text{EVSI} = \begin{bmatrix} \text{Expected value of the} \\ \text{optimal decision } \textit{without} \\ \text{sample information} \end{bmatrix} - \begin{bmatrix} \text{Expected value of the} \\ \text{optimal decision } \textit{with} \\ \text{sample information} \end{bmatrix}.$$

fastest; however, Rona pointed out that traffic jams on the expressway sometimes led to long delays. Rona preferred the somewhat longer, but more consistent, Queen City Avenue. While Jane still preferred the expressway, she agreed with Rona that they should take Queen City Avenue if the expressway had a traffic jam. Unfortunately, they did not know the state of the expressway ahead of time. The following payoff table provides the one-way time estimates (in minutes) for traveling to or from work:

		States of Nature	
		Expressway Open (s_1)	*Expressway Jammed* (s_2)
Expressway	(d_1)	25	45
Queen City Avenue	(d_2)	30	30

After driving to work on the expressway for 1 month (20 days), they found the expressway jammed three times. Assume that these days were representative of future days. Should they continue to use the expressway for traveling to work? Explain. Would it make sense not to adopt the expected value criterion for this particular problem? Explain.

22. In Exercise 21, suppose that Rona and Jane wished to determine the best way to return home in the evenings. In 20 days of traveling home on the expressway they found the expressway jammed six times. Use the travel time table shown in Exercise 21. What route would you recommend they take on their way home in the evening? If they had perfect information about the traffic condition of the expressway, what would be their savings in terms of expected travel time?

23. A firm produces a perishable food product at a cost of $10 per case. The product sells for $15 per case. For planning purposes the company is considering possible demands of 100, 200, or 300 cases. If the demand is less than production, the excess production is lost. If demand is more than production, the firm, in an attempt to maintain a good service image, will satisfy the excess demand with a special production run at a cost of $18 per case. The product, however, always sells at the $15 per case price.

a. Set up the payoff table for this problem.

b. If $P(100) = .2$, $P(200) = .2$, and $P(300) = .6$, use the expected opportunity loss criterion to determine the solution.

c. What is the EVPI?

24. The Kremer Chemical Company has a contract with one of its customers to supply a unique liquid chemical product that will be used by the customer in the manufacturing of a lubricant for airplane engines. Because of the chemical process used by the Kremer Company batch sizes for the liquid chemical product must be 1,000 pounds. The customer has agreed to adjust manufacturing to the full batch quantities and will order either one, two or three batches every 6 months. Since an aging process of 2 months exists for the product, Kremer will have to make its production (how much to manufacture) decision before its customer places an order. Thus Kremer can list the product demand alternatives of 1,000, 2,000, or 3,000 pounds, but the exact demand is unknown.

Kremer's manufacturing costs are $15 per pound, and the product sells at the fixed contract price of $20 per pound. If the customer orders more than Kremer has produced, Kremer has agreed to absorb the added cost of filling the order by purchasing a higher-quality substitute product from another chemical firm. The substitute product, including transportation expenses, will cost Kremer $24 per pound. Since the product cannot be stored more than 4 months without spoilage, Kremer cannot inventory excess production until the customer's next 6 month order. If the customer's current order is less than Kremer has produced, the excess production will be reprocessed and is valued at $5 per pound.

The inventory decision in this problem is how much Kremer should produce given the above costs and the possible demands of 1,000, 2,000, or 3,000 pounds. From historical data and an analysis of the customer's future demands Kremer has assessed the following probability distribution for demand:

Demand	*Probability*
1,000	.3
2,000	.5
3,000	.2
	1.0

a. Develop a payoff table for the Kremer problem.
b. How many batches should Kremer produce every 6 months?
c. How much of a discount should Kremer be willing to allow the customer for specifying in advance exactly how many batches will be purchased?

25. A quality control procedure involves 100% inspection of parts received from a supplier. Historical records show that the following defective rates have been observed:

Number Defective	*Probability*
0	.15
1	.25
2	.40
3	.20

The cost for the quality control 100% inspection is $250 for each shipment of 500 parts. If the shipment is not 100% inspected, defective parts will cause rework problems later in the production process. The rework cost is $25 for each defective part.

a. Complete the following payoff table, where the entries represent the total cost of inspection and reworking:

	Percentage Defective			
	0%	*1%*	*2%*	*3%*
100% inspection	$250	$250	$250	$250
No inspection	?	?	?	?

b. The plant manager is considering eliminating the inspection process in order to save the $250 inspection cost per shipment. Do you support this action? Use EMV to justify your answer.
c. Show the decision tree for this problem.

26. Milford Trucking, located in Chicago, has requests to haul two shipments: one to St. Louis and one to Detroit. Because of a scheduling problem, Milford will be able to select only one of these assignments. The St. Louis customer has guaranteed a return shipment, but the Detroit customer has not. Thus if Milford accepts the Detroit shipment and cannot find a Detroit-to-Chicago return shipment, the truck will return to Chicago empty. The payoff table showing profit is as follows:

	Return Shipment from Detroit (s_1)	*No Return Shipment from Detroit* (s_2)
St. Louis (d_1)	2,000	2,000
Detroit (d_2)	2,500	1,000

a. If the probability of Detroit return shipment is .4, what should Milford do?
b. What is the expected value of information that would tell Milford whether or not Detroit had a return shipment?

27. The payoff table for Martin's Service Station (Exercises 9 and 19) is as follows:

	Snowfall		
	Heavy (s_1)	*Moderate* (s_2)	*Light* (s_3)
Purchase snowplow (d_1)	7,000	2,000	−9,000
Do not invest (d_2)	0	0	0
Purchase blade for service truck (d_3)	3,500	1,000	−1,500
Probabilities of states of nature	.4	.3	.3

Suppose that Martin decides to wait to check the September temperature pattern before making a final decision. Estimates of the probabilities associated with an unseasonably cold September (I) are as follows: $P(I \mid s_1) = .30$; $P(I \mid s_2) = .20$; $P(I \mid s_3) = .05$. If Martin observes an unseasonably cold September, what is the recommended decision? If Martin does not observe an unseasonably cold September, what is the recommended decision?

28. The Gorman Manufacturing Company (Exercise 20) has the following payoff table for a make-or-buy decision:

		Demand		
		Low (s_1)	*Medium* (s_2)	*High* (s_3)
Manufacture component	(d_1)	−20	40	100
Purchase component	(d_2)	10	45	70
Probabilities		.35	.35	.30

A test market study of the potential demand for the product is expected to report either a favorable (I_1) or unfavorable (I_2) condition. The relevant conditional probabilities are as follows:

$$P(I_1 \mid s_1) = .10, \qquad P(I_2 \mid s_1) = .90,$$
$$P(I_1 \mid s_2) = .40, \qquad P(I_2 \mid s_2) = .60,$$
$$P(I_1 \mid s_3) = .60, \qquad P(I_2 \mid s_3) = .40.$$

a. What is the probability that the market research report will be favorable?
b. What is Gorman's optimal decision strategy?
c. What is the expected value of the market research information?
d. What is the efficiency of the information?

29. The traveling time to work for Rona and Jane has the following time payoff table (see Exercise 21):

		States of Nature	
		Expressway Open (s_1)	*Expressway Jammed* (s_2)
Expressway	(d_1)	25	45
Queen City Avenue	(d_2)	30	30
Probability of states of nature		.85	.15

After some time Rona and Jane noted that the weather seemed to affect the traffic conditions on the expressway. They identified three weather conditions (indicators) with the following conditional probabilities:

$$I_1 = \text{clear}, \quad I_2 = \text{overcast}; \quad I_3 = \text{rain};$$
$$P(I_1 \mid s_1) = .8, \quad P(I_2 \mid s_1) = .2, \quad P(I_3 \mid s_1) = 0,$$
$$P(I_1 \mid s_2) = .1, \quad P(I_2 \mid s_2) = .3, \quad P(I_3 \mid s_2) = .6.$$

a. Show the decision tree for the problem of traveling to work.
b. What are the optimal decision strategy and the expected travel time?
c. What is the efficiency of the weather information?

30. The research and development (R&D) manager for Beck Company is trying to decide whether or not to fund a project to develop a new lubricant. It is assumed that the project will be a major technical success, a minor technical success, or a failure. The company has estimated that the value of a major technical success is $150,000, since the lubricant can be used in a number of products that the company is making. If the project is a minor technical success, its value is $10,000, since Beck feels that the knowledge gained will benefit some other ongoing projects. If the project is a failure, it will cost the company $100,000. According to the opinion of the scientists involved and the manager's own subjective assessment, the assigned prior probabilities are as follows:

$$P(\text{major success}) = .15$$
$$P(\text{minor success}) = .45$$
$$P(\text{failure}) = .40$$

a. With the expected monetary value criterion, should the project be funded?
b. Suppose that a group of expert scientists from a research institute could be hired as consultants to study the project and make a recommendation. If this study will cost $30,000, should the Beck Company consider hiring the consultants?

31. Consider again the problem faced by the R&D manager of Beck Company (Exercise 30). Suppose that an experiment can be conducted to shed some light on the technical feasibility of the project. There are three possible outcomes for the experiment:

I_1 = prototype lubricant works well at all temperatures,
I_2 = prototype lubricant works well only at temperatures above 10°,
I_3 = prototype lubricant does not work well at any temperature.

Suppose that we can determine the following conditional probabilities:

$$P(I_1 \mid \text{major success}) = .70$$
$$P(I_1 \mid \text{minor success}) = .10$$
$$P(I_1 \mid \text{failure}) = .10$$

$$P(I_2 \mid \text{major success}) = .25$$
$$P(I_2 \mid \text{minor success}) = .70$$
$$P(I_2 \mid \text{failure}) = .30$$

$$P(I_3 \mid \text{major success}) = .05$$
$$P(I_3 \mid \text{minor success}) = .20$$
$$P(I_3 \mid \text{failure}) = .60$$

a. If the experiment is conducted and the prototype lubricant works well at all temperatures, should the project be funded?
b. If the experiment is conducted and the prototype lubricant works well only at temperatures above 10°, should the project be funded?

c. Develop a decision strategy that Beck's R&D manager can use to recommend a funding decision based on the outcome of the experiment.
d. Find the EVSI for the experiment. How efficient is the information in the experiment?

32. The payoff table for the Kremer Chemical Company (Exercise 24) is as follows:

Production Quantity	*Demand* 1,000 (s_1)	2,000 (s_2)	3,000 (s_3)
1,000 (d_1)	5,000	1,000	−3,000
2,000 (d_2)	−5,000	10,000	6,000
3,000 (d_3)	−15,000	0	15,000
Probabilities	.30	.50	.20

Kremer has identified a pattern in the demand for the product based on the customer's previous order quantity. Let

I_1 = customer's last order was 1,000 pounds,
I_2 = customer's last order was 2,000 pounds,
I_3 = customer's last order was 3,000 pounds.

The conditional probabilities are as follows:

$$P(I_1 \mid s_1) = .10, \quad P(I_2 \mid s_1) = .30, \quad P(I_3 \mid s_1) = .60,$$
$$P(I_1 \mid s_2) = .30, \quad P(I_2 \mid s_2) = .30, \quad P(I_3 \mid s_2) = .40,$$
$$P(I_1 \mid s_3) = .80, \quad P(I_2 \mid s_3) = .20, \quad P(I_3 \mid s_3) = .00.$$

a. Develop an optimal decision strategy for Kremer.
b. What is the EVSI?
c. What is the efficiency of the information for the most recent order?

33. Milford Trucking Co. (Exercise 26) has the following payoff table:

	Return Shipment from Detroit (s_1)	*No Return Shipment from Detroit* (s_2)
St. Louis (d_1)	2,000	2,000
Detroit (d_2)	2,500	1,000
Probabilities	.40	.60

a. Milford can phone a Detroit truck dispatch center and determine if the general Detroit shipping activity is busy (I_1) or slow (I_2). If the report is "busy," the chances of obtaining a return shipment will increase. Suppose that the following conditional probabilities are given:

$$P(I_1 \mid s_1) = .6, \quad P(I_2 \mid s_1) = .4,$$
$$P(I_1 \mid s_2) = .3, \quad P(I_2 \mid s_2) = .7.$$

What should Milford do?
b. If the Detroit report is "busy" (I_1), what is the probability that Milford obtains a return shipment if the trip to Detroit is made?
c. What is the efficiency of the phone information?

34. The quality control inspection process (Exercise 25) has the following payoff table:

	Percentage Defective			
	0% (s_1)	*1%* (s_2)	*2%* (s_3)	*3%* (s_4)
100% inspection (d_1)	250	250	250	250
No inspection (d_2)	0	125	250	375
Probabilities	*.15*	*.25*	*.40*	*.20*

Suppose that a sample of five parts is selected from the shipment and one defective part is found.

a. Let I = one defective part in a sample of five. Use the binomial probability distribution to compute $P(I \mid s_1)$, $P(I \mid s_2)$, $P(I \mid s_3)$, and $P(I \mid s_4)$, where the state of nature identifies the value for p.

b. If I occurs, what are the revised probabilities for the states of nature?

c. Should the entire shipment be 100% inspected whenever one defective part is found in a sample of five?

d. What is the cost savings associated with the sample information?

35. A food processor considers daily production runs of 100, 200, or 300 cases. Possible demands for the product are 100, 200, or 300 cases. The payoff table is as follows:

	Demand (cases)		
Production	*100* (s_1)	*200* (s_2)	*300* (s_3)
100 (d_1)	500	200	−100
200 (d_2)	−400	800	700
300 (d_3)	−1,000	−200	1,600

a. If $P(s_1) = .20$, $P(s_2) = .20$, and $P(s_3) = .60$, what is your recommended production quantity?

b. On some days the firm receives phone calls for advance orders and on some days it does not. Let I_1 = advance orders are received and I_2 = no advance orders are received. If $P(I_2 \mid s_1) = .80$, $P(I_2 \mid s_2) = .40$, and $P(I_2 \mid s_3) = .10$, what is your recommended production quantity for days that the company does not receive any advance orders?

Application

Ohio Edison Company*

Akron, Ohio

Ohio Edison Company is an investor-owned electric utility headquartered in Northeastern Ohio. Ohio Edison and a Pennsylvania subsidiary provide electrical service to over 2 million people. Most of this electricity is generated by coal-fired power plants. In order to meet evolving air quality standards, Ohio Edison embarked on a program to replace existing pollution control equipment on most of its generating plants with more efficient equipment. The combination of this program to upgrade air quality control equipment with the continuing need to construct new generating plants to meet future power requirements has resulted in a large capital investment program.

A DECISION ANALYSIS APPLICATION

The flue gas emitted by coal-fired power plants contains small ash particles and sulfur dioxide (SO_2). Federal and state regulatory agencies have established emission limits for both particulates and sulfur dioxide. Recently, Ohio Edison developed a plan to comply with new air quality standards at one of its largest power plants. This plant consists of seven coal-fired units and constitutes about one-third of the generating capacity of Ohio Edison and the subsidiary company. Most of these units had been constructed in the 1960s. Although all of the units had initially been constructed with equipment to control particulate emissions, that equipment was not capable of meeting new particulate emission requirements.

A decision had already been made to burn low sulfur coal in four of the smaller units (Units 1–4) at the plant in order to meet SO_2 emission standards. Fabric filters were to be installed on these units to control particulate emissions. Fabric filters, also known as baghouses, use thousands of fabric bags to filter out the particulates; they function in much the same way as a household vacuum cleaner.

It was considered likely, although not certain, that the three larger units (Units 5, 6, and 7) at this plant would burn medium to high sulfur coal. A method of controlling particulate emissions at these units had not yet been selected. Preliminary studies had narrowed the particulate control equipment choice to a decision between fabric filters and electrostatic precipitators (which remove particulates suspended in the flue gas as charged particles by passing the flue gas through a strong electric field). This decision

*The authors are indebted to Thomas J. Madden and M. S. Hyrnick of Ohio Edison Company, Akron, Ohio, for providing this application.

Decision analysis aided in the selection of particulate control equipment being installed at Ohio Edison's W.H. Sammis Electric Generating Plant

was affected by a number of uncertainties, including the following:

Uncertainty in the way some air quality laws and regulations might be interpreted. Certain interpretations could require that either low sulfur coal or high sulfur Ohio coal (or neither) be burned in Units 5, 6, and 7.

Potential future changes in air quality laws and regulations.

An overall plant reliability improvement program was underway at this plant. The outcome of this program would affect the operating costs of whichever pollution control technology was installed in these units.

Construction costs of the equipment were uncertain, particularly since limited space at the plant site made it necessary to install the equipment on a massive bridge deck over a four lane highway immediately adjacent to the power plant.

The costs associated with replacing the electrical power required to operate the particulate control equipment were uncertain.

Various uncertain factors, including potential accidents and chronic operating problems, which could increase the costs of operating the generating units were identified. The degree to which each of these factors affected operating costs varied with the choice of technology and with the sulfur content of the coal.

DECISION ANALYSIS

The decision to be made involved a choice between two types of particulate control equipment (fabric filters or electrostatic precipitators) for Units 5, 6, and 7. Because of the complexity of the problem, the high degree of uncertainty associated with factors affecting the decision, and the importance (due to potential reliability and cost impact on Ohio Edison) of the choice, decision analysis was used in the selection process.

The decision measure used to evaluate the outcomes of the particulate technology decision analysis was the annual revenue requirements for the three large units over their remaining lifetime. Revenue requirements are the monies that would have to be collected from the utility customers in order to recover costs that are a result of the decision. They include not only direct costs but also the cost of capital and return on investment.

A decision tree was constructed to represent the particulate control decision, its uncertainties and costs. A simplified version of this decision tree is shown in Figure 19A.1. The decision and state-of-nature nodes are indicated. Note that to conserve space a type of shorthand notation is used. The coal sulfur content state-of-nature node should actually be located at the end of each branch of the capital cost state-of-nature node, as the dotted lines indicate. Each of the indicated state-of-nature nodes actually represents several probabilistic cost models or submodels. The total revenue requirements calculated are the sum of the revenue requirements for capital and operating costs. Costs associated with these models were obtained from engineering calculations or estimates. Probabilities were obtained from existing data or the subjective assessments of knowledgeable persons.

RESULTS

A decision tree similar to that shown in Figure 19A.1 was used to generate cumulative probability distributions for the annual revenue requirements outcomes calculated for each of the two particulate control alternatives. Careful study of these results led to the following conclusions:

The expected value of annual revenue requirements for the electrostatic precipitator technology was approximately $1 million lower than that for the fabric filters.

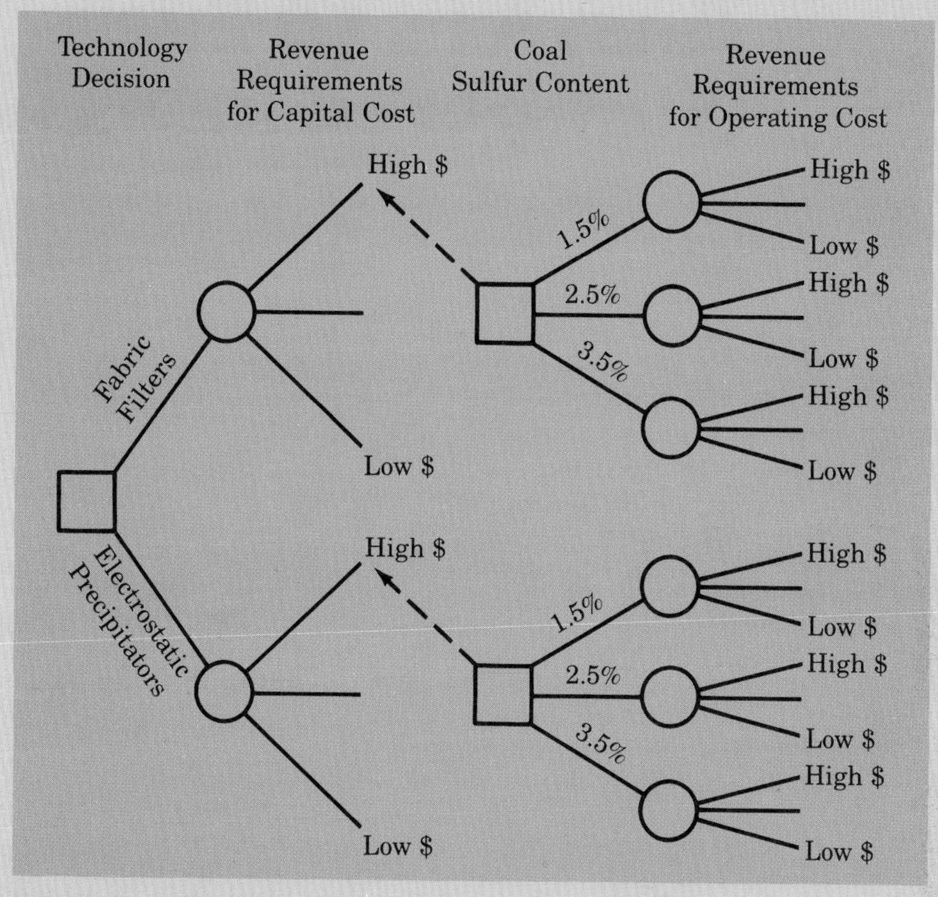

FIGURE 19A.1 Simplified Particulate Control Equipment Decision Tree

The fabric filter alternative had a higher "upside risk"—that is, a higher probability of high revenue requirements—than did the precipitator alternative.

The precipitator technology had nearly an 80% probability of having lower annual revenue requirements than the fabric filters.

Although the capital cost of the fabric filter equipment (the cost of installing the equipment) was lower than for the precipitator, this was more than offset by the higher operating costs associated with the fabric filter.

These results led Ohio Edison to select the electrostatic precipitator technology for the generating units in question. Had the decision analysis not been performed, the particulate control decision might have been based chiefly on capital cost, a decision measure which would have favored the fabric filter equipment. Decision analysis offers a means for effectively analyzing the uncertainties involved in a decision. Because of this, it is felt that the use of decision analysis methodology in this application resulted in a decision which yielded both lower expected revenue requirements and lower risk.

Answers to Even-Numbered Exercises

Chapter 2

2.

	Relative Frequency			
	(a) *Accounting*	*(b)* *Finance*	*(c)* *Marketing*	*(d)* *Management*
1,300 but less than 1,400	.037	.000	.125	.200
1,400 but less than 1,500	.111	.167	.188	.120
1,500 but less than 1,600	.259	.333	.125	.160
1,600 but less than 1,700	.185	.250	.312	.240
1,700 but less than 1,800	.148	.083	.125	.120
1,800 but less than 1,900	.111	.083	.063	.040
1,900 but less than 2,000	.037	.000	.063	.080
2,000 but less than 2,100	.111	.083	.000	.040

4.

	(a)	*(b)*	*(c)*	*(d)*
15 but less than 20	3	.150	3	.150
20 but less than 25	5	.250	8	.400
25 but less than 30	5	.250	13	.650
30 but less than 35	6	.300	19	.950
35 but less than 40	1	.050	20	1.000
	20	1.000		

6. a. .34
b. 85
c. Frequencies are as follows: 25, 55, 85, 45, 40

10. d. .60

12.

Time (minutes)	*Frequency*	*Relative Frequency*
2 but less than 4	5	.25
4 but less than 6	9	.45
6 but less than 8	4	.20
8 but less than 10	0	.00
10 but less than 12	2	.10
	20	1.00

14.

Sales (1,000,000's)	*(a)*	*(b)*	*(c)*	*(d)*
0 but less than 5	7	.175	7	.175
5 but less than 10	11	.275	18	.450
10 but less than 15	4	.100	22	.550
15 but less than 20	4	.100	26	.650
20 but less than 25	4	.100	30	.750
25 but less than 30	1	.025	31	.775
30 but less than 35	1	.025	32	.800
35 but less than 40	2	.050	34	.850
40 but less than 45	5	.125	39	.975
45 but less than 50	1	.025	40	1.000
	40	1.000		

16.

Mortgage ($1,000's)	*Frequency*	*Relative Frequency*
10 but less than 15	2	.067
15 but less than 20	0	.000
20 but less than 25	2	.067
25 but less than 30	6	.200
30 but less than 35	8	.266
35 but less than 40	5	.167
40 but less than 45	4	.133
45 but less than 50	1	.033
50 but less than 55	2	.067
	30	1.000

18.

Closing Price	*Frequency*	*Relative Frequency*	*Cumulative Frequency*	*Cumulative Relative Frequency*
5 but less than 15	18	.450	18	.450
15 but less than 25	13	.325	31	.775
25 but less than 35	2	.050	33	.825
35 but less than 45	4	.100	37	.925
45 but less than 55	1	.025	38	.950
55 but less than 65	2	.050	40	1.000
	40	1.000		

20.

Grade Point Average	*Relative Frequency*	*Cumulative Relative Frequency*
1.6 but less than 1.9	.034	.034
1.9 but less than 2.2	.100	.134
2.2 but less than 2.5	.200	.334
2.5 but less than 2.8	.333	.667
2.8 but less than 3.1	.200	.867
3.1 but less than 3.4	.100	.967
3.4 but less than 3.7	.033	1.000
3.7 but less than 4.0	.000	1.000
	1.000	

22.

Days	*Frequency*	*Relative Frequency*	*Cumulative Relative Frequency*
3 but less than 7	5	.25	.25
7 but less than 11	9	.45	.70
11 but less than 15	1	.05	.75
15 but less than 19	3	.15	.90
19 but less than 23	2	.10	1.00
		1.00	

24.

Time (minutes)	*Frequency*	*Relative Frequency*
3 but less than 8	4	.20
8 but less than 13	8	.40
13 but less than 18	5	.25
18 but less than 23	2	.10
23 but less than 28	1	.05
	20	1.00

Chapter 3

2. 1,900; 2,000; 1,850

4. a. 178

b. 178

c. Do not report a mode; not appropriate here

d. 184

6. 17.4; 18; 18; 18

8. b. City: 15.58, 15.9, 15.3; country: 18.92, 18.7, (18.6 and 19.4); better mileage in country

10. 250; 15.81

12. Dawson: .74, 2; Clark: 2.59, 8; Dawson shows less dispersion and is more consistent

14. 7; 4.97; 2.23

16. $s^2 = .002$; do not shut down

18. a. at least 56%
b. at least 84%
c. approximately 26 to 474

20. 1,638.75; 1,621.05; 1,650; 35,694.62; 188.93

22. 74.52; 75.45; 75, 158.19; 12.58 (variance and standard deviation shown assumes sample data)

24. 307.27; 317.14; 350; 12,505.63; 111.83

26. 32,940; 33,250; modes at 30,500 and 33,500

28. a. 18.62; 19; modes at 18 and 20
b. 17; 21
c. 18; 4
d. 4.57; 24.54%

30. 1,583.33; 1,550; 1,500; 800; 69,666.67; 263.94.

32. a. 5.55
b. 5.4
c. 4.8
d. 8.4
e. 4.9353
f. 2.221

34. a. at least 75%
b. at least 84%
c. 800 ± 141.4

36. 12.8; 12.57; 14; 28.8; 5.37

38. a. 56.22
b. 56.07; 57.5
c. 31.94; 5.65
d. 63.5

Chapter 4

2. b. 8 sample points
c. 16 sample points

4. a. 9
c. 5
d. 1

6. a. 1,000,000
b. 5,760,000
c. More letters

8. a. 6
b. Relative frequency based on the historical data
c. .12; .24; .30; .20; .10; .04

10. a. (4,6); (4,7); (4,8)
b. .30
c. (2,8); (3,8); (4,8)

d. .25
e. .15

12. a. 0; .05
b. 4, 5; .20
c. 0, 1, 2; .55

14. Revise probability estimates such that the sum = 1

16. a. .40; .50; .60
b. $\{E_1, E_2, E_4, E_6, E_7\}$; .65
c. $\{E_4\}$; .25
d. Yes
e. $\overline{B} = \{E_1, E_3, E_5, E_6\}$; .50

18. a. No; no
b. Yes; .45
c. .80

20. a. $\{E_1, E_2, E_3, E_4, E_5, E_8\}$
b. $\{E_2, E_5\}$
c. $\{E_4, E_6, E_7, E_8\}$
d. No

22. a.

	Single	*Married*	*Total*
Under 30	.55	.10	.65
30 or over	.20	.15	.35
Total	.75	.25	1.00

b. 65% under 30, 35% 30 or over
c. 75% single, 25% married
d. 55
e. .8462
f. No, since P (single | under 30) $\neq$ P (single)

24. a. Let A = event Ms. Smith gets the first job
B = event Ms. Smith gets the second job
$P(A) = .5$, $P(B) = .6$, $P(A \cap B) = .15$
b. .30
c. .95
d. .05
e. No, since $P(B \mid A) = .30 \neq P(B)$

26. a. .12
b. .24
c. No, since P (productive type | A) $\neq$ P(productive)

28. a. 0
b. 0
c. No

30. a. .10; .20; .09
b. .51
c. .26; .51; .23

32. a. .21
b. Yes: the probability of default is greater than .20

34. .6007

36. a. $\{E_1, E_2, E_3, E_4\}$
b. $\{E_1, E_5, E_6, E_7, E_8\}$
c. $\{E_2\}$
d. $\{E_7, E_8\}$
e. no sample points
f. $\{E_4, E_5, E_6, E_7, E_8\}$
g. $\{E_1, E_2, E_3, E_4\}$
h. $\{E_1, E_2, E_3, E_4\}$
i. $\{E_1, E_2, E_3,\}$
j. No, $A \cap B = \{E_2\}$
k. Yes, $B \cap C = \varnothing$

38. a. No, since $P(S \cap R) \neq 0$
b. No
c. .5

40. a. $P(A \cup B) = .35$; $P(A \mid B) = .8$; $P(B \mid A) = .67$
b. No, $P(A \mid B) = .8 \neq P(A) = .30$

42. a. .90, 0
b. No

44. d. .333
e. .114
f. No
g. Smoking increases the probability of heart disease

46. a. .25
b. Yes, since $P(A \cap B) = .20 = P(A)\,P(B) = (.80)(.25)$
c. No, it doesn't seem to have any effect on purchase behavior

48. a. .25, .40, .10
b. .25
c. independent; program appears to have no effect

50. a. .20
b. .35

52. a. .21; should call back

54. .0344

56. a. .12
b. .625; check adjustment
c. .305; either check adjustment now, or continue sampling until probabilities show a clear-cut decision

Chapter 5

2. a. Discrete, values: 0, 1, . . . , 20
b. Discrete, values: 0, 1, 2, . . .
c. Discrete, values: 0, 1, 2, . . .
d. Continuous, values: $0 \leq x \leq 8$
e. Continuous, values: $x \geq 0$

4. a. Satisfies (5.1) and (5.2)

b.

x	$F(x)$
148,000	.2
150,000	.6
152,000	1.0

c. $F(150{,}000) = .6$

6. a. .05; This is the probability of a $200,000 profit.

b. .70

c. .40

d. .10, .30, .60, .85, .95, 1.00

8. a. 1.25

b. 1.29

c. 1.13

10. a. $83

b. $−47

12. a. 445

b. $1,250 loss

14. a. E(medium) = 145; E(large) = 140; medium-scale project

b. Var (medium) = 2,725; Var (large) = 12,400; medium-scale less risk

16. a. $1,140

b. Variance = 2.1275; standard deviation = 1.4586

c. Variance = 340,400; standard deviation = $583.44

18. a. $3,500

b. $500

20. a. No; for example, $f(0,0) = .52 \neq f_x(0)\,f_y(0) = .48$

b. x: .20, .40; y: .40, .49

c. Yes; $f_x(1) = .20; f(x = 1 \mid y = 1) = .12/.40 = .30.$

22. a.

$$f(x) = \begin{cases} 1/4 & \text{for } 3 \le x \le 7 \\ 0 & \text{elsewhere} \end{cases}$$

b. 0

c. .50

d. 5, 1.15

24. a. 1.6

b. $32

26. y is acceptable; x and z are not

28. a.

x	$f(x)$
0	.10
1	.40
2	.30
3	.15
4	.05

b. $E(x) = 1.65$; $\text{Var}(x) = 1.0275$

30. a.

x	$f(x)$
9	.30
10	.20
11	.25
12	.05
13	.20

b. 10.65
c. 2.1275
d. Looks good, expected profit 1.35 million

32. a.

y	$f(y)$	x	$f(x)$
−5.00	.4	−2.50	.4
0.00	.2	+2.50	.2
+5.00	.4	+2.50	.4

b. $E(y) = 0$; $E(x) = 1.00$; stock one unit

34. a.

$$f(x) = \begin{cases} 1/5 & \text{for } 0 \le x \le 5 \\ 0 & \text{elsewhere} \end{cases}$$

b. .30
c. .15
d. .40
e. 2.50

36. a.

$$f(x) = \begin{cases} 1/200 & \text{for } 3{,}900 \le x \le 4{,}100 \\ 0 & \text{elsewhere} \end{cases}$$

b. .25

Chapter 6

2. a. .2060
b. .2182
c. .5886

4. a. 0
b. .1422
c. .3012
d. .2642

6. a. .0324; .2952; .6724
b. .1936; .4928; .3136
c. Yes, foul the player with the worst free-throw percentage

8. a. .0498
b. .1465

c. .3528
d. .0009

10. a. .000045
b. .010245
c. .0821
d. .9179

12. a. .2241
b. .4422

14. a. 4.75%
b. .99%
c. 79.38%
d. $796.75

16. a. .0228
b. .2857
c. 9.522

18. a. .4592
b. .0301
c. 3,976

20. a. 54
b. 24.84
c. .1841

22. a. .5905
b. .1937
c. .6082

24. .5886; $20,000

26. a. .1488
b. .1869
c. 43.05%
d. .8 per day; 4 over a 5 day week

28. .1249

30. .0189

32. a. .1394
b. .1952

34. a. .0918
b. 12,468

36. a. .0228
b. $50

38. a. .3830, or 38.3%
b. 96.41% do worse
c. .3821, or 38.21%

40. 19.23 ounces

42. a. approximately 1
b. .0228

Chapter 7

2. 283, 610, 39, 254, 568, 353, 602, 421, 638, 164

4. 364, 702, 782, 263, 281, 243, 493, 337, 525, 825

6. a. Finite
b. Infinite
c. Infinite
d. Infinite
e. Finite

8. a. 93
b. 5.385

10. .333

12. a. 17, 13, 11, 15, 16, 14, 18, 10, 14, 12
b.

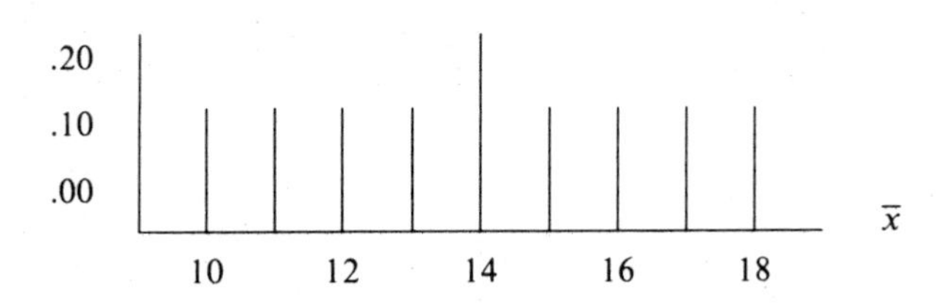

c. Sampling distribution of $\overline{x}$

14. a. 123, 124, 125, 134, 135, 145, 234, 235, 245, 345
b. 3.00, 3.67, 3.00, 3.33, 2.67, 3.33, 4.00, 3.33, 4.00, 3.67

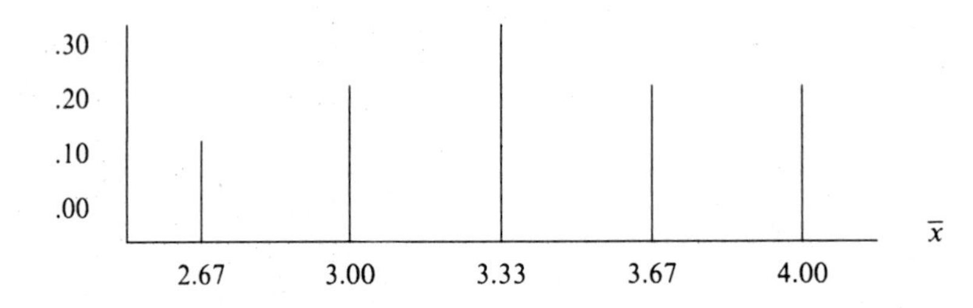

16. a. 1.6; 1.02
b. 123, 124, 125, 134, 135, 145, 234, 235, 245, 345
c. 1.00, 1.67, 2.00, 1.33, 1.67, 2.33, 1.00, 1.33, 2.00, 1.67

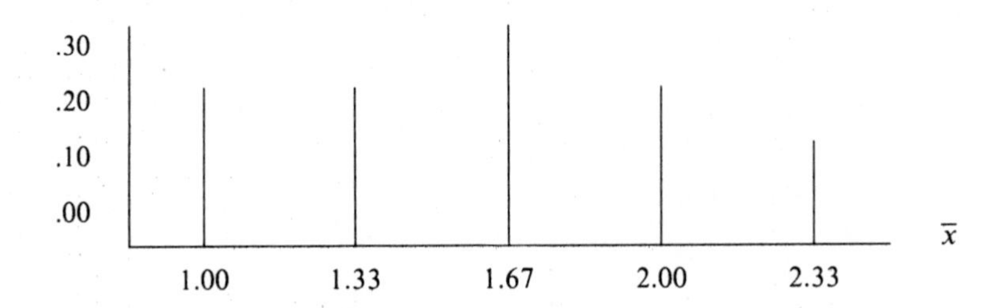

d. Mean of $\overline{x}$ = 1.6; standard deviation of $\overline{x}$ = .416
e. $E(\overline{x})$ = 1.6; $\sigma_{\overline{x}}$ = .416
f. Same results, the approach of part e is easier

18. a. .6680
b. .7888

20. b. $E(\bar{x}) = 22, \sigma_{\bar{x}} = .99$, normal
c. .99
d. $E(\bar{x}) = 22, \sigma_{\bar{x}} = .70$, normal

22. a. No: 40/4000 = .01
b. 1.290 with, 1.297 without

24. a. $E(\bar{x}) = 320, \sigma_{\bar{x}} = 13.693$, normal
b. $\sigma_{\bar{x}} = 13.693$
c. .8558
d. .3557

26. a. $E(\bar{p}) = .09, \sigma_{\bar{p}} = .0101$, normal
b. .8389

28. a. $E(\bar{p}) = .15, \sigma_{\bar{p}} = .0505$, normal
b. .4448
c. .8389

30. 263, 1763, 1199, 1628, 401

32. a. 185; 45.72
c. 185; 18.66

34. .9864; .0384

36. a. 20.11; 20.26; 20.31
b. Approximately .78 for all three firms

38. 2.0; .33

40. b. $E(\bar{p}) = .35, \sigma_{\bar{p}} = .0533$, normal
c. $E(\bar{p}) = .35, \sigma_{\bar{p}} = .0337$, normal

42. .9525

44. .4714, $E(\bar{p}) = .50, \sigma_{\bar{p}} = .079$, normal

Chapter 8

2. a. 3.96 or less
b. 68.04 to 75.96
c. 4.72 or less; 67.28 to 76.72

4. 12.865 to 14.335

6. 229.53 to 250.47

8. 75.90 to 84.10; 75.03 to 84.97

10. a. 21.15 to 23.65
b. 21.12 to 23.68
c. Essentially same interval for both z and t

12. a. 5.55
b. 4.51 to 6.59

14. a. 50, 89, 200
b. Only if absolutely necessary to have $E = 1$

16. a. 49
b. 30 to 34

18. .556 to .744

20. a. .30
b. .22 to .38

22. a. 494
b. 543

24. 82 to 90

26. a. 7.74 to 10.06
b. 7.08 to 10.72
28. 9.12 to 20.88
30. 12, 9.2 to 14.8
32. 468
34. 37
36. .028 to .062
38. 4,626
40. a. .481 to .619
b. 381

Chapter 9

2. H_0: $\mu = 2$; do not shut down
H_1: $\mu \neq 2$; shut down
4. H_0: $\mu \geq 220$; stay with current
H_1: $\mu < 220$; change to new
6. $c = 45.80$; accept H_0; president's claim cannot be rejected
8. $c = 16.58$; reject H_0; charge premium rate
10. Use t distribution with $t_{.05} = 1.895$; $c = 57.82$; reject H_0; claim work week exceeds 55 hours
12. a. Accept H_0 if $15.71 \leq \overline{x} \leq 16.29$
Reject H_0 if $\overline{x} < 15.71$ or if $\overline{x} > 16.29$
b. Reject H_0; shut line down
c. Accept H_0; continue production
14. $c_1 = 13.14$, $c_2 = 17.26$; accept H_0; no evidence to conclude a change
16. a. 12.236 to 16.364
b. Accept H_0, since 15.20 is in the interval
18. a. \$8.66; 95% confident that μ is \$8.66 or less
b. Reject H_0; conclude μ is less than \$9.00
20. Accept H_0 if $-z_{\alpha/2} \leq (\overline{x} - \mu_0)/\sigma_{\overline{x}} \leq z_{\alpha/2}$
Reject H_0 otherwise
22. a. Accept $H_0(\overline{x} - 75)/s_{\overline{x}} \leq 1.645$; otherwise reject H_0
b. $(82.50 - 75)/4.74 = 1.58$; accept H_0; $\mu \leq 75$ cannot be rejected
c. p-value $= .0571$; accept H_0
24. a. Accept H_0 if $\dfrac{\overline{x} - 40}{20/\sqrt{50}} \leq 2.05$; $z = 1.77$; accept H_0
b. p-value $= .0384$; with $\alpha = .02$, accept H_0
26. a. 910.05 to 959.95; reject H_0
b. $z = 2.75$; reject H_0
c. p-value $= .006$; reject H_0
28. a. Accept H_0 if $\overline{x} \geq 23.88$; Reject H_0 if $\overline{x} < 23.88$
b. .0537
c. .5871
d. 0; Type II error cannot be made when H_0 is true
30. a. Keep line running when filling is off standard.
b. .0749

c. .9251
d. It shows the probability of stopping and adjusting the machine when filling is off standard.

32. a. Accepting $\mu \leq 100$ when it is false.
b. .484
c. .1894
d. .2296, .0268
e. Increasing n lowers the probability of making the Type II error.

34. $n = 45$

36. $n = 76$

38. a. $c_1 = .20$, $c_2 = .30$; accept H_0; magazine's claim cannot be rejected
b. .16 to .26
c. Accept H_0

40. $z = -1.20$, accept H_0

42. a. .0793
b. .6879
c. .3783

44. $c = 17{,}352$; reject H_0; manager's claim is rejected

46. $\bar{x} = 118.9$, $s = 4.93$, $c_1 = 116.47$, $c_2 = 123.53$; accept H_0; mean production of 120 cannot be rejected

48. a. p-value $= .0718$; accept H_0: $\mu = 350$
b. $z = -1.80$; accept H_0

50. a. .2912
b. .7939
c. 0; the Type II error cannot be made when H_0 is true

52. a. $n = 45$
b. For example, for $\mu = 118$, the probability of accepting H_0 is .2358

54. $c = .252$; reject H_0; manager's claim is rejected

56. $c = .835$; $\bar{p} = .845$; accept H_0; we cannot reject the station's claim; p-value $=$ .0808

Chapter 10

2. a. \$125
b. \$36.83 to \$213.17

4. a. Worker 2 has the higher rate by 1.2 units/hour
b. Worker 2 − worker 1; −1.14 units to 3.54 units; no

6. a. H_0: $\mu_1 - \mu_2 \leq 0$; H_1: $\mu_1 - \mu_2 > 0$
b. Mean $= 0$; estimated $\sigma_{\bar{x}_1 - \bar{x}_2} = .56$; distribution normal
c. Accept H_0 if $\bar{x}_1 - \bar{x}_2 \leq .92$
d. Reject H_0; go with supplier B

8. $c_1 = 4.72$, $c_2 = 5.28$; with $\bar{x}_1 - \bar{x}_2 = 4.4$, reject H_0

10. $c_1 = 7.58$, $c_2 = 12.42$; with $\bar{d} = 8.86$, accept H_0: $\mu_d = 10$; interval: 6.44 to 11.28

12. a. Difference = after − before; accept H_0 if $\bar{d} \leq .80$; with $\bar{d} = 2.2$, reject H_0; new bonus plan appears to increase mean sales
b. 1.4 to 3.0

14. .022 to .158

16. a. $c_1 = -.086$, $c_2 = .086$; with $\bar{p}_1 - \bar{p}_2 = -.15$, reject H_0; a difference exists
b. Women response over men: .064 to .236
18. \$1,354 to \$2,646
20. $n = 45$ for each section
22. $c_1 = -7.13$, $c_2 = 7.13$; with $\bar{x}_1 - \bar{x}_2 = -6$, accept H_0; no difference
24. $c = 5.42$; with $\bar{d} = 6.17$, reject H_0; program provides loss
26. a. $c_1 = -.039$, $c_2 = +.039$; with $\bar{p}_1 - \bar{p}_2 = .09$, reject H_0; difference exists
b. .051 to .129

Chapter 11

2. .22 to .71, .47 to .84
4. $\chi^2 = 36.25$, $\chi^2_{.05} = 42.56$; accept H_0
6. $\chi^2 = 23.33$, $\chi^2_{.975} = 24.43$, $\chi^2_{.025} = 59.34$; reject H_0
8. $F = 2.08$, $F_{.05} = 1.94$; reject H_0
10. $F = 2.20$, $F_{.05} = 2.03$; reject H_0; therefore cannot pool
12. 22.83 to 65.05, 4.78 to 8.07
14. a. $\chi^2 = 27.44$, $\chi^2_{.10} = 21.06$; reject H_0
b. .00012 to .00042
16. a. $n = 15$, b. 6.25 to 11.13
18. $F = 2.75$, $F_{.025} = 2.62$; reject H_0; females
20. $F = 2.12$, $F_{.05} = 3.79$; accept H_0; 63.67

Chapter 12

2. $\chi^2 = 6.24$, $\chi^2_{.10} = 4.61$; reject H_0
4. $\chi^2 = 2.31$, $\chi^2_{.05} = 5.99$; accept H_0
6. $\chi^2 = 12.39$, $\chi^2_{.01} = 11.34$; reject H_0
8. $\chi^2 = 7.96$, $\chi^2_{.05} = 9.49$; accept H_0
10. $\chi^2 = 4.32$, $\chi^2_{.05} = 5.99$; accept H_0; Poisson
12. $\chi^2 = 2.80$, $\chi^2_{.10} = 6.25$; accept H_0; normal
14. $\chi^2 = 8.04$, $\chi^2_{.05} = 7.81$; reject H_0
16. $\chi^2 = 2.21$, $\chi^2_{.05} = 7.81$; independent
18. $\chi^2 = 6.17$, $\chi^2_{.05} = 7.81$; accept H_0; binomial

Chapter 13

2. Cannot conclude there is a difference; $F = 2.54 < F_{.05} = 3.24$
4. Not the same; $F = 17.5$, $F_{.05} = 3.89$
6.

Source	*SS*	*DF*	*MS*	*F*
Treatments	70	2	35	17.5
Error	24	12	12	
Total	94	14		

8.

Source	*SS*	*DF*	*MS*	*F*
Treatments	61.64	3	20.55	17.56
Error	23.41	20	1.17	
Total	85.05	23		

Not the same; $F = 17.56$, $F_{.05} = 3.10$

10. Not the same; $F = 20.57$, $F_{.05} = 18.51$

12.

Source	*SS*	*DF*	*MS*	*F*
Treatments	310	4	77.5	17.71
Blocks	85	2	42.5	
Error	35	8	4.38	
Total	430	14		

Not the same; $F = 17.71$, $F_{.05} = 3.84$

14. No significant difference between manufacturer 1 and manufacturer 2

16. No significant difference between company A and company B

18. No significant difference

20. The ANOVA procedure is used to test whether the means of k populations are equal.

22. In order to work correctly, the ANOVA procedure must have the population variances the same.

24. MSTR is an unbiased estimate of σ^2 when all the population means are the same. When they are not the squared deviations will be larger causing MSTR to be larger.

26. Not the same; $F = 4.45$, $F_{.05} = 4.26$

28. Not the same; $F = 12.90$

30. No significant difference; $F = 1.48$, $F_{.05} = 3.35$

32. No significant difference; $F = 1.42$, $F_{.05} = 4.26$

34. Population means are not equal; $F = 5.19$, $F_{.05} = 4.26$

36. No; means were not significantly different

38. Without removing the block effect we cannot detect any significant difference between the three brands of gasoline

40. Significant difference due to language

Chapter 14

2. a. $E(y) = 2 + 3x$

c. $E(y) = 2x$

4. b. Yes

c. The graph goes through the following points:

x	$E(y) = 60x/(2 + x)$
1	20
4	40
6	45
10	50
14	52.5

Yes, the approximation is a good one.

d. Yes

e. Linear approximation: sum of absolute value of errors = 27.5
Curvilinear approximation: sum of absolute value of errors = 12.5

6. b. Scatter diagram suggests a linear relationship between price and the number of loaves sold

8. a. $\hat{y} \approx -589.19 + 621.62x$
b. $1,275.67; $1,586.48

10. a. $\hat{y} \approx 1061.82 - 1363.64x$
b. 202.73
c. 189.09; number actually sold = 190; difference = .91

12. $\hat{y} = 80 + 4x$
a. 170
b. 179.28 for $\hat{y} = 78 + 4.2x$; least squares line has smallest $\Sigma(y_i - \hat{y}_i)^2$

14. a. SSR = 285,945.95
b. SST = 340,000.00
c. 84.10%

16. a. $\hat{y} = -54.84 + 98.80x$
b. $r^2 = .99$; yes
c. 38.13 mgs

18. a. $\bar{x} = 3.8$; $\bar{y} = 23.2$
b. Yes
c. Yes; the formula for b_0 guarantees it

20. a. $\hat{y} \approx .7542 + .5088x$
b. MSE = 1.77; MSR = 5.90
c. $F = 3.33$, $F_{.05} = 10.13$; cannot reject H_0

22. $b_1/s_{b_1} = 45.68/6.58 = 6.94$; $t_{.05} = 2.353$; square footage and selling price are related

24. a. $F = 106.92$, $F_{.05} = 5.32$; reject
b. $t = 10.3401$, $t_{.025} = 2.306$; reject
c. $t^2 = 106.92$, $t^2_{.025} = 5.32$; both tests yield same conclusion

26. a. $\hat{y} \approx 25.38 + 2.32x$
b. $t = 2.9$, $t_{.05} = 2.353$; reject
d. Looks like a curvilinear relationship is present; see Section 15.9 for a better model

28. b. Assumptions seem reasonable

30. $106.88 ± 6.71

32. $\hat{y}_p = 1,275.67$, $t_{.025} = 2.776$, $s_{ind} = 128.44$; $1,275.67 ± 2.776 (128.44)

34. 135 ± 1.86 (14.52)

36. a. 9
b. $\hat{y} = 20.0 + 7.21x$

c. 1.3626
d. $F = 27.999$; reject
e. $380,500

38. a. $\hat{y} = 80.0 + 50x$
b. 30
c. $F = 83.17$; reject
d. $680,000

40. $r = 3.2(10/35) = .91$

42. .92

44. No; regression or correlation analysis can never prove that two variables are causally related

46. By the value of r^2

48. To guard against using a model for which the assumptions concerning ϵ are violated

50. a. $\hat{y} = 10.53 + .95x$
b. $F = 47.62$, $F_{.05} = 5.32$; reject, the relationship is significant
c. Model appears adequate
d. $2,865 to $4,941
e. Yes

52. a. $\hat{y} = 3{,}766.67 - 322.22x$
b. Yes
c. Yes, $r^2 = .94$
d. $2{,}477.79 \pm 361.81$

54. a. $\hat{y} = 220 + 131.67x$
b. Variables are related
c. Yes, $r^2 = .87$
d. 746.68 ± 150.96

56. a. $\hat{y} = 5.85 + .83x$
b. Relationship is significant
c. No
d. 84.7
e. 69.13 to 100.27

Chapter 15

2. a. $\hat{y} = 88.64 + 1.60x$
b. $94,240

4. a.

$$10b_0 + 1{,}510b_1 + 1{,}880b_2 = 783$$
$$1{,}510b_0 + 233{,}200b_1 + 284{,}700b_2 = 120{,}160$$
$$1{,}880b_0 + 284{,}700b_1 + 407{,}200b_2 = 133{,}040$$

b. $b_0 = 66.5176$, $b_1 = .4139$, $b_2 = -.2698$
c. 93.7126
d. b_1 is change in quantity sold when competitor increases price by $1.00 and Heller does not change price; b_2 is change in quantity sold when Heller increases price by $1.00 and competition does not change price

6. a. $F = 28.37$, $F_{.01} = 13.27$; reject, the relationship is significant
b. $t = 1.30/.3207 = 4.05$, $t_{.025} = 2.571$; reject, do not drop x_1 from model
c. $t = 2.29/.3041 = 7.53$, $t_{.025} = 2.571$; reject, do not drop x_2 from model

8. a. 2, 14
b. SSE = 18.3, MSR = 45.15, MSE = 1.525; t for x_1 is 3.13, t for x_2 is 4.66
c. $F = 29.61$, $F_{.01} = 6.93$; relationship is significant
d. $t_{.025} = 2.179$; do not drop either variable

10. a. $R^2 = .83$
b. $R_a^2 = .80$
c. Yes

12. a. $15,300
b. $56,100
c. $41,600

14. Poor fit; $r^2 = .53$

16. $\hat{y} = .9305 + .3876x_1 + 1.2627x_2$

18. Yes, R^2 is much larger

20. a. $t_{.025} = 2.069$, $t = 4.34$; reject
b. $t_{.025} = 2.069$, $t = 1.48$; cannot reject
c. $t_{.025} = 2.069$, $t = 4.46$; reject
d. x_2 can be dropped, since it is not significant

22. a. 45
b. 100
c. $F = (55/2)/(45/25) = 27.5/1.8 = 15.28$, $F_{.05} = 3.39$; reject—they do contribute significantly

24. a. $184,000
b. $191,000
c. $15,000

26. a. $b_0 = 432.51$, $b_1 = 37.43$, $b_2 = -.38$
b. $F = 73.15$, $F_{.01} = 30.82$; reject, relationship is significant
c. 1302.01

28. a. $E(y) = \beta_0 + \beta_1 x_1 + \beta_2 x_2$
b. $\hat{y} = 2.3 + 5x_1 - 2x_2$
c. $H_0: \beta_1 = \beta_2 = 0$
d. MSR/MSE = 10.64
e. Reject

30. a. $R^2 = 18051.63/19065.93 = .95$, $R_a^2 = .93$
b. $F = 4512.9075/78.023 = 57.84$, $F_{.01} = 5.21$; reject, relationship is significant

32. a. 3.61 and 5.08
b. $t_{.025} = 2.179$; both are significant
c. Regression DF = 2, residual DF = 12, total DF = 14, SSE = 134.48, SST = 1746.48, MSR = 806, MSE = 11.21
d. .91

34. $\hat{y} = -.8021 + 6.3801x$

36. Expected increase in annual cost for energy package over standard package

Chapter 16

2. a. Selected forecasts: week 4, 19.33; week 9, 18.33; week 12, 17.83
b. MSE = 11.49

4. a. $F_{13} = .2Y_{12} + .16Y_{11} + .128Y_{10} + .1024Y_9 + .08192Y_8 + .32768F_8$
b. The weights sum to 1 and decrease as we use more remote data values; moving averages weights the data values equally

6. Selected forecasts: month 5, 112.5, month 10, 116.95, month 12, 99.24; MSE = 540.54 for $\alpha = .5$, MSE = 510.28 for $\alpha = .3$; $\alpha = .3$ provides better forecasts
8. $T_t = 28{,}800 + 421t$
10. a. Linear trend appears to exist
b. $T_t = 19.99 + 1.77t$; average cost increase is $1.77
12. Selected seasonal factors: month 4, .966; month 8, 1.225; month 12, .787
14. 20.26
16. $T_t = 197.71 + 6.82t$; $T_8 = 252.27$ and $T_9 = 259.09$
18. a. Yes
b. $T_t = 22.86 + 15.54t$
c. $T_8 = 147.18$
20. a. Selected seasonal factors: quarter 2, .6138; quarter 4, 1.6163
b. Largest effect in quarter 4
22. a. $T_t = -.345 + .995t$
b. 20.55, 21.55, 22.54, 23.54
c. 26.10, 13.15, 11.27, 38.13

Chapter 17

2. $I = 105$
4. a. 158, 113, 96
b. 120
6. 109, 105, 102; $I = 105$
8. a. 155, 155, 158, 162, 164
b. 155, 158, 162, 165, 168
c. PPI
10. 99
12. a. 151, 197, 143, 178
b. 170
14. January: 96; March: 92
16. 101
18. 11,500; 12,727; 11,218; 10,044; 10,771
20. 143

Chapter 18

2. $c_1 = 83.2$, $c_2 = 126.8$; W = 130; reject H_0; conclude salaries differ
4. One-tailed test: rejection if posttest scores higher; $c = -27.8$; T = −33; reject H_0; conclude test scores differ in favor of posttest scores
6. $c_1 = -50$, $c_2 = +50$, T = −66; reject H_0; conclude perceptions of prices differ
8. Eight "+" and two "−"; $c_1 = 1.9$, $c_2 = 8.1$; accept H_0; no difference between methods
10. a. +1.0
b. −1.0
12. $r_s = .76$; significant
14. $c_1 = 88.4$, $c_2 = 151.6$; W = 116; accept H_0; unable to reject assumption of same prices in the two cities
16. $c_1 = 90$, $c_2 = 140$; W = 70; reject H_0; difference exists
18. $c_1 = 70$, $c_2 = 95$; reject H_0; preferences differ
20. −.92; significant and negative
22. .64; not significant

Chapter 19

2. a. d_2 or d_3
b. d_1
c. d_2

4. a. d_1 (11.3)
b. d_4 (9.5)

6. d_2

8. d_2; EMV(d_2) = 500

10. EVPI = 25

12. EVPI = 195

14. a. d_1
b. $P(s_1 \mid I) = .516, P(s_2 \mid I) = .484; d_3$

16. a. d_2
b. EVSI = 12
c. EVPI = 60
d. 20%

18. a. If I_1 then d_2; if I_2 then d_2; if I_3 then d_3
b. EVSI = $38,400
c. EVPI = $195,000; efficiency = 19.7%

20. a. d_2 (purchase)
b. $9,000

22. d_2; EVPI = 3.5 minutes

24. a.

Order Quantities	*Demands* 1,000	*2,000*	*3,000*
1,000	5,000	1,000	−3,000
2,000	−5,000	10,000	6,000
3,000	−15,000	0	15,000

b. d_2: 2,000 units
c. $4,800

26. a. d_1 (St. Louis)
b. $200

28. a. .355
b. If favorable, manufacture; if unfavorable, purchase
c. $3,710
d. 41.2%

30. a. Do not fund
b. $27,000

32. a. Always d_2
b. EVSI = $0
c. 0%

34. a. $P(I \mid s_1) = 0; P(I \mid s_2) = 0.048; P(I \mid s_3) = .092; P(I \mid s_4) = .133$
b. $P(s_1 \mid I) = 0; P(s_2 \mid I) = .159; P(s_3 \mid I) = .488; P(s_4 \mid I) = .353$
c. Yes

Appendices

A. References and Bibliography

GENERAL

Freund, J. E., and R. E. Walpole, *Mathematical Statistics,* 3rd ed., Englewood Cliffs, N.J., Prentice-Hall, 1980.

Hogg, R. V., and A. T. Craig, *Introduction to Mathematical Statistics,* 4th ed., New York, Macmillan, 1978.

Mood, A. M., F. A. Graybill, and D. C. Boes, *Introduction to the Theory of Statistics,* 3rd ed., New York, McGraw-Hill, 1974.

Neter, J., W. Wasserman, and G. A. Whitmore, *Applied Statistics,* 2nd ed., Boston, Allyn & Bacon, 1982.

Winkler, R. L., and W. L. Hays, *Statistics: Probability, Inference, and Decision,* 2nd ed., New York, Holt, Rinehart & Winston, 1975.

PROBABILITY

Feller, W., *An Introduction to Probability Theory and Its Applications,* Vol. I, 3rd ed., New York, John Wiley & Sons, 1968.

Feller, W., *An Introduction to Probability Theory and Its Applications,* Vol. II, 2nd ed., New York, John Wiley & Sons, 1971.

Hoel, P. G., S. C. Port, and C. J. Stone, *Introduction to Probability Theory,* Boston, Houghton Mifflin, 1971.

Parzen, E., *Modern Probability Theory and Its Applications,* New York, John Wiley & Sons, Inc., 1960.

Wadsworth, G. P., and J. G. Bryan, *Applications of Probability and Random Variables,* 2nd ed., New York, McGraw-Hill, 1974.

Zehna, P. W., *Probability Distributions and Statistics,* Boston, Allyn & Bacon, 1970.

SAMPLING METHODS

Cochran, W. G., *Sampling Techniques,* 3rd ed., New York, John Wiley & Sons, 1977.

Kish, L., *Survey Sampling,* New York, John Wiley & Sons, 1965.

Scheaffer, R. L., W. Mendenhall, and L. Ott, *Elementary Survey Sampling,* 2nd ed., North Scituate, Mass., Duxbury Press, 1979.

Williams, B., *A Sampler on Sampling,* New York, John Wiley & Sons, 1978.

REGRESSION ANALYSIS

Chatterjee, S., and B. Price, *Regression Analysis by Example,* New York, John Wiley & Sons, Inc., 1978.

Draper, N. R., and H. Smith, *Applied Regression Analysis,* 2nd ed., New York, John Wiley & Sons, 1981.

Gunst, R. F. and R. L. Mason, *Regression Analysis and Its Applications,* New York, Dekker, 1980.

Kleinbaum, D. G., and L. L. Kupper, *Applied Regression Analysis and Other Multivariable Methods,* North Scituate, Mass., Duxbury Press, 1978.

Mendenhall, W., *Introduction to Linear Models and the Design and Analysis of Experiments,* Belmont, Calif., Wadsworth Publishing, 1968.

Mosteller, F., and J. W. Tukey, *Data Analysis and Regression: A Second Course in Statistics,* Reading, Mass., Addison-Wesley, 1977.

Neter, J., and W. Wasserman, *Applied Linear Statistical Models,* Homewood, Ill., Richard D. Irwin, 1974.

Wesolowsky, G. O., *Multiple Regression and Analysis of Variance,* New York, John Wiley & Sons, 1976.

ANALYSIS OF VARIANCE AND EXPERIMENTAL DESIGN

Anderson, V. L., and R. A. McLean, *Design of Experiments: A Realistic Approach,* New York, Marcel Dekker, 1974.

Cochran, W. G., and G. M. Cox, *Experimental Designs,* 2nd ed., New York, John Wiley & Sons, 1957.

Mendenhall, W., *Introduction to Linear Models and the Design and Analysis of Experiments,* Belmont, Calif., Duxbury Press, 1968.

Montgomery, D. C., *Design and Analysis of Experiments,* New York, John Wiley & Sons, 1976.

NONPARAMETRIC METHODS

Conover, W. J., *Practical Nonparametric Statistics,* 2nd ed., New York, John Wiley & Sons, 1980.

Gibbons, J. D., *Nonparametric Statistical Inference,* New York, McGraw-Hill, 1971.

Gibbons, J. D., I. Olkin, and M. Sobel, *Selecting and Ordering Populations: A New Statistical Methodology,* New York, John Wiley & Sons, 1977.

Lehmann, E. L., *Nonparametrics: Statistical Methods Based on Ranks,* San Francisco, Holden-Day, 1975.

Mosteller, F., and R. E. K. Rourke, *Sturdy Statistics,* Reading, Mass., Addison-Wesley, 1973.

Siegel, S., *Nonparametric Statistics for the Behavioral Sciences,* New York, McGraw-Hill, 1956.

FORECASTING AND TIME SERIES

Bowerman, B. L., and R. T. O'Connell, *Time Series and Forecasting: An Applied Approach,* North Scituate, Mass., Duxbury Press, 1979.

Box, G. E. P., and G. M. Jenkins, *Time Series Analysis, Forecasting and Control,* rev. ed., San Francisco, Holden-Day, 1976.

Brillinger, D.R., *Time Series,* San Francisco, Holden-Day, 1981.

Gilchrist, W. G., *Statistical Forecasting,* New York, John Wiley & Sons, 1976.

Makridakis, S., and S. C. Wheelwright, *Forecasting: Methods and Applications,* New York, John Wiley & Sons, 1978.
Nelson, C. R., *Applied Time Series Analysis for Managerial Forecasting,* San Francisco, Holden-Day, 1973.
Thomopoulos, N. T., *Applied Forecasting Methods,* Englewood Cliffs, N.J., Prentice-Hall, 1980.
Wheelwright, S. C., and S. Makridakis, *Forecasting Methods for Management,* 3rd ed., New York, John Wiley & Sons, 1980.

DECISION THEORY

Aitchison, J., *Choice Against Chance,* Reading, Mass., Addison-Wesley, 1970.
Berger, J., *Statistical Decision Theory, Foundations, Concepts, and Methods,* New York, Springer-Verlag, 1980.
Brown, R. V., A. S. Kahn, and C. Peterson, *Decision Analysis for the Manager,* New York, Holt, Rinehart & Winston, 1974.
Hadley, G., *Introduction to Probability and Statistical Decision Theory,* San Francisco, Holden-Day, 1967.
Kwon, I., *Statistical Decision Theory with Business and Economic Applications: A Bayesian Approach,* New York, Petrocelli/Charter, 1978.
Luce, R. D., and H. Raiffa, *Games and Decisions: Introduction and Critical Survey,* New York, John Wiley & Sons, 1957.
Raiffa, H., *Decision Analysis: Introductory Lectures on Choices under Uncertainty,* Reading, Mass., Addison-Wesley, 1968.
Winkler, R. L., *An Introduction to Bayesian Inference and Decision,* New York, Holt, Rinehart & Winston, 1972.

B.
Tables

TABLE 1 Standard Normal Distribution

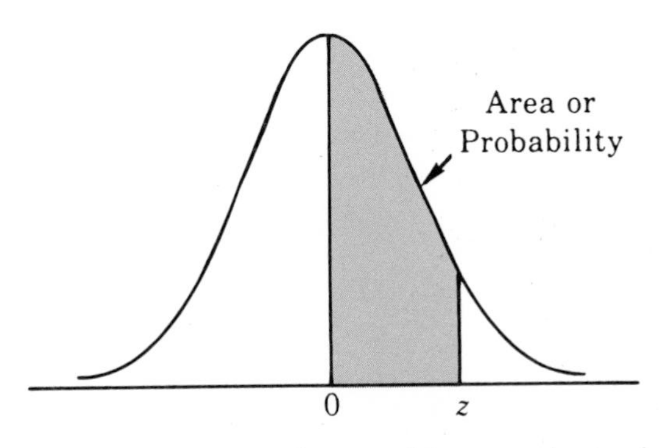

Entries in the table give the area under the curve between the mean and z standard deviations above the mean. For example, for $z = 1.25$ the area under the curve between the mean and z is .3944.

z	.00	.01	.02	.03	.04	.05	.06	.07	.08	.09
.0	.0000	.0040	.0080	.0120	.0160	.0199	.0239	.0279	.0319	.0359
.1	.0398	.0438	.0478	.0517	.0557	.0596	.0636	.0675	.0714	.0753
.2	.0793	.0832	.0871	.0910	.0948	.0987	.1026	.1064	.1103	.1141
.3	.1179	.1217	.1255	.1293	.1331	.1368	.1406	.1443	.1480	.1517
.4	.1554	.1591	.1628	.1664	.1700	.1736	.1772	.1808	.1844	.1879
.5	.1915	.1950	.1985	.2019	.2054	.2088	.2123	.2157	.2190	.2224
.6	.2257	.2291	.2324	.2357	.2389	.2422	.2454	.2486	.2518	.2549
.7	.2580	.2612	.2642	.2673	.2704	.2734	.2764	.2794	.2823	.2852
.8	.2881	.2910	.2939	.2967	.2995	.3023	.3051	.3078	.3106	.3133
.9	.3159	.3186	.3212	.3238	.3264	.3289	.3315	.3340	.3365	.3389
1.0	.3413	.3438	.3461	.3485	.3508	.3531	.3554	.3577	.3599	.3621
1.1	.3643	.3665	.3686	.3708	.3729	.3749	.3770	.3790	.3810	.3830
1.2	.3849	.3869	.3888	.3907	.3925	.3944	.3962	.3980	.3997	.4015
1.3	.4032	.4049	.4066	.4082	.4099	.4115	.4131	.4147	.4162	.4177
1.4	.4192	.4207	.4222	.4236	.4251	.4265	.4279	.4292	.4306	.4319
1.5	.4332	.4345	.4357	.4370	.4382	.4394	.4406	.4418	.4429	.4441
1.6	.4452	.4463	.4474	.4484	.4495	.4505	.4515	.4525	.4535	.4545
1.7	.4554	.4564	.4573	.4582	.4591	.4599	.4608	.4616	.4625	.4633
1.8	.4641	.4649	.4656	.4664	.4671	.4678	.4686	.4693	.4699	.4706
1.9	.4713	.4719	.4726	.4732	.4738	.4744	.4750	.4756	.4761	.4767
2.0	.4772	.4778	.4783	.4788	.4793	.4798	.4803	.4808	.4812	.4817
2.1	.4821	.4826	.4830	.4834	.4838	.4842	.4846	.4850	.4854	.4857
2.2	.4861	.4864	.4868	.4871	.4875	.4878	.4881	.4884	.4887	.4890
2.3	.4893	.4896	.4898	.4901	.4904	.4906	.4909	.4911	.4913	.4916
2.4	.4918	.4920	.4922	.4925	.4927	.4929	.4931	.4932	.4934	.4936
2.5	.4938	.4940	.4941	.4943	.4945	.4946	.4948	.4949	.4951	.4952
2.6	.4953	.4955	.4956	.4957	.4959	.4960	.4961	.4962	.4963	.4964
2.7	.4965	.4966	.4967	.4968	.4969	.4970	.4971	.4972	.4973	.4974
2.8	.4974	.4975	.4976	.4977	.4977	.4978	.4979	.4979	.4980	.4981
2.9	.4981	.4982	.4982	.4983	.4984	.4984	.4985	.4985	.4986	.4986
3.0	.4986	.4987	.4987	.4988	.4988	.4989	.4989	.4989	.4990	.4990

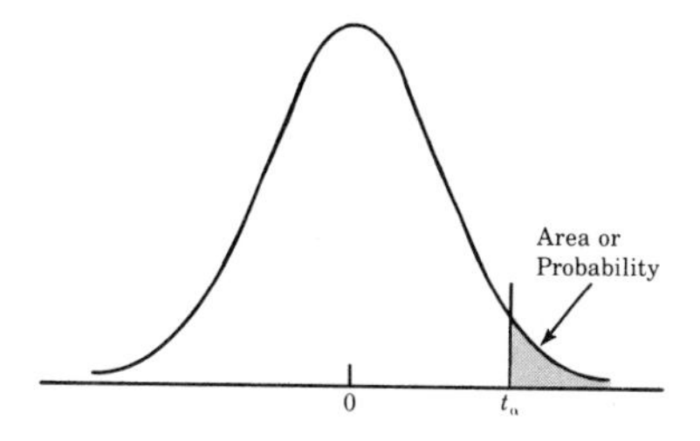

Entries in the table give t_α values, where α is the area or probability in the upper tail of the t distribution. For example, with 10 degrees of freedom and a .05 area in the upper tail, $t_{.05} = 1.812$.

Degrees of Freedom	*Area in Upper Tail*				
	.10	*.05*	*.025*	*.01*	*.005*
1	3.078	6.314	12.706	31.821	63.657
2	1.886	2.920	4.303	6.965	9.925
3	1.638	2.353	3.182	4.541	5.841
4	1.533	2.132	2.776	3.747	4.604
5	1.476	2.015	2.571	3.365	4.032
6	1.440	1.943	2.447	3.143	3.707
7	1.415	1.895	2.365	2.998	3.499
8	1.397	1.860	2.306	2.896	3.355
9	1.383	1.833	2.262	2.821	3.250
10	1.372	1.812	2.228	2.764	3.169
11	1.363	1.796	2.201	2.718	3.106
12	1.356	1.782	2.179	2.681	3.055
13	1.350	1.771	2.160	2.650	3.012
14	1.345	1.761	2.145	2.624	2.977
15	1.341	1.753	2.131	2.602	2.947
16	1.337	1.746	2.120	2.583	2.921
17	1.333	1.740	2.110	2.567	2.898
18	1.330	1.734	2.101	2.552	2.878
19	1.328	1.729	2.093	2.539	2.861
20	1.325	1.725	2.086	2.528	2.845
21	1.323	1.721	2.080	2.518	2.831
22	1.321	1.717	2.074	2.508	2.819
23	1.319	1.714	2.069	2.500	2.807
24	1.318	1.711	2.064	2.492	2.797
25	1.316	1.708	2.060	2.485	2.787
26	1.315	1.706	2.056	2.479	2.779
27	1.314	1.703	2.052	2.473	2.771
28	1.313	1.701	2.048	2.467	2.763
29	1.311	1.699	2.045	2.462	2.756
30	1.310	1.697	2.042	2.457	2.750
40	1.303	1.684	2.021	2.423	2.704
60	1.296	1.671	2.000	2.390	2.660
120	1.289	1.658	1.980	2.358	2.617
∞	1.282	1.645	1.960	2.326	2.576

TABLE 3 Chi-Square Distribution

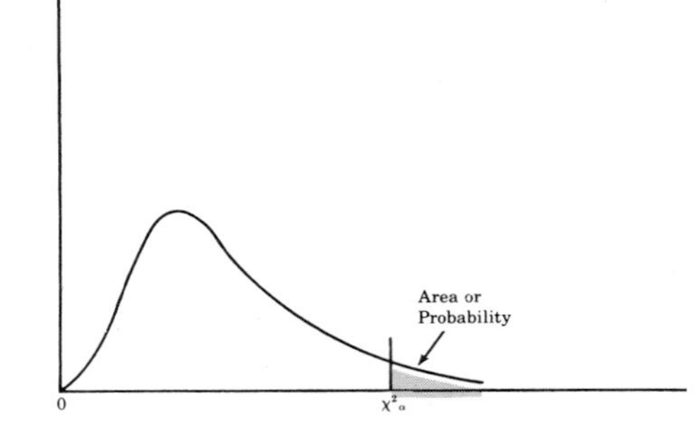

Entries in the table give χ^2_α values, where α is the area or probability in the upper tail of the chi-square distribution. For example, with 10 degrees of freedom and a .01 area in the upper tail, $\chi^2_{.01} = 23.2093$.

Degrees of Freedom					*Area in Upper Tail*					
	.995	*.99*	*.975*	*.95*	*.90*	*.10*	*.05*	*.025*	*.01*	*.005*
1	$392{,}704 \times 10^{-10}$	$157{,}088 \times 10^{-9}$	$982{,}069 \times 10^{-9}$	$393{,}214 \times 10^{-8}$	.0157908	2.70554	3.84146	5.02389	6.63490	7.87944
2	.0100251	.0201007	.0506356	.102587	.210720	4.60517	5.99147	7.37776	9.21034	10.5966
3	.0717212	.114832	.215795	.351846	.584375	6.25139	7.81473	9.34840	11.3449	12.8381
4	.206990	.297110	.484419	.710721	1.063623	7.77944	9.48773	11.1433	13.2767	14.8602
5	.411740	.554300	.831211	1.145476	1.61031	9.23635	11.0705	12.8325	15.0863	16.7496
6	.675727	.872085	1.237347	1.63539	2.20413	10.6446	12.5916	14.4494	16.8119	18.5476
7	.989265	1.239043	1.68987	2.16735	2.83311	12.0170	14.0671	16.0128	18.4753	20.2777
8	1.344419	1.646482	2.17973	2.73264	3.48954	13.3616	15.5073	17.5346	20.0902	21.9550
9	1.734926	2.087912	2.70039	3.32511	4.16816	14.6837	16.9190	19.0228	21.6660	23.5893
10	2.15585	2.55821	3.24697	3.94030	4.86518	15.9871	18.3070	20.4831	23.2093	25.1882
11	2.60321	3.05347	3.81575	4.57481	5.57779	17.2750	19.6751	21.9200	24.7250	26.7569
12	3.07382	3.57056	4.40379	5.22603	6.30380	18.5494	21.0261	23.3367	26.2170	28.2995
13	3.56503	4.10691	5.00874	5.89186	7.04150	19.8119	22.3621	24.7356	27.6883	29.8194
14	4.07468	4.66043	5.62872	6.57063	7.78953	21.0642	23.6848	26.1190	29.1413	31.3193
15	4.60094	5.22935	6.26214	7.26094	8.54675	22.3072	24.9958	27.4884	30.5779	32.8013
16	5.14224	5.81221	6.90766	7.96164	9.31223	23.5418	26.2962	28.8454	31.9999	34.2672
17	5.69724	6.40776	7.56418	8.67176	10.0852	24.7690	27.5871	30.1910	33.4087	35.7185
18	6.26481	7.01491	8.23075	9.39046	10.8649	25.9894	28.8693	31.5264	34.8053	37.1564
19	6.84398	7.63273	8.90655	10.1170	11.6509	27.2036	30.1435	32.8523	36.1908	38.5822

20	7.43386	8.26040	9.59083	10.8508	12.4426	28.4120	31.4104	34.1696	37.5662	39.9968
21	8.03366	8.89720	10.28293	11.5913	13.2396	29.6151	32.6705	35.4789	38.9321	41.4010
22	8.64272	9.54249	10.9823	12.3380	14.0415	30.8133	33.9244	36.7807	40.2894	42.7958
23	9.26042	10.19567	11.6885	13.0905	14.8479	32.0069	35.1725	38.0757	41.6384	44.1813
24	9.88623	10.8564	12.4011	13.8484	15.6587	33.1963	36.4151	39.3641	42.9798	45.5585
25	10.5197	11.5240	13.1197	14.6114	16.4734	34.3816	37.6525	40.6465	44.3141	46.9278
26	11.1603	12.1981	13.8439	15.3791	17.2919	35.5631	38.8852	41.9232	45.6417	48.2899
27	11.8076	12.8786	14.5733	16.1513	18.1138	36.7412	40.1133	43.1944	46.9630	49.6449
28	12.4613	13.5648	15.3079	16.9279	18.9392	37.9159	41.3372	44.4607	48.2782	50.9933
29	13.1211	14.2565	16.0471	17.7083	19.7677	39.0875	42.5569	45.7222	49.5879	52.3356
30	13.7867	14.9535	16.7908	18.4926	20.5992	40.2560	43.7729	46.9792	50.8922	53.6720
40	20.7065	22.1643	24.4331	26.5093	29.0505	51.8050	55.7585	59.3417	63.6907	66.7659
50	27.9907	29.7067	32.3574	34.7642	37.6886	63.1671	67.5048	71.4202	76.1539	79.4900
60	35.5346	37.4848	40.4817	43.1879	46.4589	74.3970	79.0819	83.2976	88.3794	91.9517
70	43.2752	45.4418	48.7576	51.7393	55.3290	85.5271	90.5312	95.0231	100.425	104.215
80	51.1720	53.5400	57.1532	60.3915	64.2778	96.5782	101.879	106.629	112.329	116.321
90	59.1963	61.7541	65.6466	69.1260	73.2912	107.565	113.145	118.136	124.116	128.299
100	67.3276	70.0648	74.2219	77.9295	82.3581	118.498	124.342	129.561	135.807	140.169

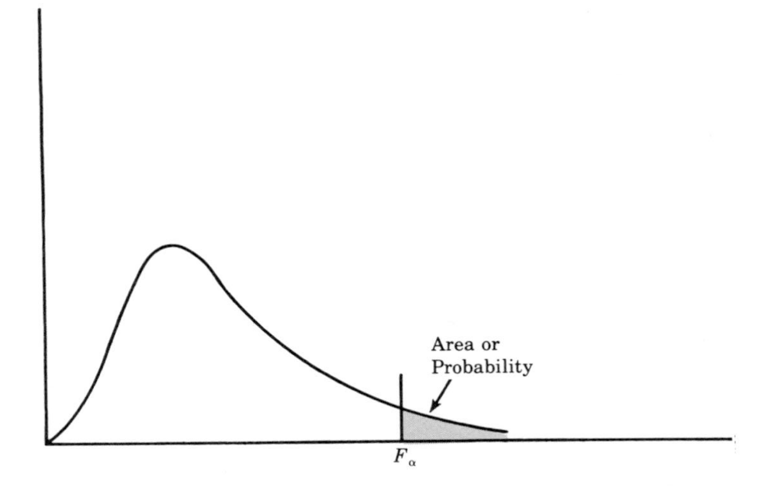

Entries in the table give F_α values, where α is the area or probability in the upper tail of the F distribution. For example, with 12 numerator degrees of freedom, 15 denominator degrees of freedom, and a .05 area in the upper tail, $F_{.05} = 2.48$.

Table of $F_{.05}$ Values

Denominator Degrees of Freedom	Numerator Degrees of Freedom 1	2	3	4	5	6	7	8	9	10	12	15	20	24	30	40	60	120	∞
1	161.4	199.5	215.7	224.6	230.2	234.0	236.8	238.9	240.5	241.9	243.9	245.9	248.0	249.1	250.1	251.1	252.2	253.3	254.3
2	18.51	19.00	19.16	19.25	19.30	19.33	19.35	19.37	19.38	19.40	19.41	19.43	19.45	19.45	19.46	19.47	19.48	19.49	19.50
3	10.13	9.55	9.28	9.12	9.01	8.94	8.89	8.85	8.81	8.79	8.74	8.70	8.66	8.64	8.62	8.59	8.57	8.55	8.53
4	7.71	6.94	6.59	6.39	6.26	6.16	6.09	6.04	6.00	5.96	5.91	5.86	5.80	5.77	5.75	5.72	5.69	5.66	5.63
5	6.61	5.79	5.41	5.19	5.05	4.95	4.88	4.82	4.77	4.74	4.68	4.62	4.56	4.53	4.50	4.46	4.43	4.40	4.36
6	5.99	5.14	4.76	4.53	4.39	4.28	4.21	4.15	4.10	4.06	4.00	3.94	3.87	3.84	3.81	3.77	3.74	3.70	3.67
7	5.59	4.74	4.35	4.12	3.97	3.87	3.79	3.73	3.68	3.64	3.57	3.51	3.44	3.41	3.38	3.34	3.30	3.27	3.23
8	5.32	4.46	4.07	3.84	3.69	3.58	3.50	3.44	3.39	3.35	3.28	3.22	3.15	3.12	3.08	3.04	3.01	2.97	2.93
9	5.12	4.26	3.86	3.63	3.48	3.37	3.29	3.23	3.18	3.14	3.07	3.01	2.94	2.90	2.86	2.83	2.79	2.75	2.71
10	4.96	4.10	3.71	3.48	3.33	3.22	3.14	3.07	3.02	2.98	2.91	2.85	2.77	2.74	2.70	2.66	2.62	2.58	2.54
11	4.84	3.98	3.59	3.36	3.20	3.09	3.01	2.95	2.90	2.85	2.79	2.72	2.65	2.61	2.57	2.53	2.49	2.45	2.40
12	4.75	3.89	3.49	3.26	3.11	3.00	2.91	2.85	2.80	2.75	2.69	2.62	2.54	2.51	2.47	2.43	2.38	2.34	2.30
13	4.67	3.81	3.41	3.18	3.03	2.92	2.83	2.77	2.71	2.67	2.60	2.53	2.46	2.42	2.38	2.34	2.30	2.25	2.21
14	4.60	3.74	3.34	3.11	2.96	2.85	2.76	2.70	2.65	2.60	2.53	2.46	2.39	2.35	2.31	2.27	2.22	2.18	2.13

15	4.54	3.68	3.29	3.06	2.90	2.79	2.71	2.64	2.59	2.54	2.48	2.40	2.33	2.29	2.25	2.20	2.16	2.11	2.07
16	4.49	3.63	3.24	3.01	2.85	2.74	2.66	2.59	2.54	2.49	2.42	2.35	2.28	2.24	2.19	2.15	2.11	2.06	2.01
17	4.45	3.59	3.20	2.96	2.81	2.70	2.61	2.55	2.49	2.45	2.38	2.31	2.23	2.19	2.15	2.10	2.06	2.01	1.96
18	4.41	3.55	3.16	2.93	2.77	2.66	2.58	2.51	2.46	2.41	2.34	2.27	2.19	2.15	2.11	2.06	2.02	1.97	1.92
19	4.38	3.52	3.13	2.90	2.74	2.63	2.54	2.48	2.42	2.38	2.31	2.23	2.16	2.11	2.07	2.03	1.98	1.93	1.88
20	4.35	3.49	3.10	2.87	2.71	2.60	2.51	2.45	2.39	2.35	2.28	2.20	2.12	2.08	2.04	1.99	1.95	1.90	1.84
21	4.32	3.47	3.07	2.84	2.68	2.57	2.49	2.42	2.37	2.32	2.25	2.18	2.10	2.05	2.01	1.96	1.92	1.87	1.81
22	4.30	3.44	3.05	2.82	2.66	2.55	2.46	2.40	2.34	2.30	2.23	2.15	2.07	2.03	1.98	1.94	1.89	1.84	1.78
23	4.28	3.42	3.03	2.80	2.64	2.53	2.44	2.37	2.32	2.27	2.20	2.13	2.05	2.01	1.96	1.91	1.86	1.81	1.76
24	4.26	3.40	3.01	2.78	2.62	2.51	2.42	2.36	2.30	2.25	2.18	2.11	2.03	1.98	1.94	1.89	1.84	1.79	1.73
25	4.24	3.39	2.99	2.76	2.60	2.49	2.40	2.34	2.28	2.24	2.16	2.09	2.01	1.96	1.92	1.87	1.82	1.77	1.71
26	4.23	3.37	2.98	2.74	2.59	2.47	2.39	2.32	2.27	2.22	2.15	2.07	1.99	1.95	1.90	1.85	1.80	1.75	1.69
27	4.21	3.35	2.96	2.73	2.57	2.46	2.37	2.31	2.25	2.20	2.13	2.06	1.97	1.93	1.88	1.84	1.79	1.73	1.67
28	4.20	3.34	2.95	2.71	2.56	2.45	2.36	2.29	2.24	2.19	2.12	2.04	1.96	1.91	1.87	1.82	1.77	1.71	1.65
29	4.18	3.33	2.93	2.70	2.55	2.43	2.35	2.28	2.22	2.18	2.10	2.03	1.94	1.90	1.85	1.81	1.75	1.70	1.64
30	4.17	3.32	2.92	2.69	2.53	2.42	2.33	2.27	2.21	2.16	2.09	2.01	1.93	1.89	1.84	1.79	1.74	1.68	1.62
40	4.08	3.23	2.84	2.61	2.45	2.34	2.25	2.18	2.12	2.08	2.00	1.92	1.84	1.79	1.74	1.69	1.64	1.58	1.51
60	4.00	3.15	2.76	2.53	2.37	2.25	2.17	2.10	2.04	1.99	1.92	1.84	1.75	1.70	1.65	1.59	1.53	1.47	1.39
120	3.92	3.07	2.68	2.45	2.29	2.17	2.09	2.02	1.96	1.91	1.83	1.75	1.66	1.61	1.55	1.50	1.43	1.35	1.25
∞	3.84	3.00	2.60	2.37	2.21	2.10	2.01	1.94	1.88	1.83	1.75	1.67	1.57	1.52	1.46	1.39	1.32	1.22	1.00

Table of $F_{.025}$ Values

Denominator Degrees of Freedom	Numerator Degrees of Freedom 1	2	3	4	5	6	7	8	9	10	12	15	20	24	30	40	60	120	∞
1	647.8	799.5	864.2	899.6	921.8	937.1	948.2	956.7	963.3	968.6	976.7	984.9	993.1	997.2	1,001	1,006	1,010	1,014	1,018
2	38.51	39.00	39.17	39.25	39.30	39.33	39.36	39.37	39.39	39.40	39.41	39.43	39.45	39.46	39.46	39.47	39.48	39.49	39.50
3	17.44	16.04	15.44	15.10	14.88	14.73	14.62	14.54	14.47	14.42	14.34	14.25	14.17	14.12	14.08	14.04	13.99	13.95	13.90
4	12.22	10.65	9.98	9.60	9.36	9.20	9.07	8.98	8.90	8.84	8.75	8.66	8.56	8.51	8.46	8.41	8.36	8.31	8.26
5	10.01	8.43	7.76	7.39	7.15	6.98	6.85	6.76	6.68	6.62	6.52	6.43	6.33	6.28	6.23	6.18	6.12	6.07	6.02
6	8.81	7.26	6.60	6.23	5.99	5.82	5.70	5.60	5.52	5.46	5.37	5.27	5.17	5.12	5.07	5.01	4.96	4.90	4.85
7	8.07	6.54	5.89	5.52	5.29	5.21	4.99	4.90	4.82	4.76	4.67	4.57	4.47	4.42	4.36	4.31	4.25	4.20	4.14
8	7.57	6.06	5.42	5.05	4.82	4.65	4.53	4.43	4.36	4.30	4.20	4.10	4.00	3.95	3.89	3.84	3.78	3.73	3.67
9	7.21	5.71	5.08	4.72	4.48	4.32	4.20	4.10	4.03	3.96	3.87	3.77	3.67	3.61	3.56	3.51	3.45	3.39	3.33
10	6.94	5.46	4.83	4.47	4.24	4.07	3.95	3.85	3.78	3.72	3.62	3.52	3.42	3.37	3.31	3.26	3.20	3.14	3.08
11	6.72	5.26	4.63	4.28	4.04	3.88	3.76	3.66	3.59	3.53	3.43	3.33	3.23	3.17	3.12	3.06	3.00	2.94	2.88
12	6.55	5.10	4.47	4.12	3.89	3.73	3.61	3.51	3.44	3.37	3.28	3.18	3.07	3.02	2.96	2.91	2.85	2.79	2.72
13	6.41	4.97	4.35	4.00	3.77	3.60	3.48	3.39	3.31	3.25	3.15	3.05	2.95	2.89	2.84	2.78	2.72	2.66	2.60
14	6.30	4.86	4.24	3.89	3.66	3.50	3.38	3.29	3.21	3.15	3.05	2.95	2.84	2.79	2.73	2.67	2.61	2.55	2.49
15	6.20	4.77	4.15	3.80	3.58	3.41	3.29	3.20	3.12	3.06	2.96	2.86	2.76	2.70	2.64	2.59	2.52	2.46	2.40
16	6.12	4.69	4.08	3.73	3.50	3.34	3.22	3.12	3.05	2.99	2.89	2.79	2.68	2.63	2.57	2.51	2.45	2.38	2.32
17	6.04	4.62	4.01	3.66	3.44	3.28	3.16	3.06	2.98	2.92	2.82	2.72	2.62	2.56	2.50	2.44	2.38	2.32	2.25
18	5.98	4.56	3.95	3.61	3.38	3.22	3.10	3.01	2.93	2.87	2.77	2.67	2.56	2.50	2.44	2.38	2.32	2.26	2.19
19	5.92	4.51	3.90	3.56	3.33	3.17	3.05	2.96	2.88	2.82	2.72	2.62	2.51	2.45	2.39	2.33	2.27	2.20	2.13
20	5.87	4.46	3.86	3.51	3.29	3.13	3.01	2.91	2.84	2.77	2.68	2.57	2.46	2.41	2.35	2.29	2.22	2.16	2.09
21	5.83	4.42	3.82	3.48	3.25	3.09	2.97	2.87	2.80	2.73	2.64	2.53	2.42	2.37	2.31	2.25	2.18	2.11	2.04
22	5.79	4.38	3.78	3.44	3.22	3.05	2.93	2.84	2.76	2.70	2.60	2.50	2.39	2.33	2.27	2.21	2.14	2.08	2.00
23	5.75	4.35	3.75	3.41	3.18	3.02	2.90	2.81	2.73	2.67	2.57	2.47	2.36	2.30	2.24	2.18	2.11	2.04	1.97
24	5.72	4.32	3.72	3.38	3.15	2.99	2.87	2.78	2.70	2.64	2.54	2.44	2.33	2.27	2.21	2.15	2.08	2.01	1.94
25	5.69	4.29	3.69	3.35	3.13	2.97	2.85	2.75	2.68	2.61	2.51	2.41	2.30	2.24	2.18	2.12	2.05	1.98	1.91
26	5.66	4.27	3.67	3.33	3.10	2.94	2.82	2.73	2.65	2.59	2.49	2.39	2.28	2.22	2.16	2.09	2.03	1.95	1.88
27	5.63	4.24	3.65	3.31	3.08	2.92	2.80	2.71	2.63	2.57	2.47	2.36	2.25	2.19	2.13	2.07	2.00	1.93	1.85
28	5.61	4.22	3.63	3.29	3.06	2.90	2.78	2.69	2.61	2.55	2.45	2.34	2.23	2.17	2.11	2.05	1.98	1.91	1.83
29	5.59	4.20	3.61	3.27	3.04	2.88	2.76	2.67	2.59	2.53	2.43	2.32	2.21	2.15	2.09	2.03	1.96	1.89	1.81
30	5.57	4.18	3.59	3.25	3.03	2.87	2.75	2.65	2.57	2.51	2.41	2.31	2.20	2.14	2.07	2.01	1.94	1.87	1.79
40	5.42	4.05	3.46	3.13	2.90	2.74	2.62	2.53	2.45	2.39	2.29	2.18	2.07	2.01	1.94	1.88	1.80	1.72	1.64
60	5.29	3.93	3.34	3.01	2.79	2.63	2.51	2.41	2.33	2.27	2.17	2.06	1.94	1.88	1.82	1.74	1.67	1.58	1.48
120	5.15	3.80	3.23	2.89	2.67	2.52	2.39	2.30	2.22	2.16	2.05	1.94	1.82	1.76	1.69	1.61	1.53	1.43	1.31
∞	5.02	3.69	3.12	2.79	2.57	2.41	2.29	2.19	2.11	2.05	1.94	1.83	1.71	1.64	1.57	1.48	1.39	1.27	1.00

Table of $F_{.01}$ Values

Denominator Degrees of Freedom	Numerator Degrees of Freedom 1	2	3	4	5	6	7	8	9	10	12	15	20	24	30	40	60	120	∞
1	4,052	4,999.5	5,403	5,625	5,764	5,859	5,928	5,982	6,022	6,056	6,106	6,157	6,209	6,235	6,261	6,287	6,313	6,339	6,366
2	98.50	99.00	99.17	99.25	99.30	99.33	99.36	99.37	99.39	99.40	99.42	99.43	99.45	99.46	99.47	99.47	99.48	99.49	99.50
3	34.12	30.82	29.46	28.71	28.24	27.91	27.67	27.49	27.35	27.23	27.05	26.87	26.69	26.60	26.50	26.41	26.32	26.22	26.13
4	21.20	18.00	16.69	15.98	15.52	15.21	14.98	14.80	14.66	14.55	14.37	14.20	14.02	13.93	13.84	13.75	13.65	13.56	13.46
5	16.26	13.27	12.06	11.39	10.97	10.67	10.46	10.29	10.16	10.05	9.89	9.72	9.55	9.47	9.38	9.29	9.20	9.11	9.06
6	13.75	10.92	9.78	9.15	8.75	8.47	8.26	8.10	7.98	7.87	7.72	7.56	7.40	7.31	7.23	7.14	7.06	6.97	6.88
7	12.25	9.55	8.45	7.85	7.46	7.19	6.99	6.84	6.72	6.62	6.47	6.31	6.16	6.07	5.99	5.91	5.82	5.74	5.65
8	11.26	8.65	7.59	7.01	6.63	6.37	6.18	6.03	5.91	5.81	5.67	5.52	5.36	5.28	5.20	5.12	5.03	4.95	4.86
9	10.56	8.02	6.99	6.42	6.06	5.80	5.61	5.47	5.35	5.26	5.11	4.96	4.81	4.73	4.65	4.57	4.48	4.40	4.31
10	10.04	7.56	6.55	5.99	5.64	5.39	5.20	5.06	4.94	4.85	4.71	4.56	4.41	4.33	4.25	4.17	4.08	4.00	3.91
11	9.65	7.21	6.22	5.67	5.32	5.07	4.89	4.74	4.63	4.54	4.40	4.25	4.10	4.02	3.94	3.86	3.78	3.69	3.60
12	9.33	6.93	5.95	5.41	5.06	4.82	4.64	4.50	4.39	4.30	4.16	4.01	3.86	3.78	3.70	3.62	3.54	3.45	3.36
13	9.07	6.70	5.74	5.21	4.86	4.62	4.44	4.30	4.19	4.10	3.96	3.82	3.66	3.59	3.51	3.43	3.34	3.25	3.17
14	8.86	6.51	5.56	5.04	4.69	4.46	4.28	4.14	4.03	3.94	3.80	3.66	3.51	3.43	3.35	3.27	3.18	3.09	3.00
15	8.68	6.36	5.42	4.89	4.56	4.32	4.14	4.00	3.89	3.80	3.67	3.52	3.37	3.29	3.21	3.13	3.05	2.96	2.87
16	8.53	6.23	5.29	4.77	4.44	4.20	4.03	3.89	3.78	3.69	3.55	3.41	3.26	3.18	3.10	3.02	2.93	2.84	2.75
17	8.40	6.11	5.18	4.67	4.34	4.10	3.93	3.79	3.68	3.59	3.46	3.31	3.16	3.08	3.00	2.92	2.83	2.75	2.65
18	8.29	6.01	5.09	4.58	4.25	4.01	3.84	3.71	3.60	3.51	3.37	3.23	3.08	3.00	2.92	2.84	2.75	2.66	2.57
19	8.18	5.93	5.01	4.50	4.17	3.94	3.77	3.63	3.52	3.43	3.30	3.15	3.00	2.92	2.84	2.76	2.67	2.58	2.49
20	8.10	5.85	4.94	4.43	4.10	3.87	3.70	3.56	3.46	3.37	3.23	3.09	2.94	2.86	2.78	2.69	2.61	2.52	2.42
21	8.02	5.78	4.87	4.37	4.04	3.81	3.64	3.51	3.40	3.31	3.17	3.03	2.88	2.80	2.72	2.64	2.55	2.46	2.36
22	7.95	5.72	4.82	4.31	3.99	3.76	3.59	3.45	3.35	3.26	3.12	2.98	2.83	2.75	2.67	2.58	2.50	2.40	2.31
23	7.88	5.66	4.76	4.26	3.94	3.71	3.54	3.41	3.30	3.21	3.07	2.93	2.78	2.70	2.62	2.54	2.45	2.35	2.26
24	7.82	5.61	4.72	4.22	3.90	3.67	3.50	3.36	3.26	3.17	3.03	2.89	2.74	2.66	2.58	2.49	2.40	2.31	2.21
25	7.77	5.57	4.68	4.18	3.85	3.63	3.46	3.32	3.22	3.13	2.99	2.85	2.70	2.62	2.54	2.45	2.36	2.27	2.17
26	7.72	5.53	4.64	4.14	3.82	3.59	3.42	3.29	3.18	3.09	2.96	2.81	2.66	2.58	2.50	2.42	2.33	2.23	2.13
27	7.68	5.49	4.60	4.11	3.78	3.56	3.39	3.26	3.15	3.06	2.93	2.78	2.63	2.55	2.47	2.38	2.29	2.20	2.10
28	7.64	5.45	4.57	4.07	3.75	3.53	3.36	3.23	3.12	3.03	2.90	2.75	2.60	2.52	2.44	2.35	2.26	2.17	2.06
29	7.60	5.42	4.54	4.04	3.73	3.50	3.33	3.20	3.09	3.00	2.87	2.73	2.57	2.49	2.41	2.33	2.23	2.14	2.03
30	7.56	5.39	4.51	4.02	3.70	3.47	3.30	3.17	3.07	2.98	2.84	2.70	2.55	2.47	2.39	2.30	2.21	2.11	2.01
40	7.31	5.18	4.31	3.83	3.51	3.29	3.12	2.99	2.89	2.80	2.66	2.52	2.37	2.29	2.20	2.11	2.02	1.92	1.80
60	7.08	4.98	4.13	3.65	3.34	3.12	2.95	2.82	2.72	2.63	2.50	2.35	2.20	2.12	2.03	1.94	1.84	1.73	1.60
120	6.85	4.79	3.95	3.48	3.17	2.96	2.79	2.66	2.56	2.47	2.34	2.19	2.03	1.95	1.86	1.76	1.66	1.53	1.38
∞	6.63	4.61	3.78	3.32	3.02	2.80	2.64	2.51	2.41	2.32	2.18	2.04	1.88	1.79	1.70	1.59	1.47	1.32	1.00

TABLE 5 Binomial Probabilities

Entries in the table give the probability of x successes in n trials of a binomial experiment, where p is the probability of a success on one trial. For example, with six trials and $p = .40$, the probability of two successes is .3110.

						p					
n	*x*	*.05*	*.10*	*.15*	*.20*	*.25*	*.30*	*.35*	*.40*	*.45*	*.50*
1	0	.9500	.9000	.8500	.8000	.7500	.7000	.6500	.6000	.5500	.5000
	1	.0500	.1000	.1500	.2000	.2500	.3000	.3500	.4000	.4500	.5000
2	0	.9025	.8100	.7225	.6400	.5625	.4900	.4225	.3600	.3025	.2500
	1	.0950	.1800	.2550	.3200	.3750	.4200	.4550	.4800	.4950	.5000
	2	.0025	.0100	.0225	.0400	.0625	.0900	.1225	.1600	.2025	.2500
3	0	.8574	.7290	.6141	.5120	.4219	.3430	.2746	.2160	.1664	.1250
	1	.1354	.2430	.3251	.3840	.4219	.4410	.4436	.4320	.4084	.3750
	2	.0071	.0270	.0574	.0960	.1406	.1890	.2389	.2880	.3341	.3750
	3	.0001	.0010	.0034	.0080	.0156	.0270	.0429	.0640	.0911	.1250
4	0	.8145	.6561	.5220	.4096	.3164	.2401	.1785	.1296	.0915	.0625
	1	.1715	.2916	.3685	.4096	.4219	.4116	.3845	.3456	.2995	.2500
	2	.0135	.0486	.0975	.1536	.2109	.2646	.3105	.3456	.3675	.3750
	3	.0005	.0036	.0115	.0256	.0469	.0756	.1115	.1536	.2005	.2500
	4	.0000	.0001	.0005	.0016	.0039	.0081	.0150	.0256	.0410	.0625
5	0	.7738	.5905	.4437	.3277	.2373	.1681	.1160	.0778	.0503	.0312
	1	.2036	.3280	.3915	.4096	.3955	.3602	.3124	.2592	.2059	.1562
	2	.0214	.0729	.1382	.2048	.2637	.3087	.3364	.3456	.3369	.3125
	3	.0011	.0081	.0244	.0512	.0879	.1323	.1811	.2304	.2757	.3125
	4	.0000	.0004	.0022	.0064	.0146	.0284	.0488	.0768	.1128	.1562
	5	.0000	.0000	.0001	.0003	.0010	.0024	.0053	.0102	.0185	.0312
6	0	.7351	.5314	.3771	.2621	.1780	.1176	.0754	.0467	.0277	.0156
	1	.2321	.3543	.3993	.3932	.3560	.3025	.2437	.1866	.1359	.0938
	2	.0305	.0984	.1762	.2458	.2966	.3241	.3280	.3110	.2780	.2344
	3	.0021	.0146	.0415	.0819	.1318	.1852	.2355	.2765	.3032	.3125
	4	.0001	.0012	.0055	.0154	.0330	.0595	.0951	.1382	.1861	.2344
	5	.0000	.0001	.0004	.0015	.0044	.0102	.0205	.0369	.0609	.0938
	6	.0000	.0000	.0000	.0001	.0002	.0007	.0018	.0041	.0083	.0156
7	0	.6983	.4783	.3206	.2097	.1335	.0824	.0490	.0280	.0152	.0078
	1	.2573	.3720	.3960	.3670	.3115	.2471	.1848	.1306	.0872	.0547
	2	.0406	.1240	.2097	.2753	.3115	.3177	.2985	.2613	.2140	.1641
	3	.0036	.0230	.0617	.1147	.1730	.2269	.2679	.2903	.2918	.2734
	4	.0002	.0026	.0109	.0287	.0577	.0972	.1442	.1935	.2388	.2734
	5	.0000	.0002	.0012	.0043	.0115	.0250	.0466	.0774	.1172	.1641
	6	.0000	.0000	.0001	.0004	.0013	.0036	.0084	.0172	.0320	.0547
	7	.0000	.0000	.0000	.0000	.0001	.0002	.0006	.0016	.0037	.0078

TABLE 5 (*Continued*)

n	x	.05	.10	.15	.20	p .25	.30	.35	.40	.45	.50
8	0	.6634	.4305	.2725	.1678	.1001	.0576	.0319	.0168	.0084	.0039
	1	.2793	.3826	.3847	.3355	.2670	.1977	.1373	.0896	.0548	.0312
	2	.0515	.1488	.2376	.2936	.3115	.2965	.2587	.2090	.1569	.1094
	3	.0054	.0331	.0839	.1468	.2076	.2541	.2786	.2787	.2568	.2188
	4	.0004	.0046	.0185	.0459	.0865	.1361	.1875	.2322	.2627	.2734
	5	.0000	.0004	.0026	.0092	.0231	.0467	.0808	.1239	.1719	.2188
	6	.0000	.0000	.0002	.0011	.0038	.0100	.0217	.0413	.0703	.1094
	7	.0000	.0000	.0000	.0001	.0004	.0012	.0033	.0079	.0164	.0312
	8	.0000	.0000	.0000	.0000	.0000	.0001	.0002	.0007	.0017	.0039
9	0	.6302	.3874	.2316	.1342	.0751	.0404	.0207	.0101	.0046	.0020
	1	.2985	.3874	.3679	.3020	.2253	.1556	.1004	.0605	.0339	.0176
	2	.0629	.1722	.2597	.3020	.3003	.2668	.2162	.1612	.1110	.0703
	3	.0077	.0446	.1069	.1762	.2336	.2668	.2716	.2508	.2119	.1641
	4	.0006	.0074	.0283	.0661	.1168	.1715	.2194	.2508	.2600	.2461
	5	.0000	.0008	.0050	.0165	.0389	.0735	.1181	.1672	.2128	.2461
	6	.0000	.0001	.0006	.0028	.0087	.0210	.0424	.0743	.1160	.1641
	7	.0000	.0000	.0000	.0003	.0012	.0039	.0098	.0212	.0407	.0703
	8	.0000	.0000	.0000	.0000	.0001	.0004	.0013	.0035	.0083	.0176
	9	.0000	.0000	.0000	.0000	.0000	.0000	.0001	.0003	.0008	.0020
10	0	.5987	.3487	.1969	.1074	.0563	.0282	.0135	.0060	.0025	.0010
	1	.3151	.3874	.3474	.2684	.1877	.1211	.0725	.0403	.0207	.0098
	2	.0746	.1937	.2759	.3020	.2816	.2335	.1757	.1209	.0763	.0439
	3	.0105	.0574	.1298	.2013	.2503	.2668	.2522	.2150	.1665	.1172
	4	.0010	.0112	.0401	.0881	.1460	.2001	.2377	.2508	.2384	.2051
	5	.0001	.0015	.0085	.0264	.0584	.1029	.1536	.2007	.2340	.2461
	6	.0000	.0001	.0012	.0055	.0162	.0368	.0689	.1115	.1596	.2051
	7	.0000	.0000	.0001	.0008	.0031	.0090	.0212	.0425	.0746	.1172
	8	.0000	.0000	.0000	.0001	.0004	.0014	.0043	.0106	.0229	.0439
	9	.0000	.0000	.0000	.0000	.0000	.0001	.0005	.0016	.0042	.0098
	10	.0000	.0000	.0000	.0000	.0000	.0000	.0000	.0001	.0003	.0010
11	0	.5688	.3138	.1673	.0859	.0422	.0198	.0088	.0036	.0014	.0005
	1	.3293	.3835	.3248	.2362	.1549	.0932	.0518	.0266	.0125	.0054
	2	.0867	.2131	.2866	.2953	.2581	.1998	.1395	.0887	.0513	.0269
	3	.0137	.0710	.1517	.2215	.2581	.2568	.2254	.1774	.1259	.0806
	4	.0014	.0158	.0536	.1107	.1721	.2201	.2428	.2365	.2060	.1611
	5	.0001	.0025	.0132	.0388	.0803	.1321	.1830	.2207	.2360	.2256
	6	.0000	.0003	.0023	.0097	.0268	.0566	.0985	.1471	.1931	.2256
	7	.0000	.0000	.0003	.0017	.0064	.0173	.0379	.0701	.1128	.1611
	8	.0000	.0000	.0000	.0002	.0011	.0037	.0102	.0234	.0462	.0806
	9	.0000	.0000	.0000	.0000	.0001	.0005	.0018	.0052	.0126	.0269
	10	.0000	.0000	.0000	.0000	.0000	.0000	.0002	.0007	.0021	.0054
	11	.0000	.0000	.0000	.0000	.0000	.0000	.0000	.0000	.0002	.0005

TABLE 5 (*Continued*)

n	*x*	.05	.10	.15	.20	*p* .25	.30	.35	.40	.45	.50
12	0	.5404	.2824	.1422	.0687	.0317	.0138	.0057	.0022	.0008	.0002
	1	.3413	.3766	.3012	.2062	.1267	.0712	.0368	.0174	.0075	.0029
	2	.0988	.2301	.2924	.2835	.2323	.1678	.1088	.0639	.0339	.0161
	3	.0173	.0853	.1720	.2362	.2581	.2397	.1954	.1419	.0923	.0537
	4	.0021	.0213	.0683	.1329	.1936	.2311	.2367	.2128	.1700	.1208
	5	.0002	.0038	.0193	.0532	.1032	.1585	.2039	.2270	.2225	.1934
	6	.0000	.0005	.0040	.0155	.0401	.0792	.1281	.1766	.2124	.2256
	7	.0000	.0000	.0006	.0033	.0115	.0291	.0591	.1009	.1489	.1934
	8	.0000	.0000	.0001	.0005	.0024	.0078	.0199	.0420	.0762	.1208
	9	.0000	.0000	.0000	.0001	.0004	.0015	.0048	.0125	.0277	.0537
	10	.0000	.0000	.0000	.0000	.0000	.0002	.0008	.0025	.0068	.0161
	11	.0000	.0000	.0000	.0000	.0000	.0000	.0001	.0003	.0010	.0029
	12	.0000	.0000	.0000	.0000	.0000	.0000	.0000	.0000	.0001	.0002
13	0	.5133	.2542	.1209	.0550	.0238	.0097	.0037	.0013	.0004	.0001
	1	.3512	.3672	.2774	.1787	.1029	.0540	.0259	.0113	.0045	.0016
	2	.1109	.2448	.2937	.2680	.2059	.1388	.0836	.0453	.0220	.0095
	3	.0214	.0997	.1900	.2457	.2517	.2181	.1651	.1107	.0660	.0349
	4	.0028	.0277	.0838	.1535	.2097	.2337	.2222	.1845	.1350	.0873
	5	.0003	.0055	.0266	.0691	.1258	.1803	.2154	.2214	.1989	.1571
	6	.0000	.0008	.0063	.0230	.0559	.1030	.1546	.1968	.2169	.2095
	7	.0000	.0001	.0011	.0058	.0186	.0442	.0833	.1312	.1775	.2095
	8	.0000	.0000	.0001	.0011	.0047	.0142	.0336	.0656	.1089	.1571
	9	.0000	.0000	.0000	.0001	.0009	.0034	.0101	.0243	.0495	.0873
	10	.0000	.0000	.0000	.0000	.0001	.0006	.0022	.0065	.0162	.0349
	11	.0000	.0000	.0000	.0000	.0000	.0001	.0003	.0012	.0036	.0095
	12	.0000	.0000	.0000	.0000	.0000	.0000	.0000	.0001	.0005	.0016
	13	.0000	.0000	.0000	.0000	.0000	.0000	.0000	.0000	.0000	.0001
14	0	.4877	.2288	.1028	.0440	.0178	.0068	.0024	.0008	.0002	.0001
	1	.3593	.3559	.2539	.1539	.0832	.0407	.0181	.0073	.0027	.0009
	2	.1229	.2570	.2912	.2501	.1802	.1134	.0634	.0317	.0141	.0056
	3	.0259	.1142	.2056	.2501	.2402	.1943	.1366	.0845	.0462	.0222
	4	.0037	.0349	.0998	.1720	.2202	.2290	.2022	.1549	.1040	.0611
	5	.0004	.0078	.0352	.0860	.1468	.1963	.2178	.2066	.1701	.1222
	6	.0000	.0013	.0093	.0322	.0734	.1262	.1759	.2066	.2088	.1833
	7	.0000	.0002	.0019	.0092	.0280	.0618	.1082	.1574	.1952	.2095
	8	.0000	.0000	.0003	.0020	.0082	.0232	.0510	.0918	.1398	.1833
	9	.0000	.0000	.0000	.0003	.0018	.0066	.0183	.0408	.0762	.1222
	10	.0000	.0000	.0000	.0000	.0003	.0014	.0049	.0136	.0312	.0611
	11	.0000	.0000	.0000	.0000	.0000	.0002	.0010	.0033	.0093	.0222
	12	.0000	.0000	.0000	.0000	.0000	.0000	.0001	.0005	.0019	.0056
	13	.0000	.0000	.0000	.0000	.0000	.0000	.0000	.0001	.0002	.0009
	14	.0000	.0000	.0000	.0000	.0000	.0000	.0000	.0000	.0000	.0001

TABLE 5 *(Continued)*

						p					
n	*x*	*.05*	*.10*	*.15*	*.20*	*.25*	*.30*	*.35*	*.40*	*.45*	*.50*
15	0	.4633	.2059	.0874	.0352	.0134	.0047	.0016	.0005	.0001	.0000
	1	.3658	.3432	.2312	.1319	.0668	.0305	.0126	.0047	.0016	.0005
	2	.1348	.2669	.2856	.2309	.1559	.0916	.0476	.0219	.0090	.0032
	3	.0307	.1285	.2184	.2501	.2252	.1700	.1110	.0634	.0318	.0139
	4	.0049	.0428	.1156	.1876	.2252	.2186	.1792	.1268	.0780	.0417
	5	.0006	.0105	.0449	.1032	.1651	.2061	.2123	.1859	.1404	.0916
	6	.0000	.0019	.0132	.0430	.0917	.1472	.1906	.2066	.1914	.1527
	7	.0000	.0003	.0030	.0138	.0393	.0811	.1319	.1771	.2013	.1964
	8	.0000	.0000	.0005	.0035	.0131	.0348	.0710	.1181	.1647	.1964
	9	.0000	.0000	.0001	.0007	.0034	.0116	.0298	.0612	.1048	.1527
	10	.0000	.0000	.0000	.0001	.0007	.0030	.0096	.0245	.0515	.0916
	11	.0000	.0000	.0000	.0000	.0001	.0006	.0024	.0074	.0191	.0417
	12	.0000	.0000	.0000	.0000	.0000	.0001	.0004	.0016	.0052	.0139
	13	.0000	.0000	.0000	.0000	.0000	.0000	.0001	.0003	.0010	.0032
	14	.0000	.0000	.0000	.0000	.0000	.0000	.0000	.0000	.0001	.0005
	15	.0000	.0000	.0000	.0000	.0000	.0000	.0000	.0000	.0000	.0000
16	0	.4401	.1853	.0743	.0281	.0100	.0033	.0010	.0003	.0001	.0000
	1	.3706	.3294	.2097	.1126	.0535	.0228	.0087	.0030	.0009	.0002
	2	.1463	.2745	.2775	.2111	.1336	.0732	.0353	.0150	.0056	.0018
	3	.0359	.1423	.2285	.2463	.2079	.1465	.0888	.0468	.0215	.0085
	4	.0061	.0514	.1311	.2001	.2252	.2040	.1553	.1014	.0572	.0278
	5	.0008	.0137	.0555	.1201	.1802	.2099	.2008	.1623	.1123	.0667
	6	.0001	.0028	.0180	.0550	.1101	.1649	.1982	.1983	.1684	.1222
	7	.0000	.0004	.0045	.0197	.0524	.1010	.1524	.1889	.1969	.1746
	8	.0000	.0001	.0009	.0055	.0197	.0487	.0923	.1417	.1812	.1964
	9	.0000	.0000	.0001	.0012	.0058	.0185	.0442	.0840	.1318	.1746
	10	.0000	.0000	.0000	.0002	.0014	.0056	.0167	.0392	.0755	.1222
	11	.0000	.0000	.0000	.0000	.0002	.0013	.0049	.0142	.0337	.0667
	12	.0000	.0000	.0000	.0000	.0000	.0002	.0011	.0040	.0115	.0278
	13	.0000	.0000	.0000	.0000	.0000	.0000	.0002	.0008	.0029	.0085
	14	.0000	.0000	.0000	.0000	.0000	.0000	.0000	.0001	.0005	.0018
	15	.0000	.0000	.0000	.0000	.0000	.0000	.0000	.0000	.0001	.0002
	16	.0000	.0000	.0000	.0000	.0000	.0000	.0000	.0000	.0000	.0000
17	0	.4181	.1668	.0631	.0225	.0075	.0023	.0007	.0002	.0000	.0000
	1	.3741	.3150	.1893	.0957	.0426	.0169	.0060	.0019	.0005	.0001
	2	.1575	.2800	.2673	.1914	.1136	.0581	.0260	.0102	.0035	.0010
	3	.0415	.1556	.2359	.2393	.1893	.1245	.0701	.0341	.0144	.0052
	4	.0076	.0605	.1457	.2093	.2209	.1868	.1320	.0796	.0411	.0182
	5	.0010	.0175	.0668	.1361	.1914	.2081	.1849	.1379	.0875	.0472
	6	.0001	.0039	.0236	.0680	.1276	.1784	.1991	.1839	.1432	.0944
	7	.0000	.0007	.0065	.0267	.0668	.1201	.1685	.1927	.1841	.1484
	8	.0000	.0001	.0014	.0084	.0279	.0644	.1134	.1606	.1883	.1855
	9	.0000	.0000	.0003	.0021	.0093	.0276	.0611	.1070	.1540	.1855

TABLE 5 (*Continued*)

n	x	.05	.10	.15	.20	p .25	.30	.35	.40	.45	.50
17	10	.0000	.0000	.0000	.0004	.0025	.0095	.0263	.0571	.1008	.1484
	11	.0000	.0000	.0000	.0001	.0005	.0026	.0090	.0242	.0525	.0944
	12	.0000	.0000	.0000	.0000	.0001	.0006	.0024	.0081	.0215	.0472
	13	.0000	.0000	.0000	.0000	.0000	.0001	.0005	.0021	.0068	.0182
	14	.0000	.0000	.0000	.0000	.0000	.0000	.0001	.0004	.0016	.0052
	15	.0000	.0000	.0000	.0000	.0000	.0000	.0000	.0001	.0003	.0010
	16	.0000	.0000	.0000	.0000	.0000	.0000	.0000	.0000	.0000	.0001
	17	.0000	.0000	.0000	.0000	.0000	.0000	.0000	.0000	.0000	.0000
18	0	.3972	.1501	.0536	.0180	.0056	.0016	.0004	.0001	.0000	.0000
	1	.3763	.3002	.1704	.0811	.0338	.0126	.0042	.0012	.0003	.0001
	2	.1683	.2835	.2556	.1723	.0958	.0458	.0190	.0069	.0022	.0006
	3	.0473	.1680	.2406	.2297	.1704	.1046	.0547	.0246	.0095	.0031
	4	.0093	.0700	.1592	.2153	.2130	.1681	.1104	.0614	.0291	.0117
	5	.0014	.0218	.0787	.1507	.1988	.2017	.1664	.1146	.0666	.0327
	6	.0002	.0052	.0301	.0816	.1436	.1873	.1941	.1655	.1181	.0708
	7	.0000	.0010	.0091	.0350	.0820	.1376	.1792	.1892	.1657	.1214
	8	.0000	.0002	.0022	.0120	.0376	.0811	.1327	.1734	.1864	.1669
	9	.0000	.0000	.0004	.0033	.0139	.0386	.0794	.1284	.1694	.1855
	10	.0000	.0000	.0001	.0008	.0042	.0149	.0385	.0771	.1248	.1669
	11	.0000	.0000	.0000	.0001	.0010	.0046	.0151	.0374	.0742	.1214
	12	.0000	.0000	.0000	.0000	.0002	.0012	.0047	.0145	.0354	.0708
	13	.0000	.0000	.0000	.0000	.0000	.0002	.0012	.0045	.0134	.0327
	14	.0000	.0000	.0000	.0000	.0000	.0000	.0002	.0011	.0039	.0117
	15	.0000	.0000	.0000	.0000	.0000	.0000	.0000	.0002	.0009	.0031
	16	.0000	.0000	.0000	.0000	.0000	.0000	.0000	.0000	.0001	.0006
	17	.0000	.0000	.0000	.0000	.0000	.0000	.0000	.0000	.0000	.0001
	18	.0000	.0000	.0000	.0000	.0000	.0000	.0000	.0000	.0000	.0000
19	0	.3774	.1351	.0456	.0144	.0042	.0011	.0003	.0001	.0000	.0000
	1	.3774	.2852	.1529	.0685	.0268	.0093	.0029	.0008	.0002	.0000
	2	.1787	.2852	.2428	.1540	.0803	.0358	.0138	.0046	.0013	.0003
	3	.0533	.1796	.2428	.2182	.1517	.0869	.0422	.0175	.0062	.0018
	4	.0112	.0798	.1714	.2182	.2023	.1491	.0909	.0467	.0203	.0074
	5	.0018	.0266	.0907	.1636	.2023	.1916	.1468	.0933	.0497	.0222
	6	.0002	.0069	.0374	.0955	.1574	.1916	.1844	.1451	.0949	.0518
	7	.0000	.0014	.0122	.0443	.0974	.1525	.1844	.1797	.1443	.0961
	8	.0000	.0002	.0032	.0166	.0487	.0981	.1489	.1797	.1771	.1442
	9	.0000	.0000	.0007	.0051	.0198	.0514	.0980	.1464	.1771	.1762
	10	.0000	.0000	.0001	.0013	.0066	.0220	.0528	.0976	.1449	.1762
	11	.0000	.0000	.0000	.0003	.0018	.0077	.0233	.0532	.0970	.1442
	12	.0000	.0000	.0000	.0000	.0004	.0022	.0083	.0237	.0529	.0961
	13	.0000	.0000	.0000	.0000	.0001	.0005	.0024	.0085	.0233	.0518
	14	.0000	.0000	.0000	.0000	.0000	.0001	.0006	.0024	.0082	.0222

TABLE 5 *(Continued)*

n	*x*	.05	.10	.15	.20	*p* .25	.30	.35	.40	.45	.50
19	15	.0000	.0000	.0000	.0000	.0000	.0000	.0001	.0005	.0022	.0074
	16	.0000	.0000	.0000	.0000	.0000	.0000	.0000	.0001	.0005	.0018
	17	.0000	.0000	.0000	.0000	.0000	.0000	.0000	.0000	.0001	.0003
	18	.0000	.0000	.0000	.0000	.0000	.0000	.0000	.0000	.0000	.0000
	19	.0000	.0000	.0000	.0000	.0000	.0000	.0000	.0000	.0000	.0000
20	0	.3585	.1216	.0388	.0115	.0032	.0008	.0002	.0000	.0000	.0000
	1	.3774	.2702	.1368	.0576	.0211	.0068	.0020	.0005	.0001	.0000
	2	.1887	.2852	.2293	.1369	.0669	.0278	.0100	.0031	.0008	.0002
	3	.0596	.1901	.2428	.2054	.1339	.0716	.0323	.0123	.0040	.0011
	4	.0133	.0898	.1821	.2182	.1897	.1304	.0738	.0350	.0139	.0046
	5	.0022	.0319	.1028	.1746	.2023	.1789	.1272	.0746	.0365	.0148
	6	.0003	.0089	.0454	.1091	.1686	.1916	.1712	.1244	.0746	.0370
	7	.0000	.0020	.0160	.0545	.1124	.1643	.1844	.1659	.1221	.0739
	8	.0000	.0004	.0046	.0222	.0609	.1144	.1614	.1797	.1623	.1201
	9	.0000	.0001	.0011	.0074	.0271	.0654	.1158	.1597	.1771	.1602
	10	.0000	.0000	.0002	.0020	.0099	.0308	.0686	.1171	.1593	.1762
	11	.0000	.0000	.0000	.0005	.0030	.0120	.0336	.0710	.1185	.1602
	12	.0000	.0000	.0000	.0001	.0008	.0039	.0136	.0355	.0727	.1201
	13	.0000	.0000	.0000	.0000	.0002	.0010	.0045	.0146	.0366	.0739
	14	.0000	.0000	.0000	.0000	.0000	.0002	.0012	.0049	.0150	.0370
	15	.0000	.0000	.0000	.0000	.0000	.0000	.0003	.0013	.0049	.0148
	16	.0000	.0000	.0000	.0000	.0000	.0000	.0000	.0003	.0013	.0046
	17	.0000	.0000	.0000	.0000	.0000	.0000	.0000	.0000	.0002	.0011
	18	.0000	.0000	.0000	.0000	.0000	.0000	.0000	.0000	.0000	.0002
	19	.0000	.0000	.0000	.0000	.0000	.0000	.0000	.0000	.0000	.0000
	20	.0000	.0000	.0000	.0000	.0000	.0000	.0000	.0000	.0000	.0000

TABLE 6 Values of $e^{-\mu}$

μ	$e^{-\mu}$	μ	$e^{-\mu}$	μ	$e^{-\mu}$
.0	1.0000	3.1	.0450	8.0	.000335
.1	.9048	3.2	.0408	9.0	.000123
.2	.8187	3.3	.0369	10.0	.000045
.3	.7408	3.4	.0334		
.4	.6703	3.5	.0302		
.5	.6065	3.6	.0273		
.6	.5488	3.7	.0247		
.7	.4966	3.8	.0224		
.8	.4493	3.9	.0202		
.9	.4066	4.0	.0183		
1.0	.3679	4.1	.0166		
1.1	.3329	4.2	.0150		
1.2	.3012	4.3	.0136		
1.3	.2725	4.4	.0123		
1.4	.2466	4.5	.0111		
1.5	.2231	4.6	.0101		
1.6	.2019	4.7	.0091		
1.7	.1827	4.8	.0082		
1.8	.1653	4.9	.0074		
1.9	.1496	5.0	.0067		
2.0	.1353	5.1	.0061		
2.1	.1225	5.2	.0055		
2.2	.1108	5.3	.0050		
2.3	.1003	5.4	.0045		
2.4	.0907	5.5	.0041		
2.5	.0821	5.6	.0037		
2.6	.0743	5.7	.0033		
2.7	.0672	5.8	.0030		
2.8	.0608	5.9	.0027		
2.9	.0550	6.0	.0025		
3.0	.0498	7.0	.0009		

TABLE 7 Poisson Probabilities

Entries in the table give the probability of x occurrences for a Poisson process with a mean μ. For example, when $\mu = 2.5$, the probability of four occurrences is .1336.

x	μ 0.1	0.2	0.3	0.4	0.5	0.6	0.7	0.8	0.9	1.0
0	.9048	.8187	.7408	.6703	.6065	.5488	.4966	.4493	.4066	.3679
1	.0905	.1637	.2222	.2681	.3033	.3293	.3476	.3595	.3659	.3679
2	.0045	.0164	.0333	.0536	.0758	.0988	.1217	.1438	.1647	.1839
3	.0002	.0011	.0033	.0072	.0126	.0198	.0284	.0383	.0494	.0613
4	.0000	.0001	.0002	.0007	.0016	.0030	.0050	.0077	.0111	.0153
5	.0000	.0000	.0000	.0001	.0002	.0004	.0007	.0012	.0020	.0031
6	.0000	.0000	.0000	.0000	.0000	.0000	.0001	.0002	.0003	.0005
7	.0000	.0000	.0000	.0000	.0000	.0000	.0000	.0000	.0000	.0001

x	μ 1.1	1.2	1.3	1.4	1.5	1.6	1.7	1.8	1.9	2.0
0	.3329	.3012	.2725	.2466	.2231	.2019	.1827	.1653	.1496	.1353
1	.3662	.3614	.3543	.3452	.3347	.3230	.3106	.2975	.2842	.2707
2	.2014	.2169	.2303	.2417	.2510	.2584	.2640	.2678	.2700	.2707
3	.0738	.0867	.0998	.1128	.1255	.1378	.1496	.1607	.1710	.1804
4	.0203	.0260	.0324	.0395	.0471	.0551	.0636	.0723	.0812	.0902
5	.0045	.0062	.0084	.0111	.0141	.0176	.0216	.0260	.0309	.0361
6	.0008	.0012	.0018	.0026	.0035	.0047	.0061	.0078	.0098	.0120
7	.0001	.0002	.0003	.0005	.0008	.0011	.0015	.0020	.0027	.0034
8	.0000	.0000	.0001	.0001	.0001	.0002	.0003	.0005	.0006	.0009
9	.0000	.0000	.0000	.0000	.0000	.0000	.0001	.0001	.0001	.0002

x	μ 2.1	2.2	2.3	2.4	2.5	2.6	2.7	2.8	2.9	3.0
0	.1225	.1108	.1003	.0907	.0821	.0743	.0672	.0608	.0550	.0498
1	.2572	.2438	.2306	.2177	.2052	.1931	.1815	.1703	.1596	.1494
2	.2700	.2681	.2652	.2613	.2565	.2510	.2450	.2384	.2314	.2240
3	.1890	.1966	.2033	.2090	.2138	.2176	.2205	.2225	.2237	.2240
4	.0992	.1082	.1169	.1254	.1336	.1414	.1488	.1557	.1622	.1680
5	.0417	.0476	.0538	.0602	.0668	.0735	.0804	.0872	.0940	.1008
6	.0146	.0174	.0206	.0241	.0278	.0319	.0362	.0407	.0455	.0540
7	.0044	.0055	.0068	.0083	.0099	.0118	.0139	.0163	.0188	.0216
8	.0011	.0015	.0019	.0025	.0031	.0038	.0047	.0057	.0068	.0081
9	.0003	.0004	.0005	.0007	.0009	.0011	.0014	.0018	.0022	.0027
10	.0001	.0001	.0001	.0002	.0002	.0003	.0004	.0005	.0006	.0008
11	.0000	.0000	.0000	.0000	.0000	.0001	.0001	.0001	.0002	.0002
12	.0000	.0000	.0000	.0000	.0000	.0000	.0000	.0000	.0000	.0001

This table is reproduced by permission from R. S. Burington and D. C. May, *Handbook of Probability and Statistics with Tables*. New York: McGraw-Hill Book Company, 1953.

TABLE 7 (*Continued*)

x	μ 3.1	3.2	3.3	3.4	3.5	3.6	3.7	3.8	3.9	4.0
0	.0450	.0408	.0369	.0344	.0302	.0273	.0247	.0224	.0202	.0183
1	.1397	.1304	.1217	.1135	.1057	.0984	.0915	.0850	.0789	.0733
2	.2165	.2087	.2008	.1929	.1850	.1771	.1692	.1615	.1539	.1465
3	.2237	.2226	.2209	.2186	.2158	.2125	.2087	.2046	.2001	.1954
4	.1734	.1781	.1823	.1858	.1888	.1912	.1931	.1944	.1951	.1954
5	.1075	.1140	.1203	.1264	.1322	.1377	.1429	.1477	.1522	.1563
6	.0555	.0608	.0662	.0716	.0771	.0826	.0881	.0936	.0989	.1042
7	.0246	.0278	.0312	.0348	.0385	.0425	.0466	.0508	.0551	.0595
8	.0095	.0111	.0129	.0148	.0169	.0191	.0215	.0241	.0269	.0298
9	.0033	.0040	.0047	.0056	.0066	.0076	.0089	.0102	.0116	.0132
10	.0010	.0013	.0016	.0019	.0023	.0028	.0033	.0039	.0045	.0053
11	.0003	.0004	.0005	.0006	.0007	.0009	.0011	.0013	.0016	.0019
12	.0001	.0001	.0001	.0002	.0002	.0003	.0003	.0004	.0005	.0006
13	.0000	.0000	.0000	.0000	.0001	.0001	.0001	.0001	.0002	.0002
14	.0000	.0000	.0000	.0000	.0000	.0000	.0000	.0000	.0000	.0001

x	μ 4.1	4.2	4.3	4.4	4.5	4.6	4.7	4.8	4.9	5.0
0	.0166	.0150	.0136	.0123	.0111	.0101	.0091	.0082	.0074	.0067
1	.0679	.0630	.0583	.0540	.0500	.0462	.0427	.0395	.0365	.0337
2	.1393	.1323	.1254	.1188	.1125	.1063	.1005	.0948	.0894	.0842
3	.1904	.1852	.1798	.1743	.1687	.1631	.1574	.1517	.1460	.1404
4	.1951	.1944	.1933	.1917	.1898	.1875	.1849	.1820	.1789	.1755
5	.1600	.1633	.1662	.1687	.1708	.1725	.1738	.1747	.1753	.1755
6	.1093	.1143	.1191	.1237	.1281	.1323	.1362	.1398	.1432	.1462
7	.0640	.0686	.0732	.0778	.0824	.0869	.0914	.0959	.1002	.1044
8	.0328	.0360	.0393	.0428	.0463	.0500	.0537	.0575	.0614	.0653
9	.0150	.0168	.0188	.0209	.0232	.0255	.0280	.0307	.0334	.0363
10	.0061	.0071	.0081	.0092	.0104	.0118	.0132	.0147	.0164	.0181
11	.0023	.0027	.0032	.0037	.0043	.0049	.0056	.0064	.0073	.0082
12	.0008	.0009	.0011	.0014	.0016	.0019	.0022	.0026	.0030	.0034
13	.0002	.0003	.0004	.0005	.0006	.0007	.0008	.0009	.0011	.0013
14	.0001	.0001	.0001	.0001	.0002	.0002	.0003	.0003	.0004	.0005
15	.0000	.0000	.0000	.0000	.0001	.0001	.0001	.0001	.0001	.0002

x	μ 5.1	5.2	5.3	5.4	5.5	5.6	5.7	5.8	5.9	6.0
0	.0061	.0055	.0050	.0045	.0041	.0037	.0033	.0030	.0027	.0025
1	.0311	.0287	.0265	.0244	.0225	.0207	.0191	.0176	.0162	.0149
2	.0793	.0746	.0701	.0659	.0618	.0580	.0544	.0509	.0477	.0446
3	.1348	.1293	.1239	.1185	.1133	.1082	.1033	.0985	.0938	.0892
4	.1719	.1681	.1641	.1600	.1558	.1515	.1472	.1428	.1383	.1339

(*continues*)

TABLE 7 (*Continued*)

5	.1753	.1748	.1740	.1728	.1714	.1697	.1678	.1656	.1632	.1606
6	.1490	.1515	.1537	.1555	.1571	.1584	.1594	.1601	.1605	.1606
7	.1086	.1125	.1163	.1200	.1234	.1267	.1298	.1326	.1353	.1377
8	.0692	.0731	.0771	.0810	.0849	.0887	.0925	.0962	.0998	.1033
9	.0392	.0423	.0454	.0486	.0519	.0552	.0586	.0620	.0654	.0688
10	.0200	.0220	.0241	.0262	.0285	.0309	.0334	.0359	.0386	.0413
11	.0093	.0104	.0116	.0129	.0143	.0157	.0173	.0190	.0207	.0225
12	.0039	.0045	.0051	.0058	.0065	.0073	.0082	.0092	.0102	.0113
13	.0015	.0018	.0021	.0024	.0028	.0032	.0036	.0041	.0046	.0052
14	.0006	.0007	.0008	.0009	.0011	.0013	.0015	.0017	.0019	.0022
15	.0002	.0002	.0003	.0003	.0004	.0005	.0006	.0007	.0008	.0009
16	.0001	.0001	.0001	.0001	.0001	.0002	.0002	.0002	.0003	.0003
17	.0000	.0000	.0000	.0000	.0000	.0001	.0001	.0001	.0001	.0001

					μ					
x	*6.1*	*6.2*	*6.3*	*6.4*	*6.5*	*6.6*	*6.7*	*6.8*	*6.9*	*7.0*
0	.0022	.0020	.0018	.0017	.0015	.0014	.0012	.0011	.0010	.0009
1	.0137	.0126	.0116	.0106	.0098	.0090	.0082	.0076	.0070	.0064
2	.0417	.0390	.0364	.0340	.0318	.0296	.0276	.0258	.0240	.0223
3	.0848	.0806	.0765	.0726	.0688	.0652	.0617	.0584	.0552	.0521
4	.1294	.1249	.1205	.1162	.1118	.1076	.1034	.0992	.0952	.0912
5	.1579	.1549	.1519	.1487	.1454	.1420	.1385	.1349	.1314	.1277
6	.1605	.1601	.1595	.1586	.1575	.1562	.1546	.1529	.1511	.1490
7	.1399	.1418	.1435	.1450	.1462	.1472	.1480	.1486	.1489	.1490
8	.1066	.1099	.1130	.1160	.1188	.1215	.1240	.1263	.1284	.1304
9	.0723	.0757	.0791	.0825	.0858	.0891	.0923	.0954	.0985	.1014
10	.0441	.0469	.0498	.0528	.0558	.0588	.0618	.0649	.0679	.0710
11	.0245	.0265	.0285	.0307	.0330	.0353	.0377	.0401	.0426	.0452
12	.0124	.0137	.0150	.0164	.0179	.0194	.0210	.0227	.0245	.0264
13	.0058	.0065	.0073	.0081	.0089	.0098	.0108	.0119	.0130	.0142
14	.0025	.0029	.0033	.0037	.0041	.0046	.0052	.0058	.0064	.0071
15	.0010	.0012	.0014	.0016	.0018	.0020	.0023	.0026	.0029	.0033
16	.0004	.0005	.0005	.0006	.0007	.0008	.0010	.0011	.0013	.0014
17	.0001	.0002	.0002	.0002	.0003	.0003	.0004	.0004	.0005	.0006
18	.0000	.0001	.0001	.0001	.0001	.0001	.0001	.0002	.0002	.0002
19	.0000	.0000	.0000	.0000	.0000	.0000	.0000	.0001	.0001	.0001

					μ					
x	*7.1*	*7.2*	*7.3*	*7.4*	*7.5*	*7.6*	*7.7*	*7.8*	*7.9*	*8.0*
0	.0008	.0007	.0007	.0006	.0006	.0005	.0005	.0004	.0004	.0003
1	.0059	.0054	.0049	.0045	.0041	.0038	.0035	.0032	.0029	.0027
2	.0208	.0194	.0180	.0167	.0156	.0145	.0134	.0125	.0116	.0107
3	.0492	.0464	.0438	.0413	.0389	.0366	.0345	.0324	.0305	.0286
4	.0874	.0836	.0799	.0764	.0729	.0696	.0663	.0632	.0602	.0573

(*continues*)

TABLE 7 *(Continued)*

5	.1241	.1204	.1167	.1130	.1094	.1057	.1021	.0986	.0951	.0916
6	.1468	.1445	.1420	.1394	.1367	.1339	.1311	.1282	.1252	.1221
7	.1489	.1486	.1481	.1474	.1465	.1454	.1442	.1428	.1413	.1396
8	.1321	.1337	.1351	.1363	.1373	.1382	.1388	.1392	.1395	.1396
9	.1042	.1070	.1096	.1121	.1144	.1167	.1187	.1207	.1224	.1241
10	.0740	.0770	.0800	.0829	.0858	.0887	.0914	.0941	.0967	.0993
11	.0478	.0504	.0531	.0558	.0585	.0613	.0640	.0667	.0695	.0722
12	.0283	.0303	.0323	.0344	.0366	.0388	.0411	.0434	.0457	.0481
13	.0154	.0168	.0181	.0196	.0211	.0227	.0243	.0260	.0278	.0296
14	.0078	.0086	.0095	.0104	.0113	.0123	.0134	.0145	.0157	.0169
15	.0037	.0041	.0046	.0051	.0057	.0062	.0069	.0075	.0083	.0090
16	.0016	.0019	.0021	.0024	.0026	.0030	.0033	.0037	.0041	.0045
17	.0007	.0008	.0009	.0010	.0012	.0013	.0015	.0017	.0019	.0021
18	.0003	.0003	.0004	.0004	.0005	.0006	.0006	.0007	.0008	.0009
19	.0001	.0001	.0001	.0002	.0002	.0002	.0003	.0003	.0003	.0004
20	.0000	.0000	.0001	.0001	.0001	.0001	.0001	.0001	.0001	.0002
21	.0000	.0000	.0000	.0000	.0000	.0000	.0000	.0000	.0001	.0001

					μ					
x	*8.1*	*8.2*	*8.3*	*8.4*	*8.5*	*8.6*	*8.7*	*8.8*	*8.9*	*9.0*
0	.0003	.0003	.0002	.0002	.0002	.0002	.0002	.0002	.0001	.0001
1	.0025	.0023	.0021	.0019	.0017	.0016	.0014	.0013	.0012	.0011
2	.0100	.0092	.0086	.0079	.0074	.0068	.0063	.0058	.0054	.0050
3	.0269	.0252	.0237	.0222	.0208	.0195	.0183	.0171	.0160	.0150
4	.0544	.0517	.0491	.0466	.0443	.0420	.0398	.0377	.0357	.0337
5	.0882	.0849	.0816	.0784	.0752	.0722	.0692	.0663	.0635	.0607
6	.1191	.1160	.1128	.1097	.1066	.1034	.1003	.0972	.0941	.0911
7	.1378	.1358	.1338	.1317	.1294	.1271	.1247	.1222	.1197	.1171
8	.1395	.1392	.1388	.1382	.1375	.1366	.1356	.1344	.1332	.1318
9	.1256	.1269	.1280	.1290	.1299	.1306	.1311	.1315	.1317	.1318
10	.1017	.1040	.1063	.1084	.1104	.1123	.1140	.1157	.1172	.1186
11	.0749	.0776	.0802	.0828	.0853	.0878	.0902	.0925	.0948	.0970
12	.0505	.0530	.0555	.0579	.0604	.0629	.0654	.0679	.0703	.0728
13	.0315	.0334	.0354	.0374	.0395	.0416	.0438	.0459	.0481	.0504
14	.0182	.0196	.0210	.0225	.0240	.0256	.0272	.0289	.0306	.0324
15	.0098	.0107	.0116	.0126	.0136	.0147	.0158	.0169	.0182	.0194
16	.0050	.0055	.0060	.0066	.0072	.0079	.0086	.0093	.0101	.0109
17	.0024	.0026	.0029	.0033	.0036	.0040	.0044	.0048	.0053	.0058
18	.0011	.0012	.0014	.0015	.0017	.0019	.0021	.0024	.0026	.0029
19	.0005	.0005	.0006	.0007	.0008	.0009	.0010	.0011	.0012	.0014
20	.0002	.0002	.0002	.0003	.0003	.0004	.0004	.0005	.0005	.0006
21	.0001	.0001	.0001	.0001	.0001	.0002	.0002	.0002	.0002	.0003
22	.0000	.0000	.0000	.0000	.0001	.0001	.0001	.0001	.0001	.0001

TABLE 7 (*Continued*)

x	μ 9.1	9.2	9.3	9.4	9.5	9.6	9.7	9.8	9.9	10
0	.0001	.0001	.0001	.0001	.0001	.0001	.0001	.0001	.0001	.0000
1	.0010	.0009	.0009	.0008	.0007	.0007	.0006	.0005	.0005	.0005
2	.0046	.0043	.0040	.0037	.0034	.0031	.0029	.0027	.0025	.0023
3	.0140	.0131	.0123	.0115	.0107	.0100	.0093	.0087	.0081	.0076
4	.0319	.0302	.0285	.0269	.0254	.0240	.0226	.0213	.0201	.0189
5	.0581	.0555	.0530	.0506	.0483	.0460	.0439	.0418	.0398	.0378
6	.0881	.0851	.0822	.0793	.0764	.0736	.0709	.0682	.0656	.0631
7	.1145	.1118	.1091	.1064	.1037	.1010	.0982	.0955	.0928	.0901
8	.1302	.1286	.1269	.1251	.1232	.1212	.1191	.1170	.1148	.1126
9	.1317	.1315	.1311	.1306	.1300	.1293	.1284	.1274	.1263	.1251
10	.1198	.1210	.1219	.1228	.1235	.1241	.1245	.1249	.1250	.1251
11	.0991	.1012	.1031	.1049	.1067	.1083	.1098	.1112	.1125	.1137
12	.0752	.0776	.0799	.0822	.0844	.0866	.0888	.0908	.0928	.0948
13	.0526	.0549	.0572	.0594	.0617	.0640	.0662	.0685	.0707	.0729
14	.0342	.0361	.0380	.0399	.0419	.0439	.0459	.0479	.0500	.0521
15	.0208	.0221	.0235	.0250	.0265	.0281	.0297	.0313	.0330	.0347
16	.0118	.0127	.0137	.0147	.0157	.0168	.0180	.0192	.0204	.0217
17	.0063	.0069	.0075	.0081	.0088	.0095	.0103	.0111	.0119	.0128
18	.0032	.0035	.0039	.0042	.0046	.0051	.0055	.0060	.0065	.0071
19	.0015	.0017	.0019	.0021	.0023	.0026	.0028	.0031	.0034	.0037
20	.0007	.0008	.0009	.0010	.0011	.0012	.0014	.0015	.0017	.0019
21	.0003	.0003	.0004	.0004	.0005	.0006	.0006	.0007	.0008	.0009
22	.0001	.0001	.0002	.0002	.0002	.0002	.0003	.0003	.0004	.0004
23	.0000	.0001	.0001	.0001	.0001	.0001	.0001	.0001	.0002	.0002
24	.0000	.0000	.0000	.0000	.0000	.0000	.0000	.0001	.0001	.0001

x	μ 11	12	13	14	15	16	17	18	19	20
0	.0000	.0000	.0000	.0000	.0000	.0000	.0000	.0000	.0000	.0000
1	.0002	.0001	.0000	.0000	.0000	.0000	.0000	.0000	.0000	.0000
2	.0010	.0004	.0002	.0001	.0000	.0000	.0000	.0000	.0000	.0000
3	.0037	.0018	.0008	.0004	.0002	.0001	.0000	.0000	.0000	.0000
4	.0102	.0053	.0027	.0013	.0006	.0003	.0001	.0001	.0000	.0000
5	.0224	.0127	.0070	.0037	.0019	.0010	.0005	.0002	.0001	.0001
6	.0411	.0255	.0152	.0087	.0048	.0026	.0014	.0007	.0004	.0002
7	.0646	.0437	.0281	.0174	.0104	.0060	.0034	.0018	.0010	.0005
8	.0888	.0655	.0457	.0304	.0194	.0120	.0072	.0042	.0024	.0013
9	.1085	.0874	.0661	.0473	.0324	.0213	.0135	.0083	.0050	.0029
10	.1194	.1048	.0859	.0663	.0486	.0341	.0230	.0150	.0095	.0058
11	.1194	.1144	.1015	.0844	.0663	.0496	.0355	.0245	.0164	.0106
12	.1094	.1144	.1099	.0984	.0829	.0661	.0504	.0368	.0259	.0176
13	.0926	.1056	.1099	.1060	.0956	.0814	.0658	.0509	.0378	.0271
14	.0728	.0905	.1021	.1060	.1024	.0930	.0800	.0655	.0514	.0387

(continues)

TABLE 7 (*Continued*)

15	.0534	.0724	.0885	.0989	.1024	.0992	.0906	.0786	.0650	.0516
16	.0367	.0543	.0719	.0866	.0960	.0992	.0963	.0884	.0772	.0646
17	.0237	.0383	.0550	.0713	.0847	.0934	.0963	.0936	.0863	.0760
18	.0145	.0256	.0397	.0554	.0706	.0830	.0909	.0936	.0911	.0844
19	.0084	.0161	.0272	.0409	.0557	.0699	.0814	.0887	.0911	.0888
20	.0046	.0097	.0177	.0286	.0418	.0559	.0692	.0798	.0866	.0888
21	.0024	.0055	.0109	.0191	.0299	.0426	.0560	.0684	.0783	.0846
22	.0012	.0030	.0065	.0121	.0204	.0310	.0433	.0560	.0676	.0769
23	.0006	.0016	.0037	.0074	.0133	.0216	.0320	.0438	.0559	.0669
24	.0003	.0008	.0020	.0043	.0083	.0144	.0226	.0328	.0442	.0557
25	.0001	.0004	.0010	.0024	.0050	.0092	.0154	.0237	.0336	.0446
26	.0000	.0002	.0005	.0013	.0029	.0057	.0101	.0164	.0246	.0343
27	.0000	.0001	.0002	.0007	.0016	.0034	.0063	.0109	.0173	.0254
28	.0000	.0000	.0001	.0003	.0009	.0019	.0038	.0070	.0117	.0181
29	.0000	.0000	.0001	.0002	.0004	.0011	.0023	.0044	.0077	.0125
30	.0000	.0000	.0000	.0001	.0002	.0006	.0013	.0026	.0049	.0083
31	.0000	.0000	.0000	.0000	.0001	.0003	.0007	.0015	.0030	.0054
32	.0000	.0000	.0000	.0000	.0001	.0001	.0004	.0009	.0018	.0034
33	.0000	.0000	.0000	.0000	.0000	.0001	.0002	.0005	.0010	.0020
34	.0000	.0000	.0000	.0000	.0000	.0000	.0001	.0002	.0006	.0012
35	.0000	.0000	.0000	.0000	.0000	.0000	.0000	.0001	.0003	.0007
36	.0000	.0000	.0000	.0000	.0000	.0000	.0000	.0001	.0002	.0004
37	.0000	.0000	.0000	.0000	.0000	.0000	.0000	.0000	.0001	.0002
38	.0000	.0000	.0000	.0000	.0000	.0000	.0000	.0000	.0000	.0001
39	.0000	.0000	.0000	.0000	.0000	.0000	.0000	.0000	.0000	.0001

TABLE 8 Random Digits

63271	59986	71744	51102	15141	80714	58683	93108	13554	79945
88547	09896	95436	79115	08303	01041	20030	63754	08459	28364
55957	57243	83865	09911	19761	66535	40102	26646	60147	15702
46276	87453	44790	67122	45573	84358	21625	16999	13385	22782
55363	07449	34835	15290	76616	67191	12777	21861	68689	03263
69393	92785	49902	58447	42048	30378	87618	26933	40640	16281
13186	29431	88190	04588	38733	81290	89541	70290	40113	08243
17726	28652	56836	78351	47327	18518	92222	55201	27340	10493
36520	64465	05550	30157	82242	29520	69753	72602	23756	54935
81628	36100	39254	56835	37636	02421	98063	89641	64953	99337
84649	48968	75215	75498	49539	74240	03466	49292	36401	45525
63291	11618	12613	75055	43915	26488	41116	64531	56827	30825
70502	53225	03655	05915	37140	57051	48393	91322	25653	06543
06426	24771	59935	49801	11082	66762	94477	02494	88215	27191
20711	55609	29430	70165	45406	78484	31639	52009	18873	96927
41990	70538	77191	25860	55204	73417	83920	69468	74972	38712
72452	36618	76298	26678	89334	33938	95567	29380	75906	91807
37042	40318	57099	10528	09925	89773	41335	96244	29002	46453
53766	52875	15987	46962	67342	77592	57651	95508	80033	69828
90585	58955	53122	16025	84299	53310	67380	84249	25348	04332
32001	96293	37203	64516	51530	37069	40261	61374	05815	06714
62606	64324	46354	72157	67248	20135	49804	09226	64419	29457
10078	28073	85389	50324	14500	15562	64165	06125	71353	77669
91561	46145	24177	15294	10061	98124	75732	00815	83452	97355
13091	98112	53959	79607	52244	63303	10413	63839	74762	50289
73864	83014	72457	22682	03033	61714	88173	90835	00634	85169
66668	25467	48894	51043	02365	91726	09365	63167	95264	45643
84745	41042	29493	01836	09044	51926	43630	63470	76508	14194
48068	26805	94595	47907	13357	38412	33318	26098	82782	42851
54310	96175	97594	88616	42035	38093	36745	56702	40644	83514
14877	33095	10924	58013	61439	21882	42059	24177	58739	60170
78295	23179	02771	43464	59061	71411	05697	67194	30495	21157
67524	02865	39593	54278	04237	92441	26602	63835	38032	94770
58268	57219	68124	73455	83236	08710	04284	55005	84171	42596
97158	28672	50685	01181	24262	19427	52106	34308	73685	74246
04230	16831	69085	30802	65559	09205	71829	06489	85650	38707
94879	56606	30401	02602	57658	70091	54986	41394	60437	03195
71446	15232	66715	26385	91518	70566	02888	79941	39684	54315
32886	05644	79316	09819	00813	88407	17461	73925	53037	91904
62048	33711	25290	21526	02223	75947	66466	06232	10913	75336
84534	42351	21628	53669	81352	95152	08107	98814	72743	12849
84707	15885	84710	35866	06446	86311	32648	88141	73902	69981
19409	40868	64220	80861	13860	68493	52908	26374	63297	45052
57978	48015	25973	66777	45924	56144	24742	96702	88200	66162
57295	98298	11199	96510	75228	41600	47192	43267	35973	23152

(continues)

TABLE 8 (*Continued*)

94044	83785	93388	07833	38216	31413	70555	03023	54147	06647
30014	25879	71763	96679	90603	99396	74557	74224	18211	91637
07265	69563	64268	88802	72264	66540	01782	08396	19251	83613
84404	88642	30263	80310	11522	57810	27627	78376	36240	48952
21778	02085	27762	46097	43324	34354	09369	14966	10158	76089

Used by permission from *A Million Random Digits with 100,000 Normal Deviates*. CA, The Rand Corporation, 1955.

C. Summation Notation

SUMMATIONS

Definition

$$\sum_{i=1}^{n} x_i = x_1 + x_2 + \cdots + x_n. \tag{C.1}$$

Example: $x_1 = 5, x_2 = 8, x_3 = 14$:

$$\begin{aligned}\sum_{i=1}^{3} x_i &= x_1 + x_2 + x_3 \\ &= 5 + 8 + 14 \\ &= 27.\end{aligned}$$

Result 1

For a constant c:

$$\sum_{i=1}^{n} c = (\underbrace{c + c + \cdots + c}_{n \text{ times}}) = nc. \tag{C.2}$$

Example: $c = 5, n = 10$:

$$\sum_{i=1}^{10} 5 = 10(5) = 50.$$

Example: $c = \bar{x}$:

$$\sum_{i=1}^{n} \bar{x} = n\bar{x}.$$

Result 2

$$\begin{aligned}\sum_{i=1}^{n} cx_i &= cx_1 + cx_2 + \cdots + cx_n \\ &= c(x_1 + x_2 + \cdots + x_n) = c\sum_{i=1}^{n} x_i.\end{aligned} \tag{C.3}$$

Example: $x_1 = 5, x_2 = 8, x_3 = 14, c = 2$:

$$\sum_{i=1}^{3} 2x_i = 2\sum_{i=1}^{3} x_i = 2(27) = 54.$$

Result 3

$$\sum_{i=1}^{n} (ax_i + by_i) = a\sum_{i=1}^{n} x_i + b\sum_{i=1}^{n} y_i. \qquad (C.4)$$

Example: $x_1 = 5, x_2 = 8, x_3 = 14, a = 2, y_1 = 7, y_2 = 3, y_3 = 8, b = 4$:

$$\begin{aligned}\sum_{i=1}^{3} (2x_i + 4y_i) &= 2\sum_{i=1}^{3} x_i + 4\sum_{i=1}^{3} y_i \\ &= 2(27) + 4(18) \\ &= 54 + 72 \\ &= 126.\end{aligned}$$

DOUBLE SUMMATIONS

Consider the following data involving the variable x_{ij}, where i is the subscript denoting the row position and j is the subscript denoting the column position:

		Column 1	Column 2	Column 3
Row	1	$x_{11} = 10$	$x_{12} = 8$	$x_{13} = 6$
	2	$x_{21} = 7$	$x_{22} = 4$	$x_{23} = 12$

Definition

$$\sum_{i=1}^{n}\sum_{j=1}^{m} x_{ij} = (x_{11} + x_{12} + \cdots + x_{1m}) + (x_{21} + x_{22} + \cdots + x_{2m})$$

$$+ (x_{31} + x_{32} + \cdots + x_{3m}) + \cdots + (x_{n1} + x_{n2} + \cdots + x_{nm}). \qquad (C.5)$$

Example:

$$\begin{aligned}\sum_{i=1}^{2}\sum_{j=1}^{3} x_{ij} &= x_{11} + x_{12} + x_{13} + x_{21} + x_{22} + x_{23} \\ &= 10 + 8 + 6 + 7 + 4 + 12 \\ &= 47.\end{aligned}$$

Definition

$$\sum_{i=1}^{n} x_{ij} = x_{1j} + x_{2j} + \cdots + x_{nj}. \qquad (C.6)$$

Example:

$$\begin{aligned}\sum_{i=1}^{2} x_{i2} &= x_{12} + x_{22} \\ &= 8 + 4 \\ &= 12.\end{aligned}$$

SHORTHAND NOTATION

Sometimes when a summation is for all values of the subscript, we use the following shorthand notations:

$$\sum_{i=1}^{n} x_i = \sum_{i} x_i, \tag{C.7}$$

$$\sum_{i=1}^{n} x_i = \sum x_i, \tag{C.8}$$

$$\sum_{i=1}^{n} \sum_{j=1}^{m} x_{ij} = \sum_{i} \sum_{j} x_{ij}, \tag{C.9}$$

$$\sum_{i=1}^{n} x_{ij} = \sum_{i} x_{ij}. \tag{C.10}$$

Index

†

t Distribution

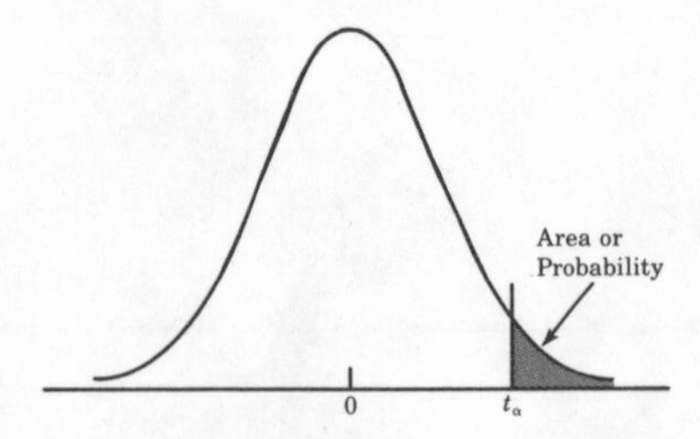

Entries in the table give t_α values, where α is the area or probability in the upper tail of the t distribution. For example, with 10 degrees of freedom and a .05 area in the upper tail, $t_{.05} = 1.812$.

Degrees of Freedom	*Area in Upper Tail* .10	.05	.025	.01	.005
1	3.078	6.314	12.706	31.821	63.657
2	1.886	2.920	4.303	6.965	9.925
3	1.638	2.353	3.182	4.541	5.841
4	1.533	2.132	2.776	3.747	4.604
5	1.476	2.015	2.571	3.365	4.032
6	1.440	1.943	2.447	3.143	3.707
7	1.415	1.895	2.365	2.998	3.499
8	1.397	1.860	2.306	2.896	3.355
9	1.383	1.833	2.262	2.821	3.250
10	1.372	1.812	2.228	2.764	3.169
11	1.363	1.796	2.201	2.718	3.106
12	1.356	1.782	2.179	2.681	3.055
13	1.350	1.771	2.160	2.650	3.012
14	1.345	1.761	2.145	2.624	2.977
15	1.341	1.753	2.131	2.602	2.947
16	1.337	1.746	2.120	2.583	2.921
17	1.333	1.740	2.110	2.567	2.898
18	1.330	1.734	2.101	2.552	2.878
19	1.328	1.729	2.093	2.539	2.861
20	1.325	1.725	2.086	2.528	2.845
21	1.323	1.721	2.080	2.518	2.831
22	1.321	1.717	2.074	2.508	2.819
23	1.319	1.714	2.069	2.500	2.807
24	1.318	1.711	2.064	2.492	2.797
25	1.316	1.708	2.060	2.485	2.787
26	1.315	1.706	2.056	2.479	2.779
27	1.314	1.703	2.052	2.473	2.771
28	1.313	1.701	2.048	2.467	2.763
29	1.311	1.699	2.045	2.462	2.756
30	1.310	1.697	2.042	2.457	2.750
40	1.303	1.684	2.021	2.423	2.704
60	1.296	1.671	2.000	2.390	2.660
120	1.289	1.658	1.980	2.358	2.617
∞	1.282	1.645	1.960	2.326	2.576